Vegetation Fires and Pollution in Asia

Krishna Prasad Vadrevu · Toshimasa Ohara ·
Chris Justice
Editors

Vegetation Fires and Pollution in Asia

Springer

Editors
Krishna Prasad Vadrevu
NASA Marshall Space Flight Center
Huntsville, AL, USA

Toshimasa Ohara
Center for Environmental Science
in Saitama
Kazo, Saitama, Japan

Chris Justice
University of Maryland
College Park, MD, USA

ISBN 978-3-031-29915-5 ISBN 978-3-031-29916-2 (eBook)
https://doi.org/10.1007/978-3-031-29916-2

This Springer imprint is published by the registered company Springer Nature Switzerland AG
The registered company address is: Gewerbestrasse 11, 6330 Cham, Switzerland

Foreword

It is my pleasure to write a Foreword to the book on *Vegetation Fires and Air Pollution in Asia*, edited by Drs. Krishna Prasad Vadrevu, Toshimasa Ohara, and Chris Justice. I wrote the forewords to their earlier books: (1) *Land-Atmosphere Interactions in South/Southeast Asia* (2018) and (2) *Biomass Burning in South and Southeast Asia* (2022) that drew wide attention in the research community. Wildfires are among the most serious disasters over the globe and in South and Southeast Asian countries, in particular. Transboundary pollution from vegetation fires is a recurrent environmental problem in Asian countries. Addressing fire-related impacts on ecosystems and the atmosphere requires robust information on their spatiotemporal characteristics, including impact assessment and uncertainties. Remote sensing and geospatial technologies offer unique opportunities for mapping and monitoring fires and emissions estimation, and impacts assessment needed for land management and air pollution control.

The current book is a compilation of articles on various fire-related topics from renowned researchers in the USA and South/Southeast Asia. The contributions in this book result from several international workshops organized in Asia in the framework of the South/Southeast Asia Research Initiative (SARI, http://sari.umd.edu), funded under the NASA Land-Cover/Land-Use Change Program (http://lcluc.hq.nasa.gov), which I manage at NASA Headquarters. Recognizing the pervasive land-use changes in Asia driven by rapid population growth and economic development, the NASA LCLUC Program launched SARI in 2015, with Dr. Krishna Prasad Vadrevu (NASA Marshall Space Flight Center) as its Project Scientist. Since then, several SARI workshops have been organized, each meeting attended by more than 100 researchers. There, important regional science issues were discussed, with research needs and priorities identified, and thus vegetation fires have been recognized as the most recurrent problem in the region that needs attention.

The NASA Land-Cover/Land-Use Change Program, which I have been leading as a Program Manager for over 23 years at NASA Headquarters, is highly interdisciplinary and integrates physical and socioeconomic sciences to address environmental and societal issues. The program aims to develop and use NASA space and airborne remote sensing technologies, relying on US and non-US satellite data sources, to improve our understanding of human interactions with the environment. One of the critical questions addressed in the LCLUC program is "What are the causes and consequences of LCLUC?". Several decades of research in the SARI region have revealed the drivers of extensive modifications of the natural environment and environmental impacts of LCLUC, such as changes in atmospheric composition and air quality, biomass burning being a significant contributor.

In my research career, I was involved in mapping and studying wildfires globally. In 2000, I published a paper on 1997–1998 El Niño year fires in Indonesia (Gutman et al. 2000) in a meteorological journal to attract the attention of meteorologists and climatologists to the richness of the NOAA Advanced Very High-Resolution Radiometer (AVHRR) observations for land-climate applications. In that paper, I demonstrated how many land and atmospheric products can be derived from AVHRR alone and highlighted the compound effect of fires from slash-and-burn agriculture and one of the strongest El Niño event on atmosphere (smoke and cloud microphysics) and land (surface reflectance, temperature, and greenness).

I am happy to see this book published with significant contributions from regional researchers from Asia, with over 125 authors contributing to the book chapters. Each article is unique. They highlight various aspects, such as mapping and monitoring fires and the resulting aerosol and greenhouse gas emission characteristics, including transport and impacts on the local climate. The authors integrate ground-based measurements and satellite data to study the region's fire regime characteristics and biomass burning effects. In addition, a separate section on modeling and decision support system includes several articles useful to address air pollution impacts, mitigation, and control.

The material in this book will help researchers working on the remote sensing of fires, air pollution, modeling, and decision support systems useful for fire management. Overall, the scientific results will be of interest to students, researchers, fire managers, policymakers, and practicing professionals in the field. With contributions from several international scholars and professionals and edited by three eminent scientists, this book is a significant addition to the literature on fire research and applications and biomass burning in South/Southeast Asia. I am impressed with the book's content and the rich contributions from regional scientists. The book is a timely contribution to NASA LCLUC SARI science and reflects the current status in this research field.

The editors of this book are highly active and reputed researchers who have published multiple books earlier. I commend all the team members, particularly the Chief Editor, Dr. Krishna Prasad Vadrevu, SARI Lead, on accomplishing another big task of compiling rich material. I am confident it will motivate new, exciting research on fire and emissions and trigger innovative ideas. I welcome more such contributions from the US and regional scientists and wish you all good reading.

Dr. Garik Gutman
Land-Cover/Land-Use Change
Program Manager
NASA Headquarters
Washington DC, USA

Reference

Gutman, G., I. Csiszar, and P. Romanov. 2000. Using NOAA/AVHRR products to monitor El Nino impacts: Focus on Indonesia in 1997–98. *Bulletin of the American Meteorological Society* 81: 1189–1205.

Preface

Vegetation fires are pervasive in South/Southeast Asian (S/SEA) countries. The most common type of fire in S/SEA is due to land clearing and land-use conversion. For example, several indigenous people in S/SEA practice slash-and-burn agriculture. In addition, fires are widely used as an affordable management tool to remove crop residues after harvest, pests, disease, rejuvenate grazing lands, etc. Recently, most of the fires in SEA have been attributed to large-scale land-clearing activities to expand the agricultural frontier for oil palm, rubber, and soybean cultivation in Indonesia, Laos, Thailand, and Myanmar. The impact of such fires on landscapes can be significant as they can destroy the vegetation cover, resulting in biodiversity loss and disruption in the nutrient cycles and hydrology. Most importantly, vegetation fires can release large volumes of radiatively active gases, aerosols, and other chemically active species, impacting the climate at multiple spatial scales. Also, the smoke particles and gases released during the burning can be harmful to human health. Thus, mapping and monitoring fires, including quantifying the impacts of biomass burning on the environment, gain significance. However, addressing fire impacts at multiple spatial scales can be challenging as fires show considerable spatial and temporal variability due to varying fuel loads, moisture content, temperature, topography, humidity, wind speed, etc. Due to these variations, the impacts of fires on the environment can vary significantly. Thus, it is essential to integrate various datasets, both top-down and bottom-up approaches, and technologies to understand vegetation fires and their impacts for effective mitigation.

The current book, *Vegetation Fires and Pollution in Asia*, is a collection of papers from several South/Southeast Asia Research Initiative (SARI, sari.umd.edu) meetings organized in Asia since 2015. SARI is a NASA Land-Cover/Land-Use Change Program (lcluc.umd.edu) funded research activity. SARI aims to develop innovative regional research, education, and capacity-building programs involving state-of-the-art remote sensing, natural sciences, engineering, and social sciences to enrich LCLUC science in S/SEA. To address LCLUC science, SARI has been utilizing a systems approach to problem-solving that examines biophysical and socioeconomic aspects of land systems, including the interactions between land use and climate and the interrelationships among policy, governance, and land use. Several SARI meetings identified fires and biomass burning as an important environmental and policy issue that need immediate attention. This book on vegetation fires and pollution was planned to meet both the research and applications community needs.

The book is divided into three parts: (a) *Mapping, Monitoring, and Modeling of Vegetation Fires*; (b) *Greenhouse Gas Emissions and Air Pollution*; (c) *Air Pollution Modeling and Decision Support Systems.* These parts are preceded by an introductory chapter by the editors, which details the latest satellite-derived fire statistics and the current fire situation in S/SEA. The authors also highlight important research needs and priorities on fire mapping and monitoring, greenhouse gas emissions estimation, including decision support systems useful for fire control, management and mitigation.

The first part, *Mapping, Monitoring, and Modeling of Vegetation Fires*, includes 13 chapters. The second chapter "Wildfire Monitoring Using Infrared Bands and Spatial Resolution Effects" by Wei Zheng et al. uses data from the Chinese FY3A meteorological satellite with 1 km infrared data and HJ-IB environmental disaster reduction satellite with 150 m and 300 m infrared data for active fire monitoring. Using a novel mixed-pixel decomposition method, they demonstrate the importance of infrared bands for small wildfire detection and fire intensity evaluation. Chapters "Status and Drivers of Forest Fires in Myanmar"–"Burnt Area Signal Variations in Agriculture and Forested Landscapes of India—A Case Study Using Sentinel 1A/B Synthetic Aperture Radar" focus on fires in South Asian countries. Chapters "Status and Drivers of Forest Fires in Myanmar" and "Vegetation Fires and Entropy Variations in Myanmar" focus on fires in Myanmar. In the third chapter "Status and Drivers of Forest Fires in Myanmar", Biswas and Vadrevu provide an overview of the status and drivers of fires using VIIRS (375 m) and a new national-level forest-type map of Myanmar (20 m). They report that the highest number of fires occur in mixed deciduous forests, followed by bamboo forests. The next follow-on the fourth chapter "Vegetation Fires and Entropy Variations in Myanmar" by Vadrevu et al. uses the entropy technique to assess spatial variations of fires and report higher entropies for forest fires than in agriculture and show hotspots of fire variability useful for fire management. The fifth chapter "Crop Residue Burning and Forest Fire Emissions in Nepal" by Bhupendra et al. reports on Nepal's crop residue burning and forest fire emissions using bottom-up and top-down approaches. Their results suggest ten times more CO_2 emissions from forest fires than crop residues. Authors provide mitigation options such as redesigning combine harvesters to collect crop residues while harvesting, including integrated livestock farming to reduce crop residue burning. The sixth chapter "Firewood Burning Dynamics by the Sri Lankan Households: Trends, Patterns, and Implications" by Pallegedara and Kumara focuses on firewood use in Sri Lanka using survey data from >57,000 households and infers that biofuel use for cooking has significant harmful effects on both health and the environment. Authors note that education and income levels are the country's primary governing factors of biofuel use and stress the need for alternative energy sources such as electric, solar, wind, and biogas to reduce indoor air pollution and associated health risks.

Chapters "Burnt Area Signal Variations in Agriculture and Forested Landscapes of India—A Case Study Using Sentinel 1A/B Synthetic Aperture Radar" and "Application of Interferometry SAR for Monitoring of Peatland Area—Case Studies in Indonesia" are novel as they use Synthetic Aperture Radar and Sentinel 1A/B data to characterize burnt area signals. In the seventh chapter "Burnt Area Signal Variations in Agriculture and Forested Landscapes of India—A Case Study Using Sentinel 1A/B Synthetic Aperture Radar", Vadrevu et al. report contrasting results using SAR data in burnt area sites, i.e., a decrease in backscatter signal in crop residue burnt areas, whereas an increase in the forest burnt sites and attribute these differences to site conditions. Authors suggest a caution toward using a universal algorithm that can capture burnt areas globally using SAR data. In the eighth chapter "Application of Interferometry SAR for Monitoring of Peatland Area—Case Studies in Indonesia", Arvelyna et al. use a differential interferometry SAR (DInSAR) analysis technique using ALOS-2 PALSAR data for peatland monitoring in Central Kalimantan and Riau Province, Indonesia. Their results suggest the potential of DInSAR in capturing peatland's surface height variabilities affected by groundwater fluctuations and fires.

Chapters "Active Fire Monitoring of Thailand and Upper ASEAN by Earth Observation Data: Benefits, Lessons Learned, and What Still Needs to Be Known" and "Detecting Vegetation Regrowth After Fires in Small Watershed Settings Using Remotely Sensed Data and Local Community Participation Approach" focus on fire mapping and monitoring in Thailand; however, the approaches presented can be adopted elsewhere for effective results. In the ninth chapter "Active Fire Monitoring of Thailand and Upper ASEAN by Earth Observation Data: Benefits, Lessons Learned, and What Still Needs to Be Known", Tanpipat et al. highlight integrating advanced geostationary fire data to enable effective fire management, whereas, in the tenth chapter "Detecting Vegetation Regrowth After Fires in Small Watershed Settings Using Remotely Sensed Data and Local Community Participation Approach", Onpraphai et al. present an operational framework to detect vegetation regrowth after fires using the THEOS satellite data. Their framework combines a local community participation process to generate income from alternative options.

Chapter "Long-Term Spatiotemporal Distribution of Fire Over Maritime Continent and Their Responses to Climate Anomalies" by Khoir et al. focuses on the climate drivers of long-term spatial and temporal distribution fires over the Maritime Continent. Using empirical orthogonal function (EOF), fast Fourier transform (FFT), and correlation method, authors infer that El Niño and positive Dipole Mode events significantly increased the total number of fire hotspots. In contrast, the Madden–Julian Oscillation (MJO) events increased the negative fire event anomalies during the study period. In the twelfth chapter "Vegetation Fires in Laos—An Overview", Vadrevu et al. provide satellite-derived fire statistics in Laos and highlight hotspot regions where fire mitigation is urgently needed and highlight including important drivers of fires.

Fire behavior aspects are highlighted in Chapters "Vegetation Fires, Fire Radiative Power, and Intermediate Fire Occurrence Intensity (IFOI) Hypothesis Testing in Myanmar, Laos, and Cambodia" and "Analyzing Fire Behavior and Calibrating a Fire Growth Model in a Seasonally Dry Tropical Forest Area". In the thirteenth chapter "Vegetation Fires, Fire Radiative Power, and Intermediate Fire Occurrence Intensity (IFOI) Hypothesis Testing in Myanmar, Laos, and Cambodia", Vadrevu et al. evaluate the Intermediate Fire Occurrence Intensity (IFOI) Hypothesis in Myanmar, Laos, and Cambodia. The IFOI states that the fire occurrence in units of time per unit of an area increases with fire intensity up until a threshold is reached above which occurrence decreases with increasing intensity, in essence, a humped relationship. Using various statistical models, the authors conclude that the IFOI hypothesis does not hold up well in Southeast Asian countries and attribute this to differences in topography, fuels, climate, and anthropogenic drivers. In the fourteenth chapter "Analyzing Fire Behavior and Calibrating a Fire Growth Model in a Seasonally Dry Tropical Forest Area", Ruecker et al. demonstrate the utility of a fire behavior model parameterized for seasonal dry tropical forests in Thailand. They suggest that widely used global databases might overestimate the fuel consumption and fire emissions for this forest type.

The second part focuses on *Greenhouse Gas Emissions and Air Pollution* and comprises 12 chapters. Chapters "Spatiotemporally Resolved Pollutant Emissions from Biomass Burning in Asia" and "Twenty-Year (2000–2019) Variations of Aerosol Optical Depth Over Asia in Relation to Anthropogenic and Biomass Burning Emissions" focus on Asia as a whole. In the fifteenth chapter "Spatiotemporally Resolved Pollutant Emissions from Biomass Burning in Asia", Xiong et al. provide spatiotemporally resolved pollutant emissions from biomass burning in Asia. The authors also provide emission estimates for Asia's $PM_{2.5}$, OC, BC, CO, and SO_2 and report substantial spatiotemporal differences in emissions. In the sixteenth chapter "Twenty-Year (2000–2019) Variations of Aerosol Optical Depth Over Asia in Relation to Anthropogenic and Biomass Burning Emissions", Itahashi et al. focus on quantifying twenty-year (2000–2019) variations in Aerosol Optical Depth (AOD) over Asia in relation to anthropogenic and biomass burning emissions. Over East Asia (China, the Republic of Korea, and Japan), authors report a gradual decline in AOD due to the reduction of anthropogenic emissions, mainly in China. Over North and Southeast Asia, AOD showed large inter-annual variability, which the authors attribute to biomass burning emissions. These results highlight anthropogenic and biomass burning emission variations useful for regional air quality and climate change studies in Asia.

Chapters "Light Absorption Properties of Biomass Burning Emissions in Bangladesh: Current State of Knowledge"–"Estimation of Ultrafine Particulate Matter Emissions from Biomass Burning Using Satellite Imaging and Burn Severity" focus on biomass burning emissions in South Asia. Zaman et al. (Chapter "Light Absorption Properties of Biomass Burning Emissions in Bangladesh: Current State of Knowledge") review the light absorption properties of biomass burning emissions in Bangladesh. They report elevated concentrations of trace elements, including the dominance of $PM_{2.5}$ from biomass burning. In the eighteenth chapter "Remote Sensing of Greenhouse Gases and Aerosols from Agricultural Residue Burning Over Pakistan", Tariq et al. report elevated AOD (>2.0) from agricultural residue burning season in addition to high concentrations of CO and CO_2, NO_2, SO_2, and O_3. Finally, a comparative study of energy, emissions, and economic efficiency of various cookstoves in Nepal is provided in the nineteenth chapter "A Comparative Study of Energy, Emissions, and Economic Efficiency of Various Cookstoves in Nepal" by Adhikari et al. Authors conclude that electric stoves as the best solution for conserving energy and, avoiding emissions; however, they also note the lack of a reliable electricity supply in Nepal as a major hindrance for this technology transition. Thus, the authors identify the need for further research on alternate cook stoves in Nepal.

Emissions from Southeast Asia are highlighted in the twentieth chapter "Estimation of Ultrafine Particulate Matter Emissions from Biomass Burning Using Satellite Imaging and Burn Severity" for Thailand, "Characteristics of Transboundary Haze and General Aerosol Over Pulau Pinang, Malaysia" for Malaysia, and "Measurements of Atmospheric Carbon Dioxide Emissions from Fire-Prone Peatlands in Central Kalimantan, Indonesia, Using Ground-Based Instruments"–"Forest Fire Emissions in Equatorial Asia and Their Recent Delay Anomaly in the Dry Season" for Indonesia. Tekasakul et al. (Chapter "Estimation of Ultrafine Particulate Matter Emissions from Biomass Burning Using Satellite Imaging and Burn Severity") identify rice residue burning as the most significant contributor, followed by sugarcane and maize for the ultrafine particles, with most of the burning hotspots in northeastern Thailand. In the twenty-first chapter "Characteristics of Transboundary Haze and General Aerosol Over Pulau Pinang, Malaysia", Lim et al., using the AERONET data from Pulau Pinang, Malaysia, report that the dominant source of pollutants reaching Malaysia is from biomass burning occurring in Indonesia. The authors also identify weather as the most influential parameter impacting transboundary haze in the region.

Chapters "Measurements of Atmospheric Carbon Dioxide Emissions from Fire-Prone Peatlands in Central Kalimantan, Indonesia, Using Ground-Based Instruments"–"Forest Fire Emissions in Equatorial Asia and Their Recent Delay Anomaly in the Dry Season" discuss greenhouse gas (GHG) and air pollutant emissions in Indonesia. In the twenty-second chapter "Measurements of Atmospheric Carbon Dioxide Emissions from Fire-Prone Peatlands in Central Kalimantan, Indonesia, Using Ground-Based Instruments", Ohashi et al. provide ground-based measurements of atmospheric CO_2 emissions from peatlands in Central Kalimantan, Indonesia. They report a mean increment in <ΔXCO2> of 7.8 ppm during the fire season from September to November and also significantly higher XCO_2 values in the north (Palangka Raya) than in the south (Banjarbaru) from peatland burning. In another study, Hayasaka et al. (Chapter "Air Pollution Caused by Peatland Fires in Central Kalimantan") report that peak concentrations of PM_{10}, SO_2, CO, and O_3 reached 1905, 85.8, 38.3, and 1003×10^{-6} gm^{-3}, respectively, when the groundwater level (GWL) was less than -1000 mm in a study of Central

Kalimantan. They also report that underground peat fires occurred when the GWL dropped below 550 mm. In the twenty-fourth chapter "Chemical Speciation of PM_{10} Emissions from Peat Burning Emission in Central Kalimantan, Indonesia", Lestari et al. report elevated PM_{10}, OC, and EC of 4141.5 $\mu g\ m^{-3}$, 2133.2 $\mu g\ m^{-3}$, and 50.9 $\mu g\ m^{-3}$, respectively, in the same study area. Their results also indicate the dominance of organic carbon (OC), ions (Cl-, $SO4^{2-}$, K^+), and metals such as Al and Zn in biomass burning plumes. Finally, in the twenty-fifth chapter "GHG Emissions' Estimation from Peatland Fires in Indonesia—Review and Importance of Combustion Factor", Saharjo highlights the importance of combustion factors while estimating GHG emissions from peatlands in Indonesia and the need for more ground-based measurements, including emission factors and site conditions, for accurate emissions estimation. The role of climate on peatland fire emissions is highlighted in Brasika (Chapter "Forest Fire Emissions in Equatorial Asia and Their Recent Delay Anomaly in the Dry Season"). The author infers that from 2009 to 2021, the peatlands had regained their ability to hold water, so carbon emissions became lower and delayed. However, they point out that more studies are needed for validation.

The third part focuses on *Air Pollution Modeling and Decision Support Systems.* Chapters "Impact of Vegetation Fires on Regional Aerosol Black Carbon Over South and East Asia" and "Detection and Modeling of South Asian Biomass Burning Aerosols from both Macro- and Micro-perspective" focus on Asia as a whole. In the twenty-seventh chapter "Impact of Vegetation Fires on Regional Aerosol Black Carbon Over South and East Asia", Kant et al. report an average increase in fire anomalies by 27.3%, 11.1%, and 12.6%, and AOD by 24.5%, 4.7%, and 7.1% in South, East, and East Asia, respectively. Their results from Potential Source Contribution Function (PSCF) and Concentrated Weighted Trajectories (CWT) suggest that both the local and external sources contribute to pollution in the Indo Gangetic Plains (IGP) and Eastern China region and that the pollutants are strongly influenced by synoptic meteorology. In the twenty-eighth chapter, Shi et al. "Detection and Modeling of South Asian Biomass Burning Aerosols from Both Macro- and Micro-perspective", present both the micro- to macro-perspective of detecting the aerosols emitted by biomass burning in South Asia. They note a decrease in the volume fraction of black carbon and Angstrom Exponent during the 48-hour aging process, but an increase in the Aerosol Sphere Fraction and Single Scattering Albedo. Using the HYSPLIT model, the authors report biomass burning aerosols originating from South Asia, being transported to China, and even reaching far over the Pacific Ocean (including part of the East China Sea and the South China Sea).

Chapters "Remote Sensing of Agricultural Biomass Burning Aerosols, Gaseous Compounds, Long-Distance Transport, and Impact on Air Quality" and "Agricultural Fires in Northeast China: Characteristics, Impacts, and Challenges" discuss biomass burning pollution and modeling in China. Wu et al. (Chapter "Remote Sensing of Agricultural Biomass Burning Aerosols, Gaseous Compounds, Long-Distance Transport, and Impact on Air Quality") present satellite remote sensing of agricultural fires in Northeast and Eastern China. Using the various satellite data and transport model (NRL-NAAPS), authors demonstrate the trans-Pacific transport of smoke plumes to the NE US from the CCNY-lidar and CALIPSO data and supported by the NOAA-HYSPLIT backward trajectories analysis. In the thirtieth chapter "Agricultural Fires in Northeast China: Characteristics, Impacts, and Challenges", Liu and Cheng present results on the impacts of biomass burning during a six-month-long heating season in the Harbin–Changchun (HC) metropolitan area, which is China's only national-level city cluster located in the severe cold climate region. In this region, biomass burning was found to be the major contributor to PM2.5 pollution.

In the thirty-first chapter "Air Pollution Modelling in Southeast Asia—An Overview", Amnuaylojaroen reviews some important air quality models, their potential, and their limitations, including applications in SEA. An air quality modeling case study over northern Thailand is also presented, where forest biomass burning is prevalent. In the thirty-second chapter "Trace Gases and Air Quality in Northwestern Vietnam During Recurrent Biomass Burning on the Indochina Peninsula Since 2014—Field Observations and Atmospheric Simulations", Pieber et al., using the Global Atmosphere Watch (GAW) station Pha Din (PDI) data on trace gases, air quality parameters, and the Copernicus Atmospheric Monitoring Service (CAMS) global reanalysis and FLEXPART model, authors confirm the contribution of elevated biomass burning emissions and aerosols. In the thirty-third chapter "Southeast Asian Transboundary Haze in the Southern Philippines, 2019 and Meteorological Drivers", Santos et al. showcase how transboundary air pollution from SEA biomass burning events influences the haze event experienced in the Southern Philippines using a variety of data such as from AERONET, Philippine Atmospheric, Geophysical, and Astronomical Services Administration (PAGASA) ground-based stations, ECMWF-CAMS and HIMAWARI-8 in combination with HYPLIT modeling and MERRA-2 analysis.

The last two chapters focus on fire decision support systems and prediction. In the thirty-fourth chapter "An Operational Fire Danger Rating System for Thailand and Lower Mekong Region: Development, Utilization, and Experiences", Tanpipat et al. describe the development, utilization, and experiences of implementing a fire danger rating forecast system for Thailand and Upper ASEAN. The DSS system became a part of the daily operation of forest fire, open burning, and smoke haze control and management for 17 northern provinces of Thailand. In addition, the DSS can be modified for other regions for effective fire management. In the final Chapter ("Fires Hotspot Forecasting in Indonesia Using Long Short-Term Memory Algorithm and MODIS Datasets"), Kadir et al. use a long short-term memory (LSTM) algorithm to predict the fire hotspots based on Indonesia's 2010 to 2021 fire data. The authors report the robustness of the LSTM algorithm in successfully predicting 2022 fires with high accuracy.

Overall, the articles in the book cover various topics on fires, and biomass burning pollutants, including emissions modeling. The book will serve as a valuable source of information for remote sensing scientists, geographers, ecologists, atmospheric and environmental scientists, and all who wish to advance their knowledge of vegetation fires and emissions in South/Southeast Asia. We profoundly thank the authors for their contributions and cooperation in bringing this excellent volume to fruition. Also, we duly acknowledge the tremendous effort of the chapter reviewers for valuable guidance and suggestions and Margaret Deignan, Springer, for her help in publishing this book.

We wish you all an exciting read.

Huntsville, AL, USA — Krishna Prasad Vadrevu
Kazo, Saitama, Japan — Toshimasa Ohara
College Park, Maryland, USA — Chris Justice

Contents

Air Pollution Modeling and Decision Support Systems

Editors and Contributors

About the Editors

Dr. Krishna Prasad Vadrevu is a remote sensing scientist at NASA Marshall Space Flight Center, Huntsville, Alabama, USA. His research focuses on land-cover and land-use change (LCLUC) studies, fires, and biomass burning emissions. He has more than 20 years of research experience in satellite remote sensing. He is currently serving as the Deputy Program Manager for the NASA LCLUC Program (lcluc.umd.edu) and leading the South/Southeast Research Initiative (www.sari.umd.edu).

Dr. Toshimasa Ohara is a scientist and research director at Center for Environmental Science (CESS) in Saitama, Japan. He has 33 years of research experience in air quality modeling, emission inventories, and pollution research. He is a lead developer for Regional Emission Inventory in Asia (REAS) and one of the highly cited researchers on the emissions. He is currently working on linking top-down and bottom-up approaches for emissions quantification from different sectors in Asia.

Dr. Chris Justice is a professor at the Department of Geographical Sciences, University of Maryland, College Park, USA. He has 40 years of research experience. His current research is on land-cover and land-use change and global agricultural monitoring using remote sensing. He is an authority on satellite remote sensing of fires. He serves as Project Scientist for the NASA LCLUC Program, the Land Discipline Lead for the NASA MODIS and the Suomi-NPP VIIRS Science Team. He is Co-Chair of the GEO Global Agricultural Monitoring initiative (GEOGLAM), Chief Scientist for NASA HARVEST, and Chair of the international Global Observations of Forest and Land-Use Dynamics (GOFC-GOLD) program.

Contributors

Narayan P. Adhikari Alternative Energy Promotion Center, Kathmandu, Nepal

Amal Abdullah AlMansour Department of Computer Science, King Abdul Aziz University, Jeddah, Saudi Arabia

Teerachai Amnuaylojaroen School of Energy and Environment, University of Phayao, Phayao, Thailand

Yessy Arvelyna Remote Sensing Technology Center of Japan, Tokyo, Japan

Rajasekhar Balasubramanian Department of Civil and Environmental Engineering, Faculty of Engineering, National University of Singapore, Singapore, Singapore

Krishna B. Bhujel Nepal Energy and Environment Development Services (NEEDS), Kathmandu, Nepal

Sumalika Biswas University of California, Los Angeles, USA

Ida Bagus Mandhara Brasika Department of Marine Science, Udayana University, Bali, Indonesia;
Department of Mathematics and Statistics, The University of Exeter, Exeter, UK

Chatchaya Buaniam Department of National Parks, Wildlife, and Plant Conservation, Ministry of Natural Resources and Environment, Bangkok, Thailand

Rejina M. Byanju Central Department of Environmental Science, Tribhuvan University, Kirtipur, Nepal

Mylene G. Cayetano Institute of Environmental Science and Meteorology, University of the Philippines-Diliman, Quezon City, Philippines

Prakash Chauhan Indian Institute of Remote Sensing (IIRS), ISRO Department of Space, Government of India, Dehradun, India

Jie Chen Key Laboratory of Radiometric Calibration and Validation for Environmental Satellites, National Satellite Meteorological Center (National Center for Space Weather), China Meteorological Administration, Beijing, People's Republic of China;
Innovation Center for FengYun Meteorological Satellite, Beijing, People's Republic of China

Tianhai Cheng State Key Laboratory of Remote Sensing Science, Aerospace Information Research Institute, Chinese Academy of Sciences, Beijing, China

Yuan Cheng State Key Laboratory of Urban Water Resource and Environment, School of Environment, Harbin Institute of Technology, Harbin, China

Bhupendra Das Nepal Energy and Environment Development Services (NEEDS), Kathmandu, Nepal;
Clean Air Asia, Pasig City, Philippines

Hanita Daud Department of Applied Mathematics, Universiti Teknologi Petronas, Perak, Malaysia

Diane Davies Trigg-Davies Consulting Ltd, Malvern, UK;
NASA-GSFC, Science Systems and Applications Inc., Lanham, MD, USA

William J. deGroot Natural Resources Canada, Ottawa, Canada

Racha Dejchanchaiwong Department of Chemical Engineering, Faculty of Engineering, Prince of Songkla University, Songkhla, Thailand

Aditya Eaturu University of Alabama Huntsville, Huntsville, AL, USA

Chris Elvidge Earth Observation Group, Payne Institute for Public Policy, Colorado School of Mines, Golden, CO, USA

Jinlong Fan Key Laboratory of Radiometric Calibration and Validation for Environmental Satellites, National Satellite Meteorological Center (National Center for Space Weather), China Meteorological Administration, Beijing, People's Republic of China;
Innovation Center for FengYun Meteorological Satellite, Beijing, People's Republic of China

Jun Fan Institute of Food Crops, Yunnan Academy of Agricultural Sciences, Kunming, China

Robert D. Field Department of Applied Physics and Applied Mathematics, NASA Goddard Institute for Space Studies, Columbia University, New York, USA

Xingfa Gu State Key Laboratory of Remote Sensing Science, Aerospace Information Research Institute, Chinese Academy of Sciences, Beijing, China

Yong Han School of Atmospheric Sciences, Sun Yat-Sen University, Zhuhai, China

Hiroshi Hayasaka Hokkaido University, Sapporo, Japan

Stephan Henne Empa, Swiss Federal Laboratories for Materials Science and Technology, Dübendorf, Switzerland

Brent N. Holben NASA Goddard Space Flight Center, Greenbelt, MD, USA

Windy Iriana Center for Environmental Studies, Bandung Institute of Technology, Bandung, Indonesia;
Graduate School of Frontier Sciences, The University of Tokyo, Kashiwa, Japan

Keiko Ishi Remote Sensing Technology Center of Japan, Tokyo, Japan

Md Safiqul Islam Department of Chemistry, University of Dhaka, Dhaka, Bangladesh

Syuichi Itahashi Sustainable System Research Laboratory (SSRL), Central Research Institute of Electric Power Industry (CRIEPI), Abiko, Chiba, Japan

Narin Jaroonrattanapak Department of National Parks, Wildlife, and Plant Conservation, Ministry of Natural Resources and Environment, Bangkok, Thailand

Farah Jeba Department of Chemistry, University of Dhaka, Dhaka, Bangladesh

Attachai Jintrawet Center for Agricultural Resources System Research, Faculty of Agriculture, Chiang Mai University, Chiang Mai, Thailand;
Institute of Food Crops, Yunnan Academy of Agricultural Sciences, Kunming, China

Chris Justice University of Maryland College Park, College Park, USA

Febri Juwita Faculty of Civil and Environmental Engineering, Bandung Institute of Technology, Bandung, Indonesia

Evizal Abdul Kadir Department of Informatics Engineering, Universitas Islam Riau, Pekanbaru, Indonesia;
Department of Computer Science, Harvard University, Cambridge, MA, USA

Yogesh Kant Indian Institute of Remote Sensing (IIRS), ISRO Department of Space, Government of India, Dehradun, India

Masahiro Kawasaki Research Institute for Humanity and Nature, Kyoto, Japan

Bounthanh Keoboualapha Upland Agriculture Research Center, National Agriculture and Forestry Research Institution, Luang Prabang, Laos

Aulia Nisa'ul Khoir Centre of Tropical Climate Change System, Institute of Climate Change, Universiti Kebangsaan Malaysia, Bangi, Selangor, Malaysia;
Centre for Applied Climate Services, Indonesia Agency for Meteorology, Climatology and Geophysics, Kemayoran, Jakarta, Indonesia

Suprapat Khuenjai Center for Agricultural Resources System Research, Faculty of Agriculture, Chiang Mai University, Chiang Mai, Thailand

Osamu Kozan Center for Southeast Asian Studies, Kyoto University, Kyoto, Japan; Research Institute for Humanity and Nature, Kyoto, Japan

Ajantha Sisira Kumara Department of Public Administration, Faculty of Management Studies and Commerce, University of Sri Jayewardenepura, Nugegoda, Sri Lanka

Hsiang Tsung Kung Department of Computer Science, Harvard University, Cambridge, MA, USA

Junichi Kurokawa Asia Center for Air Pollution Research (ACAP), Niigata, Niigata, Japan

Puji Lestari Faculty of Civil and Environmental Engineering, Bandung Institute of Technology, Bandung, Indonesia

Xiaoyang Li State Key Laboratory of Remote Sensing Science, Aerospace Information Research Institute, Chinese Academy of Sciences, Beijing, China;
University of Chinese Academy of Sciences, Beijing, China

Yajun Li Key Laboratory of Radiometric Calibration and Validation for Environmental Satellites, National Satellite Meteorological Center (National Center for Space Weather), China Meteorological Administration, Beijing, People's Republic of China

Cheng Liu Key Laboratory of Radiometric Calibration and Validation for Environmental Satellites, National Satellite Meteorological Center (National Center for Space Weather), China Meteorological Administration, Beijing, People's Republic of China;
Innovation Center for FengYun Meteorological Satellite, Beijing, People's Republic of China

Jiumeng Liu State Key Laboratory of Urban Water Resource and Environment, School of Environment, Harbin Institute of Technology, Harbin, China

Pong Loungmoon Highland Research and Training Center, Faculty of Agriculture, Chiang Mai University, Chiang Mai, Thailand

Bijaya Maharjan Nepal Energy and Environment Development Services (NEEDS), Kathmandu, Nepal

Ezekiel Kaura Makama Department of Physics, University of Jos, Jos, Nigeria

Rachane Malinee Air Pollution and Health Effect Research Center, Prince of Songkla University, Songkhla, Thailand;
Energy Technology Program, Department of Specialized Engineering, Faculty of Engineering, Prince of Songkla University, Songkhla, Thailand

Kasemsan Manomaiphiboon The Joint Graduate School of Energy and Environment, King Mongkut's University of Technology Thonburi, Bangkok, Thailand

Ajay Mathema School of Environmental Science and Management (SchEMS), Pokhara University, Lekhnath, Nepal

Jessica L. McCarty Department of Geography and Geospatial Analysis Center, Miami University, Oxford, OH, USA

Usman Mehmood Remote Sensing, GIS and Climatic Research Lab (National Center of GIS and Space Applications), Centre for Remote Sensing, University of the Punjab, Lahore, Pakistan;
Department of Political Science, University of Management and Technology, Lahore, Pakistan

Yatai Men College of Urban and Environmental Sciences, Peking University, Beijing, China

Debashis Mitra Indian Institute of Remote Sensing (IIRS), ISRO Department of Space, Government of India, Dehradun, India

John Morris School of Industrial Education and Technology, King Mongkut's Institute of Technology Ladkrabang, Bangkok, Thailand

Fred Moshary Optical Remote Sensing Lab, The City College of New York, NY, USA

Nur Nazmi Liyana Binti Mohd Napi Centre of Tropical Climate Change System, Institute of Climate Change, Universiti Kebangsaan Malaysia, Bangi, Selangor, Malaysia

Arbi Haza Nasution Department of Informatics Engineering, Universitas Islam Riau, Pekanbaru, Indonesia

Aryan Natwariya Indian Institute of Remote Sensing (IIRS), ISRO Department of Space, Government of India, Dehradun, India

Hasan Nawaz Remote Sensing, GIS and Climatic Research Lab (National Center of GIS and Space Applications), Centre for Remote Sensing, University of the Punjab, Lahore, Pakistan;
Centre for Atmospheric Chemistry, School of Earth, Atmospheric and Life Sciences, University of Wollongong, Wollongong, Australia

Dhurba Neupane University of Nevada, Reno, USA

Dac-Loc Nguyen Institut Für Umweltmedizin, Helmholtz Zentrum München, Munich, Germany;
Chair of Analytical Chemistry, University of Rostock, Rostock, Germany

Nhat Anh Nguyen Hydro-meteorological Observation Center, Vietnam Meteorological and Hydrological Administration, Ministry of Natural Resources and Environment, Ha Noi, Vietnam

Prayoonyong Nhuchaiya Department of National Parks, Wildlife, and Plant Conservation, Ministry of Natural Resources and Environment, Bangkok, Thailand

Prisco D. Nilo Institute of Environmental Science and Meteorology, University of the Philippines-Diliman, Quezon City, Philippines

Narissara Nuthammachot Faculty of Environmental Management, Prince of Songkla University, Songkhla, Thailand

Toshimasa Ohara Center for Environmental Science in Saitama (CESS), Kazo, Saitama, Japan

Masafumi Ohashi Graduate School of Science and Engineering, Kagoshima University, Kagoshima, Japan

Thaworn Onpraphai Department of Highland Agriculture and Natural Resources, Faculty of Agriculture, Chiang Mai University, Chiang Mai, Thailand

Maggie Chel Gee Ooi Centre of Tropical Climate Change System, Institute of Climate Change, Universiti Kebangsaan Malaysia, Bangi, Selangor, Malaysia

Mahmod Othman Department of Applied Mathematics, Universiti Teknologi Petronas, Perak, Malaysia

Asankha Pallegedara Department of Industrial Management, Faculty of Applied Sciences, Wayamba University of Sri Lanka, Kuliyapitiya, Sri Lanka

Chittana Phompila Faculty of Forest Sciences (FFS), National University of Laos (NUoL), Vientiane, Laos

Simone M. Pieber Empa, Swiss Federal Laboratories for Materials Science and Technology, Dübendorf, Switzerland;
AirUCI (Atmospheric Integrated Research), University of California, Irvine, USA

Suyesh Prajapati MinErgy Private Limited, Kathmandu, Nepal

Prayoto Pranoto Riau Environmental and Forestry Office, Pekanbaru, Indonesia

Siva Praveen Puppala NSW Department of Planning and Environment, Lidcombe, Australia

Jeffrey S. Reid Marine Meteorology Division, Naval Research Laboratory, Monterey, CA, USA

Sri Listia Rosa Department of Informatics Engineering, Universitas Islam Riau, Pekanbaru, Indonesia

Shatabdi Roy Department of Chemistry, University of Dhaka, Dhaka, Bangladesh

Gernot Ruecker ZEBRIS Geo-IT GmbH, Munich, Germany

Bambang Hero Saharjo Forest Fire Laboratory, Division of Forest Protection, Department of Silviculture, Faculty of Forestry and Environment, IPB University, Bogor, Indonesia

Abdus Salam Department of Chemistry, University of Dhaka, Dhaka, Bangladesh

Pranith Salikineedi University of Alabama, Huntsville, USA

Lim Hwee San School of Physics, Universiti Sains Malaysia, Pulau Pinang, Malaysia

Krishna E. Santos Institute of Environmental Science and Meteorology, University of the Philippines-Diliman, Quezon City, Philippines

Wilfrid Schroeder NOAA-NESDIS-OSPO Satellite Analysis Branch, College Park, MD, USA

Prajwal R. Shakya Institute of Engineering, Tribhuwan University, Kathmandu, Nepal

Guofeng Shen College of Urban and Environmental Sciences, Peking University, Beijing, China

Shuaiyi Shi State Key Laboratory of Remote Sensing Science, Aerospace Information Research Institute, Chinese Academy of Sciences, Beijing, China

Shubha Laxmi Shrestha Alternative Energy Promotion Center, Kathmandu, Nepal

Angkana Somsak Center for Agricultural Resources System Research, Faculty of Agriculture, Chiang Mai University, Chiang Mai, Thailand

Thav Sopheak Royal University of Agriculture, Phnom Penh, Cambodia

Martin Steinbacher Empa, Swiss Federal Laboratories for Materials Science and Technology, Dübendorf, Switzerland

Veerachai Tanpipat Upper ASEAN Wildland Fire Special Research Unit, Forestry Research Center, Faculty of Forestry, Kasetsart University, Bangkok, Thailand

Salman Tariq Remote Sensing, GIS and Climatic Research Lab (National Center of GIS and Space Applications), Centre for Remote Sensing, University of the Punjab, Lahore, Pakistan;
Department of Space Science, University of the Punjab, Lahore, Pakistan

Perapong Tekasakul Air Pollution and Health Effect Research Center, Prince of Songkla University, Songkhla, Thailand;
Department of Mechanical and Mechatronics Engineering, Faculty of Engineering, Prince of Songkla University, Songkhla, Thailand

Kenichi Tonokura Graduate School of Frontier Sciences, The University of Tokyo, Kashiwa, Japan

Zia Ul-Haq Remote Sensing, GIS and Climatic Research Lab (National Center of GIS and Space Applications), Centre for Remote Sensing, University of the Punjab, Lahore, Pakistan

Aswin Usup Palangka Raya University, Palangka Raya, Indonesia

Isna Utami Faculty of Civil and Environmental Engineering, Bandung Institute of Technology, Bandung, Indonesia

Krishna Prasad Vadrevu NASA Marshall Space Flight Center, Huntsville, AL, USA

Siheng Wang Institute of Remote Sensing Satellite, China Academy of Space Technology, Beijing, China

Yuyang Wang Institute of Remote Sensing Satellite, China Academy of Space Technology, Beijing, China

Kobsak Wanthongchai Upper ASEAN Wildland Fire Special Research Unit, Faculty of Forestry, Forestry Research Center, Kasetsart University, Bangkok, Thailand

Yonghua Wu Optical Remote Sensing Lab, The City College of New York, NY, USA

Yu Wu State Key Laboratory of Remote Sensing Science, Aerospace Information Research Institute, Chinese Academy of Sciences, Beijing, China

Rui Xiong College of Urban and Environmental Sciences, Peking University, Beijing, China

Jittisak Yodcum Royal Forest Department, Ministry of Natural Resources and Environment, Bangkok, Thailand

Mohamad Farid Izzat Bin Zahari School of Physics, Universiti Sains Malaysia, Pulau Pinang, Malaysia

Shahid Uz Zaman Department of Chemistry, University of Dhaka, Dhaka, Bangladesh;
Department of Chemistry, Faculty of Science, Bangladesh University of Engineering and Technology, Dhaka, Bangladesh

Wei Zheng Key Laboratory of Radiometric Calibration and Validation for Environmental Satellites, National Satellite Meteorological Center (National Center for Space Weather), China Meteorological Administration, Beijing, People's Republic of China;
Innovation Center for FengYun Meteorological Satellite, Beijing, People's Republic of China

Vegetation Fires and Biomass Burning in South/Southeast Asia—An Overview

Krishna Prasad Vadrevu, Toshimasa Ohara, and Chris Justice

Abstract Vegetation fires are important emission sources of greenhouse gases, pollutants, and aerosols in many regions, including South/Southeast Asia (S/SEA). Fire frequency, intensity, and emissions are driven by climate and land use practices. The air pollutants released from fires can circulate regionally and get deposited far from their sources, as in the case of the Indonesian fire emissions reaching Singapore, Malaysia, Brunei, and southern Thailand. Due to such significant transboundary nature of pollutants impacting air quality and health, vegetation fires are attracting international attention. Managing fires requires synoptic information at multiple scales on where fires occur, their spatial and temporal characteristics, including drivers, and impacts. This introductory chapter provides an overview of the satellite-derived fire statistics in S/SEA and some critical needs and priorities for fire science and management and related pollution mitigation. In this study, we specifically discuss three crucial aspects: mapping, monitoring, and modeling of vegetation fires, integrating top-down and bottom-up approaches, effective emissions monitoring, and impacts assessment, including the need for the development of robust decision support systems (DSS) useful for fire management, mitigation, and air pollution control.

Keywords Vegetation fires · Satellite data · Greenhouse gas (GHG) emissions · Decision support systems (DSS) · South/Southeast Asia · Air pollution

K. P. Vadrevu (✉)
NASA Marshall Space Flight Center, Huntsville, AL, USA
e-mail: krishna.p.vadrevu@nasa.gov

T. Ohara
Center for Environmental Science in Saitama, Kazo, Japan

C. Justice
University of Maryland College Park, College Park, USA

K. P. Vadrevu et al. (eds.), *Vegetation Fires and Pollution in Asia*,
https://doi.org/10.1007/978-3-031-29916-2_1

1 Introduction

Vegetation fires are common in several regions, including South/Southeast Asian (S/SEA) countries (Fig. 1). Fire occurrence and spread are governed by fuel type, topography, climate, weather, lightning, etc. In addition to these natural factors, most of the fires in S/SEA are human-initiated associated with land use (Vadrevu et al. 2019). Recent changes in fire regimes, i.e., frequency, intensity, duration, and extent, promote extensive habitat degradation leading to social, economic, and environmental concerns. The drivers of fires vary in S/SEA, both anthropogenic and climatic. For example, fire is often used as a land-clearing tool through slash-and-burn agriculture in several countries of S/SEA, such as Nepal (Mukul and Byg 2020), India (Sharma et al. 2022), Bangladesh (Chowdhury et al. 2022), Myanmar (Biswas et al. 2015a, b), Malaysia (de Neergaard et al. 2008), Philippines (Gabriel et al. 2020), Indonesia (Murdjoko et al. 2022) Thailand (Arunrat et al. 2022), Laos (Phompila et al. 2022), Cambodia (Mellac 2021), and Vietnam (Trang et al. 2022). More recently, fires have been extensively used for clearing land for rubber and oil palm expansion for example in Indonesia (Dhandapani and Evers 2020). Also, most of the countries in S/SEA are agrarian, where farmers use fire to burn crop residues after harves to plant the next crop (Vadrevu and Lasko 2018). These biomass burning practices alter landscape structure and function at various spatial scales. For example, slash and burn can result in the loss of forests and biodiversity, disrupting biogeochemical cycles (Bruun et al. 2009). After slash and burn, secondary vegetation can result in mixed landscape patches with invasive species (Palm et al. 2005). In the case of cropland fires, rather than burning, incorporating crop residues after harvest into the soil can benefit soil health (Magdoff 2018). The burning of biomass can release significant amounts of greenhouse gases, pollutants and aerosols such as CO_2, CO, NO_x, CH_4, non-methane hydrocarbons, and other chemical species, which can impact the radiative budget, air quality, and health at both local and regional scales (Crutzen and Andreae 1990; Penner et al. 1992; Sigsgaard et al. 2015; Gupta et al. 2001). In addition, biomass burning pollutants can be transported over long distances impacting regional climate. Thus, quantifying vegetation fires and biomass burning impacts for different regions of the world, including S/SEA countries, gains significance. In this study, we provide an overview of the satellite remote sensing of fire-related statistics in S/SEA and briefly highlight some of the critical needs and priorities to address the region's vegetation fires and biomass burning issues.

2 Mapping, Monitoring, and Modeling of Vegetation Fires

Three important fire products are generated using satellite remote sensing data: (A) active fire products that rely on middle infrared and thermal infrared (usually around 3.7–11 mm) satellite bands (Dozier 1981) for detecting fires (Giglio et al. 2009; Schroeder et al. 2014); (B) burned area products that rely on visible and short-wave

Fig. 1 Fires and thermal anomalies (Day and Night, 375 m) Suomi NPP/VIIRS data, April 2, 2022

infrared wavelengths for detecting changes in land surface reflectance after burning (Roy et al. 2005; Giglio et al. 2009); and (C) fire radiative power (FRP) products that represent instantaneous fire radiative power measured from middle infrared satellite bands, to characterize the intensity and which can be further used to estimate the amount of biomass burned (Wooster et al. 2005). Satellite fire characteristics derived from these products in South/Southeast Asia are given below:

The averaged annual fire counts obtained from the VIIRS 375 m resolution active fire product for South and Southeast Asia from 2012 to 2019 are shown in Fig. 2a, b, respectively. The VIIRS algorithm builds on the well-established MODIS Fire and Thermal Anomalies product, using a contextual approach to detect fires (Schroeder et al. 2014). Due to its higher spatial resolution, the VIIRS 375 m active fire product captures more fire pixels than the MODIS MCDML 1 km product. In South Asia, India had the highest number of fires, with more than 350,000 fire counts per year, followed by Pakistan (37,513), Nepal (19,259), with the least occurring in Afghanistan (714) (Fig. 2a). In the case of Southeast Asia, Myanmar had the most fires (348074), followed by Indonesia (275064), Cambodia (177631), Thailand (168250), Laos (150051), etc., with the least in East Timor (4473) (Fig. 2b).

Figures 3 and 4 depicts the spatial/temporal variation in fires for different years, including seasonality for South/Southeast Asian countries, respectively. For example, the peak fire season for the South Asian countries of India, Nepal, Bhutan, and Bangladesh runs from March to May. In contrast, in Pakistan, it is from April to June, in Afghanistan from June to August, and in Sri Lanka from July to September. In

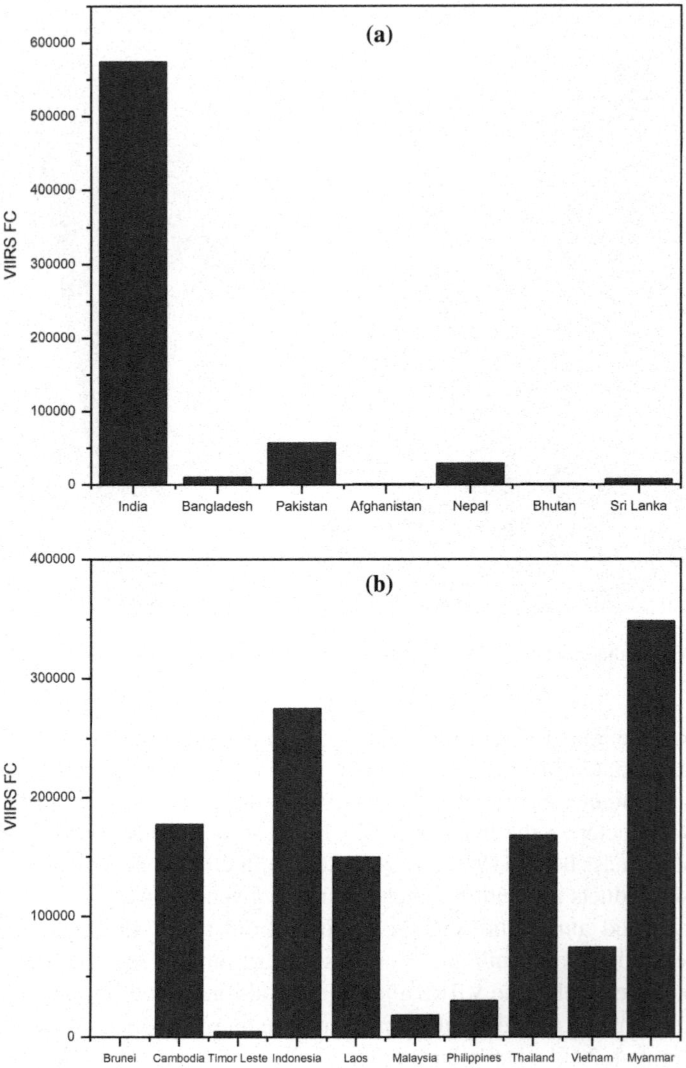

Fig. 2 VIIRS (375 m) active fire counts data per year (averaged from 2012 to 2022) for South Asia (**a**) and Southeast Asian (**b**) countries

Southeast Asia, Myanmar, Laos, Vietnam, Thailand, Cambodia, and the Philippines had a peak fire season from February to April, whereas in Malaysia, the fire season is from July to September, in Indonesia from August to October, and in East Timor from September to November. The peak fire season in S/SEA coincides with the dry season timing associated with the monsoon cycles.

The burnt area temporal variations derived from the MODIS (MCD64A1) product for South and Southeast Asian countries are shown in Fig. 5a, b. The burnt area

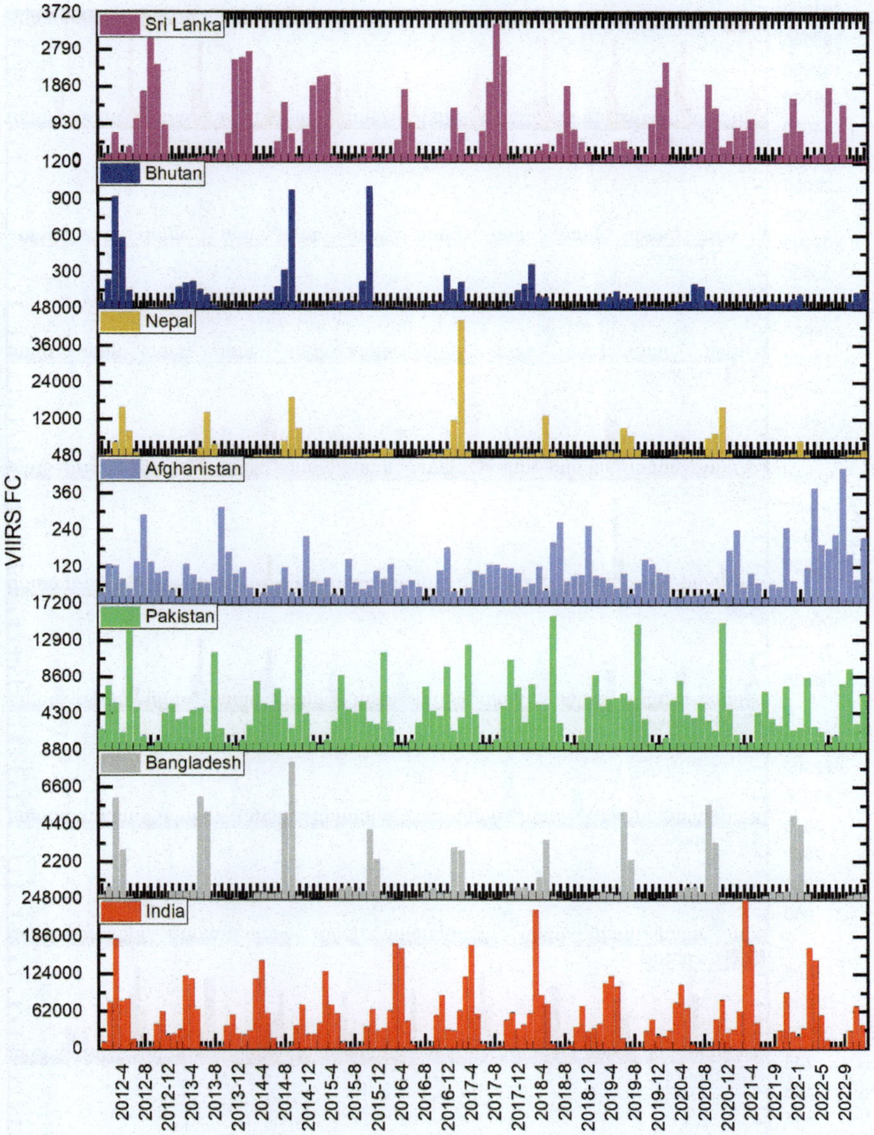

Fig. 3 Vegetation fires (VIIRS Day and Night, 375 m) Suomi NPP/VIIRS data, 2012–2020 for different South Asian countries

detection is based on an automated method using 500 m MODIS imagery and 1 km active fire observations. A hybrid approach is used in several steps for mapping post-fire burned areas. A time series of daily band 1 (0.620–0.670 μm), 5 (1.230–1.250 μm), and 7 (2.105–2.155 μm) reflectances for each pixel are used to compute a daily burn-sensitive vegetation index (VI = (B5 − B7)/(B5 + B7)), to detect a

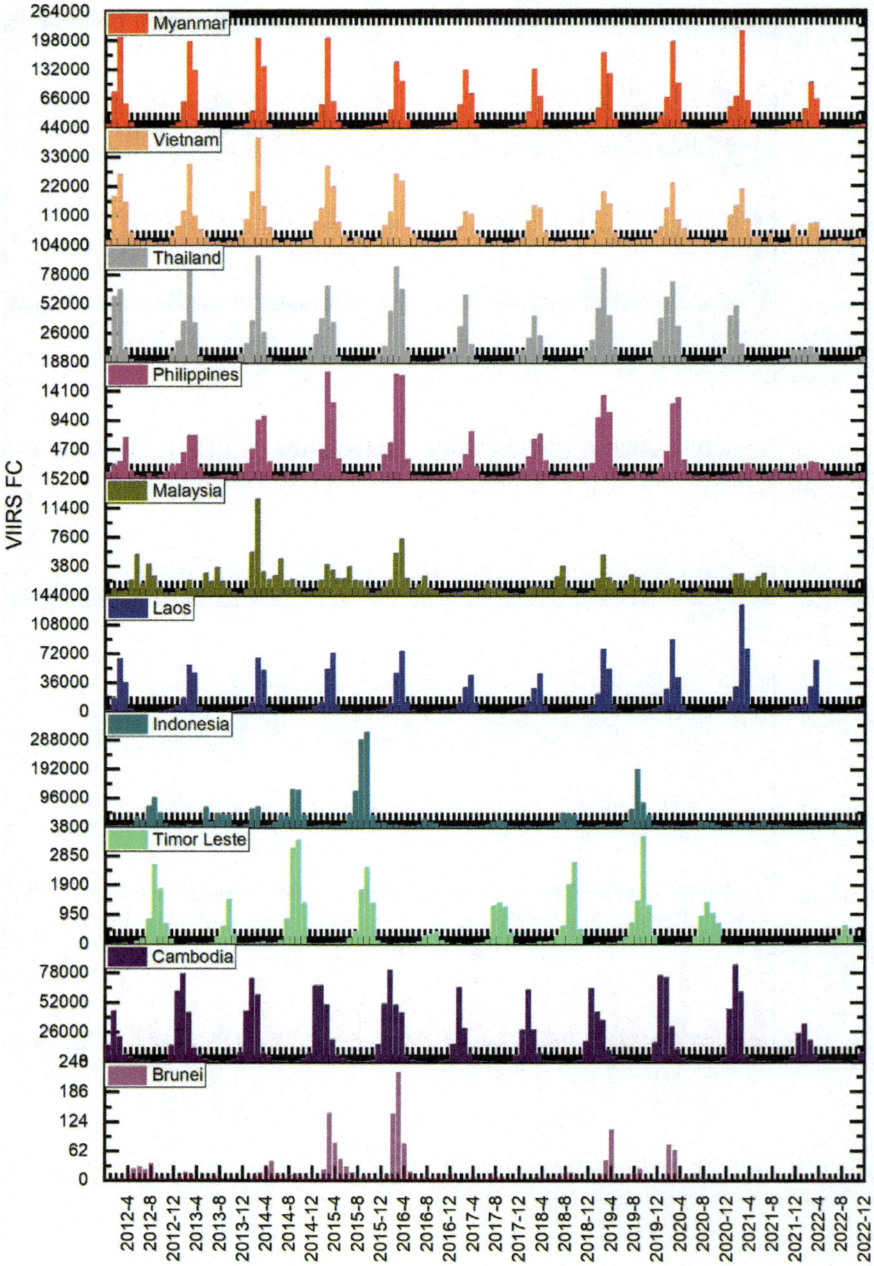

Fig. 4 Vegetation fires (VIIRS Day and Night, 375 m) Suomi NPP/VIIRS data, 2012–2020 for different Southeast Asian countries

decrease in the Vegetation Index signal to detect burned areas. A mask with active fires also sorts burned/unburned pixels. The MODIS-derived annual burnt areas (km^2) averaged from 2012 to 2022 suggested India with the highest per year (39,079.0 km^2) followed by Pakistan (1819 km^2), Nepal (1706 km^2), and least for Brunei (1.78 km^2). In Southeast Asia, Myanmar had the highest burned area (32,764 km^2), followed by Cambodia (25,551.0 km^2), Thailand (14,424.13), Indonesia (10,178.7 km^2), and least in Timor Leste (175.83) (Fig. 5b).

Fire radiative power (FRP), which is a measure of intensity derived from the VIIRS data (VNP14IMG) for the entire S/SEA at 0.5 degree (30-minute intervals), is shown in Fig. 6. The FRP is calculated using a combination of both VIIRS 375 m and 750 m data. The former is used to identify fire-affected, cloud, water, and valid background pixels, then the co-located M13 channel radiance data at 750 m coinciding with fire pixels is used for estimating the FRP. Valid background pixels are also used in the FRP calculation. More details about the product can be found in Schroeder et al. (2014). The FRP values in S/SEA vary from 0 to 56.2 (MW) per 0.5-degree grid cells. The highest values can be seen in Northeast India, Myanmar, Laos, and Kalimantan in Indonesia.

In our earlier papers, we discussed important needs and priorities of fire research and applications in S/SEA useful for fire mitigation and management (Vadrevu and Justice 2011; Vadrevu et al. 2014, 2017). In this paper, we provide additional new points. We broadly infer the following as priorities for fire mapping, monitoring, and modeling studies: (a) calibration and validation of satellite-derived fire datasets; (b) harmonized multi-sensor fire datasets that integrate polar and geostationary data and at a higher resolution for effective fire monitoring and characterizing fire regimes; (c) meeting the requirements of the Global Climate Observing System (GCOS) World Meteorological Organization (WMO) Fire Essential Climate Variables (ECV) for active fires, burnt areas, and FRP products (GCOS-WMO 2022); (d) quantifying and characterizing small fires from crop residue burning; (e) quantifying the important drivers of fires and making the linkages with policy, for effective mitigation; (f) quantifying the impacts of low-intensity fires on biodiversity and ecosystem functions; (g) monitoring smoldering fires such as from peatlands using latest remote sensing techniques and novel algorithms; (h) integrating satellite data and ground-based measurements for characterizing fire behavior modeling input parameters (fuel loads, vegetation type, canopy height, topography, meteorology, soil and fuel moisture, and other site conditions); (i) using the latest artificial intelligence and deep learning techniques for fire behavior modeling and prediction; and (j) building decision support systems (DSS) useful for mapping, monitoring, and quantifying the impacts of fires on the environment. Some of the above goals are implemented as a part of Global Observations of Forest and Land Use Dynamics (GOFC-GOLD) and Global Wildfire Information Systems (GWIS) activities (gofcgold.org; https://gwis.jrc.ec.europa.eu/).

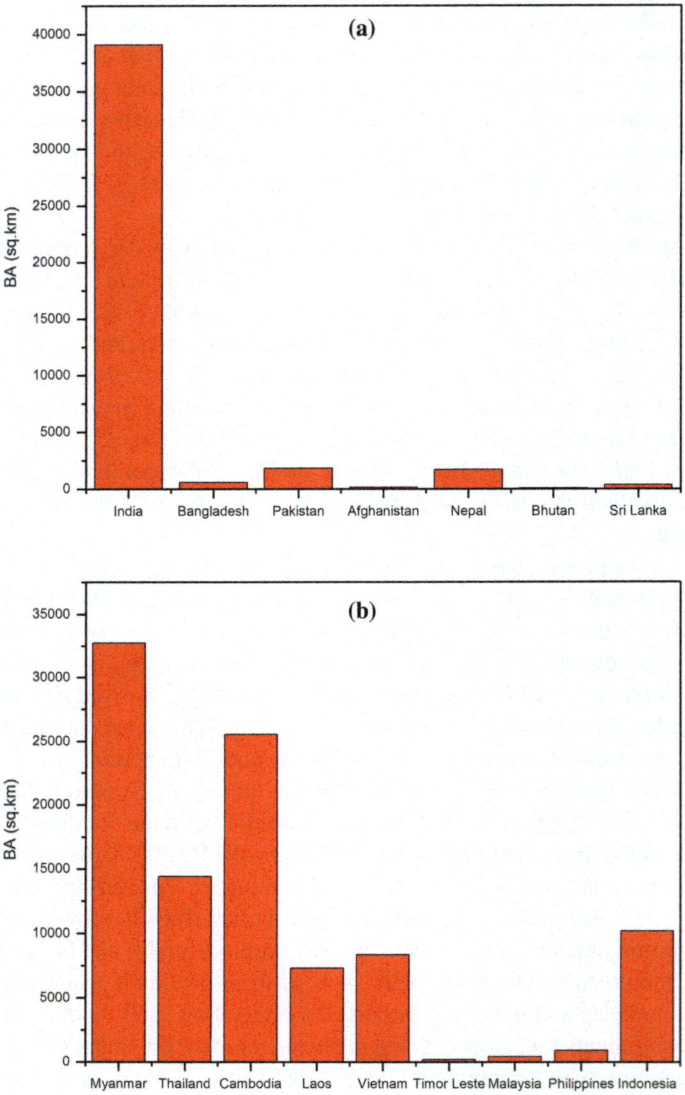

Fig. 5 MODIS (MCD64A1) averaged (2002–2022) annual burnt areas (km^2) in South Asia (**a**) and Southeast Asian countries (**b**)

3 GHG Emissions and Air Pollution

Due to the devastating fire impacts worldwide, decision makers are interested in understanding the risks of increasing emissions and the opportunities for mitigation. In particular, vegetation fires are one of the largest challenges for emissions accounting. A major reason is that the fires exhibit considerable spatial and temporal

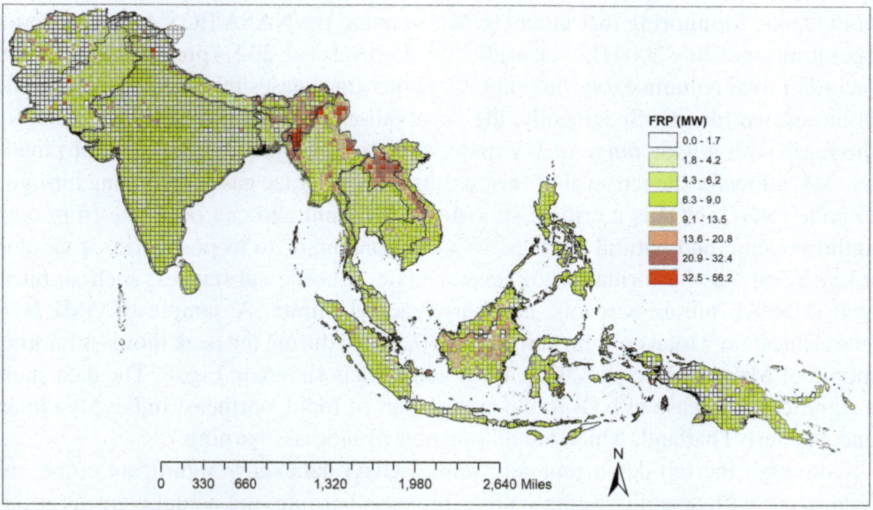

Fig. 6 VIIRS fire radiative power (FRP in MW) for South/Southeast Asian countries. The data has been averaged from 2012 to 2022 to retrieve annual FRP at 0.5 degree grid cells

variability, including drivers. The inputs to calculate biomass burning emissions have significant uncertainties, such as the combustion efficiency, emission factors, and trace gases produced in these fires. It is still unclear how these factors govern the emissions due to changing meteorology and on-site conditions. Although the measurements made by satellites or aircrafts can be used to estimate areas burned or ratios of specific emissions, ground-based measurements are needed to estimate the amount of fuel consumed, the fuel consumption rate, heat release, and the amount of carbon released. Rapid progress has been made concerning quantifying emissions in general. For example, now a variety of satellite sensors measure GHG's such as water vapor, methane, carbon dioxide, nitrous oxide, and ozone, using different wavelengths and passive and active sensors. New instruments (e.g., TROPOMI, OCO-2 and 3, GOSAT) and other related sensors are providing excellent data to advance the mapping of greenhouse gas emissions. These datasets can be related to fire events to characterize emissions directly.

For example, the Ozone Mapping and Profiler Suite (OMPS) onboard the joint NASA/NOAA Suomi National Polar-orbiting Partnership (Suomi NPP) satellite, launched in October 2011, delivers several products such as total column ozone, total column sulfur dioxide (SO_2), vertical ozone profile swath, and aerosol index (AI). A sample of single-date data for the AI product for the peak biomass burning month of March (March 27, 2021), for the entire South/Southeast Asia, is shown in Fig. 7. AI indicates the presence of ultraviolet (UV)-absorbing particles in the air (aerosols), such as from black carbon and desert dust. The black carbon source is mainly from biomass burning. The higher the AI, the more polluted the aerosols in the atmospheric environment. In Fig. 7, a significant enhancement of AI from biomass burning aerosols in northeast India and Myanmar can be seen. The Dutch Finnish built Ozone Monitoring Instrument (OMI) onboard the NASA EOS Aura spacecraft operating since July 2004 (Levelt et al. 2018; Lamsal et al. 2021) provides a long-term record of total column ozone and monitors other trace gases relevant to tropospheric pollution worldwide. Specifically, the observations of sunlight backscattered from the Earth over a wide range of UV and visible wavelengths ($\sim$ 260–500 nm) made by OMI allow for the retrieval of various atmospheric trace gases, including nitrogen dioxide (NO_2). NO_2 is a critical short-lived pollutant and can originate from both anthropogenic and natural sources. NO_2 is a precursor to tropospheric ozone and a key agent for the formation of several toxic airborne substances, such as nitric acid (HNO_3), nitrate aerosols, and peroxyacetyl nitrate. A sample of OMI NO_2 (molecules/cm^2) total column daily data at 0.25 deg during the peak biomass burning month of March (March 8, 2022), for South Asia is shown in Fig. 8. The data show a significant increase of NO_2 in the eastern part of India, northeast India, Myanmar, and northern Thailand, which are all hotspots of biomass burning.

Although the top-down remotely sensed GHG data show significant emissions coinciding with specific events such as biomass burning, the actual concentrations can be mixed due to atmospheric diffusion and transport. For example, biomass burning aerosols and pollutants in South Asia often get mixed with dust aerosols. How much of the total pollutants are from dust versus biomass burning is unclear. The atmosphere mixes and integrates with other surface fluxes that can vary spatially and temporally. The top-down satellite retrievals rely on observations of atmospheric gas concentrations, and it is not easy to separate mixed concentrations. In contrast,

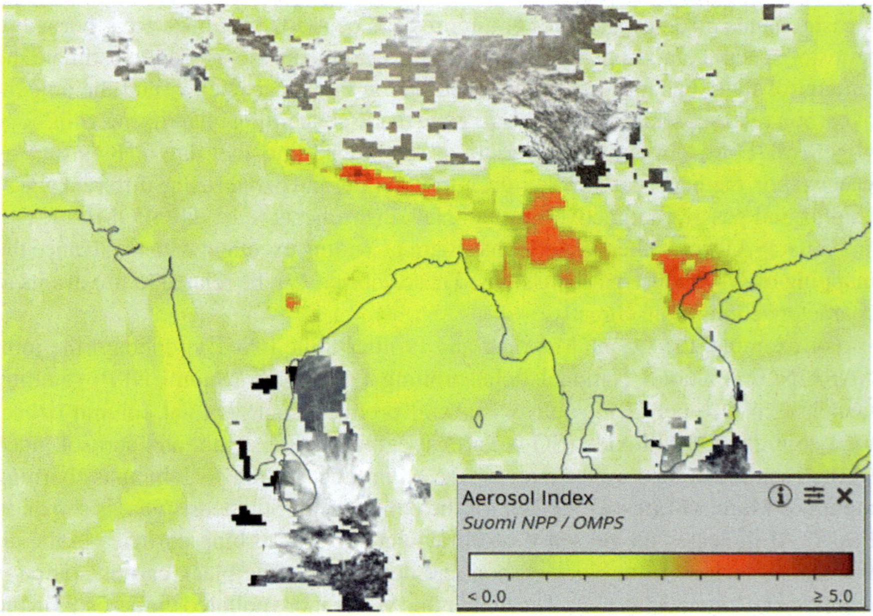

Fig. 7 Suomi NPP/OMPS aerosol index, March 27th, 2021, South/Southeast Asia

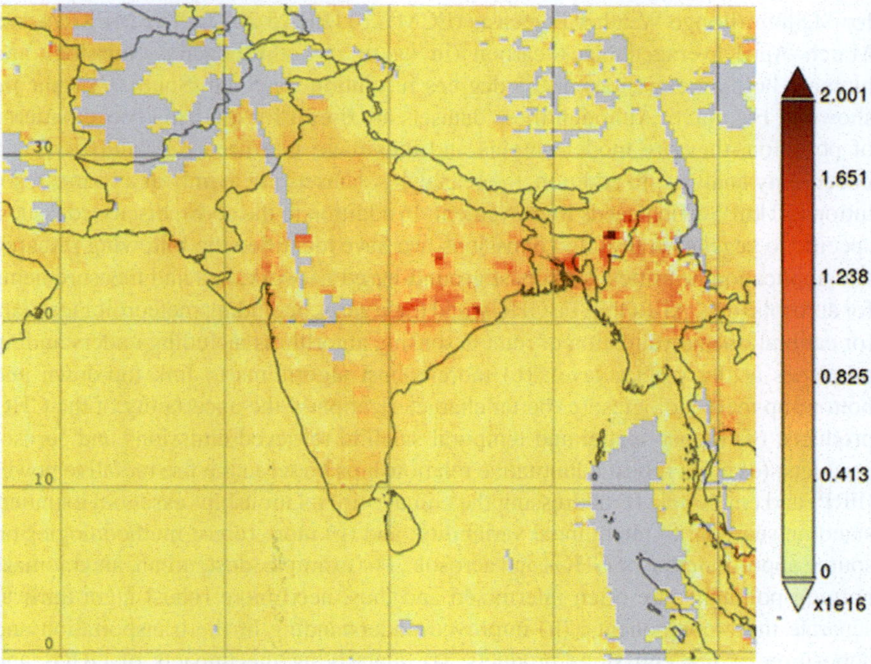

Fig. 8 OMI NO_2 (OMI OM02d v003, molecules/cm^2) total column daily data at 0.25 deg, March 8th, 2022 in South Asia

bottom-up accounting relies on point-based observations of GHGs or aerosols or net fluxes at the Earth's surface and infers the changes in the atmosphere. The key challenge with point-based measurements lies in spatial and temporal upscaling of short-term point observations to estimate large-scale emissions and quantifying the uncertainty associated with the upscaling. Similar issues arise in the case of top-down approaches while downscaling the emissions to specific sources. Thus, it is crucial and necessary to reconcile the top-down and bottom-up approaches and integrate them robustly to resolve any differences and uncertainties in GHG emissions (Guevara et al. 2017; Kondo et al. 2020; Elguindi et al. 2020).

Atmospheric inversion methodologies have proven useful in linking top-down approaches with bottom-up inventory data (Kurokawa et al. 2009; Itahashi et al. 2019). Inversions allow the mismatch between the modeled and observed concentrations to be minimized allowing measurement and model errors to be accounted for (Müller and Stavrakou 2005). In addition, the reanalysis datasets extend over several decades and cover the entire globe from the Earth's surface to well above the stratosphere. The reanalysis products integrate a variety of numerical models, satellite, ground-based, airborne data, etc., useful for understanding climate, weather, and pollution phenomena. One of the best examples of reanalysis products is from the Modern-Era Retrospective analysis for Research and Application (MERRA-2) (MERRA 2022) and the European ReAnalysis (ERA5) from the European Centre for Medium-Range Weather Forecasts (ECMWF) (ERA5 2022). The MERRA data March–April averaged black carbon (in kg m^{-2}) anthropogenic emissions and biomass burning emissions at 0.5 degrees resolution for South/Southeast Asia are shown in Fig. 9a, b. Although these datasets are useful for inferring broad patterns of pollution, they are model outputs and thus there is a need for validation, and uncertainty analysis. In addition, the reanalysis datasets are mainly at a coarse resolution (50 km), which needs improvement. In addition to the above discussed issues, specific to vegetation fires and emissions, we have identified the following research and application priorities: (a) robust ground-based network data and measurements for aerosols (e.g., AERONET), GHGs, and trace gases, including meteorological data for calibration and validation of remote sensing algorithms including understanding processes at the surface level; (b) more robust algorithms to link top-down and bottom-up methodologies; (c) better characterization of the uncertainty of the GHG products; (d) higher spatial and temporal satellite retrieved emissions and aerosol products; (e) more robust quantitative relationships between the fire radiative power (FRP) and emissions; (f) addressing the climate-fire relationships essential to understanding seasonal to interannual variability; and (g) more robust methodologies on source apportionment of GHGs and aerosols. For example, dust, urban, and biomass burning pollutants are often intermixed and thus, need more robust algorithms to separate these components; (h) improved understanding of the transportation and deposition of fire emissions products; (i) quantifying the impacts of GHGs and aerosols on the environment and human health; (j) focused international regional biomass burning field campaigns in South/Southeast Asia to address pollution and climate impacts; and (k) strengthening remote sensing and atmospheric modeling community in South/Southeast Asia through capacity building and training.

4 Fire Prediction, Air Pollution Modeling, and Decision Support Systems (DSS)

Specific to fire prediction, an accurate estimation of fire behavior is required to analyze potential impact and risk. Over the past 30 years, some countries, such as Canada, have realized the potential of building fire behavior models integrated into decision support systems for operational use. For example, the Canadian forest fire management agencies have demonstrated that information technology is essential in tracking fire weather, predicting fire behavior, and monitoring fire occurrence (Tymstra et al. 2020). Turning the data into information in near real-time useful for decision-making is the strength of the DSS. The best example is the Canadian Forest Fire Danger Rating System (CFFDRS 2022). It is a comprehensive system of tools that synthesize the environmental wildfire influencing factors such as fire's ignition, spread, and behavior to inform the ground fire management decisions. In addition, the CFFDRS is combined with other data on Canada's Wildfire Information Systems (CWFIS 2022), which provides information on fire weather, fire behavior

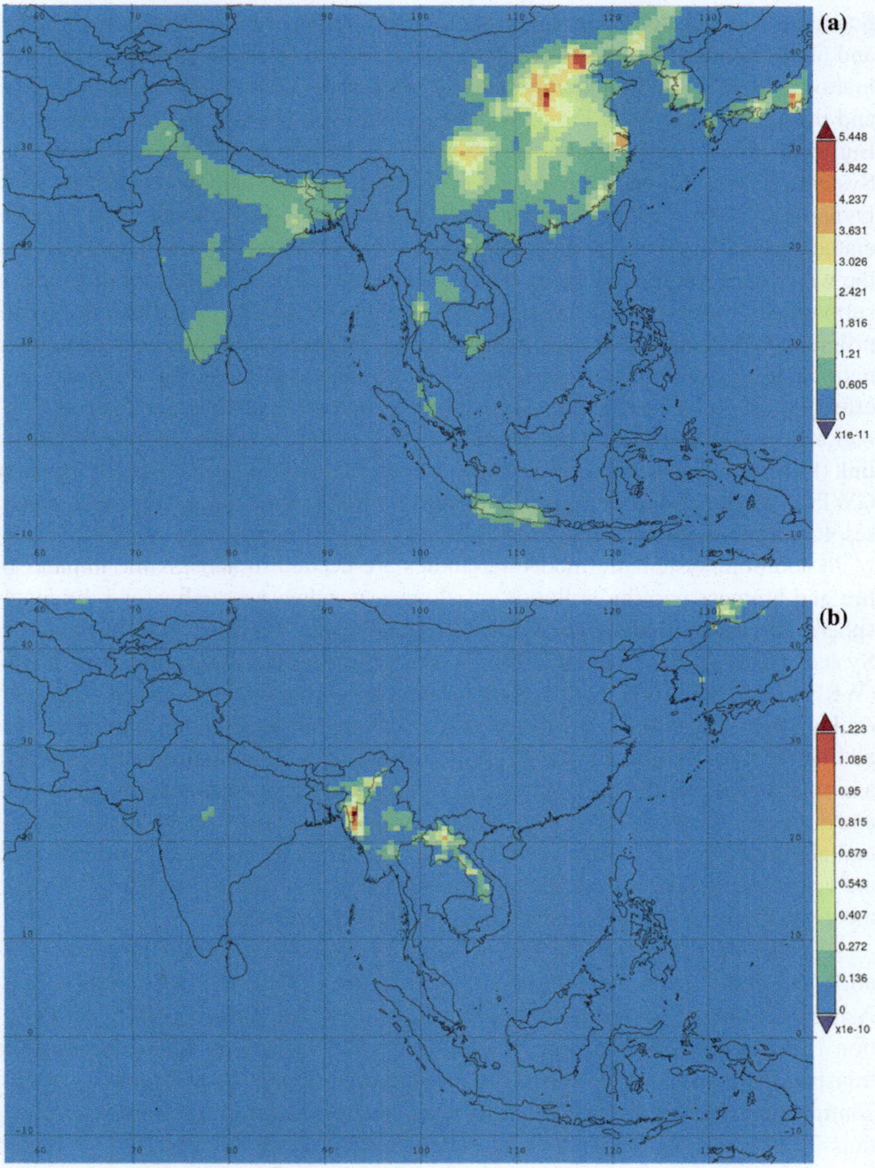

Fig. 9 March–April averaged black carbon (in kg m^{-2}) anthropogenic emissions (**a**) derived from MERRA reanalysis data at 0.5 degrees for South/Southeast Asia; **b** March–April averaged black carbon (in kg m^{-2}) biomass burning emissions derived from MERRA reanalysis data at 0.5 degrees for South/Southeast Asia

potential, and selected upper atmospheric conditions used to produce fire weather and fire behavior maps. A similar approach is adopted by the Global Wildfire Information System (GWIS), a joint initiative of the Group on Earth Observations (GEO) and the Copernicus Work Programs. GWIS builds on the ongoing activities of the European Forest Fire Information System (EFFIS), the Global Terrestrial Observing System (GTOS), Global Observation of Forest Cover-Global Observation of Land Dynamics (GOFC-GOLD), Fire Implementation Team (GOFC Fire IT), and the associated regional networks, complementing existing activities that are ongoing around the world concerning wildfire information gathering. The development of GWIS is supported by partner organizations and space agencies. In the new GEO GWIS work program, GWIS aims to bring together existing information sources at regional and national levels to provide a comprehensive view and evaluation of fire regimes and effects at the global level and tools to support operational wildfire management from national to global scales. A prototype of GWIS can be visualized at the following link (https://gwis.jrc.ec.europa.eu); however, there is a need to significantly enhance GWIS at regional and local scales in terms of both higher temporal and spatial resolution for operational use.

In addition, systematic modeling studies are needed to address the impacts of fire and biomass burning pollution on climate, weather, air quality, and the atmosphere. Air quality models such as The Community Multi-scale Air Quality Modeling System (CMAQ 2022) and Weather Research and Forecasting/Chemistry model (WRF/Chem) (UCAR 2022) that combine emission inventory information with meteorology and atmospheric chemistry and transport mechanisms are one option for addressing the pollution impacts. Also, there is a need to explore the potential of data assimilation and modeling systems that can integrate ground-based emissions data and airborne and satellite data to quantify pollutant exposure estimates to populations and aid in health risk assessments. However, such studies in South/Southeast Asia are very meager, and thus need attention. Also, regional capacity building activities are a must to train scientists and address pollution problems.

It is also clear that addressing transboundary pollution requires stronger cooperation and multi-lateral agreements among countries, especially in South/Southeast Asia. There are successful examples, such as Long-Range Transboundary Air Pollution (LRTAP), which has been active since 1983 in Europe for addressing the transboundary pollution problem. For more than 36 years, the protocols have been continuously revised to meet essential needs and priorities of air pollution mitigation. The protocols are based on the latest updated research. The LRTAP is also a legal instrument of air pollution control in Europe. Both South and Southeast Asian countries could develop a similar framework to LRTP for regional pollution mitigation. Although all countries that are part of ASEAN have ratified a form of the ASEAN Agreement on Transboundary Haze Pollution (ATHP) to reduce the amount of air pollution in Southeast Asia, it is not legally binding (ASEAN 2023). The success of the ASEAN protocols requires commitments on implementation from the participating countries. In contrast to the LRTAP, the ASEAN is also hindered by other factors, such as funding, institutional capacities, and manpower (Umar and

Kurrahman 2022). For addressing pollution in general and biomass burning in particular in S/SEA, more stringent multi-lateral agreements are needed. To strengthen research and its applications in addressing fires and biomass burning pollution, there is a need for more collaboration among regional researchers from the South/Southeast Asian countries. This should also include capacity building and training efforts.

Acknowledgements We are grateful to several authors from the USA and South/Southeast Asia who contributed to the book. This work is a part of the South/Southeast Asia Research Initiative (SARI) funded by the NASA Land Cover/Land Use Change Program.

References

Arunrat, N., S. Sereenonchai, and R. Hatano. 2022. Effects of fire on soil organic carbon, soil total nitrogen, and soil properties under rotational shifting cultivation in northern Thailand. *Journal of Environmental Management* 302: 113978.

ASEAN. 2023. https://asean.org/wp-content/uploads/2021/01/ASEANAgreementonTransboundaryHazePollution-1.pdf.

Biswas, S., K.D. Lasko, and K.P. Vadrevu. 2015a. Fire disturbance in tropical forests of Myanmar—Analysis using MODIS satellite datasets. *IEEE Journal of Selected Topics in Applied Earth Observations and Remote Sensing* 8 (5): 2273–2281.

Biswas, S., K.P. Vadrevu, Z.M. Lwin, K. Lasko, and C.O. Justice. 2015b. Factors controlling vegetation fires in protected and non-protected areas of Myanmar. *PLoS One* 10 (4): e0124346.

Bruun, T.B., A. De Neergaard, D. Lawrence, and A.D. Ziegler. 2009. Environmental consequences of the demise in swidden cultivation in Southeast Asia: Carbon storage and soil quality. *Human Ecology* 37 (3): 375–388.

CFFDRS. 2022. https://www.nwcg.gov/publications/pms437/cffdrs/overview.

Chowdhury, F.I., I. Barua, A.I. Chowdhury, V. Resco de Dios, and M.S. Alam. 2022. Agroforestry shows higher potential than reforestation for soil restoration after slash-and-burn: A case study from Bangladesh. *Geology, Ecology, and Landscapes* 6 (1): 48–54.

CMAQ. 2022. https://www.epa.gov/cmaq.

Crutzen, P.J., and M.O. Andreae. 1990. Biomass burning in the tropics: Impact on atmospheric chemistry and biogeochemical cycles. *Science* 250 (4988): 1669–1678.

CWFIS. 2022. https://cwfis.cfs.nrcan.gc.ca/home.

de Neergaard, A., J. Magid, and O. Mertz. 2008. Soil erosion from shifting cultivation and other smallholder land use in Sarawak, Malaysia. *Agriculture, Ecosystems & Environment* 125 (1–4): 182–190.

Dhandapani, S., and S. Evers. 2020. Oil palm 'slash-and-burn' practice increases post-fire greenhouse gas emissions and nutrient concentrations in burnt regions of an agricultural tropical peatland. *Science of the Total Environment* 742: 140648.

Dozier, J. 1981. A method for satellite identification of surface temperature fields of subpixel resolution. *Remote Sensing of the Environment* 11: 221–229.

Elguindi, N., C. Granier, T. Stavrakou, S. Darras, M. Bauwens, H. Cao, C. Chen, H.A.C. Denier van der Gon, O. Dubovik, T.M. Fu, and D.K. Henze. 2020. Intercomparison of magnitudes and trends in anthropogenic surface emissions from bottom-up inventories, top-down estimates, and emission scenarios. *Earth's Future* 8 (8): e2020EF001520.

ERA5. 2022. https://www.ecmwf.int/en/forecasts/datasets/reanalysis-datasets/era5.

Gabriel, A.G., M. De Vera, and B. Antonio. 2020. Roles of indigenous women in forest conservation: A comparative analysis of two indigenous communities in the Philippines. *Cogent Social Sciences* 6(1): 1720564.

GCOS-WMO. 2022. https://gcos.wmo.int/en/essential-climate-variables/fire.
Giglio, L., T. Loboda, D.P. Roy, B. Quayle, and C.O. Justice. 2009. An active-fire based burned area mapping algorithm for the MODIS sensor. *Remote Sensing of Environment* 113: 408–420.
Guevara, M., S. Lopez-Aparicio, C. Cuvelier, L. Tarrason, A. Clappier, and P. Thunis. 2017. A benchmarking tool to screen and compare bottom-up and top-down atmospheric emission inventories. *Air Quality, Atmosphere & Health* 10 (5): 627–642.
Gupta, P.K., V.K. Prasad, Y. Kant, C. Sharma, A.B. Ghosh, M.C. Sharma, A.K. Sarkar, S.L. Jain, O.P. Tripathi, R.C. Sharma, and K.V.S. Badarinath. 2001. Study of trace gases and aerosol emissions due to biomass burning at shifting cultivation sites in East Godavari District (Andhra Pradesh) during INDOEX IFP-99. *Current Science* 186–196.
Itahashi, S., K. Yumimoto, J.I. Kurokawa, Y. Morino, T. Nagashima, K. Miyazaki, T. Maki, and T. Ohara. 2019. Inverse estimation of NO_x emissions over China and India 2005–2016: Contrasting recent trends and future perspectives. *Environmental Research Letters* 14 (12): 124020.
Kondo, M., P.K. Patra, S. Sitch, P. Friedlingstein, B. Poulter, F. Chevallier, P. Ciais, J.G. Canadell, A. Bastos, R. Lauerwald, and L. Calle. 2020. State of the science in reconciling top-down and bottom-up approaches for terrestrial CO_2 budget. *Global Change Biology* 26 (3): 1068–1084.
Kurokawa, J.I., K. Yumimoto, I. Uno, and T. Ohara. 2009. Adjoint inverse modeling of NO_x emissions over eastern China using satellite observations of NO_2 vertical column densities. *Atmospheric Environment* 43 (11): 1878–1887.
Lamsal, L.N., N.A. Krotkov, A. Vasilkov, S. Marchenko, W. Qin, E.S. Yang, Z. Fasnacht, J. Joiner, S. Choi, D. Haffner, and W.H. Swartz. 2021. Ozone monitoring instrument (OMI) aura nitrogen dioxide standard product version 4.0 with improved surface and cloud treatments. *Atmospheric Measurement Techniques* 14 (1): 455–479.
Levelt, P.F., J. Joiner, J. Tamminen, J.P. Veefkind, P.K. Bhartia, D.C. Stein Zweers, B.N. Duncan, D.G. Streets, H. Eskes, R. van der A, C. McLinden, V. Fioletov, S. Carn, J. de Laat, M. DeLand, S. Marchenko, R. McPeters, J. Ziemke, D. Fu, X. Liu, K. Pickering, A. Apituley, G. González Abad, A. Arola, F. Boersma, C. Chan Miller, K. Chance, M. de Graaf, J. Hakkarainen, S. Hassinen, I. Ialongo, Q. Kleipool, N. Krotkov, C. Li, L. Lamsal, P. Newman, C. Nowlan, R. Suleiman, L.G. Tilstra, O. Torres, H. Wang, and K. Wargan. 2018. The ozone monitoring instrument: Overview of 14 years in space. *Atmospheric Chemistry and Physics* 18: 5699–5745. http://doi.org/10.5194/acp-18-5699-2018.
Magdoff, F. 2018. Soil quality and management. In *Agroecology*, 349–364. Boca Raton: CRC Press.
Mellac, M. 2021. When land and sky turn French: Is there any place left for the Bunong when private land property disrupts customary spaces along the Vietnamese-Cambodian border? In *Spaces of law and custom*, 141–159. London: Routledge.
MERRA. 2022. https://gmao.gsfc.nasa.gov/reanalysis/MERRA/.
Mukul, S.A., and A. Byg. 2020. What determines indigenous Chepang farmers' swidden land-use decisions in the central hill districts of Nepal? *Sustainability* 12 (13): 5326.
Müller, J.F., and T. Stavrakou. 2005. Inversion of CO and NO_x emissions using the adjoint of the IMAGES model. *Atmospheric Chemistry and Physics* 5 (5): 1157–1186.
Murdjoko, A., F.Q. Brearley, A. Ungirwalu, D.A. Djitmau, and N.M. Benu. 2022. Secondary succession after slash-and-burn cultivation in Papuan lowland forest, Indonesia. *Forests* 13 (3): 434.
Palm, C., S.A. Vosti, P.A. Sanchez, and P.J. Ericksen (eds.). 2005. *Slash-and-burn agriculture: The search for alternatives*. New York: Columbia University Press.
Penner, J.E., R.E. Dickinson, and C.A. O'Neill. 1992. Effects of aerosol from biomass burning on the global radiation budget. *Science* 256 (5062): 1432–1434.
Phompila, C., V. Sihapanya, B. Chanthavong, S. Thongmanivong, P. Van Khoa, K. Noda, and K.P. Vadrevu. 2022. Agricultural use/cover changes in the Vientiane. In *Remote sensing of agriculture and land cover/land use changes in South and Southeast Asian countries*, 573–586. Cham: Springer.

Roy, D.P., Y. Jin, P.E. Lewis, and C.O. Justice. 2005. Prototyping a global algorithm for systematic fire-affected area mapping using MODIS time series data. *Remote Sensing of Environment* 97 (2): 137–162.

Schroeder, W., P. Oliva, L. Giglio, and I.A. Csiszar. 2014. The new VIIRS 375 m active fire detection data product: Algorithm description and initial assessment. *Remote Sensing of Environment* 143: 85–96.

Sharma, S.B., S. Kumar, E.Y. Ovung, and B. Konsam. 2022. Vegetation dynamics and soil nutrients across different shifting cultivation fallows in Montane Subtropical Forest of Mizoram, NE India. *Acta Oecologica* 115: 103833.

Sigsgaard, T., B. Forsberg, I. Annesi-Maesano, A. Blomberg, A. Bølling, C. Boman, J. Bønløkke, M. Brauer, N. Bruce, M.E. Héroux, and M.R. Hirvonen. 2015. Health impacts of anthropogenic biomass burning in the developed world. *European Respiratory Journal* 46 (6): 1577–1588.

Trang, P.T., M.E. Andrew, T. Chu, and N.J. Enright. 2022. Forest fire and its key drivers in the tropical forests of northern Vietnam. *International Journal of Wildland Fire* 31 (3): 213–229.

Tymstra, C., B.J. Stocks, X. Cai, and M.D. Flannigan. 2020. Wildfire management in Canada: Review, challenges and opportunities. *Progress in Disaster Science* 5: 100045.

UCAR. 2022. https://www2.acom.ucar.edu/wrf-chem.

Umar, W., and T. Kurrahman. 2022. Possibility to adopt LRTAP against transboundary haze pollution: What should ASEAN look for? *Indonesian Comparative Law Review* 5 (1): 12–222.

Vadrevu, K.P., and C.O. Justice. 2011. Vegetation fires in the Asian region: Satellite observational needs and priorities. *Global Environmental Research* 15 (1): 65–76.

Vadrevu, K., and K. Lasko. 2018. Intercomparison of MODIS AQUA and VIIRS I-Band fires and emissions in an agricultural landscape—Implications for air pollution research. *Remote Sensing* 10 (7): 978.

Vadrevu, K.P., T. Ohara, and C. Justice. 2014. Air pollution in Asia. *Environmental Pollution (Barking, Essex: 1987)* 195: 233–235.

Vadrevu, K., T. Ohara, and C. Justice. 2017. Land cover, land use changes and air pollution in Asia: A synthesis. *Environmental Research Letters* 12 (12): 120201.

Vadrevu, K.P., K. Lasko, L. Giglio, W. Schroeder, S. Biswas, and C. Justice. 2019. Trends in vegetation fires in South and Southeast Asian countries. *Scientific Reports* 9 (1): 1–13.

Wooster, M.J., G. Roberts, G. L. W. Perry, and Y.J. Kaufman. 2005. Retrieval of biomass combustion rates and totals from fire radiative power observations: FRP derivation and calibration relationships between biomass consumption and fire radiative energy release. *Journal of Geophysical Research: Atmospheres* 110(D24).

Mapping, Monitoring, and Modeling of Vegetation Fires

Wildfire Monitoring Using Infrared Bands and Spatial Resolution Effects

Wei Zheng, Jie Chen, Jinlong Fan, Yajun Li, and Cheng Liu

Abstract In the past, the application of wildfire monitoring by satellite remote sensing mainly used the mid-infrared band data with kilometer resolution, and there needs to be more quantitative research on the fire detection sensitivity of other infrared bands, like far infrared and short infrared. This study uses the mixed-pixel decomposition method to quantitatively analyze the effect of various spatial resolutions, i.e., 150 m, 300 m, and 1 km resolution infrared bands on fire monitoring. The results show that the fire detection sensitivity of the mid-infrared channel with 150 m resolution is about 30 times higher than that of the 1 km channel. The far-infrared channel with 300 m resolution can detect a fire in hundreds of square meters level; the short-infrared channel with 150 m resolution has an obvious response to the high-intensity flame fire area. This study also uses FY3A meteorological satellite 1 km infrared data and HJ-IB environmental disaster reduction satellite 150 m and 300 m infrared data to verify the above analysis by monitoring individual forest fire cases in Heilongjiang Province in spring and a straw-burning fire in Anhui Province in summer of 2009. The results show that improving the resolution of the infrared band will significantly improve the application ability of satellite remote sensing in small wildfire detection, fire field dynamic monitoring, and fire intensity evaluation.

Keywords Fire monitoring · Infrared bands · Spatial resolution · Detecting sensitivity · Quantitative analyzing

W. Zheng (✉) · J. Chen · J. Fan · Y. Li · C. Liu
Key Laboratory of Radiometric Calibration and Validation for Environmental Satellites, National Satellite Meteorological Center (National Center for Space Weather), China Meteorological Administration, Beijing, People's Republic of China
e-mail: zhengw@cma.gov.cn

W. Zheng · J. Chen · J. Fan · C. Liu
Innovation Center for FengYun Meteorological Satellite, Beijing, People's Republic of China

K. P. Vadrevu et al. (eds.), *Vegetation Fires and Pollution in Asia*,
https://doi.org/10.1007/978-3-031-29916-2_2

1 Introduction

Meteorological satellites were used for wildfire monitoring since the 1980s (Matson and Schneider 1984) and often took an important role in forest fire prevention (Zhang et al. 1989b). The major channel used for wildfire monitoring applications is the mid-infrared channel with a wavelength of 3.5–4 μm, which is sensitive to high-temperature targets like wildfire (Kaufman et al. 1998; Justice et al. 2002). Previous research on remote sensing of wildfire monitoring focused on using mid-infrared channel data, and very few studies used the other infrared channels. This is because most early meteorological satellites had a low IR resolution of ~ 1 km (He et al. 2011). Although the mid-infrared channel is sensitive to the high-temperature heat source, it has some limitations in fire monitoring. For example, the brightness temperature may be saturated due to the large fires, affecting fire intensity evaluation (Elvidge et al. 2021). In addition, it is easy to be contaminated by the reflection of solar radiation at the cloud edge, especially in the flare area. The far-infrared and short-infrared channels can compensate for these two sides. However, due to the spatial resolution limitation, it is challenging to realize the role of these two channels in wildfire monitoring at a 1 km resolution data. Therefore, few studies focused on applying far-infrared and short-infrared channels in wildfire monitoring.

The environmental disaster mitigation satellite B infrared camera (HJ-IB/IRS) launched by China has mid-infrared and short-infrared channels with 150 m resolution and a far-infrared channel with a 300 m, which provides data conditions for studying the impact of spatial resolution on wildfire monitoring in each infrared channel. There are some earlier articles that introduced the ability of HJ-IB/IRS for wildfire monitoring, but they still only use mid-infrared channel and not analyze the ability of other infrared channel in wild fire monitoring (Zhang et al. 1989b; Qin et al. 2010; Peng et al. 2011).

This study uses the 150 m and 300 m data of HJ-1B/IRS and 1 km data of FY3A meteorological satellite as data sources to quantitatively analyze the difference in fire detection sensitivity of short-infrared, mid-infrared, and far-infrared channels with different spatial resolution by using the mixed-pixel linear decomposition method. The influence of infrared channels' spatial resolution on fire monitoring applications is also addressed. The results are verified by monitoring the forest fires in Heilongjiang Province in the spring and straw-burning fires in Anhui Province in the summer of 2009. The results show that improving the spatial resolution of infrared channels, including mid-infrared, far infrared, and short infrared, can help in the sensitivity of fire detection, dynamic monitoring of fire fields, and fire intensity assessment by using satellite remote sensing. Our study provides a quantitative assessment of the effect for improving the resolution of mid-infrared, far-infrared and short-infrared channels on wildfire detection and fire condition assessment for the first time, which will provide reference information for promoting the comprehensive application for forest, grassland fire and straw-burning fire monitoring using multi-infrared band data.

2 Analyzing the Effect of Spatial Resolution of Infrared Band on Fire Monitoring

2.1 Principle of Fire Detection Using Infirared Channel

According to Wien's displacement law: $T * \lambda_{max} = 2897.8$ (K, μm) which describes the relationship between black body temperature and radiation peak wavelength, Blackbody temperature T and radiation peak wavelength λ_{Max} is inversely proportional, that is, the higher the temperature, the smaller the radiation peak wavelength.

The peak wavelength of the general surface temperature (about 300 K) is close to the far-infrared channel wavelength (11 μm). The temperature of wildfire is usually around 600–1200 K, and the peak wavelength of thermal radiation is close to the mid-infrared (3.5–4 μm). Therefore, the mid-infrared channel (3.5–4 μm) is very sensitive to the forest, grassland, and straw-burning fires. Satellite remote sensing instruments that can be used for fire monitoring generally have medium infrared, far-infrared, and short-wave infrared channels. HJ-1B/IRS has four channels, including near infrared, short infrared, medium infrared, and far infrared. Their wavelengths are similar to those of the infrared channels in FY3A/VIRR (Polar-orbiting meteorological satellite/Visible and Infrared Radiometer) with a 1 km resolution. Table 1 lists the major relevant parameters of HJ-1 B/IRS and FY3A/VIRR instrument.

It can be seen from the blackbody emissivity curve that at different temperatures and spectral wavelengths calculated according to the Planck blackbody radiation equation (see Fig. 1), the peak wavelength of general surface temperature (about 300 K) radiation is about the wavelength range of HJ/IRS channel 4 (11.58 μm). When the temperature increases, especially above 600 K, the peak wavelength of radiation quickly moves to the wavelength range of channel 3 (3.72 μm). When the temperature exceeds 900 K, the radiation peak wavelength is close to the wavelength range of HJ/IRS channel 2 (1.65 μm). It can also be seen that when the temperature increases, the emissivity corresponding to each band also increases with the increase of the total emission power.

Table 1 Major parameters of HJ-1B/IRS and FY3A/VIRR channels

HJ-1B/IRS			FY3A/VIRR		
Channel No.	Central wavelength (μm)	Spatial resolution (m)	Channel No.	Central wavelength (μm)	Spatial resolution (m)
1	0.9	150	2	0.865	1000
2	1.65	150	6	1.61	1000
3	3.72	150	3	3.7	1000
4	11.58	300	4	10.8	1000

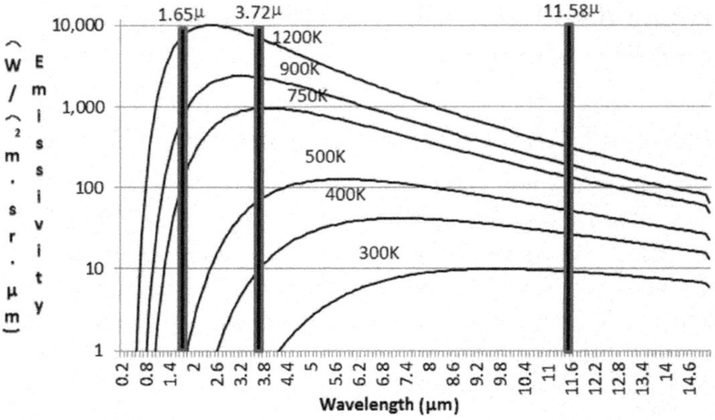

Fig. 1 Planck blackbody emissivity curve

The pixels covering the fire fields can be regarded as mixed pixels containing the fire area. According to the mixed pixel linear decomposition method, the radiation energy of the field of view of the infrared channel is the linear combination of the radiation energy of each part of the object in the pixel. That is, the radiation of a pixel is the weighted average of the radiance of all kinds of ground objects in the pixel, and its area accounts for the proportion of the pixel, as shown in Eq. (1).

When a wildfire occurs, the temperature of the fire field is generally 600–1200 K or above. Therefore, the pixel containing the fire can be regarded as a mixed pixel composed of only two types of end elements in the fire area and no fire area. Dozier (1982) proposed that the radiance of mixed pixels containing fire area can be expressed by Eq. (2):

$$L_t = \left(\sum_{i=1}^{n} \Delta S_i L_{T_i}\right)/S \tag{1}$$

$$\begin{aligned} L_{\text{mix}} &= P * L_{T_{\text{ft}}} + (1-P) * L_{T_{\text{bg}}} \\ &= P * \frac{C_1 \lambda_i^{-5}}{\pi * \left(e^{C_2/\lambda_i T_{\text{ft}}} - 1\right)} + (1-P) * \frac{C_1 \lambda_i^{-5}}{\pi * \left(e^{C_2/\lambda_i T_{\text{bg}}} - 1\right)} \end{aligned} \tag{2}$$

In Eq. (1), L_t is the radiance of the pixel observed by the satellite, t is the equivalent blackbody temperature corresponding to the radiance L_t, ΔS_i is the area of the ith sub-region in the pixel, L_{T_i} is the emissivity of the sub-region (W m^{-2} sr^{-1} μm^{-1}), T_i is the temperature of the sub-area i, and S is the total size of the pixel.

In Eq. (2), L_{mix} is the radiance of mixed pixel of HJ infrared channel, P is the percentage of sub-pixel fire size in the pixel area, $L_{T_{\text{ft}}}$ is the sub-pixel fire emissivity, $L_{T_{\text{bg}}}$ is the background emissivity around the fire point, T_{ft} is the sub-pixel fire temperature, and T_{bg} is the background temperature, λ_i is the central wavelength, $C_1 = 1.191043 * 10^8$ W μm^4 m^{-2} sr^{-1}, and $C_2 = 1.438768 * 10^4$ μm K.

The main principle to discern the fire pixel is whether the brightness temperature of the detected pixel is abnormally higher than that of the surrounding pixels. Assuming that the brightness temperature of the mixed pixel containing the fire field is T_{mix}, the brightness temperature of the background pixel is T_{bg}, the temperature of the fire field is T_{ft}, and the brightness temperature difference between the mixed pixel of channel i and the background pixel is ΔT_i, then:

$$\Delta T_i = T_{i\text{mix}} - T_{i\text{bg}} = \frac{C_2}{\lambda_1 \text{Ln}\left(1 + \frac{C_1}{\pi * \lambda_i^5 * L_{i\text{mix}}}\right)} - \frac{C_2}{\lambda_i \text{Ln}\left(1 + \frac{C_1}{\pi^* \lambda_1^5 * L_{\text{bg}}}\right)} \tag{3}$$

Figure 2a shows the curve with variations of the brightness temperature increase of 3.72 μm and 11.58 μm channel mixed pixels with the proportion of sub-pixel fire size produced by Eq. (3), in which the fire area temperature is 600 K, 800 K, and 1000 K respectively, and the background temperature is 300 K. It can be seen from Fig. 2a that even if the fire spot area is very small, it will cause the brightness temperature of mid-infrared pixels containing fire spots to rise rapidly, while the temperature rise of 11.58 μm channel (far infrared) is much lower than that of the mid-infrared channel. This phenomenon shows the difference between the brightness temperature of the mid-infrared channel and the surrounding and the subtraction between the brightness temperature of the mid-infrared and far infrared and the surrounding difference can be used to detect fire information.

Further, the subtraction between the brightness temperature of the mid-infrared and far infrared and the surrounding difference can be used to detect the fire information.

The short-infrared band with a wavelength of 1.65 μm mainly reflects solar radiation. Figure 2b shows the radiation emittance of the blackbody temperature of 300–1200 K at the wavelength of 1.65 μm and the solar irradiance at the top of the atmosphere at the wavelength of 1.65 μm calculated by the Planck function. It can be seen from Fig. 2b that when the temperature is 900 K, 1000 K, and 1200 K,

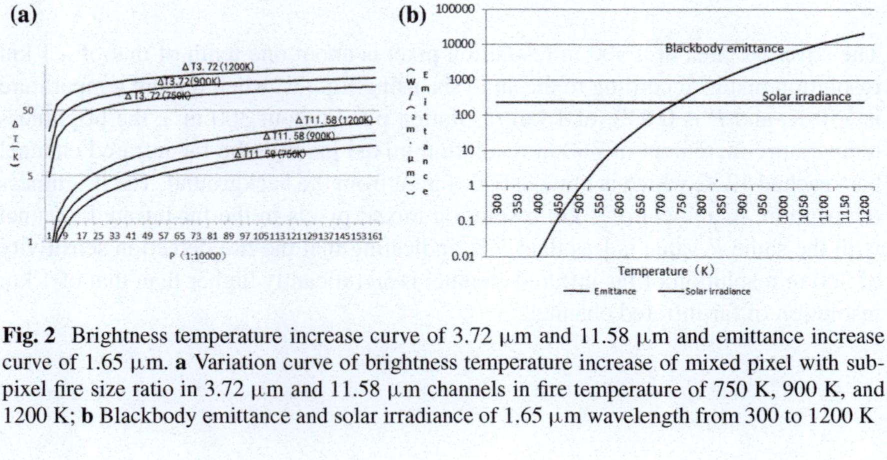

Fig. 2 Brightness temperature increase curve of 3.72 μm and 11.58 μm and emittance increase curve of 1.65 μm. **a** Variation curve of brightness temperature increase of mixed pixel with sub-pixel fire size ratio in 3.72 μm and 11.58 μm channels in fire temperature of 750 K, 900 K, and 1200 K; **b** Blackbody emittance and solar irradiance of 1.65 μm wavelength from 300 to 1200 K

the emittance is 8 times, 22 times, and 94 times of the solar irradiance at 1.65 μm, respectively. Therefore, strong flaming fires may cause the reflectivity of the 1.65 μm channel of the fire pixel to increase rapidly, which is significantly different from the surrounding pixels.

2.2 Analysis on the Influence of Infrared Channels with Different Resolution on Fire Monitoring

Equations (3) and (2) are used to quantitatively analyze the influence of infrared channels with different resolutions on fire monitoring.

2.2.1 Sensitivity Analysis of Fire Detection in 150 m and 1 km Mid-Infrared Channels

The coverage area of 150 m resolution pixel is about one-fortieth of that of 1 km resolution pixel. According to Eq. (3), when the fire temperature is 750 K, and P is 0.0005% of 1 km resolution pixel (about 5 m^2), the brightness temperature increase of 150 m resolution mixed pixel is up to 10 K, reaching the fire identification threshold. The P of 1 km resolution pixel is up to 0.017%, then the brightness temperature increment reaches 8 K, which is the fire detection threshold, and the area difference between 150 m and 1 km resolution for fire detection is about 34 times. Therefore, it can be seen that the sensitivity of 150 m resolution mid-infrared channel to fire identification is about 30 times than that of 1 km resolution data, and then the flame fire with an area of square meters level can be detected.

2.2.2 Sensitivity Analysis of Fire Detection of 300 m and 1 km Far-Infrared Channels

The coverage area of a 300 m resolution pixel is about one-tenth of that of a 1 km resolution pixel. According to the analysis, using Eq. (3), when the fire temperature is 750 K, and P is 0.03% of 1 km resolution pixel (about 300 m^2), the brightness temperature increment of 300 m resolution mixed pixel in the far-infrared channel has reached 12 K, which is obviously different from the background. The brightness temperature increment of 1 km resolution mixed pixels in the far-infrared channel with the same P value is less than 1 K, indicating that the fire detection sensitivity of 300 m resolution of far-infrared channel is significantly higher than that of 1 km resolution of far-infrared channel.

2.2.3 Sensitivity Analysis of Fire Detection in 150 m Short-Infrared Channel

Figure 2b shows that high-intensity flaming fire may cause the reflectivity of short-infrared channels to increase significantly. Equation (4) has been used to calculate the difference between the reflectance of fire pixels and the background of 1.65 μm short-infrared channel established by using the mixed-pixel linear decomposition method:

$$\begin{aligned}\Delta R_{1.65} &= R_{1.65\_\min} - R_{1.65\_\mathrm{bg}} \\ &= \pi \cdot \left(P \cdot L_{1.65\_\mathrm{ft}} + (1-P) \cdot L_{1.65\_\mathrm{bg}}\right) \cdot d^2/E_0 \cdot \cos\theta_s \\ &\quad - R_{1.65\_\mathrm{bg}}\end{aligned} \tag{4}$$

where $\Delta R_{1.65}$ is the reflectivity difference between mixed pixel and background in 1.65 μm channel; $R_{1.65\_\mathrm{mix}}$ is the reflectance of 1.65 μm channel mixed pixel; $R_{1.65\_\mathrm{bg}}$ is the background reflectance of 1.65 μm channel, which is set to 0.2 here; $L_{1.65\_\mathrm{ft}}$ is the radiance of the fire pixel of the 1.65 μm channel, ft is the fire temperature, and $L_{1.65\_\mathrm{bg}}$ is the radiance corresponding to the background reflectance of 1.65 μm channel. By substituting Eq. (2) into Eq. (4), it can be seen that when the sub-pixel fire temperature (T_{ft}) is 750 K and P increases from 0.01 to 5.21%, the reflectance increment of mixed pixels is only 7%. When T_{ft} is 900 K, and P reaches 3%, the reflectance of fire mixed pixel increases by about 26%; When T_{ft} is 1200 K and P is about 0.4%, the reflectance increases about 40%, and when P is about 0.8%, the reflectance increases about 80%. If the coverage area of pixel is calculated in 150 m resolution, when the temperature of the flame fire area is 1200 K and the area is 90 m^2, the reflectance increases about 40%, which will be significantly different from the surrounding pixels and easy to be identified as fire pixel.

3 Case Study of Fire Monitoring Using Infrared Channels with Different Resolutions

3.1 Fire Detection

In this study, we used a contextual method to identify the fire pixel, i.e., when the detected pixel meets the following conditions, it is identified as the fire pixel (Eq. 5):

$$\begin{aligned}&T_{\mathrm{MIR}} > T_{\mathrm{MIR\_bg}} + a\delta T_{\mathrm{MR\_\ bg}} \text{ and } T_{\mathrm{MIR\text{-}FIR}} > T_{\mathrm{MIR\text{-}FIR\_bg}} + a\delta T_{\mathrm{MRR\text{-}FIR\_bg}}; \text{ or} \\ &T_{\mathrm{MIR}} > 360\,\mathrm{K}\end{aligned} \tag{5}$$

where T_{MIR}, $T_{\mathrm{MIR\_bg}}$, and $\delta T_{\mathrm{MIR\_bg}}$ are the brightness temperature of the mid-infrared channel, the background brightness temperature of the mid-infrared channel, and the standard deviation of the background brightness temperature of the detected pixel, respectively; $T_{\text{MIR-FIR}}$, $T_{\text{MIR-FIR\_bg}}$, and $\delta T_{\text{MIR-FIR\_bg}}$ are the brightness temperature difference between mid-infrared and far-infrared channels, the background brightness temperature difference between mid-infrared and far-infrared channels, and the standard deviation of background brightness temperature difference, respectively; a is the coefficient, the 1 km data is set as 4, and 150 m data is set as 5.

3.2 Evaluation of the Size of Sub-pixel Fire

A single-channel method has been used to evaluate the sub-pixel fire area. The estimation equation is shown in Eq. (6). When the mid-infrared channel is unsaturated, the mid-infrared channel is used for estimation, as shown in Eq. (7); when mid-infrared channel is saturated, the far-infrared channel is used for estimation, as shown in Eq. (8):

$$S_f = P \times S \tag{6}$$

$$P = \left(L_{\text{MIR\_mix}}\left(T_{\text{MIR\_mix}}\right) - L_{\text{MIR\_bg}}\left(T_{\text{MIR\_bg}}\right)\right) / \left(L_{\text{MIR}}(T_f) - L_{\text{MRR\_bg}}\left(T_{\text{MIR\_bg}}\right)\right) \tag{7}$$

$$P = \left(L_{\text{FIR\_ mix}}\left(T_{\text{FIR\_mix}}\right) - L_{\text{FIR\_bg}}\left(T_{\text{FIR\_bg}}\right)\right) / \left(L_{\text{FIR}}(T_f) - L_{\text{FIR\_bg}}\left(T_{\text{FIR\_bg}}\right)\right) \tag{8}$$

where S_f is the sub-pixel fire size, P is the proportion of sub-pixel fire size, and S is the pixel area. T_f is the sub-pixel fire temperature, set as 750 K. $L_{\text{MIR\_mix}}$, $L_{\text{FAR\_mix}}$ is the emissivity of mixed pixels in mid-infrared channel and far-infrared channel, $L_{\text{MIR\_bg}}$, $L_{\text{FAR\_bg}}$ is the background pixel emissivity of mid-infrared channel and far-infrared channel, and $T_{\text{MIR\_mix}}$, $T_{\text{FAR\_mix}}$ is the brightness temperature of mid-infrared and far-infrared mixed pixels, respectively. $T_{\text{mid\_bg}}$, $T_{\text{far\_bg}}$ is the brightness temperature of background pixels in mid-infrared and far-infrared channels, respectively.

3.3 A Case Study of Straw-Burning Fire Monitoring

Using HJ/IRS (150 m) at 11:10 on June 4, 2009 (Beijing time) and FY3A meteorological satellite (1 km) at 10:45 (Beijing time) on the same day, the information of straw-burning fire in some areas of Anhui Province was derived. The geographical range is 116.5° to 117° E and 32.5° to 33° N. Figure 3a shows the RGB image composited using HJ/IRS channels 3, 1, and 2. Figure 3b shows the RGB image composited by using FY3A/VIRR channels 3, 2, and 1. Figure 3c, d show the fire spot detected from Fig. 3a, b, respectively.

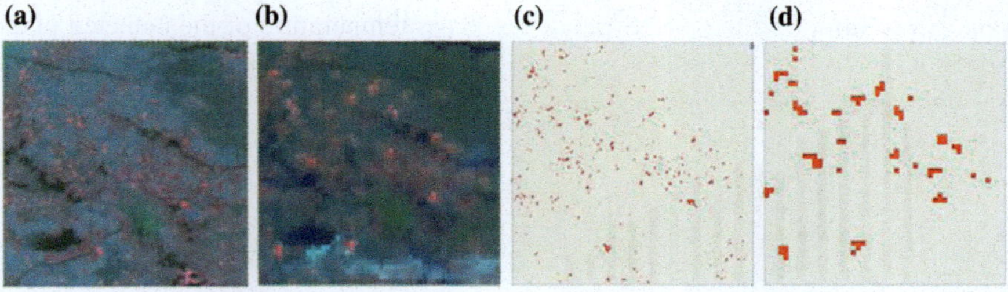

Fig. 3 Fire image and fire spot of HJ/IRS and meteorological satellite. **a** HJ/IRS fire image; **b** meteorological satellite fire image; **c** HJ/IRS fire spot; **d** meteorological satellite fire spot

Table 2 Statistics of fire information identified by HJ/IRS and meteorological satellite

Satellite	Number of fire region	Number of pixels	Satellite	Number of fire region	Number of pixels
HJ/IRS	210	837	Meteorological sat	26	78

It can be seen from the comparison that in the same area, the fire information identified by HJ/IRS is significantly more than that of meteorological satellites, especially in many areas where HJ/IRS identifies small fire, and meteorological satellites do not identify fire.

Table 2 lists the statistics of the number of sub-pixel fire size and fire pixels identified by HJ/IRS and meteorological satellites in the region. It can be seen from the table that the number of fire regions and fire pixels detected by HJ-1B/IRS are 7 times and 11 times higher than that of meteorological satellites respectively.

Figure 4a, b show the statistics of sub-pixel fire size and relevant number of pixels in the above areas by HJ/IRS and meteorological satellite using Eq. (6) respectively. It can be seen from the figure that the sub-pixel fire size estimated by HJ/IRS ranges from square meters to nearly 100 m^2, and the sub-pixel fire size of a considerable number of pixels is less than 10 m^2, while the sub-pixel fire size estimated by meteorological satellite ranges from 100 to 1000 m^2, mostly from 300 to 400 m^2.

3.4 A Case Study of Forest Fire Monitoring

In late April 2009, a forest fire occurred in Heilongjiang Province of China. The fire information was analyzed using HJ/IRS data at 02:40 (World time) on April 29, 2009, and compared with the meteorological satellite image at 02:15 (World time) on the same day to analyze the fire monitoring characteristics of HJ/IRS and meteorological satellite.

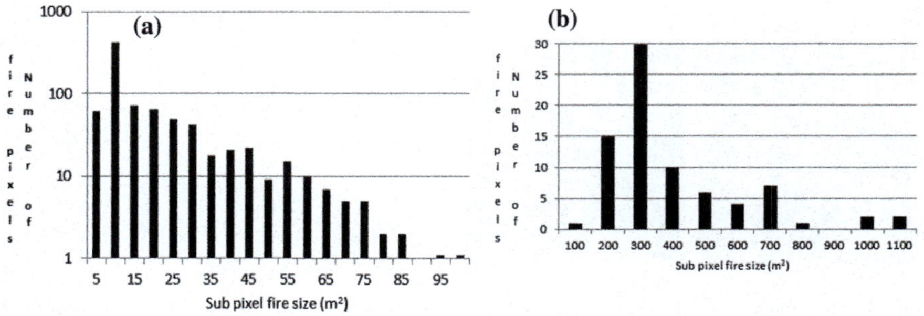

Fig. 4 Statistics of pixel numbers with different sub-pixel fire size. **a** Number of pixels with different sub pixels size in HJ/IRS; **b** Number of pixels with different sub-pixel size in meteorological satellite

3.4.1 The Response of Far-Infrared Channel (300 m) of HJ/IRS to Strong Fire

Figure 5a shows the fire field RGB image composed of 3.72 μm, 11.58 μm, and 11.58 μm of HJ/IRS in Heilongjiang Province of China. The red patches in the figure are the fire pixels detected by the mid-infrared channel, and the white lines are the fire pixels detected by both the mid-infrared and far-infrared channel. It can be seen from the figure that because the sensitivity of the mid-infrared channel to the fire is much higher than that of the far infrared, the far-infrared channel does not reflect at most of the fire fields detected by the mid-infrared channel, but only reacts in some local areas of the fire site, indicating that the fire is very strong there.

Figure 5b shows the brightness temperature profile of 3.72 μm and 11.58 μm when passing through the fire field from point A (128.221° E, 48.61° N) to point B (128.666° E, 48.61° N) in Fig. 5a. It can be seen from the figure that the 3.72 μm channel has an obvious temperature increase when passing through the fire site

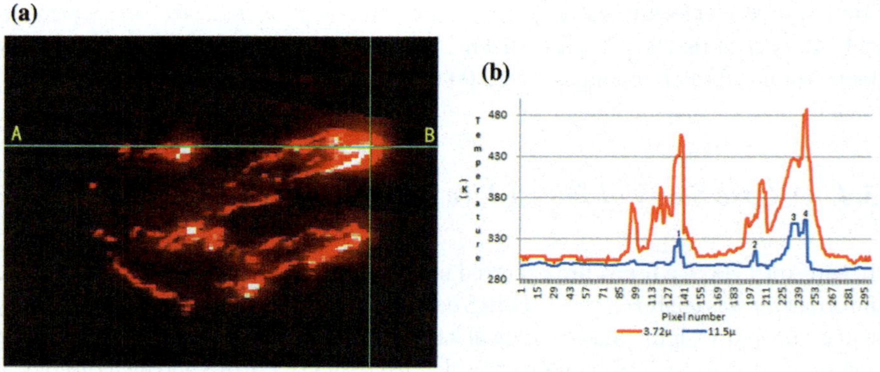

Fig. 5 HJ/IRS RGB fire image and profile of 3.72 μm, 11.58 μm data across the fire field. **a** HJ/IRS RGB fire image, 3.72 μm, 11.58 μm, 11.58 μm, 2009/0429/02:40 (GT); **b** HJ/IRS 3.72 μm, 11.58 μm profile. A: 128.221 E, 48.61 N; B: 128.666 E, 48.61 N

Table 3 Brightness temperature increment of 11.58 μm and 3.72 μm of HJ/IRS in four peaks of the profile where 11.58 μm has obvious temperature increase

Pixel No.	11.58 μm BT (K)	3.72 μm BT (K)
1	34	121
2	21	49
3	54	116
4	59	170

with multiple spikes or high platforms. In comparison, the 11.58 μm channel has an obvious temperature increase only when the 3.72 μm channel has a significant temperature increase.

Table 3 lists the brightness temperature increment of 11.58 μm and 3.72 μm channels, where the four peaks of 11.58 μm channels in Fig. 5b have obvious temperature increases. It can be seen from the table that the response of the 11.58 μm channel to strong fire does not increase linearly with the increment of the 3.72 μm channel.

3.4.2 The Response of Short-Infrared Channel (150 m) to High-Intensity Fire

Figure 6a shows the fire field RGB image composed of 3.72 μm, 1.65 μm, and 1.65 μm of HJ/IRS in Heilongjiang Province of China. The red patches in the figure are the fire pixels detected by the 3.72 μm channel, and the bright white spots or lines are the fire pixels detected by both 3.72 μm and 1.65 μm channels. The mid-infrared channel is very sensitive to high-temperature targets. In contrast, the 1.65 μm channel only reflects strong high-temperature targets and does not reflect most fire pixels detected by the 3.72 μm channel. It can be seen from the figure that the fire pixels detected by the 1.65 μm channel are generally located in the front of the fire field, indicating that there is a high-intensity flaming fire, which may be a canopy fire or a rapid surface fire. Figure 6b shows the profile of 3.72 μm bright temperature and 1.65 μm channel reflectivity when passing through the fire field from point A (128.221° E, 48.61° N) to point B (128.666° E, 48.61° N) in Fig. 6a. It can be seen from the figure that the reflectivity of the 1.65 μm channel increases significantly only when the 3.72 μm channel has a significant temperature increase.

Table 4 lists the reflectivity increment of the 1.65 μm channel at four peaks and the brightness temperature increment of the 3.72 μm channel in Fig. 6b. It can be seen that only when the brightness temperature increment of the 3.72 μm channel reaches more than hundreds K, the reflectivity of the 1.65 μm channel increases significantly, indicating that the 1.65 μm channel only has an obvious response to the high-intensity flame fire area, so it can reflect the fire area with the strongest fire in the large fire field.

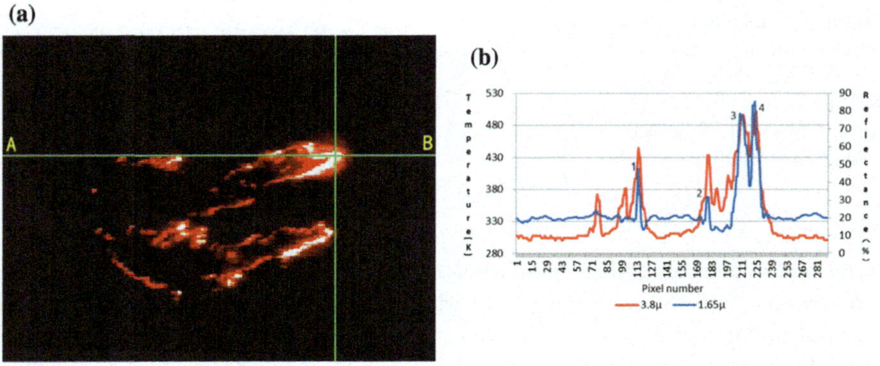

Fig. 6 HJ/IRS RGB fire image and profile of 3.72 μm, 1.65 μm data across the fire field. **a** HJ/IRS RGB fire image, 3.72 μm, 1.65 μm, 1.65 μm, 2009/0429/02:40 (GT); **b** HJ/IRS 3.72 μm, 1.65 μm profile. A: 128.221 E, 48.61 N; B: 128.666 E, 48.61 N

Table 4 Reflectance increase of 1.65 μm and brightness temperature increase of 3.72 μm of HJ/IRS in four peaks of the profile where 1.65 μm has obvious reflectance increase

Peak number	1.65 μm increase (%)	3.72 μm increase (K)
1	30	134
2	13	117
3	62	182
4	65	190

4 Conclusion

Through the quantitative analysis of the spatial and temperature response characteristics of various infrared bands to high-temperature targets such as wildfires with different resolutions of 150 m, 300 m, and 1 km, and the verification of forest fire and straw-burning fire monitoring examples, we show that improving the resolution of infrared channel will significantly improve the application ability of satellite remote sensing in small fire detection and assessment. As shown in the case studies, the mid-infrared channel with 150 m resolution can detect the small fire with the size of square meters level, which has great significance in forest and grassland fire monitoring. The far-infrared channel with 300 m resolution can detect fire with the size of hundreds square meters level and will be not affected by the reflection interference of solar radiation in the cloud area and the solar flare area; thus, can be used to evaluate the fire areas. The short-infrared channel with 150 m resolution has obvious response to the high-intensity flame fire area, which can provide a reference for judging the area with high-intensity flame fire (such as rapid spreading surface fire, and canopy fire) in the fire field. Our results demonstrate that satellites having infrared bands such as FY3D/MERSI-II with far-infrared channels (250 m resolution) and NPP with infrared channels with a resolution of 375 m. Using the variety of infrared bands will further improve the satellite remote sensing of wildfire mapping and monitoring.

Forest and grassland fires are complex and dynamic. In the forest and grassland fire prevention work, besides finding the fire source in time, it is also necessary to know the type and intensity of fire. Especially for large-scale fires with the area of tens or even hundreds of square kilometers, it is necessary to know the spatial distribution and dynamic development of different types of fires and intensity. Satellite remote sensing is the only way to obtain the continuous observation information of large-scale fires. Specifically, use of infrared bands can further improve the satellite remote sensing of wildfire mapping and monitoring.

Acknowledgements This paper supported by the National Key R&D Program of China (2021YFC3000300).

References

Dozier, J. 1982. A method for satellite identification of surface temperature fields of sub-pixel resolution. *Remote Sensing of Environment* 11: 221–229.

Elvidge, C.D., M. Zhizhin, K. Baugh, and F.C. Hsu. 2021. Identification of smoldering peatland fires in Indonesia via triple-phase temperature analysis of VIIRS nighttime data. In *Biomass burning in South and Southeast Asia*, 25–38. Boca Raton: CRC Press.

He, Bao-Hua, Liang-Fu Chen, and Jin-Hua Tao. 2011. A contextual fire detection algorithm based on observation geometry for HJ-IB-IRS. *Journal of Infrared and Millimeter Waves* 30 (2): 104–108.

Justice, C.O., L. Giglio, S. Korontzi, J. Owens, J.T. Morisette, D. Roy, J. Descloitres, S. Alleaume, F. Petitcolin, and Y. Kaufman. 2002. The MODIS fire products. *Remote Sensing of Environment* 83 (1–2): 244–262.

Kaufman, Y.J., C.O. Justice, L.P. Flynn, J.D. Kendall, E.M. Prins, L. Giglio, D.E. Ward, W.P. Menzel, and A.W. Setzer. 1998. Potential global fire monitoring from EOS-MODIS. *Journal of Geophysical Research: Atmospheres* 103 (D24): 32215–32238.

Matson, M., and S.R. Schneider. 1984. *Fire detection using the NOAA-series satellite*. NOAA Technical Report NESDIS 7.

Peng, Guangxiong, Wei Shen, and Jifa Guo. 2011. Method on fire point automatic detection using HJ satellite. *Infrared and Laser Engineering* 40 (9): 1618–1623.

Qin, Xianlin, Zihui Zhang, and Zengyuan Li. 2010. An automatic forest fire identification method using HJ-IB data. *Remote Sensing Technology and Application* 25 (5): 700–706.

Zhang, Jijia, Qingshan Zhang, and Changgong Dian. 1989b. Detection of forest fire in Da Hinggan Ling region by meteorological satellite. *Acta Meteorological Sinica* 3 (4): 562–568.

Status and Drivers of Forest Fires in Myanmar

Sumalika Biswas and Krishna Prasad Vadrevu

Abstract The spatial relationship of fire and forest types at different scales provides insight into the underlying drivers of fires. In this study, we investigate the status of fires by forest type in Myanmar using two high-resolution datasets, active fire data from VIIRS (375 m), and a new national-level forest-type map of Myanmar (20 m). We assessed fire variations by forest type at the national, state, and landscape levels. At the national level, forest fires were the highest in Mixed Deciduous forests, followed by Bamboo, Upland Evergreen, Dry Deciduous, and Lowland Evergreen forests. During the dry season, fires occur expectedly in seasonally dry forests (Mixed Deciduous, Bamboo, Dry Deciduous) due to their low moisture and high fuel content. However, fires in Evergreen forests were also observed despite high moisture content. Fires in Upland Evergreen forests in Myanmar are mostly attributed to shifting agriculture, but the reason for fires in Lowland Evergreen forests remains unknown. Fire in Lowland Evergreen forests was most dominant in Tanintharyi. Analysis of the landscape-level fires in the Lowland Evergreen forests of Tanintharyi helped us understand the drivers of the fires.

Keywords Fire · Forest type · Lowland evergreen · Tanintharyi · Myanmar · Forest fires · Fire variations · Fire drivers

1 Introduction

Vegetation fires occur globally and have been detected at all times of the year (Dwyer et al. 2000; Giglio et al. 2006). They are common in moisture-deficient, fuel-laden dry forests (Otterstrom et al. 2006; Stephens et al. 2020) and are an integral part of fire-dependent ecosystems (Fava and Colombo 2017; Ratnam et al. 2019). In

S. Biswas (✉)
University of California, Los Angeles, USA
e-mail: sumalikabiswas@ucla.edu

K. P. Vadrevu
NASA Marshall Space Flight Center, Huntsville, AL, USA

K. P. Vadrevu et al. (eds.), *Vegetation Fires and Pollution in Asia*,
https://doi.org/10.1007/978-3-031-29916-2_3

moist tropical forests, fires are commonly used by humans to clear vegetated lands (Ketterings et al. 1999; Stolle et al. 2003). In lowland areas, forests are generally permanently converted for agriculture, pasture, urban, and infrastructure development (Curran et al. 2004; Pacheco 2006; Müller et al. 2012; Carter et al. 2017). In contrast, forests are temporarily converted in mountainous regions to practice shifting agriculture (Fox et al. 2000; Heinimann et al. 2017). Both forms of forest conversion are known to play an important role in regional and global climate change (Van der Werf et al. 2009; Carter et al. 2017; Pendrill et al. 2019). The other anthropogenic reasons of forest clearing include oil palm expansion, and agricultural residue or waste burning for clearing of land for the next crop, promoting grass growth for cattle grazing, accidental fires due to human negligence, etc., including policy decisions (Krishna Prasad and Badarinath 2006; Badarinath and Prasad 2011; Badarinath et al. 2007, 2008, 2009; Prasad et al. 2001a, b, 2002a; Albar et al. 2018; Biswas et al. 2021; Hayasaka et al. 2021; Lasko et al. 2017, 2018; Perez et al. 2021; Vadrevu 2008, 2015; Vadrevu and Badarinath 2009; Vadrevu and Justice 2011; Vadrevu et al. 2008a, b). At a regional and local scales, the burning of biomass from these activities can result in the disruption of forest structure and functions such as biogeochemical cycles, release of large amounts of radiatively active gases, aerosols, and other chemically active species that significantly alter the Earth's radiation balance and atmospheric chemistry (Choi et al. 2008; Kant et al. 2000; Prasad et al. 2002b, 2003a, b, 2004, 2005a, b; Prasad and Badarinath 2004; Kharol et al. 2012; Lasko and Vadrevu 2018; Vadrevu and Choi 2011; Vadrevu and Lasko 2015; Vadrevu et al. 2012, 2013, 2014, 2017, 2018, 2019a, b, c 2021a, b, 2022; Vay et al. 2011; Adam and Balasubramanian 2021; Phairuang 2021; Lam and Roy 2021; Park and Takeuchi 2021; Lee and Wang 2021; Uranishi et al. 2021; Syaufina and Maulana 2021; ul Haz and Tariq 2021).

Increased anthropogenic activity and climate change have greatly impacted the existing fire regimes, making the fire–climate–anthropogenic relationship more complex. Remote sensing has significant potential in mapping and monitoring of fires including emissions estimation (Justice et al. 2015; Wooster et al. 2021; Elvidge et al. 2021; Eaturu and Vadrevu 2021; Vadrevu 2021; Vadrevu et al. 2020, 2021a, b, 2022). The earliest quantitative study of fire dynamics on a global scale, based on National Oceanographic and Atmospheric Administration (NOAA) Advanced Very High Resolution Radiometer (AVHRR) satellite observations, revealed that 80% of the fires occur in the tropics (Dwyer et al. 1998). The potential of satellite remote sensing for mapping and monitoring of fires in Asia has been documented in Justice et al. (2015); Petropoulos et al. (2013). In mainland Southeast Asia, fires are common in all countries, with Myanmar leading in the number of fires observed (Vadrevu et al. 2019a, b, c). Recently available high-resolution VIIRS active fire data show that five times more VIIRS fires were observed in a shorter period in Myanmar compared to MODIS (Vadrevu et al. 2019a, b, c), including variations during COVID and pre-pandemic (Vadrevu et al. 2022).

Myanmar is a regional fire hotspot due to its high percentage of recurring fires (Vadrevu et al. 2019a, b, c). Most fires occur in March over forests, especially deciduous forests (Biswas et al. 2015a, b). Central Myanmar, which is dominated by Mixed Deciduous forests (Biswas et al. 2020), has been identified as a significant fire hotspot within the country (Biswas et al. 2015a, b). Previous studies on forest fires in Myanmar were conducted using coarse-resolution continental scale (300 m) global land cover maps or moderate-resolution global forest cover (30 m) products because, till recently, Myanmar lacked a national-level forest-type map. A new forest-type map developed by Biswas et al. (2021) addressed this need. The new map shows the extent and distribution of the diverse forest types in Myanmar for 2020 at 20 m resolution. The current study advances the previous efforts by using the newly developed national-level forest-type map to study forest fires at different scales in Myanmar. We answered the following questions,

Q1. What are the major forest types associated with fire in Myanmar at the national scale?
Q2. Which forest types do the fires dominate at the state level?
Q3. What drives the fires in the Lowland Evergreen forests of Tanintharyi?

2 Study Area

Myanmar lies between South and Southeast Asia in the Indo-Burma biodiversity hotspot. It extends from 9° 32′ N to 28° 31′ N and 92° 10′ E and 101° 11′ E, covering an area of 676,580 km^2. It neighbors include India and Bangladesh on the west, China on the north, and Laos and Thailand on the East. Myanmar comprises seven states (Chin, Kachin, Kayah, Kayin, Rakhine, Shan, Mon), seven regions (Ayeyarwady, Bago, Magway, Mandalay, Sagaing, Tanintharyi, Yangon), and the Union territory of Nay Pyi Taw, which is also its capital. Myanmar is one of the most forested countries in mainland Southeast Asia. Most of its forests are on mountains along the international borders and in the central hills of Bago Yoma. The large extents of highly biodiverse forests have been recently mapped (Fig. 1).

3 Data and Methods

3.1 Data

We used the highest resolution fire and forest-type dataset available for Myanmar.

Active fire data developed from the Visible Infrared Imaging Radiometer Suite (VIIRS) I band was used (Schroeder et al. 2014). The product has a spatial resolution of 375 m. Due to its high spatial resolution, it captures fires in relatively small areas (Vadrevu et al. 2019a, b, c). We expect this characteristic to help capture fires in the small, fragmented, dry forests in Central Myanmar. VIIRS active fire dataset for the year 2020 was obtained from Fire Information for Resource Management System (FIRMS) (https://firms.modaps.eosdis.nasa.gov/download/). For each fire observation, the location of the fire in latitude, longitude, and confidence level was given, among other properties. Active fires with high confidence were selected for analysis.

Fig. 1 Map of study area

More details about the VIIRS active fire data can be found at https://earthdata.nasa.gov/earth-observation-data/near-real-time/firms/viirs-i-band-active-fire-data.

Forest-type information was derived from a recent forest-type map of Myanmar (Biswas et al., 2021). The forest-type map was developed using optical (Sentinel-2) and radar (Sentinel-1, PALSAR) satellite images, training data on forest types and machine learning (Random Forest) methods. It has a spatial resolution of 20 m and represents the forest types in Myanmar in 2020. The major forest types mapped include Mangroves, Upland Evergreen forests, Lowland Evergreen forests, Mixed Deciduous forests, Dry Deciduous forests, Thorn forests, Bamboo, Swamp, and Plantations.

3.2 Methods

The VIIRS active fire data is available in point geometry and shapefile format. Each point observation represents the center of a 375 m pixel as the dataset has a spatial resolution of 375 m. A 375 m^2 buffer was created around each active fire point location to represent the active fire pixel. All active fire pixels were overlaid on the national-level forest-type map. For national-level analysis, the dominant forest type for each active fire pixel was computed using the majority function of zonal statistics in ArcGIS. This step reveals the dominant forest type associated with each fire observation. Next, the active fire pixels with associated dominant forest-type information were analyzed to determine each state's major forest type affected by fires. Finally, for landscape-level analysis, we used high-resolution imagery on Google Earth to investigate the clusters of fire pixels within the significant forest type affected to understand the drivers of the fires.

4 Results and Discussion

4.1 National Level

For the year 2020, we detected 17,190 high-confidence VIIRS active fire pixels at the national level in Myanmar. Of this, 77.14% (13,260) of fires occurred in forests. Most forest fires (13,257, 77.12%) occurred between January and May and between November and December (Fig. 2a). March was the peak forest fire month, with the highest number of forest fires (8510) observed (Fig. 2a). The fires (13,156, 76.53%) predominantly occurred between 5 and 8 am (Fig. 2b).

The highest number of fires was observed in the Mixed Deciduous forests (3529), followed by Bamboo (3485), Upland Evergreen (3166), and Dry Deciduous (1360) forests (Table 1). During the dry season, fires are a common occurrence in the seasonally dry forests (Mixed Deciduous, Bamboo, and Dry Deciduous forests) due to

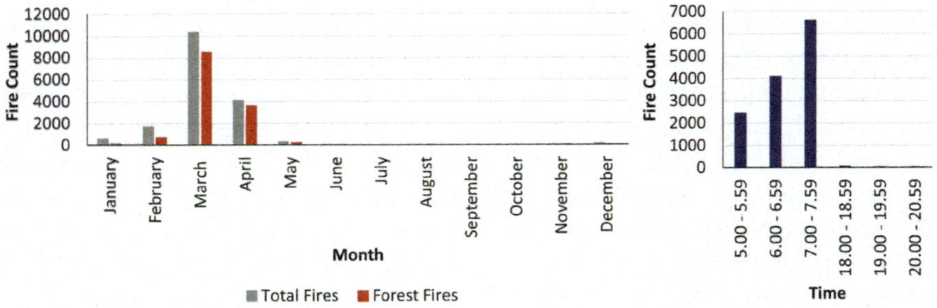

Fig. 2 **a** Monthly distribution of forest fire in Myanmar. **b** Distribution of forest fire in Myanmar by time of the day

their high fuel and low moisture content (Stott et al. 1990; Giri and Shrestha 2000; Baker et al. 2008; Baker and Bunyavejchewin 2009). Central Myanmar, which is dominated by Mixed Deciduous, Bamboo, and Dry Deciduous forests (Biswas et al. 2020), is a well-known fire hotspot within Myanmar (Biswas et al. 2015a, b). In Bago Yoma, in Central Myanmar, overexploitation of hardwood forests, especially globally renowned Teak forests, has led to considerable degradation (Mon et al. 2012), leading to the establishment of patches of Bamboo. Outside Rakhine state, Bamboo occurs as secondary vegetation in degraded forests or recovering fallow lands and is often considered a sign of forest degradation. The mixture of Mixed Deciduous forests and Bamboo patches makes these forests highly susceptible to forest fires (Stott et al. 1990; Giri and Shrestha 2000; Biswas et al. 2015a, b). In areas where Bamboo occurs in recovering fallow lands, fires indicate shifting cultivation practices after the fallow period. Bamboo brakes also occur in Rakhine and are considered a natural part of the ecosystem (Davis 1964). Fires in Bamboo breaks in Rakhine may be associated with Bamboo flowering events (Platt et al. 2010; Fava and Colombo 2017).

It is unexpected to find fires in Evergreen forests with high moisture content all year. Among the Upland and Lowland Evergreen forests, a large percentage (23.88%) of the total forest fires occurred in Upland Evergreen forests (Table 1). Fires in Upland Evergreen forests in montane mainland Southeast Asia are generally indicative of shifting cultivation (Fox et al. 2000; Heinimann et al. 2017), including upland areas of Myanmar (Biswas et al. 2015a, b; Shimizu et al. 2018). Though a smaller percentage of fires (5.28%) were observed in the Lowland Evergreen forests, it is of extreme importance as fires in lowland tropical forests often indicate land clearing resulting in deforestation (Langner and Siegert 2009; Van der Werf et al. 2009).

Fires in the remaining forest types (Thorn, Mangrove, and Swamp) were less than 1%. Though fires in high moisture forest ecosystems like Mangroves and Swamp are a clear sign of deforestation, Thorn forests are known to be cleared for agriculture using fire. In this study, we will focus on fires in the top five forest types. Among other classes studied, Plantations had 6.76% of the total fires. Setting fires in Plantations after harvest is standard practice for clearing land for the next set of commercial

Table 1 Number of fires by dominant forest type at national level

S. No.	Dominant forest type	Fire counts	Percent of total fires
1	Mixed Deciduous	3529	26.61
2	Bamboo	3485	26.28
3	Upland Evergreen	3166	23.88
4	Dry Deciduous	1360	10.26
5	Plantation	897	6.76
6	Lowland Evergreen	700	5.28
7	Thorn	59	0.44
8	Mangrove	35	0.26
9	Swamp	29	0.22

crops, similar to agriculture (Simorangkir 2007; Thumaty et al. 2015). Plantations are not considered a part of forests; hence not discussed further.

4.2 State Level

In Kachin, Chin, and Rakhine, fires were dominant in the Bamboo forests, while in Shan and Kayin in eastern Myanmar, fires were dominant in Upland Evergreen forests (Fig. 3). All the states lie on mountains inhabited mainly by ethnic minorities who practice the regional tradition of shifting cultivation to grow food to support themselves (Biswas et al. 2015a, b). In the method of shifting cultivation, native vegetation on hills or mountains, generally Upland Evergreen forests, is cleared for agriculture using fire, followed by planting food crops. After the harvest of food crops, the land is left barren and allowed to recover naturally. Bamboo, a fast-growing grass species, often grows as secondary vegetation in formerly fallow lands. Since Bamboo is not the native vegetation in these regions, the presence of Bamboo is considered a sign of degradation. The inhabitants re-cultivate the formerly abandoned lands after a specific time called the fallow period. Hence, the fires in Upland Evergreen and Bamboo forests could be due to land clearing for shifting agriculture. In addition, as discussed before, Bamboo brakes in Rakhine are considered a part of the natural forest ecosystem, and fires are an integral part of its ecology (Platt et al. 2010; Fava and Colombo 2017).

Fires dominated the Mixed Deciduous forests in Kayah, while forest fires dominated Sagaing, Magway, and Mandalay in the Dry Deciduous forests. As discussed before, seasonally dry forests (Mixed and Dry Deciduous forests) are naturally fire-prone due to their low moisture content. Further, the Dry Deciduous forests in the north of Central Myanmar, where these states are located, have been historically

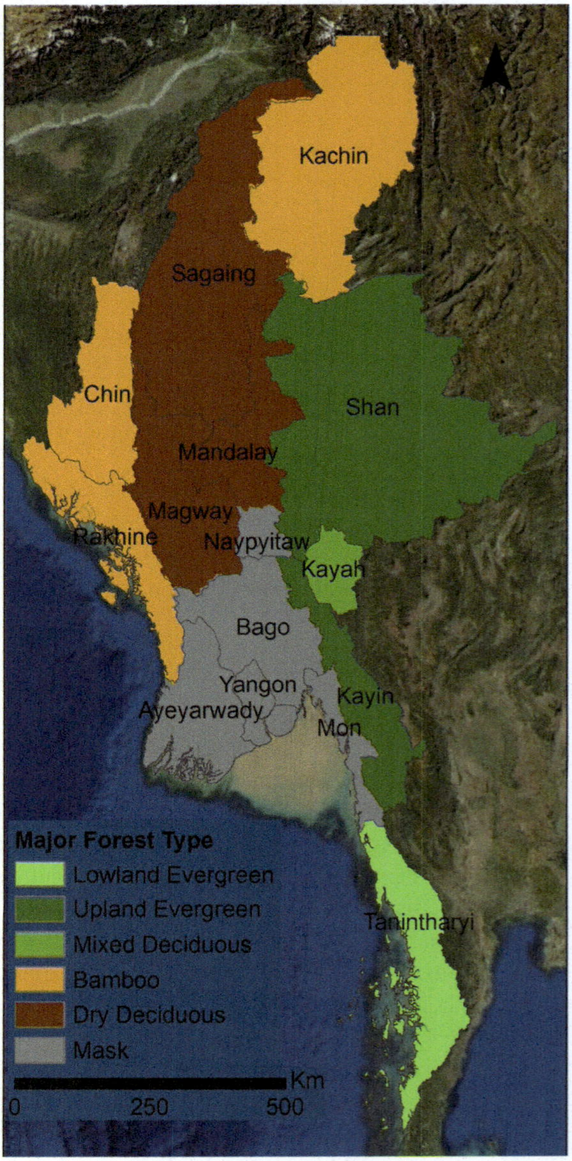

Fig. 3 Major forest types associated with fire at state level

deforested for conversion to agriculture (Leimgruber et al. 2005; Songer 2006), so the fires observed could be due to natural fire ecology or forest conversion by humans.

Most of the fires in Ayeyarwady, Bago, Nay Pyi Taw, Mon, and Yangon occurred in masked areas. The masked areas include the non-forest classes like agriculture, urban, and water. Investigation using high-resolution images showed that the fire locations coincided with agricultural lands, so we attribute the fires in masked areas to agricultural fires (Biswas et al. 2015a, b; Vadrevu et al. 2019a, b, c).

Tanintharyi, the southernmost state of Myanmar, was the only state where fires were predominant in Lowland Evergreen forests. The Lowland Evergreen forests in Tanintharyi constitute the Tanintharyi Sundaic Lowland Evergreen forests (Murray et al. 2020). They are highly diverse in flora and fauna and are located between the transition zone of the Indochinese and Sundaic regions (Donald et al. 2009). These are among the last significant stretches of remaining lowland forests in the Indochinese and Sundaic regions (Lambert and Collar 2002; Aung et al. 2017; De Alban et al. 2018). Other Lowland Evergreen forests in Thailand and Malaysia have been cleared for agriculture, notable conversion to oil palm and rubber (Aratrakorn et al. 2006). The forests in Tanintharyi were protected till 2011, mainly due to isolation caused by political instability in the region (De Alban et al. 2018). However, post-2011, a huge number of forest conversions to Plantations (oil palm and rubber) was reported in this region (Donald et al. 2015; Nomura et al. 2019; Biswas et al. 2021). Lowland Evergreen forests are naturally not prone to fire due to their high moisture content year-round, so the presence of fire observations implies forest conversion, as previously reported.

4.3 Landscape Level

Although Tanintharyi was dominated by fires in Lowland Evergreen forests at state-level analysis, the spatial pattern of forest fire distribution is not uniform. It had a distinct spatial distribution of forest fires at the landscape level (Fig. 4). Fires in northern Tanintharyi occurred mostly in Upland Evergreen forests (Fig. 4a), while most fires in southern Tanintharyi occurred in Lowland Evergreen forests (Fig. 4b). The fires in Upland Evergreen forests occurred in remote regions, deep inside the forests, and far away from transportation (road, river) networks (Fig. 5a). In contrast, fires in the Lowland Evergreen forests occurred near the roads along the edges of the mountains or rivers (Fig. 5b–d). The presence of fires which is commonly used as a land-clearing tool in the tropics (Ketterings et al. 1999; Stolle et al. 2003) near transportation networks is a clear indication of temporary or permanent forest cover change.

Using high-resolution images, we investigated three clusters of fires in Lowland Evergreen forests (Fig. 5b–d) to determine the drivers of the fires.

The first cluster is located near the town of Mawdaung, close to the Myanmar–Thailand border (Fig. 5b). As seen from high-resolution images, a huge oil palm Plantation lies at the center of the cluster along with a road. The road leads to Singkhon Pass, a nearby border checkpoint on the Myanmar–Thailand border on the eastern side, and Myeik, the largest city in Tanintharyi on the western side, which is distant. Analysis of the date and timings of the fires in the cluster show that on March 9, 2020, nine fires were observed between 5.56 and 7.18 am. It is reasonable to assume that the fires were set to clear the nearby Lowland Evergreen forests for Plantation expansion. The proximity of the Plantation to the Thai border and the ease of transportation of the products along the paved road to the international markets in

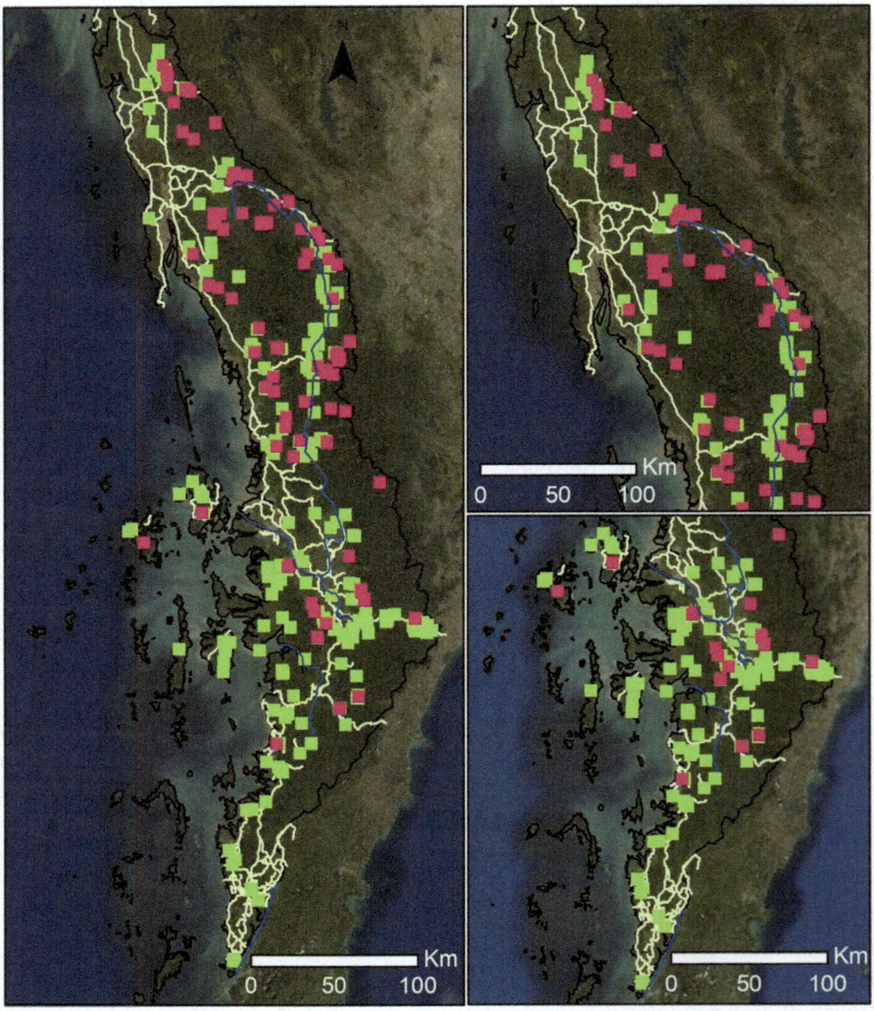

Fig. 4 Spatial distribution of forest fires in Tanintharyi. **a** Shows dominant presence of forest fires in Upland Evergreen forests (magenta). **b** Shows dominant presence of forest fires in Lowland Evergreen forests (light green)

neighboring Thailand fuel the expansion. Thus, in this cluster, Plantation expansion is the direct driver of fires.

The next cluster is located near the coast in Bokpyin township of Tanintharyi (Fig. 5c). This cluster lies along a tertiary road near the boundary of an oil palm Plantation and a Lowland Evergreen forest. Most fires in this cluster were observed on March 8, 2020, between 5.56 and 7.36 am. Like the cluster before, it is reasonable to assume that these fires were set to clear land for Plantation expansion, given its proximity to existing Plantations and the transportation network. These clusters are

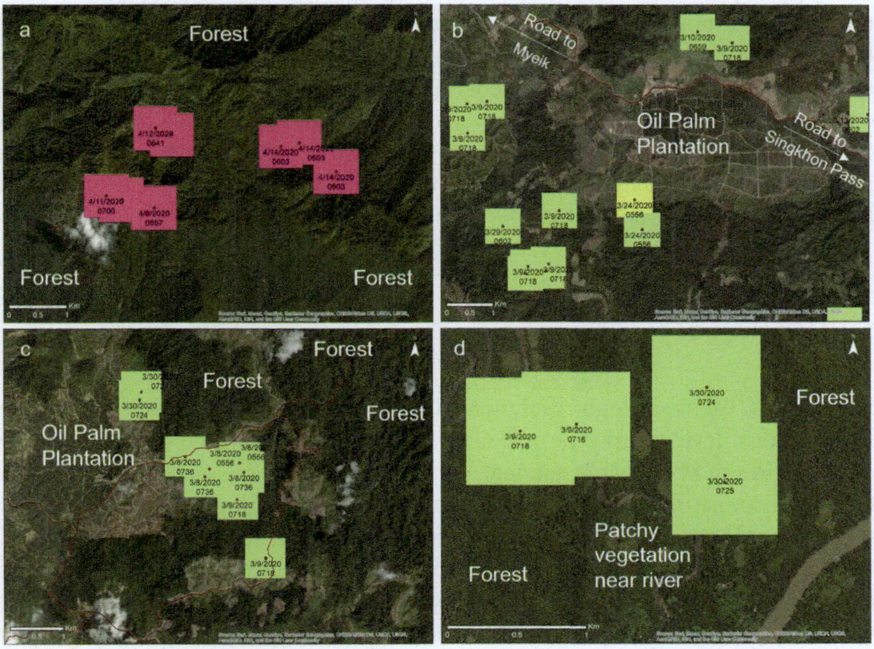

Fig. 5 High-resolution images of forest fire clusters in Tanintharyi. **a** Typical cluster in Upland Evergreen forest, characterized by remote location. **b–d** Typical cluster in Lowland Evergreen forests characterized by closeness to transportation (roads, river) networks

textbook examples of how gradual Plantation expansions encroach on neighboring forests. Thus, even in this cluster, fire is directly driven by Plantation expansion.

The final cluster is in Tenasserim in central Tanintharyi (Fig. 5d). This cluster lies along the tributary of the Tanintharyi River. Three fires in this cluster were observed on March 30, 2020, at approximately around 7.24 am. In contrast to the two previous examples, this cluster is far away from Plantations and roads. The river provides transportation to this site. Given the patchy nature of the surrounding landscape, a possible reason for the fire is shifting cultivation. Thus, the driver of fire in this cluster is shifting cultivation.

Our analysis of the fire locations, transportation networks (roads, rivers), and neighboring land use in all three clusters indicate that the fires were used to clear existing Lowland Evergreen forests for conversion to agriculture. In the first two clusters, it was to expand oil palm Plantations, i.e., forest conversion for large-scale commercial agriculture, and the last cluster was for shifting cultivation, which may be considered a form of subsistence agriculture. Thus, the main driver of fire in the Lowland Evergreen forests of Tanintharyi was conversion to agriculture. Agriculture is well recognized as the leading driver of tropical forest loss (DeFries et al. 2010; Carter et al. 2017; Pendrill et al. 2019). In the Asian region, the expansion of commercial agriculture to meet the demands of the global market is a well-known driver of tropical deforestation in Lowland Evergreen forests (Donald et al. 2015), while the traditional practice of shifting cultivation in the upland mountains for subsistence continues to drive temporary forest loss (Heinimann et al. 2017). However, shifting cultivation is gradually being replaced by more permanent forms of agriculture (Ziegler et al. 2011). The conversion of the last large tracts of the invaluable Sundaic Lowland forests to agriculture is of grave concern. If left unchecked, the global community faces the risk of losing the highly biodiverse Sundaic lowland forests and the biodiversity it supports forever.

5 Conclusion

In this study, we used a high-resolution VIIRS (375 m) Active Fire dataset and a new 20 m national-level forest-type map of Myanmar to investigate the status and drivers of forest fires in Myanmar at national, regional, and landscape scales. 77.14% of the fires in Myanmar occur in forest areas, in the morning, during the dry season, with peak fires in March. The top five forest types dominated by fire include Mixed Deciduous, Bamboo, Dry Deciduous, Upland Evergreen, and Lowland Evergreen forests. Previous research shows that fires may be expected in seasonal forests (Mixed Deciduous, Bamboo, Dry Deciduous), which are naturally fire-prone in the dry season, while shifting agriculture practice is known to drive fires in the Upland Evergreen forests in mountainous regions of Asia. The reason for fires in Lowland Evergreen forests in Tanintharyi was unknown. We investigated three different clusters of forest fires in the Lowland Evergreen forests of Tanintharyi and found conversion to agriculture (commercial and subsistence) to be a significant driver of forest fires in this biodiverse region. We also highlighted the concern in Tanintharyi, where the last large tract of highly biodiverse Sundaic Lowland Evergreen forests is being converted to agriculture. If left unchecked, these invaluable forests and the biodiversity it supports could be lost forever.

References

Adam, M.G., and R. Balasubramanian. 2021. Black carbon emissions from biomass burning in Southeast Asia—A review. In *Biomass burning in South and Southeast Asia*, 95–108.

Albar, I., I. Jaya, B.H. Saharjo, B. Kuncahyo, and K.P. Vadrevu. 2018. Spatio-temporal analysis of land and forest fires in Indonesia using MODIS active fire dataset. In *Land-atmospheric research applications in South and Southeast Asia*, 105–127. Cham: Springer.

Aratrakorn, S., S. Thunhikorn, and P.F. Donald. 2006. Changes in bird communities following conversion of lowland forest to oil palm and rubber plantations in southern Thailand. *Bird Conservation International* 16 (1): 71–82.

Aung, S.S., et al. 2017. Surveys in southern Myanmar indicate global importance for tigers and biodiversity. *Oryx* 51 (1): 13.

Badarinath, K.V.S., and K.V. Prasad. 2011. Carbon dioxide emissions from forest biomass burning in India. *Global Environmental Research* 15: 45–52.

Badarinath, K.V.S., S.K. Kharol, K.M. Latha, T.K. Chand, V.K. Prasad, A.N. Jyothsna, and K. Samatha. 2007. Multiyear ground-based and satellite observations of aerosol properties over a tropical urban area in India. *Atmospheric Science Letters* 8 (1): 7–13.

Badarinath, K.V.S., S. Kumar Kharol, V. Krishna Prasad, A. Rani Sharma, E.U.B. Reddi, H.D. Kambezidis, and D.G. Kaskaoutis. 2008. Influence of natural and anthropogenic activities on UV index variations—A study over tropical urban region using ground based observations and satellite data. *Journal of Atmospheric Chemistry* 59 (3): 219–236.

Badarinath, K.V.S., S.K. Kharol, A.R. Sharma, and V.K. Prasad. 2009. Analysis of aerosol and carbon monoxide characteristics over Arabian Sea during crop residue burning period in the Indo-Gangetic Plains using multi-satellite remote sensing datasets. *Journal of Atmospheric and Solar-Terrestrial Physics* 71 (12): 1267–1276.

Baker, P.J., and S. Bunyavejchewin. 2009. Fire behavior and fire effects across the forest landscape of continental Southeast Asia. In *Tropical fire ecology: Climate change, land use, and ecosystem dynamics*, ed. by M.A. Cochrane, 311–334. Berlin, Heidelberg: Springer (Springer Praxis Books). http://doi.org/10.1007/978-3-540-77381-8_11.

Baker, P.J., S. Bunyavejchewin, and A.P. Robinson. 2008. The impacts of large-scale, low-intensity fires on the forests of continental South-East Asia. *International Journal of Wildland Fire* 17 (6): 782–792.

Biswas, S., K.D. Lasko, and K.P. Vadrevu. 2015a. Fire disturbance in tropical forests of Myanmar—Analysis using MODIS satellite datasets. *IEEE Journal of Selected Topics in Applied Earth Observations and Remote Sensing* 8 (5): 2273–2281.

Biswas, S., K.P. Vadrevu, Z.M. Lwin, K. Lasko, and C.O. Justice. 2015b. Factors controlling vegetation fires in protected and non-protected areas of Myanmar. *PLoS One* 10 (4): e0124346.

Biswas, S., et al. 2020. A multi sensor approach to forest type mapping for advancing monitoring of sustainable development goals (SDG) in Myanmar. *Remote Sensing* 12 (19): 3220.

Biswas, S., K.P. Vadrevu, M.S. Mon, and C. Justice. 2021. Contemporary forest loss in Myanmar: Effect of democratic transition and subsequent timber bans on landscape structure and composition. *Ambio* 50 (4): 914–928.

Biswas, S., Q. Huang, M. Aung, F. Arnold, M.S. Mon and P. Leimgruber. 2021. Myanmar forest type map 2020. National Zoo and Smithsonian Conservation Biology Institute. Figure. https://doi.org/10.25573/data.16613818.v1

Carter, S., et al. 2017. Agriculture-driven deforestation in the tropics from 1990–2015: Emissions, trends and uncertainties. *Environmental Research Letters* 13 (1): 014002.

Choi, Y., S.A. Vay, K.P. Vadrevu, A.J. Soja, J.H. Woo, S.R. Nolf, G.W. Sachse, G.S. Diskin, D.R. Blake, N.J. Blake, and H.B. Singh. 2008. Characteristics of the atmospheric CO_2 signal as observed over the conterminous United States during INTEX-NA. *Journal of Geophysical Research: Atmospheres* 113 (D7).

Curran, L.M., et al. 2004. Lowland forest loss in protected areas of Indonesian Borneo. *Science* 303 (5660): 1000–1003.

Davis, J. H. (1964) *The forests of Burma*. New York: New York Botanical Garden.

De Alban, J.D.T., et al. 2018. Combined Landsat and L-band SAR data improves land cover classification and change detection in dynamic tropical landscapes. *Remote Sensing* 10 (2): 306. https://doi.org/10.3390/rs10020306.

DeFries, R.S., et al. 2010. Deforestation driven by urban population growth and agricultural trade in the twenty-first century. *Nature Geoscience* 3 (3): 178–181.

Donald, P.F., et al. 2009. Population, distribution, habitat use and breeding of Gurney's Pitta Pitta gurneyi in Myanmar and Thailand. *Bird Conservation International* 19 (4): 353–366.

Donald, P.F., et al. 2015. Social reform and a growing crisis for southern Myanmar's unique forests. *Conservation Biology* 29 (5): 1485–1488.

Dwyer, E., J.M. Gregoire, and J.P. Malingreau. 1998. A global analysis of vegetation fires using satellite images: Spatial and temporal dynamics. *Ambio* 27. Available at: https://www.osti.gov/etdeweb/biblio/612117. Accessed: July 30, 2021.

Dwyer, E., et al. 2000. Global spatial and temporal distribution of vegetation fire as determined from satellite observations. *International Journal of Remote Sensing* 21 (6–7): 1289–1302. https://doi.org/10.1080/014311600210182.

Eaturu, A., and K.P. Vadrevu. 2021. Evaluation of Sentinel-3 SLSTR data for mapping fires in forests, peatlands, and croplands—A case study over Australia, Indonesia, and India. In *Biomass burning in South and Southeast Asia*, 39–59. Boca Raton: CRC Press.

Elvidge, C.D., M. Zhizhin, K. Baugh, and F.C. Hsu. 2021. Identification of smoldering peatland fires in Indonesia via triple-phase temperature analysis of VIIRS nighttime data. In *Biomass burning in South and Southeast Asia*, 25–38. Boca Raton: CRC Press.

Fava, F., and R. Colombo. 2017. Remote sensing-based assessment of the 2005–2011 bamboo reproductive event in the Arakan mountain range and its relation with wildfires. *Remote Sensing* 9 (1): 85. https://doi.org/10.3390/rs9010085.

Fox, J., et al. 2000. Shifting cultivation: A new old paradigm for managing tropical forests. *BioScience* 50 (6): 521–528.

Giglio, L., I. Csiszar, and C.O. Justice. 2006. Global distribution and seasonality of active fires as observed with the terra and aqua moderate resolution imaging spectroradiometer (MODIS) sensors. *Journal of Geophysical Research: Biogeosciences* 111 (G2). http://doi.org/10.1029/2005JG000142.

Giri, C., and S. Shrestha. 2000. Forest fire mapping in Huay Kha Khaeng wildlife sanctuary, Thailand.

Hayasaka, H., A. Sepriando, A. Usup, and N. Daisuke. 2021. Air pollution conditions near peat fire-prone areas during El Niño in Central Kalimantan, Indonesia. In *Biomass burning in South and Southeast Asia*, 233–247. Boca Raton: CRC Press.

Heinimann, A., et al. 2017. A global view of shifting cultivation: Recent, current, and future extent. *PLoS One* 12 (9): e0184479.

Justice, C., G. Gutman, and K.P. Vadrevu. 2015. NASA land cover and land use change (LCLUC): An interdisciplinary research program. *Journal of Environmental Management* 148: 4–9.

Kant, Y., V.K. Prasad, and K.V.S. Badarinath. 2000. Algorithm for detection of active fire zones using NOAA AVHRR data. *Infrared Physics & Technology* 41 (1): 29–34.

Ketterings, Q.M., et al. 1999. Farmers' perspectives on slash-and-burn as a land clearing method for small-scale rubber producers in Sepunggur, Jambi Province, Sumatra, Indonesia. *Forest Ecology and Management* 120 (1–3): 157–169.

Kharol, S.K., K.V.S. Badarinath, A.R. Sharma, D.V. Mahalakshmi, D. Singh, and V.K. Prasad. 2012. Black carbon aerosol variations over Patiala city, Punjab, India—A study during agriculture crop residue burning period using ground measurements and satellite data. *Journal of Atmospheric and Solar-Terrestrial Physics* 84: 45–51.

Krishna Prasad, V., and K.V.S. Badarinath. 2006. Soil surface nitrogen losses from agriculture in India: A regional inventory within agroecological zones (2000–2001). *The International Journal of Sustainable Development and World Ecology* 13 (3): 173–182.

Lam, Y.F., and S. Roy. 2021. Impact of biomass burning on local air quality in South China. In *Biomass burning in South and Southeast Asia*, 39–56. Boca Raton: CRC Press.

Lambert, F.R., and N.J. Collar. 2002. The future for Sundaic lowland forest birds: Long-term effects of commercial logging and fragmentation. *Forktail* 127–146.

Langner, A., and F. Siegert. 2009. Spatiotemporal fire occurrence in Borneo over a period of 10 years. *Global Change Biology* 15 (1): 48–62.

Lasko, K., and K. Vadrevu. 2018. Improved rice residue burning emissions estimates: Accounting for practice-specific emission factors in air pollution assessments of Vietnam. *Environmental Pollution* 236: 795–806.

Lasko, K., K.P. Vadrevu, V.T. Tran, E. Ellicott, T.T. Nguyen, H.Q. Bui, and C. Justice. 2017. Satellites may underestimate rice residue and associated burning emissions in Vietnam. *Environmental Research Letters* 12 (8): 085006.

Lasko, K., K.P. Vadrevu, and T.T.N. Nguyen. 2018. Analysis of air pollution over Hanoi, Vietnam using multi-satellite and MERRA reanalysis datasets. *PLoS One* 13 (5): e0196629.

Lee, H.H., and C. Wang. 2021. Impacts of biomass burning aerosols on air quality and convective systems in Southeast Asia. In *Biomass burning in South and Southeast Asia*, 267–280. Boca Raton: CRC Press.

Leimgruber, P., et al. 2005. Forest cover change patterns in Myanmar (Burma) 1990–2000. *Environmental Conservation* 356–364.

Mon, M.S., et al. 2012. Factors affecting deforestation and forest degradation in selectively logged production forest: A case study in Myanmar. *Forest Ecology and Management* 267: 190–198.

Müller, R., et al. 2012. Proximate causes of deforestation in the Bolivian lowlands: An analysis of spatial dynamics. *Regional Environmental Change* 12 (3): 445–459.

Murray, N.J., et al. 2020. Threatened ecosystems of Myanmar. An IUCN red list of ecosystems assessment. Version 1.0.

Nomura, K., et al. 2019. Oil palm concessions in southern Myanmar consist mostly of unconverted forest. *Scientific Reports* 9 (1): 11931. https://doi.org/10.1038/s41598-019-48443-3.

Otterstrom, S.M., M.W. Schwartz, and I. Velázquez-Rocha. 2006. Responses to fire in selected tropical dry forest trees 1. *Biotropica* 38 (5): 592–598.

Pacheco, P. 2006. Agricultural expansion and deforestation in lowland Bolivia: The import substitution versus the structural adjustment model. *Land Use Policy* 23 (3): 205–225.

Park, H., and W. Takeuchi. 2021. Satellite-based estimation of global CO_2 emissions from biomass burning. In *Biomass burning in South and Southeast Asia*, 109–125. Boca Raton: CRC Press.

Pendrill, F., et al. 2019. Agricultural and forestry trade drives large share of tropical deforestation emissions. *Global Environmental Change* 56: 1–10.

Perez, G.J., J.C. Comiso, and M.G. Cayetano. 2021. Swidden agriculture and biomass burning in the Philippines. In *Biomass burning in South and Southeast Asia*, 183–198. Boca Raton: CRC Press.

Petropoulos, G.P., K.P. Vadrevu, and C. Kalaitzidis. 2013. Spectral angle mapper and object-based classification combined with hyperspectral remote sensing imagery for obtaining land use/cover mapping in a Mediterranean region. *Geocarto International* 28 (2): 114–129.

Phairuang, W. 2021. Biomass burning and their impacts on air quality in Thailand. In *Biomass burning in South and Southeast Asia*, 21–38. Boca Raton: CRC Press.

Platt, S.G., et al. 2010. Notes on Melocanna baccifera and bamboo brakes in the Rakhine Hills of western Myanmar. *Bamboo Science & Culture* 23 (1).

Prasad, V.K., and K.V.S. Badarinth. 2004. Land use changes and trends in human appropriation of above ground net primary production (HANPP) in India (1961–98). *Geographical Journal* 170 (1): 51–63.

Prasad, V.K., Y. Kant, and K.V.S. Badarinath. 2001a. CENTURY ecosystem model application for quantifying vegetation dynamics in shifting cultivation areas: A case study from Rampa Forests, Eastern Ghats (India). *Ecological Research* 16 (3): 497–507.

Prasad, V.K., Y. Kant, P.K. Gupta, C. Sharma, A.A. Mitra, and K.V.S. Badarinath. 2001b. Biomass and combustion characteristics of secondary mixed deciduous forests in Eastern Ghats of India. *Atmospheric Environment* 35 (18): 3085–3095.

Prasad, V.K., Y. Kant, and K.V.S. Badarinath. 2002a. Land use changes and modeling carbon fluxes from satellite data. *Advances in Space Research* 30 (11): 2511–2516.

Prasad, V.K., Y. Kant, P.K. Gupta, C. Elvidge, and K.V.S. Badarinath. 2002b. Biomass burning and related trace gas emissions from tropical dry deciduous forests of India: A study using DMSP-OLS data and ground-based measurements. *International Journal of Remote Sensing* 23 (14): 2837–2851.

Prasad, V.K., K.V.S. Badarinath, H. Tsuruta, S. Sudo, S. Yonemura, J. Cardina, B. Stinner, R. Moore, D. Stinner, and C. Hoy. 2003a. Implications of land use changes on carbon dynamics and sequestration—Evaluation from forestry datasets, India. *The Environmentalist* 23 (2): 175–187.

Prasad, V.K., M. Lata, and K.V.S. Badarinath. 2003b. Trace gas emissions from biomass burning from northeast region in India—Estimates from satellite remote sensing data and GIS. *The Environmentalist* 23 (3): 229–236.

Prasad, V.K., K.V.S. Badarinath, S. Yonemura, and H. Tsuruta. 2004. Regional inventory of soil surface nitrogen balances in Indian agriculture (2000–2001). *Journal of Environmental Management* 73 (3): 209–218.
Prasad, V.K., E. Anuradha, and K.V.S. Badarinath. 2005a. Climatic controls of vegetation vigor in four contrasting forest types of India—Evaluation from National Oceanic and Atmospheric Administration's advanced very high resolution radiometer datasets (1990–2000). *International Journal of Biometeorology* 50 (1): 6–16.
Prasad, V.K., A. Ortiz, B. Stinner, D. Mccartney, J. Parker, D. Hudgins, C. Hoy, and R. Moore. 2005b. Exploring the relationship between hydrologic parameters and nutrient loads using digital elevation model and GIS—A case study from Sugarcreek headwaters, Ohio, USA. *Environmental Monitoring and Assessment* 110 (1): 141–169.
Ratnam, J., et al. 2019. Functional traits of trees from dry deciduous "forests" of southern India suggest seasonal drought and fire are important drivers. *Frontiers in Ecology and Evolution* 7: 8.
Schroeder, W., et al. 2014. The new VIIRS 375 m active fire detection data product: Algorithm description and initial assessment. *Remote Sensing of Environment* 143: 85–96.
Shimizu, K., et al. 2018. Patch-based assessments of shifting cultivation detected by Landsat time series images in Myanmar. *Sustainability* 10 (9): 3350.
Simorangkir, D. 2007. Fire use: Is it really the cheaper land preparation method for large-scale plantations? *Mitigation and Adaptation Strategies for Global Change* 12 (1): 147–164. https://doi.org/10.1007/s11027-006-9049-2.
Songer, M.A. 2006. Endangered dry deciduous forests of Upper Myanmar (Burma): A multi-scale approach for research and conservation. Ph.D. thesis.
Stephens, S.L., et al. 2020. Fire and climate change: Conserving seasonally dry forests is still possible. *Frontiers in Ecology and the Environment* 18 (6): 354–360.
Stolle, F., et al. 2003. Land use and vegetation fires in Jambi Province, Sumatra, Indonesia. *Forest Ecology and Management* 179 (1–3): 277–292.
Stott, P.A., J.G. Goldammer, and W.L. Werner. 1990. The role of fire in the tropical lowland deciduous forests of Asia. In *Fire in the tropical biota*, 32–44. Berlin: Springer.
Syaufina, L., and S.I. Maulana. 2021. Biomass burning emissions in Indonesia and policy measures—An overview. In *Biomass burning in South and Southeast Asia*, 135–148.
Thumaty, K.C., et al. 2015. Spatio-temporal characterization of agriculture residue burning in Punjab and Haryana, India, using MODIS and Suomi NPP VIIRS data. *Current Science* 109 (10): 1850.
ul Haq, Z., and S. Tariq. 2021. Impact of biomass burning on surface-level carbon monoxide over Lahore and Karachi and their comparison with South Asian megacities. In *Biomass burning in South and Southeast Asia*, 57–75. Boca Raton: CRC Press.
Uranishi, K., H. Shimadera, and A. Kondo. 2021. Biomass burning influence on PM2.5 regional and long-range transport in Northeast Asia. In *Biomass burning in South and Southeast Asia*, 281–295. Boca Raton: CRC Press.
Vadrevu, K.P. 2008. Analysis of fire events and controlling factors in eastern India using spatial scan and multivariate statistics. *Geografiska Annaler: Series A, Physical Geography* 90 (4): 315–328.
Vadrevu, K. 2015. International regional science meeting of NASA-LCLUC: Land cover/land use change (LC/LUC) and environmental impacts in South Asia, Coimbatore, India, 19–23 January 2013. *Journal of Environmental Management* 148: 1–163.
Vadrevu, K.P. 2021. Fire danger indices and methods: An appraisal. In *Biomass burning in South and Southeast Asia*, 201–232. Boca Raton: CRC Press.
Vadrevu, K.P., and K.V.S. Badarinath. 2009. Spatial pattern analysis of fire events in Central India—A case study. *Geocarto International* 24 (2): 115–131.
Vadrevu, K.P., and Y. Choi. 2011. Wavelet analysis of airborne CO_2 measurements and related meteorological parameters over heterogeneous landscapes. *Atmospheric Research* 102 (1–2): 77–90.

Vadrevu, K.P., and C.O. Justice. 2011. Vegetation fires in the Asian region: Satellite observational needs and priorities. *Global Environmental Research* 15 (1): 65–76.

Vadrevu, K., and K. Lasko. 2015. Fire regimes and potential bioenergy loss from agricultural lands in the Indo-Gangetic Plains. *Journal of Environmental Management* 148: 10–20.

Vadrevu, K.P., K.V.S. Badarinath, and E. Anuradha. 2008a. Spatial patterns in vegetation fires in the Indian region. *Environmental Monitoring and Assessment* 147 (1): 1–13.

Vadrevu, K.P., J. Cardina, F. Hitzhusen, I. Bayoh, R. Moore, J. Parker, B. Stinner, D. Stinner, and C. Hoy. 2008b. Case study of an integrated framework for quantifying agroecosystem health. *Ecosystems* 11 (2): 283–306.

Vadrevu, K.P., L. Giglio, and C. Justice. 2013. Satellite based analysis of fire–carbon monoxide relationships from forest and agricultural residue burning (2003–2011). *Atmospheric Environment* 64: 179–191.

Vadrevu, K.P., T. Ohara, and C. Justice. 2014. Air pollution in Asia. *Environmental Pollution (Barking, Essex: 1987)* 195: 233–235.

Vadrevu, K., T. Ohara, and C. Justice. 2017. Land cover, land use changes and air pollution in Asia: A synthesis. *Environmental Research Letters* 12 (12): 120201.

Vadrevu, K.P., T. Ohara, and C. Justice (eds.). 2018. *Land-atmospheric research applications in South and Southeast Asia.*

Vadrevu, K.P., et al. 2019a. Trends in vegetation fires in South and Southeast Asian countries. *Scientific Reports* 9 (1): 7422. https://doi.org/10.1038/s41598-019-43940-x.

Vadrevu, K., A. Heinimann, G. Gutman, and C. Justice. 2019b. Remote sensing of land use/cover changes in South and Southeast Asian countries. *International Journal of Digital Earth* 12 (10): 1099–1102.

Vadrevu, K.P., V.K. Dadhwal, G. Gutman, and C. Justice. 2019c. Remote sensing of agriculture—South/Southeast Asia research initiative special issue. *International Journal of Remote Sensing* 40 (21): 8071–8075.

Vadrevu, K.P., E. Ellicott, L. Giglio, K.V.S. Badarinath, E. Vermote, C. Justice, and W.K. Lau. 2012. Vegetation fires in the himalayan region–Aerosol load, black carbon emissions and smoke plume heights. *Atmospheric Environment* 47: 241–251.

Vadrevu, K.P., A. Eaturu, S. Biswas, K. Lasko, S. Sahu, J.K. Garg, and C. Justice. 2020. Spatial and temporal variations of air pollution over 41 cities of India during the COVID-19 lockdown period. *Scientific Reports* 10 (1): 1–15.

Vadrevu, K.P., T. Ohara, and C. Justice (eds.). 2021a. *Biomass burning in South and Southeast Asia: Impacts on the biosphere*, vol. 2. Boca Raton: CRC Press.

Vadrevu, K.P., T. Ohara, and C. Justice (eds.). 2021b. *Biomass burning in South and Southeast Asia: Mapping and monitoring*, vol. 1. Boca Raton: CRC Press.

Vadrevu, K.P., A. Eaturu, E. Casadaban, and S. Biswas. 2022. Agricultural fires in South Asian countries and implications. In *remote sensing of agriculture and land cover/land use changes in South and Southeast Asian countries*, 501–516. Cham: Springer.

Van der Werf, G.R., et al. 2009. Estimates of fire emissions from an active deforestation region in the southern Amazon based on satellite data and biogeochemical modelling. *Biogeosciences* 6 (2): 235–249.

Vay, S.A., Y. Choi, K.P. Vadrevu, D.R. Blake, S.C. Tyler, A. Wisthaler, A. Hecobian, Y. Kondo, G.S. Diskin, G.W. Sachse, and J.H. Woo. 2011. Patterns of CO_2 and radiocarbon across high northern latitudes during International Polar Year 2008. *Journal of Geophysical Research: Atmospheres* 116 (D14).

Wooster, M.J., G.J. Roberts, L. Giglio, D.P. Roy, P.H. Freeborn, L. Boschetti, C. Justice, C. Ichoku, W. Schroeder, D. Davies, and A.M. Smith. 2021. Satellite remote sensing of active fires: History and current status, applications and future requirements. *Remote Sensing of Environment* 267: 112694.

Ziegler, A.D., et al. 2011. Recognizing contemporary roles of swidden agriculture in transforming landscapes of Southeast Asia. *Conservation Biology* 25 (4): 846–848.

Vegetation Fires and Entropy Variations in Myanmar

Krishna Prasad Vadrevu, Pranith Salikineedi, Aditya Eaturu, and Sumalika Biswas

Abstract Vegetation fires are most common in South/Southeast Asian countries. For effective mitigation and control of fires, it is essential to quantify the spatial and temporal patterns, including the variability. In this study, we use Shannon's entropy measure to quantify the heterogeneity or a degree of randomness in fires in Myanmar. We used VIIRS 375 m and MODIS 1 km satellite data to quantify the spatial and temporal variations in fire counts (FC) and burnt areas (BA). VIIRS fire analysis suggested the mean FC from 2012 to 2020 for Myanmar with 34,7930 FC per year, with the highest 403,292 in 2013 and a minimum of 254,106 in 2018. Most of the fires with high intensity (FRP) occurred in the dry season (February-March-April), with the highest intensity in March (~ 203,897MW). Temporal variations in BA suggested an average of 655,296.85 (km^2) BA per year, with the highest in forests (411,125.75 km^2), followed by croplands (159,908.93 km^2) and grasslands (84,262.16 km^2). Results suggested forest fires with higher entropies than agricultural fires. Specifically, the forest fires in central Myanmar bordering the southern Sagaing, Shan north, and Mandalay and Magway regions had higher entropies. Further, entropy values did not show significant variations with the elevation, except in northern Kachin, Shan (east), northwestern Sagaing, and forest lands in the Chin state and along Magway and Rakhine regions. The entropy index indicates variability and is a measure of disorder or a degree of uncertainty, such as randomness, unevenness, irregularity, and complexity. Thus, controlling the fires in regions with high entropy can be challenging. The results also identify hotspots of vegetation fires with less entropy for effective fire control and management.

Keywords Fire variability · Satellite data · Entropy · Myanmar

K. P. Vadrevu (✉)
NASA Marshall Space Flight Center, Huntsville, AL, USA
e-mail: Krishna.p.vadrevu@nasa.gov

P. Salikineedi · A. Eaturu
University of Alabama, Huntsville, USA

S. Biswas
University of California, Los Angeles, USA

K. P. Vadrevu et al. (eds.), *Vegetation Fires and Pollution in Asia*,
https://doi.org/10.1007/978-3-031-29916-2_4

1 Introduction

Fire is a natural phenomenon in most ecosystems that have shaped them over time (Chuvieco 2009; Bond and Keeley 2005). In addition to the natural drivers such as lightning fires or spontaneous combustion of dried leaves or vegetation due to intense droughts, fires in tropical ecosystems, in-particular South/Southeast Asia (S/SEA), are driven by humans (Stott 2000). Fire is used as a land clearing tool through slash and burns in S/SEA countries (Biswas et al. 2015a, b, 2021; Prasad et al. 2002a; Yadav 2013). In addition, in several agricultural ecosystems, fires are used to dispose of crop residues and clearing of land for the next crop. Also, fires are used for hunting, maintaining grasslands, controlling pests, facilitating cattle migration, etc. (Hough 1993; Albar et al. 2018; Lasko and Vadrevu 2018; Lasko et al. 2017, 2018). Regardless of the reasons for the ignition, fires can cause devastating damage to both environment and humans. Fires can result in a total or partial loss of vegetation resulting in changes in vegetation composition, structure, and function, including the disruption of biogeochemical cycles. For example, repeated burning in forested ecosystems can result in pyrodenitrification (Crutzen and Andreae 1990). Also, the landscapes that are affected by recurrent burning are characterized by soils of low nutrient status. The other significant effect of fires and biomass burning relates to the release of greenhouse gas emissions and aerosols, which can have both direct and indirect effects on the climate (Badarinath et al. 2007, 2008, 2009; Badarinath and Prasad 2011; Choi et al. 2008; Kant et al. 2000; Kharol et al. 2012; Lata et al. 2001; Vadrevu et al. 2008, 2013, 2014a, 2022a, b; Vay et al. 2011). The impacts of biomass burning, especially the pollutants released, can depend on various factors such as the type of vegetation and the amount burnt, including the frequency, intensity of the fires, and the burning efficiency (Eva and Lambin 2000; Vadrevu et al. 2017, 2018, 2019a, b, 2022).

Fighting vegetation fires and addressing the resulting impacts on the environment is highly challenging due to their spatial and temporal variability caused due to climate, vegetation, topography, and several other land-use/cover change drivers, including wide-ranging ecological, environmental, and social impacts (Prasad et al. 2001a, b, 2002a, b, 2003, 2004, 2005; Prasad and Badarinath 2004, 2006; Vadrevu 2015; Vadrevu and Lasko 2015; Vadrevu et al. 2019a, b). Mapping and monitoring vegetation fires are critical to mitigating fires and any impacts. Typical ground-based surveys to monitor fire occurrences are complex as they can occur in remote areas that are not easily accessible. In contrast, frequent updates concerning the progress of vegetation fires are essential for effective mitigation and control. In recent decades, rapid developments in remote sensing and geospatial technologies led to the introduction of a new generation of solutions for the early detection, mapping, and monitoring of fires (Justice et al. 2015; Wooster et al. 2021). Remote sensing data has proven indispensable for fire mapping and monitoring due to its multi-spectral, multi-temporal, synoptic, and repetitive coverage capabilities (Vadrevu and Justice 2011).

Combined with the robust algorithms, the remote sensing data can be effectively used to map and monitor spatial and temporal variability in fires and impacts (Vadrevu et al. 2021a, b), including post-fire land cover changes (Petropoulos et al. 2013) useful for fire control in different regions of the world.

One of the critical characteristics of fires that make it challenging to monitor is spatial variability (Vadrevu 2008; Vadrevu et al. 2008). Fires can vary spatially and temporally, and it is important to capture the variations for designing appropriate prevention measures and allocating firefighting resources for effective management and control (Vadrevu and Badarinath 2009). The present study aims to capture the spatial fire variability using the entropy index in Myanmar based on VIIRS satellite-derived fires. We focused on Myanmar, which stands first in the entire S/SEA for total fire counts (Vadrevu et al. 2019a, b). The spatial fire variability has been mapped to quantify the degree of randomness or irregularity in different regions of the country. We analyzed fire variability for the entire country and specific vegetation types, i.e., agricultural and forest fires, separately for entropy and relative entropy variations. Further, fire intensity variations were analyzed in addition to burnt areas and temporal variations. Furthermore, we used topography as an indicator of fire entropy variations. A lack of clustering among fires would indicate that local-scale mechanisms such as vegetation diversity and structure, topography, and soil variability, including different land-use practices, have played a more significant role in determining spatial fire variability. The discussion focuses on the entropy of fires and the implications of fire heterogeneity for fire mitigation and control, including post-fire vegetation response and restoration efforts.

2 Study Area

We focused our study on Myanmar. The country is formerly known as Burma and lies between latitudes 9° and 29° N, and longitudes 92° and 102° E. It is the largest country in mainland Southeast Asia. It is bordered by Bangladesh and India to its northwest, China to its northeast, Laos and Thailand to its east and southeast, and the Andaman Sea and the Bay of Bengal to its south and southwest. The country's capital city is Naypyidaw, and its largest city is Yangon. The country's total area is 678,500 km^2 (262,000 mi^2). Much of the country lies between the Tropic of Cancer and the Equator. The country lies in the monsoon region of Asia, with its coastal regions receiving over 5000 mm (196.9 in.) of rain annually. Annual rainfall in the delta region is approximately 2500 mm (98.4 in.), while average annual rainfall in the dry zone in central Myanmar is less than 1000 mm (39.4 in.). The northern regions of Myanmar are the coolest, with average temperatures of 21 °C (70 °F). Coastal and delta regions have an average maximum temperature of 32 °C (89.6 °F). The country is divided into seven states and seven regions, formerly called divisions (Fig. 1). Regions are predominantly inhabited by Bamar–Myanmar's dominant ethnic

Fig. 1 Myanmar with seven states and seven regions. Kachin, Kayah, Kayin, Chin, Mon, Rakhine, and Shan are states, whereas Sagaing, Tanintharyi, Bago, Magaway, Mandalay, and Ayeyarwady are regions

group. States, in essence, are regions that are home to particular ethnic minorities. The administrative divisions are further subdivided into districts, which are further subdivided into townships, wards, and villages. The elevation varies from 0 to 3917 (m) (Fig. 2).

3 Datasets

3.1 VIIRS Fires

We used the 375 m active fire product derived from the VIIRS instruments onboard the Suomi National Polar-orbiting Partnership (S-NPP) and NOAA-20 satellites. In contrast to other coarser resolution satellite fire detection products such as MODIS ($\geq$ 1 km), the improved 375 m data provide increased detection of smaller fires and improved mapping of large fire perimeters (Schroeder et al. 2014). The VIIRS 375 m

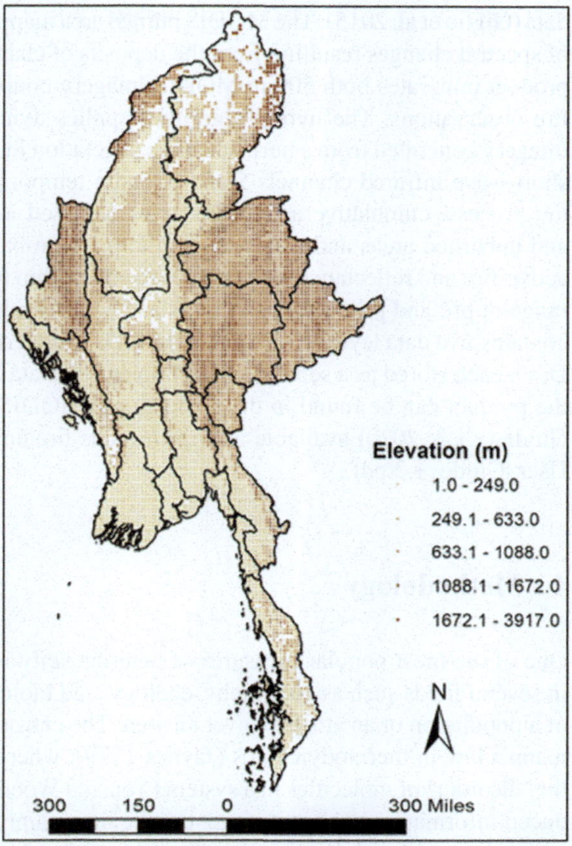

Fig. 2 Elevation of Myanmar derived from GTOPO-30 data at 1 km resolution

fire product builds on the earlier MODIS fire product heritage (Giglio et al. 2003, 2018; Kaufman et al. 1998), using a multi-spectral contextual algorithm to identify sub-pixel fire activity and other thermal anomalies in the Level 1 (swath) input data. The algorithm uses all five 375 m VIIRS channels to detect fires and separate land, water, and cloud pixels in the image (Schroeder et al. 2014). Near real-time data are available in various formats, including the TXT, SHP, KML, and WMS from https://earthdata.nasa.gov/active-fire-data. We gridded the VIIRS fire data at 5-arcminutes, i.e., 9.297 km intervals.

3.2 *MODIS Burnt Areas*

We used the latest MCD64A1 burned area product (Collection 6) for our study. The product is generated using an improved version of the MCD64 burned area mapping algorithm based on the Collection 6 surface reflectance and active fire input data (Giglio et al. 2015). The MODIS burned area mapping algorithm takes advantage of spectral changes resulting from the deposits of charcoal and ash. The burned area product integrates both 500 m MODIS imagery coupled with 1 km MODIS active fire observations. The hybrid algorithm applies dynamic thresholds to composite imagery generated from a burn-sensitive vegetation index (VI) derived from MODIS short-wave infrared channels 5 and 7, and a temporal texture measure. As part of the process, cumulative active fire maps are used as a guide to filter the burned and unburned areas and to guide specific prior probabilities. The combined use of active-fire and reflectance data enables the algorithm to adapt regionally over a wide range of pre-and post-burn conditions and across multiple ecosystems. The product contains five data layers (Burn Date, Burn Date Uncertainty, QA, First Day, and Last Day), each stored as a separate HDF4 Scientific Data Set (SDS). More details about the product can be found in the Collection 6 MODIS Burned Area Product User's Guide (v.1.3, 2020) available at https://modis-fire.umd.edu/files/MODIS_C6_BA_User_Guide_1.3.pdf.

4 Methodology

One of the most popular measures of heterogeneity is Shannon' entropy. It is used in several fields such as geography, ecology, and biology to assess the heterogeneity of a population or an attribute over an area. The entropy concept goes back to Boltzmann's law in thermodynamics (Jaynes 1979), where it is defined as a measure of the 'disorder' of molecules in a system (You and Wood 2005). Shannon (1948) introduced information entropy to measure the uncertainty of the expected information, which later gave birth to a new discipline called information theory. For a given probability distribution $\{p1, p2, \ldots, pk\}$, Shannon's information entropy (amount of information) for a given a categorical variable X with I possible outcomes, the entropy is defined as (Altieri et al. 2018):

$$H(X) = \sum_{i=1}^{I} p(x_i) \log\left(\frac{1}{p(x_i)}\right) \tag{1}$$

where $p(x_i)$ is the probability of the ith outcome and $\log(1 = p(x_i))$ is the information function, which measures the information brought by outcome x_i (Cover and Thomas 2006). Entropy is a non-negative quantity, which measures the average 'information' or 'surprise' concerning an outcome of X, fires in our case. The more the categories of X are equally likely, the higher the entropy; if a category of X is far more likely than others, the entropy is low, as one can predict the behavior of X, and data do not carry much information (Altieri et al. 2018). Thus, entropy synthesizes the heterogeneity of X outcomes in a single number; data with very different spatial configurations but the same probability mass function for X share the same entropy. In the context of vegetation fires, the more the entropy, the more the randomness, which is not desirable as it is more difficult to control or manage the fires. We computed the entropy of vegetation fires in Myanmar to capture randomness. In addition, we also hypothesized that among the forest and agricultural fires, the former will have relatively more entropy than the latter and that topography can govern these variations. To test the hypothesis, we separated the forest fires from agricultural fires using a Worldcover vegetation mask (Worldcover 2020). The fire data was then subset and gridded at 5-min intervals for forests and agriculture separately. The values of Shannon's entropy range between 0 and log(I), I being the number of categories of the variable under study. We also computed the relative version of Shannon's entropy, i.e., the entropy divided by log(I); this index is specifically useful for comparison across datasets with different I. Our results highlight the spatial variations in entropy in fires in Myanmar useful for fire management and mitigation.

5 Results and Discussion

Vegetation fire analysis from the VIIRS satellite data suggested significant spatial variations (Fig. 3). The gridded fire data at 5-min intervals suggested 1-430 fire counts (FC), with the highest in Shan (East), Shan (South), Kayah, Kayin, Chin, central and western Sagaing, western Magway, and other regions. VIIRS fire analysis suggested the mean FC from 2012 to 2020 for the entire country with 34,7930 FC per year, with the highest 403,292 in 2013 and a minimum of 254,106 in 2018. A typical monthly fire cycle for 2015 with 341,781 FC is depicted in Fig. 4. Most of the fires with high intensity (FRP) occur in the dry season (February-March-April), with the highest Sum of FRP in March (~ 203,897MW). VIIRS-derived annual FC for different years is shown in Fig. 5. Of the different years, 2020 and 2013 had the highest FC, with the least in 2018. The mean FRP (MW) for the 5-min interval data varied from 0.3 to 191.6 (MW) (Fig. 6). Also, spatial variations in FRP (Fig. 6) suggested the highest mean FRP in Shan (north), northern Kayin, northern Sagaing, eastern and southern Chin, and borderlands of northern Kayah and southern Shan regions.

The burnt area statistics derived from the MODIS data for different years are shown in Fig. 7. An average BA of 31,204.61 (km^2) per year was recorded. Further, of the different years, 2004 and 2010 had the highest BA with the least in 2002. Temporal variations in MODIS-derived burnt areas for different vegetation types are shown in Fig. 8, which suggested an average of 655,296.85 (km^2) BA per year with the highest BA in forests (411,125.75 km^2), followed by croplands (159,908.93 km^2) and grasslands (84,262.16 km^2). The percentage variations in BA for different vegetation types are shown in Fig. 9. Figure 10a, b presents the spatial distribution of the entropy values of for forest and agricultural fires, respectively. Results suggested forest fires with higher entropies than agricultural fires (Fig. 11), with the former varying from 0.01 to 0.32 (Fig. 10a) and the latter with 0.01–0.14 (Fig. 10b) at 5-min grid intervals. Also, the relative entropy for forests was higher (0.836) than the agriculture (0.753). The highest entropies (> 0.2) of forest fires were found in central Myanmar bordering the southern Sagaing, Shan north, and Mandalay and

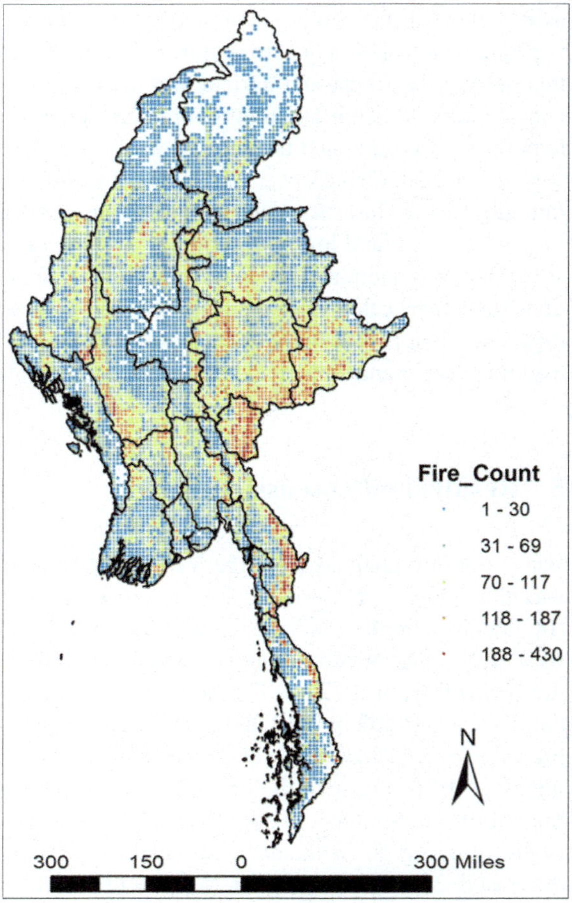

Fig. 3 Fire counts (FC) derived from VIIRS 375 (m) resolution data and gridded at 5-min intervals, 2020

Magway regions. In addition, the northern Sagaing and northern Kachin, including southern Tanintharyi, also had relatively higher entropies. In contrast, agricultural fires' entropies were highly scattered throughout the country. Considerable overlaps of agriculture and forest fires were seen in central Myanmar. The elevation for the gridded data varied from 1 to 3917 (m) (Fig. 2); however, the spatial distribution of the fire entropy values for the forest data did not show significant variations with the elevation, except in northern Kachin, Shan (east), northwestern Sagaing, and forest lands in the Chin stage and along Magway and Rakhine regions.

Understanding the spatiotemporal variability of vegetation fires is important for fire control, mitigation, and management, including impact assessment. In recent years, several researchers used information theory-based measures such as entropy to capture heterogeneity or variability such as evapotranspiration, precipitation, runoff, and discharge (Cui et al. 2018). However, the use of such entropy measures for capturing fire variability is rarely studied, especially in S/SEA countries, which is attempted in this study. Entropy measure is advantageous as it does not make any

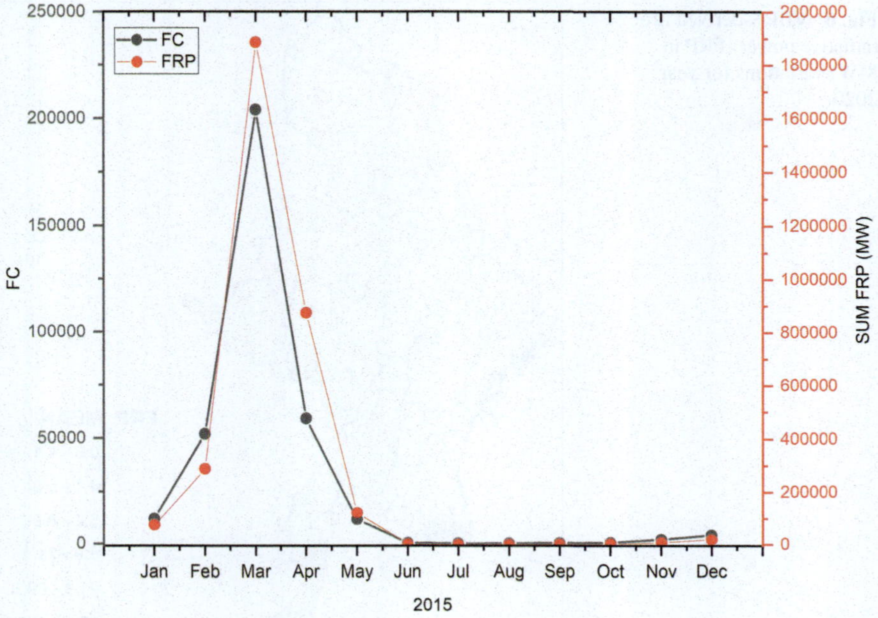

Fig. 4 Fire counts (FC) and FRP (MW) variations for year 2015

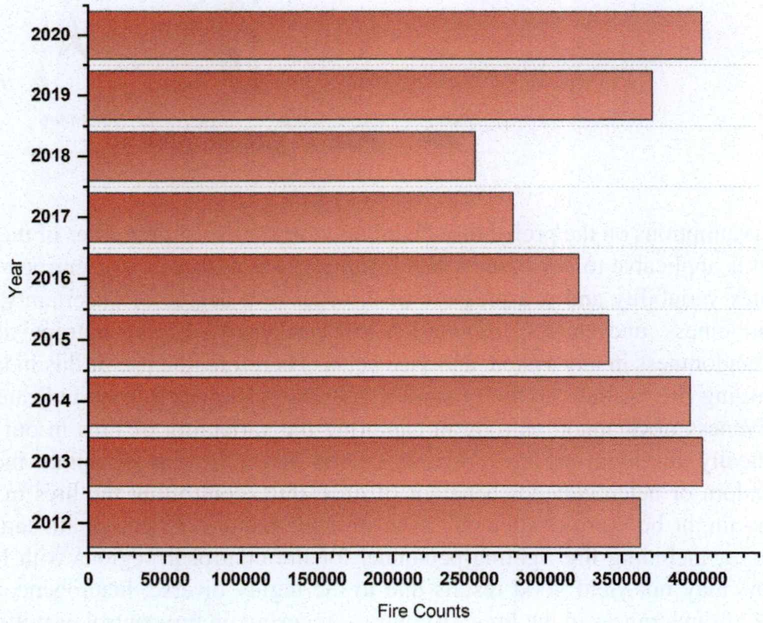

Fig. 5 VIIRS-derived temporal fire counts (FC) for different years

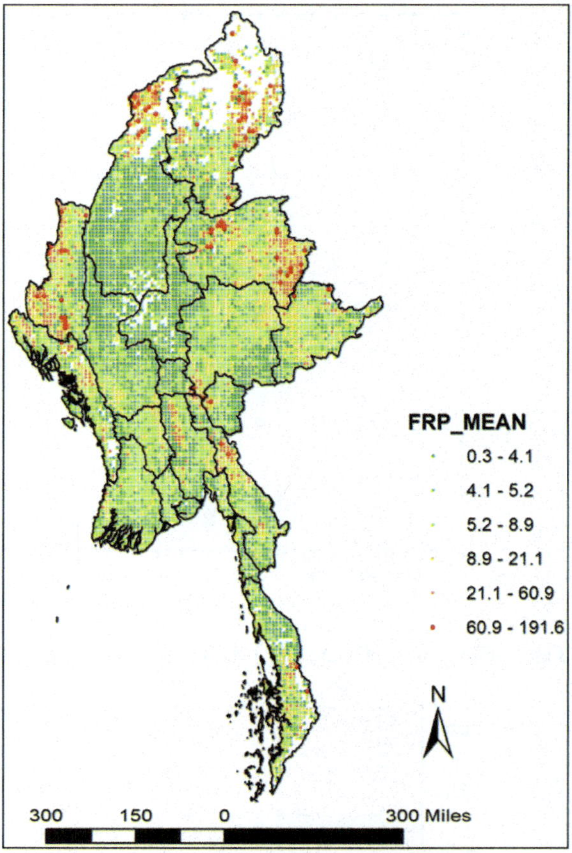

Fig. 6 VIIRS-derived fire radiative power (FRP in MW) variations for year 2020

prior assumptions on the probability distribution or statistical properties of the data; thus, it is applicable to any type of data (Maruyama et al. 2005). The entropy index indicates variability and is a measure of disorder or a degree of uncertainty, such as randomness, unevenness, irregularity, and complexity. Higher entropy reflects more randomness in any system and vice versa. Therefore, the use of this index for quantifying fire spatial variability studies is highly justified. Results indicated the effectiveness of Shannon entropy in capturing the variability of fires in our case. Specifically, the high entropy values in central and northwest Myanmar indicate the random or heterogeneous behavior of fires; thus, controlling the fires in these regions might be more challenging than in other regions. Effective allocation of resources, including fire fighting personnel for fire control in regions with higher entropy, may not yield good results due to the highly diverse, heterogeneous, or degree of randomness in the fires. In contrast, focusing on fire control or mitigation measures where fires are high (Fig. 3) and have less entropy (Fig. 10a), such as in Kayah, eastern Kayin, northern Shan, and western Magway regions, might yield more beneficial results. Specific to agriculture, it was obvious that agricultural fires

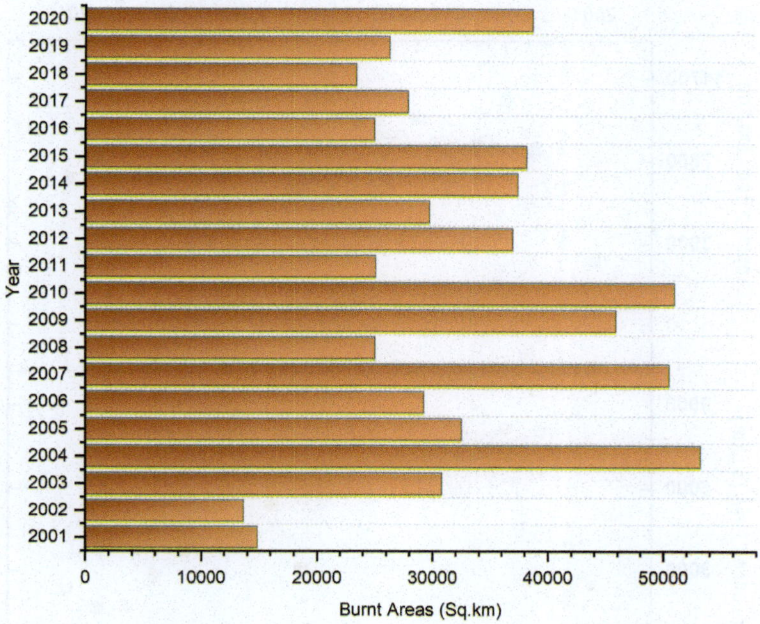

Fig. 7 MODIS-derived burnt areas (km^2) derived from the MODIS data for different years

showed relatively less entropy (Fig. 10b) than the forest fires as most of these fires are human initiated for clearing of crop residues for the next crop cycle, compared to forest fires which might be due to other factors such as climate or other human drivers. Our hypothesis relating to the topography as a governing factor of forest fire entropy did not yield good results from the typical geographical information system overlay and spatial visualization analysis. A more robust statistical analysis of entropy and elevation gradients might be needed to get meaningful results. Among the fires and entropy in agricultural landscapes, the relatively higher entropy in some regions might be due to the small fields that are highly dispersed such as in the western Chin due to shifting cultivation and agriculture. In contrast, in central Mandalay, Mandalay–Magway border, the clustering is attributed to the small rice paddy fields that are close to each other and related crop residue burning. In addition, the clustering or diverse nature of fires in agricultural landscapes might also be due to varied agricultural practices, type of crop grown, and burnt, including field size variations. Although the current study captured the spatial variations in fire-related entropy spatial variations, the drivers of entropy behavior are not addressed. Such an in-depth analysis requires correlating the entropy behavior with various biophysical and socioeconomic indices, including ground truth verification. Moreover, our study focused on a single year (2020), and the temporal analysis might also yield more robust results on variations and drivers of spatial fire variability. Our future studies will focus on addressing these limitations. Despite these limitations, our initial results highlight typical fire behavior

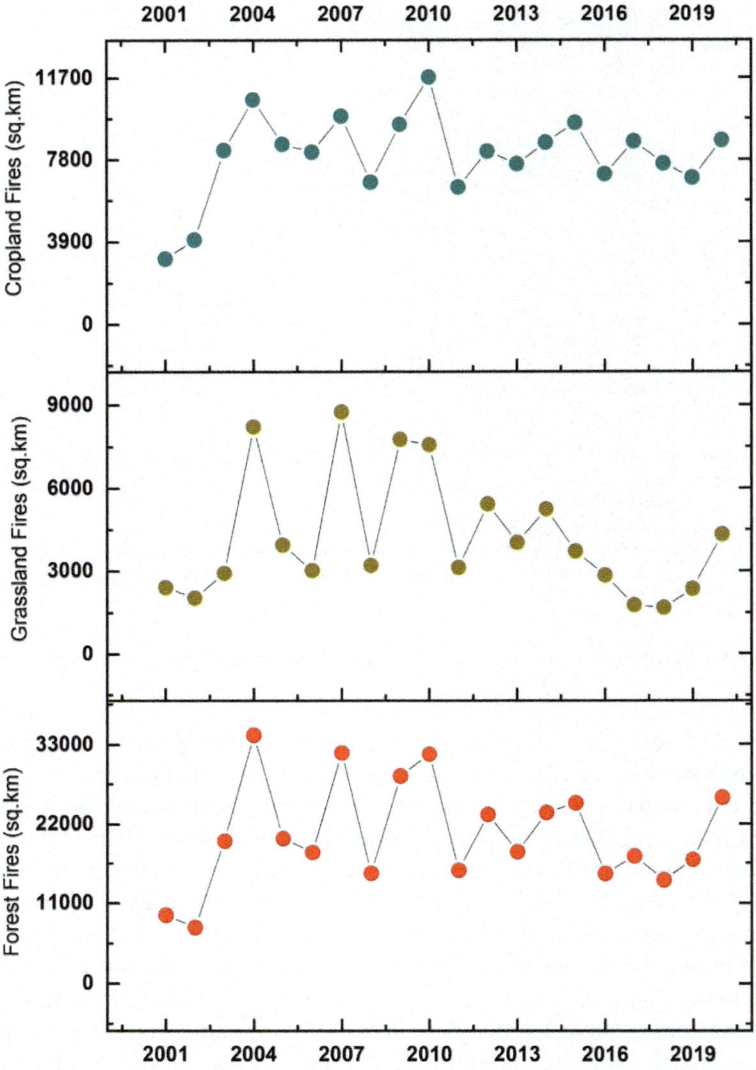

Fig. 8 MODIS-derived burnt areas (km^2) in different vegetation types for different years

and variability in Myanmar as captured by entropy useful for fire management and mitigation.

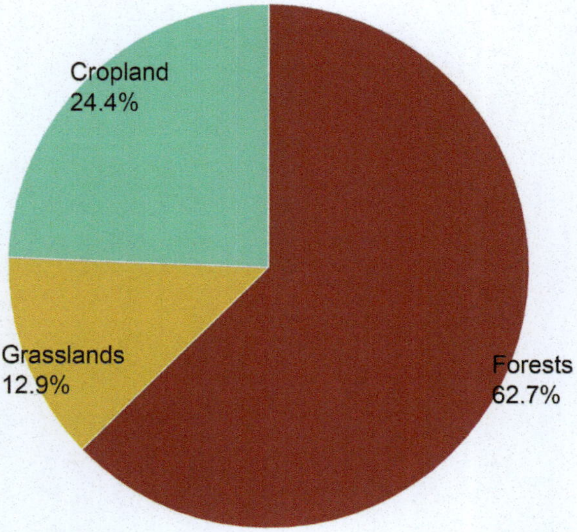

Fig. 9 Burnt area percentages for different vegetation classes derived from MODIS data. The data has been averaged from 2012 to 2020

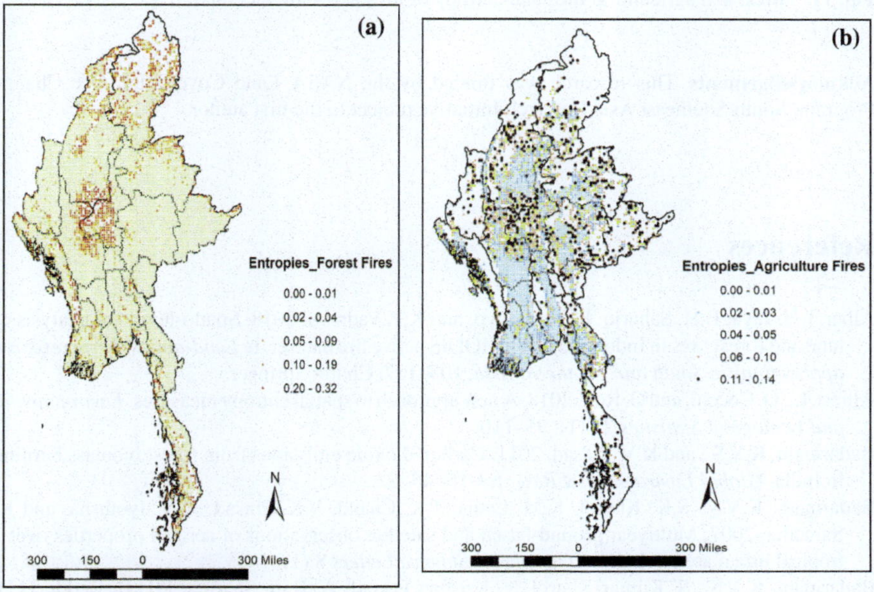

Fig. 10 **a**, **b** Forest and agricultural fires and entropy of fires at 5-min intervals

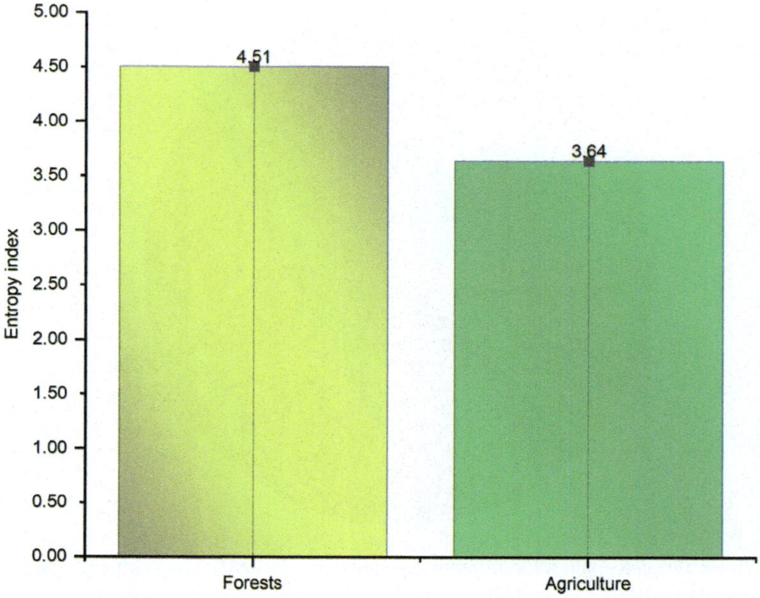

Fig. 11 Forest and agricultural fires and entropy of fires at 5-min intervals

Acknowledgements This research was funded by the NASA Land Cover/Land Use Change Program, South/Southeast Asia Research Initiative project to the first author.

References

Albar, I., I. Jaya, B.H. Saharjo, B. Kuncahyo, and K.P. Vadrevu. 2018. Spatio-temporal analysis of land and forest fires in Indonesia using MODIS active fire dataset. In *Land-atmospheric research applications in South and Southeast Asia*, 105–127. Cham: Springer.

Altieri, L., D. Cocchi, and G. Roli. 2018. A new approach to spatial entropy measures. *Environmental and Ecological Statistics* 25 (1): 95–110.

Badarinath, K.V.S., and K.V. Prasad. 2011. Carbon dioxide emissions from forest biomass burning in India. *Global Environmental Research* 15: 45–52.

Badarinath, K.V.S., S.K. Kharol, K.M. Latha, T.K. Chand, V.K. Prasad, A.N. Jyothsna, and K. Samatha. 2007. Multiyear ground-based and satellite observations of aerosol properties over a tropical urban area in India. *Atmospheric Science Letters* 8 (1): 7–13.

Badarinath, K.V.S., S. Kumar Kharol, V. Krishna Prasad, A. Rani Sharma, E.U.B. Reddi, H.D. Kambezidis, and D.G. Kaskaoutis. 2008. Influence of natural and anthropogenic activities on UV index variations—A study over tropical urban region using ground based observations and satellite data. *Journal of Atmospheric Chemistry* 59 (3): 219–236.

Badarinath, K.V.S., S.K. Kharol, A.R. Sharma, and V.K. Prasad. 2009. Analysis of aerosol and carbon monoxide characteristics over Arabian Sea during crop residue burning period in the Indo-Gangetic Plains using multi-satellite remote sensing datasets. *Journal of Atmospheric and Solar-Terrestrial Physics* 71 (12): 1267–1276.

Biswas, S., K.D. Lasko, and K.P. Vadrevu. 2015a. Fire disturbance in tropical forests of Myanmar—Analysis using MODIS satellite datasets. *IEEE Journal of Selected Topics in Applied Earth Observations and Remote Sensing* 8 (5): 2273–2281.

Biswas, S., K.P. Vadrevu, Z.M. Lwin, K. Lasko, and C.O. Justice. 2015b. Factors controlling vegetation fires in protected and non-protected areas of Myanmar. *PLoS One* 10 (4): e0124346.

Biswas, S., K.P. Vadrevu, M.S. Mon, and C. Justice. 2021. Contemporary forest loss in Myanmar: Effect of democratic transition and subsequent timber bans on landscape structure and composition. *Ambio* 50 (4): 914–928.

Bond, W.J., and J.E. Keeley. 2005. Fire as a global 'herbivore': The ecology and evolution of flammable ecosystems. *Trends in Ecology & Evolution* 20 (7): 387–394.

Choi, Y., S.A. Vay, K.P. Vadrevu, A.J. Soja, J.H. Woo, S.R. Nolf, G.W. Sachse, G.S. Diskin, D.R. Blake, N.J. Blake, and H.B. Singh. 2008. Characteristics of the atmospheric CO_2 signal as observed over the conterminous United States during INTEX-NA. *Journal of Geophysical Research: Atmospheres* 113 (D7).

Chuvieco, E. 2009. *Earth observation of wildland fires in Mediterranean ecosystems*. Berlin: Springer.

Cover, T., and J. Thomas. 2006. *Elements of information theory*, 2nd ed. Hoboken: Wiley.

Crutzen, P.J., and M.O. Andreae. 1990. Biomass burning in the tropics: Impact on atmospheric chemistry and biogeochemical cycles. *Science* 250 (4988): 1669–1678.

Cui, H., B. Sivakumar, and V.P. Singh. 2018. Entropy applications in environmental and water engineering. *Entropy* 20 (8): 598.

Eva, H., and E.F. Lambin. 2000. Fires and land-cover change in the tropics: A remote sensing analysis at the landscape scale. *Journal of Biogeography* 27 (3): 765–776.

Giglio, L., J. Descloitres, C.O. Justice, and Y.J. Kaufman. 2003. An enhanced contextual fire detection algorithm for MODIS. *Remote Sensing of Environment* 87 (2–3): 273–282.

Giglio, L., C. Justice, L. Boschetti, D. Roy. 2015. MCD64A1 MODIS/Terra+Aqua Burned Area Monthly L3 Global 500m SIN Grid V006. distributed by NASA EOSDIS Land Processes DAAC, https://doi.org/10.5067/MODIS/MCD64A1.006. Accessed 2023-04-03.

Giglio, L., L. Boschetti, D.P. Roy, M.L. Humber, and C.O. Justice. 2018. The collection 6 MODIS burned area mapping algorithm and product. *Remote Sensing of Environment* 217: 72–85.

Hough, J.L. 1993. Why burn the bush? Social approaches to bush-fire management in West African national parks. *Biological Conservation* 65 (1): 23–28.

Jaynes, E.T. 1979. Concentration of distributions at entropy maxima. In *ET Jaynes: Papers on probability, statistics and statistical physics*, 315.

Justice, C., G. Gutman, and K.P. Vadrevu. 2015. NASA land cover and land use change (LCLUC): An interdisciplinary research program. *Journal of Environmental Management* 148: 4–9.

Kant, Y., V.K. Prasad, and K.V.S. Badarinath. 2000. Algorithm for detection of active fire zones using NOAA AVHRR data. *Infrared Physics & Technology* 41 (1): 29–34.

Kaufman, Y.J., C.O. Justice, L.P. Flynn, J.D. Kendall, E.M. Prins, L. Giglio, D.E. Ward, W.P. Menzel, and A.W. Setzer. 1998. Potential global fire monitoring from EOS-MODIS. *Journal of Geophysical Research: Atmospheres* 103 (D24): 32215–32238.

Kharol, S.K., K.V.S. Badarinath, A.R. Sharma, D.V. Mahalakshmi, D. Singh, and V.K. Prasad. 2012. Black carbon aerosol variations over Patiala city, Punjab, India—A study during agriculture crop residue burning period using ground measurements and satellite data. *Journal of Atmospheric and Solar-Terrestrial Physics* 84: 45–51.

Lasko, K., and K. Vadrevu. 2018. Improved rice residue burning emissions estimates: Accounting for practice-specific emission factors in air pollution assessments of Vietnam. *Environmental Pollution* 236: 795–806.

Lasko, K., K.P. Vadrevu, V.T. Tran, E. Ellicott, T.T. Nguyen, H.Q. Bui, and C. Justice. 2017. Satellites may underestimate rice residue and associated burning emissions in Vietnam. *Environmental Research Letters* 12 (8): 085006.

Lasko, K., K.P. Vadrevu, and T.T.N. Nguyen. 2018. Analysis of air pollution over Hanoi, Vietnam using multi-satellite and MERRA reanalysis datasets. *PLoS One* 13 (5): e0196629.

Lata, K.M., C.H. Sankar Rao, V. Krishna Prasad, K.V.S. Badrinath, and V. Raghavaswamy. 2001. Measuring urban sprawl: A case study of Hyderabad. *GIS Development* 5 (12): 22–25.

Maruyama, T., T. Kawachi, and V.P. Singh. 2005. Entropy-based assessment and clustering of potential water resources availability. *Journal of Hydrology* 309 (1–4): 104–113.

Petropoulos, G.P., K.P. Vadrevu, and C. Kalaitzidis. 2013. Spectral angle mapper and object-based classification combined with hyperspectral remote sensing imagery for obtaining land use/cover mapping in a Mediterranean region. *Geocarto International* 28 (2): 114–129.

Prasad, V.K., and K.V.S. Badarinth. 2004. Land use changes and trends in human appropriation of above ground net primary production (HANPP) in India (1961–98). *Geographical Journal* 170 (1): 51–63.

Prasad, K.V., and K.V.S. Badarinath. 2006. Soil surface nitrogen losses from agriculture in India: A regional inventory within agroecological zones (2000–2001). *The International Journal of Sustainable Development and World Ecology* 13 (3): 173–182.

Prasad, V.K., Y. Kant, P.K. Gupta, C. Sharma, A.A. Mitra, and K.V.S. Badarinath. 2001a. Biomass and combustion characteristics of secondary mixed deciduous forests in Eastern Ghats of India. *Atmospheric Environment* 35 (18): 3085–3095.

Prasad, V.K., Y. Kant, and K.V.S. Badarinath. 2001b. CENTURY ecosystem model application for quantifying vegetation dynamics in shifting cultivation areas: A case study from Rampa Forests, Eastern Ghats (India). *Ecological Research* 16 (3): 497–507.

Prasad, V.K., Y. Kant, P.K. Gupta, C. Elvidge, and K.V.S. Badarinath. 2002a. Biomass burning and related trace gas emissions from tropical dry deciduous forests of India: A study using DMSP-OLS data and ground-based measurements. *International Journal of Remote Sensing* 23 (14): 2837–2851.

Prasad, V.K., Y. Kant, and K.V.S. Badarinath. 2002b. Land use changes and modeling carbon fluxes from satellite data. *Advances in Space Research* 30 (11): 2511–2516.

Prasad, V.K., M. Lata, and K.V.S. Badarinath. 2003. Trace gas emissions from biomass burning from northeast region in India—Estimates from satellite remote sensing data and GIS. *The Environmentalist* 23 (3): 229–236.

Prasad, V.K., K.V.S. Badarinath, S. Yonemura, and H. Tsuruta. 2004. Regional inventory of soil surface nitrogen balances in Indian agriculture (2000–2001). *Journal of Environmental Management* 73 (3): 209–218.

Prasad, V.K., E. Anuradha, and K.V.S. Badarinath. 2005. Climatic controls of vegetation vigor in four contrasting forest types of India—Evaluation from National Oceanic and Atmospheric Administration's advanced very high resolution radiometer datasets (1990–2000). *International Journal of Biometeorology* 50 (1): 6–16.

Schroeder, W., P. Oliva, L. Giglio, and I.A. Csiszar. 2014. The new VIIRS 375 m active fire detection data product: Algorithm description and initial assessment. *Remote Sensing of Environment* 143: 85–96.

Shannon, C.E. 1948. A mathematical theory of communication. *The Bell System Technical Journal* 27 (3): 379–423.

Stott, P. 2000. Combustion in tropical biomass fires: A critical review. *Progress in Physical Geography* 24 (3): 355–377.

Vadrevu, K.P. 2008. Analysis of fire events and controlling factors in eastern India using spatial scan and multivariate statistics. *Geografiska Annaler: Series A, Physical Geography* 90 (4): 315–328.

Vadrevu, K. 2015. International regional science meeting of NASA-LCLUC: Land cover/land use change (LC/LUC) and environmental impacts in South Asia, Coimbatore, India, 19–23 January 2013. *Journal of Environmental Management* 148: 1–163.

Vadrevu, K.P., and K.V.S. Badarinath. 2009. Spatial pattern analysis of fire events in Central India—A case study. *Geocarto International* 24 (2): 115–131.

Vadrevu, K.P., and C.O. Justice. 2011. Vegetation fires in the Asian region: Satellite observational needs and priorities. *Global Environmental Research* 15 (1): 65–76.

Vadrevu, K., and K. Lasko. 2015. Fire regimes and potential bioenergy loss from agricultural lands in the Indo-Gangetic Plains. *Journal of Environmental Management* 148: 10–20.

Vadrevu, K.P., K.V.S. Badarinath, and E. Anuradha. 2008. Spatial patterns in vegetation fires in the Indian region. *Environmental Monitoring and Assessment* 147 (1): 1–13.

Vadrevu, K.P., L. Giglio, and C. Justice. 2013. Satellite based analysis of fire–carbon monoxide relationships from forest and agricultural residue burning (2003–2011). *Atmospheric Environment* 64: 179–191.

Vadrevu, K.P., T. Ohara, and C. Justice. 2014. Air pollution in Asia. *Environmental Pollution (Barking, Essex: 1987)* 195: 233–235.

Vadrevu, K., T. Ohara, and C. Justice. 2017. Land cover, land use changes and air pollution in Asia: A synthesis. *Environmental Research Letters* 12 (12): 120201.

Vadrevu, K.P., T. Ohara, and C. Justice (eds.). 2018. *Land-atmospheric research applications in South and Southeast Asia*. Berlin: Springer.

Vadrevu, K., A. Heinimann, G. Gutman, and C. Justice. 2019a. Remote sensing of land use/cover changes in South and Southeast Asian Countries. *International Journal of Digital Earth* 12 (10): 1099–1102.

Vadrevu, K.P., K. Lasko, L. Giglio, W. Schroeder, S. Biswas, and C. Justice. 2019b. Trends in vegetation fires in South and Southeast Asian countries. *Scientific Reports* 9 (1): 1–13.

Vadrevu, K.P., T. Ohara, and C. Justice (eds.). 2021a. *Biomass burning in South and Southeast Asia: Impacts on the biosphere*, vol. 2. Boca Raton: CRC Press.

Vadrevu, K.P., T. Ohara, and C. Justice (eds.). 2021b. *Biomass burning in South and Southeast Asia: Mapping and monitoring*, vol. 1. Boca Raton: CRC Press.

Vadrevu, K.P., T. Le Toan, S.S. Ray, and C. Justice. 2022a. *Remote sensing of agriculture and land cover/land use changes in South and Southeast Asian countries*. Cham: Springer.

Vadrevu, K., A. Eaturu, E. Casadaban, K. Lasko, W. Schroeder, S. Biswas, L. Giglio, and C. Justice. 2022b. Spatial variations in vegetation fires and emissions in South and Southeast Asia during COVID-19 and pre-pandemic. *Scientific Reports* 12 (1): 1–21.

Vay, S.A., Y. Choi, K.P. Vadrevu, D.R. Blake, S.C. Tyler, A. Wisthaler, A. Hecobian, Y. Kondo, G.S. Diskin, G.W. Sachse, and J.H. Woo. 2011. Patterns of CO_2 and radiocarbon across high northern latitudes during International Polar Year 2008. *Journal of Geophysical Research: Atmospheres* 116 (D14).

Wooster, M.J., G.J. Roberts, L. Giglio, D.P. Roy, P.H. Freeborn, L. Boschetti, C. Justice, C. Ichoku, W. Schroeder, D. Davies, and A.M. Smith. 2021. Satellite remote sensing of active fires: History and current status, applications and future requirements. *Remote Sensing of Environment* 267: 112694.

Worldcover. 2020. https://worldcover2020.esa.int/data/docs/WorldCover_PUM_V1.1.pdf.

Yadav, P.K. 2013. Slash-and-burn agriculture in north-east India. *Journal of Expert Opinion on Environmental Biology* 2: 2–5.

You, L., and S. Wood. 2005. Assessing the spatial distribution of crop areas using a cross-entropy method. *International Journal of Applied Earth Observation and Geoinformation* 7 (4): 310–323.

Crop Residue Burning and Forest Fire Emissions in Nepal

Bhupendra Das, Siva Praveen Puppala, Bijaya Maharjan, Krishna B. Bhujel, Ajay Mathema, Dhurba Neupane, and Rejina M. Byanju

Abstract Open burning of biomass is a significant source of air pollution in Nepal. Both the bottom-up and top-down methods were used in this study for estimating the emission of pollutants from crop residue open burning and forest fires. The range of uncertainties has been estimated through Monte Carlo simulations. Through the bottom-up approach, in 2016/17, the greenhouse gases from CROB were estimated to be 4140 Gg for CO_2 and 6.5 Gg for CH_4. Likewise, air pollutants' emission from forest fire in Nepal in the year 2008/09 was estimated to be SO_2 16.4, NO_x (70.3), CO (2985.2), NMVOC (232.5), NH_3 (37.3), $PM_{2.5}$ (261.2), PM_{10} (301.4), BC (18.9), OC (149.3), CO_2 (45,352), and CH_4 (195.2), all in the unit of Gg. Addressing emissions from crop residue burning and forest fires in Nepal will require a participatory approach involving locals.

Keywords Crop residue open burning · Forest fires · High-resolution · Emission factors · Emission inventory

---

B. Das (✉) · B. Maharjan · K. B. Bhujel
Nepal Energy and Environment Development Services (NEEDS), Kathmandu, Nepal
e-mail: bhupenids@gmail.com

B. Das
Clean Air Asia, Pasig City, Philippines

A. Mathema
School of Environmental Science and Management (SchEMS), Pokhara University, Lekhnath, Nepal

S. P. Puppala
NSW Department of Planning and Environment, Lidcombe, Australia

D. Neupane
University of Nevada, Reno, USA

R. M. Byanju
Central Department of Environmental Science, Tribhuvan University, Kirtipur, Nepal

K. P. Vadrevu et al. (eds.), *Vegetation Fires and Pollution in Asia*,
https://doi.org/10.1007/978-3-031-29916-2_5

1 Introduction

In recent years, biomass burning has been globally threatening ecosystems and the environment. Crop residue open burning (CROB) and forest fires are the major sources of biomass burning globally, including Nepal. Open burning of this sector releases greenhouse gases (CO_2, CH_4), aerosols ($PM_{2.5}$, PM_{10}, BC, OC, EC), and trace gases (NO_X, CO, NMVOC, NH_3, SO_2) (Kant et al. 2000a; Andreae and Merlet 2001; Prasad et al. 2001, 2002, 2003; Das et al. 2020; Park and Takeuchi 2021). Following their release into the atmosphere, air pollutants undergo additional chemical and physical transformations and can disrupt the Earth's radiation balance and negatively impact air quality (Permadi and Oanh 2013; Jethva et al. 2019) and human health (Tipayarom and Oanh 2007; Zhang et al. 2017). Unlike many industrial point sources, the impact of biomass burning can be widespread (Vadrevu et al. 2017, 2021a, b). The open burning can impact weather in the form of fog, haze, and smog that can last for a couple of weeks, leading to reduced atmospheric visibility (Chauhan and Singh 2017), including serious health outcomes associated with biomass burning such as asthma attacks among children, premature deaths, respiratory disease, and cardiovascular disease (Sigsgaard et al. 2015).

Bottom-up and top-down approaches are two different methods of estimating biomass burning and emissions, both of which have a range of uncertainties. The bottom-up approach relies on the activity data at the ground level (i.e., survey-based, secondary sources). In top-down approaches, data from remote sensing are mainly used to assess biomass burning (Vadrevu and Justice 2011; Chang et al. 2013). When employing a satellite-based dataset, there are crucial caveats and constraints to consider, such as the area of biomass land and active fire patches being much smaller than the geographical scale of satellites with 1 km^2 resolution, resulting in mixed pixel difficulties (Chang et al. 2013; Elvidge et al. 2021). Furthermore, satellite data are unable to identify short-lived fires, and active fire data may not be accounted for in such cases, resulting in a significant level of uncertainty (Kant et al. 2000b; Kanabkaew and Oanh 2011; Vander-Werf et al. 2010; Benali et al. 2016; Lasko et al. 2017; Lasko and Vadrevu 2018). In addition, prolonged cloud cover and many fire occurrences within a single fire perimeter can add to the fire detection uncertainty. Despite being a widely used emissions inventory, the Global Fire Emissions Database (GFED) cannot capture emissions from smaller fires (Zhang et al. 2018), adding to the uncertainty. The main objective of this study is to provide an estimate of forest and crop residue burning emissions from Nepal. More specifically, the air quality in Nepal during April 2021 has been the worst on record. Thus, it is essential to review the critical sources of biomass burning pollution, including methods useful for emissions estimation. In this study, we review both the top-down and bottom-up methods useful for emissions estimation.

2 Study Area

The country of Nepal is located on the southern slopes of the Himalayan Mountain ranges. It is a landlocked country having an area of 147,930.15 km^2, bordered by India on the east, south, and west, and the Tibet Autonomous Region of China on the north. The country is divided into 77 districts for administrative purposes (CBS 2019). For both the crop residue and forest fire emission inventories, the study area includes all 77 districts of Nepal (Figs. 1 and 2). Altogether 300,721 fire spots of biomass burning have been detected from 2012 to 2021 in Nepal through VIIRS (Tables 1 and 2).

3 Data and Methods

3.1 Crop Residue Open Burning

The Ministry of Agriculture Development (MoAD) Annual Reports provide Nepal's district-level crop production data from 2003/04 to 2016/17 (tons/year). Paddy, maize, millet, wheat, barley, oil crops, potato, sugarcane, jute, and pulses were considered in the study. Lentils, chickpeas, pigeon peas, black gram, grass peas, horse peas, soybeans, and other crops are all examples of pulses. The national crop production is approximated by summing the crop production in all districts of Nepal. The residue-to-crop ratio is used to assess the amount of residue generated during crop production. Likewise, dry matter-to-crop residue ratio and burn efficiency are reported in the literature (e.g., Das et al. 2020).

3.2 Forest Fire

Forest cover is the dominant land cover with 6.67 million ha, i.e., about 41.69%, followed by 24.21% cropland and 13.27% grassland in Nepal. Around 20% of land cover is covered by snow, bare rock, glacier, riverbed, built-up water bodies, and soil (FRTC 2022). According to land cover mapping, forest occupies a total of 6.67 million ha (41.69%) and 0.53 million ha (3.62%) of the total area of the country and altogether represent 45.31% of the total forest area of the country (FRTC 2022).

The University of Maryland developed the Fire Information for Resource Management System (FIRMS) in 2007 with funding from NASA's Applied Sciences Program and the United Nations Food and Agriculture Organization (UN FAO) to provide near real-time active fire locations to natural resource managers who faced challenges to get timely satellite-derived fire information. NASA's Moderate Resolution Imaging Spectroradiometer (MODIS) aboard the Terra and Aqua satellites, as well as NASA's Visible Infrared Imaging Radiometer Suite (VIIRS) aboard the joint NASA/NOAA

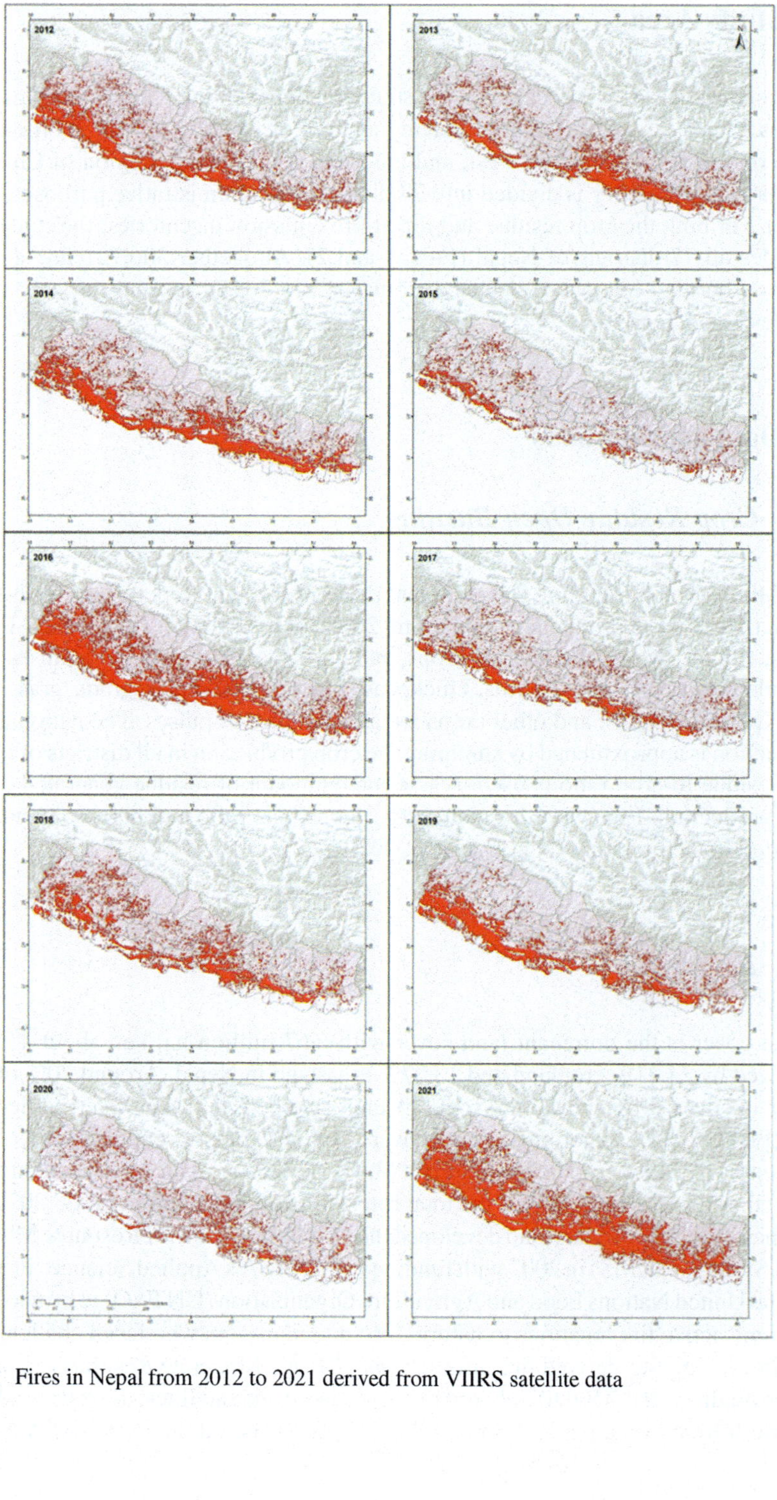

Fig. 1 Fires in Nepal from 2012 to 2021 derived from VIIRS satellite data

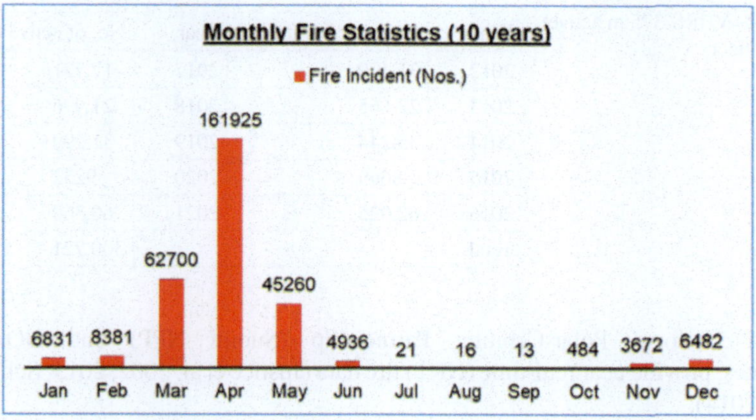

Fig. 2 Seasonal variations in fires in Nepal showing the peak fires in April

Table 1 Attribute fields for NRT VIIRS 375 m active fire data distributed by FIRMS

Attribute	Short description	Long description
Latitude	Latitude	Center of nominal 375 m fire pixel
Longitude	Longitude	Center of nominal 375 m fire pixel
Bright_ti4	Brightness temperature I-4	VIIRS I-4 channel brightness temperature of the fire pixel measured in Kelvin
Scan	Along scan pixel size	The algorithm produces approximately 375 m pixels at nadir. Scan and track reflect actual pixel size
Track	Along track pixel size	The algorithm produces approximately 375 m pixels at nadir. Scan and track reflect actual pixel size
Acq_Date	Acquisition date	Date of VIIRS acquisition
Acq_Time	Acquisition time	Time of acquisition/overpass of the satellite (in UTC)
Satellite	Satellite	N = Suomi National Polar-orbiting Partnership (Suomi NPP), 1 = NOAA-20 (designated JPSS-1 prior to launch)
Confidence	Confidence	It is intended to help users gauge the quality of individual hotspot/fire pixels. Confidence values are set to low, nominal and high
Version	Version (Collection and source)	Version identifies the collection and source of data processing or standard processing
Bright_ti5	Brightness temperature I-5	I-5 Channel brightness temperature of the fire pixel measured in Kelvin
FRP	Fire radiative power	FRP depicts the pixel-integrated fire radiative power in MW (megawatts)
DayNight	Day or night	D = Daytime fire, N = Nighttime fire

Table 2 VIIRS-375 m active fire product

Year	No. of active fire	Year	No. of active fire
2012	31,490	2017	17,721
2013	22,665	2018	21,996
2014	35,244	2019	31,299
2015	8669	2020	9252
2016	62,025	2021	60,360
Total			300,721

Suomi National Polar-Orbiting Partnership (Suomi NPP) and NOAA-20 satellites, provide near real-time (NRT) fire data (Justice et al. 2002, 2015; Schroeder et al. 2014).

The MODIS was the first sensor to provide global mapping capability for both fire locations and direct burned area mapping. MODIS detects fires within one square kilometer as a single fire pixel, with each active fire location being the center of a 1 km pixel, i.e., a spatial resolution of 1 km. As a result, MODIS data deliver a higher frequency of fire data with a lesser spatial resolution. Since 2012, the Visible Infrared Imaging Radiometer Suite (VIIRS), a newly built intermediate resolution sensor, has been providing daily global active fire products with a finer spatial resolution of 375 m and high fire sensitivity. As a result, the VIIRS-375 m fire product can detect smaller and colder fires (Schroeder et al. 2014).

3.3 Approach

3.3.1 Crop Residue Open Burning

Bottom-Up Approach

After multiplying the crop production amount by residue to crop ratio and dry matter-to-crop residue ratio, the total dry matter is calculated using Eq. 1 (Das et al. 2020; Kanabkaew and Oanh 2011; Shrestha 2018; IPCC 1996). The amount of crop residue available for open burning (Eq. 2) following its consumption pattern (i.e., fuel energy, roof thatching, and animal food) is calculated following Das et al. (2020). Crop residue for open burning has been assumed to be zero in districts with a residue consumption deficit. The mass of residue burned (M_1) in the agricultural field is calculated by multiplying B_l by η (burn efficiency fraction) (Eq. 3) (Kanabkaew and Oanh 2011; Shrestha 2018; IPCC 1996).

$$C_R = P_l \times N_l \times D_l \tag{1}$$

where

C_R is the total dry matter/dry residue; P_l is the crop production amount for crop type l; N_l is the residue-to-crop ratio for crop type l; and D_l is the dry matter-to-crop residue ratio for crop type l.

$$B_l = C_R - (C_F + C_T + C_A) \tag{2}$$

where

B_l is the amount of crop residue available for open burning; C_F is the crop residue used as fuel energy; C_T is the crop residue used in thatching the roof; and C_A is the crop residue used as animal fodder.

$$M_l = B_l \times \eta \tag{3}$$

Using Eq. (3), emissions '*E*' for pollutant type I and crop type 'l' are calculated, where 'M_1' from Eq. (3) is multiplied by the emission factor '$EF_{i,l}$' unique to pollutant type I from crop type 'l' (Kanabkaew and Oanh 2011; Shrestha et al. 2013; Shrestha 2018; IPCC 2006).

$$E_{I,l} = \sum_l M_l \times EF_{I,l} \tag{4}$$

Top-Down Approach

The total dry matter burned from a crop residue could be calculated by referring to the below equation:

$$M_l = N_f \times G_a \times \rho_l \times \eta_l \tag{5}$$

where M_l is the mass of dry matter burned (tons/year) of crop residue; N_f is the number of fire spots captured by the satellites; G_a is the grid area of crop fire actually burned (ha); ρ_l is dry matter density (tons/ha) of crops, and η_l is burning efficiency (oxidized in the combustion).

Using Eq. (6), emissions '*E*' for pollutant type I and crop type 'l' are calculated, where 'M_1' from Eq. (5) is multiplied by the emission factor '$EF_{i,l}$' unique to pollutant type I from crop type 'l' (Kanabkaew and Oanh 2011; Shrestha et al. 2013; Shrestha 2018; IPCC 2006).

$$E_{I,l} = \sum_l M_l \times EF_{I,l} \tag{6}$$

3.3.2 Forest Fire

Bottom-Up Approach

The total dry matter burned from a forest fire is calculated using Eq. (7) (Shrestha 2018).

$$M_f = A_{ba} \times \rho_f \times \eta_f \tag{7}$$

where M_f is mass of dry matter burned (tons/year); A_{ba} is the actual burned area (ha); ρ_f is dry matter density (tons/ha), and η_f is the burning efficiency (oxidized in the combustion).

Equation (8) is used to compute the actual burned area (A_{ba}) (Shrestha 2018):

$$A_{ba} = A_l \times f_l \tag{8}$$

where A_l is area of land cover type l (ha); and f_l is the fraction of the total area of land cover type l burned annually.

Further, using the below Eq. (9), emissions 'E' for pollutant type I and land cover type 'l' are calculated, where 'M_f' from Eq. (7) is multiplied by emission factor '$EF_{I,l}$' unique to pollutant type I from land cover type 'l' (Shrestha 2018; IPCC 2006).

$$E_{I,l} = \sum_{l} M_f \times EF_{I,l} \tag{9}$$

Top-Down Approach

The total dry matter burned from forest biomass has been using Eq. (10):

$$M_l = N_f \times G_a \times \rho_f \times \eta_f \tag{10}$$

where M_l is the mass of dry matter burned (tons/year) of forest biomass; N_f is the number of fire spots captured by the satellites; G_a is the grid area of forest fire burned (ha); ρ_f is dry matter density (tons/ha) of forest biomass, and η_f is burning efficiency (oxidized in the combustion).

Using Eq. (11), emissions 'E' for pollutant type I and land cover type 'l' are calculated, where 'M_l' from Eq. (10) is multiplied by the emission factor '$EF_{I,l}$' unique to pollutant type I from land cover type 'l' (Shrestha 2018; Chang and Song 2009).

$$E_{I,l} = \sum_{l} M_l \times EF_{I,l} \tag{11}$$

The VIIRS/MODIS active fire data for various years have been obtained from Fire Information for Resource Management System (FIRMS) for estimating the biomass burnt area. Moreover, the shapefile of forest types of Nepal has been obtained from the Department of Forests, Government of Nepal (DoF 2002). From the shapefile, burnt areas were delineated using the ArcGIS software. The accuracy of the burnt area has been validated by direct field observations and also compared with the general accuracy statement of the VIIRS/MODIS product performance.

Uncertainty Estimations

The Monte Carlo approach has been used to examine the uncertainties in the emission inventory (e.g., Das et al. 2020; Ni et al. 2015; Zhao et al. 2011; Zhang et al. 2019). The coefficient of variation (CV), mean, and SD were calculated for the activity data. Then, averaging 20,000 Monte Carlo simulations with a 95% confidence interval and transforming to a percentage of the mean, the range of residue burnt, EFs, and emissions were calculated, which are reported comprehensively in Das et al. (2020).

4 Results and Discussion

For the entire country, 1159.65 million tons (194.51 t ha^{-1}) of aboveground air-dried biomass and 1054.97 million tons (176.95 t ha^{-1}) of carbon stock have been estimated (DFRS 2015). Through the bottom-up approach, on average, dry matter (DM) generation from crop residue is estimated to be 13,500 Gg in the year 2016/17 in Nepal. Of it, the total open burning was estimated to be 2908 Gg. The highest emissions from CROB occurred in the Terai region (91%), followed by the hills (6%) and the mountains (3%). Likewise, ~ 400,000 ha of forest has been reported to burn in Nepal (Bajracharya 2002). The total number of stems with a Diameter at Breast Height (DBH) $\geq$ 10 cm estimated in the Forest of Nepal is 2563.27 million (429.93 ha^{-1}). The total stem volume is 982.33 million m^3 (164.76 m^3 ha^{-1}). The total aboveground air-dried biomass in the Forest of Nepal is 1159.65 million tons (194.51 t ha^{-1}) (DFRS 2015). Parajuli et al. (2015) have reported increasing forest fire incidences in Nepal. Moreover, Khanal (2015) reported 375,000 ha of forests burnt from 2001 to 2015. According to Matin et al. (2017), 18 districts of Nepal, including Nawalparasi, are at very high risk of fire. Altogether 300,721 fire spots of biomass burning have been detected from 2012 to 2021 in Nepal through VIIRS-375 m resolution (Table 2) (Earth Data 2022).

April is the peak burning time, whereas, in July–August, the monsoon season, active biomass burning fires number were the lowest (Fig. 2) (Earth Data 2022).

Through the bottom-up approach, in 2016/17, the greenhouse gases from CROB were estimated to be 4140 Gg CO_2 and 6.5 Gg CH_4. The other harmful pollutants were 154 CO, 1.2 SO_2, 24.5 PM2.5, 8.6 OC, 2.2 BC, 7 NO_x, 22.5 NMVOC, and 2.7 NH_3 Gg, respectively (Das et al. 2020). From February to May, open burning of crop residue produced more than 80% of all air pollutants. Likewise, air pollutants' emission from the forest fire in Nepal in the year 2008/09 was estimated to be SO_2 16.4, NO_x (70.3), CO (2985.2), NMVOC (232.5), NH_3 (37.3), PM2.5 (261.2), PM10 (301.4), BC (18.9), OC (149.3), CO_2 (45,352), and CH_4 (195.2) Gg (Shrestha 2018).

In addition to the negative impacts on the climate, harmful pollutants from the open burning of biomass can cause severe health effects such as respiratory and cardiovascular diseases, including allergies, and premature deaths (Zhang et al. 2017) as CO_2, CH_4, N_2O, and BC have proven global warming potential (Gupta et al. 2001; Badarinath et al. 2007, 2009; Kharol et al. 2012; Das et al. 2020; Lasko and Vadrevu 2018). Crop residue open burning is primarily caused by labor shortages and combined harvesters. The study by Gupta (2014) and Gupta et al. (2004) suggests that combine harvesters can be redesigned to collect crop residues separately while harvesting. Open burning of crop residues is less likely to occur where farmers have more cattle per hectare, (Gupta 2014). As a result, rearing cattle for additional money by local farmers could be a mitigating measure to prevent open burning. In addition, animal feed, alternative energy production (e.g., bio-briquette), and raw materials for the industry can all benefit from chopped crop waste (e.g., mushroom cultivation, paper production, and brick kilns). Likewise, enough fire lines and conservation ponds, as well as capacity-building and awareness-raising programs for local communities and forest fire watch groups, are necessary for curbing air pollution from the forest fire.

5 Conclusion

This study used bottom-up and top-down approaches to calculate pollutant emissions from agricultural residue burning and forest fires. The emissions were calculated by integrating crop-related statistics, ground-based methods, satellite data, and empirical methods for the entire country. In addition, Monte Carlo simulations were used to assess the range of uncertainties. The results suggested GHG emissions from crop residue burning about 4140 Gg CO_2 and 6.5 Gg CH_4. From February to May, open burning of crop residue produced more than 80% of all air pollutants. Likewise, air pollutants' emission from the forest fire in Nepal in the year 2008/09 was estimated to be SO_2 16.4, NO_x (70.3), CO (2985.2), NMVOC (232.5), NH_3 (37.3), $PM_{2.5}$ (261.2), PM_{10} (301.4), BC (18.9), OC (149.3), CO_2 (45,352), and CH_4 (195.2) Gg. Some mitigation options to curb biomass burning pollution in Nepal include redesigning combined harvesters to collect residues while harvesting, encouraging integrated farming practices such as livestock with cropping, alternative energy production measures from the crop residues, and to capacity-building and awareness-raising programs for local communities.

Acknowledgements The authors acknowledge the support from the Nepal Energy and Environment Development Services (NEEDS), Kathmandu, Nepal. A special thanks to Dr. Prakash V. Bhave (Duke University, USA) and Dr. Maheswar Rupakheti (IASS, Potsdam, Germany) for their kind suggestions, especially in the open-burning crop residue emissions estimation.

References

Andreae, M.O., and P. Merlet. 2001. Emission of trace gases and aerosols from biomass burning. *Global Biogeochemical Cycles* 15 (4): 955–966.

Badarinath, K.V.S., S.K. Kharol, K.M. Latha, T.K. Chand, V.K. Prasad, A.N. Jyothsna, and K. Samatha. 2007. Multiyear ground-based and satellite observations of aerosol properties over a tropical urban area in India. *Atmospheric Science Letters* 8 (1): 7–13.

Badarinath, K.V.S., S.K. Kharol, A.R. Sharma, and V.K. Prasad. 2009. Analysis of aerosol and carbon monoxide characteristics over Arabian Sea during crop residue burning period in the Indo-Gangetic Plains using multi-satellite remote sensing datasets. *Journal of Atmospheric and Solar-Terrestrial Physics* 71 (12): 1267–1276.

Bajracharya, K.M. 2002. Forest fire situation in Nepal. *International Forest Fire News* 26: 84–86.

Benali, A., A. Russo, A.C.L. Sá, R.M.S. Pinto, O. Price, N. Koutsias, and J.M.C. Pereira. 2016. Determining fire dates and locating ignition points with satellite data. *Remote Sensing* 8 (326): 1–20. http://doi.org/10.3390/rs8040326.

Central Bureau of Statistics (CBS). 2019. *Nepal statistical year book*. Government of Nepal.

Chang, D., and Y. Song. 2009. Estimates of biomass burning emissions in tropical Asia based on satellite-derived data. *Atmospheric Chemistry and Physics* 10 (5): 2335–2351. https://doi.org/10.5194/acp-10-2335-2010.

Chang, C.H., C.C. Liu, and P.Y. Tseng. 2013. Emissions inventory for rice straw open burning in Taiwan based on burned area classification and mapping using Formosat-2 satellite imagery. *Aerosol and Air Quality Research* 13: 474–487. https://doi.org/10.4209/aaqr.2012.06.0150.

Chauhan, A., and R.P. Singh. 2017. Poor air quality and dense haze/smog during 2016 in the indo-gangetic plains associated with the crop residue burning and diwali festival. In *IEEE international geoscience and remote sensing symposium (IGARSS)* 6048–6051, Fort Worth, TX, USA. https://doi.org/10.1109/IGARSS.2017.8128389.

Das, B., P.V. Bhave, S.P. Puppala, K. Shakya, B. Maharjan, and R.M. Byanju. 2020. A model-ready emission inventory for crop residue open burning in the context of Nepal. *Environmental Pollution* 266: 1–10. https://doi.org/10.1016/j.envpol.2020.115069.

DFRS. 2015. *State of Nepal's forests*. Forest resource assessment Nepal project. Kathmandu, Nepal: Department of Forest Research and Survey, Government of Nepal.

DoF. 2002. *Forest and vegetation types of Nepal. Tree improvement and silviculture component*. Document Series No. 105. Department of Forests, His Majesty's Government of Nepal.

Earth Data. 2022. Retrieved from https://earthdata.nasa.gov/earth-observation-data/near-real-time/firms; https://earthdata.nasa.gov/earth-observation-data/near-real-time/firms/viirs-i-band-active-fire-data

Elvidge, C.D., M. Zhizhin, K. Baugh, and F.C. Hsu. 2021. Identification of smoldering peatland fires in Indonesia via triple-phase temperature analysis of VIIRS nighttime data. In *Biomass burning in South and Southeast Asia*, 25–38. Boca Raton: CRC Press.

Forest Research and Training Centre (FRTC). 2022. Government of Nepal.

Gupta, R. 2014. Low-hanging fruit in black carbon mitigation: Crop residue burning in South Asia. *Climate Change Economics* 5 (4): 1450012. https://doi.org/10.1142/S2010007814500122.

Gupta, P.K., V.K. Prasad, C. Sharma, A.K. Sarkar, Y. Kant, K.V.S. Badarinath, and A.P. Mitra. 2001. CH_4 emissions from biomass burning of shifting cultivation areas of tropical deciduous forests—Experimental results from ground-based measurements. *Chemosphere-Global Change Science* 3 (2): 133–143.

Gupta, P.K., S. Sahai, N. Singh, C.K. Dixit, D.P. Singh, C. Sharma, M.K. Tiwari, R.K. Gupta, and S.C. Garg. 2004. Residue burning in rice-wheat cropping system: Causes and implications. *Current Science* 87 (12): 1713–1717.

Intergovernmental Panel on Climate Change, IPCC. 1996. *Revised IPCC guidelines for national greenhouse gas inventories*. Reference manual, vol. 3. Bracknell, USA: Intergovernmental Panel on Climate Change (IPCC).

Intergovernmental Panel on Climate Change, IPCC. 2006. IPCC guidelines for national greenhouse gas inventories, prepared by the Johnson EA, Larsen CPS (1991) climatically induced change in fire frequency in the Southern Canadian Rockies. *Ecology* 72: 194–201.

Jethva, H., O. Torres, R.D. Field, A. Lyapustin, R. Gautam, and V. Kayetha. 2019. Connecting crop productivity, residue fires, and air quality over Northern India. *Scientific Reports* 9 (16594): 1–11. https://doi.org/10.1038/s41598-019-52799-x.

Justice, C.O., L. Giglio, S. Korontzi, J. Owens, J.T. Morisette, D. Roy, J. Descloitres, S. Alleaume, F. Petitcolin, and Y. Kaufman. 2002. The MODIS fire products. *Remote Sensing of Environment* 83 (1–2): 244–262.

Justice, C., G. Gutman, and K.P. Vadrevu. 2015. NASA land cover and land use change (LCLUC): An interdisciplinary research program. *Journal of Environmental Management* 148: 4–9.

Kanabkaew, T., and N.T.K. Oanh. 2011. Development of spatial and temporal emission inventory for crop residue field burning. *Environmental Modeling & Assessment* 16: 453–464. https://doi.org/10.1007/s10666-010-9244-0.

Kant, Y., A.B. Ghosh, M.C. Sharma, P.K. Gupta, V.K. Prasad, K.V.S. Badarinath, and A.P. Mitra. 2000a. Studies on aerosol optical depth in biomass burning areas using satellite and ground-based observations. *Infrared Physics & Technology* 41 (1): 21–28.

Kant, Y., V.K. Prasad, and K.V.S. Badarinath. 2000b. Algorithm for detection of active fire zones using NOAA AVHRR data. *Infrared Physics & Technology* 41 (1): 29–34.

Khanal, S. 2015. Wildfire trends in Nepal based on MODIS burnt-area data. *BankoJanakari* 25 (1): 76.

Kharol, S.K., K.V.S. Badarinath, A.R. Sharma, D.V. Mahalakshmi, D. Singh, and V.K. Prasad. 2012. Black carbon aerosol variations over Patiala city, Punjab, India—A study during agriculture crop residue burning period using ground measurements and satellite data. *Journal of Atmospheric and Solar-Terrestrial Physics* 84: 45–51.

Lasko, K., and K. Vadrevu. 2018. Improved rice residue burning emissions estimates: Accounting for practice-specific emission factors in air pollution assessments of Vietnam. *Environmental Pollution* 236: 795–806.

Lasko, K., K.P. Vadrevu, V.T. Tran, E. Ellicott, T.T. Nguyen, H.Q. Bui, and C. Justice. 2017. Satellites may underestimate rice residue and associated burning emissions in Vietnam. *Environmental Research Letters* 12 (8): 085006.

Matin, M.A., V.S. Chitale, M.S.R. Murthy, K. Uddin, B. Bajracharya, and S. Pradhan. 2017. Understanding forest fire patterns and risk in Nepal using remote sensing, geographic information system and historical fire data. *International Journal of Wildland Fire* 26 (4): 276–286.

Ministry of Agriculture Development, MoAD. 2003/04–2016/17. *Statistical information on Nepalese agriculture*. Singha Durbar, Kathmandu: Agri-Business Promotion and Statistics Division, Statistics Section, Government of Nepal.

Ni, H., Y. Han, J. Cao, L.-W.A. Chen, J. Tian, X. Wang, J.C. Chow, J.G. Watson, Q. Wang, P. Wang, H. Li, and R.-J. Huang. 2015. Emission characteristics of carbonaceous particles and trace gases from open burning of crop residues in China. *Atmospheric Environment* 123: 399–406. https://doi.org/10.1016/j.atmosenv.2015.05.007.

Parajuli, A., D.B. Chand, B. Rayamajhi, R. Khanal, S. Baral, Y. Malla, and S. Poudel. 2015. *Spatial and temporal distribution of forest fires in Nepal*. Available at http://www.forestrynepal.org/images/publications/wfc2015_forestfirenepal.pdf. Retrieved 28 Feb 2016.

Park, H., and W. Takeuchi. 2021. Satellite-based estimation of global CO_2 emissions from biomass burning. In *Biomass burning in South and Southeast Asia*, 109–125. Boca Raton: CRC Press.

Permadi, D.A., and N.T.K. Oanh. 2013. Assessment of biomass open burning emissions in Indonesia and potential climate forcing impact. *Atmospheric Environment* 78: 250–258. https://doi.org/10.1016/j.atmosenv.2012.10.016.

Prasad, V.K., Y. Kant, P.K. Gupta, C. Sharma, A.A. Mitra, and K.V.S. Badarinath. 2001. Biomass and combustion characteristics of secondary mixed deciduous forests in Eastern Ghats of India. *Atmospheric Environment* 35 (18): 3085–3095.

Prasad, V.K., Y. Kant, P.K. Gupta, C. Elvidge, and K.V.S. Badarinath. 2002. Biomass burning and related trace gas emissions from tropical dry deciduous forests of India: A study using DMSP-OLS data and ground-based measurements. *International Journal of Remote Sensing* 23 (14): 2837–2851.

Prasad, V.K., M. Lata, and K.V.S. Badarinath. 2003. Trace gas emissions from biomass burning from northeast region in India—Estimates from satellite remote sensing data and GIS. *The Environmentalist* 23 (3): 229–236.

Schroeder, W., P. Oliva, L. Giglio, and I.A. Csiszar. 2014. The new VIIRS 375 m active fire detection data product: Algorithm description and initial assessment. *Remote Sensing of Environment* 143: 85–96.

Shrestha, R.M. 2018. Emissions of air pollutants and greenhouse gases in Nepal, an integrated inventory. In *A study carried out for Regional Resource Centre for Asia and the Pacific (RRC.AP) as a part of activities of Project Atmospheric Brown Cloud, Regional Energy Resources Information Center (RERIC)*. Thailand: Asian Institute of Technology.

Shrestha, R.M., N.T. Kim Oanh, R.P. Shrestha, M. Rupakheti, S. Rajbhandari, D.A. Permadi, T. Kanabkaew, and M. Iyngararasan. 2013. Atmospheric brown clouds (ABC) emission inventory manual. United Nations Environment Programme, Nairobi, Kenya.

Sigsgaard, T., B. Forsberg, I. Annesi-Maesano, A. Blomberg, A. Bølling, C. Boman, J. Bønløkke, M. Brauer, N. Bruce, M.E. Héroux, and M.R. Hirvonen. 2015. Health impacts of anthropogenic biomass burning in the developed world. *European Respiratory Journal* 46 (6): 1577–1588.

Tipayarom, D., and N.T.K. Oanh. 2007. Effects from open rice straw burning emission on air quality in the Bangkok Metropolitan Region. *ScienceAsia* 33: 339–345. http://doi.org/10.2306/scienceasia1513-1874.2007.33.339.

Vadrevu, K.P., and C.O. Justice. 2011. Vegetation fires in the Asian region: Satellite observational needs and priorities. *Global Environmental Research* 15 (1): 65–76.

Vadrevu, K., T. Ohara, and C. Justice. 2017. Land cover, land use changes and air pollution in Asia: A synthesis. *Environmental Research Letters* 12 (12): 120201.

Vadrevu, K.P., T. Ohara, and C. Justice (eds.). 2021a. *Biomass burning in South and Southeast Asia: Impacts on the biosphere*, vol. 2. Boca Raton: CRC Press.

Vadrevu, K.P., T. Ohara, and C. Justice (eds.). 2021b. *Biomass burning in South and Southeast Asia: Mapping and monitoring*, vol. 1. Boca Raton: CRC Press.

Vander-Werf, G.R., J.T. Randerson, L. Giglio, G.J. Collatz, M. Mu, P.S. Kasibhatla, D.C. Morton, R.S. Defries, Y. Jin, and T.T. van Leeuwen. 2010. Global fire emissions and the contribution of deforestation, savanna, forest, agricultural, and peat fires (1997–2009). *Atmospheric Chemistry and Physics* 10: 11707–11735. https://doi.org/10.5194/acp-10-11707-2010.

Zhang, H., J. Hu, Y. Qi, C. Li, J. Chen, X. Wang, J. He, S. Wang, J. Hao, L. Zhang, L. Zhang, Y. Zhang, R. Li, S. Wang, and F. Chai. 2017. Emission characterization, environmental impact, and control measure of PM2.5 emitted from agricultural crop residue burning in China. *Journal of Cleaner Production* 149: 629–635. https://doi.org/10.1016/j.jclepro.2017.02.092.

Zhang, T., M.J. Wooster, M.C. de Jong, and W. Xu. 2018. How well does the 'Small Fire Boost' methodology used within the GFED4.1s fire emissions database represent the timing, location and magnitude of agricultural burning? *Remote Sensing* 10 (823): 1–18. http://doi.org/10.3390/rs10060823.

Zhang, X., Y. Lu, Q. Wang, and X. Qian. 2019. A high-resolution inventory of air pollutant emissions from crop residue burning in China. *Atmospheric Environment* 213: 207–214. https://doi.org/10.1016/j.atmosenv.2019.06.009.

Zhao, Y., C.P. Nielsen, Y. Lei, M.B. McElroy, and J. Hao. 2011. Quantifying the uncertainties of a bottom-up emission inventory of anthropogenic atmospheric pollutants in China. *Atmospheric Chemistry and Physics* 11 (5): 2295–2308. https://doi.org/10.5194/acp-11-2295-2011.

Firewood Burning Dynamics by the Sri Lankan Households: Trends, Patterns, and Implications

Asankha Pallegedara and Ajantha Sisira Kumara

Abstract Many Sri Lankans rely on firewood for cooking despite confirmed issues of burning firewood in households for cooking purposes. This paper explores the trends and patterns of firewood burning for cooking by Sri Lankan households using data collected from more than 57,000 households. Analyzing three waves of nationwide Household Income and Expenditure Survey (HIES) data collected in 1990–1991, 2002, and 2016, this study found that, on average, although reliance on firewood for cooking has been declining over time, more than 78% rural households still use firewood for cooking. Relatively poor-income households are more likely to still use firewood for cooking, while relatively higher-income households are expected to significantly reduce their firewood usage over time. The education level of the household head is another major factor determining the use of firewood for cooking purposes. Based on these results, the study suggests promoting education and awareness campaigns to understand the adverse effects of using firewood targeting rural households and generating alternative energy sources such as solar power, wind power, and biogas for domestic purposes. Moreover, the policy implications are linked with promoting safe cooking stoves and conducive kitchen designs, even for urban dwellings, to mitigate the health consequences of using firewood.

Keywords Firewood burning · Households · Energy · Air pollution · Health impacts

A. Pallegedara (✉)
Department of Industrial Management, Faculty of Applied Sciences, Wayamba University of Sri Lanka, Kuliyapitiya, Sri Lanka
e-mail: asankap@wyb.ac.lk

A. S. Kumara
Department of Public Administration, Faculty of Management Studies and Commerce, University of Sri Jayewardenepura, Nugegoda, Sri Lanka
e-mail: mhasisira@sjp.ac.lk

K. P. Vadrevu et al. (eds.), *Vegetation Fires and Pollution in Asia*,
https://doi.org/10.1007/978-3-031-29916-2_6

1 Introduction

Sri Lankan households depend mainly on firewood, liquefied petroleum gas (LPG), kerosene, and electricity for cooking. Nonetheless, approximately 72% of households still use firewood for cooking (Department of Census and Statistics 2018) for various reasons. For instance, most Sri Lankans live in rural sectors and are in a good position to collect the required firewood from their vicinity without incurring any financial cost. Moreover, according to Sri Lankan food culture, people strongly believe that food gets tastier though quality may not differ when cooked with firewood. In particular, the food cooked with firewood is considered a part of Sri Lankan traditional meals, characterized by a unique flavor and authenticity. This special flavor is believed to be added by the slow-cooked heating process of firewood stoves. Further, firewood prices are relatively lower compared to the prices of other energy sources used for cooking. For instance, the retail unit price of a bundle of 5 kg of firewood is LKR 100 (0.49 USD) on average.

Recently, Sri Lankan households have faced several issues concerning the supply and pricing of LPG and the safety of gas cylinders. Nearly 800 incidents of cooking-gas-related explosions have been recorded from January 1, 2021 to December 31, 2021. The LPG supply in Sri Lanka is being done with a system of re-filling using gas cylinders. Also, the quality assurance in filling LPG has several issues. Thus, Sri Lankan households are now moving to firewood, believing that firewood is safer than LPG. Consequently, firewood usage for cooking purposes may have substantially increased, and this study examines its dynamics to understand how trends in using firewood have changed over time until the present.

Using biomass and firewood for cooking and heating is categorized as one of the dirty energy sources as there are confirmed issues of burning firewood in households for cooking and heating purposes (Vadrevu and Lasko 2015). *First*, burning wood generates a mixture of gases, liquids, and solid particles, reducing oxygen while increasing the background temperature. Exposure to wood smoke is found to be associated with several adverse health effects. Using firewood as an energy source comes under the dirty fuel category due to the emission of pollutants, including carbon dioxide, carbon monoxide, nitrogen dioxide, aerosols, and volatile organic compounds, which are sources of indoor air pollution (Gupta et al. 2001a, b; Kant et al. 2000; Prasad et al. 2000, 2001, 2002; Badarinath et al. 2007, 2008, 2009; Lasko et al. 2021; Vadrevu et al. 2014, 2018; 2021a, b).

They include toxic substances which negatively affect the respiratory health of the individuals exposed to wood smoke. As examined by Pallegedara and Kumara (2022), there is a significant positive association between burning firewood for cooking and the respiratory health of people in Sri Lanka. Asthma prevalence is significantly higher among those who burn firewood for cooking. A systematic review by Bede-Ojimadu and Orisakwe (2020) also confirms that high levels of exposure to wood smoke cause respiratory health issues for people living in sub-Saharan Africa.

Second, there are also environmental consequences of using firewood for cooking. For instance, increased demand for firewood can only be fulfilled by cutting down trees in the forest, and as a consequence, the forest would be destroyed. Sri Lanka is endowed with a natural forest cover of approximately two million hectares of land area or a forest cover of 20.7% (Sri Lanka UN-REDD 2017), which is the leading provider of fuel wood. However, for various reasons, Sri Lankan forest cover has reduced to 16.5% in 2019. Extracting trees for firewood production has been one of the main factors contributing to Sri Lanka's higher rate of deforestation. In addition, selling firewood as a business endeavor has substantially increased in Sri Lanka due to increased demand triggered by the issues on LPG supply.

Third, even though firewood has been a low-cost alternative for cooking, its accessibility is not universal. In particular, the houses in urban sectors are not designed to use firewood. Therefore, in the absence of LPG, urban households face various issues relating to sourcing and using firewood for cooking.

With the safety issue of LPG, Sri Lankan policymakers in the energy sector are searching for alternatives. Policymakers consider introducing suitable wood-fired cooking stoves. However, we need to ensure a proper firewood supply mechanism, particularly in urban areas. Replanting trees used for firewood like Gliricidia is highlighted much in policy debates to ensure an adequate supply of dried and processed firewood. The health hazards associated with firewood burning are more likely in poorly ventilated and confined spaces. However, kitchens with properly built chimneys would solve the issue to a greater extent. Moreover, a policy evaluation effort is required to assess the possibility of using electricity for cooking without burdening the national electricity grid. Policymakers are evaluating the potential of using solar rooftop systems to generate the electricity required for cooking.

This paper contributes to the ongoing policy debate on using firewood as a source of energy for cooking by examining its trends and dynamics. First, the paper provides empirical evidence on the dynamics of using firewood for cooking from 1990 to the present and how they differ across different household types in terms of living sector, income levels, nature of household head, household size, housing information, and livelihood-related factors. Second, this paper discusses the probable implications of firewood burning by Sri Lankan households. Accordingly, the paper contains vital information that may be useful to design and popularize safe firewood stoves, design projects of replanting trees, and ventilated kitchens and understand probable subpopulations for solar system installation.

The paper is structured as follows: Sect. 2 reviews the previous related literature. Then, Sect. 3 describes the data and methods applied, while Sect. 4 presents the main results of the analysis and discusses it. Finally, Sect. 5 concludes by explaining policy implications.

2 Literature Review

The United Nations (UN) identified the importance of using clean energy sources by including Sustainable Development Goal (SDG) goal 7: "Affordable and clean energy" as a critical challenge to achieve by 2030 (UNDP 2018). However, as many as 2.6 billion people, especially in the developing world, still use dangerous and inefficient cooking systems, which cause adverse health impacts (UN 2021). Many households use direct firewood or sometimes charcoal derived from wood for cooking their food. Although firewood can be considered a renewable resource technically, the forest destruction rate from firewood consumption is often more significant than the forest replacement rate. Thus, firewood burning for cooking can be considered one of the critical sources of forest degradation.

Earlier studies illustrated the links between firewood consumption and forest degradation (Baland et al. 2010; Heltberg et al. 2000; Kirubi et al. 2000; Negi et al. 2018; Specht et al. 2015; Trossero 2002). For example, Specht et al. (2015) found that, on average, Brazilian rural households consume 961 kg/person/year, which causes significant forest degradation in the Brazilian Atlantic forests. It was estimated that the annual demand for firewood for cooking from 210 municipalities might reach 303,793 tons, equivalent to 1.2–2.1 thousand hectares of tropical forest. Many households in developed and developing countries also burn firewood for heating purposes, especially in cold regions (Schueftan et al. 2016; Negi et al. 2018; Zhu et al. 2020). For instance, 25.6% of Italian households still use firewood for domestic heating, increasing to 38.7% for mountain regions (Negi et al. 2018). Therefore, using low-efficient firewood as a heating energy source could also increase forest degradation.

Firewood consumption for domestic purposes depends significantly on households' socio-economic and demographic characteristics. The income and wealth status of the households are significantly associated with firewood consumption patterns (Heltberg et al. 2000; Rahut et al. 2016). Poor-income households tend to consume more firewood collected from forests, while relatively high-income households consume less firewood because of the higher opportunity cost of collecting firewood from forests. Studies found that the education level of the household heads and family members seems to reduce firewood usage in homes (Mottaleb et al. 2017; Pandey and Chaubal 2011). Education may bring about awareness of the adverse health impacts of air pollution caused by burning firewood. Demographic factors such as family size and composition are positively associated with using firewood for domestic purposes (Dewees 1989; Rao and Reddy 2007). Large families are expected to provide additional labor resources for collecting firewood from nearby forests.

Studies on firewood consumption and its impacts in the Sri Lankan context are scarce due to difficulty in obtaining recorded data. Previous studies primarily relied on household survey data to estimate the usage of firewood consumption (Rajmohan and Weerahewa 2010; Nandasena et al. 2012; Wickramasinghe 2011; Wijayatunga and Attalage 2002; Pallegedara et al. 2021). Wickramasinghe (2011) and Nandasena et al. (2012) found that despite improving income level and quality of life, most households still use firewood as the primary energy source for cooking, particularly in rural areas. However, there is evidence of energy switching from dirty energy usage to cleaner energy usage over time (Rajmohan and Weerahewa 2010; Pallegedara et al. 2021). There are significant adverse health impacts from firewood usage for cooking reported in Sri Lanka (Pallegedara and Kumara 2022; Lall et al. 2021) and environmental impacts through greenhouse gas emissions (Perera and Sugathapala 2002). However, replacing traditional cooking stoves with improved ones could reduce harmful gas emissions.

3 Data and Methods

This paper uses data from the Household Income and Expenditure Survey (HIES) compiled by the Department of Census and Statistics (DCS) of Sri Lanka. Three waves of HIES data collected in 1990/1991, 2002, and 2016 will be used to analyze the household firewood burning dynamics over time. HIES is a nationwide cross-sectional household survey intended to cover all districts in Sri Lanka and collect income, expenditure, consumption, and demographic details of Sri Lankan households. DCS collects data using a two-stage stratified random sampling method. First, DCS selects census blocks as the primary sampling units, which are proportionate to the number of housing units in Sri Lanka (DCS 2015). Second, DCS selects housing units as the secondary sampling units from the primary sampling units selected in the first stage. DCS collects and records income and demographic data at the individual level, while it collects consumption and expenditure data at the household level. In addition, detailed household consumption expenditure data for all food and non-food expenditures are collected weekly or monthly. This study primarily uses household energy consumption expenditures on firewood for domestic cooking purposes for the analysis.

Descriptive statistics are used to describe and uncover the trends and patterns of household firewood burning over time. Frequency distribution graphs by different socio-economic categories of households will be depicted. To precisely understand the central tendency and the variability of firewood burning by households, mean values and standard deviation will be calculated.

4 Results and Discussion

Table 1 shows the background information of sampled households. It shows that 18,462, 16,924, and 21,748 households were surveyed in HIES 1990/91, HIES 2002, and HIES 2016. The average household size seems to decline over the years. For example, there were five members in a household in 1990/91, but it decreased to around four members in 2016. The average age of the household head was about

Table 1 Background information on the sampled households by survey period

Variables	Survey period		
	1990/91	2002	2016
No. of observations	18,462	16,924	21,748
Household size (members)	5.07 (2.14)	4.36 (1.77)	3.81 (1.59)
Age of the household head (HH) (years)	49.57 (14.28)	49.21 (13.88)	52.70 (14.06)
Urban households (%)	33.70 (47.27)	19.15 (39.35)	15.77 (36.44)
HH only attended up to primary school (%)	44.35 (49.68)	33.82 (47.31)	26.23 (43.99)
HH attended up to secondary school (%)	53.46 (49.88)	63.83 (48.05)	70.80 (45.47)
HH attended tertiary education (%)	1.77 (13.19)	2.36 (15.17)	2.97 (16.96)
Sinhalese households (%)	82.66 (37.85)	83.89 (36.77)	72.57 (44.34)
Tamil households (%)	9.44 (28.96)	9.34 (29.10)	18.70 (38.76)

Note Standard deviations are in parentheses
Source Authors' calculation based on HIES 1990/91, HIES 2002, HIES 2016

50 years old in 1990/91 and 2002, but it increased to about 53 years in 2016. About 34%, 19%, and 16% of households in urban areas in 1990/91, 2002, and 2016, respectively. In the 1990/91 survey, the DCS oversampled the urban sector in each district to allocate roughly one-third of the total population to urban areas. Table 1 also shows that the education level of the household head tends to increase over the years. For example, on average, 44% of household heads had completed up to a primary school level in 1990/91, but this percentage decreased to 34% in 2002 and 26% in 2016. In contrast, the percentage of household heads with education up to secondary school level increased from 53% in 1990/91 to 64% in 2002 and 71% in 2016. More than 82% of surveyed households were ethnic Sinhalese in 1990/91 and 2002 survey rounds, while it was about 73% in 2016. In contrast, about 9% of surveyed households were ethnic Tamils in 1990/91 and 2002 survey rounds, while it was about 19% in 2016. It should be mentioned that compared to HIES 1990/91 and HIES 2002 surveys, the percentage of ethnic Tamil households increased because HIES 1990/91 and HIES 2002 surveys could not collect information from the households in North and Eastern parts of Sri Lanka where most Tamil households reside due to civil war ended only in 2009.

Figure 1 illustrates the distribution of households that used firewood for cooking purposes across the survey years. It shows that the use of firewood for cooking has decreased over time. For instance, over 89% of households used firewood for cooking in 1990/91, which declined to about 80% in 2002 and to about 70% in 2016. It is expected that the percentage of households that used firewood for cooking declined because of the increased income level of the households and the expansion of middle-class people over the years. For example, the per capita gross domestic product (GDP) of Sri Lanka has increased from about US$ 465 in 1990 to US$ 3800 in 2015 (World Bank 2017).

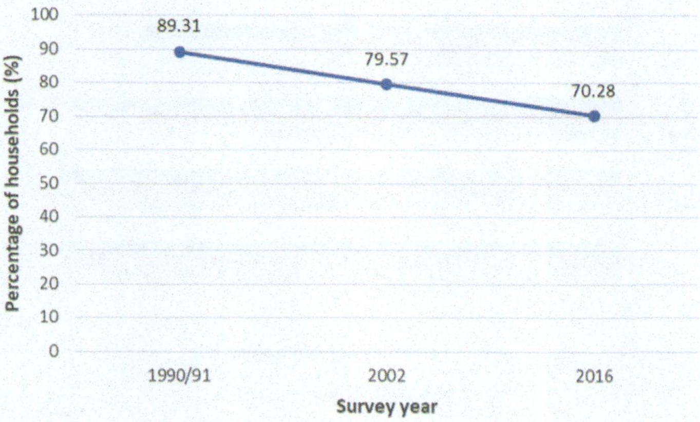

Fig. 1 Distribution of households using firewood for cooking by survey period. *Source* Authors' calculation based on HIES 1990/91, HIES 2002, and HIES 2016

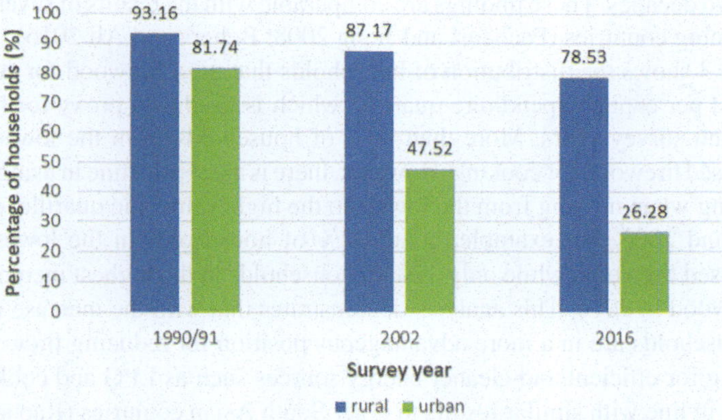

Fig. 2 Distribution of households using firewood for cooking by survey period and living sector of households. *Source* Authors' calculation based on HIES 1990/91, HIES 2002, and HIES 2016

Figure 2 presents the distribution of households that used firewood for cooking by survey years and the location of where the household resides. According to that, more rural households used firewood for cooking purposes compared to urban households. However, for rural and urban households, firewood usage has decreased over time. For example, about 93% of rural households used firewood in 1990/91, declining to 87% in 2002 and 79% in 2016. In contrast, about 82% of urban households used firewood in 1990/91, which decreased to about 48% in 2002 and 26% in 2016. Thus, urban households significantly reduce firewood usage over time. Since firewood is not easily available in urban areas and other clean energy sources such as liquefied petroleum gas (LPG) and electricity are easily available in urban areas, urban households are more likely to switch from firewood to other clean energy sources for cooking during

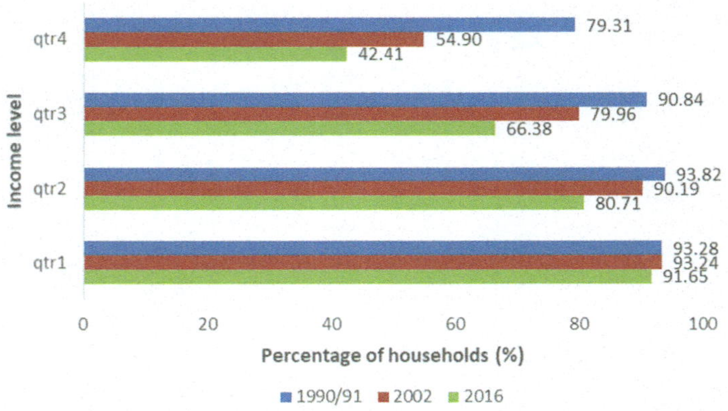

Fig. 3 Distribution of households using firewood for cooking by survey period and income quartile of households. *Source* Authors' calculation based on HIES 1990/91, HIES 2002, and HIES 2016

the last two decades. These findings are comparable with the results of several studies in developing countries (Pachauri and Jiang 2008; Behera and Ali 2016).

Figure 3 shows the distribution of households that used firewood for cooking by household per capita expenditure quartile, which is used as a proxy for household income and survey years. More than 90% of households from the lowest income quartile used firewood for cooking. However, there is a steep decline in using firewood for cooking when moving from the lowest to the highest income quartile, especially in 2016 and 2002. For example, about 92% of households in the lowest income quartile used firewood, while only 42% of households in the highest income quartile used firewood in 2016. This analysis demonstrates that with the increase in income level, households are in a more advantageous position for reducing firewood usage and opting for efficient and cleaner energy sources such as LPG and cooking. This finding is in line with similar results in other South Asian countries (Rao and Reddy 2007; Farsi et al. 2007; Rahut et al. 2020).

Figure 4 depicts the relationship between the households that used firewood for cooking and the education level of the household head. There is a steep downward trend in using firewood for cooking when the education level of the household head improved from below primary school education to more than tertiary education. This negative linkage between the level of education and firewood usage is more visible in recent survey data. For example, in 2016, about 87% of households that used firewood for cooking were headed by a primary school-educated household head, while this percentage decreased to 66% when a secondary school-educated household head headed households and further decreased to 25% when a tertiary-level educated household head headed households. This is because more educated household heads know the adverse health and environmental impacts of using firewood for cooking and increasingly switch to a more efficient and cleaner energy source like LPG. In

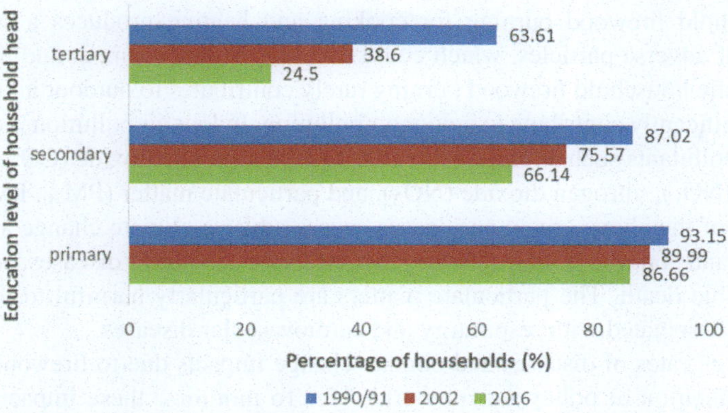

Fig. 4 Distribution of households that used firewood for cooking by survey period and education level of the household head. *Source* Authors' calculation based on HIES 1990/91, HIES 2002, and HIES 2016

addition, educated household heads tend to use less firewood for cooking because the opportunity cost of time involved in collecting firewood from the neighborhood forests may be more valuable for them. Previous studies also confirmed the negative relationship between the education level and firewood usage in other countries (Nepal et al. 2011; Heltberg 2004; Israel 2002).

5 Conclusion and Policy Implications

This study examined the trends and patterns of household firewood usage for cooking purposes in Sri Lanka using three waves of nationwide household survey data covering 1990–2016. The descriptive analysis of trends and patterns of household firewood use for cooking shows that the use of firewood for cooking has decreased over time. However, rural households are more likely to still use firewood for cooking purposes compared to urban households. In addition, relatively poor-income households tend to stick to using firewood for cooking, while relatively higher-income households significantly reduce their firewood usage. This study further confirms that educated household heads seem to significantly reduce the use of firewood domestically during the survey period from 1990 to 2016.

Firewood burning for domestic purposes such as household cooking and heating has environmental and health implications. Extensive firewood collection may cause forest degradation. Extracting trees for firewood production is one of the main factors contributing to Sri Lanka's higher rate of deforestation over time. If the forests are cut down in an unsustainable way to obtain firewood in rural areas, it may cause lands to be unproductive, resulting in adverse economic impacts on rural farming communities in Sri Lanka.

Household firewood burning for cooking and heating produces a substantial amount of adverse particles, which could damage the environment and the human body. While household firewood burning rarely contributes to outdoor air pollution, it can significantly contribute to indoor air pollution. Indoor air pollution is caused by harmful pollutants such as carbon dioxide (CO_2), carbon monoxide (CO), oxides of nitrogen (NO_x), nitrogen dioxide (NO_2), and particulate matter (PM_{10}, $PM_{2.5}$). The emission of the above greenhouse gases causes adverse climate change and global warming and contributes to respiratory diseases, cardiovascular diseases, headaches, and possible death. The particulate matters are particularly harmful to the human body and associated with respiratory and cardiovascular diseases.

The high rates of diseases and climate change impacts due to firewood burning call the attention of policymakers worldwide. To minimize these impacts, policymakers should focus on increasing educational and awareness programs targeted primarily at rural households. It can increase the understanding of the adverse consequences of using firewood for cooking and provide an incentive to use clean energy sources. Since income is directly related to the use of firewood (usually collected free from the forests), new public policies should favor access to cleaner energy sources such as LPG for relatively poor-income households. In addition, the implementation of alternative energy sources such as solar power, wind power, and biogas should be promoted. In order to reduce indoor air pollution and associated health risks, the reduction of the use of traditional firewood cooking stoves and the adoption of improved modern cooking stoves should also be considered.

Firewood burning at the household level for cooking has become health-hazardous, particularly in urban dwellings in Sri Lanka. Urban houses are generally designed for cleaner energy, like LPG for cooking, with modern kitchen structures with no traditional chimney to emit hazardous elements safely. Thus, firewood burning in the urban dwelling may not be advisable, and a chimney with an adequate height may be recommended for house construction projects in the country's urban sector. Moreover, respiratory health risks induced by firewood usage have been gendered, as females mainly do household-level cooking in Sri Lanka (Pallegedara and Kumara 2022). However, this fact needs to be taken seriously in the discussions on women's empowerment. Thus, promoting cleaner energy, safe cooking stoves, and safe kitchen designs need to be prioritized when setting an agenda for women's empowerment in countries.

Acknowledgements The authors thank the Sri Lanka Department of Census and Statistics for providing household income and expenditure data. All remaining errors are attributed to the authors.

References

Available at SSRN: https://ssrn.com/abstract=3834319 or http://doi.org/10.2139/ssrn.3834319.

Badarinath, K.V.S., S.K. Kharol, K.M. Latha, T.K. Chand, V.K. Prasad, A.N. Jyothsna, and K. Samatha. 2007. Multiyear ground-based and satellite observations of aerosol properties over a tropical urban area in India. *Atmospheric Science Letters* 8 (1): 7–13.

Badarinath, K.V.S., S. Kumar Kharol, V. Krishna Prasad, A. Rani Sharma, E.U.B. Reddi, H.D. Kambezidis, and D.G. Kaskaoutis. 2008. Influence of natural and anthropogenic activities on UV index variations—A study over tropical urban region using ground based observations and satellite data. *Journal of Atmospheric Chemistry* 59 (3): 219–236.

Badarinath, K.V.S., S.K. Kharol, A.R. Sharma, and V.K. Prasad. 2009. Analysis of aerosol and carbon monoxide characteristics over Arabian Sea during crop residue burning period in the Indo-Gangetic Plains using multi-satellite remote sensing datasets. *Journal of Atmospheric and Solar-Terrestrial Physics* 71 (12): 1267–1276.

Baland, J.M., P. Bardhan, S. Das, and D. Mookherjee. 2010. Forests to the people: Decentralization and forest degradation in the Indian Himalayas. *World Development* 38 (11): 1642–1656.

Bede-Ojimadu, O., and O.E. Orisakwe. 2020. Exposure to wood smoke and associated health effects in Sub-Saharan Africa: A systematic review. *Annals of Global Health* 86 (1).

Behera, B., and A. Ali. 2016. Household energy choice and consumption intensity: Empirical evidence from Bhutan. *Renewable and Sustainable Energy Reviews* 53: 993–1009.

Department of Census and Statistics (DCS). 2018. *Final report. Household income and expenditure survey-2016*. Colombo: Department of Census and Statistics.

Department of Census and Statistics of Sri Lanka (DCS). 2015. *Household income and expenditure survey, 2012/2013 final report*. Colombo: Department of Census and Statistics of Sri Lanka.

Dewees, P.A. 1989. The woodfuel crisis reconsidered: Observations on the dynamics of abundance and scarcity. *World Development* 17 (8): 1159–1172.

Farsi, M., M. Filippini, and S. Pachauri. 2007. Fuel choices in urban Indian households. *Environment and Development Economics* 12 (6): 757–774.

Gupta, P.K., V.K. Prasad, Y. Kant, C. Sharma, A.B. Ghosh, M.C. Sharma, A.K. Sarkar, S.L. Jain, O.P. Tripathi, R.C. Sharma, and K.V.S. Badarinath. 2001a. Study of trace gases and aerosol emissions due to biomass burning at shifting cultivation sites in East Godavari District (Andhra Pradesh) during INDOEX IFP-99. *Current Science* 186–196.

Gupta, P.K., V.K. Prasad, C. Sharma, A.K. Sarkar, Y. Kant, K.V.S. Badarinath, and A.P. Mitra. 2001b. CH_4 emissions from biomass burning of shifting cultivation areas of tropical deciduous forests—Experimental results from ground-based measurements. *Chemosphere-Global Change Science* 3 (2): 133–143.

Heltberg, R. 2004. Fuel switching: Evidence from eight developing countries. *Energy Economics* 26 (5): 869–887.

Heltberg, R., T.C. Arndt, and N.U. Sekhar. 2000. Fuelwood consumption and forest degradation: A household model for domestic energy substitution in rural India. *Land Economics* 213–232.

Israel, D. 2002. Fuel choice in developing countries: Evidence from Bolivia. *Economic Development and Cultural Change* 50 (4): 865–890.

Kant, Y., A.B. Ghosh, M.C. Sharma, P.K. Gupta, V.K. Prasad, K.V.S. Badarinath, and A.P. Mitra. 2000. Studies on aerosol optical depth in biomass burning areas using satellite and ground-based observations. *Infrared Physics & Technology* 41 (1): 21–28.

Kirubi, C., W.N. Wamicha, and J.K. Laichena. 2000. The effects of woodfuel consumption in the ASAL areas of Kenya: The case of Marsabit forest. *African Journal of Ecology* 38 (1): 47–52.

Lall, O., G. Bowatte, S. Dharmaratne, A. Lowe, A. Vakalopoulos, I. Ambrose, P. Jayasinghe, D. Yasaratne, J. Heyworth, and S. Dharmage. 2021. Household use of biomass fuel, especially traditional stove, is associated with childhood wheeze and eczema: A cross sectional study of rural communities in Kandy, Sri Lanka. http://doi.org/10.2139/ssrn.3834319.

Lasko, K., K.P. Vadrevu, V. Bandaru, T.N.T. Nguyen, and H.Q. Bui. 2021. PM2.5 emissions from biomass burning in South/Southeast Asia—Uncertainties and trade-offs. In *Biomass burning in South and Southeast Asia*, 149–169. Boca Raton: CRC Press.

Mottaleb, K., D. Rahut, and A. Ali. 2017. An exploration into the household energy choice and expenditure in Bangladesh. *Energy* 135: 767–776.

Nandasena, S., A.R. Wickremasinghe, and N. Sathiakumar. 2012. Biomass fuel use for cooking in Sri Lanka: Analysis of data from national demographic health surveys. *American Journal of Industrial Medicine* 55 (12): 1122–1128.

Negi, V.S., B.C. Joshi, R. Pathak, R.S. Rawal, and K.C. Sekar. 2018. Assessment of fuelwood diversity and consumption patterns in cold desert part of Indian Himalaya: Implication for conservation and quality of life. *Journal of Cleaner Production* 196: 23–31.

Nepal, M., A. Nepal, and K. Grimsrud. 2011. Unbelievable but improved cookstoves are not helpful in reducing firewood demand in Nepal. *Environment and Development Economics* 16 (1): 1–23.

Pachauri, S., and L. Jiang. 2008. The household energy transition in India and China. *Energy Policy* 36 (11): 4022–4035.

Pallegedara, A., and A.S. Kumara. 2022. Impacts of firewood burning for cooking on respiratory health and healthcare utilisation: Empirical evidence from Sri Lankan micro-data. *International Journal of Health Planning and Management* 37 (1): 465–485.

Pallegedara, A., K.A. Mottaleb, and D.B. Rahut. 2021. Exploring choice and expenditure on energy for domestic works by the Sri Lankan households: Implications for policy. *Energy* 222: 119899.

Pandey, V.L., and A. Chaubal. 2011. Comprehending household cooking energy choice in rural India. *Biomass and Bioenergy* 35 (11): 4724–4731.

Perera, K.K.C.K., and A.G.T. Sugathapala. 2002. Fuelwood-fired cookstoves in Sri Lanka and related issues. *Energy for Sustainable Development* 6 (1): 85–94.

Prasad, V.K., P.K. Gupta, C. Sharma, A.K. Sarkar, Y. Kant, K.V.S. Badarinath, T. Rajagopal, and A.P. Mitra. 2000. NO_x emissions from biomass burning of shifting cultivation areas from tropical deciduous forests of India—Estimates from ground-based measurements. *Atmospheric Environment* 34 (20): 3271–3280.

Prasad, V.K., Y. Kant, P.K. Gupta, C. Sharma, A.A. Mitra, and K.V.S. Badarinath. 2001. Biomass and combustion characteristics of secondary mixed deciduous forests in Eastern Ghats of India. *Atmospheric Environment* 35 (18): 3085–3095.

Prasad, V.K., Y. Kant, P.K. Gupta, C. Elvidge, and K.V.S. Badarinath. 2002. Biomass burning and related trace gas emissions from tropical dry deciduous forests of India: A study using DMSP-OLS data and ground-based measurements. *International Journal of Remote Sensing* 23 (14): 2837–2851.

Rahut, D., B. Behera, and A. Ali. 2016. Patterns and determinants of household use of fuels for cooking: Empirical evidence from sub-Saharan Africa. *Energy* 117 (P1): 93–104.

Rahut, D.B., A. Ali, K.A. Mottaleb, and J.P. Aryal. 2020. Understanding households' choice of cooking fuels: Evidence from urban households in Pakistan. *Asian Development Review* 37 (1): 185–212.

Rajmohan, K., and J. Weerahewa. 2010. Household energy consumption patterns in Sri Lanka. *Sri Lankan Journal of Agricultural Economics* 9. http://doi.org/10.4038/sjae.v9i0.1833.

Rao, M.N., and B.S. Reddy. 2007. Variations in energy use by Indian households: An analysis of micro level data. *Energy* 32 (2): 143–153.

Schueftan, A., J. Sommerhoff, and A.D. González. 2016. Firewood demand and energy policy in south-central Chile. *Energy for Sustainable Development* 33: 26–35.

Specht, M.J., S.R.R. Pinto, U.P. Albuquerque, M. Tabarelli, and F.P. Melo. 2015. Burning biodiversity: Fuelwood harvesting causes forest degradation in human-dominated tropical landscapes. *Global Ecology and Conservation* 3: 200–209.

Sri Lanka UN-REDD. 2017. Sri Lanka's forest reference level submission to the UNFCCC.

Trossero, M.A. 2002. Wood energy: The way ahead. *Unasylva (English Ed.)* 53 (211): 3–12.

UNDP. 2018. *Sustainable development goals goal 7: Affordable and clean energy*. http://www.undp.org/content/undp/en/home/sustainable-development-goals/goal-7-affordable-and-clean-energy.html.

United Nations (UN). 2021. *SDG goal 7*. Department of Economic and Social Affairs. https://sdgs.un.org/goals/goal7.

Vadrevu, K., and K. Lasko. 2015. Fire regimes and potential bioenergy loss from agricultural lands in the Indo-Gangetic Plains. *Journal of Environmental Management* 148: 10–20.

Vadrevu, K.P., T. Ohara, and C. Justice. 2014. Air pollution in Asia. *Environmental Pollution (Barking, Essex: 1987)* 195: 233–235.

Vadrevu, K.P., T. Ohara, and C. Justice (eds.). 2018. *Land-atmospheric research applications in South and Southeast Asia.*

Vadrevu, K.P., T. Ohara, and C. Justice (eds.). 2021a. *Biomass burning in South and Southeast Asia: Impacts on the biosphere*, vol. 2. Boca Raton: CRC Press.

Vadrevu, K.P., T. Ohara, and C. Justice (eds.). 2021b. *Biomass burning in South and Southeast Asia: Mapping and monitoring*, vol. 1. Boca Raton: CRC Press.

Wickramasinghe, A. 2011. Energy access and transition to cleaner cooking fuels and technologies in Sri Lanka: Issues and policy limitations. *Energy Policy* 39 (12): 7567–7574.

Wijayatunga, P.D., and R.A. Attalage. 2002. Analysis of household cooking energy demand and its environmental impact in Sri Lanka. *Energy Conversion and Management* 43 (16): 2213–2223.

World Bank. 2017. *World development indicator*. Washington DC, United States: World Bank. http://data.worldbank.org/data-catalog/worlddevelopment-indicators.

Zhu, L., H. Liao, B. Hou, L. Cheng, and H. Li. 2020. The status of household heating in northern China: A field survey in towns and villages. *Environmental Science and Pollution Research* 27 (14): 16145–16158.

Burnt Area Signal Variations in Agriculture and Forested Landscapes of India—A Case Study Using Sentinel-1A/B Synthetic Aperture Radar

Krishna Prasad Vadrevu, Aditya Eaturu, and Sumalika Biswas

Abstract In this study, we report contrasting burnt area signals in Sentinel-1A Synthetic Aperture Radar (SAR) VV polarized data versus Sentinel-2 optical data for agriculture and forested ecosystems. A clear decrease in the Sentinel-1 SAR backscatter has been observed in agricultural burnt areas, whereas an increase in the forest burnt areas. Furthermore, in contrast to the SAR data, the optical data showed a clear decrease in the normalized difference vegetation index (NDVI) in agricultural and forest burnt areas. Our results suggest a caution toward using a universal algorithm that can capture burnt areas at a global scale using the SAR data.

Keywords Sentinel-1A SAR data · NDVI · Burnt signal variations · India

1 Introduction

Biomass burning is prevalent in the tropics, especially in South/Southeast Asian (S/SEA) countries. In S/SEA, fire is widely used a land clearing tool through the slash and burn agriculture, clearing agricultural crop residues for planting the next crop, hunting, pest control, etc. (Stolle et al. 2004; Suyanto 2006; Albar et al. 2018; Badarinath et al. 2007, 2008, 2009; Badarinath and Prasad 2011; Biswas et al. 2015a, b, 2021). In addition, factors such as lightning or negligence too can lead to the ignition of vegetation fires. Furthermore, both prescribed burns and accidentally ignited fires can easily spread to neighboring landscapes in dry conditions and develop into uncontrolled wildfires. The impacts of the fires on the ecosystems can vary widely based on the type of vegetation burnt, amount, timing, etc. (Prasad et al. 2001a, b, 2002a, b, 2003, 2004, 2005; Prasad and Badarinath 2004, 2006; Vadrevu 2008;

K. P. Vadrevu (✉)
NASA Marshall Space Flight Center, Huntsville, AL, USA
e-mail: Krishna.p.vadrevu@nasa.gov

A. Eaturu
University of Alabama, Huntsville, USA

S. Biswas
University of California, Los Angeles, USA

K. P. Vadrevu et al. (eds.), *Vegetation Fires and Pollution in Asia*,
https://doi.org/10.1007/978-3-031-29916-2_7

Vadrevu and Badarinath 2009; Justice et al. 2015; Vadrevu et al. 2022). The impacts also depend on how well the ecosystem is resistant and adapted to the fires. Typically, partial destruction of vegetation cover by fire can impact species composition, vegetation structure, and nutrient fluxes in a variety of ways (Dwyer et al. 2000). In addition, another important effect of fires or biomass burning is the release of greenhouse gas emissions and aerosols which can affect the atmospheric environment and atmospheric chemistry (Choi et al. 2008; Lata et al. 2001; Levine 2000; Kant et al. 2000; Kharol et al. 2012; Lasko and Vadrevu 2018; Lasko et al. 2017, 2018; Vadrevu et al. 2019a, b). Over the 2002–2019 period, vegetation fires have affected worldwide, causing average carbon emissions from 2.1 $\pm$ 0.2 ($\pm$ 1σ interannual variability) Pg C yr^{-1} (van Wees et al. 2022). Considering these local to regional scale impacts, mapping, and monitoring fires and burnt areas, including emissions in diverse landscapes, are very much needed.

The most popular satellite derived methods of fire detection include active fire detection and burnt area mapping. Mostly, active fire detection is based on the mid or thermal infrared radiation (TIR) relevant bands, whereas the detection of the burnt area is based on the visible channels (Eva and Lambin 1998; Petropoulos et al. 2013). Satellite derived burnt area methods aim at detecting and delineating burn scars using the spectral signature of the vegetation and burnt areas before and after fires and landscape-related change conditions. In conjunction with other ancillary data, such as vegetation and biomass information, the burnt scar information can be used to estimate the effects, including the impacts of fires in any ecosystems. The most popular emissions estimation from biomass burning following Seiler and Crutzen (1980) follows the equation: $G(x) = A \times B \times C \times E$, where $G(X)$ is the total amount of trace gas released, A is the amount of the area burned, B is the biomass or fuel load, C is the combustion or burning efficiency, i.e., percentage of biomass consumed, and E is the amount of trace gas released per unit of dry matter burned. The satellite derived parameter in the above equation is the amount of the area burned, which is an important parameter for estimating the emissions. Specific to the burnt area mapping methods, most rely on polar satellites and use automatic or semi-automatic processing using medium spatial and temporal resolution satellite data (Giglio et al. 2018). Using these methods, the burnt areas can be estimated; however, the overall results can be underestimated especially in tropical areas, due to the cloud cover, which prohibits fire detection and burnt area mapping when using polar satellites. In addition, satellite resolution too can play an important role in burnt scar delineation. Despite these limitations, due to the great need for information on the biomass burning impacts on ecosystems (Vadrevu and Justice 2011; Vadrevu and Lasko 2015; Vadrevu 2015; Vadrevu et al. 2008a, b, 2013, 2014, 2017, 2018, 2019a, b, 2021b, c; Vay et al. 2011; Wooster et al. 2021), polar satellites are frequently used for both active fire and burnt area mapping at local and regional scales (Giglio et al. 2006) and at the regional level in different parts of the world (Stolle et al. 2004; Miettinen et al. 2013). However, these studies report widely divergent estimates.

In contrast to the polar satellites, Synthetic Aperture Radar (SAR) data is well suited for land cover mapping and monitoring as the data is insensitive to cloud and haze (Engelbrecht et al. 2017). SAR sensors acquire day-and-night data using microwave radiation at various frequencies and incidence angles (Moreira et al. 2013). The resulting backscatter depends on the characteristics of the sensor, such as frequency and incidence angle, but also on the scatterers' size, structure, and dielectric properties (Wallington and Woodhouse 2006). SAR data potential for burnt area delineation lies in the sensitivity of SAR backscatter to vegetation structure and biomass, which can vary based on the SAR data polarizations, including the incident angles. In particular, the SAR backscatter can vary after the fire due to vegetation removal. Then, various approaches, such as single-pass or multi-temporal analysis approaches, can be used to detect these backscatter changes (Jolly et al. 2015; Menges et al. 2004). For example, Siegert and Hoffmann (2000) used multi-temporal data acquired by the single-polarization (VV) C-band SAR onboard ESA's European Remote Sensing (ERS-2) satellite to map the extensive forest fires in East Kalimantan in 1998. The multi-temporal SAR data (before and after the fire) was subjected to principal component analysis and visually interpreted to map burnt areas, but no information is provided about the accuracy of the method. In Australia, Menges et al. (2004) used the JPL AIRSAR multi-frequency (C-, L-, and P-band) data to discriminate savanna fires and found that the C-band data provided some degree of separability between burnt and unburnt areas, whereas L- and P-band were ineffective in detecting changes due to the low intensity fires. Tanase et al. (2010) found that in Mediterranean Pine forests, the backscatter increased with burn severity for X- and C-bands, whereas it decreased for L-band. Further, cross-polarized (HV) backscatter decreased with burn severity for all frequencies. Lohberger et al. (2018) used Sentinel-1 imagery to map the burnt areas in Indonesia in the 2015 fire season with 84% accuracy. In general, SAR backscatter is sensitive to many factors, including moisture conditions, surface roughness, and biomass (Leblon et al. 2002). These conditions can be highly site specific; thus, several researchers report increased or decreased SAR backscatter associated with burnt conditions (Menges et al. 2004; Polychronaki et al. 2013). Belenguer-Plomer et al. (2019) found that Sentinel-1C backscatter and accuracy of burnt area delineation were most influenced by the dual-pass SAR data availability, topography, and soil moisture and that most accurate detections were observed in over forests and least over grasslands. In particular, delineating burnt areas in agricultural systems, such as after the crop residue burning, can be difficult as the signal can be relatively temporary due to various management practices by the farmers, in contrast to burnt area delineation in forested ecosystems. Therefore, it is difficult to delineate the burnt areas using a universal backscatter-based algorithm. Another challenge for burnt area extraction using SAR data is that SAR-based change detection is generally considered challenging due to speckle effects, textural differences, and heterogeneous image characteristics (Polychronaki et al. 2013). To address these challenges, region-based and object-orientated approaches based on image segmentation were recommended (Yang et al. 2016).

Compared with several studies conducted on other regions of the world, only a few studies have focused on burnt area delineation using the SAR data in S/SEA countries. S/SEA countries are unique, with varied landscapes that make fire monitoring particularly challenging due to persistent cloud cover in some regions and diverse geographic and climatic gradients. Thus, it is essential to explore methods using SAR data that can penetrate clouds on fire mapping and monitoring. In this study, we use Sentinel-1A/B SAR data in conjunction with the very high-resolution PlanetScope to infer BA signal variations from agricultural and forested landscapes of India. In India, agricultural fires are most common and are routinely observed after crop harvest. Farmers burn agricultural residues to clear the fields for the next crop. Specific to forest fires, most of them are surface fires, in contrast to canopy fires that are most common in the North American or Russian continents. Irrespective of the type and nature of fires, biomass burning pollution caused due to fires is a significant issue in India. Thus, mapping and monitoring of fires, including burnt areas in the region, is needed to aid in air pollution mitigation efforts, including forest conservation and restoration.

2 Study Areas

We chose two specific states and study sites, in Punjab, representing agricultural fires, and in Uttarakhand, representing forest fires. The state of Punjab is in northern India (Fig. 1) with border states of Jammu and Kashmir to the north, Himachal Pradesh to the east, Haryana to the south and southeast, Rajasthan to the southwest, and the Pakistan province of Punjab to the west (https://en.wikipedia.org/wiki/Punjab,India). Ninety-nine percent of the net sown area is under irrigated agriculture through canals, tube wells, and other water sources (http://agripb.gov.in). On average, rice (37.15%) and wheat (48.76%) together constitute 85.91% of grown crops. Crops such as maize, jowar, cotton, and others constitute only 14% of the total cropped area (http://agripb.gov.in/home.php?page=astat). In this region, rice is usually grown in the wet summer season (sown in July–August and harvested in October–November) and wheat in the dry winter season (sown in November–December and harvested in April–May). The summer wheat residue burning season is during April and May, and winter rice residue burning is during October and November.

Uttarakhand is located in northern India and is mostly hilly, with thirteen different districts (Fig. 2) occupying an area of 53,483 km^2 (20,650 mi^2), of which 86% is mountainous, and 65% is covered by forest. The following vegetation type dominates at different altitudes: alpine shrubs (3000–5000 m); subalpine conifer forests (3000–2600 m); broadleaf forests (2600–1500 m); subtropical pine forests (1500 m), and moist-deciduous forests below 1500 m. Uttarakhand witnessed a massive wildfire in May 2019, during which more than 900ha of forest was burned. Incidents of fire have been reported from all 13 districts of Uttarakhand, the worst affected districts being Nainital and Almora, which are mostly comprised of oak and pine trees. The primary driver of fires was anthropogenic and due to negligence. We used Sentinel-1A/B data in conjunction with the high-resolution PlanetScope data in Nainital (Fig. 2 bottom) to characterize the burnt area signal variations in these two study areas.

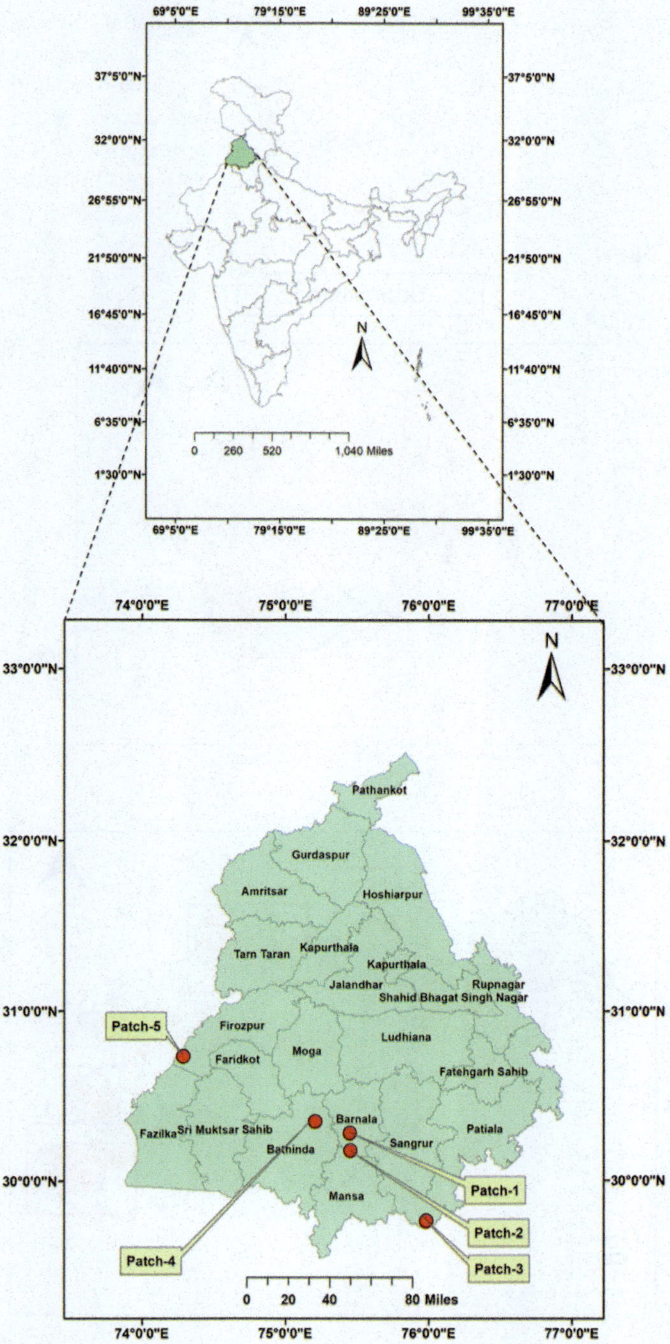

Fig. 1 Punjab, India with study area location sites

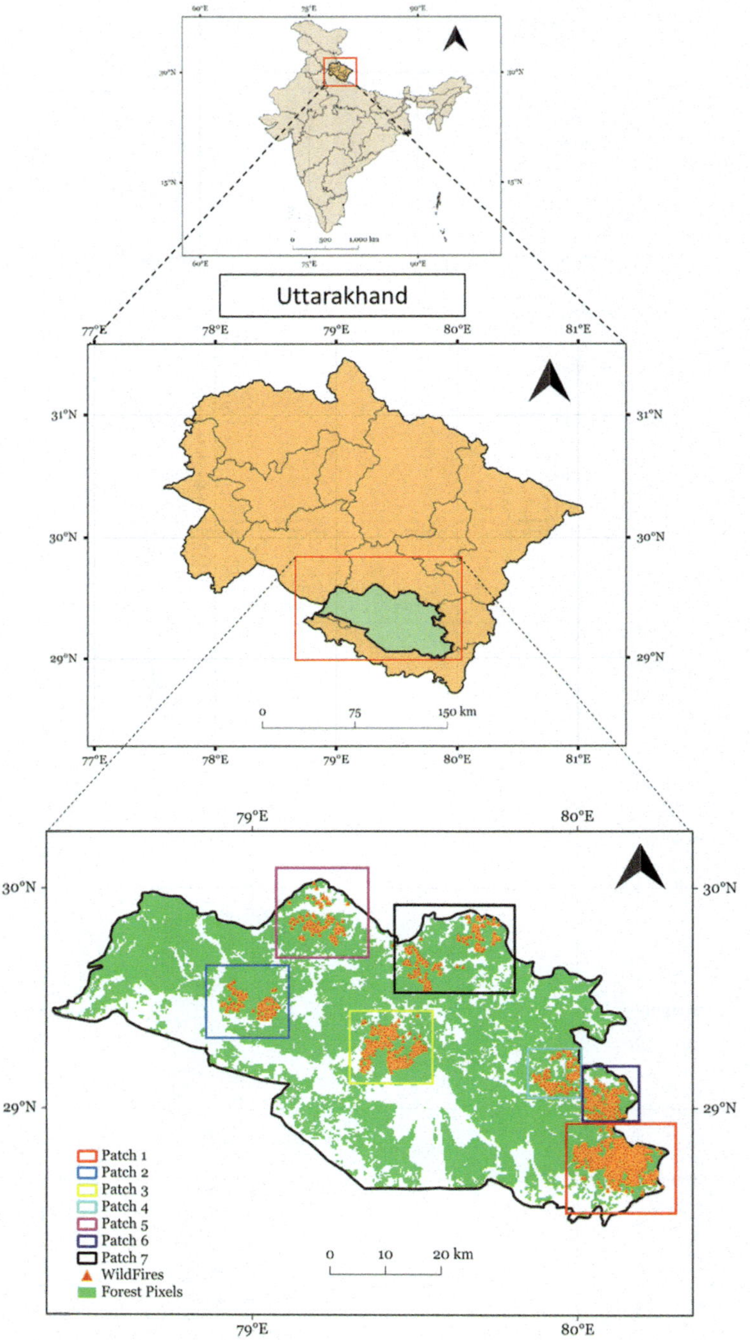

Fig. 2 Nainital, India with study area location sites

3 Datasets

3.1 Sentinel-1A/B Datasets

The Sentinel-1A C-band SAR was the first of a series of operational Earth Observation Satellites from the European Space Agency Copernicus Program (Butler 2014). Sentinel-1A and 1B are in the same orbit, but 180° apart and were launched in April 2014 and April 2016, respectively. Together, they provided global coverage every 12 days in dual-polarization (VV + VH). Interferometric Wide swath (IW) mode is at 20 m (range) × 22 m (azimuth) ground resolution depolarization (VV + VH) single-look complex data. Sentinel-1B malfunctioned on December 23, 2021, and has not provided data since then. S-1A IW data is acquired with a 250 km swath, using three sub-swaths with the Terrain Observation with Progressive Scans SAR (TOPSAR) technique (De Zan and Guarnieri 2006) with an incidence angle range of 29° to 46°.

4 Methodology

The Sentinel-1 imagery is provided as dual-polarized Interferometric Wide swath (IW) data with vertical transmit, vertical receive (VV), and vertical transmit, horizontal receive (VH) polarizations. Each polarization is at a nominal spatial resolution of 5 m × 20 m before preprocessing. The Level-1 Sentinel-1A/B ground range detected, descending mode, IW imagery was processed using the free and open-source Sentinel-1 toolbox. The ground range detected images were processed following guidelines, including applying restituted orbit files, multi-look azimuthal compressions to 20 m, terrain correction using SRTM 30 m version 4DEM, radiometric calibration adjustments to correct for viewing geometry effects, and refined lee speckle filter to reduce constructive and destructive interference, all resulting in sigma-nought backscatter data logarithmically scaled in decibel (Lasko et al. 2018). The final data had VV, including incidence angle data for 2019. Using these data, we assessed the backscatter and amplitude signal variations for different burnt area patches in Punjab and Nainital, Uttarakhand, India districts.

Since the Sentinel SAR data is at a nominal resolution of 5 m × 20 m, to avoid confusion on the burnt area delineation from SAR, we used the PlanetScope data at 3 m resolution. Each PlanetScope image covers an area of 24.6 × 16.4 km^2. Specifically, we used the PlanetScope Analytic Ortho Scene surface reflectance (SR). Level 3B data has four spectral bands (blue, green, red, and near-infrared). All PlanetScope data was acquired for May 2019 peak biomass burning month with less than 50% cloud cover. Although we used surface reflectance products, we had to do preprocessing to correct the hazy images. We used a cumulative distribution function (CDF)-based normalization method, to account for atmospheric haze as below:

$$h(v) = \text{round}\left(\frac{\text{cdf}(v) - \text{cdf}_{\text{min}}}{(M \times N) - \text{cdf}_{\text{min}}} \times (L - 1)\right)$$

h denotes equalized, v is pixel intensity, cdf_{min} is the minimum nonzero value of the cumulative distribution function, $M \times N$ gives the image's number of pixels width and height, and L is the number of gray levels. After the image correction, a minimum of 30 burnt area points per PlanetScope scene were created through visual interpretation of the data and exported as shapefiles. We then used the exported shapefiles to extract the VV backscatter signal and the amplitude from the Sentinel-1 data for different seasons. We used only the VV signal as VH signal data was unavailable for different seasons and dates compared to VV, which was consistently available.

5 Results and Discussion

To delineate the burnt patches, the PlanetScope imagery was quite useful; however, it needed additional preprocessing. For example, although the surface reflectance images were procured, there was a significant haze in the images, especially in the Punjab region, which hindered burnt area delineation. The images are provided as a 16-bit GeoTIFF image with reflectance values scaled by 10,000. The haze in the PlanetScope surface reflectance product is attributed to the aerosol model used in the atmospheric correction algorithm, i.e., continental type of aerosol representation used uniformly across the world, whereas the Punjab region, India, has a mixed aerosol from urban pollution, dust including biomass burning. In addition, the effects of haze and thin cirrus clouds are not corrected in the Planet Surface Reflectance data, including adjacency effects; thus, the images can be noisy. Therefore, fine-tuning atmospheric correction parameters are needed when using the PlanetScope images. In our case, CDF normalization of the data could help remove some noise and delineate the burnt patches well (Fig. 3). The burnt patches appeared dark blackish on the crop residue burning sites on the PlanetScope imagery compared to the light grayish ones, which mostly represented sites after the signal was lost due to various other management practices.

Using the Sentinel-1A/B data earlier, our research has shown a clear distinction between the VV and VH backscatter signal with the varying incident angles, i.e., VV had a much higher backscatter than VH at all angles. Further, during March, the VV signal for burnt areas was in the range of −7 to −10 dB, whereas it was −13 to −16.5 dB for VH centered around 37.5–39.5 incident angles. As the incident angle increased beyond 40°, the data was highly scattered for VV and VH during March (Vadrevu et al. 2021a). The results obtained from the VV backscatter signal for different seasons, dates, and months including amplitude for different agricultural burnt area patches are shown in Figs. 4, 5 and 6. The seasonal results suggested an apparent decrease in the backscatter and amplitude during the peak March–April–May (Summer) followed by October (Winter) months. The reduction in vegetation greenness was also evident in the monthly normalized difference vegetation index

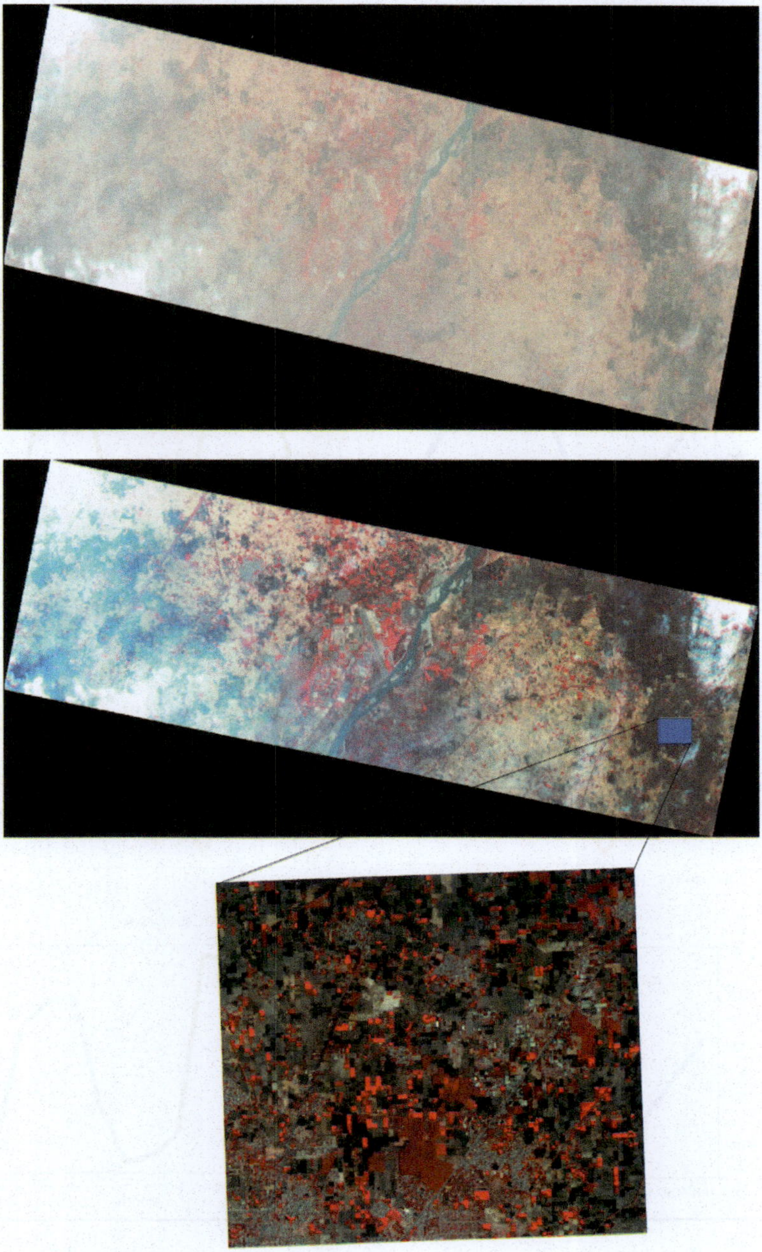

Fig. 3 Planetscope data over Barnala, Punjab, India. The top image is for typical burnt area image during May 18th, 2019 surface reflectance data; the middle one after the cumulative distribution function (cdf) normalization and the bottom image shows zoomed in areas with burnt patches in black color

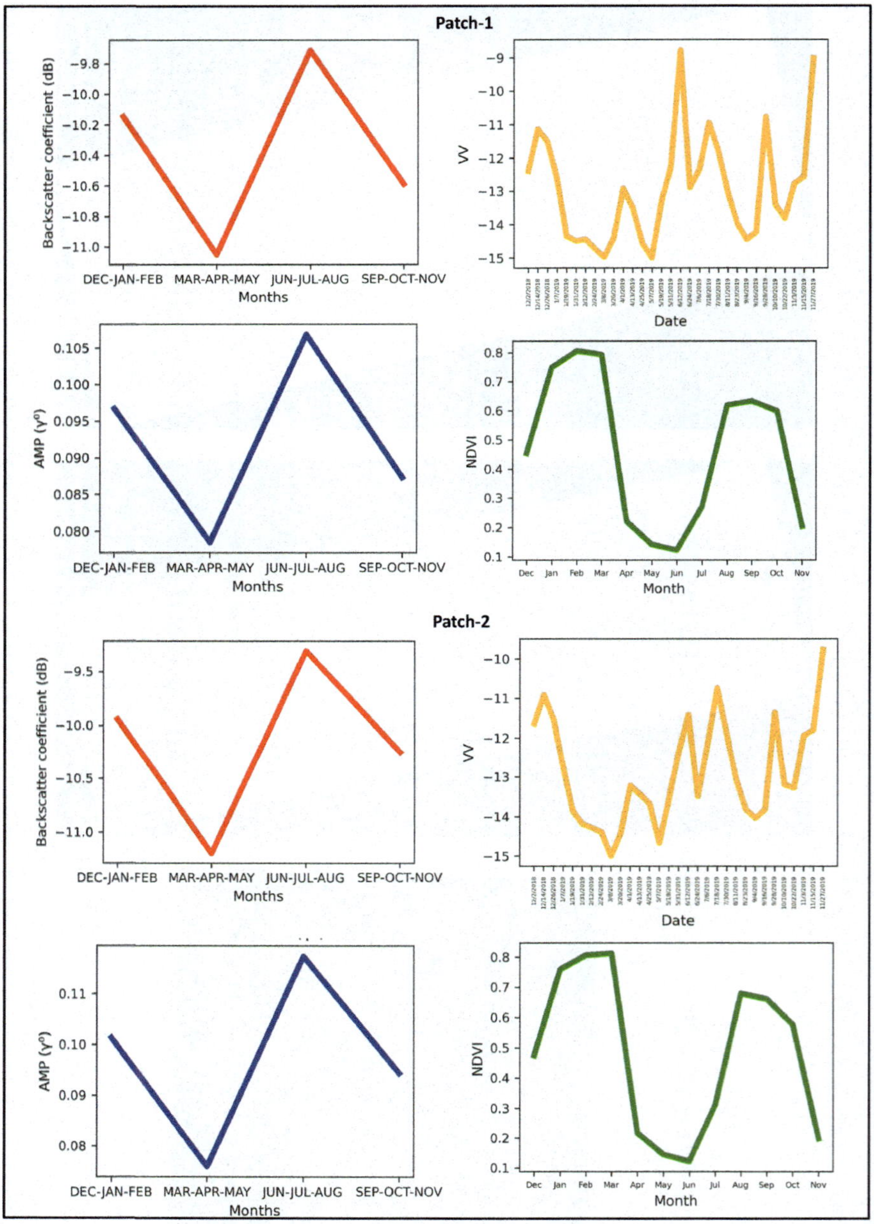

Fig. 4 Sentinel 1A/B derived backscatter and amplitude signal in Patch-1 and 2 burnt area locations in Barnala, Punjab, 2020. Temporal variations in VV backscatter for different dates are also shown in addition to monthly NDVI values from Sentinel-2 data

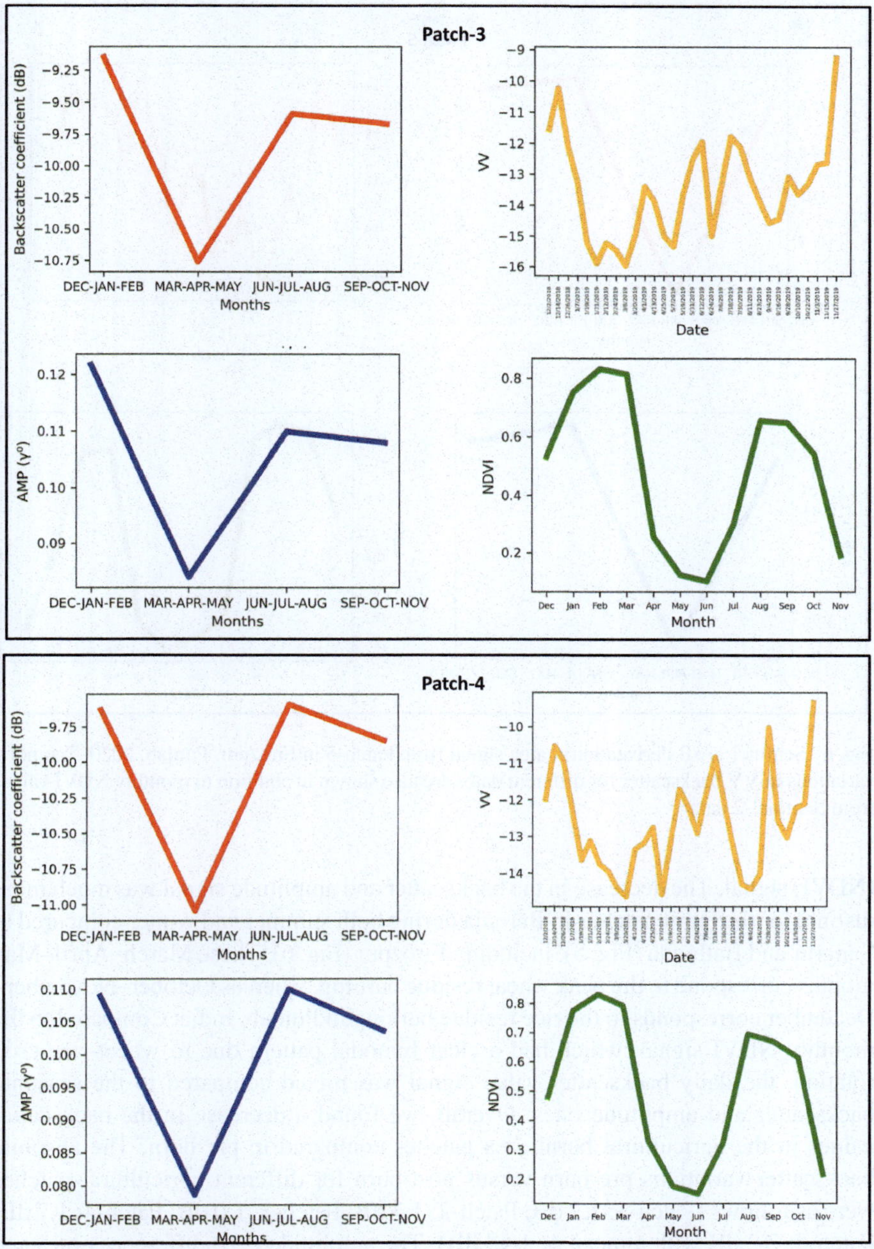

Fig. 5 Sentinel 1A/B derived burnt area signal from Patch-3 (Southern Sangrur) and Patch-4 (Bathinda, Punjab, 2020). Temporal variations in VV backscatter for different dates are also shown in addition to monthly NDVI values from Sentinel-2 data

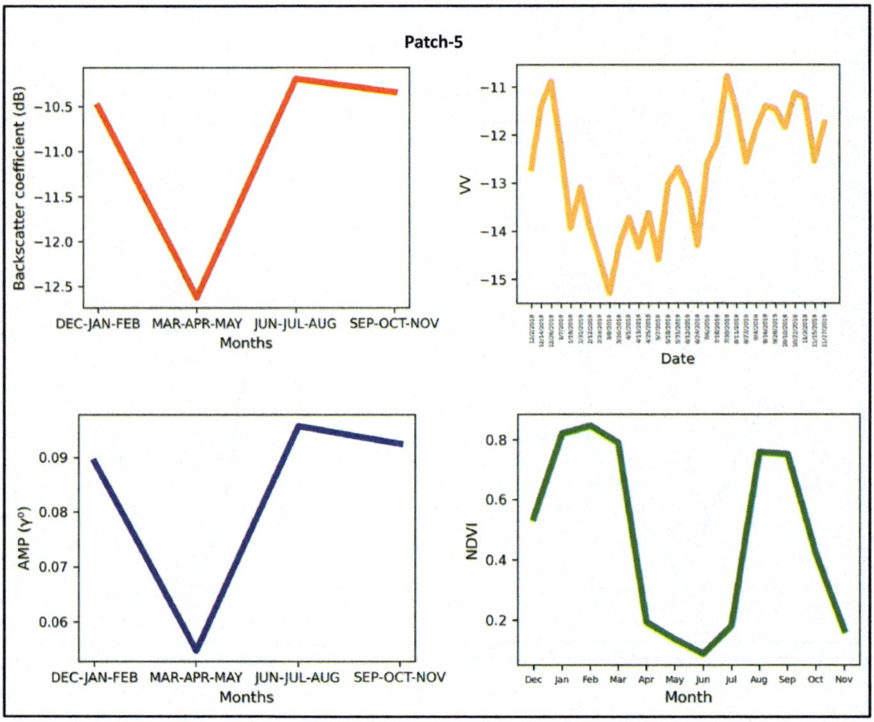

Fig. 6 Sentinel 1A/B derived burnt area signal from Patch-5 in Firozpur, Punjab, 2020. Temporal variations in VV backscatter for different dates are also shown in addition to monthly NDVI values from Sentinel-2 data

(NDVI) signal. The decrease in the backscatter and amplitude signal was much more distinct for the burnt patches in Barnala during both summer and winter compared to Sangrur and Bathinda (Fig. 5), including Firozpur (Fig. 6). These March–April–May months correspond to the peak wheat residue burning, whereas October–November–December corresponds to the rice residue burning in Punjab, India. Compared to the monthly NDVI signal, which had a clear bimodal pattern due to wheat-rice crop rotation, the daily backscatter value signal was mixed compared to the seasonal backscatter and amplitude ones. Overall, we found a decrease in the backscatter values in the agricultural burnt area patches compared to pre-burn. The absolute backscatter variations pre-burn versus post-burn for different agricultural patches were as follows (Patch-1: 1.2 dB; Patch-2: 1.3 dB; Patch-3: 1.0 dB; Patch-4: 1.7 dB; Patch-5: 2.2 dB with a mean of 1.48 dB). The amplitude ($\gamma 0$) differences pre- and post-burn for different agricultural patches were as follows (Patch-1: 0.020; Patch-2: 0.025; Patch-3: 0.040; Patch-4: 0.030; and Patch-5: 0.040 with a mean of 0.031) (Figs. 3, 4 and 5).

The burnt area patches over Nainital, Uttarakhand, India, in the PlanetScope data for May 4, 2019, are shown in Fig. 7, which look dark black in color. Compared

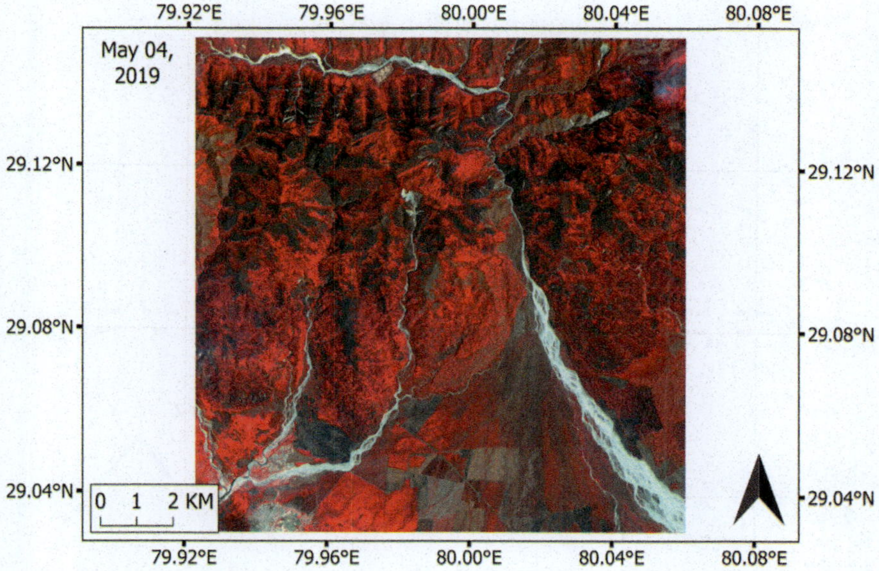

Fig. 7 Planetscope data over Nainital, Uttarakhand, India with burnt patches in black color

to the backscatter values in the agricultural sites, the absolute backscatter values in the forest burnt patches of Nainital for Patch-1 and Patch-2 (Figs. 2 and 8) from December–January–February and March–April–May were 0.01 dB and 0.1 dB, and 0 and 0.01 for amplitude ($\gamma 0$) differences, respectively (Fig. 9). These results suggest relatively higher backscatter differences for agricultural burnt area patches for pre- and post-burn seasons than the forest ones. In addition, the results for several other forest burnt areas patches 3–7 (Fig. 9) suggested an increase in backscatter during March–April–May, peak burning season compared to pre-burn December–January–February. These results are contrasting to agricultural burnt area patches where a decreased backscatter has been noted (Figs. 5 and 6). In addition, the Sentinel-2-derived NDVI showed a clear decline in the forested and agricultural areas during the peak burning seasons. We note that the amplitude differences in the burnt area patches can also depend on the burnt intensity as severe burns can result in loss of vegetation specific to height, thus more amplitude differences. Therefore, specific fire types, such as canopy fires versus surface fires, can impact amplitude variations. In the case of agriculture, as the crop residues are more or less of the same height, after burning, the differences may not be distinct in the same crop unless it is a different crop. However, these inferences need ground verification in both the agriculture and forest sites.

Most studies on SAR-derived BA mapping were done in Mediterranean regions, and very few focused on Asian countries, including India. Earlier studies suggested that SAR-derived backscattering observations are dependent on multiple factors such as radar parameters, geometry characteristics such as surface roughness, topography, and structure, dielectric properties of the vegetation, including the timing of SAR

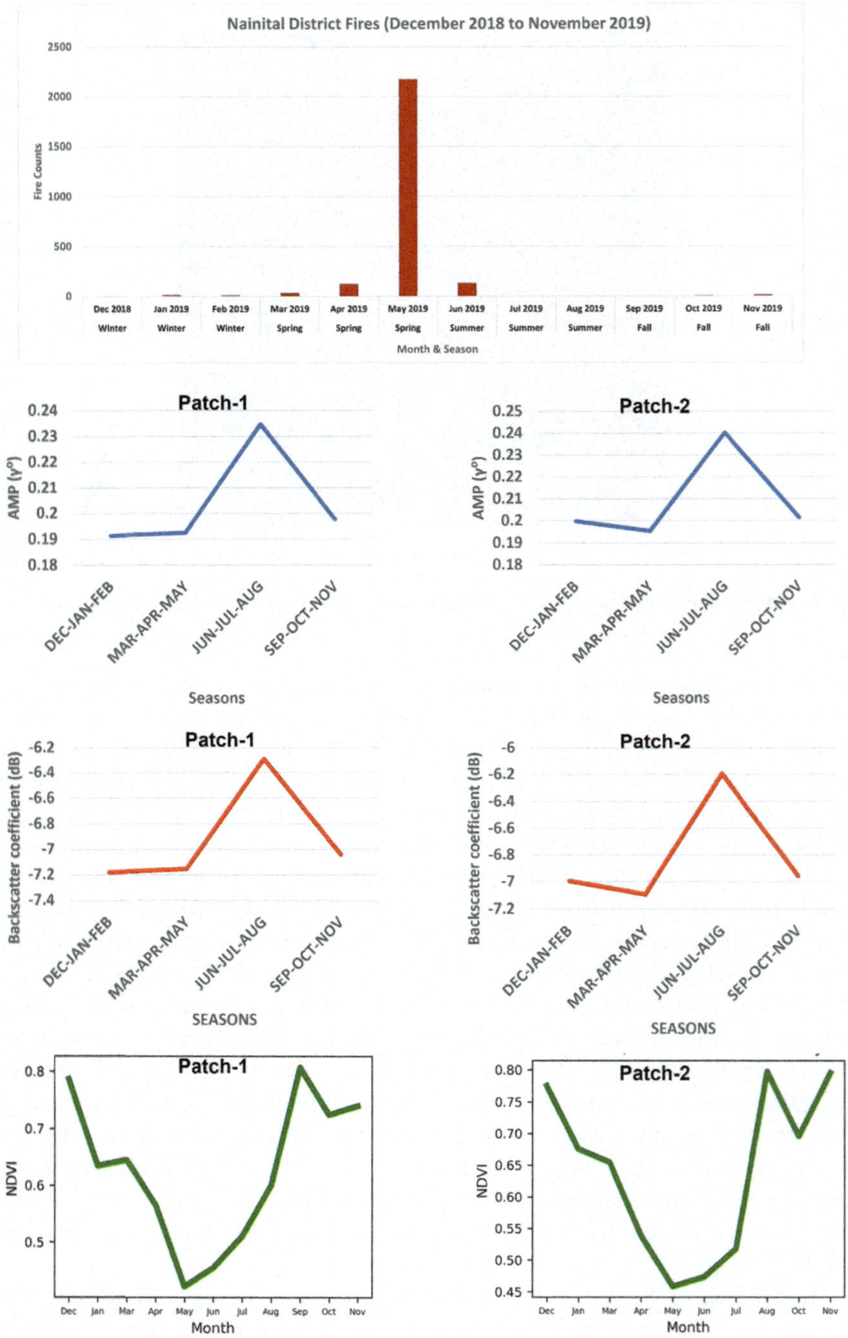

Fig. 8 Sentinel-1A VV-polarized amplitude, backscatter and Sentinel-2 derived NDVI signal variations for two different burnt patches in Nainital, Uttarakhand, India. The monthly fire counts from VIIRS 375 m data are shown on the top. See Fig. 9 for data on other burnt patches

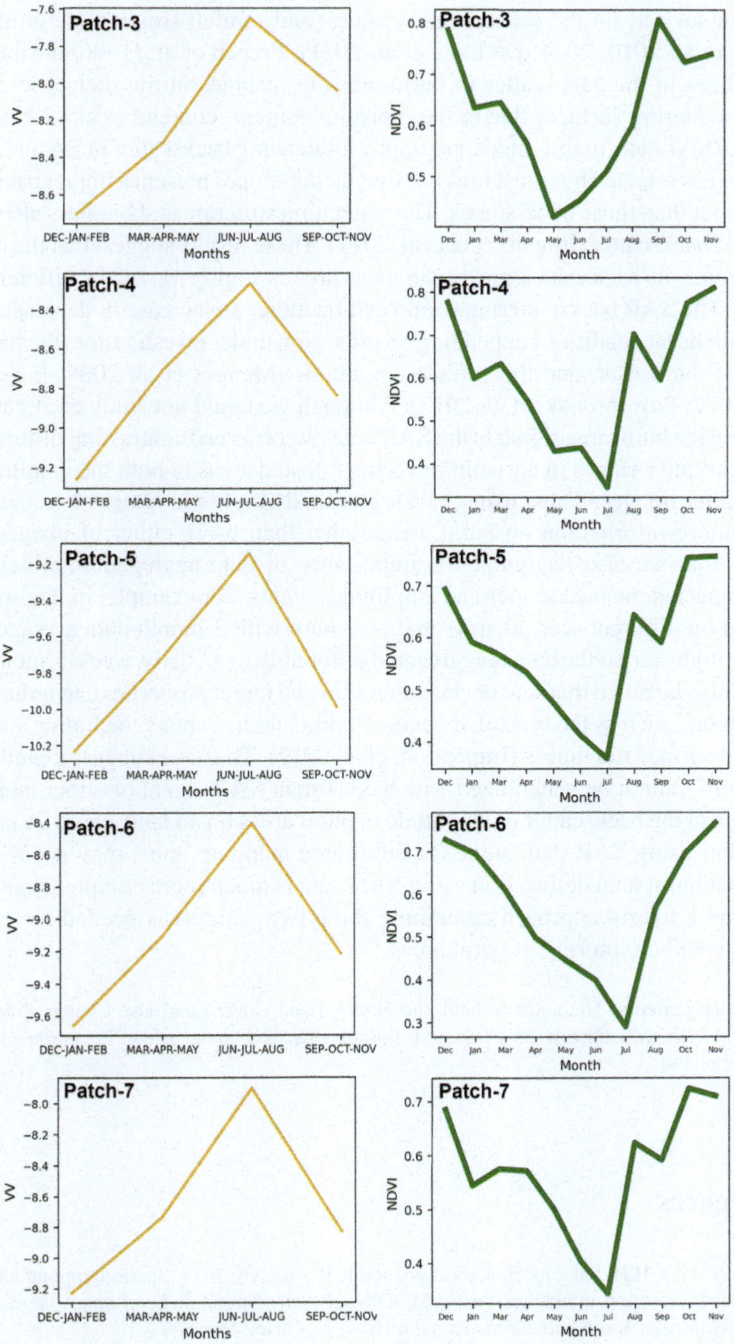

Fig. 9 Sentinel-1A VV-polarized backscatter and Sentinel-2 derived NDVI signal variations for Patches-3–5 (see Fig. 2 for patch locations) in Nainital, Uttarakhand, India. Sentinel-2 NDVI showed a clear decline whereas Sentiel-1A backscatter signal didn't decrease during the March–May

imaging such as on the season, temperature, and rainfall (Imperatore et al. 2017; Tanase et al. 2010, 2014; De Luca et al. 2021). French et al. (1996) attributed the differences in the backscatter in burnt areas to changes in the dielectric constant of the scattering surfaces due to the moisture content. Gimeno et al. (2004), using RADARSAT data, highlighted topography influencing backscatter in BA and, specifically, areas affected by forest fires on front facing-slopes present a higher backscatter coefficient than those back-slopes. The vegetation structure and biomass also impact the SAR backscatter (Engelbrecht et al. 2017). These results suggest that the increase or decrease in backscatter pre-versus post-fires is highly varied in different landscapes. The SAR backscattering would exhibit either an increase or decrease associated with burnt conditions depending on the region under investigation, the incidence angle of the sensor, and the surface conditions (Menges et al. 2004; Engelbrecht et al. 2017; Polychronaki et al. 2013). Although we could not study each parameter affecting the burnt area signal in the SAR data, we observed contrasting differences in the backscatter values in agriculture versus forested areas in both the amplitude and backscatter images. Thus, using backscatter and amplitude images together might reveal more information on burnt areas rather than using either of images alone. In addition, we also recognize the importance of data aggregation, which might have impacted the backscatter and amplitude signals. For example, in this study, we focused on different seasons (pre- and post-burn with 3-month data averages). The results might vary if the data is aggregated differently (e.g., daily, weekly, monthly). It should also be noted that sensor characteristics and target properties can influence the SAR signal, such as the type of vegetation burnt and its canopy, including soil moisture and surface roughness (Imperatore et al. 2017). Thus, we suggest a caution, and the results cannot be generalized with a consistent SAR signal of either increase or decrease in the backscatter or amplitude in burnt areas on all landscapes. A universal algorithm using SAR data alone for burnt area mapping, thus, may not work. We infer that burnt area delineation using SAR data is much more complex than optical data, and a hybrid approach combining these two datasets is needed for effective mapping and monitoring of burnt areas.

Acknowledgements The authors thank the NASA Land Cover/Land Use Change Program for funding the South/Southeast Asia Research Initiative (SARI), under which the current study has been carried out.

References

Albar, I., I. Jaya, B.H. Saharjo, B. Kuncahyo, and K.P. Vadrevu. 2018. Spatio-temporal analysis of land and forest fires in Indonesia using MODIS active fire dataset. In *Land-atmospheric research applications in South and Southeast Asia*, 105–127. Cham: Springer.

Badarinath, K.V.S., and K.V. Prasad. 2011. Carbon dioxide emissions from forest biomass burning in India. *Global Environmental Research* 15: 45–52.

Badarinath, K.V.S., S.K. Kharol, K.M. Latha, T.K. Chand, V.K. Prasad, A.N. Jyothsna, and K. Samatha. 2007. Multiyear ground-based and satellite observations of aerosol properties over a tropical urban area in India. *Atmospheric Science Letters* 8 (1): 7–13.

Badarinath, K.V.S., S. Kumar Kharol, V. Krishna Prasad, A. Rani Sharma, E.U.B. Reddi, H.D. Kambezidis, and D.G. Kaskaoutis. 2008. Influence of natural and anthropogenic activities on UV index variations—A study over tropical urban region using ground based observations and satellite data. *Journal of Atmospheric Chemistry* 59 (3): 219–236.

Badarinath, K.V.S., S.K. Kharol, A.R. Sharma, and V.K. Prasad. 2009. Analysis of aerosol and carbon monoxide characteristics over Arabian Sea during crop residue burning period in the Indo-Gangetic Plains using multi-satellite remote sensing datasets. *Journal of Atmospheric and Solar-Terrestrial Physics* 71 (12): 1267–1276.

Belenguer-Plomer, M.A., M.A. Tanase, A. Fernandez-Carrillo, and E. Chuvieco. 2019. Burned area detection and mapping using Sentinel-1 backscatter coefficient and thermal anomalies. *Remote Sensing of Environment* 233: 111345.

Biswas, S., K.D. Lasko, and K.P. Vadrevu. 2015a. Fire disturbance in tropical forests of Myanmar—Analysis using MODIS satellite datasets. *IEEE Journal of Selected Topics in Applied Earth Observations and Remote Sensing* 8 (5): 2273–2281.

Biswas, S., K.P. Vadrevu, Z.M. Lwin, K. Lasko, and C.O. Justice. 2015b. Factors controlling vegetation fires in protected and non-protected areas of Myanmar. *PLoS One* 10 (4): e0124346.

Biswas, S., K.P. Vadrevu, M.S. Mon, and C. Justice. 2021. Contemporary forest loss in Myanmar: Effect of democratic transition and subsequent timber bans on landscape structure and composition. *Ambio* 50 (4): 914–928.

Butler, D. 2014. Earth observation enters next phase. *Nature* 508 (7495): 160–161.

Choi, Y., S.A. Vay, K.P. Vadrevu, A.J. Soja, J.H. Woo, S.R. Nolf, G.W. Sachse, G.S. Diskin, D.R. Blake, N.J. Blake, and H.B. Singh. 2008. Characteristics of the atmospheric CO_2 signal as observed over the conterminous United States during INTEX-NA. *Journal of Geophysical Research: Atmospheres* 113 (D7).

De Luca, G., J.M. Silva, and G. Modica. 2021. A workflow based on Sentinel-1 S.A.R. data and open-source algorithms for unsupervised burned area detection in Mediterranean ecosystems. *GIScience & Remote Sensing* 58 (4): 516–541.

De Zan, F., and A.M. Guarnieri. 2006. TOPSAR: Terrain observation by progressive scans. *IEEE Transactions on Geoscience and Remote Sensing* 44 (9): 2352–2360.

Dwyer, E., S. Pinnock, J.M. Grégoire, and J.M.C. Pereira. 2000. Global spatial and temporal distribution of vegetation fire as determined from satellite observations. *International Journal of Remote Sensing* 21 (6–7): 1289–1302.

Engelbrecht, J., A. Theron, L. Vhengani, and J. Kemp. 2017. A simple normalized difference approach to burnt area mapping using multi-polarisation C-band SAR. *Remote Sensing* 9 (8): 764.

Eva, H., and E.F. Lambin. 1998. Remote sensing of biomass burning in tropical regions: Sampling issues and multisensor approach. *Remote Sensing of Environment* 64 (3): 292–315.

French, N.H.F., E.S. Kasischke, L.L. Bourgeau-Chavez, and P.A. Harrell. 1996. Sensitivity of ERS-1 SAR to variations in soil water in fire-disturbed boreal forest ecosystems. *International Journal of Remote Sensing* 17 (15): 3037–3053.

Giglio, L., I. Csiszar, and C.O. Justice. 2006. Global distribution and seasonality of active fires as observed with the terra and aqua moderate resolution imaging spectroradiometer (MODIS) sensors. *Journal of Geophysical Research: Biogeosciences* 111 (G2).

Giglio, L., L. Boschetti, D.P. Roy, M.L. Humber, and C.O. Justice. 2018. The collection 6 MODIS burned area mapping algorithm and product. *Remote Sensing of Environment* 217: 72–85.

Gimeno, M., J. San-Miguel-Ayanz, and G. Schmuck. 2004. Identification of burnt areas in Mediterranean forest environments from ERS-2 SAR time series. *International Journal of Remote Sensing* 25 (22): 4873–4888.

Imperatore, P., R. Azar, F. Calo, D. Stroppiana, P.A. Brivio, R. Lanari, and A. Pepe. 2017. Effect of the vegetation fire on backscattering: An investigation based on Sentinel-1 observations.

IEEE Journal of Selected Topics in Applied Earth Observations and Remote Sensing 10 (10): 4478–4492.
Jolly, W.M., M.A. Cochrane, P.H. Freeborn, Z.A. Holden, T.J. Brown, G.J. Williamson, and D.M.J.S. Bowman. 2015. Climate-induced variations in global wildfire danger from 1979 to 2013. *Nature Communications* 6: 7537.
Justice, C., G. Gutman, and K.P. Vadrevu. 2015. NASA land cover and land use change (LCLUC): An interdisciplinary research program. *Journal of Environmental Management* 148: 4–9.
Kant, Y., V.K. Prasad, and K.V.S. Badarinath. 2000. Algorithm for detection of active fire zones using NOAA AVHRR data. *Infrared Physics & Technology* 41 (1): 29–34.
Kharol, S.K., K.V.S. Badarinath, A.R. Sharma, D.V. Mahalakshmi, D. Singh, and V.K. Prasad. 2012. Black carbon aerosol variations over Patiala city, Punjab, India—A study during agriculture crop residue burning period using ground measurements and satellite data. *Journal of Atmospheric and Solar-Terrestrial Physics* 84: 45–51.
Lasko, K., and K. Vadrevu. 2018. Improved rice residue burning emissions estimates: Accounting for practice-specific emission factors in air pollution assessments of Vietnam. *Environmental Pollution* 236: 795–806.
Lasko, K., K.P. Vadrevu, V.T. Tran, E. Ellicott, T.T. Nguyen, H.Q. Bui, and C. Justice. 2017. Satellites may underestimate rice residue and associated burning emissions in Vietnam. *Environmental Research Letters* 12 (8): 085006.
Lasko, K., K.P. Vadrevu, and T.T.N. Nguyen. 2018. Analysis of air pollution over Hanoi, Vietnam using multi-satellite and MERRA reanalysis datasets. *PLoS One* 13 (5): e0196629.
Lata, K.M., C.H. Sankar Rao, V. Krishna Prasad, K.V.S. Badrinath, and V. Raghavaswamy. 2001. Measuring urban sprawl: A case study of Hyderabad. *GIS Development* 5 (12): 22–25
Leblon, B., E. Kasischke, M. Alexander, M. Doyle, and M. Abbott. 2002. Fire danger monitoring using ERS-1 SAR images in the case of northern boreal forests. *Natural Hazards* 27 (3): 231–255.
Levine, J.S. 2000. Global biomass burning: A case study of the gaseous and particulate emissions released to the atmosphere during the 1997 fires in Kalimantan and Sumatra, Indonesia. In *Biomass burning and its inter-relationships with the climate system*, 15–31. Dordrecht: Springer.
Lohberger, S., M. Stängel, E.C. Atwood, and F. Siegert. 2018. Spatial evaluation of Indonesia's 2015 fire-affected area and estimated carbon emissions using Sentinel-1. *Global Change Biology* 24 (2): 644–654.
Menges, C.H., R.E. Bartolo, D. Bell, and G.E. Hill. 2004. The effect of savanna fires on SAR backscatter in northern Australia. *International Journal of Remote Sensing* 25 (22): 4857–4871.
Miettinen, J., E. Hyer, A.S. Chia, L.K. Kwoh, and S.C. Liew. 2013. Detection of vegetation fires and burnt areas by remote sensing in insular Southeast Asian conditions: Current status of knowledge and future challenges. *International Journal of Remote Sensing* 34 (12): 4344–4366.
Moreira, A., P. Prats-Iraola, M. Younis, G. Krieger, I. Hajnsek, and K.P. Papathanassiou. 2013. A tutorial on synthetic aperture radar. *IEEE Geoscience and Remote Sensing Magazine* 1 (1): 6–43.
Petropoulos, G.P., K.P. Vadrevu, and C. Kalaitzidis. 2013. Spectral angle mapper and object-based classification combined with hyperspectral remote sensing imagery for obtaining land use/cover mapping in a Mediterranean region. *Geocarto International* 28 (2): 114–129.
Polychronaki, A., I.Z. Gitas, S. Veraverbeke, and A. Debien. 2013. Evaluation of ALOS PALSAR imagery for burned area mapping in Greece using object-based classification. *Remote Sensing* 5: 5680–5701.
Prasad, V.K., and K.V.S. Badarinth. 2004. Land use changes and trends in human appropriation of above ground net primary production (HANPP) in India (1961–98). *Geographical Journal* 170 (1): 51–63.
Prasad, K.V., and K.V.S. Badarinath. 2006. Soil surface nitrogen losses from agriculture in India: A regional inventory within agroecological zones (2000–2001). *The International Journal of Sustainable Development and World Ecology* 13 (3): 173–182.

Prasad, V.K., Y. Kant, and K.V.S. Badarinath. 2001a. CENTURY ecosystem model application for quantifying vegetation dynamics in shifting cultivation areas: A case study from Rampa Forests, Eastern Ghats (India). *Ecological Research* 16 (3): 497–507.

Prasad, V.K., Y. Kant, P.K. Gupta, C. Sharma, A.A. Mitra, and K.V.S. Badarinath. 2001b. Biomass and combustion characteristics of secondary mixed deciduous forests in Eastern Ghats of India. *Atmospheric Environment* 35 (18): 3085–3095.

Prasad, V.K., Y. Kant, and K.V.S. Badarinath. 2002a. Land use changes and modeling carbon fluxes from satellite data. *Advances in Space Research* 30 (11): 2511–2516.

Prasad, V.K., Y. Kant, P.K. Gupta, C. Elvidge, and K.V.S. Badarinath. 2002b. Biomass burning and related trace gas emissions from tropical dry deciduous forests of India: A study using DMSP-OLS data and ground-based measurements. *International Journal of Remote Sensing* 23 (14): 2837–2851.

Prasad, V.K., M. Lata, and K.V.S. Badarinath. 2003. Trace gas emissions from biomass burning from northeast region in India—Estimates from satellite remote sensing data and GIS. *Environmentalist* 23 (3): 229–236.

Prasad, V.K., K.V.S. Badarinath, S. Yonemura, and H. Tsuruta. 2004. Regional inventory of soil surface nitrogen balances in Indian agriculture (2000–2001). *Journal of Environmental Management* 73 (3): 209–218.

Prasad, V.K., E. Anuradha, and K.V.S. Badarinath. 2005. Climatic controls of vegetation vigor in four contrasting forest types of India—Evaluation from National Oceanic and Atmospheric Administration's advanced very high resolution radiometer datasets (1990–2000). *International Journal of Biometeorology* 50 (1): 6–16.

Seiler, W., and P.J. Crutzen. 1980. Estimates of gross and net fluxes of carbon between the biosphere and the atmosphere from biomass burning. *Climatic change* 2(3): 207–247.

Siegert, F., and A.A. Hoffmann. 2000. The 1998 forest fires in East Kalimantan (Indonesia): A quantitative evaluation using high resolution, multitemporal ERS-2 SAR images and NOAA-AVHRR hotspot data. *Remote Sensing of Environment* 72 (1): 64–77.

Stolle, F., R.A. Dennis, I. Kurniwan, and E.F. Lambin. 2004. Evaluation of remote sensing-based active fire datasets in Indonesia. *International Journal of Remote Sensing* 20: 471–479.

Suyanto, S. 2006. Underlying cause of fire: Different form of land tenure conflicts in Sumatra. *Mitigation and Adaptation Strategies for Global Change* 12: 67–74.

Tanase, M.A., M. Santoro, J. de La Riva, P. Fernando, and T. Le Toan. 2010. Sensitivity of X-, C-, and L-band S.A.R. backscatter to burn severity in Mediterranean pine forests. *IEEE Transactions on Geoscience and Remote Sensing* 48 (10): 3663–3675.

Tanase, M.A., R. Panciera, K. Lowell, S. Tian, J.M. Hacker, and J.P. Walker. 2014. Airborne multi-temporal L-band polarimetric SAR data for biomass estimation in semi-arid forests. *Remote Sensing of Environment* 145: 93–104.

Vadrevu, K.P. 2008. Analysis of fire events and controlling factors in eastern India using spatial scan and multivariate statistics. *Geografiska Annaler: Series A, Physical Geography* 90 (4): 315–328.

Vadrevu, K. 2015. International regional science meeting of NASA-LCLUC: Land cover/land use change (LC/LUC) and environmental impacts in South Asia, Coimbatore, India, 19–23 January 2013. *Journal of Environmental Management* 148: 1–163.

Vadrevu, K.P., and K.V.S. Badarinath. 2009. Spatial pattern analysis of fire events in Central India—A case study. *Geocarto International* 24 (2): 115–131.

Vadrevu, K.P., and C.O. Justice. 2011. Vegetation fires in the Asian region: Satellite observational needs and priorities. *Global Environmental Research* 15 (1): 65–76.

Vadrevu, K., and K. Lasko. 2015. Fire regimes and potential bioenergy loss from agricultural lands in the Indo-Gangetic Plains. *Journal of Environmental Management* 148: 10–20.

Vadrevu, K.P., J. Cardina, F. Hitzhusen, I. Bayoh, R. Moore, J. Parker, B. Stinner, D. Stinner, and C. Hoy. 2008a. Case study of an integrated framework for quantifying agroecosystem health. *Ecosystems* 11 (2): 283–306.

Vadrevu, K.P., K.V.S. Badarinath, and E. Anuradha. 2008b. Spatial patterns in vegetation fires in the Indian region. *Environmental Monitoring and Assessment* 147 (1): 1–13.

Vadrevu, K.P., L. Giglio, and C. Justice. 2013. Satellite based analysis of fire–carbon monoxide relationships from forest and agricultural residue burning (2003–2011). *Atmospheric Environment* 64: 179–191.

Vadrevu, K.P., T. Ohara, and C. Justice. 2014. Air pollution in Asia. *Environmental Pollution (Barking, Essex: 1987)* 195: 233–235.

Vadrevu, K., T. Ohara, and C. Justice. 2017. Land cover, land use changes and air pollution in Asia: A synthesis. *Environmental Research Letters* 12 (12): 120201.

Vadrevu, K.P., T. Ohara, and C. Justice (eds.). 2018. *Land-atmospheric research applications in South and Southeast Asia.*

Vadrevu, K., A. Heinimann, G. Gutman, and C. Justice. 2019a. Remote sensing of land use/cover changes in South and Southeast Asian countries. *International Journal of Digital Earth* 12 (10): 1099–1102.

Vadrevu, K.P., V.K. Dadhwal, G. Gutman, and C. Justice. 2019b. Remote sensing of agriculture–South/Southeast Asia research initiative special issue. *International Journal of Remote Sensing* 40 (21): 8071–8075.

Vadrevu, K.P., A. Eaturu, S. Biswas, and C. Phompila. 2021a. An assessment of burnt area signal variations in Laos using Sentinel-1A&B datasets. In *Biomass burning in South and Southeast Asia*, 61–76. Boca Raton: CRC Press.

Vadrevu, K.P., T. Ohara, and C. Justice (eds.). 2021b. *Biomass burning in South and Southeast Asia: Impacts on the biosphere*, vol. 2. Boca Raton: CRC Press.

Vadrevu, K.P., T. Ohara, and C. Justice (eds.). 2021c. *Biomass burning in South and Southeast Asia: Mapping and monitoring*, vol. 1. Boca Raton: CRC Press.

Vadrevu, K., A. Eaturu, E. Casadaban, K. Lasko, W. Schroeder, S. Biswas, L. Giglio, and C. Justice. 2022. Spatial variations in vegetation fires and emissions in South and Southeast Asia during COVID-19 and pre-pandemic. *Scientific Reports* 12 (1): 1–21.

van Wees, Dave, Guido R. van der Werf, James T. Randerson, Brendan M. Rogers, Yang Chen, Sander Veraverbeke, Louis Giglio, and Douglas C. Morton. 2022. Global biomass burning fuel consumption and emissions at 500-m spatial resolution based on the Global Fire Emissions Database (GFED). *Geoscientific Model Development Discussions* 1–46.

Vay, S.A., Y. Choi, K.P. Vadrevu, D.R. Blake, S.C. Tyler, A. Wisthaler, A. Hecobian, Y. Kondo, G.S. Diskin, G.W. Sachse, and J.H. Woo. 2011. Patterns of CO2 and radiocarbon across high northern latitudes during International Polar Year 2008. *Journal of Geophysical Research: Atmospheres* 116(D14).

Wallington, E.D., and I.H. Woodhouse. 2006. Forest height retrieval from commercial X-band SAR products. *IEEE Transactions on Geoscience and Remote Sensing* 44 (4): 863–870.

Wooster, M.J., G.J. Roberts, L. Giglio, D.P. Roy, P.H. Freeborn, L. Boschetti, C. Justice, C. Ichoku, W. Schroeder, D. Davies, and A.M. Smith. 2021. Satellite remote sensing of active fires: History and current status, applications and future requirements. *Remote Sensing of Environment* 267: 112694.

Yang, W., X. Yang, T. Yan, H. Song, and G.S. Xia. 2016. Region-based change detection for polarimetric SAR images using Wishart mixture models. *IEEE Transactions on Geoscience and Remote Sensing* 54: 6746–6756.

Application of Interferometry SAR for Monitoring of Peatland Area—Case Studies in Indonesia

Yessy Arvelyna, Prayoto Pranoto, Keiko Ishi, and Krishna Prasad Vadrevu

Abstract Fires are a significant threat to peat-rich biomes resulting in extensive carbon loss. The peat fires are dominated by smoldering combustion and can persist even in wet conditions. As a result, the peatland surface height can change considerably based on the water levels beneath the peat. Use of field-based methods to monitor peatland surface height changes is challenging as they can cover vast areas. Thus, remote sensing methods are the only way to capture the subsidence-related changes in peatlands at a large spatial scale. In this study, we used the differential interferometry SAR (DInSAR) analysis technique using ALOS-2 PALSAR data for peatland monitoring in Central Kalimantan and Riau Province, Indonesia. The results suggest that this technique can capture peatland's surface height variabilities affected by groundwater fluctuation, peatland fires, etc. A case study is presented which demonstrates the SAR data potential for monitoring peatland surface height changes in Indonesia.

Keywords Peatland fires · Central Kalimantan · SAR · DInSAR · Indonesia

1 Introduction

Biomass burning is most prevalent in South/Southeast Asian countries as fire is often used as a land-clearing tool (Albar et al. 2018; Badarinath et al. 2007, 2008, 2009; Badarinath and Prasad 2011; Biswas et al. 2015a, b, 2021; Eaturu and Vadrevu 2021; Justice et al. 2015). However, such biomass burning events can cause significant economic losses, including adverse effects on the ecosystem functions such as loss of biodiversity, biomass, nutrients, and soil disturbance (Prasad et al. 2001a, b, 2002a, b,

Y. Arvelyna (✉) · K. Ishi
Remote Sensing Technology Center of Japan, Tokyo, Japan
e-mail: arvelyna_yessy@restec.or.jp

P. Pranoto
Riau Environmental and Forestry Office, Pekanbaru, Indonesia

K. P. Vadrevu
NASA Marshall Space Flight Center, Huntsville, USA

K. P. Vadrevu et al. (eds.), *Vegetation Fires and Pollution in Asia*,
https://doi.org/10.1007/978-3-031-29916-2_8

2003, 2004, 2005; Prasad and Badarinath 2004, 2006; Vadrevu 2008, 2015; Vadrevu and Badarinath 2009; Vadrevu and Justice 2011; Vadrevu and Lasko 2015a, b). Furthermore, several researchers also identified biomass burning as a significant source of air pollution impacting human health (Kant et al. 2000; Kharol et al. 2012; Lasko and Vadrevu 2018; Lasko et al. 2017, 2018, 2021). In addition, the greenhouse gas emissions and aerosols released from biomass burning can impact local as well as regional climate (Prasad et al. 2000, 2002a, b, 2003, 2004; Kant et al. 2000; Gupta et al. 2001; Badarinath et al. 2008; Vadrevu 2008; Vadrevu et al. 2015, 2018, 2019; Lasko and Vadrevu 2018). Thus, characterizing the burnt areas and the impact of biomass burning in different regions of the world gain significance as the results can aid in fire management and mitigation efforts (Choi et al. 2008; Vay et al. 2011; Vadrevu et al. 2008, 2013, 2014a, b, c, 2017, 2020, 2021a, b, 2022a, b).

Remote sensing has significant potential for mapping and monitoring fires, including burnt area assessment (Justice et al. 2015; Wooster et al. 2021). Due to its synoptic coverage and multi-temporal, multispectral, and repetitive coverage capabilities, remote sensing data can be effectively used in fire and biomass-burning emissions research (Vadrevu and Justice 2011). Specifically, the burnt areas can be detected from the satellites as the spectral response of the land surface is changed after fires due to the loss of vegetation, greenness, and water content or changes in the moisture content, soil color, etc., which can be detected by the satellites (Vadrevu et al. 2006, 2008). Mainly, such changes are detected as a decrease in spectral reflectance in the visible-near-infrared and an increase in the mid-infrared wavelengths of the spectrum (Van Wagtendonk et al. 2004). Thus, one of the most widely used indices for burnt area mapping is normalized burn ratio (NBR) (Key and Benson 2003), which combines the reflectances in the near-infrared (ρNIR) and mid-infrared bands (ρMIR). However, factors like fire regime characteristics such as fire frequency, intensity, and temporal distribution can impact the fire signal (Vadrevu et al. 2012), including the NBR. In addition, different fire regimes can result in varying burnt scar patterns that might create signal differences from the remote sensing satellites in a given region.

However, the peatlands of Indonesia are unique as they are covered with the water beneath (Hayasaka et al. 2014; Albar et al. 2018; Nurbaya et al. 2020). The trigger factor of forest and peatland fires ranges from natural drivers such as drought due to the El Niño–Southern Oscillation (ENSO) in Southeast Asia and Indonesia and man-made drivers such as slash-and-burn agriculture and deforestation for commercial plantation and urban development (Nurbaya et al. 2020; Vadrevu et al. 2014a, b, c, 2020). The organic content of the peat layer at the surface consists of organic content of at least 30% (Joosten and Clarke 2002). The peat deposits act as the primary fuels that increase the rate of fire and create smoldering wildfires that can release several GHGs (Davies et al. 2008). In Kalimantan and Sumatra, Indonesia, peatland areas are wide and cover about 8.4 million ha and 9.6 million ha, respectively (Nurbaya et al. 2020).

Since 2016, the Ministry of Forestry and Environment of Indonesia has implemented national measures for preventing, monitoring, and responding to the forest and peatland fires, such as community-based peatland conservation and restoration and monitoring of forest resources and hotspots using medium and high-resolution optical satellite images, i.e., Landsat-7/8 and SPOT-6/7, respectively (Nurbaya et al. 2020). The utilization of SAR image data has shown its effectiveness in detecting burnt forest and peatland areas under cloud cover and smoke released from burning (Arvelyna et al. 2015, 2018; Lasko et al. 2018). Furthermore, the differential interferometry SAR (DInSAR) technique using ALOS-2 PALSAR-2 data is capable of measuring ground surface movement of peatlands affected by groundwater level changes at a good resolution (up to centimeter) (Arvelyna et al. 2018) and useful for peatland fire monitoring, and analysis of the fire incidents has been demonstrated by Arvelyna et al. (2021a, b). This study proposes applying time series DInSAR analysis using the Small Baseline Subset (SBAS) method on PALSAR-2 data for monitoring the forest and peatland fire area. We implemented the analysis in Central Kalimantan and Riau Province, Indonesia, where peatland areas are widely distributed and often affected by fires.

2 Study Area

2.1 Central Kalimantan Province

In Central Kalimantan Province, we selected study areas where hotspots occurred extensively during 2014–2021 at Kalampangan District, Pulang Pisau District, and Mantangai District in the southeast region of the Palangka Raya city (Fig. 1, right box). The distribution of land cover at observation areas varies from secondary swamp forest (T6, T12), swamp shrub (T1–T5, T7), dryland agriculture, and barren land (T8–T11) (SIGAP KLHK 2020) with peat depth 0.5–3 m (PRIMS 2022). Peat depth from field observation at Kalampangan District was about 3.72–6 m (Indrayani et al. 2011). In Pulang Pisau District, groundwater level (GWL) and ground surface level (GSL) data from Station Taka-1 to Kalteng-1, etc., were used to analyze the ground surface movement of the peatland area (Takahashi et al. 2018; Arvelyna 2022).

2.2 Kampar Peninsula, Riau Province

Kampar Peninsula is Indonesia's largest peat swamp forest, consisting of 18,705.82 ha of Adat forests (Nurbaya et al. 2020). The Kampar Peninsula area is located between the Kampar River in the south and the Siak River in the north. The Kampar Peninsula is managed by the KPHP Model Tasik Besar Serkap. The stipulation of the Tasik Besar Serkap Model KPHP based on the Ministry of Forestry Decree No. 509/Menhut-VII/2010 covers an area of 513,276 ha consisting of 2660 ha of limited production forest, 491,768 ha of permanent production forest, and 18,848 ha of convertible production forest (FMU/REDD+ 2016; Prayoto et al. 2017). The study

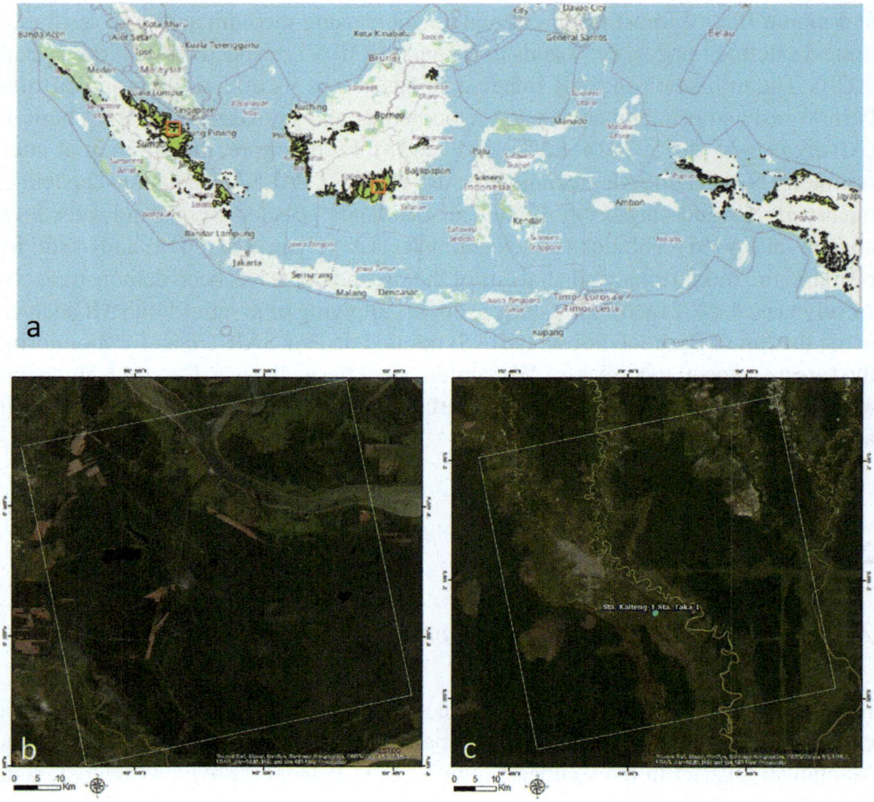

Fig. 1 **a** Peatland map and study area; **b** Central Kalimantan and **c** Kampar, Riau Province. *Source* KLHK, ESRI

area is located in Siak District and Pelalawan District (Fig. 1, left box). The land cover in this region is mostly secondary swamp forests (SIGAP KLHK 2020), with a peat depth of > 5 m (PRIMS 2022). The fire hotspots in 2015–2020 were distributed in an area with land cover (and peat depth) at plantation forest (3 to < 5 m), swamp shrub (2 to < 3 m), shrub-mixed dryland farm (2 to < 3 m), and barren land (1 to < 2 m) (SIGAP KLHK 2020; PRIMS 2022). The peat depth from a field survey at Siak District held in 2012 is distributed between 6 and 10.5 m (Arvelyna et al. 2019), while peat dome existed with peat depth > 12 m (Karyanto et al. 2015).

3 Data and Methods

3.1 Data

The fire hotspot data in the study areas were retrieved from the FIRMS NASA website for the MODIS NRT dataset and the LAPAN Fire Hotspot website for MODIS Aqua and Terra and SNPP VIIRS data (Roswintiarti 2016). Land cover data are acquired from the KLHK Land Cover map retrieved from the KLHK website. Generally, the study area's peat depth is retrieved from BRGM's PRIMS website. The field data (Fig. 2a, b) are retrieved from the Riau Province government. The peat depth (Fig. 3) at the study area is retrieved from BRGM's PRIMS website.

3.2 Satellite Data Used

We obtained the ALOS-2 PALSAR-2 L1.1 dataset over study areas in Central Kalimantan and Riau Province, mainly in the dry season (between July and September) during 2015–2021 (Table 1) through the second Earth Observation Collaborative Research from JAXA. PALSAR-2 L1.1 datasets were downloaded from JAXA AUIG2 and G-Portal platform (Table 1), then processed using the DInSAR method and time series DInSAR SBAS processing to retrieve deformation of peatland surface height. In addition, the other datasets, i.e., ASTER, Sentinel-2, Landsat, KLHK's Land Cover map, LAPAN Fire Hotspot map, and BMKG's rainfall data, were also gathered for the study.

3.3 Approach

A pair of PALSAR-2 datasets were processed using the DInSAR method, and time series of deformation processing were computed using the Small Baseline Subset (SBAS) DInSAR method of GAMMA software using UNIX programming. The flowchart is shown in Fig. 4. In addition, comparative studies are done between the derived DInSAR data and field data of the GWL and the GSL data from the SESAME project (Arvelyna et al. 2021a, b) in Central Kalimantan and bore field data at the study area in Siak District, Riau Province.

3.3.1 DInSAR Processing

We applied DInSAR processing on pairs of PALSAR-2 L1.1 datasets to observe peatland surface height deformation due to extensive peatland fire and other phenomena such as subsidence. For all processing, we used the GAMMA software. The DInSAR

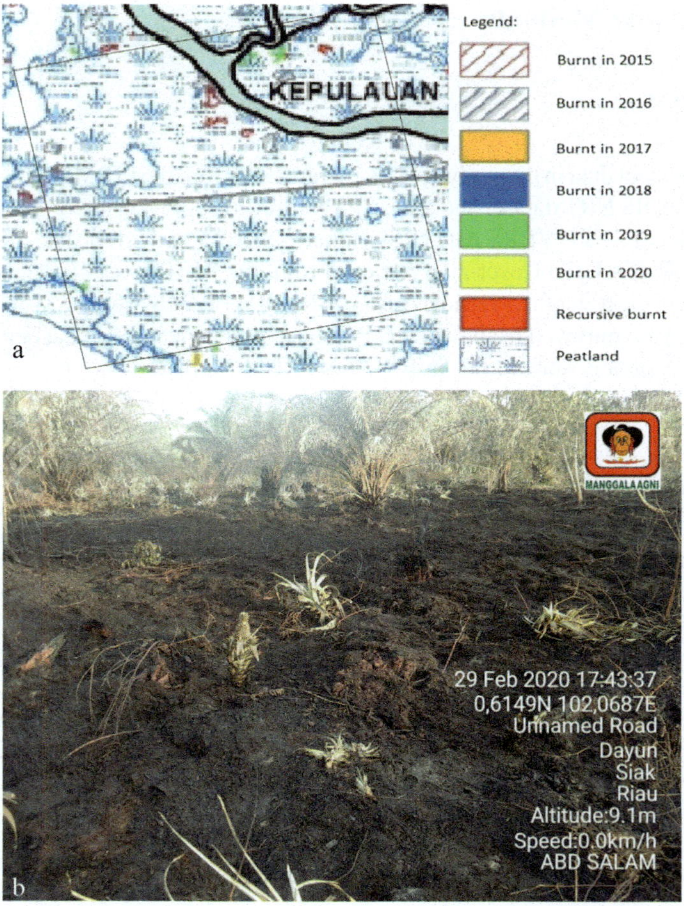

Fig. 2 **a** Peatland fire map in Riau Province 2015–2020; **b** Field photo of peatland fire incident in Kampar Peninsula, Riau Province, Indonesia on 2020/02/29, 17:43 local time (data provided by: KLHKBPPIKHL Manggala Agni Daops Siak)

processing of PALSAR-2 data involved utilizing interferometric SAR processor (ISP) and DIFF-GEO packages, including MLI data generation, radiometric calibration, band filtering in range and azimuth, and estimation of baseline. First, the interferogram, including the topographic and displacement phase, is generated, then the interferometric coherence is calculated. Next, phase unwrapping is generated using the Minimum Cost Flow (MCF) algorithm to obtain the unwrapped phase of the elevation map using the baseline refinement. Next, the inversion is done from the unwrapped phase to generate the height map. Then, a differential interferogram and displacement map are generated (GAMMA Remote Sensing 2013).

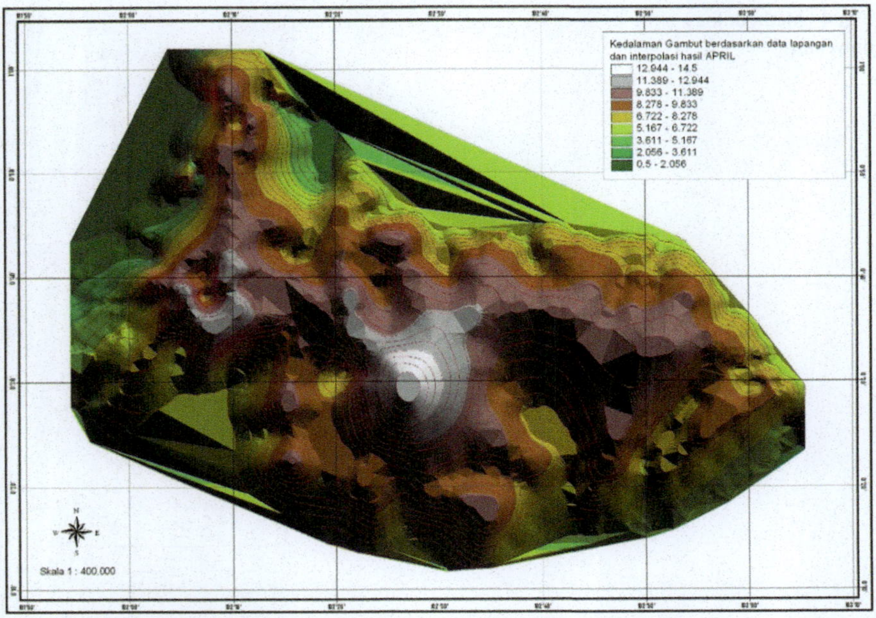

Fig. 3 Peatland depth (m) in Riau Province. *Source* BRGM's PRIMS website

Table 1 PALSAR data used in the study

No.	Date	Mode	Direction
Central Kalimantan Province			
1	10/8/2015	SM3	Ascending
2	10/6/2016		
3	10/5/2017		
4	9/6/2018		
5	9/5/2019		
6	9/3/2020		
7	9/2/2021		
Kampar Peninsula, Riau Province			
1	2/14/2015	SM3	Ascending
2	9/26/2015		
3	7/16/2016		
4	9/24/2016		
5	12/3/2016		
6	7/15/2017		
7	8/25/2018		
8	8/24/2019		
9	8/22/2020		
10	8/21/2021		

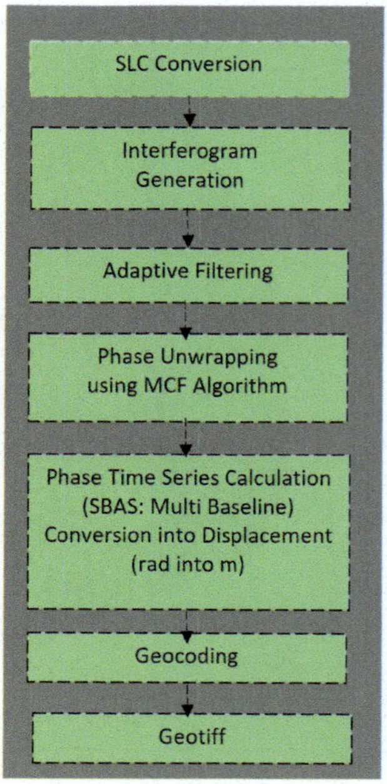

Fig. 4 Methodology flowchart

3.3.2 Time Series DInSAR SBAS Processing Method

The time series DInSAR SBAS (Small Baseline Subset) processing method is applied to process a stack of PALSAR-2 data using the combination of the ISP package, DIFF-GEO package, and Interferometric Point Target Analysis (IPTA) package of the GAMMA software (GAMMA Remote Sensing 2013). The IPTA package calculates the temporal and spatial characteristics of interferometric signatures of the data stack to improve the model parameters by achieving an optical match of interferometric phases. The surface map deformation is calculated for selected points from multi-looked interferogram data, thus increasing processing and data storage efficiency. Multi-base line processing is calculated within a stack of SAR data, and then the criteria of baseline less than 500 m are applied to select pairs for interferometry processing.

4 Results and Discussion

4.1 *The Observation of Peatland Surface Height Variability*

Peatland surface height variability for study sites in Central Kalimantan has been studied using differential SAR interferometry (DInSAR) analysis of PALSAR-2 data during the 2015–2021 period (Fig. 5). The comparative study with the GWL and GSL data from observed stations showed a good correlation with the fluctuation direction of the GWL/GSL of field data from the SESAME project (Takahashi et al. 2018). The variability of peatland surface height affected due to the groundwater fluctuations was captured by DInSAR analysis. The difference in peatland surface height variability between the DInSAR analyses with the GSL data is about $\pm$ 2.6 cm for SM2 mode and ~ $\pm$ 5.5 cm for SM3 mode (Table 2). The results were correlated with a previous study that revealed the GWL data follow the GSL data in peatland areas (Arvelyna et al. 2021a, b).

4.2 *Peatland Drought Analysis in ENSO year*

The El Niño–Southern Oscillation (ENSO) event in the Pacific Ocean and atmosphere involves extremely warm events for about two years and generates a warm and dry climate in Southeast Asia. The ENSO event in 2015 correlated well with the forest fire incidents in peatland areas in Indonesia. The smoke released from the fires affected Indonesia and neighboring countries, causing severe visibility issues and ill health. The DInSAR analysis of PALSAR-2 data over Central Kalimantan during the 2015–2020 period showed downward vertical movement from DInSAR data analysis for the observed stations in the ENSO year 2015–2016 (−2.8 to −4.2 cm) and 2018–2019 (−4.3 to −7.5 cm) during August–October period (dry season). At the same time, the surface height fluctuations of the presumed stable areas were below −2 cm. A comparative study with the field data revealed that the GWL data at the observed station marked the lowest level of about −1.44 m on 2019/10/1 (Fig. 6a, b). Although the lowest downward movement of the ground surface of the peatland area is shown in 2019, the forest fire incidents this year occurred 0.5 times compared to 2015, which may indicate the impact of implementing new regulations on peatland management since 2016. The study results revealed that the fluctuation of DInSAR data derived from PALSAR-2 data correlated well with the ENSO cycle year in Indonesia, which occurs every two years, and peaks every four years. Thus, the proposed analysis methods help monitor the surface height fluctuations in peatlands affected by fires that amplify during the ENSO years.

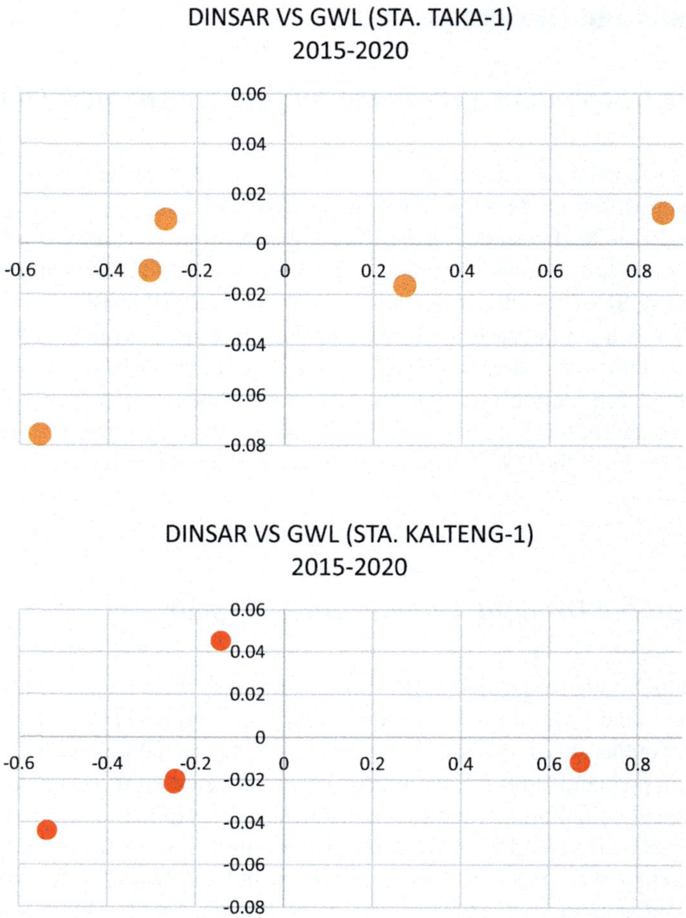

Fig. 5 Comparison of ground surface variability of peatland from DInSAR data and field data (GWL) at **a** Sta. Taka-1 and **b** Sta. Kalteng-1

Table 2 Results from the comparative study between the DINSAR versus ground surface level (cm)

Mode	Date (DMY)	DINSAR versus GSL (cm)
Asc-SM2	15/04/09	−2.6702
Asc-SM3	17/10/05	2.1236
Asc-SM3	18/09/06	0.1189
Asc-SM3	19/09/05	−5.3484
Asc-SM3	20/09/03	0.0599

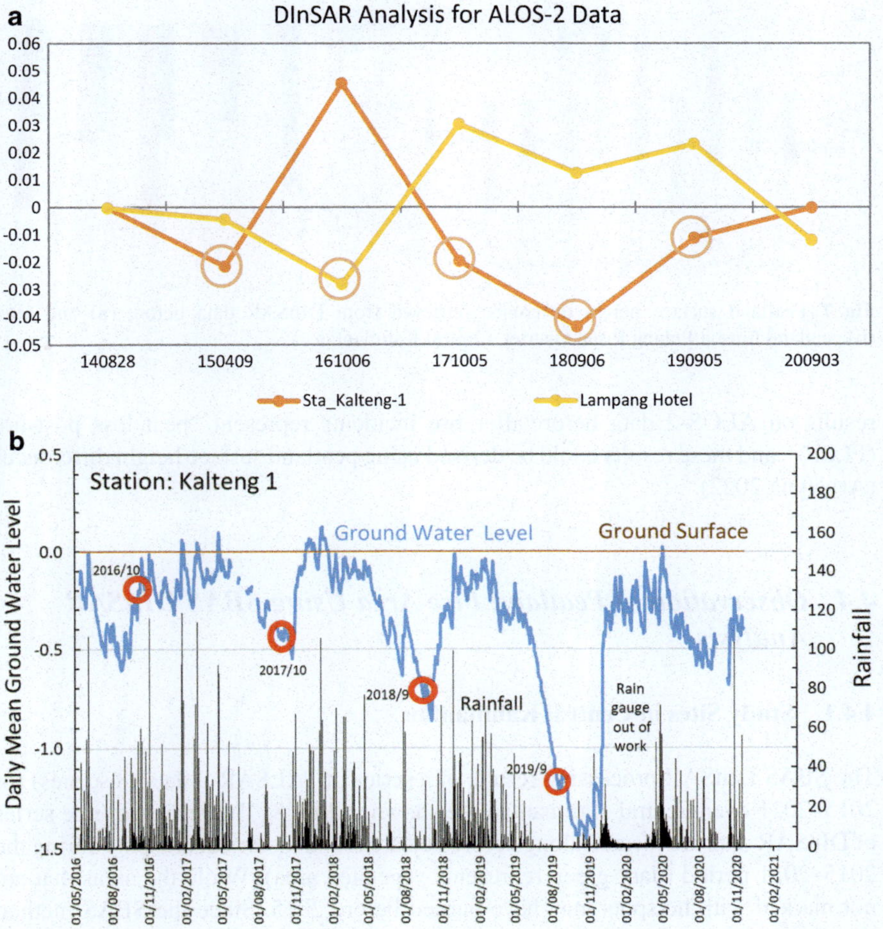

Fig. 6 **a**, **b** DInSAR data of peatland area at Station Kalteng-1 (top figure) and the GWL data (bottom figure). The *y*-axis is in meters (m)

4.3 *Peatland Surface Loss Due to Fires*

DInSAR processing is applied on PALSAR-2 data pairs prior to and post-fire incidents for selected fire hotspots with 80% of coincidence during 2015–2018. The DInSAR data analysis (Fig. 7a–d) shows that the fluctuation of the maximum height difference of peatland surface for T1–T9 hotspots in the downward direction is about −2.9 cm before fire incidents and −23.5 cm after fire incidents suggesting the possibility of peat loss after fires (Fig. 7). Peat loss is higher at a location around fire hotspots in the barren land (Site 2). Higher rainfall data affected the fluctuation of peatland surface due to more water absorption, shown by smaller downward DInSAR data for data pair at swamp shrub and secondary swamp forest. DInSAR analysis

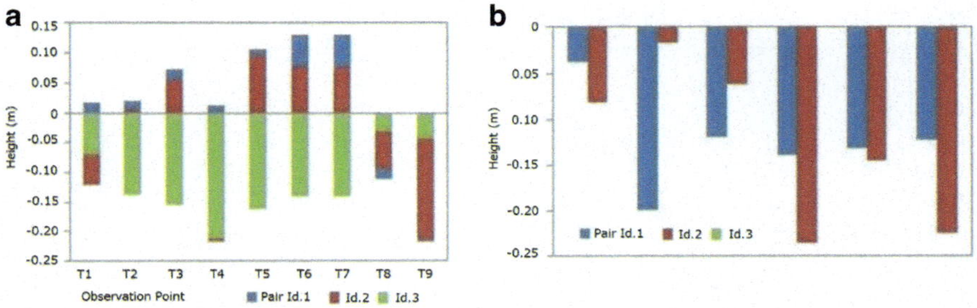

Fig. 7 Peatland surface height difference retrieved from DInSAR data before (**a**) and after (**b**) peatland fires at Pulang Pisau District, Central Kalimantan

results on ALOS-2 data before/after fire incidents represent "peat loss post-fire (PLPF)," and these results could be derived using peatland surface height differences (Arvelyna 2022).

4.4 Observation of Peatland Fire Area Using SBAS DInSAR Analysis

4.4.1 Study Sites in Central Kalimantan

The SBAS DInSAR processing results of a series of PALSAR-2 data (8 scenes) for 2015–2021 over Central Kalimantan are shown in Fig. 8. The retrieved time series of DInSAR data are located along deforested areas and historical hotspots during the 2015–2021 period (dark green represents vegetated area). While the areas that are not marked with hotspots may have burned before 2015. Since the SBAS method computed differential interferometry for the whole data pairs (each pair is defined with a small baseline, i.e., less than 500 m), high coherence data have been obtained from areas where surface deformation mainly occurred. Thus, the time series SBAS DInSAR method can delineate critical areas such as deforested and peatland fires. For the hotspots with 80% probability (red marker in the middle of Fig. 6), the preliminary result for the average peatland surface fluctuation using the SBAS DInSAR method is about −1.89 cm/year.

4.4.2 Study Sites in Kampar, Riau

The processing results of the SBAS DInSAR method of PALSAR-2 over the study area in Kampar Peninsula, Riau Province, are shown in Fig. 9. The area with burnt incidents, such as incidents in 2019 (green polygon in Fig. 2) and recursive burnt (red polygon), was delineated as the deformed area of yellowish-greenish color on the image.

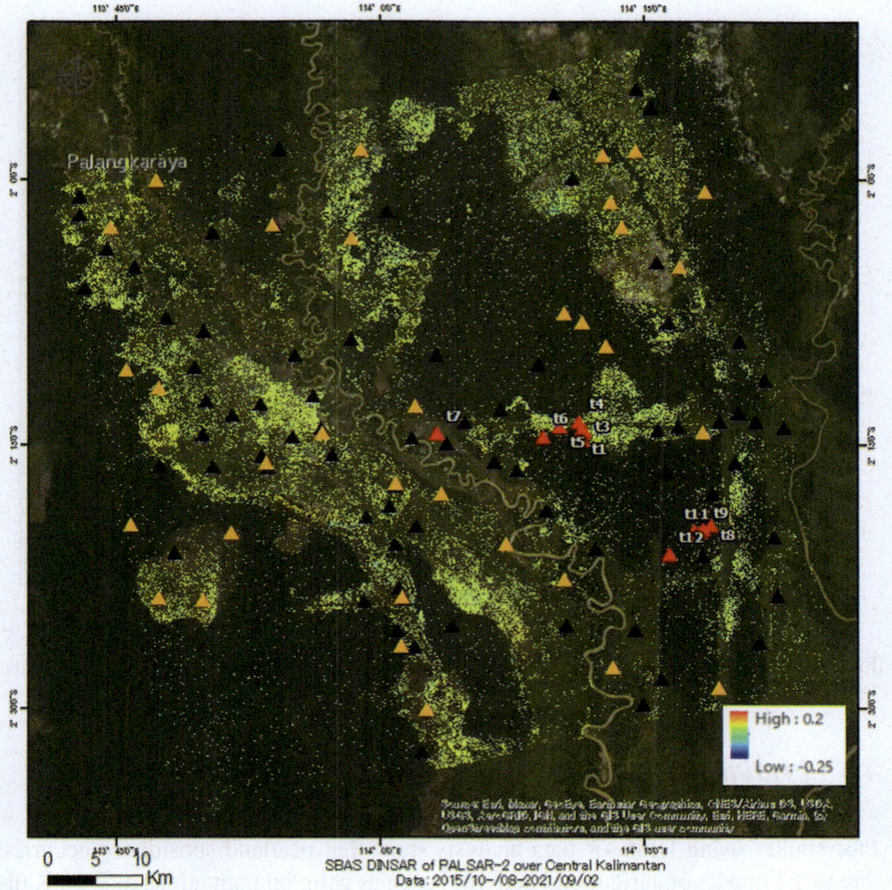

Fig. 8 SBAS DInSAR results for PALSAR-2 data (in m) with hotspots—black triangle representing medium probability; orange representing high probability and red representing 80% probability for period 2015–2021 in Central Kalimantan

4.5 *Peat Dome Analysis*

The analysis of the multi-temporal intensity of PALSAR-2 data during the 2015–2018 period over Kampar Peninsula, Riau Province (Fig. 10), shows that peat dome areas have lower backscattering of about 0.067–0.22 dB along the vertical cross line ($X1$–$X2$ and $Y1$–$Y2$). The maximum peat depth for these areas is about 14 m, while the bore data for the non-peat dome area are about 6 m. DInSAR analysis results for PALSAR-2 data show large vertical fluctuations (~ −40.5 cm) in peat dome areas larger than non-peat dome areas (−27.8 cm), possibly due to higher moisture and bearing capacity (Arvelyna et al. 2019).

Fig. 9 SBAS DInSAR processing result (in m) for PALSAR-2 data over study area in Kampar Peninsula, Riau Province

4.6 Peatland Subsidence Analysis

Past studies using DInSAR data analysis show that peatland subsidence occurred due to the impact of agricultural activities such as palm oil plantations in West Kalimantan. The subsided areas were low in organic matter from laboratory tests (Lost on Ignition method). It is considered to be affected by the decomposition process (Arvelyna et al. 2015). The application of the SBAS DInSAR method on PALSAR-2 data over the study area in Kampar Peninsula, Riau Province, showed the tendency of peatland subsidence at the areas with shallow peat thickness along the coast, river, and canal (Fig. 11a, b), possibly due to decomposition by water sedimentation transport. From 17 processed interferogram data for Kampar Peninsula, the average deformation (subsidence) rate at study sites along the coast retrieved is about 8.5 cm/year (Fig. 10, box).

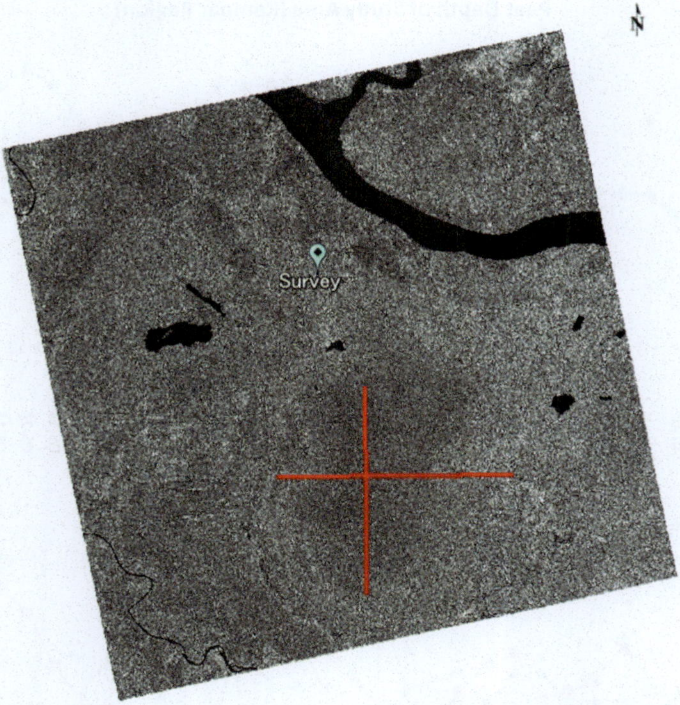

Fig. 10 Multi temporal intensity image resulted from ALOS-2 PALSAR-2 data pair 2017/7/15 and 2018/8/25 (©JAXA). Crosslines are drawn on observed peat domes

5 Conclusions

The study demonstrates the usefulness of SAR interferometry to map and monitor peatland changes due to the fires, deep peat assessment, and peatland subsidence characterization. DInSAR analysis results on PALSAR-2 data before/after fire incidents showed that peat loss after fire incidents could be derived using peatland surface height difference analysis. The proposed methods can increase the accuracy of existing monitoring methods using optical satellite images, and the SAR data can retrieve information even when clouds and smoke cover persist.

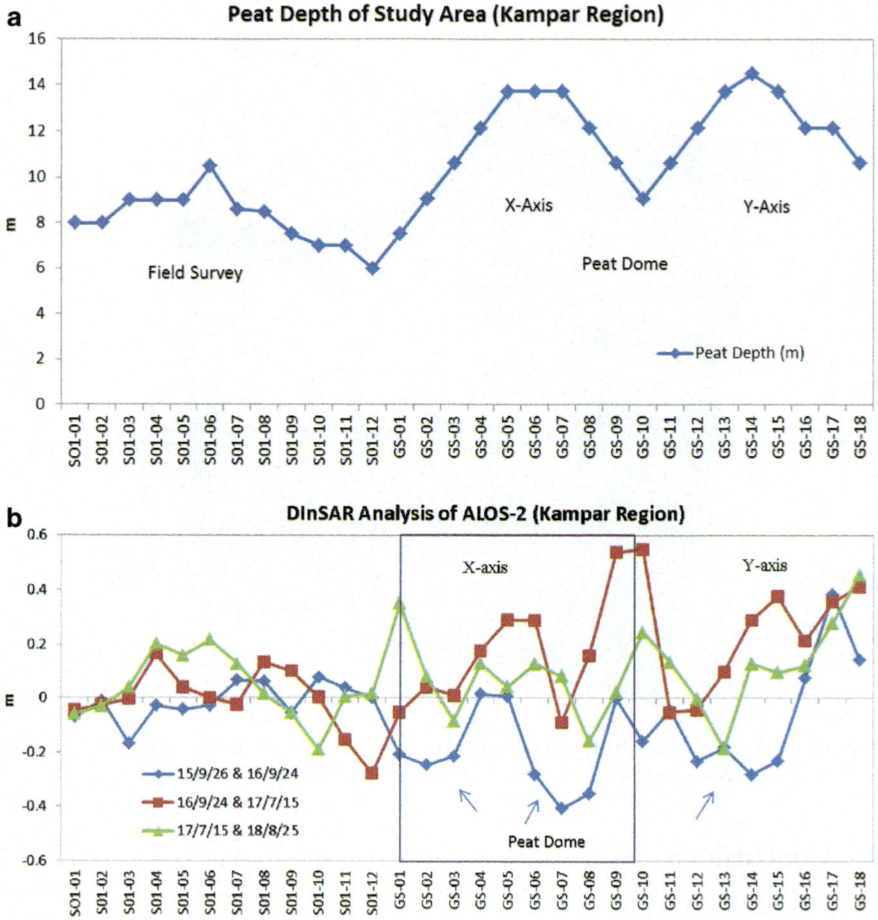

Fig. 11 **a** Peat depth in study area and peat dome shown on *X* and *Y* axes and their **b** ground surface fluctuation retrieved from DInSAR processing

Acknowledgements The authors thank JAXA for providing the ALOS-2 PALSAR-2 image data through the second Earth Observation Collaborative Research (PI: ER2A2N201).

References

Albar, I., I. Jaya, B.H. Saharjo, B. Kuncahyo, and K.P. Vadrevu. 2018. Spatio-temporal analysis of land and forest fires in Indonesia using MODIS active fire dataset. In *Land-atmospheric research applications in South and Southeast Asia*, 105–127. Cham: Springer.

Arvelyna, Y. 2022. The achievement of the observation of peatland in Indonesia using interferometry SAR. JAXA PI Meeting, Tokyo, Japan.

Arvelyna, Y., K. Hirose, G. Anshari, T. Kunihiro, H. Koichi, and R. Yanto. 2015. The application of interferometry SAR data analysis for peatland and land use monitoring in West Kalimantan. In *The 59th Japan Remote Sensing Conference Proceeding, Tokyo, Japan*, 207–208.

Arvelyna, Y., H. Takahashi, M. Osaki, H. Hirose, and T. Sato. 2018. Monitoring of peatland surface height variability using DInSAR analysis on ALOS-2 PALSAR data and field observation of ground water/surface level data. In *The 64th Japan Remote Sensing Conference Proceeding*, 111–112, Tokyo, Japan.

Arvelyna, Y., Prayoto, and R. Maryani. 2019. Observation of peat dome and subsidence using DInSAR analysis on ALOS-2 PALSAR data (Study case: Kampar Peninsula, Central Sumatra). In *The 66th Japan Remote Sensing Conference Proceeding*, 86–88.

Arvelyna, Y., H. Takahashi, L. Indrayanti, H. Hayasaka, K.P. Vadrevu, R. Maryani, M. Osaki, and K. Hirose. 2021a. Peatland surface loss due to fires in Central Kalimantan, Indonesia—A case study using differential interferometry SAR (DInSAR). In *Biomass burning mapping and monitoring*, vol. I, 77–90, ed. by P.K. Vadrevu, T. Ohara and C. Justice.

Arvelyna, Y., H. Takahashi, R. Maryani, and M. Osaki. 2021b. Peatland monitoring using DInSAR of PALSAR-2 data and field observation of ground water/surface level data during the ENSO year (Study case: Central Kalimantan, Indonesia). In *The 71th Japan Remote Sensing Conference Proceeding*, Tokyo, Japan.

Badarinath, K.V.S., and K.V. Prasad. 2011. Carbon dioxide emissions from forest biomass burning in India. *Global Environmental Research* 15: 45–52.

Badarinath, K.V.S., S.K. Kharol, K.M. Latha, T.K. Chand, V.K. Prasad, A.N. Jyothsna, and K. Samatha. 2007. Multiyear ground-based and satellite observations of aerosol properties over a tropical urban area in India. *Atmospheric Science Letters* 8 (1): 7–13.

Badarinath, K.V.S., S. Kumar Kharol, V. Krishna Prasad, A. Rani Sharma, E.U.B. Reddi, H.D. Kambezidis, and D.G. Kaskaoutis. 2008. Influence of natural and anthropogenic activities on UV index variations—A study over tropical urban region using ground based observations and satellite data. *Journal of Atmospheric Chemistry* 59 (3): 219–236.

Badarinath, K.V.S., S.K. Kharol, A.R. Sharma, and V.K. Prasad. 2009. Analysis of aerosol and carbon monoxide characteristics over Arabian Sea during crop residue burning period in the Indo-Gangetic Plains using multi-satellite remote sensing datasets. *Journal of Atmospheric and Solar-Terrestrial Physics* 71 (12): 1267–1276.

Biswas, S., K.D. Lasko, and K.P. Vadrevu. 2015a. Fire disturbance in tropical forests of Myanmar—Analysis using MODIS satellite datasets. *IEEE Journal of Selected Topics in Applied Earth Observations and Remote Sensing* 8 (5): 2273–2281.

Biswas, S., K.P. Vadrevu, Z.M. Lwin, K. Lasko, and C.O. Justice. 2015b. Factors controlling vegetation fires in protected and non-protected areas of Myanmar. *PLoS One* 10 (4): e0124346.

Biswas, S., K.P. Vadrevu, M.S. Mon, and C. Justice. 2021. Contemporary forest loss in Myanmar: Effect of democratic transition and subsequent timber bans on landscape structure and composition. *Ambio* 50 (4): 914–928.

Choi, Y., S.A. Vay, K.P. Vadrevu, A.J. Soja, J.H. Woo, S.R. Nolf, G.W. Sachse, G.S. Diskin, D.R. Blake, N.J. Blake, and H.B. Singh. 2008. Characteristics of the atmospheric CO_2 signal as observed over the conterminous United States during INTEX-NA. *Journal of Geophysical Research: Atmospheres* 113 (D7).

Davies, D.K., S. Ilavajhala, M.M. Wong, and C.O. Justice. 2008. Fire information for resource management system: archiving and distributing MODIS active fire data. *IEEE Transactions on Geoscience and Remote Sensing* 47(1):72–79.

Eaturu, A., and K.P. Vadrevu. 2021. Evaluation of Sentinel-3 SLSTR data for mapping fires in forests, peatlands, and croplands—A case study over Australia, Indonesia, and India. In *Biomass burning in South and Southeast Asia*, 39–59. Boca Raton: CRC Press.

FMU/REDD+. 2016. *Research service for acquiring carbon credits from REDD+*, 272, 425 p. Project report. Indonesia: Ministry of Environment and Forestry (in Indonesian).

GAMMA Remote Sensing. 2013. *GAMMA software introduction*. Documentation—User's guide, 4, 9–12, 14 p

Gupta, P.K., V.K. Prasad, Y. Kant, C. Sharma, A.B. Ghosh, M.C. Sharma, A.K. Sarkar, S.L. Jain, O.P. Tripathi, R.C. Sharma, and K.V.S. Badarinath. 2001. Study of trace gases and aerosol emissions due to biomass burning at shifting cultivation sites in East Godavari District (Andhra Pradesh) during INDOEX IFP-99. *Current Science* 186–196.

Hayasaka, H., I. Noguchi, E.I. Putra, N. Yulianti and K. Vadrevu. 2014. Peat-fire-related air pollution in Central Kalimantan, Indonesia. *Environmental pollution* 195: 257–266.

Indrayani, L., S.N. Marsoem, T.A. Prayitno, H. Supriyo, and B. Radjagukguk. 2011. Distribusi ketebalan gambut dan sifat-sifat tanah di hutan rawa gambut Kalampangan, Kalimantan Tengah. *Wanatropika* 56–72.

Joosten, H., and D. Clarke. 2002. *Wise use of mires and peatlands*, 24, 304 p. Finland: International Mire Conservation Group and International Peat Society.

Justice, C., G. Gutman, and K.P. Vadrevu. 2015. NASA land cover and land use change (LCLUC): An interdisciplinary research program. *Journal of Environmental Management* 148: 4–9.

Kant, Y., V.K. Prasad, and K.V.S. Badarinath. 2000. Algorithm for detection of active fire zones using NOAA AVHRR data. *Infrared Physics & Technology* 41 (1): 29–34.

Karyanto, O., A. Susanti, N. Qomar, W. Wardhana, D. Soeprijadi, J. Sartohadi, and E. Faridah. 2015. *Developing carbon-incentive mechanisms for forest resource management toward the emission reduction in Kampar Peninsular peatlands: Establishing baseline activities*, 37, 97 p. Final report. http://doi.org/10.13140/RG.2.1.3108.9763.

Key, C.H., and N.C. Benson. 2003. *The normalized burn ratio, a Landsat TM radiometric index for burn severity*. Available online at: http://nrmsc.usgs.gov/research/nbr.htm.

Kharol, S.K., K.V.S. Badarinath, A.R. Sharma, D.V. Mahalakshmi, D. Singh, and V.K. Prasad. 2012. Black carbon aerosol variations over Patiala city, Punjab, India—A study during agriculture crop residue burning period using ground measurements and satellite data. *Journal of Atmospheric and Solar-Terrestrial Physics* 84: 45–51.

LAPAN Fire Hotspot. http://103.51.131.166/.

Lasko, K., and K. Vadrevu. 2018. Improved rice residue burning emissions estimates: Accounting for practice-specific emission factors in air pollution assessments of Vietnam. *Environmental Pollution* 236: 795–806.

Lasko, K., K.P. Vadrevu, V.T. Tran, E. Ellicott, T.T. Nguyen, H.Q. Bui, and C. Justice. 2017. Satellites may underestimate rice residue and associated burning emissions in Vietnam. *Environmental Research Letters* 12 (8): 085006.

Lasko, K., K.P. Vadrevu, and T.T.N. Nguyen. 2018. Analysis of air pollution over Hanoi, Vietnam using multi-satellite and MERRA reanalysis datasets. *PLoS One* 13 (5): e0196629.

Lasko, K., K.P. Vadrevu, V. Bandaru, T.N.T. Nguyen and H.Q. Bui. 2021. PM2. 5 emissions from biomass burning in South/Southeast Asia–Uncertainties and trade-offs. *Biomass burning in south and Southeast Asia* 149–169.

Nurbaya, S., Efransjah, S. Murniningtyas, Erwinsyah, and E.K. Damayanti (eds.). 2020. *The state of Indonesia's forests*, 36; 110, 234 p. Jakarta: Ministry of Environment and Forestry, Republic of Indonesia.

Prasad, V.K., and K.V.S. Badarinth. 2004. Land use changes and trends in human appropriation of above ground net primary production (HANPP) in India (1961–98). *Geographical Journal* 170 (1): 51–63.

Prasad, K.V., and K.V.S. Badarinath. 2006. Soil surface nitrogen losses from agriculture in India: A regional inventory within agroecological zones (2000–2001). *The International Journal of Sustainable Development and World Ecology* 13 (3): 173–182.

Prasad, K.V., Y. Kant, and K.V.S. Badarinath. 2000. Quantifying Short-Term Carbon Dynamics from Land Use Changes Using Satellite Data-A Case Study from Rampa Forests (Eastern Ghats) India. *Geocarto International* 15 (2): 71–78.

Prasad, V.K., Y. Kant, and K.V.S. Badarinath. 2001a. CENTURY ecosystem model application for quantifying vegetation dynamics in shifting cultivation areas: A case study from Rampa Forests, Eastern Ghats (India). *Ecological Research* 16 (3): 497–507.

Prasad, V.K., Y. Kant, P.K. Gupta, C. Sharma, A.A. Mitra, and K.V.S. Badarinath. 2001b. Biomass and combustion characteristics of secondary mixed deciduous forests in Eastern Ghats of India. *Atmospheric Environment* 35 (18): 3085–3095.

Prasad, V.K., Y. Kant, and K.V.S. Badarinath. 2002a. Land use changes and modeling carbon fluxes from satellite data. *Advances in Space Research* 30 (11): 2511–2516.

Prasad, V.K., Y. Kant, P.K. Gupta, C. Elvidge, and K.V.S. Badarinath. 2002b. Biomass burning and related trace gas emissions from tropical dry deciduous forests of India: A study using DMSP-OLS data and ground-based measurements. *International Journal of Remote Sensing* 23 (14): 2837–2851.

Prasad, V.K., M. Lata, and K.V.S. Badarinath. 2003. Trace gas emissions from biomass burning from northeast region in India—Estimates from satellite remote sensing data and GIS. *The Environmentalist* 23 (3): 229–236.

Prasad, V.K., K.V.S. Badarinath, S. Yonemura, and H. Tsuruta. 2004. Regional inventory of soil surface nitrogen balances in Indian agriculture (2000–2001). *Journal of Environmental Management* 73 (3): 209–218.

Prasad, V.K., E. Anuradha, and K.V.S. Badarinath. 2005. Climatic controls of vegetation vigor in four contrasting forest types of India—Evaluation from National Oceanic and Atmospheric Administration's advanced very high resolution radiometer datasets (1990–2000). *International Journal of Biometeorology* 50 (1): 6–16.

Prayoto, M.I. Ishihara, R. Firdaus, and N. Nakagoshi. 2017. Peatland fires in Riau, Indonesia, in relation to land cover type, land management, landholder, and spatial management. *Journal of Environmental Protection* 8: 1312–1332. https://doi.org/10.4236/jep.2017.811081.

PRIMS (Peatland Restoration Information and Management System). 2022. *Peat depth map*. Indonesia: Peatland Restoration and Mangrove Agency's (BRGM). https://en.prims.brg.go.id/.

Roswintiarti, O. 2016. *Informasi titik panas (Hotspot) kebakaran hutan/lahan*. Guide V.01. LAPAN. http://pusfatja.lapan.go.id/files_uploads_ebook/publikasi/Panduan_hotspot_2016%20versi%20draft%201_LAPAN.pdf.

SIGAP KLHK. 2020. *Interactive land cover map 2020*. Jakarta: Ministry of Environment and Forestry, Republic of Indonesia. http://sigap.menlhk.go.id/sigap/peta_interaktif.php.

Takahashi, H., Y. Shigenaga, Y. Hamada, M. Osaki, and B. Setiadi. 2018. Near-real time remote monitoring of ground-water/surface levels and warning to carbon loss from tropical peatlands in Indonesia. In *The proceedings of the International Workshop on Forest Ecological Resources Security for Next Generation: Development and Routine Utilization of Forest Ecological Resources and their Domestication*, 80–90, ed. by S. Shiodera, M. Fujita and S. Kobayashi. Kyoto: Center for Southeast Asian Studies, Research Center, Kyoto University.

Vadrevu, K.P. 2008. Analysis of fire events and controlling factors in eastern India using spatial scan and multivariate statistics. *Geografiska Annaler: Series A, Physical Geography* 90 (4): 315–328.

Vadrevu, K. 2015. International regional science meeting of NASA-LCLUC: Land cover/land use change (LC/LUC) and environmental impacts in South Asia, Coimbatore, India, 19–23 January 2013. *Journal of Environmental Management* 148: 1–163.

Vadrevu, K.P., and K.V.S. Badarinath. 2009. Spatial pattern analysis of fire events in Central India—A case study. *Geocarto International* 24 (2): 115–131.

Vadrevu, K.P., and C.O. Justice. 2011. Vegetation fires in the Asian region: Satellite observational needs and priorities. *Global Environmental Research* 15 (1): 65–76.

Vadrevu, K.P. and K. Lasko. 2015a. Satellite-derived nitrogen dioxide variations from biomass burning in a subtropical evergreen forest, Northeast India. In *Remote Sensing Handbook* (pp. 545–559). CRC Press.

Vadrevu, K., and K. Lasko. 2015b. Fire regimes and potential bioenergy loss from agricultural lands in the Indo-Gangetic Plains. *Journal of Environmental Management* 148: 10–20.

Vadrevu, K.P., A. Eaturu, and K.V.S. Badarinath. 2006. Spatial distribution of forest fires and controlling factors in Andhra Pradesh, India using spot satellite datasets. *Environmental Monitoring and Assessment* 123: 75–96.

Vadrevu, K.P., K.V.S. Badarinath, and E. Anuradha. 2008. Spatial patterns in vegetation fires in the Indian region. *Environmental Monitoring and Assessment* 147 (1): 1–13.

Vadrevu, K.P., L. Giglio, and C. Justice. 2013. Satellite based analysis of fire–carbon monoxide relationships from forest and agricultural residue burning (2003–2011). *Atmospheric Environment* 64: 179–191.

Vadrevu, K.P., K. Lasko, L. Giglio, and C. Justice. 2014a. Analysis of Southeast Asian pollution episode during June 2013 using satellite remote sensing datasets. *Environmental Pollution* 195: 245–256.

Vadrevu, K.P., T. Ohara, and C. Justice. 2014b. Air pollution in Asia. *Environmental Pollution* 12: 233–235.

Vadrevu, K.P., T. Ohara, and C. Justice. 2014c. Air pollution in Asia. *Environmental Pollution (Barking, Essex: 1987)* 195: 233–235.

Vadrevu, K., T. Ohara, and C. Justice. 2017. Land cover, land use changes and air pollution in Asia: A synthesis. *Environmental Research Letters* 12 (12): 120201.

Vadrevu, K.P., T. Ohara, and C. Justice (eds.). 2018. *Land-atmospheric research applications in South and Southeast Asia.*

Vadrevu, K., A. Heinimann, G. Gutman, and C. Justice. 2019. Remote sensing of land use/cover changes in South and Southeast Asian Countries. *International Journal of Digital Earth* 12 (10): 1099–1102.

Vadrevu, K.P., E. Ellicott, L. Giglio, K.V.S. Badarinath, E. Vermote, C. Justice, and W.K. Lau. 2012. Vegetation fires in the himalayan region–Aerosol load, black carbon emissions and smoke plume heights. *Atmospheric Environment* 47: 241–251.

Vadrevu, K.P., A. Eaturu, S. Biswas, K. Lasko, S. Sahu, J.K. Garg, and C. Justice. 2020. Spatial and temporal variations of air pollution over 41 cities of India during the COVID-19 lockdown period. *Scientific Reports* 10 (1): 1–15.

Vadrevu, K.P., T. Ohara, and C. Justice (eds.). 2021a. *Biomass burning in South and Southeast Asia: Impacts on the biosphere*, vol. 2. Boca Raton: CRC Press.

Vadrevu, K.P., T. Ohara, and C. Justice (eds.). 2021b. *Biomass burning in South and Southeast Asia: Mapping and monitoring*, vol. 1. Boca Raton: CRC Press.

Vadrevu, T. Le Toan, S.S. Ray, and C. Justice. 2022a. *Remote sensing of agriculture and land cover/land use changes in South and Southeast Asian Countries.* Cham: Springer.

Vadrevu, K., A. Eaturu, E. Casadaban, K. Lasko, W. Schroeder, S. Biswas, L. Giglio, and C. Justice. 2022b. Spatial variations in vegetation fires and emissions in South and Southeast Asia during COVID-19 and pre-pandemic. *Scientific Reports* 12 (1): 1–21.

Van Wagtendonk, J.W., R.R. Root, and C.H. Key. 2004. Comparison of AVIRIS and Landsat ETM+ detection capabilities for burn severity. *Remote Sensing of Environment* 92 (3): 397–408.

Vay, S.A., Y. Choi, K.P. Vadrevu, D.R. Blake, S.C. Tyler, A. Wisthaler, A. Hecobian, Y. Kondo, G.S. Diskin, G.W. Sachse, and J.H. Woo. 2011. Patterns of CO_2 and radiocarbon across high northern latitudes during International Polar Year 2008. *Journal of Geophysical Research: Atmospheres* 116 (D14).

WEBGIS of Ministry of Environment and Forestry Indonesia. http://webgis.menlhk.go.id:8080/kemenhut/index.php/en/map/interactive-map-2.

Wooster, M.J., G.J. Roberts, L. Giglio, D.P. Roy, P.H. Freeborn, L. Boschetti, C. Justice, C. Ichoku, W. Schroeder, D. Davies, and A.M. Smith. 2021. Satellite remote sensing of active fires: History and current status, applications and future requirements. *Remote Sensing of Environment* 267: 112694.

Active Fire Monitoring of Thailand and Upper ASEAN by Earth Observation Data: Benefits, Lessons Learned, and What Still Needs to Be Known

Veerachai Tanpipat, Jessica L. McCarty, Diane Davies, Wilfrid Schroeder, and Chris Elvidge

Abstract This paper describes the history of active fire monitoring of Thailand and Upper ASEAN (Cambodia, Lao PDR, Myanmar, and Vietnam) by Earth Observation Data, including benefits, lessons learned, and the unknowns. Satellite remote sensing offers a unique potential for mapping and monitoring fires at large spatial scales. Active fire information from satellites is one of the fundamental data in mitigating the smoke haze effect; therefore, active fire monitoring has become an essential tool for daily fires and smoke haze control, management, and mitigation in Thailand. The three critical sources of active fire hotspot information used in Thailand and elsewhere include (a) NASA's Fire Information for Resource Management System (FIRMS); (b) the National Environmental Satellite, Data, and Information Service (NESDIS), Active Fire Alert System; and (c) the Colorado School of Mines VIIRS NightFire Alert System. On their own, satellite-derived active fire information is insufficient, but the combination of satellite-derived data and traditional fire monitoring from ground-based methods can work well. Adding advanced geostationary fire detection is needed as a next step to enable fire suppression teams to reach fires as quickly as possible, resulting in much more efficient and safer fire control.

Keywords Active fire · Hotspot · Thailand · Upper ASEAN · NASA-FIRMS · VIIRS · NOAA · NightFire

---

V. Tanpipat (✉)
Upper ASEAN Wildland Fire Special Research Unit, Kasetsart University, Bangkok, Thailand
e-mail: iamtanpipat@hotmail.com

J. L. McCarty
Department of Geography and Geospatial Analysis Center, Miami University, Oxford, OH, USA

D. Davies
Trigg-Davies Consulting Ltd, Malvern, UK

NASA-GSFC, Science Systems and Applications Inc., Lanham, MD, USA

W. Schroeder
NOAA-NESDIS-OSPO Satellite Analysis Branch, College Park, MD, USA

C. Elvidge
Earth Observation Group, Payne Institute for Public Policy, Colorado School of Mines, Golden, CO, USA

K. P. Vadrevu et al. (eds.), *Vegetation Fires and Pollution in Asia*,
https://doi.org/10.1007/978-3-031-29916-2_9

1 Introduction

In recent years, forest fires and associated haze have become a major environmental problem in the tropical ecosystems of the upper Association of Southeast Asian Nations (ASEAN). Upper ASEAN, or Lower Mekong River Region, consists of five countries: Cambodia, Lao PDR, Myanmar, Thailand, and Vietnam (Biswas et al. 2015; Lasko and Vadrevu 2018; Yu et al. 2017; Albar et al. 2018; Inoue 2018; Hayasaka and Sepriando 2018). Forest fires, open burning, and smoke haze became common environmental issues in this region, specifically, a Lower Mekong River Region that comprises northern Thailand, northeastern-eastern Myanmar, and northern Laos. Smoke haze has long stood out as a significant yearly environmental concern for the region, with its severity varying year-to-year (Adam and Balasubramanian 2021). The term "smoke haze" means atmospheric haze mainly due to smoke from fires or open burning, particularly in forests and agricultural lands. Incomplete burning releases a significant amount of smoke. In Thailand, smoke haze has become a conflict issue between rural and urban people and government authorities in northern Thailand (Marks and Miller 2022).

Moreover, smoke haze affects the tourism industry and causes more respiratory illness, resulting in losing money in all sectors. Humans are known to be the sole player in fires in the region. Farmers want to clear land and eliminate weeds and the remains of previous-cycle crops or seek new ground and resort to burning as an inexpensive tool (Oanh et al. 2018). For forest fires, causes include slash and burn practice, hunting-gathering, and arguably land or personal conflicts (Saharjo and Yungan 2018; Vadrevu et al. 2021a, b). Smoke haze commonly contains tiny and unhealthy particles called Particulate Matter 2.5 or PM2.5 (Phairuang 2021). The smoke haze transboundary issue exists, but there needs to be scientifically sound research and studies to confirm the smoke haze movement and behavior on a small scale. Simply looking only at active fire hotspot information will not address the smoke haze transboundary issue, as the transport depends on various parameters. Active fire information from satellites is one of the fundamental data in mitigating the smoke haze effect (Elvidge et al. 2021); therefore, active fire monitoring has become an essential monitoring tool for daily fires and smoke haze control, management, and mitigation in Thailand and Upper ASEAN.

2 Study Area

Upper ASEAN is located in the Indochina Peninsula at 9.0 and 25.0 north latitudes, 92.5 and 110 east longitudes. Thailand is centrally located in the Indochina Peninsula between 5 40 and 20 30 north latitudes and 90 70 and 105 45 east longitudes, covering an area of 513,115 km^2 characterized by a monsoon climate (Fig. 1). It has two dominant vegetation types: evergreen and deciduous. The evergreen forest contains a significant proportion of non-leaf shedding species and covers about 40% of the area. It can be further classified into tropical rainforests, dry evergreen, hill evergreen, coniferous, mangrove, swamp, and beach forests. The remaining 60% belongs to the deciduous forest, which comprises species with leafless periods. Trees growing in this latter type of forest tend to develop growth or annual rings that cannot be found in the evergreen forest species. The deciduous forest, prone to surface fires during dry seasons, can be categorized into mixed deciduous, dry dipterocarp, and savanna. Though fires in the mixed deciduous forest have great potential to be severe and damaging, if bamboo constitutes a majority of this forest, they play a vital role in its regeneration (Charuppat 2001).

Forest fires are common in Thailand as wildland fires and are caused by humans (Akaakara 2001). Forest fires fall into six general categories in Thailand: surface, crown, ground fires, semi-crown, and semi-ground fires. First, a surface fire burns

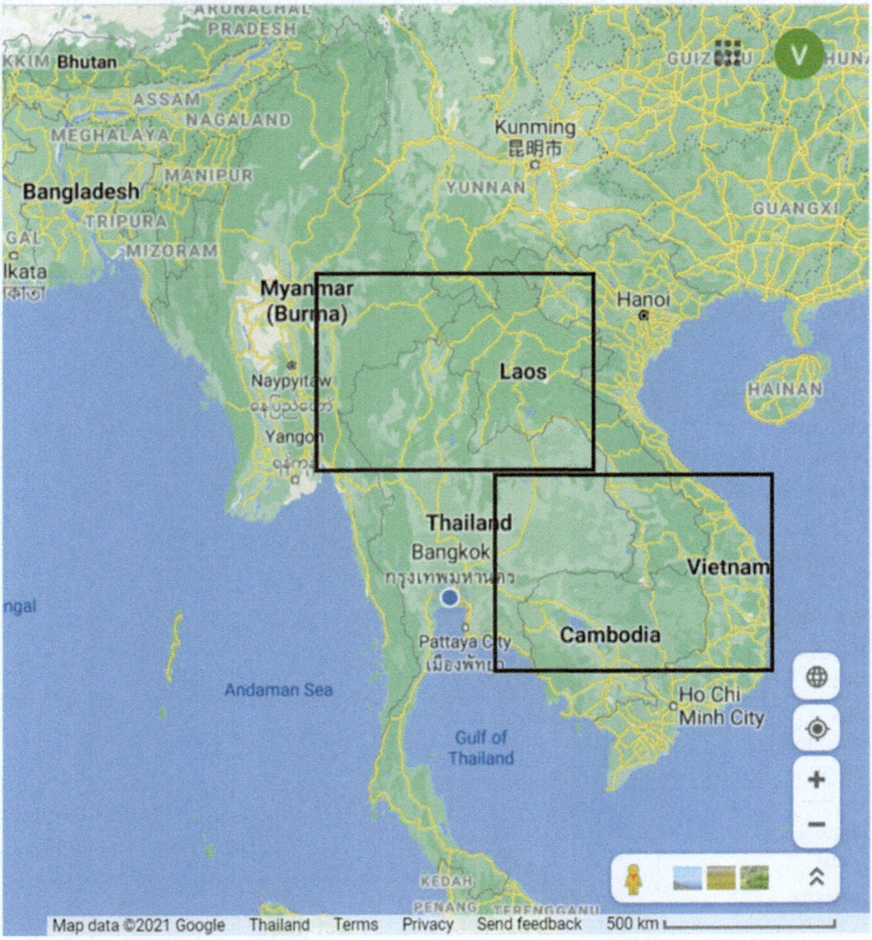

Fig. 1 Upper ASEAN and Thailand (https://www.google.co.th/)

organic materials in soil layers and often the surface litter, loose derbies such as leaves and fallen branches, low-growing vegetation, and other fire fuels on the forest floor. Surface fires occur in the dry dipterocarp, mixed deciduous plantations, dry evergreen, hill evergreen, and some parts of the tropical rainforests. Second, a crown fire is a fire that burns primarily in the leaves and needles of trees, spreading from tree to tree above the ground. It occurs in the coniferous forest and pine plantations in the country's northern region. Third, a ground fire is a fire that burns primarily under the forest floor, spreading underground from tree to tree and usually lasting a long time. It occurs only in the peat forest in the southern region of Thailand (Plodpail et al. 1987; Akaakara 2001). Three additional types of fires occur in the transition, which are semi-crown fires, semi-ground fires, and spot fires which recently happened in southern Thailand in a degraded peat forest where paperbark trees dominated. Surface fire is a common fire type in Thailand and this region. Forest fires in Thailand occur annually during the dry season from late December to early May, with the most occurring in March (Akaakara 2001; Tanpipat et al. 2009). The relationships between the forest fires and the forest types are summarized in Table 1.

Table 1 Relationships between the forest fires and the forest types in Thailand

	Surface fire	Ground fire	Crown fire	Semi-ground fire	Semi-ground fire	Spot fire
Tropical rainforest	X					
Dry evergreen forest	X					
Hill evergreen forest	X					
Coniferous forest	X	X	X	X	X	X
Mangrove forest	X					
Peat forest	X	X	X	X	X	X
Beach forest	X					
Mixed deciduous forest	X					
Dry dipterocarp forest	X					
Savanna	X					X

The forest fire situation in Upper ASEAN has been widely reported in the media, with increasing access to information through the rapid development of cyber-technologies. Satellite monitoring of active fires has led to greater transparency, and the number of fires detected since 2003 has decreased (Copernicus 2020), but the forest fire, open burning, and smoke haze continue to be a problem every dry season. From AFoCO's Forest Fire Management Training Course's discussions among AFoCO's member states in 2019 and 2020, just for Upper ASEAN participants, the causes and needs can be listed. Those causes are all related to human activities by the country in Table 2 (AFoCO 2019, 2020).

Fires are essential for tree regeneration in dry dipterocarp and deciduous forest and usually occur annually during the dry season. There can be natural causes, such as lightning strikes, but not in this region. In the case of Thailand or the Upper ASEAN or tropical zone, lighting occurs during thunderstorms (wet lighting), so such a fire does not spread from where it occurs. Unfortunately, the major causes of forest fires are related to activities of those who live in rural areas, not the natural; they are incendiary fires ignited by people for the gathering of forest non-timber products, agricultural land preparation, political conflict, hunting, timber, cattle grazing, agricultural land expansion, and carelessness to say a few. The officially recorded statistics collected since 1980 by Thailand's Forest Fires Control Division of the National Park, Wildlife, and Plant Conservation Department, Ministry of Natural Resources and Environment (a former division in the Thai Royal Forest Department) clearly showed a very insignificantly small number of fires were of natural causes and they were not spread

Table 2 Fire causes of Upper ASEAN countries

	Cambodia	Lao PDR	Myanmar	Thailand	Vietnam
Firewood utilization	X	X	X	X	
Slash and burn	X	X	X	X	
Cash mono-crop		X	X	X	X
Hunting		X	X	X	
Cattle			X	X	
Plantation project		X			
Harvesting honey					X
Carelessness	X				X
Resin collection	X				
Concessions	X				
Burn after logging	X				
New settlement	X				
Conflicts				X	
Timber				X	
Land encroachment				X	

Source AFoCO's Country Report during Forest Fire Management Short Training Course's discussions among AFoCO's member states in 2019 and 2020

out (DNP 2020). Therefore, in this context, it is safe to conclude that all forest fires in Thailand are man-made. And their major causes are directly related to the activities of the inhabitants in the rural areas who live near or within the forests.

Smog and haze from wildfires can have a profound impact on health and everyday life. It is a transboundary issue that will require an international effort to solve. The latest examples are peat fires at South Sumatra in 2015 and 2019 (Ismi 2019).

Active fire information is the most common fire information that has been used in Upper Southeast Asia and Thailand in almost the past 2 decades since the AVHRR World Fire Web, DMSP-OLS NOAA System, FIRMS University of Maryland, etc. These systems gradually become an important part and much needed information for wildland fire and smoke haze control and management from the top to the bottom and vice versa in Thailand. The information on the active fires has become the most important information for addressing fire and smoke haze for daily operations.

3 History of Active Fires Monitoring in Thailand and Upper ASEAN

Satellite-derived active fire information has been used in Upper ASEAN or Southeast Asia and Thailand for over two decades. In 1999, the Asian Center for Research on Remote Sensing (ACRoRS), Asia Institute Technology (AIT) began using The Advanced Very-High-Resolution Radiometer (AVHRR) World Fire Web, which was started in 1998 (GFMC 2020). With informal Committee of Earth Observation System (CEOS) Working Group on Information Systems and Services (WGISS) meeting in Bangkok, Thailand, in 1999 and WGISS/GOFC (Global Observation of Forest Cover) Demonstration in 2000 meetings (Cahoon et al. 2000), the Global Observation Information Network (GOIN) 1999 by NASA Research and Education Network was demonstrated to show the Global Internet capabilities through Asia Pacific Advanced Network (APAN) and the possibility of utilizing Earth Observation Satellite (EOS) information to support a daily fire control was feasible through NOAA-AVHRR's fire products. In 1999, launching the Terra Satellite and later the Afternoon Train proved useful. EOS later became essential information supporting fire control and smoke haze management. The expansion of active fire detection from the NOAA-AVHRR-World Fire Web system (Li et al. 2000) by The Defense Meteorological Program (DMSP) Operational Line-Scan System (OLS) NOAA System was set up at Asian Center for Research on Remote Sensing (ACRoRS), Asian Institute of Technology (AIT) in 2003. Later in 2006, the Department of National Park Wildlife and Plants Conservation requested support from the Fire Information for Resource Management System (FIRMS) at the Department of Geography, University of Maryland; it has been using FIRMS (http://earthdata.nasa.gov/firms) ever since. Initially, FIRMS provided data from MODIS onboard the Aqua and Terra satellites; VIIRS 375 m active fire data (Schroeder et al. 2014) were added when it became available from Suomi-NPP and NOAA-20. FIRMS was transitioned to

NASA's Land and Atmosphere Near-real-time Capability for EOS (LANCE) in 2012. In 2017, the collaboration with NOAA-National Centers for Environmental Information [NCEI, Formal National Geophysical Data Center (NGDC)] was started, and the VIIRS NightFire alert system for Southeast Asia was established. Later in 2019, the system was moved to the Colorado School of Mines. In March 2019, following Thailand's National Forest Fire and Smoke Haze meeting, the Royal Forest Department requested official support from NOAA-National Environmental Satellite, Data and Information Service (NESDIS) to obtain quicker VIIRS active fire products to Thailand and later the whole Southeast Asia.

Active fire information has become a key component of forest fires, open burning, and smoke haze control monitoring programs in Thailand. All levels of management routinely use it for daily fire and smoke haze operations across the country. During the 2007 fire season, satellite-derived active fire data became part of the National Forest Fire Control Operation through the Forest Fire Control Division, Forest Protection, and Fire Control Office, Department of National Park, Wildlife and Plants Conservation, and Ministry of Natural Resources and Environment to the National Ad hoc Committees under the Deputy Prime Minister. To build confidence in the satellite-derived fire products, the Forest Fire Control Division carried out a field validation campaign to determine the accuracy of the MODIS active fire product; the results showed good accuracy, where burned areas were found near active fire hotspots reported at 91.84%, 95.60%, and 97.53% for 2007, 2008, and 2009 (Tanpipat et al. 2009). The Forest Fire Control Division continues this field validation task until today, and the information is used for the routine suppression task. Later in 2016, the VIIRS active fire product had 98% verification accuracy with the smallest burned area found at four m^2 (unpublished DNP report 2016), which is the same burned area's size as what found by Schroeder et al. (2014) in Brazil. With very high accuracy results, it later became a Key Performance Index (KPI) of many government agencies until today. Nowadays, the number of active fires detected inside and outside protected areas comparison has been a key statistic for people involved in daily fires and smoke haze control and management. A reliance on statistics in the past few years has led to conflicts between local people and government authorities, resulting in increasing numbers of arsonists setting fires in the deeper forested and mountainous areas where firefighters hardly reach to control them. Moreover, during the past five years of The Asian Forest Cooperation Organization's (AFoCO) 15 member states training and Disaster Mitigation Working Group of Asia Pacific Advanced Network, active fire information is always mentioned to educate participants about existing technology that can be used to a part of their forest fire control within their country.

Since 2003, there has been a gradual reduction in active fires detected using MODIS data. However, air quality is still a major concern. According to the Pollution Control Department of Thailand, the number of bad-quality air days increased from 60 days in 2017 to 112 days in 2020. The smoke haze from fires has become a national issue and is pushing the People's Clean Air Act into Thailand Parliament, hopefully again by the end of 2021 because the Prime Minister of Thailand just rejected the first "People's Clean Air Act" proposal in late March 2021 resulting in another rolling motion to collect 10,000 Thai's signatures to put it through the Thai parliament again.

At a regional level, the use of active fire information systems is taught as one of the Forest Fire Management topics in an annual five-day training course organized by the Asian Forest Cooperation Organization (AFoCO, http://afocosec.org/) for its member states. The three active fire information systems have been introduced to participants, so they can be used in their countries. For example, Cambodia used active fire information from NASA-FIRMS after the training in 2015.

All three systems have an email alert with a KML file which can be easily used on a smartphone. Representatives of forestry government agencies from Thailand, Lao PDR, Cambodia, Myanmar, and Vietnam who participated in AFoCO's Forest Fire Management training course are the people who receive those email alerts and distribute them to responsible people in the country. Only Thailand has used such information extensively, as mentioned. Web Map Service of FIRMS is also commonly used by active Thai citizens and NGOs who would like to follow fire situations closely.

4 Description of Main Existing Fire Monitoring Systems and Sources Using in Upper ASEAN

4.1 FIRMS NASA

NASA's Fire Information for Resource Management System (FIRMS, Figs. 2 and 3) delivers global active fire information, derived from MODIS (Aqua and Terra) and VIIRS (S-NPP and NOAA-20), in easy-to-use formats, within 3 h of satellite observation, as well as the full archive of active fire data. FIRMS has users in over 160 countries with over 12,000 users registered for near-real-time, daily or weekly, active fire email alerts. The Fire Map interface enables users to view and query active fires by time since detection, as well as view corrected reflectance imagery (from MODIS and VIIRS), Aerosol Index from the Ozone Mapping Profiler Suite (OMPS) and historic burned areas (derived from MODIS). Originally developed by the University of Maryland, with funds from NASA's Applied Sciences and the United Nations Food and Agriculture Organization (UN FAO), FIRMS is now part of NASA's Land, Atmosphere Near-real-time Capability for Earth Observing System (EOS) (LANCE, Fig. 4), which is managed by NASA's Earth Observing System Data and Information System (EOSDIS).

4.2 VIIRS NOAA-NESDIS

The ground segment serving VIIRS onboard S-NPP and JPSS (e.g., NOAA-20) spacecrafts includes primary 300 Mbps Ka-band receiving stations located in the

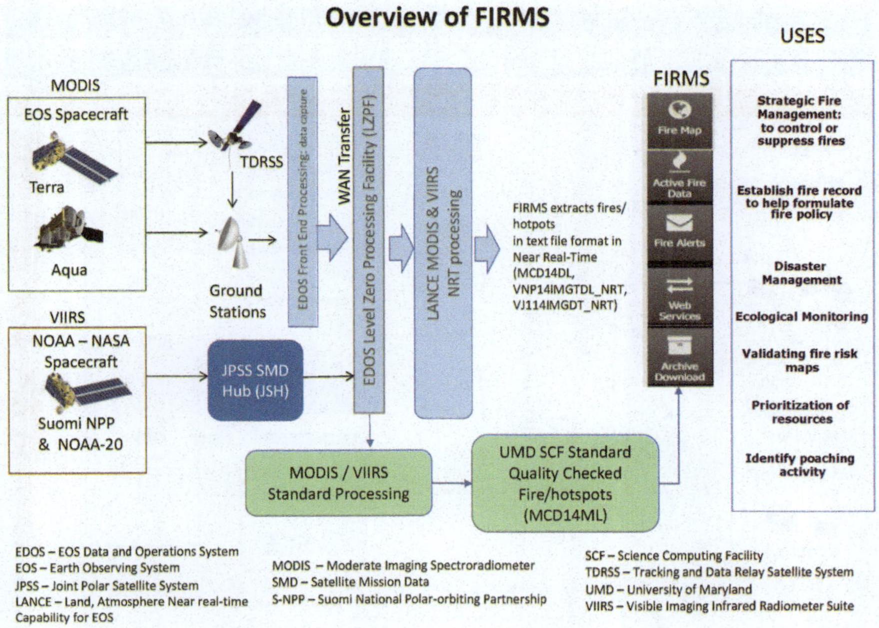

Fig. 2 Overview of FIRMS (Diane Davies, NASA-LANCE Operation Manager)

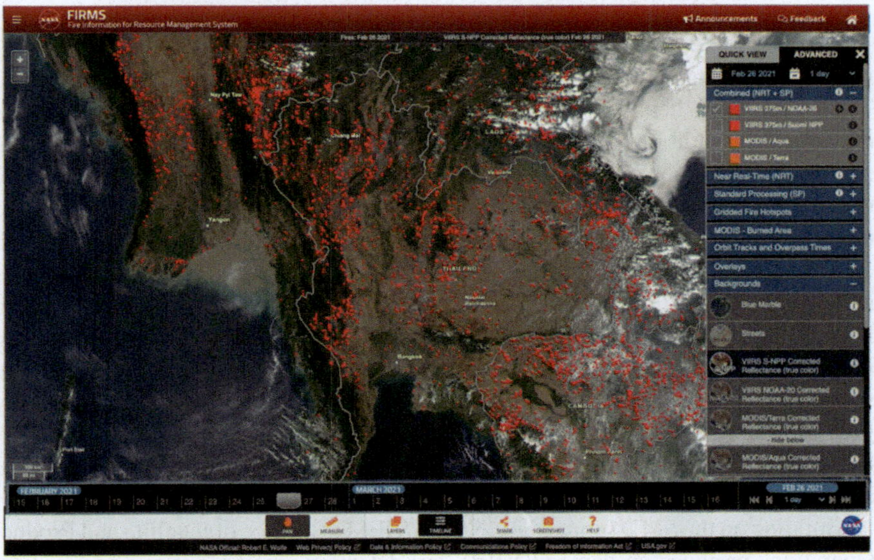

Fig. 3 FIRMS fire map showing VIIRS 375 m active fire detections (in red) from NOAA-20 (https://go.nasa.gov/3sRsP3b)

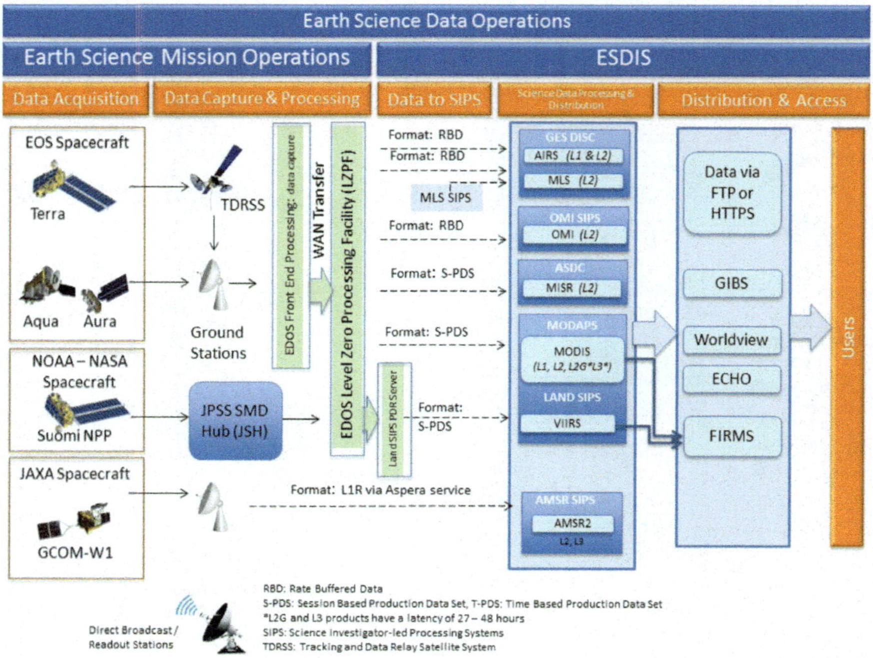

Fig. 4 Overview of LANCE (Diane Davies, NASA-LANCE Operation Manager)

northern (Svalbard/Norway and Fairbanks/Alaska) and southern (McMurdo and Troll/Antarctica) hemispheres, providing Stored Mission Data (SMD) downlink capabilities in addition to telemetry and commanding operations. From the receiving stations, data are relayed to a primary processing node located at the NOAA Satellite Operations Facility (NSOF) in Suitland/Maryland—USA, where mission data are converted into raw, sensor, and environmental products for distribution and archival. A more stringent 80-min data latency makes the JPSS mission particularly well-suited for near-real-time forest fire detection applications. In comparison, S-NPP data latency requirements are set at a somewhat higher 140-min. Data quality monitoring and calibration are performed continuously ensuring that products remain within specifications throughout the mission's lifetime. In addition to NOAA's primary ground system, a network of 15 Mbps X-band ground receiving stations found in all five continents provides access to direct broadcasting data used in support of local and regional applications.

Building on the SMD source data, NOAA/NESDIS Office of Satellite and Product Operations provides on-demand VIIRS fire detection email alerts to various users in Thailand as well as neighboring Upper ASEAN member states. Users receive country-specific fire detection data every 12 h or less, including location, timing, and intensity (fire radiative power [FRP]) information packaged in a GIS-friendly format (comma-separated value [csv], KML). Both S-NPP and NOAA-20 data were being served at the time of writing.

4.3 VIIRS NightFires (VNF)

VIIRS NightFires (VNF) is a multispectral global fire detection product developed by the Earth Observation Group (EOG) in 2012 (Elvidge et al. 2013). EOG produces VNF at Colorado School Mines, and nightly global data are made accessible in near real time for nightly global detection of infrared emitters such as wildfire and natural gas flaring. In 2012, NASA and NOAA began collecting data with the Visible Infrared Imaging Radiometer Suite (VIIRS) from a polar orbit. The VIIRS collects a complete set of orbital strips covering the globe daytime and nighttime at near 1 km^2 resolution every 24 h over a wide range of spectral channels. The sensor specifications came from the meteorological community. Traditional satellite fire detection is performed with mid-wave infrared (MWIR) channels, and VIIRS collects that in two channels. EOG reviewed nighttime VIIRS data in April 2012 were surprised by the large number of infrared emitter detections present in daytime channels designed for measurement of reflected sunlight. This includes bands M7 and M8 in the near infrared and M10 in the shortwave infrared. These spectral bands are typical for earth observation sensors, but usually the nighttime data are not collected. The decision to collect the NIR and SWIR at night originated with a VIIRS design engineer who added them to the nighttime collection mode to mitigate for mid-wave infrared (MWIR) signal saturation anticipated for large fires. EOG realized that with the detections spread across the NIR, SWIR, and MWIR, it may be possible to analyze fire temperatures via Planck curve fitting. The original development of VNF was funded by NOAA's Joint Polar Satellite System (JPSS) proving ground program. For pixels with IR emitter detection in two or more channels, the VNF algorithm calculates temperature, source size, and radiant heat using Planck's Law and its derivatives, including Wien's displacement law and the Stefan–Boltzmann law. VNF traces its roots back to methods pioneered by Dozier (1981).

The importance of the NIR and SWIR data in VNF's success turns out to be crucial. With sunlight eliminated, the VIIRS NIR and SWIR band data act as super-detectors for sub-pixel infrared emitters. These wavelengths have some advantage over mid-wave (MWIR) and long-wave infrared (LWIR) channels used in that all the energy measured in the NIR and SWIR at night can be attributed to emitting sources present on the ground. In addition, the SWIR detection limits are quite low due to the low levels of incoming solar radiation and the signal-to-noise requirements for daytime imaging. It turns out that the SWIR data are particularly important for the detection and monitoring of gas flares as they coincide with the peak of radiant emissions from flares due to their high temperature. It has been shown that the inclusion of the shorter wavelength channels extends the IR emitter detection limits as compared to the MWIR fire products (Elvidge et al. 2019). EOG provides near-real-time VNF detection alerts for the Upper ASEAN region by email.

5 Benefits, Lessons Learned, and What Still Needs to Be Known

5.1 Benefits

The active fire monitoring systems described above are useful for daily fire and smoke haze control, management, and mitigation. They provide an overview, in near real time, of where fires occur, and the information is being used to empower citizens to monitor the government's work in fire control and funded low-cost air quality networks, which are expanding due to public concern about air quality. As mentioned earlier, air quality is a key concern that has led to the 2021 People's Clean Air Act being taken to the House of Representatives in the Thai Government.

5.2 Troubles

Transboundary smoke and haze can be seen using satellite data under the right conditions; however, as MODIS and VIIRS are both optical instruments, they cannot penetrate clouds or thick smoke, so it is not always possible to determine which fires cause the problem. There are also issues with simply looking at the number of active fire detectors without considering underlying vegetation types and other factors that affect smoke behavior. Apportioning blame can be politically sensitive. Slow delivery of active fire information to the public needs to be improved as well.

5.3 Lessons Learned

Users sometimes need help understanding the caveats and limitations of using satellite-derived active fire information. The MODIS and VIIRS instruments are on polar-orbiting satellites, meaning active fires are only detected as the satellite passes overhead. Because they are optical instruments, clouds, smoke, and thick haze can obscure active fire detection, so using several active fire hotspots detected as a Key Performance Indicator (KPI) will not take into account fires ignited when the satellites are not passing over in the early evening until late night, which will result in creating more smoke haze as fuel on the ground has higher moisture during that time period. To this end, the number of active fire hotspots detected does not reflect the total picture of smoke haze, which affects air quality. More active fires detected in other countries mean additional sources of smoke haze that can be transported long distances to another country. Factors such as wind and air pressure are involved in the movement of smoke haze. Users do not understand what an active fire 'hotspot' is and are sometimes too obsessed with active fire hotspot information, so they depend on it too much. The current active fire delivery platforms need to add reliable geostationary products. So far, The Japan Aerospace Exploration Agency (JAXA) Himawari AHI (Advanced Himawari Imager) active fire products science team is still working on that. There also needs to be culturally competent training extended to mobile applications, including multiple fires and smoke haze products and information ready to use anywhere. The linking to fire emissions such as PM2.5, CO, smoke transport, and modeling in Thailand is not robust yet. There is an ongoing collaboration among the Royal Forest Department, Office of Information Technology Administration for Educational Development (UniNet), Upper ASEAN Wildland Fire Special Research Unit (WFSRU), Forestry Research Center, Faculty of Forestry, Kasetsart University, Webster University Thailand and Chulalongkorn University to deliver fire emissions data from Copernicus Atmosphere Monitoring Service (CAMS) Global Fire Assimilation System (GFAS) to the public in easier to understand ways. The EO APIs need to be improved. Addressing transboundary fire emissions needs a better monitoring system and validation of fire emissions (fuel type, emission variables). Monitoring platforms need to include ancillary data (land cover, land use type, winds, soil moisture, fire weather, etc.). Models need to be integrated into the existing platforms for holistic fire and smoke management. Also, there needs to be a learning process from and expanding on for-profit and non-profit partnerships to address the problem.

5.4 What Still Needs to Be Known

Quicker delivery time of fire information is fundamental to fire control. In the future, additional satellites in the Earth Observing System (EOS) better than VIIRS's capabilities with faster, within 10 min, downlink time onboard are needed for more efficient forest fire control. Faster delivery time of high-quality fire products can help fire managers address the problem quickly with less damage. Improvements in geostationary thermal detection sensors and systems are needed. Further, the estimated size of burned areas (by both optical and microwave data) needs to be delivered faster than the current rate. Fire spread direction, including intensity, is needed for more efficient fire control and suppression. Online analysis tools should be developed where all the statistics of active fire, burned areas, fire emissions, and other related products can be analyzed. Also, the information should be disseminated faster on fire emissions, useful for smoke haze control and air quality management. More international collaborations on forest fire control and smoke haze management can help address the problem effectively.

6 Conclusion

Monitoring wildland fires and smoke haze using Earth Observation Satellite (EOS) has become an integral tool for monitoring active fires and smoke haze in Upper ASEAN. The control, management, and mitigation measures used in Thailand are now expanding to all countries in Upper ASEAN (Lower Mekong River Region). Access to free high-quality active fire information from NASA-FIRMS, VIIRS NightFire Alert System, and VIIRS NOAA Alert System has changed how fires are managed in the region. On their own, satellite-derived active fire information is insufficient, but the combination of satellite-derived data and traditional fire monitoring and detection methods works well. Adding advanced geostationary fire detection is needed as a next step to enable fire control teams to reach fires as quickly as possible.

References

Adam, M.G., and R. Balasubramanian. 2021. Black carbon emissions from biomass burning in Southeast Asia—A review. In *Biomass burning in South and Southeast Asia*, 95–108.

Akaakara, S. 2001. *Forest fire control in Thailand*. Bangkok, Thailand: Forest Fire Control Office, Royal Forest Department.

Albar, I., I. Jaya, B.H. Saharjo, B. Kuncahyo, and K.P. Vadrevu. 2018. Spatio-temporal analysis of land and forest fires in Indonesia using MODIS active fire dataset. In *Land-atmospheric research applications in South and Southeast Asia*, 105–127. Cham: Springer.

Asian Forest Cooperation Organization (AFoCO). 2019. *Forest fire management short training course's country reports*. Hmawbi Township, Myanmar: The AFoCO Regional Education and Training Center (RETC) (Unpublished).

Asian Forest Cooperation Organization (AFoCO). 2020. *Forest fire management short training course's country reports*. Hmawbi Township, Myanmar: The AFoCO Regional Education and Training Center (RETC) (Unpublished).

Biswas, S., K.P. Vadrevu, Z.M. Lwin, K. Lasko, and C.O. Justice. 2015. Factors controlling vegetation fires in protected and non-protected areas of Myanmar. *PLoS One* 10 (4): e0124346.

Cahoon, D.R. Jr., B.J. Stocks, M.E. Alexander, B.A. Baum, and J.G. Goldammer. 2000. Wildland fire detection from space: Theory and application. In *Biomass burning and its inter-relationships with the climate system*, ed. by J.L. Innes, M.M. Verstraete and M. Beniston, 151–169. Advances in Global Change Research Series, ed. by M. Beniston. Dordrecht and Boston: Kluwer Academic Publishers.

Charuppat, T. 2001. *Application of remote sensing for forest fires monitoring in Thailand*. Bangkok, Thailand: Royal Forest Department.

Copernicus Atmospheric Monitoring Services (CAMS). 2020. *Tropical fire season in the northern hemisphere: How did 2020 compare to previous years*? https://atmosphere.copernicus.eu/tropical-fire-season-2020.

Department of National Park, Wildlife and Plants Conservation (DNP). 2016. *Thailand's VIIRS active fire field validation report*. Bangkok, Thailand (Unpublished).

Department of National Park, Wildlife and Plants Conservation (DNP). 2020. *Forest fire annual report*. Bangkok, Thailand (in Thai). https://www.dnp.go.th/forestfire/.

Dozier, J. 1981. A method for satellite identification of surface temperature fields of subpixel resolution. *Remote Sensing of Environment* 11: 221–229.

Elvidge, C.D., M. Zhizhin, F.C. Hsu, and K.E. Baugh. 2013. VIIRS nightfire: Satellite pyrometry at night. *Remote Sensing* 5 (9): 4423–4449.

Elvidge, C.D., M. Zhizhin, K. Baugh, F.C. Hsu, and T. Ghosh. 2019. Extending nighttime combustion source detection limits with short wavelength VIIRS data. *Remote Sensing* 11 (4): 395–406.

Elvidge, C.D., M. Zhizhin, K. Baugh, and F.C. Hsu. 2021. Identification of smoldering peatland fires in Indonesia via triple-phase temperature analysis of VIIRS nighttime data. In *Biomass burning in South and Southeast Asia*, 25–38. Boca Raton: CRC Press.

Global Fire Monitoring Center. 2020. *GFMC selected examples for imagery information sources: An imagery interpretation aid*. https://www2.fire.uni-freiburg.de/photos/satex/sele_ex_2.htm.

Hayasaka, H., and A. Sepriando. 2018. Severe air pollution due to peat fires during 2015 super El Nino in Central Kalimantan, Indonesia. In *Land-atmospheric research applications in South and Southeast Asia*, 105–127. Cham: Springer.

Inoue, Y. 2018. Ecosystem carbon stock, atmosphere, and food security in slash-and-burn land use: A geospatial study in mountainous region of Laos. In *Land-atmospheric research applications in South and Southeast Asia*, 641–665. Cham: Springer.

Ismi, N. 2019. *Peatland fires rage through Indonesia's Sumatra Island*. https://news.mongabay.com/2019/11/peat-forest-fires-indonesia-sumatra-photos/.

Lasko, K., and K. Vadrevu. 2018. Improved rice residue burning emissions estimates: Accounting for practice-specific emission factors in air pollution assessments of Vietnam. *Environmental Pollution* 236: 795–806.

Li, Z., Y.J. Kaufman, C. Ichoku, R. Fraser, A. Trishchenko, L. Giglio, J. Jin, and X. Yu. 2000. *A review of AVHRR-based active fire detection algorithms: Principles, limitations, and recommendations*. https://gofcgold.umd.edu/sites/default/files/docs/fire_ov.pdf.

Marks, D., and M.A. Miller. 2022. A transboundary political ecology of air pollution: Slow violence on Thailand's margins. *Environmental Policy and Governance*. https://doi.org/10.1002/eet.1976.

Oanh, K.N.T., D.A. Permadi, N.P. Dong, and D.A. Nguyet. 2018. Emission of toxic air pollutants and greenhouse gases from crop residue open burning in Southeast Asia. In *Land-atmospheric research applications in South and Southeast Asia*, 47–66. Cham: Springer.

Phairuang, W. 2021. Biomass burning and their impacts on air quality in Thailand. In *Biomass burning in South and Southeast Asia*, 21–38. Boca Raton: CRC Press.

Plodpail, A., S. Akaakara, B. Manirat, W. Parnnakapitak, and N. Songporn. 1987. *The management of forest fire control in Thailand*. Bangkok, Thailand: Natural Disaster Office, Royal Forest Department.

Saharjo, B.H., and A. Yungan. 2018. Forest and land fires in Riau Province: A case study in fire prevention policy implementation with local concession holders. In *Land-atmospheric research applications in South and Southeast Asia*, 143–169. Cham: Springer.

Schroeder, W., P. Oliva, L. Giglio, and I.A. Csiszar. 2014. The New VIIRS 375 m active fire detection data product: Algorithm description and initial assessment. *Remote Sensing of Environment* 143: 85–96. https://doi.org/10.1016/j.rse.2013.12.008.

Tanpipat, V., K. Honda, and P. Nuchaiya. 2009. MODIS hotspot validation over Thailand. *Remote Sensing* 1: 1043–1054. ISSN 2072-4292. http://doi.org/10.3390/rs1041043.

Vadrevu, K.P., T. Ohara, and C. Justice (eds.). 2021a. *Biomass burning in South and Southeast Asia: Impacts on the biosphere*, vol. 2. Boca Raton: CRC Press.

Vadrevu, K.P., T. Ohara, and C. Justice (eds.). 2021b. *Biomass burning in South and Southeast Asia: Mapping and monitoring*, vol. 1. Boca Raton: CRC Press.

Yu, B., F. Chen, B. Li, L. Wang, and M. Wu. 2017. Fire risk prediction using remote sensed products: A case of Cambodia. *Photogrammetric Engineering & Remote Sensing* 83 (1): 19–25.

Detecting Vegetation Regrowth After Fires in Small Watershed Settings Using Remotely Sensed Data and Local Community Participation Approach

Thaworn Onpraphai, Attachai Jintrawet, Angkana Somsak, Suprapat Khuenjai, Pong Loungmoon, Bounthanh Keoboualapha, and Jun Fan

Abstract The purpose of this paper is to present an operational framework to detect vegetation regrowth after fires and implement alternative income options in small watershed settings. The framework combined remotely sensed (RS) data sets to provide spatial and temporal resolution observations and local community participation (PAR) process to detect vegetation regrowth and generate income from alternative options. THEOS satellite image was spatially analyzed with the *K*-Nearest Neighbor classification method to identify various land use types, including growing maize areas in the highlands, in 2016. The imagery data from the Sentinel-2 satellite, integrated with unmanned aerial vehicle (UAV), were also spatially explored to classify the land use types in 2019 with the techniques of *K*-Nearest Neighbor classification and visual interpretation to monitor the regrowth of vegetation on highlands. The classification accuracy was assessed by sampling coordinated positions for various land use types in 2016 and 2019, with overall accuracies and kappa statistics of 93.7%, 0.924, and 94.8%, 0.938, respectively. Thirty farmers implemented selected alternative income options to increase vegetation regrowth and eliminated all hotspots in the highlands. Compared to income from maize production in the highland, all options have significantly increased the average farm incomes from

T. Onpraphai
Department of Highland Agriculture and Natural Resources, Faculty of Agriculture, Chiang Mai University, Chiang Mai 50200, Thailand

A. Jintrawet (✉) · A. Somsak · S. Khuenjai
Center for Agricultural Resources System Research, Faculty of Agriculture, Chiang Mai University, Chiang Mai 50200, Thailand
e-mail: attachai.j@cmu.ac.th

P. Loungmoon
Highland Research and Training Center, Faculty of Agriculture, Chiang Mai University, Chiang Mai 50200, Thailand

B. Keoboualapha
Upland Agriculture Research Center, National Agriculture and Forestry Research Institution, Luang Prabang 0600, Laos

A. Jintrawet · J. Fan
Institute of Food Crops, Yunnan Academy of Agricultural Sciences, Kunming 650200, China

K. P. Vadrevu et al. (eds.), *Vegetation Fires and Pollution in Asia*,
https://doi.org/10.1007/978-3-031-29916-2_10

1159 to 2368 USD per farm per year in the community areas. Our operational framework could be practically implemented in other small watersheds in Thailand and Mekong River Basin (MRB).

Keywords Vegetation fire · Vegetation regrowth · Maize · Highland · Remotely sensed data · Community participation · Land use change

1 Introduction

Population growth has increased the pressure on natural resources of small watersheds in the Mekong River Basin (MRB) and has caused improper utilization of natural resources in different areas worldwide (Eskandari and Moradi 2020; Pech and Sunada 2008). The efforts at the national and local levels to provide and implement alternative options for better utilization of resources, with evidence of the reduction in hotspots from vegetation fires and the increase in forest regrowth, are being monitored and detected by various national agencies (Justice et al. 2013, 2015). The current monitoring and detecting method uses remote sensing data sets from various satellites (GISTDA 2016; ESA 2019). These methods provide periodic historical data sets of land uses and vegetation regrowth. The additional data sets from the participation of local communities can provide a better picture of alternative income options to reduce pressure, especially in maize producing watersheds (Prem and Mohan 1996). In addition, the success or failure design of public policies for reducing damages from fires requires an understanding of their local drivers, i.e., farmers' knowledge and livelihood opportunities and networks/institutions that meet their priorities and needs (Prasad and Badarinth 2004; Biswas et al. 2015a, b; Lasko et al. 2017; Phompila et al. 2022; Temudo et al. 2020; Thapa et al. 2021; Vadrevu et al. 2019a, b; 2021a, b, c, d, e, 2022).

This paper presents an operational framework that uses remote sensing (RS) technology with community participatory (PAR) process to better classify the land uses and detect the vegetation regrowth resulting from the intervention of public policies and measures to stop maize-growing areas on highlands. Technically, the spatial data sets from various sources and unmanned aerial vehicles (UAV) were analyzed and utilized spatially for planning and implementation with local communities. The approach of resolving vegetation and forest fire problems was explicitly conducted in a selected small watershed using the process of community participation. The new alternative income options were collaboratively identified and implemented to eliminate highland vegetation fires, formerly used for growing maize, and generate income in community areas. This framework allowed local communities and implementing agencies to collaborate and evaluate various options to prevent and mitigate vegetation fires, detect vegetation regrowth, and maintain small farms' income and livelihood.

2 Study Area

We selected our study area, a small watershed named "Yang watershed", located in Pa Leaw Luang sub-district, San-ti Suk district, Nan province (Fig. 1). The watershed was selected due to its shared border with MRB and its importance for the vegetation fire and regrowth issue. The Yang watershed covers an area of approximately 4681.6 ha. Its geography includes mountainous areas with sloping lands and mostly highlands. The elevation ranges between 257 and 737 m above mean sea level (MSL). During the past three decades, the watershed's land use has been tremendously transformed from natural forests to agricultural lands, particularly field crops, with mainly the monoculture of industrial maize (Greenpeace Southeast Asia 2020). The research site, "Yang watershed", represents a collaborative effort toward monitoring land uses and detecting vegetation regrowth using combined approaches of remotely sensed data and community participation.

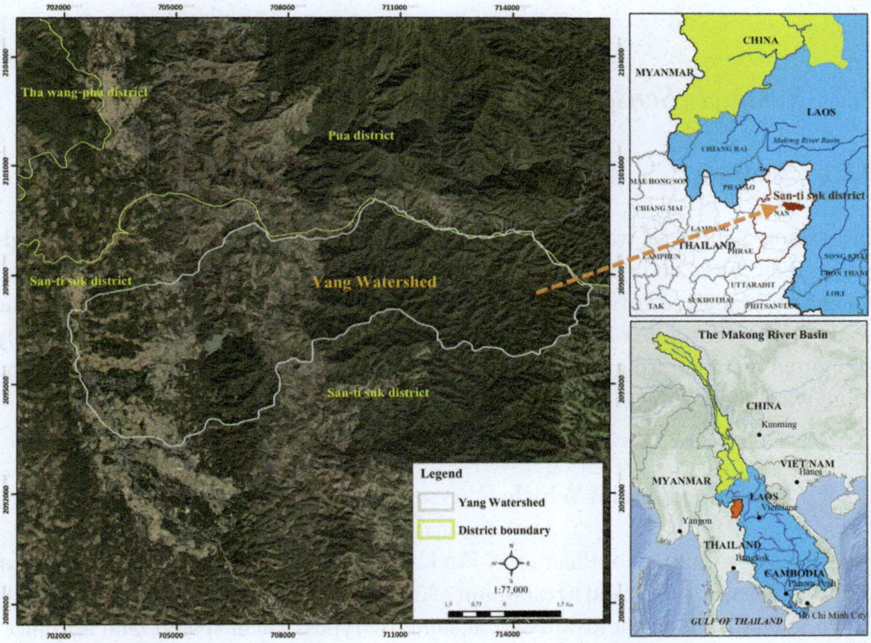

Fig. 1 "Yang watershed" in San-ti Suk district of Nan province, which it shares border with Mekong River Basin (MRB)

3 Data and Methods

3.1 *Remotely Sensed Data Used in 2016*

3.1.1 THEOS (in Thai Pronunciation "Thaichote")

THEOS satellite images, with multispectral bands (red, green, blue, and near-infrared) and a resolution of 15 m received from GISTDA (2016), were classified to identify land use types in the Yang watershed in November 2016. We classified THEOS images using the supervised classification, known as the *K*-Nearest Neighbor image processing method (Harrison 2018; Srivastava 2018). This task was intended to detect how the lands in the study area were spatially utilized by farmers and the community. Moreover, the classified land use types provided crucial information to farmers and the community for a better understanding of their performance and activities concerning natural resources and agricultural lands.

3.2 *Remotely Sensed Data Used in 2019*

3.2.1 Sentinel-2

In addition, Sentinel-2 was also a source of satellite images, with multispectral bands (red, green, blue, and near-infrared) and a resolution of 10.0 m (ESA 2019). The images were analyzed to identify various land use types and vegetation regrowth in the study area in November 2019. The information on land use types was academically supported by monitoring and comparing the land use and vegetation regrowth in the Yang watershed with collaborative activities under the process of PAR.

3.2.2 Unmanned Aerial Vehicle (UAV or Drone)

In this study, a multi-rotor Phantom 4 Pro UAV, with multispectral bands (red, green, blue, and near-infrared) and a resolution of 2.0 m (DJI 2019), was also systematically flown to take a set of photos of various land use types. Then, spatial land use information obtained from the UAV was geographically integrated to confirm precisely the 30 collaborative farms and land use types from the Sentinel-2 satellite images as the whole land use information of the watershed in 2019.

3.3 Approach

3.3.1 In 2016

With the RS data sets and the PAR process, the actors were collaboratively planning using 2016 land use types data sets from the RS procedure to design new alternative income options to be implemented in highlands and community areas (Fig. 2). The options on the highlands were based on agroforestry principles, including reforestation and new income alternatives. In the community areas, options were implemented to achieve food security and generate additional income to reduce burning activities in the highlands.

Based on the prevention of vegetation fires policy and regulations, the PAR process was inaugurated and implemented by Pa Leaw Luang sub-district Local Administrative Organization (LAO). The PAR process was twofold (1) to transform the former growing maize areas on highlands into suitable economic fruit trees mixed with forests, based on agroforestry principles and (2) to generate new income from alternative options in the community areas. LAO staff organized a series of meetings with the communities in the Yang watershed to obtain information from the communities (Creighton 2005). With land use types data, classified from THEOS satellite image in 2016 together with the vegetation fire records from the highlands by the LAO, the meetings have learned together about deforestation, growing maize, and the vegetation fires in the highlands. The meetings concluded with 30 representative farmers with maize plots on the highland (Fig. 3) who volunteered to implement selected options in highlands and community areas.

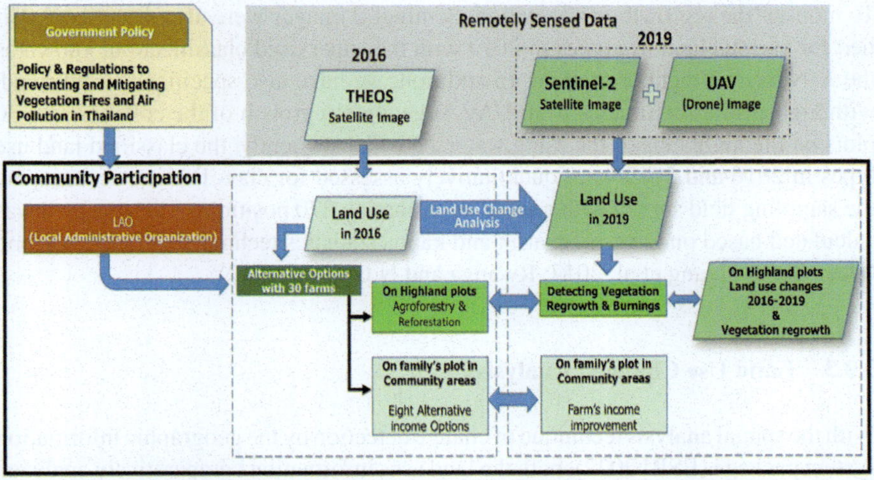

Fig. 2 Operational framework by combining remotely sensed data and community participation in detecting regrowth from 2016 to 2019 seasons

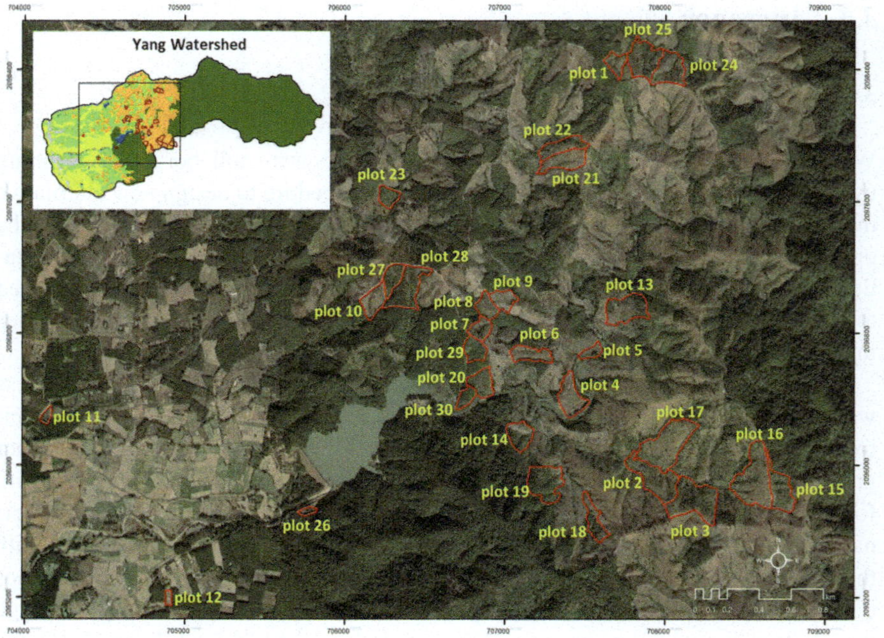

Fig. 3 Boundaries of thirty maize plots of collaborative farmers in highlands of Yang watershed designated for the implementation of agroforestry and reforestation option

3.3.2 In 2019

To monitor the regrowth of vegetation, Sentinel-2 images were also spatially classified for various land use types in 2019 with the supervised classification, known as the *K*-Nearest Neighbor method. In addition, we have also specifically confirmed, with high-resolution images from UAV, vegetation regrowth of the collaborative 30 plots on the highlands of the Yang watershed. Consequently, the classified land use types in 2016 and 2019 were quantitatively assessed for classification accuracy by the sampling field survey with a set of coordinating 270 positions. The accuracy was calculated based on the error matrix and kappa statistics techniques (Congalton and Green 2009; Huang et al. 2017; Rwanga and Ndambuki 2017).

3.3.3 Land Use Changed Analysis

With the spatial analysis technique of change detection by the geographic information system package (ESRI 2022), both the land uses information comparatively analyzed their transformations in 2016 and 2019 (Chen 2002; Zhou et al. 2011; Jeon et al. 2013; Kim 2016; Ayele et al. 2018; Tewabe and Fentahun 2020). Finally, the transition matrix was comprehensively presented and explained the land use transformations in the Yang watershed (FAO 1995; Zhang et al. 2017; Hu et al. 2019).

4 Results and Discussion

4.1 Community Participation

Our operational framework yielded 30 volunteered farms to implement selected options in highland plots and community area plots (Table 1). In 2016, land areas under maize in the highlands ranged from 0.3 to 6.3 ha per farm, with an average of 2.5 ha per farm. Various burning activities for maize field clearing and land preparation ranged from 1.0 to 13.0 burnings per plot, with an average of 5.0 burnings per plot. In terms of annual income, our PAR process in 2016 revealed that income per farm ranged from 138.9 to 2916.7 USD, with an average of 1159.0 USD from one year of maize-growing activities in their highland plots. All thirty farms had no additional income from other alternative options in community areas.

In 2019, with inputs from RS data sets and the PAR process, all thirty farms successfully implemented agroforestry options in their highland plots (Fig. 4). The selected fruit trees under the agroforestry option will yield some marketable products in the near future based on the fruit tree life cycle. The most encouraging outcome of the process was that there were no burning activities in all thirty highland plots belonging to thirty farms.

Again in 2019, the PAR process successfully implemented options to generate additional income from family plots within the urban and built-up area (Table 1). There were eight alternative income options, namely (a) organic vegetable, (b) chicken raising, (c) pig raising, (d) fish pond, (e) growing banana, (f) growing mango, (g) bamboo handicraft, and (h) food processing and preservation (Fig. 5). Organic vegetable, chicken raising, pig raising, and fish pond options were implemented by 30, 10, 10, and 22 farms. However, growing bananas, mango, bamboo handicrafts, and food processing and preservation were adopted by only five farms due to longer growing seasons and high learning curves. In 2019, we found that 2, 12, 13, and 3 farms adopted one, two, three, and four alternative income options, respectively. Additional income generated from these adopted options (1–4 options per farm) averaged 164; 895, 1121, and 3545 USD per farm, respectively.

All thirty farms generated an annual income ranging from 939 to 5867 USD per farm, with an average of 2368 USD per farm. This was approximately 200% higher than the average farm's income in 2016. On average, income from pig raising, organic vegetables, chicken, and fish ponds yielded 1957; 1079; 664; and 257 USD per farm per year. Our results underlined the potential of alternative income options and local markets to support vegetation regrowth similar to crop-livestock integration reported by Paul et al. (2022).

The policies and regulations for preventing vegetation fires, particularly in northern Thailand, were issued by the central Thai Government. The Ministry of Interior and the Ministry of Natural Resources and Environment initially proposed these policy documents. To implement the policy, the provincial-level offices under each Ministry convened a series of meetings to coordinate and plan to communicate to the local level units, i.e., Local Administrative Organizations (LAOs) (Chiang Mai

Table 1 Comparison of annual incomes from maize in 2016 and from alternative income options in 2019

	In 2016		In 2019					
	Number of burning activity (times/year)	Annual income from former growing maize on highland (US$/farm/year)	Annual income from new alternative options in community area (US$/farm/year)					
			(a)* org-veg	(b)* chicken	(c)* pig	(d)* fish	(e–h)* etc.	Total income
# 1 (1.3 ha)	3	601.9	1365.4	×	×	146.7	×	1512.1
# 2 (4.6 ha)	9	2129.6	1527.5	×	1546.7	176.0	×	3250.2
# 3 (5.6 ha)	11	2592.6	1298.3	1184.0	×	146.7	×	2629.0
# 4 (2.8 ha)	6	1296.3	1536.6	×	1600.0	176.0	1600.0	4912.6
# 5 (0.8 ha)	2	370.4	624.2	592.0	×	234.7	×	1450.9
# 6 (2.0 ha)	4	925.9	939.4	×	×	×	×	939.4
# 7 (1.4 ha)	3	648.2	775.6	×	1280.0	×	×	2055.6
# 8 (1.5 ha)	3	694.4	1150.2	296.0	×	×	×	1446.2
# 9 (1.5 ha)	3	694.4	1434.1	×	×	234.7	×	1668.8
# 10 (2.1 ha)	4	972.2	1542.4	×	1760.0	205.3	×	3507.7
# 11 (0.5 ha)	1	231.5	1171.8	×	×	293.3	×	1465.2
# 12 (0.3 ha)	1	138.9	540.2	444.0	×	176.0	×	1160.2
# 13 (3.6 ha)	7	1666.7	982.8	×	×	234.7	1200.0	2417.5
# 14 (2.1 ha)	4	972.2	1286.2	×	×	×	×	1286.2
# 15 (3.1 ha)	6	1435.2	1384.3	×	×	264.0	×	1648.3
# 16 (6.3 ha)	13	2916.7	626.3	×	2133.3	293.3	×	3053.0
# 17 (6.3 ha)	13	2916.7	1581.5	1184.0	×	352.0	×	3117.5
# 18 (2.0 ha)	4	925.9	805.3	×	×	440.0	×	1245.3

(continued)

Table 1 (continued)

	In 2016		In 2019					
	Number of burning activity (times/year)	Annual income from former growing maize on highland (US$/farm/year)	Annual income from new alternative options in community area (US$/farm/year)					
			(a)* org-veg	(b)* chicken	(c)* pig	(d)* fish	(e–h)* etc.	Total income
# 19 (3.8 ha)	8	1759.3	1418.2	444.0	×	146.7	×	2008.9
# 20 (2.0 ha)	4	925.9	750.2	×	×	264.0	×	1014.2
# 21 (2.7 ha)	5	1250.0	616.2	740.0	2506.7	205.3	×	4068.2
# 22 (2.2 ha)	4	1018.5	1018.2	666.0	×	234.7	×	1918.9
# 23 (1.2 ha)	2	555.6	530.2	592.0	×	293.3	×	1415.6
# 24 (3.1 ha)	6	1435.2	1213.5	×	×	352.0	×	1565.5
# 25 (3.6 ha)	7	1666.7	1107.0	×	2720.0	440.0	1600.0	5867.0
# 26 (0.4 ha)	1	185.2	765.8	×	1440.0	352.0	×	2557.8
# 27 (1.7 ha)	3	787.0	674.0	×	2560.0	×	×	3234.0
# 28 (3.9 ha)	8	1805.6	1494.3	500.2	×	×	×	1994.5
# 29 (1.7 ha)	3	787.0	662.4	×	2026.7	×	1400.0	4089.1
# 30 (1.0 ha)	2	463.0	1540.0	×	×	×	1000.0	2540.0

* *Remark* (a) organic vegetable option, (b) chicken option, (c) pig farm, (d) fish option, (e–h) other option; such as (e) growing banana, (f) growing mango, (g) bamboo handicraft, and (h) food processing, (×) not implemented

Agricultural Extension Office 2022). Each LAO has to plan and manage forest and agricultural lands for forest conservation and fire prevention objectives. Networks of volunteers were established to operate in a specified zone (Center of Disaster Prevention and Mitigation of Thailand 2020). LAOs in Nan province and other provinces in Northern Thailand can learn from and implement our operational framework to better plan and manage their forests and other alternative income options, as demonstrated in the Yang watershed.

Regarding the information from remotely sensed data, the 30 collaborative farmers learned that their maize-growing activities in the highland areas caused vegetation fires, as detected and reported by GISTDA (2016). Using RS data from THEOS

Fig. 4 Examples of activities under agroforestry (**a**, **b**), reforestation (**c**) options, with small water pond (**d**) to supply irrigation water for vegetation regrowth in thirty collaborative plots on highlands of Yang watershed *Photo credit* Mr. Natchaphol Phomkham (Kame)

Fig. 5 Examples of activities under new income generation options on plots within community areas implemented by thirty collaborative farms *Photo credit* Mr. Natchaphol Phomkham (Kame)

in 2016, the PAR process yielded an implementation plan for alternative income generation options. These options included agroforestry and reforestation in formerly maize-growing plots on the highlands. Moreover, many tiny water ponds were rapidly constructed to support young trees under newly established agroforestry plots. In addition, some abandoned plots on the highland had been wholly changed for vegetation regeneration by reforestation activities. These activities evidently resulted in vegetation regrowth in 2019 and could be detected using the remotely sensed data of Sentinel-2 integrated with the high-resolution UAV images.

Concerning the policy and regulation, the LAO of Pa Leaw Luang sub-district played a major supervision role in communicating with the public to understand the situations of vegetation fires and alternative income options. The situations were confirmed by the RS data sets in 2016 and 2019. In 2016, RS data sets provided common data that all participants shared and realized how to collaborate to stop the vegetation fires. In 2019, RS data sets confirmed the vegetation regrowth in the highlands. This is a practical and innovative model to prevent vegetation fires and detect regrowth with the community participation process in a small watershed.

Regarding the PAR, the thirty farmers had learned and experienced the negative effects of burning activities on maize-growing areas, i.e., deforestation, vegetation fires, and poor health from agricultural chemical uses. Therefore, PAR provided a platform for technical staff to design new income alternatives (Prem and Mohan 1996). Subsequently, the farmers have implemented and evaluated new suitable alternatives within the community areas, i.e., organic vegetables, banana tree, chicken raising, pig raising, and fish ponds. As a result, the new alternatives have significantly generated higher incomes for all thirty farms.

4.2 *Land Use and Land Use Changes*

4.2.1 Land Uses in 2016

As a result, with THEOS satellite imagery in 2016, the land use types of the Yang small watershed were spatially classified and identified into six classes. In 2016, the major land use in the small watershed was forest (F) of 2387.6 ha (51.0%), while field crop and maize (A2) and agroforestry (A3) covered 817.4 ha (17.5%) and 789.6 ha (16.9%), respectively. The rest of the land use types were paddy rice (A1), urban and build-up (U), and water body (W), with total areas of 541.7 ha (11.6%), 97.5 ha (2.1%), and 47.8 ha (1.0%), respectively (Table 2). The results of classified land use types in 2016 obtained a high overall accuracy of 93.7% and kappa statistics of 0.924.

4.2.2 Land Uses in 2019

In 2019, with Sentinel-2 satellite images integrated with UAV photos, the land use types in the Yang watershed were spatially classified into six classes. The main land use type was still forest (F), with a total area of 2715.2 ha (58.0% of the Yang watershed total area), an increase of 8% compared with 2016. The second main land use had become agroforestry (A3), with a total area of 914.0 ha (19.5%).

Table 2 Land use types of Yang watershed classified with THEOS satellite image in 2016 and Sentinel-2 satellite image integrated with UAV images in 2019

Land use types	Area in 2016 (THEOS)		Area in 2019 (Sentinel-2 and UAV)	
	(ha)	(%)	(ha)	(%)
Paddy rice (A1)	541.7	11.6	476.1	10.2
Field crop and maize (A2)	817.4	17.5	413.4	8.8
Agroforestry (A3)	789.6	16.9	914.0	19.5
Forest (F)	2387.6	51.0	2715.2	58.0
Urban and built-up (U)	97.5	2.1	98.1	2.1
Water body (W)	47.8	1.0	64.8	1.4
Total	4681.6	100.0	4681.6	100.0

Finally, the field crop and maize (A2) had significantly decreased, with a total area of 413.4 ha (8.8%). In 2019, A2 land use types decreased some 404 ha or 8.7% reduction compared to 2016. The other land use types were paddy rice (A1), urban and build-up (U), and water body (W), with areas of 476.1 (10.2%), 98.1 (2.1%), and 64.8 ha (1.4%), respectively (Table 2 and Fig. 6). In 2019, land use types classification assessment obtained high overall accuracy and kappa statistics of 94.8% and 0.938, respectively.

UAV is an excellent advantageous tool providing accurate data with imagery and photo above the crop to farmers. It simply offers digital imagery data, without the need for human scouts, particularly in remote mountainous areas, with rapidity, accuracy, and cost-effectiveness (Krishna 2018). However, applications of UAV-based RS in agriculture and related industries require consideration of various issues, including technological adaptations, high initial investment cost, adverse weather conditions,

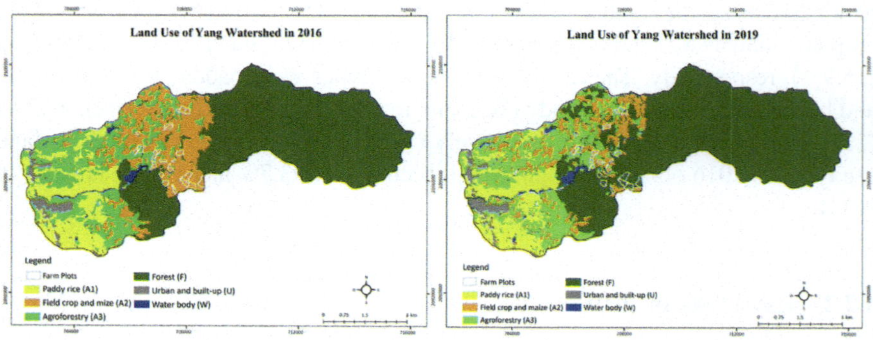

Fig. 6 Land use types and boundary of thirty collaborative plots in highland of Yang watershed classified from THEOS satellite images in 2016 and from Sentinel-2 satellite images integrated with UAV images in 2019

communication failures, and data interpretation for a better understanding of policy and regulations (Amarasingam et al. 2022).

4.2.3 Land Use Changed from 2016 to 2019

With the analysis of land use changes between 2016 and 2019, the land uses had mainly changed between forests (F) and crops, particularly field crops and maize (A2) areas. In 2019, the forests (F) had significantly recovered, with areas of 2715.2 ha (58.0%). The recovery of forests promptly increased from 2016 to 2019, with incremental areas of 327.5 ha (34.9%). The increasing forest areas had resulted from field crops and maize, agroforestry, paddy rice, water body, and urban and build-up, with the turning over areas of 332.3, 23.6, 2.2, 3.4, and 1.0 ha, respectively. Meanwhile, field crop and maize areas (A2) had collaboratively decreased from 817.4 ha (17.5%) in 2016 to 413.5 ha (8.8%) in 2019.

It showed that forest (F) areas had been converted from field crops and maize (A2), and some 332.3 ha were detected in 2019. Furthermore, areas under agroforestry (A3) had increased, mostly, from field crops and maize (A2), with an increase of 914.0 ha (19.5%). Agroforestry (A3) and forests (F) had increased from 789.6 ha (16.9%) and 2387.6 ha (51.0%) in 2016 to 914.0 ha (19.5%) and 2715.2 ha (58.0%) in 2019, respectively. The field crop and maize (A2) land use type in 2016 had also been detected as agroforestry, paddy, water body, and urban and built-up land use types in 2019, with the converted areas of 165.2, 4.1, 3.2, and 0.7 ha, respectively. Moreover, it was also found that the water body had increased relative to increasing agroforestry (A3), with areas of 47.8 ha (1.0%) in 2016 to 64.9 ha in 2019. The result of land use change was confirmed in each land use class using the transition matrix in Table 3. Selected examples of changes in vegetation regrowth of individual plots in the highlands from 2016 (a, b) to 2019 (c, d) were depicted and captured by UAV (Fig. 7). Our results agreed with previous studies in the upper Lancang-Mekong River Basin. For example, Jiang et al. (2021) showed that a light UAV remote sensing data set is an attractive choice for investigating vegetation in the Lantsang cascade reservoir system. In particular, small watershed development projects in northern Laos, where hydroelectric dams have been built (Sivongxay et al. 2017). It can be implemented to enhance international cooperation at the level of institutions or countries (Hognogi et al. 2021).

5 Conclusion

We developed an operational framework to monitor land use and vegetation regrowth changes in a small watershed. The framework combined remotely sensed data (RS) and community participation (PAR) to detect the former growing maize areas and vegetation regrowth, which significantly reduces vegetation fire risk. The framework was collaboratively implemented in Northern Thailand's Yang watershed, Nan

Table 3 Matrix of land use changed analysis of Yang watershed from 2016 to 2019 by community participation

Land use classes	Area in 2016		Changed area in 2019					
	(ha)	(%)	A1	A2	A3	F	U	W
Paddy rice (A1)	541.7	11.6	**446.3**	22.8	51.8	2.2	9.8	8.8
Field crop and maize (A2)	817.4	17.5	4.1	**311.9**	165.2	332.3	0.7	3.2
Agroforestry (A3)	789.6	16.9	16.1	65.7	**666.7**	23.6	7.9	9.7
Forest (F)	2387.6	51.0	1.2	11.1	17.0	**2352.6**	0.2	5.5
Urban and built-up (U)	97.5	2.1	5.8	1.4	9.9	1.0	**79.4**	0.1
Water body (W)	47.8	1.0	2.7	0.6	3.3	3.4	0.1	**37.6**
Total areas	4681.6	100.0	476.2	413.5	914.0	2715.2	98.1	64.9
(%)			10.2	8.8	19.5	58.0	2.1	1.4
Area changes (ha)	938.9		– 65.6	– 403.9	124.4	327.6	0.5	17.0
(%)	100.0		– 7.0	– 43.0	13.2	34.9	0.1	1.8

'Bold' face letters represent area remained as the same land use class in 2016 and 2019. For example, A1 class in 2016 was 541.7 ha and in 2019 some 446.3 ha remained as A1

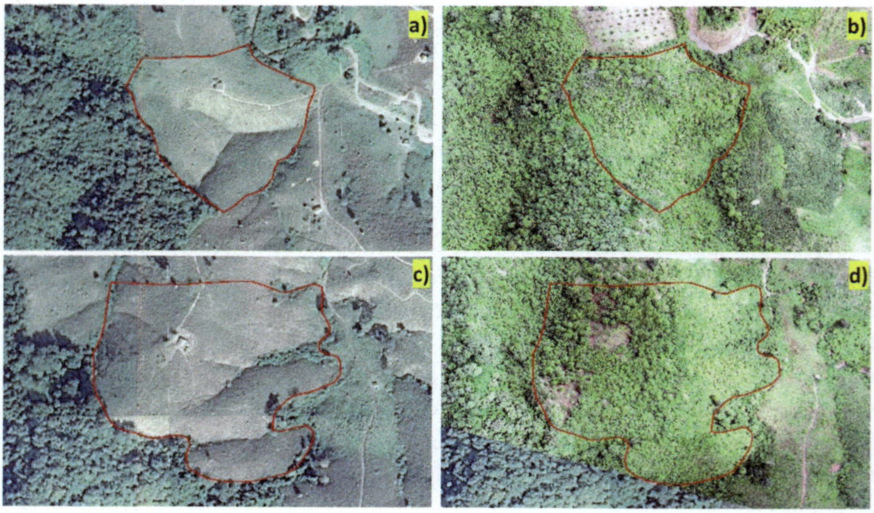

Fig. 7 Examples of vegetation regrowth monitored between 2016 and 2019 seasons of plot 14 and plot 19 in highlands of Yang watershed. **a** Plot 14 in 2016 and **b** Plot 14 in 2019. **c** Plot 19 in 2016 **d** Plot 19 in 2019. The red line indicates the plot boundary

province. Remotely sensed data allowed bird-eye views of changes over space (30 farms) and time (2016–2019 growing seasons). The PAR process supported the community in developing mainly two major transitional options: (1) transforming the former growing maize areas to new regrowth of vegetation plots based on agroforestry principles and (2) designing and practicing new alternative income options within the community areas to compensate incomes from not growing maize on highlands, such as organic vegetable farm, banana tree, chicken farm, pig farm, fish farm, and so on. As a result, the PAR process reduced burning the growing maize areas and increased vegetation regrowth on highlands and incomes from alternative options on plots in the community areas. The combined RS and PAR methods allowed actors to monitor changes in small watershed settings efficiently. They could be extended effectively to other watersheds in Northern Thailand and Mekong River Basin (MRB) to resolve the vegetation fires and deforestation situations.

Acknowledgements We appreciate the financial support from the Thailand Science Research and Innovation (TSRI), formerly Thailand Research Fund (TRF), during 2016–2019. We are very grateful to the spirit of learning and kindness of giving by 30 collaborative farmers, in particularly Mr. Natchaphol Phomkham (Kame), the leader of the farmer group during the project period.

References

Amarasingam, N., A.S. Ashan Salgadoe, K. Powell, L.F. Gonzalez, and S. Natarajan. 2022. A review of UAV platforms, sensors, and applications for monitoring of sugarcane crops. *Remote Sensing Applications: Society and Environment* 26: 100712. https://doi.org/10.1016/j.rsase.2022.100712.

Ayele, G.T., A.K. Tebeje, S.S. Demissie, M.A. Belete, M.A. Jemberrie, W.M. Teshome, D.T. Mengistu, and E.Z. Teshale. 2018. Time series land cover mapping and change detection analysis using geographic information system and remote sensing, Northern Ethiopia. *Air, Soil and Water Research* 11: 1–18. https://doi.org/10.1177/1178622117751603.

Biswas, S., K.D. Lasko, and K.P. Vadrevu. 2015a. Fire disturbance in tropical forests of Myanmar—Analysis using MODIS satellite datasets. *IEEE Journal of Selected Topics in Applied Earth Observations and Remote Sensing* 8 (5): 2273–2281.

Biswas, S., K.P. Vadrevu, Z.M. Lwin, K. Lasko, and C.O. Justice. 2015b. Factors controlling vegetation fires in protected and non-protected areas of Myanmar. *PLoS ONE* 10 (4): e0124346.

Center of Disaster Prevention and Mitigation of Thailand. 2020. *Protecting and resolving forest fires, smog and particulate matters ($PM_{2.5}$) in 2020–2021*, 12. Department of Disaster Prevention and Mitigation, Ministry of Interior, Krung Thep Maha Nakhon (Bangkok), Thailand (in Thai).

Chen, X. 2002. Using remote sensing and GIS to analyse land cover change and its impacts on regional sustainable development. *International Journal of Remote Sensing* 23 (1): 107–124. https://doi.org/10.1080/01431160010007051.

Chiang Mai Agricultural Extension Office. 2022. *The report of crop production data; Estimating agricultural residues*, 13. Chiang Mai, Thailand (in Thai).

Congalton, R.G., and K. Green. 2009. *Assessing the accuracy of remotely sensed data; Principles and practices*, 2nd ed., 105–110. Boca Raton: CRC Press; Taylor & Francis Group.

Creighton, J.L. 2005. *The public participation handbook: Making better decisions through citizen involvement*, 102–137. Jossey-Bass, A Wiley Imprint.

DJI. 2019. Phantom 4 Pro; Specs. Shenzhen DJI Sciences and Technologies Ltd. Available at: https://doi.org/10.1579/0044-7447(2008)37[219:pganpi]2.0.co;2.

ESA. 2019. *Sentinel products*. The European Space Agency (ESA); Sentinel Online. Available at: https://sentinel.esa.int/web/sentinel/sentinel-data-access/sentinel-products. Accessed: October 30, 2019.

Eskandari, S., and A. Moradi. 2020. Mapping the land uses and analysing the landscape elements in south-western Iran: Application of Landsat-7 field data and landscape metrics. *International Journal of Conservations Science* 11 (2): 557–564.

ESRI. 2022. Change detection in ArcGIS Pro. ArcGIS Pro; Resources. Available at: https://pro.arcgis.com/en/pro-app/2.8/help/analysis/image-analyst/change-detection-in-arcgis-pro.htm#ESRI_SECTION1_9FF9489173C741DD95472F21B5AD8374. Accessed: March 1, 2022.

FAO. 1995. Forest resources assessment 1990, Global Synthesis, Food and Agriculture Organization of the United Nations, Rome.

GISTDA, 2016. THEOS satellite image; our service; Satellite—Thaichote (THEOS). Data Service Center, Geo-Informatics and Space Technology Development Agency (GISTDA), Krung Thep Maha Nakhon (Bangkok), Thailand. Available at: https://www2.gistda.or.th/main/th/satellite. Accessed: January 28, 2022.

Greenpeace Southeast Asia. 2020. *Maize, land use change, and transboundary haze pollution*, 16. Available at: https://www.greenpeace.org/southeastasia/publication/4117/maize-land-use-change-and-transboundary-haze-pollution/. Accessed: January 28, 2022.

Harrison, O. 2018. Machine learning basics with the *K*-nearest neighbors algorithm. Available at: https://towardsdatascience.com/machine-learning-basics-with-the-k-nearest-neighbors-algorithm-6a6e71d01761. Accessed: January 31, 2022.

Hognogi, G.-G., A.-M. Pop, A.-C. Marian-Potra, and T. Someşfălean. 2021. The role of UAS–GIS in digital era governance. A systematic literature review. *Sustainability* 13(19): 11097. https://doi.org/10.3390/su131911097.

https://www.dji.com/phantom-4-pro/info#specs. Accessed: January 28, 2022.

https://www.fao.org/3/x5669e/x5669e00.htm#Contents. Accessed: February 26, 2022.

Hu, Y., Batunacun, L. Zhen, and D. Zhuang. 2019. Assessment of land-use and land-cover change in Guangxi. *Scientific Reports* 9(1), 2189. https://doi.org/10.1038/s41598-019-38487-w.

Huang, D., S. Xu, J. Sun, S. Liang, W. Song, and Z. Wang. 2017. Accuracy assessment model for classification result of remote sensing image based on spatial sampling. *Journal of Applied Remote Sensing* 11 (4): 48–59.

Jeon, S.B., P. Olofsson, and C.E. Woodcock. 2013. Land use change in New England: A reversal of the forest transition. *Journal of Land Use Science* 9 (1): 105–130. https://doi.org/10.1080/1747423X.2012.754962.

Jiang, W., L. Liu, H. Xiao, S. Zhu, W. Li, and Y. Liu. 2021. Composition and distribution of vegetation in the water level fluctuating zone of the Lantsang cascade reservoir system using UAV multispectral imagery. *PLoS ONE* 16 (3): e0247682. https://doi.org/10.1371/journal.pone.0247682.

Justice, C., C. Ivan, B. Luigi, K. Stefania, S. Wilfrid, G. Louis, P.V. Krishna, and R. David. 2013. Satellite monitoring and inventory of global vegetation fire. In *Vegetation fires and global change*, 261–276. A Publication of the Global Fire Monitoring Center (GFMC). Remagen: Kessel Publishing.

Justice, C., G. Gutman, and K.P. Vadrevu. 2015. NASA land cover and land use change (LCLUC): An interdisciplinary research program. *Journal of Environmental Management* 148: 4–9.

Kim, C. 2016. Land use classification and land use change analysis using satellite images in Lombok Island, Indonesia. *Forest Science and Technology* 12 (4): 183–191. https://doi.org/10.1080/21580103.2016.1147498.

Krishna, K.R. 2018. *Agricultural drones; A peaceful pursuit*, 1–91. Waretown, New Jersey, USA: Apple Academic Press, Inc.

Lasko, K., K.P. Vadrevu, V.T. Tran, E. Ellicott, T.T. Nguyen, H.Q. Bui, and C. Justice. 2017. Satellites may underestimate rice residue and associated burning emissions in Vietnam. *Environmental Research Letters* 12 (8): 085006.

Paul, B.K., C.A. Epper, D.J. Tschopp, C.T.M. Long, V. Tungani, D. Burra, L. Hok, P. Phengsavanh, and S. Douxchamps. 2022. Crop-livestock integration provides opportunities to mitigate environmental trade-offs in transitioning smallholder agricultural systems of the Greater Mekong Subregion. *Agricultural Systems* 195: 103285.

Pech, S., and K. Sunada. 2008. Population growth and natural-resources pressures in the Mekong River Basin. *Ambio* 37 (3): 219–224.

Phompila, C., V. Sihapanya, B. Chanthavong, S. Thongmanivong, P. Van Khoa, K. Noda, and K.P. Vadrevu. 2022. Agricultural use/cover changes in the Vientiane. In *Remote sensing of agriculture and land cover/land use changes in South and Southeast Asian countries*, 573–586. Cham: Springer.

Prasad, V.K., and K.V.S. Badarinth. 2004. Land use changes and trends in human appropriation of above ground net primary production (HANPP) in India (1961–98). *Geographical Journal* 170 (1): 51–63.

Prem, N.S., and P.W. Mohan. 1996. Case studies of people's participation in watershed management in Asia; Part I: Nepal, China and India. UNDP/FAO/Netherlands.

Rwanga, S.S., and J.M. Ndambuki. 2017. Accuracy assessment of land use/land cover classification using remote sensing and GIS. *International Journal of Geosciences* 4 (8): 611–622.

Sivongxay, A., R. Greiner, and S.T. Garnett. 2017. Livelihood impacts of hydropower projects on downstream communities in central Laos and mitigation measures. *Water Resources and Rural Development* 9: 46–55.

Srivastava, T. 2018. Introduction to *k*-nearest neighbors: A powerful machine learning algorithm (with implementation in Python & R). Available at: https://www.analyticsvidhya.com/blog/2018/03/introduction-k-neighbours-algorithm-clustering/. Accessed: January 31, 2022.

Temudo, M.P., J.M.C. Pereira, and D. Oom. 2020. Bio-cultural fire regions of Guinea-Bissau: Analysis combining social research and satellite remote sensing. *Applied Geography* 118. https://doi.org/10.1016/j.apgeog.2020.102203.

Tewabe, D., and T. Fentahun. 2020. Assessing land use and land cover change detection using remote sensing in the Lake Tana Basin, Northwest Ethiopia. *Cogent Environmental Science* 6 (1): 1–11. https://doi.org/10.1080/23311843.2020.1778998.

Thapa, P., K.P. Vadrevu, A. Eaturu, and S. Biswas. 2021. Vegetation fire status and management in Bhutan. In *Biomass burning in South and Southeast Asia*, 135–157. CRC Press.

Vadrevu, K., A. Heinimann, G. Gutman, and C. Justice. 2019a. Remote sensing of land use/cover changes in South and Southeast Asian Countries. *International Journal of Digital Earth* 12 (10): 1099–1102.

Vadrevu, K.P., V.K. Dadhwal, G. Gutman, and C. Justice. 2019b. Remote sensing of agriculture–South/Southeast Asia research initiative special issue. *International Journal of Remote Sensing* 40 (21): 8071–8075.

Vadrevu, K.P., A. Eaturu, S. Biswas, and C. Phompila. 2021a. An assessment of burnt area signal variations in Laos using Sentinel-1A&B datasets. In *Biomass burning in South and Southeast Asia*, 61–76. CRC Press.

Vadrevu, K.P., A. Eaturu, and S. Biswas. 2021b. Burnt area mapping in Nainital, Uttarakhand, India, using very high-resolution PlanetScope imagery. In *Biomass burning in South and Southeast Asia*, 91–103. CRC Press.

Vadrevu, K.P., T. Ohara, and C. Justice. 2021c. Biomass burning in South/Southeast Asia–Needs and priorities. In *Biomass burning in South and Southeast Asia*, 1–21. CRC Press.

Vadrevu, K.P., T. Ohara, and C. Justice, eds. 2021d. *Biomass burning in South and Southeast Asia: Impacts on the biosphere*, vol. 2. CRC Press.

Vadrevu, K.P., T. Ohara, and C. Justice, eds. 2021e. *Biomass burning in South and Southeast Asia: Mapping and monitoring*, vol. 1. CRC Press.

Vadrevu, K.P., T. Le Toan, S.S. Ray, and C. Justice. 2022. *Remote sensing of agriculture and land cover/land use changes in South and Southeast Asian Countries*. Cham: Springer. https://doi.org/10.1007/978-3-030-92365-5.

Zhang, B., Q. Zhang, C. Feng, Q. Feng, and S. Zhang. 2017. Understanding land use and land cover dynamics from 1976 to 2014 in Yellow River Delta. *Land* 6 (1): 20. https://doi.org/10.3390/land6010020.

Zhou, Q., B. Li, and Y. Chen. 2011. Remote sensing change detection and process analysis of long-term land use change and human impacts. *Ambio* 40 (7): 807–818. https://doi.org/10.1007/s13280-011-0157-1.

Long-Term Spatiotemporal Distribution of Fire Over Maritime Continent and Their Responses to Climate Anomalies

Aulia Nisa'ul Khoir, Maggie Chel Gee Ooi, and Nur Nazmi Liyana Binti Mohd Napi

Abstract Annual biomass burning (BB) through a slash-and-burn, forest fire, and open burning that occurred in the Maritime Continent (MC) has become a notable concern. The persistent occurrence of BB events has caused adverse effects not only on regional and local air quality but also negative impacts on human health and societal and environmental issues. Much research about this annual problem has been reported, particularly about its characteristics and precursor factors in specific periods or areas in MC. However, the long-term pattern and precursor factors of BB in MC have not been analyzed. Given that MC is remarkable as a region with one of the most complicated meteorology systems, this research aims to understand the spatial and temporal distribution of burning events in the long term and its relationship with climate anomalies that influence observed burning activity. Satellite-based Moderate Resolution Imaging Spectroradiometer (MODIS) products of fire hotspots and aerosol optical depth (AOD), satellite-based precipitation, and climate anomalies indices over MC were examined. Using Empirical Orthogonal Function (EOF), fast Fourier transform (FFT), and correlation method, this research assesses the spatiotemporal distributions of fire events and their response to climate anomalies over MC during 2001–2020. The result shows that the first primary EOF mode explains that the annual fire event has a homogeneous influence over a wide region of MC, which has a positive loading value. Rainfall is found to have a significant correlation and impact on fire hotspots in the same month. The El Niño and positive Dipole Mode events have increased the total hotspots significantly, while the MJO events have increased the negative fire event anomalies during the study period.

Keywords Biomass burning · Fire regimes · Rainfall · Climate anomalies

A. N. Khoir · M. C. G. Ooi (✉) · N. N. L. B. M. Napi
Centre of Tropical Climate Change System, Institute of Climate Change, Universiti Kebangsaan Malaysia, UKM, 43600 Bangi, Selangor, Malaysia
e-mail: chelgee.ooi@ukm.edu.my

A. N. Khoir
Centre for Applied Climate Services, Indonesia Agency for Meteorology, Climatology and Geophysics, Kemayoran, Jakarta, Indonesia

K. P. Vadrevu et al. (eds.), *Vegetation Fires and Pollution in Asia*,
https://doi.org/10.1007/978-3-031-29916-2_11

1 Introduction

Biomass burning (BB) emissions are often associated with human health and social and environmental issues as they produce numerous hazardous particulate matter and other gases that degrade the air quality. Worst, it has occurred annually and led to local and regional air quality problems. In the Maritime Continent (MC), biomass burning has been closely linked with agricultural activities such as slash-and-burn, wildfire, peatland, and open burning for land clearing and management purposes (Albar et al. 2018; Hayasaka et al. 2014; Saharjo and Yungan 2018; Vadrevu et al. 2021a, b). The previous finding for the year 2015 by Kiely et al. (2019) found that 53% of fire events happened over peatlands in Sumatra and Borneo. It contributed to about 3.7% of global fire carbon emissions (Hayasaka et al. 2014). A massive amount of aerosol was suspended in the air in the form of thick black smoke and managed to transport across boundaries, possibly affecting the entire MC atmosphere (Amnuaylojaroen and Anuma 2021; Latif et al. 2018; Ward et al. 2021). Moreover, it can also disrupt Earth's energy balance directly by absorbing solar radiation (solar-radiation interaction) and also modifying the cloud properties (aerosol-cloud interaction) as an indirect effect (Pavuluri and Kawamura 2018; Zhang 2020; Zhu et al. 2020). The persistence of this condition would lead to global climate change, which can modify the weather conditions to the point where they can be more vulnerable to fire events (Vadrevu et al. 2018).

Global climate change is always intertwined with weather anomalies, which modify ocean-atmospheric interaction. Weather anomaly modes such as El Nino–Southern Oscillation (ENSO), Madden–Julian oscillation (MJO), and Indian Ocean Dipole (IOD) have characterized surface temperature, precipitation efficiency, and convection in local and regional areas. Hence, it can affect fire regimes and transboundary smoke events (Islam et al. 2018). For example, El Niño years have been observed to occur every 2–8 years in the equatorial Pacific Ocean; they have introduced an increase in sea surface temperature and low intensity of precipitation over the Western Pacific (Australia, Philippines, Indonesia, and India) (Mendi et al. 2015). Meanwhile, MJO originates over the Western Indian Ocean, which scales to the east with an interval of 30–60 days. It affects the weather by disrupting daily rainfall patterns by interacting with the sea level pressure (Windayati et al. 2022). A previous study by Xian et al. (2013) found that El Niño years and MJO phases can trigger smoke events during 2003–2010 by sustaining the fire activities. This is because its emissions could be transported farther than non-El Niño years and earlier MJO phases in MC. Putra et al. (2020) mentioned that the spatial correlation of hotspots with SST indices of Niño 1.2, Niño 3, Niño 3.4, and Niño 4 had a positive correlation with coefficient values ranging from 0.1 to 0.4 in almost all regions of Indonesia during 2001–2019. Hence, both anomalies initiated drier conditions in MC, which are more susceptible to fire activity.

IOD is a couple of modes of natural ocean–atmosphere determined when there is a negative sea surface temperature in the tropical Eastern Indian Ocean and a positive sea surface temperature in the tropical Western Indian Ocean. However, positive

IOD is always linked with the presence of El Nino, resulting in a deficit of rainfall intensity, especially in Indonesia (Mareta et al. 2019). A study by Dafri et al. (2021) showed that the number of fire hotspots in Indonesia has a high relationship with ENSO and IOD climate anomalies through the Nino 3.4 index and DMI indicators.

In this study, the Niño3.4 index was used to identify the ENSO (El Niño–Southern Oscillation), which describes the fire variance during the climate anomaly index from 2001 to 2020 over the MC, constructed by the SST anomalies in the Niño3.4 region (5° N–5° S, 120° W–170° W). El Niño event is defined as a Niño3.4 value larger than 0.5, and the La Nina event is defined as a Niño3.4 value smaller than − 0.5. Meanwhile, positive Dipole Mode Index (DMI), which is equal to or more than 0.4, is used to identify Indian Ocean Dipole (IOD). IOD shows the difference in SST anomaly in the Indian Ocean between the tropical western (50°–70° E, 10° S–10° N) and the tropical southeastern (90–110° E, 10° S-equator). Active MJO phases 4 and 5 were also extracted to study the relationship between the fire activities and inter-seasonal activities climate, MJO. By phase 4 of the MJO, convection approaches MC, and by phase 5, convection enters the Pacific Ocean, and significant drying begins over the MC.

This study used fire hotspot data from the Moderate Resolution Imaging Spectroradiometer (MODIS) sensor, which is mounted on two satellites, Terra and Aqua. MODIS can detect and provide data regarding the troposphere layer's aerosol product, cloud, and water vapor profiling (Borbas et al. 2021). Moreover, it can also detect active fire that occurs worldwide by fire hotspots through the Fire Information for Resources Management System (FIRMS) provided by NASA (Giglio et al. 2016). Hantson et al. (2013) showed that MODIS hotspots are highly reliable in detecting true burned areas, assuming that the hotspot is associated with an actual fire and covers a specific burned area. However, active fires generally can be influenced by several weather conditions (Elvidge et al. 2021). Hence, the Empirical Orthogonal Functions (EOF) analysis is performed to identify the pattern of the influencing factors of the fire hotspot. It has been applied widely in climatology studies to identify weather-influencing factors contributing to vast biomass burning during specific periods. EOF has been applied to determine the variability of the fire hotspot activities with the aerosol optical depth (AOD) and the weather conditions (Khoir et al. 2021; Monahan et al. 2009). A previous study by Juneng et al. (2009) *showed that* PM_{10} concentration variability had decomposed into four dominant modes in Malaysia through the application of EOF. Recently, Huang et al. (2016) also found that EOF can reduce the variance of hotspots to identify two geographical regions for biomass-burning activities in Indo-China.

In addition, other methods, such as the fast Fourier transform (FFT) method, also have been used to identify influencing factors through the power spectrum by converting signal factors to individual spectral components from the time domain to the frequency domain. This method was applied to analyze the 6-month frequency of rainfall signals in Kalimantan by Nurdiati et al. (2021). Aldrian and Djamil (2008) also used the fast Fourier transform (FFT) method to find the climatic change of rainfall in East Java, Indonesia, from 1955 to 2005. However, since MC experiences different significant couples modes of climate anomalies in specific periods, combining both methods is more suitable for the long-term analysis to determine which weather patterns exert a more significant influence on temporal fire activity. Hence, this study aims to determine the distribution of fire hotspots in MC and their interrelationship with climate anomalies. The distribution of fire hotspots is determined with the spatial and temporal distribution of hotspot data in the MC area using EOF and FFT. Furthermore, the distribution is related to possible factors caused by hotspot fluctuations, namely AOD, and also possible driven factors, namely climate anomalies and precipitation. It is hoped that this research can understand how weather anomalies affect fire activities and provide information on fire potential as a mitigation consideration in the future.

2 Study Area

The study focused is over the Maritime Continent, defined by 10° S–10° N and 90° E–130° E (Fig. 1a). MC has tropical climate as it lies across the equatorial line and has been known as a major burning smoke aerosol source (Hansen et al. 2019; Lee et al. 2018; Xian et al. 2013). MC is a notable region for global circulation as the change of the sea surface temperature across the MC is significant to the convection in the eastern/western pole in the ascending region. The complexity of these factors also affects how fire distribution occurs. Previous research has shown that the Southeast Asia (SEA) region has its own unique total burning mass variance profile that are different from other regions in Asia (Khoir et al. 2021) where the region is divided into three modes of burning activities. Therefore, understanding on specific areas, the MC area in this case, needs to be done as it is essential to determine the characteristic of different region for appropriate mitigation to be carried out. The fluctuation of rainfall in maritime area is strongly influenced by the monsoon activities. The highest average rainfall occurs in NDJ (November–December–January) and the lowest occurs in JAS (July–August–September) (Fig. 1b). The total of the rainfall average in NDJ and JAS is 22.6 mm day^{-1} and 15 mm day^{-1}, respectively. The lowest rainfall in JAS has corresponded to the increase in the total number of fire hotspots. The previous study by Khoir et al. (2021) and Indratmoko and Rizqihandari (2019) found the maximum total count of fire hotspots in South Sumatra Province alone had already reached about 35,008 during the strong El Nino year in 2015.

3 Data and Methods

3.1 Data

The fire data used in this study are from the MOD14 (Terra) and MYD14 (Aqua) satellites provided by NASA's FIRMS system (Giglio et al. 2003). Fire data are detected

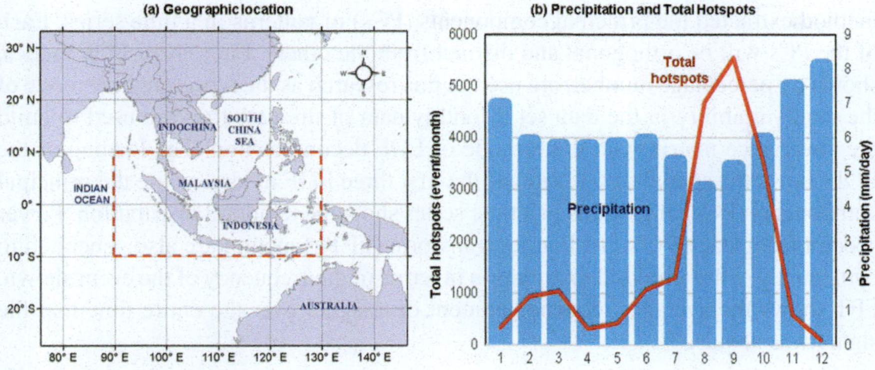

Fig. 1 **a** Geographic location of study area. **b** Fire hotspot variations in relation to precipitation

from the thermal anomalies from multiple channels of MODIS sensor (Justice et al. 2002). Historical archive data on fire activities are generated for the last two decades, starting from January 2001 to December 2020. The daily data for fire activities used in the research are combined from Terra and Aqua satellites. A high level of confidence will result in fewer false alarms (Giglio 2015). Hence, fire activity data are selected based on the confidence level and only use data with a confidence level above 80% (Giglio 2015). Hantson et al. (2013) showed that MODIS hotspots are highly reliable to detect true burned areas with assumption that detected hotspot is associated with an actual fire and that it covers a certain burned area.

In addition, rainfall data and aerosol data are also obtained from model reanalysis and satellite data, respectively. Surface rainfall data are generated from ECMWF reanalysis data set model with $0.1^\circ \times 0.1^\circ$ resolution for period of 2001–2020. While the aerosol data (AOD_550_Dark_Target_Deep_Blue_Combined) are obtained from Terra (MOD08_D3) and Aqua (MYD08_D3), MODIS L3 aerosol products include a new Scientific Data Set which represents the merged aerosol product (DTB) based on the Dark Target (DT) and Deep Blue (DB) aerosol retrieval algorithms. DTB data were retrieved with a resolution of $1^\circ \times 1^\circ$ over 2001–2020. The performance of the selected product DTB has been studied in several previous studies (Unnithan et al. 2020; Wei et al. 2019; Xie et al. 2019).

3.2 Methods

Empirical Orthogonal Function (EOF) analysis was performed to analyze the temporal and spatial distribution of fire activities in MC. EOF is a multivariate analysis technique to filter many data variables to only a few representative variables without changing most of the variance to describe the data (Hannachi et al. 2007). EOF is used in this study to reduce the total data of high dimensions fire hotspots to lower ones while retaining the essential information from the initial data. The EOF method extracted the principal components (PCs) of patterns in a time series. Each of the PCs will be orthogonal and distinct from the others. First, three EOF modes, shown by percentage number, are used in this research as the representative mode of the total variability in the data set. Monthly data of fire activities are used to build the covariance matrix with the S mode of EOF decomposition. The loading of the EOF shows the spatial distribution of the first three EOF modes, while the principal components (PCs) of patterns in a time series show the temporal distribution. Power spectra using each PC's fast Fourier transform (FFT) function are also generated to better analyze the temporal distribution in terms of the frequency of the event shown. FFT shows the average frequency content of a signal over the entire time that the signal was acquired.

4 Results and Discussion

4.1 Spatiotemporal Distribution of Total Fire Hotspot in Maritime Continent

We used the first three PCs to show the spatial and temporal distribution of the fire hotspots. The first three PC values represented the entire data as they have an eigenvalue > 1 (Dillon and Goldstein 1984) (Fig. 2). The first PC (PC1) explained 22.8% of the total variance and became the dominant pattern. PC1 has similar positive signs all over the study area, which indicates a homogeneous influence (Fig. 2). Then PC2 and PC3 are the second and third most dominant PC, which explains 6.8% and 3.7% of the total variance, respectively. These three PCs only explain less than half of the total variance because the study period took the whole data set of 20 years. The coefficients of the time series revealed that PC1 represents the annual patterns. Meanwhile, the coefficients of the PC2 and PC3 time series are hard to explain. To clarify which frequency dominates the total hotspot pattern, the FFT power spectrum of each PC is investigated. Figure 3 illustrates the power spectrum of the total hotspot pattern from the three PCs. Important timescales are indicated by a period (month/cycle). Those important periods are 6, 12, 17, 60, and 120 months/cycle, representing semi-annual, annual, and decadal patterns.

Signals captured from the power spectrum show the frequency distributions of the principal components of the dominated spatial pattern. PC1 has an annual signal which is shown by the highest power value during the 12-month/cycle period in the first PC mode plot. In addition, lower power values are recorded, namely during the 17- and 120-month/cycle periods. PC2 also shows an annual signal with slightly lower power than PC1, which is shown by the highest power value during the 12-month/cycle period. The second-highest power value is during the 60-month/cycle period. PC3 occurs in the 12 month/cycle as the highest power value and 6 and 60 month/cycles. The latter is slightly lower than the highest power, which indicates annual and semi-annual signals. The annual signal, as the signal that occurs in all

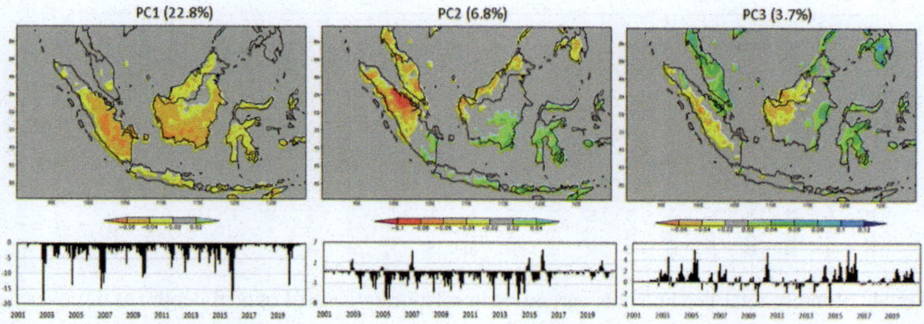

Fig. 2 Three different principal components derived from Empirical Orthogonal Function (EOF) analysis to describe fire events (see text for description)

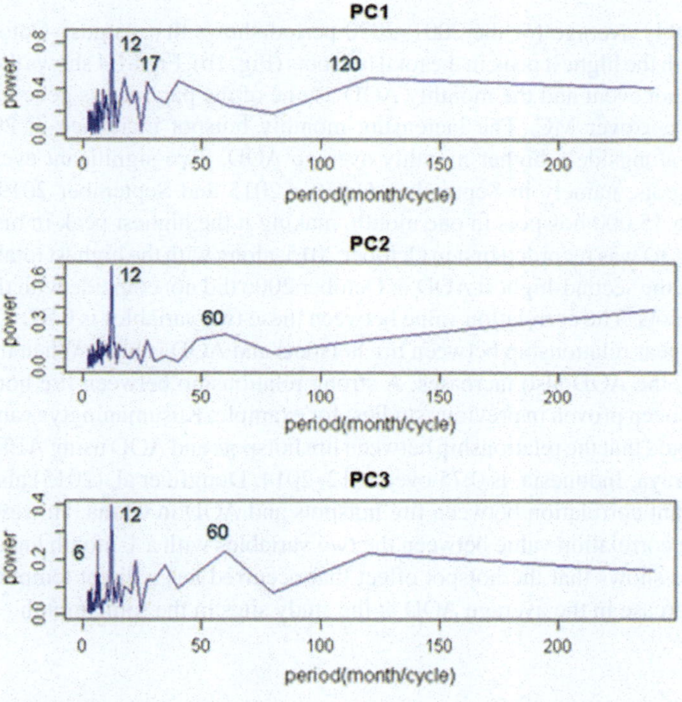

Fig. 3 Power spectrum of the total fire hotspot pattern from the three principle components (PCs)

PCs, can be related to the annual monsoon effect. Saha (2010) mentioned that MC experiences a monsoonal climate, mostly warm, and humid throughout the year. Furthermore, the 60-month/cycle period that occurs in PC2 and PC3 can be explained by the ENSO event, where previously Wang and Picaut (2004) reported that the Niño3.4 index time series has three periodic components with periods 31, 45, and 61 months over 1950–2003.

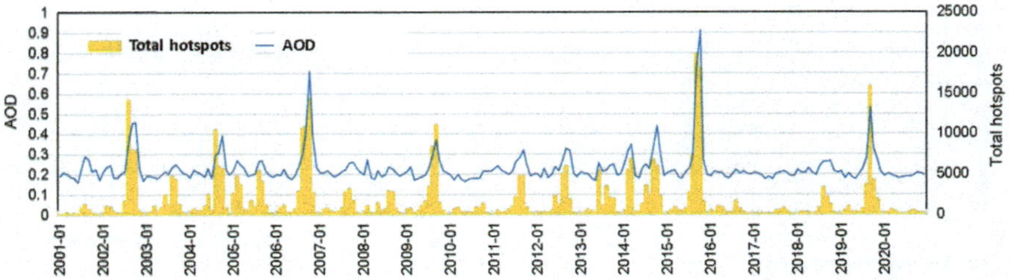

Fig. 4 Monthly total fire hotspots variations in relation to aerosol optical depth (AOD) over the maritime continent

4.2 Relationship of Total Hotspot to AOD

The monthly average for the 2001–2020 period shows that August–October is the month with the highest peak in the total hotspots (Fig. 1b). Figure 4 shows the monthly total hotspot event and the monthly AOD as one of the parameters generated by the total hotspot over MC. The increasing monthly hotspot incidence in 2001–2020 occurred alongside a higher monthly average AOD. Two significant events of the total hotspots, namely in September–October 2015 and September 2019, reached more than 15,000 hotspots in one month, making it the highest peak in history. The highest AOD was recorded first in October 2015, along with the highest total hotspots. However, the second-highest AOD in October 2006 did not coincide with the highest total hotspots. The correlation value between these two variables is 0.87. This shows that the linear relationship between fire hotspots and AOD is high. When the hotspot increases, the AOD also increases. A strong relationship between fire hotspots and AOD has been proven in previous studies; for example, Kusumaningtyas and Aldrian (2016) noted that the relationship between fire hotspots and AOD using AERONET in Palangkaraya, Indonesia, is 0.75 over 2012–2014. Dongfu et al. (2015) also showed a significant correlation between fire hotspots and AOD in China. However, in this study, the correlation value between the two variables with a 1-month lag of AOD is 0.36. This shows that the hotspot effect that occurred has a direct temporal impact on the increase in the average AOD at the study sites in the same month.

4.3 Relationship of the Total Hotspot to Its Precipitation

Possible relationships between total hotspots and the factors that influence the fluctuation of total hotspots are described in this section. Variability of rainfall is a significant parameter in MC due to the active monsoonal activity in this region (Saha 2010). The variability of rainfall can be an annual precursor factor for fire burning because of its variability, which is influenced by monsoons, as it has one peak every year. Figure 5 shows the monthly time series of total hotspots in August–September–October (ASO), where these months peak the average of total hotspots over the last

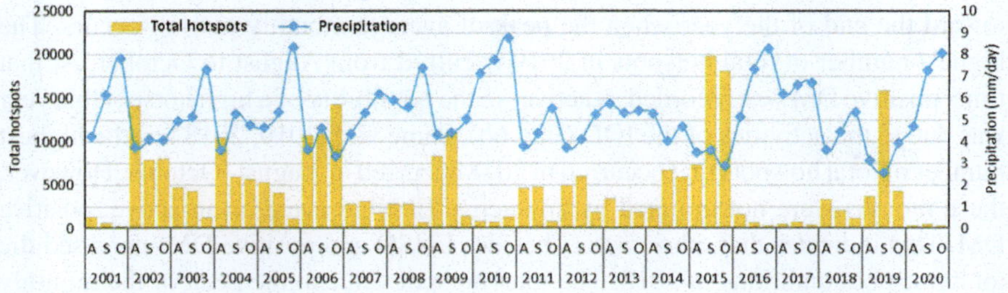

Fig. 5 Monthly time series of total fire hotspots during August–September–October (ASO), where the peak fires can be seen

two decades. Total hotspots are compared with rainfall data at the same chosen months to see how they fluctuate.

As seen from Fig. 5, the increase in precipitation in the month of ASO coincides with a decrease in the total hotspots in the same months every year. The rainfall increases, and the total hotspot value decreases in October every year, and vice versa. Rainfall also has an impact on the interannual pattern of the hotspot. The lowest average rainfall occurred in July–August–September (JAS), reaching an average of 5.03 mm/day, followed by the average months with the highest number of hotspots in ASO (Fig. 1b). While the highest average rainfall occurs in NDJ (Nov–Dec–Jan), reaching an average of 7.53 mm/day, followed by the average months with the lowest number of hotspots in NDJ. This pattern is supported by the high correlation value between the two parameters, reaching − 0.69. A strong negative correlation value indicates that an increase will follow a decrease in the amount of rainfall in the number of hotspots and vice versa. The relationship between rainfall and hotspots is proportionate with previous studies (Ceccato et al. 2010; Indra and Hasayaka 2011; Saharjo and Velicia 2018; Soro et al. 2021). As a maritime area, the rainfall variability in MC is strongly influenced by monsoon activity. The MC experiences monsoon wind systems throughout the year, with wind directions reversing between winter and summer. MC has substantial seasonal variation in wind and rainfall regimes, which consist of prevailing easterly (westerly) wind and dry (wet) conditions during the boreal summer (winter).

4.4 *Relationship of the Total Hotspot to Climate Anomalies*

Burning mass in MC appeared to be significantly influenced by major climate events. Several studies have shown how ENSO can affect hotspot occurrence, especially in the MC region. Figure 6. a shows the total hotspot fluctuates when El Niño and La Nina occur. El Niño events increased the total hotspots significantly, for example, in 2002/2003, 2004/2005, 2006/2007, 2009/2010, 2014/2015, and 2015/2016. Interestingly, El Niño in 2018/2019 and 2019/2020 did not increase total hotspot incidence. This is probably due to the rainfall, considering that El Niño in that period occurred toward the end of the year when the peak of average monthly rainfall occurs. The highest number of total hotspots in 2019 occurred from August to October. At that time, positive DM was recorded as active, so the possibility of a high hotspot this year was due to the activities of the IOD (Fig. 6b). Same with 2019, 2018 has the highest number of total hotspots that occurred in 2018 occurred in August–October. However, these months were not initiated by any active El Niño event or an active, positive DM. This indicates that other factors besides ENSO and positive DM increased the total hotspot conditions in 2018. The total hotspot value at the peak of the monthly hotspot event in 2018 was still below the average because there was no active ENSO, positive Dipole, and MJO phases 4 and 5. In 2018 and 2019, the active El Niño event experienced an anomaly delayed in the active period, which is usually onset in the middle of the year during the dry condition, to shift at the end until early the year when the wet condition occurred.

The strong El Niño event in 2015/2016, which is also mentioned as a super El Niño event in some studies (Chen et al. 2017; Hayasaka and Sepriando 2018), is associated with an increased incidence of burning haze in the MC region. For example, Islam et al. (2018) proved that the concurrence of ENSO, positive IOD, and MJO phenomenon in 2015 strongly influenced biomass burning in Southeast Asia as the factors of increasing surface air temperature and decreasing rainfall. Hayasaka and Sepriando (2018) also shared that during the 2015 super El Niño, the dry season was longer and very dry, affecting the severe burning haze and air pollution event in Palangkaraya, Indonesia.

In a year without El Niño and positive Dipole, the number of monthly hotspot anomalies in August–October was less than 5000 hotspot events than the average.

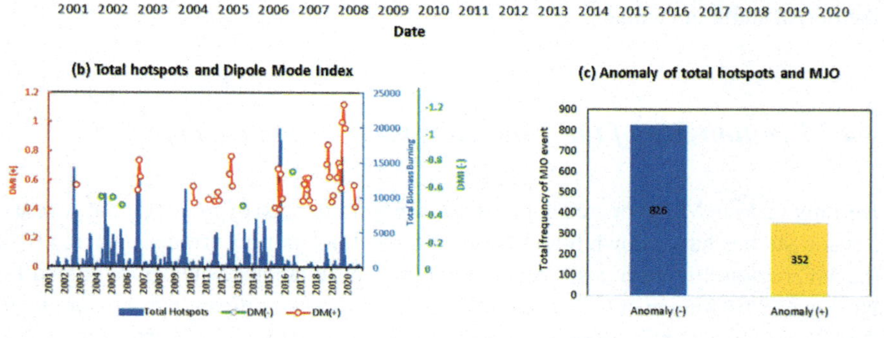

Fig. 6 **a** Monthly total fire hotspots and ENSO (2001–2020). **b** Total fire hotspots and Dipole Mode Index. **c** Anomaly of total fire hotspots and the Madden–Julian oscillation (MJO)

La Nina events that occur in August–October also reduce hotspot events compared to the average. The La Nina event that coincided with the positive Dipole Mode also reduced the monthly hotspot value from the average (Fig. 6a, b). Positive Dipole Mode in some events affects the increase in total hotspots, but when it coincides with La Nina, the effect will be reduced. Meanwhile, the occurrence of MJO affects more negative anomalies than the occurrence of total hotspots (Fig. 6c). It shows that when the MJO mainly occurs, the anomaly of the total hotspot is low. This complements the previous study in Borneo or Sumatra by Xian et al. (2013), which mentioned that most of the longest-living smoke events originated during the MJO later phases (phases 4–8). Reid et al. (2012) also reported that MJO becomes an interannual phenomenon that correlates to total seasonal burning, mainly when visible burning occurs.

5 Conclusion

The spatiotemporal distribution of the fire events in the Maritime Continent was analyzed in the long term. The EOF was performed to obtain significant spatiotemporal signals of fire events from the compressed data set over 2001–2020. The three largest PCs could explain the distribution of fire events in MC. The power spectrum results provided the frequency of occurrence of the three PCs. The strongest signal from the three PCs suggested the annual occurrence of fire hotspots. This agrees with the high correlation between fire hotspots and rainfall as one of the factors affecting fire hotspots in MC. The peak total monthly hotspots in MC occurred in the month of ASO. When there are climatic phenomena such as the onset of El Niño and La Nina in the month of ASO, the number of hotspots that occur is greatly affected. Likewise, when there is a positive Dipole Mode, there is an increase in the total number of hotspots in the MC. However, when El Niño occurs during a wet month, there is no significant increase in fire hotspots. Meanwhile, MJO phases 4 and 5 increase negative fire event anomalies during the study period.

Acknowledgements We would like to acknowledge National University of Malaysia, also known as Universiti Kebangsaan Malaysia (UKM) for the funding support of the research work (GGPM-2020-033).

References

Albar, I., I. Jaya, B.H. Saharjo, B. Kuncahyo, and K.P. Vadrevu. 2018. Spatio-temporal analysis of land and forest fires in Indonesia using MODIS active fire dataset. In *Land-atmospheric research applications in South and Southeast Asia*, 105–127. Cham: Springer.

Aldrian, E., and Y.S. Djamil. 2008. The impact of the positive indian ocean dipole on Zimbabwe droughts tropical climate is understood to be dominated by. *International Journal of Climatology* 28 (2007): 435–448. https://doi.org/10.1002/joc.1543.

Amnuaylojaroen, T., and N. Anuma. 2021. Estimation of PM_{10} concentrations from biomass burning and anthropogenic emissions using the WRF–HYSPLIT modeling system. In *Biomass burning in South and Southeast Asia*, 297–308. CRC Press.

Borbas, E., E. Weisz, C. Moeller, P.W. Menzel, and B.A. Baum. 2021. Improvement in tropospheric moisture retrievals from VIIRS through the use of infrared absorption bands constructed from VIIRS and CrIS data fusion. *Atmospheric Measurement Techniques* 14 (2): 1191–1203. https://doi.org/10.5194/amt-14-1191-2021.

Ceccato, P., I.N.S. Jaya, J.H. Qian, M.K. Tippett, A.W. Robertson, and S. Someshwar. 2010. *Early warning and response to fires in Kalimantan, Indonesia*, 23. https://iri.columbia.edu/docs/publications/TR10-14_EarlyWarningandResponse_PCeccato.pdf.

Chen, L., T. Li, B. Wang, and L. Wang. 2017. Formation mechanism for 2015/16 Super El Niño. *Scientific Reports* 7 (1). https://doi.org/10.1038/s41598-017-02926-3.

Dafri, M., S. Nurdiati, and A. Sopaheluwakan. 2021. Quantifying ENSO and IOD impact to hotspot in Indonesia based on Heterogeneous Correlation Map (HCM). *Journal of Physics: Conference Series* 1869 (1). https://doi.org/10.1088/1742-6596/1869/1/012150.

Dillon, W.R., and M. Goldstein. 1984. *Multivariate analysis: methods and applications*. Wiley.

Dongfu, F., Y. Shuyun, W. Biwen, F. Dongqing, W. Jun, J. Bo, and C. Xiaolong. 2015. Monitoring of biomass burning based on AOD. *Remote Sensing for Land & Resources* 2 (2): 133–138. https://doi.org/10.6046/gtzyyg.2015.02.21.

Elvidge, C.D., M. Zhizhin, K. Baugh, and F.C. Hsu. 2021. Identification of smoldering Peatland fires in Indonesia via triple-phase temperature analysis of VIIRS nighttime data. In *Biomass burning in South and Southeast Asia*, 25–38. CRC Press.

Giglio, L. 2015. MODIS collection 6 active fire product user's guide revision A. Department of Geographical science, University of Maryland.

Giglio, L., J. Descloitres, C.O. Justice, and Y.J. Kaufman. 2003. An enhanced contextual fire detection algorithm for MODIS. *Remote Sensing of Environment* 87 (2–3): 273–282. https://doi.org/10.1016/S0034-4257(03)00184-6.

Giglio, L., W. Schroeder, and C.O. Justice. 2016. The collection 6 MODIS active fire detection algorithm and fire products. *Remote Sensing of Environment* 178: 31–41. https://doi.org/10.1016/j.rse.2016.02.054.

Hannachi, A., I.T. Jolliffe, and D.B. Stephenson. 2007. Empirical orthogonal functions and related techniques in atmospheric science: A review. *International Journal of Climatology*: 1119–1152.

Hansen, A.B., C.S. Witham, W.M. Chong, E. Kendall, B.N. Chew, C. Gan, M.C. Hort, and S. Lee. 2019. *Haze in Singapore—Source attribution of biomass burning PM_{10} from Southeast Asia*, 5363–5385.

Hantson, S., M. Padilla, D. Corti, and E. Chuvieco. 2013. Strengths and weaknesses of MODIS hotspots to characterize global fire occurrence. *Remote Sensing of Environment* 131: 152–159. https://doi.org/10.1016/j.rse.2012.12.004.

Hayasaka, H., I. Noguchi, E.I. Putra, N. Yulianti, and K. Vadrevu. 2014. Peat-fire-related air pollution in Central Kalimantan, Indonesia. *Environmental Pollution* 195: 257–266.

Hayasaka, H., A. Sepriando, K.P. Vadrevu, T. Ohara, and C. Justice. 2018. Land-atmospheric research applications in south and southeast Asia severe air pollution due to peat fires during 2015 super El Niño in Central Kalimantan Indonesia, 129–142. Cham: Springer.

Huang, W.-R., S.-H. Wang, M.-C. Yen, N.-H. Lin, and P. Promchote. 2016. Interannual variation of springtime biomass burning in Indochina: Regional differences, associated atmospheric dynamical changes, and downwind impacts. *Journal of Geophysical Research* 121 (17): 10016–10028. https://doi.org/10.1002/2016JD025286.

Indra, P.E., and H. Hasayaka. 2011. The effect of the precipitation pattern of the dry season on peat fire occurrence in the Mega Rice Project area, Central Kalimantan, Indonesia. *Tropics* 19 (4): 145–156. https://doi.org/10.3759/tropics.19.145.

Indratmoko, S., and N. Rizqihandari. 2019. Burn area detection using Landsat 8 OLI TIRS. *IOP Conference Series: Earth and Environmental Science* 338 (1). https://doi.org/10.1088/1755-1315/338/1/012035.

Islam, M.A., A. Chan, M.J. Ashfold, C.G. Ooi, and M. Azari. 2018. Effects of El-Niño, Indian ocean dipole, and Madden-Julian oscillation on surface air temperature and rainfall anomalies over Southeast Asia in 2015. *Atmosphere* 9 (9): 1–14. https://doi.org/10.3390/atmos9090352.

Juneng, L., M.T. Latif, F.T. Tangang, and H. Mansor. 2009. Spatio-temporal characteristics of PM_{10} concentration across Malaysia. *Atmospheric Environment* 43 (30): 4584–4594. https://doi.org/10.1016/j.atmosenv.2009.06.018.

Justice, C.O., L. Giglio, S. Korontzi, J. Owens, J.T. Morisette, D. Roy, J. Descloitres, S. Alleaume, F. Petitcolin, and Y. Kaufman. 2002. The MODIS fire products. *Remote Sensing of Environment* 83 (1–2): 244–262. https://doi.org/10.1016/S0034-4257(02)00076-7.

Khoir, A.N., M.C.G. Ooi, J. Liew, Suradi, A. Kurniawan, and A. Ni'amillah. 2021. MODIS-derived fire spatial and temporal distribution during haze season in Southeast Asia using empirical orthogonal function. *IOP Conference Series: Earth and Environmental Science* 880 (1): 012003. https://doi.org/10.1088/1755-1315/880/1/012003.

Kiely, L., D. Spracklen, C. Wiedinmyer, L. Conibear, C. Reddington, S. Archer-Nicholls, D. Lowe, S. Arnold, C. Knote, M.F. Khan, M.T. Latif, M. Kuwata, S.H. Budisulistiorini, and L. Syaufina. 2019. *Revised estimate of particulate emissions from Indonesian peat fires in 2015*, 11105–11121. https://doi.org/10.5194/acp-2019-323.

Koplitz, S.N., L.J. Mickley, D.J. Jacob, M.E. Marlier, R.S. DeFries, D.L.A. Gaveau, B. Locatelli, J.S. Reid, P. Xian, and S.S. Myers. 2018. Role of the Madden-Julian oscillation in the transport of smoke from Sumatra to the Malay Peninsula during severe non-El Niño Haze Events. *Journal of Geophysical Research: Atmospheres* 123 (11): 6282–6294. https://doi.org/10.1029/2018JD028533.

Kusumaningtyas, S.D.A., and E. Aldrian. 2016. Impact of the June 2013 Riau province Sumatera smoke haze event on regional air pollution. *Environmental Research Letters* 11 (7): 075007.

Latif, M.T., M. Othman, N. Idris, L. Juneng, A.M. Abdullah, W.P. Hamzah, M.F. Khan, N.M. Nik Sulaiman, J. Jewaratnam, N. Aghamohammadi, M. Sahani, C.J. Xiang, F. Ahamad, N. Amil, M. Darus, H. Varkkey, F. Tangang, and A.B. Jaafar. 2018. Impact of regional haze towards air quality in Malaysia: A review. *Atmospheric Environment* 177 (January): 28–44. https://doi.org/10.1016/j.atmosenv.2018.01.002.

Lee, H., S. Jeong, M. Tosca, and S. Kim. 2018. Special section: Characterization of wildfire-induced aerosol emissions from the maritime continent Peatland and Central African Dry Savannah with MISR and CALIPSO aerosol products. *Journal of Geophysical Research: Atmospheres*: 3116–3125. https://doi.org/10.1002/2017JD027415.

Mareta, L., R. Hidayat, R. Hidayati, and G. Alsepan. 2019. Influence of the positive Indian ocean dipole in 2012 and El Niño-southern oscillation (ENSO) in 2015 on the Indonesian rainfall variability. *IOP Conference Series: Earth and Environmental Science* 284 (1). https://doi.org/10.1088/1755-1315/284/1/012018.

Mendi, V. 2015. El Nino: A review. *International Journal of Earth Sciences and Engineering* 8 (2): 130–137.

Monahan, A.H., J.C. Fyfe, M.H.P. Ambaum, D.B. Stephenson, and G.R. North. 2009. Empirical orthogonal functions: The medium is the message. *Journal of Climate* 22 (24): 6501–6514. https://doi.org/10.1175/2009JCLI3062.1.

Nurdiati, S., E. Khatizah, M.K. Najib, and R.R. Hidayah. 2021. Analysis of rainfall patterns in Kalimantan using fast fourier transform (FFT) and empirical orthogonal function (EOF). *IOP Conference Series: Earth and Environmental Science* 1796 (1). https://doi.org/10.1088/1742-6596/1796/1/012053.

Pavuluri, C.M., and K. Kawamura. 2018. Organic aerosols in South and East Asia: Composition and sources. In *Land-atmospheric research applications in South and Southeast Asia*, 379–408. Cham: Springer.

Putra, I.D.G.A., E. Heriyanto, A. Sopaheluwakan, R.P. Pradana, and D.E. Nuryanto. 2020. Seasonal analysis of the hotspot spatial grid in indonesia and the relationship of the hotspot grid with the Nino SST indices. In *2020 IEEE Asia-Pacific conference on geoscience, electronics and remote sensing technology (AGERS)*, 63–68.

Reid, J.S., P. Xian, E.J. Hyer, M.K. Flatau, E.M. Ramirez, F.J. Turk, C.R. Sampson, C. Zhang, E.M. Fukada, and E.D. Maloney. 2012. Multi-scale meteorological conceptual analysis of observed active fire hotspot activity and smoke optical depth in the Maritime Continent. *Atmospheric Chemistry and Physics* 12 (4): 2117–2147. https://doi.org/10.5194/acp-12-2117-2012.

Saha, K. 2010. Tropical systems and monsoons. *Springer*. https://doi.org/10.1007/978-3-642-03373-5.

Saharjo, B.H. and W.A. Velicia. 2018. Peran Curah Hujan Terhadap Penurunan Hotspot Kebakaran Hutan dan Lahan di Empat Provinsi di Indonesia pada Tahun 2015–2016. The Role of Rainfall Towards Forest and Land Fires Hotspot Reduction in Four Districs in Indonesia on 2015–2016. *Jurnal Silvikultur Tropika*, 9 (1): 24–30. https://doi.org/10.29244/j-siltrop.9.1

Saharjo, B.H., and A. Yungan. 2018. Forest and land fires in Riau Province: A case study in fire prevention policy implementation with local concession holders. In *Land-atmospheric research applications in South and Southeast Asia*, 143–169. Cham: Springer.

Soro, T.D., M. Koné, A.B. N'Dri, and E.T. N'Datchoh. 2021. Identified main fire hotspots and seasons in Côte d'Ivoire (West Africa) using MODIS fire data. *South African Journal of Science* 117(1–2): 1–13. https://doi.org/10.17159/SAJS.2021/7659

Unnithan, S.L.K., L. Gnanappazham, and S.L.K. Unnithan. 2020. Spatiotemporal mixed effects modeling for the estimation of $PM_{2.5}$ from MODIS AOD over the Indian subcontinent spatiotemporal mixed effects modeling for the estimation of $PM_{2.5}$ from MODIS AOD over the Indian subcontinent. *GIScience & Remote Sensing* 00 (00): 1–15. https://doi.org/10.1080/15481603.2020.1712101.

Vadrevu, K.P., T. Ohara, and C. Justice, eds. 2018. *Land-atmospheric research applications in South and Southeast Asia.*

Vadrevu, K.P., T. Ohara, and C. Justice, eds. 2021a. *Biomass burning in south and Southeast Asia: Impacts on the biosphere*, vol. 2. CRC Press.

Vadrevu, K.P., T. Ohara, and C. Justice, eds. 2021b. *Biomass burning in South and Southeast Asia: Mapping and monitoring*, vol. 1. CRC Press.

Wang, C., and J. Picaut. 2004. Understanding ENSO physics—A review. In *Earth's climate: The ocean–atmosphere interaction, geophysical monograph*, vol. 147, 21–48.

Ward, C., L.C. Stringer, E. Warren-Thomas, F. Agus, M. Crowson, K. Hamer, B. Hariyadi, W.D. Kartika, J. Lucey, C. McClean, N.L. Nurida, N. Petorelli, E. Pratiwi, A. Saad, R. Andriyani, T. Ariani, H. Sriwahyuni, and J.K. Hill. 2021. Smallholder perceptions of land restoration activities: rewetting tropical peatland oil palm areas in Sumatra, Indonesia. *Regional Environmental Change* 21 (1). https://doi.org/10.1007/s10113-020-01737-z.

Windayati, R., B.W. Mutaqin, M.A. Marfai, E.H. Pangaribowo, M. Helmi, and M. Rindarjono. 2022. Assessment of coral-reef ecosystem services in West Buleleng conservation Zone, Bali, Indonesia. *Journal of Coastal Conservation* 26 (5): 1–9.

Wei, J., Z. Li, L. Sun, Y. Peng, and L. Wang. 2019. An improved merge scheme for MODIS collection 6.1 dark target and deep blue combined aerosol products. *Atmospheric Environment*. https://doi.org/10.1016/j.atmosenv.2019.01.016.

Xian, P., J.S. Reid, S.A. Atwood, R.S. Johnson, E.J. Hyer, D.L. Westphal, and W. Sessions. 2013. Smoke aerosol transport patterns over the maritime continent. *Atmospheric Research* 122: 469–485. https://doi.org/10.1016/j.atmosres.2012.05.006.

Xie, G., M. Wang, J. Pan, and Y. Zhu. 2019. Spatio-temporal variations and trends of MODIS C6.1 dark target and deep blue merged aerosol optical depth over China during 2000–2017. *Atmospheric Environment*: 116846. https://doi.org/10.1016/j.atmosenv.2019.116846

Zhang, B. 2020. The effect of aerosols to climate change and society. *Journal of Geoscience and Environment Protection* 08 (08): 55–78. https://doi.org/10.4236/gep.2020.88006.

Zhu, A., H. Xu, J. Deng, J. Ma, and S. Li. 2020. ENSO effect on interannual variability of Spring aerosols over East Asia. *Atmospheric Chemistry and Physics Discussions*: 1–32. https://doi.org/10.5194/acp-2020-859.

Vegetation Fires in Laos—An Overview

Krishna Prasad Vadrevu, Chittana Phompila, and Aditya Eaturu

Abstract Of the different countries in Southeast Asia, Laos has the most recurrent fires. This study highlights the vegetation fire characteristics in Laos using various satellite datasets. The European Space Agency WorldCover (10 m) data in Laos suggested tree cover as the dominant class with 78.2%, followed by grasslands (12.8%), croplands (7.29%), and with the least percentage of built-up areas (0.31%). These land cover statistics suggest that the country still has rich tree cover, which is under threat due to various drivers of change, including fires. VIIRS satellite-derived fires over Laos for 10 years suggested a mean of 138,934 fire counts, with the highest during March–May every year. The spatial patterns suggested a mean of 0–23 and the highest of 128–338 fires per 5-min grid cells. Similarly, the fire radiative power (FRP) varied from 0–7.49 MW to 43.15–120.15 MW at a similar resolution. Also, a close analysis suggested North Western Laos provinces of Oudomxai Louang Namtha, Bokeo, and Xiagnabouri with the highest fires, and a similar pattern is shown for FRP. Results from the MODIS burnt area datasets revealed 665,923.49 ha per year burnt based on 23 years of data. Further, the burnt areas were the highest in the forest class, followed by grasslands and croplands. To infer the role of climate in regulating fires, we used the MODIS data to compute the temperature condition index (TCI) and related it to fire counts (FC). The Ordinary Least Squares (OLS) regression between TCI and FC resulted in a negative slope, and TCI could explain 29% of the variation in fires. We also used the quantile regression, and the results suggested the pseudo-R^2 of 0.35 at 0.9 quantiles. Both these regression results suggested a minor role of TCI as governing factor of fires. A literature review indicated that most of the fires in the country are caused due to humans, primarily through slash-and-burn agriculture. Therefore, we infer that effective mitigation of fires is only possible through people's participation through consultation, cooperation, and collective action.

K. P. Vadrevu (✉)
NASA Marshall Spaceflight Center, Huntsville, AL, USA
e-mail: krishna.p.vadrevu@nasa.gov

C. Phompila
Faculty of Forest Sciences (FFS), National University of Laos (NUoL), Vientiane, Laos

A. Eaturu
University of Alabama Huntsville, Huntsville, AL, USA

K. P. Vadrevu et al. (eds.), *Vegetation Fires and Pollution in Asia*,
https://doi.org/10.1007/978-3-031-29916-2_12

Keywords Fires · Satellite data · Slash and burn · Laos

1 Introduction

Vegetation fires are prevalent worldwide, including in South/Southeast Asian countries. Vegetation fires can significantly alter landscape structure and function. The impact of fires on landscape structure and composition in different ecosystems is documented in Goldammer and Price (1998). The adverse effects of the fires on landscapes include loss of forests and biodiversity, including ecosystem services such as timber, recreation, nutrients, and water retention, including bioenergy loss (Myers 1996; Justice et al. 2015; Vadrevu and Lasko 2015). Continuous burning results in loss of soil fertility (Crutzen and Andreae 1990; Prasad et al. 2004). The drivers of vegetation fires include the climate and anthropogenic factors (Albar et al. 2018; Biswas et al. 2015a, b, 2021; Eaturu and Vadrevu 2021; Hayasaka et al. 2014; Inoue 2018; Justice et al. 2015). In several countries of Asia, fires are used as a land-clearing tool through slash and burn of forests and clearing of agricultural residues after harvest (Badarinath et al. 2008; Prasad et al. 2001a, b; 2002a, b, 2003, 2004; Prasad and Badarinath 2004; Prasad et al. 2005; Prasad and Badarinth 2006; Lasko et al. 2017, 2018; Thapa et al. 2021). The burning of biomass from these activities is an important source of greenhouse gas emissions and aerosols (Badarinath et al. 2007, 2008, 2009; Badarinath and Prasad 2011; Choi et al. 2008; Goldammer and Seibert 1990; Kant et al. 2000a, b, Kharol et al. 2012; Lasko and Vadrevu 2018; Lasko et al. 2017, 2018, 2021). The fires can also emit substantial amounts of particulate matter (PM) and other pollutants into the atmosphere, which may get transported over long distances impacting both the local and regional air quality and health (Crutzen and Andreae 1990; Gupta et al. 2001a, b; Vadrevu et al. 2008; 2013, 2014a, b, 2017, 2018, 2019, 2020, 2021a, b, 2022a, b; Vay et al. 2011). Some researchers reported positive aspects of fires, including the availability of nutrients after burning for the subsequent crops, promoting vegetation growth such as grass cover for livestock in some ecosystems (Bond and Wilgen 2012). However, fires can also determine the regrowth of new vegetation, including type, structure, and composition (Dansereau and Bergeron 1993) and, thus, can affect ecological processes (Turner 1989). Considering these impacts, quantifying the spatial and temporal variations in fires can help understand their ecological and environmental impacts (Vázquez and Moreno 2001; Vadrevu 2008; Vadrevu et al. 2008; Vadrevu and Justice 2011). Specific to Laos, the country has significant recurrent fires during the dry season. Although ground-based methods are helpful at small spatial scales, monitoring fires at large spatial scales is challenging. Remote sensing technologies have been widely used in fire detection, mapping, and monitoring studies. In particular, remote sensing technology, with its multitemporal, multi-spectral, synoptic, and repetitive coverage capabilities, provides significant benefits compared to ground-based fire monitoring (Petropoulos et al. 2013; Vadrevu 2008; Vadrevu and Justice 2011). Remote sensing data can provide helpful information on the fire counts, the amount of area burned, and the type of ecosystem burned (Vadrevu et al. 2012; Petropoulos et al. 2013; Vadrevu and Badarinath 2009; Vadrevu and Lasko 2015; Vadrevu 2015). Fires emit thermal radiation with a peak in the mid-infrared region, following Planck's theory of blackbody radiation, which satellites can capture. Specifically, active fires can be detected using mid-infrared and thermal infrared (usually around 3.7–11 mm) information from satellites (Kant et al. 2000b; Wooster et al. 2021). This study uses various satellite data to characterize fires in Laos.

2 Study Area

Laos is the only landlocked country in Southeast Asia (Fig. 1). The country lies between latitudes 14° and 23° N (a small area is south of 14°) and longitudes 100° and 108° E. Myanmar and China border it to the northwest, Vietnam to the east, Cambodia to the southeast, and Thailand to the west and southwest. Its capital city is Vientiane. The total land area of the country is 23.68 million hectares. The country has rich forest landscapes with rugged mountains, the highest of which is Phou Bia at 2818 m, with some plains and plateaus. The Mekong River forms a large part of the western boundary with Thailand, where the mountains of the Annamite Range form most of the eastern border with Vietnam, and the Luang Prabang Range is the northwestern border with the Thai highlands. There are two plateaus, the Xiangkhoang in the north and the Bolaven Plateau at the southern end. The country has 17 provinces (khoeng, qwang, or khoueng) and one prefecture, the Vientiane capital city municipality (Fig. 1). The provinces are then subdivided into districts (muang) and villages (baan). Further, the provinces are grouped geographically into three strata, north (from Phongsaly to Saiyabouly, Luang Prabang, and Xiangkhoang), central (Vientiane and Bolikhamxay), and south (from Khammuane to Champasak). The climate is primarily tropical savanna and influenced by the monsoon pattern. There is a distinct rainy season from May to October, followed by a dry season from November to April. Laos's population was estimated at 7.45 million in 2020 (https://en.wikipedia.org/wiki/Laos). Slash-and-burn agriculture is most common in northern hill districts of Laos (Fig. 2a–f).

3 Datasets

3.1 Active Fires

We used the Near real-time (NRT) Suomi National Polar-orbiting Partnership (Suomi NPP) Visible Infrared Imaging Radiometer Suite (VIIRS) I Band 375 m active fire product (VNP14IMGTDL_NRT) for the study. The first VIIRS was launched in

Fig. 1 Laos country with 17 different provinces

October 2011 aboard the Suomi National Polar-orbiting Partnership (S-NPP) satellite. The VIIRS instrument carries two separate sets of multi-spectral channels providing full global coverage at both 375 and 750 m nominal resolutions every 12 h or less depending on the latitude. The VIIRS satellite incorporates fire-sensitive channels, including a dual-gain, high-saturation temperature 4 μm channel, enabling active fire detection and characterization. The active fire product, based on the 375 m (I-bands) and 750 m moderate resolution "M" bands of VIIRS, is currently generated (Schroeder et al. 2014). In this study, we specifically used the VIIRS 375 m active fire product (VNP14IMG), which builds on the well-established MODIS Fire and Thermal Anomalies product using a contextual approach to detect thermal anomalies (Schroeder et al. 2014). Due to its higher spatial resolution, the VNP14IMG active fire product captures more fire pixels than the MODIS MCDML product. Specific to the FRP, the VNP14IMG FRP is calculated using both VIIRS 375 and 750 m data. The former is used to identify fire-affected, cloud (solid blue), water (dashed blue), and valid background pixels. Then, co-located M13 channel radiance data (750 m) coinciding with fire pixels and valid background pixels are used in the FRP calculation. In contrast to other coarser resolution ($\geq$ 1 km) satellite fire detection products

Fig. 2 **a–f** Slash-and-burn agricultural sites in Northern Laos

such as MODIS, the improved 375 m data provide greater response over fires of relatively small areas, as well as improved mapping of large fire perimeters (Schroeder et al. 2014). Thus, the data are well suited for use in support of fire management, including other science applications (Vadrevu and Lasko 2018). The data are available in various formats, including the TXT, SHP, KML, and WMS, from the NASA FIRMS website (https://earthdata.nasa.gov/active-fire-data). We used these VIIRS 375 m location (x, y) datasets to infer fire variations in Laos.

3.2 MODIS Burned Areas

We also used the latest MODIS collection 6.1 burnt area product (MCD64A1) from 2000 to 2022 to assess the total areas burnt in different land cover classes, i.e., agriculture, forests, and grasslands. The MCD64A1 product contains the burning and quality information on a per-pixel basis and is produced using both MODIS Terra and AQUA surface reflectance inputs. The MODIS burnt area algorithm characterizes burnt area dynamics based on the surface reflectance data to identify changes in the landscapes and uses that information to detect the approximate date of burning, mapping the spatial extent of recent fires only (MODIS burnt area collection 6.1 User's Guide 2022). The product provides burnt information at a pixel level. The algorithm improves on previous methods by using a BRDF model-based change detection approach to handle angular variations in the data and uses a statistical measure to identify change probability from a previously observed state (MODIS 2022).

3.3 Land Cover

In this study, we used the Sentinel WorldCover 2021 (ver. 2) produced by the European Space Agency (https://worldcover2021.esa.int/) to characterize the land cover statistics for Laos (Fig. 3). The algorithm used to generate the ESA WorldCover product is based on the earlier algorithm that was used to produce the dynamic yearly Copernicus Global Land Service Land Cover (CGLS-LC) map at 100 m resolution (Buchhorn et al. 2020a, b). Both Sentinel-1 and 2 data are integrated into the WorldCover land cover product using different expert rules and are subsequently tiled into 3 $\times$ 3-degree tiles in geographic projection (EPSG:4326). More details about the algorithm can be found at https://worldcover2021.esa.int/data/docs/WorldCover_PUM_V2.0.pdf.

3.4 Temperature Condition Index (TCI)

We calculated the temperature condition index (TCI) using the MOD11A1 v006 LST data to relate fire counts to the land surface temperature variations. The MOD11A1 Version 6 product provides daily per-pixel land surface temperature and emissivity (LST&E) at a 1-km (km) spatial resolution in a 1200 by 1200-km grid. The pixel temperature value is derived from the MOD11 L2 swath product. More details about the LST product can be found at (https://lpdaac.usgs.gov/products/mod11a1v006/). TCI values give extra information about vegetation stress and determine whether stress is the result of dryness and extra wetness. TCI is calculated using the below equation (Kogan 1995)

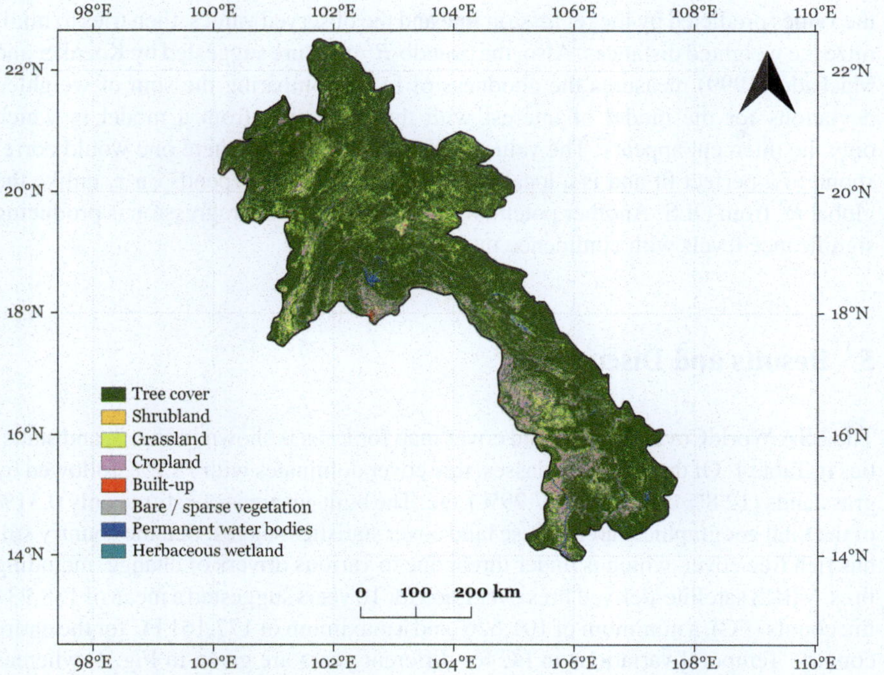

Fig. 3 Land cover map of Laos derived from WorldCover 2021 (ver. 2), European Space Agency. Tree cover class dominates with more than 75% of the total land cover

$$\mathrm{TCI} = \mathrm{LST}_{\max} - \mathrm{LST}_i / \mathrm{LST}_{\max} - \mathrm{LST}_{\min}$$

where LST_i defines LST value for a specific month and $\mathrm{LST}_{\min}$ and $\mathrm{LST}_{\max}$ denote maximum and minimum LST values. The TCI values range from 0 to 1 (or 0–100 based on scaling). The higher TCI values indicate the wet conditions, whereas the lower values indicate dry conditions.

4 Methods

To relate fires to the vegetation condition, we employed the traditional Ordinary Least Squares (OLS) regression method followed by a quantile regression. OLS focuses on the average relationship between a dependent variable and a set of explanatory variables, whereas quantile regression is relatively more advantageous than the typical OLS as it deals with heteroscedasticity by not assuming the distribution of the residuals, and it tends to resist the impact of outlying observations (Koenker and Hallock 2001; Yu et al. 2003; Hao et al. 2007). OLS regression aims to minimize the distances between the values predicted by the regression line and the observed values. In contrast, quantile regression differentially weights the distances between the values predicted by the regression line and the observed values, then tries to minimize the weighted distances. Also, the pseudo-R^2 measure suggested by Koenker and Machado (1999) measures the goodness of fit by comparing the sum of weighted deviations for the model of interest with the same sum from a model in which only the intercept appears. The values lie between 0 and 1, where one would correspond to a perfect fit and is a local measure of fit since it depends on τ, unlike the global R^2 from OLS. Another potential of using the quantile regression is producing significance levels with confidence intervals.

5 Results and Discussion

The ESA WorldCover derived land cover map for Laos is shown in Fig. 3, and statistics in Table 1. Of the different classes, tree cover dominates with 78.2%, followed by grasslands (12.8%), croplands (7.29%), etc. The built-up areas constitute only 0.31% of the total geographical area. These land cover statistics suggest that the country still has rich tree cover, which is under threat due to various drivers of change, including fires. VIIRS satellite-derived fires over Laos for 10 years suggested a mean of 138,934 fire counts (FC), a minimum of 101,626, and a maximum of 177,161 FC for the entire country. Temporal variations in FC for different years are given in Fig. 4, whereas seasonal trends in Fig. 5. The highest fires were found during the dry season, i.e., March till May, with the peak in April. The 5-min interval gridded data for FC and fire radiative power (FC) averaged across ten years is shown in Fig. 6a, b, respectively. The spatial patterns suggest a mean of 0–23 FC per cell and the highest of 128–338 FC per grid cell. Similarly, the FRP varied from a minimum of 0–7.49 MW to 43.15–120.15 MW. A close analysis of FC suggests that North Western Laos provinces of Oudomxai Louang Namtha, Bokeo, and Xiagnabouri had the highest fires in addition to the Northeastern province of Xiangkhoang, Southeastern province of Savannakhet, Western Savan, southern northern Atapu and southwestern Champasak. Similar patterns were shown for FRP (MW) with the highest in the Northern Laos provinces of Bokeo, Louang Namtha, Southern Phongsali, Louangphrabang, and Houaphan with lesser FRP (MW) intensities in southern provinces of Xekong and Saravan.

Results from the MODIS burnt area datasets revealed 665,923.49 ha per year for 23 years, with a minimum of 6568.56 ha and a maximum of 1,620,085 ha. Further, the burnt areas were the highest in the forests, with a mean of 311,529.86 per year, a minimum of 2615.2 ha/year, and a maximum of 799,795.65 ha/year. The grasslands had the second highest burnt areas with a mean of 210,810.17 ha/year, a minimum of 1484.64 ha/year, and a maximum of 532,624.68 ha/year. Finally, the croplands had a mean BA of 140,063.87 ha/year, a minimum of 2516.02 ha/year, and a maximum of 340,146.73/year. Temporal trends in burnt areas for forests, grasslands, and croplands are shown in Fig. 7.

The averaged TCI map for the entire country for a typical dry season from March–May, 2021 is shown in Fig. 8, which suggests significant variations. The Ordinary

Table 1 Land cover statistics derived from European Space Agency, WorldCover Map, 2021

Class name	Percent
Tree cover	78.22883549
Shrubland	0.018081947
Grassland	12.85098396
Cropland	7.297637842
Built-up	0.310053809
Bare/sparse vegetation	0.174861535
Permanent water bodies	1.118976344
Herbaceous wetland	0.000569075
Total	100

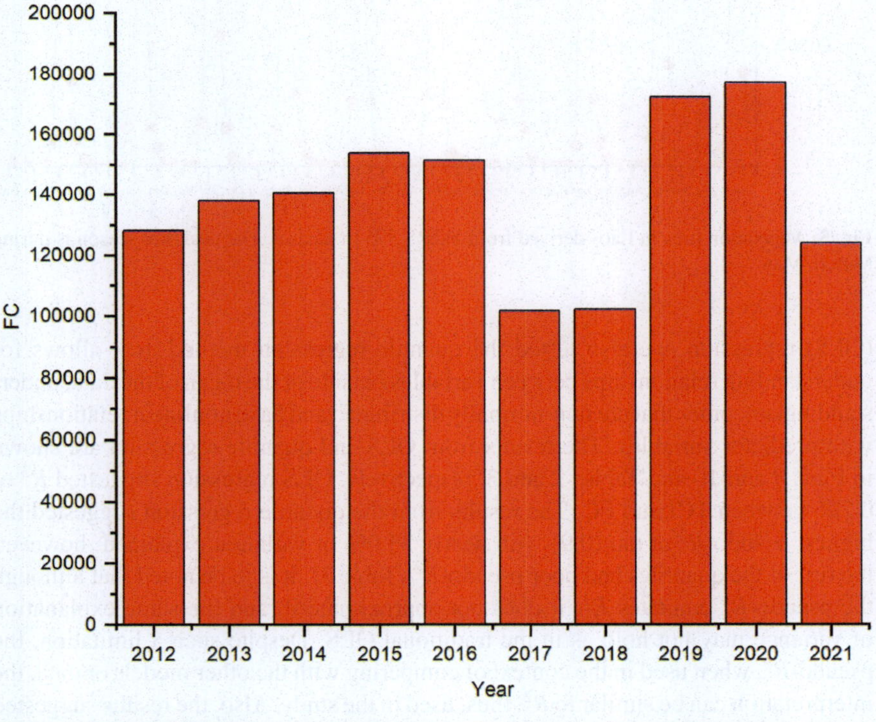

Fig. 4 Temporal variations in fire counts (FC) derived from VIIRS 375 m resolution dataset

Least Squares (OLS) regression plot with TCI as predictor and fire counts as response variable is shown in Fig. 8. TCI versus FC showed a negative slope, i.e., as the TCI increased, the FC decreased. These results are justified as the lower TCI indicates drought, and higher values suggest increased moisture. Overall, TCI could explain 28.1% of the variation in the fire datasets. In addition to the Ordinary Least Squares

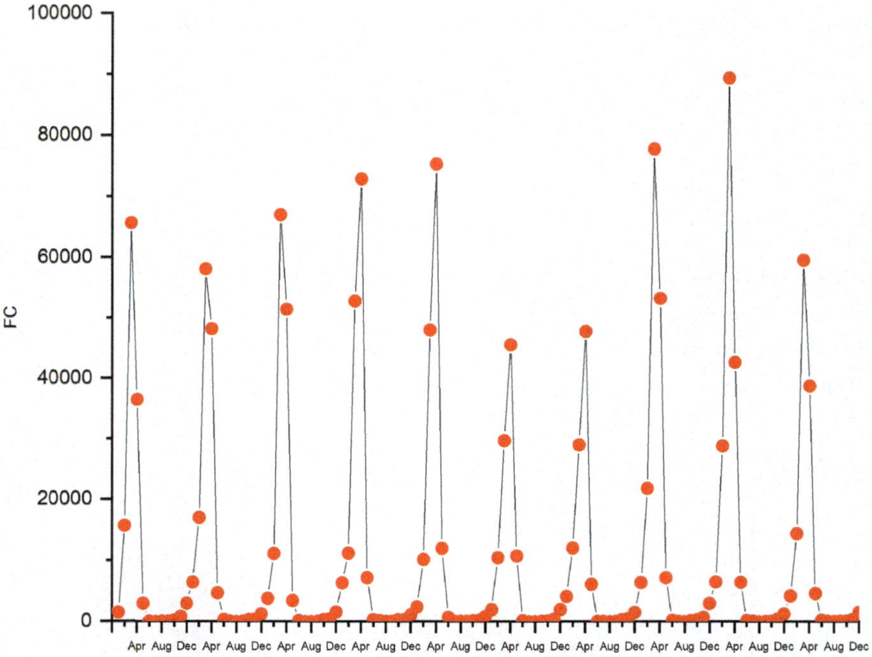

Fig. 5 Vegetation fires in Laos derived from VIIRS 375 m dataset. The peak fire season is during March–May

(OLS) regression, we also tested the quantile regression method as it allows for understanding relationships between variables outside of the mean of the data, understanding outcomes that are non-normally distributed and have nonlinear relationships with predictor variables. The results from OLS and quantile regression are shown in Figs. 9 and 10 and Tables 2 and 3, respectively. OLS regression suggested R^2 of 0.296 between TCI and FC. The results from the quantile regression suggested the highest pseudo-R^2 at tau (0.9) with nearly 35.0% of variation explained; however, the rest of the quantiles had poor pseudo-R^2 (Table 3). It is to be noted that although the pseudo-R^2 emulates R^2, it does not approximate it, and the same explanation of variance may not hold as in the traditional OLS. Despite such a limitation, the pseudo-R^2, when used in the context of comparing with the other models or data, the interpretation can be similar to R^2, thus, used in the study. Also, the results suggested that the magnitude of the impact of TCI on fires decreased as the quantiles decreased (Table 4). These results suggest that at 0.9 quantiles, TCI-FC relationship was relatively stronger than the other quantiles. Table 4 also provides coefficient differences for different quantiles, including lower and upper confidence intervals. The results suggest that the coefficients' magnitude and intensity change across the quantiles. Further, Prob > $|t|$ in the table represents the results from the p-value for the two-tailed test. A two-tailed t-test will test both if the mean is significantly greater than or less than x. The mean is considered significantly different from x if the test statistic is

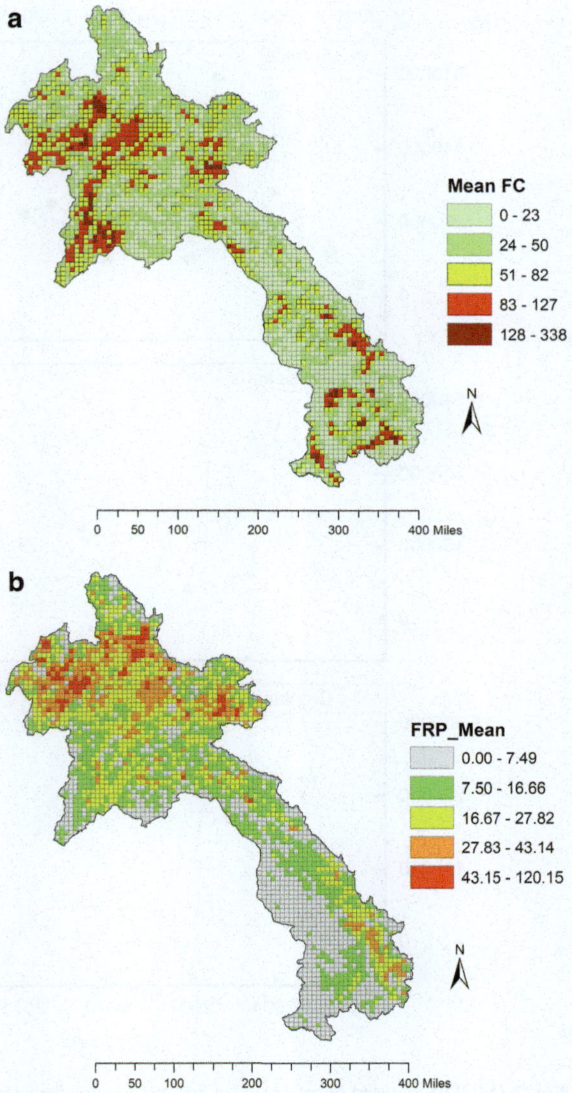

Fig. 6 **a** Vegetation fires in Laos derived from VIIRS 375 m and gridded at 5-min intervals. **b** Fire radiative power (FRP in MW) gridded data at 5-min intervals. High FC and FRP can be seen in the northern provinces of Laos

in the top 2.5% or bottom 2.5% of its probability distribution, resulting in a *p*-value less than 0.05. Thus, the *p*-value for 0.5 and 0.9 taus is significant.

The results from both the OLS and quantile regression suggest a minor role of LST as a driving factor of fires as they could only explain 29 and 35% of the variation in fire occurrence in Laos. More than the climate, previous studies highlighted that most of the fires in the country are caused due to anthropogenic interference, especially through slash-and-burn agriculture (Fig. 2a–f). In Laos, especially in northern provinces where fires are prevalent, including the slash and burn, important drivers

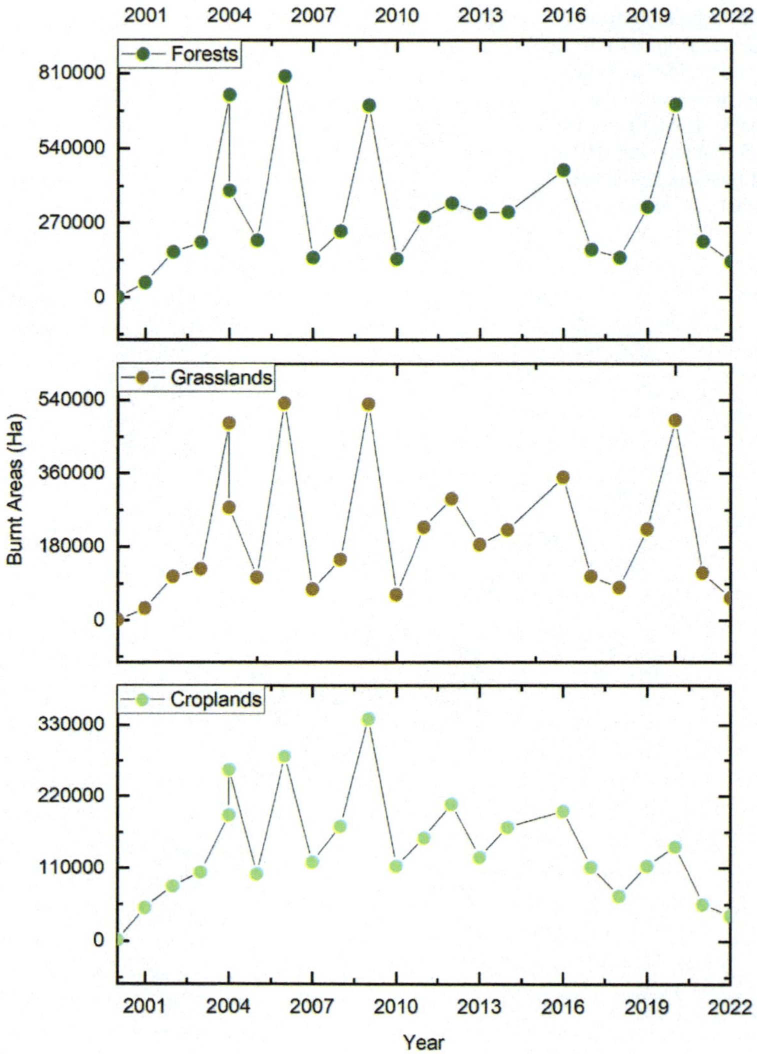

Fig. 7 MODIS derived burnt areas (in ha) in Laos for forests, grasslands, and croplands. Forests had the highest annual burnt areas followed by grasslands and croplands

include subsistence farming, economic and political factors, and government policies, including recent market forces. The region's land use changes are driven by the demand from the neighboring economies of China, Thailand, and Vietnam through trade and transportation (Hurni et al. 2013). Slash-and-burn agriculture is common in several northern provinces of Laos, where the terrain is hilly and mountainous. Mostly, upload rice is grown in the slash-and-burn systems. Forests are cleared in January and February and burnt during the dry months of March and April, where the peak fires are also seen (Fig. 4). Traditional rice is planted using a dibble stick in

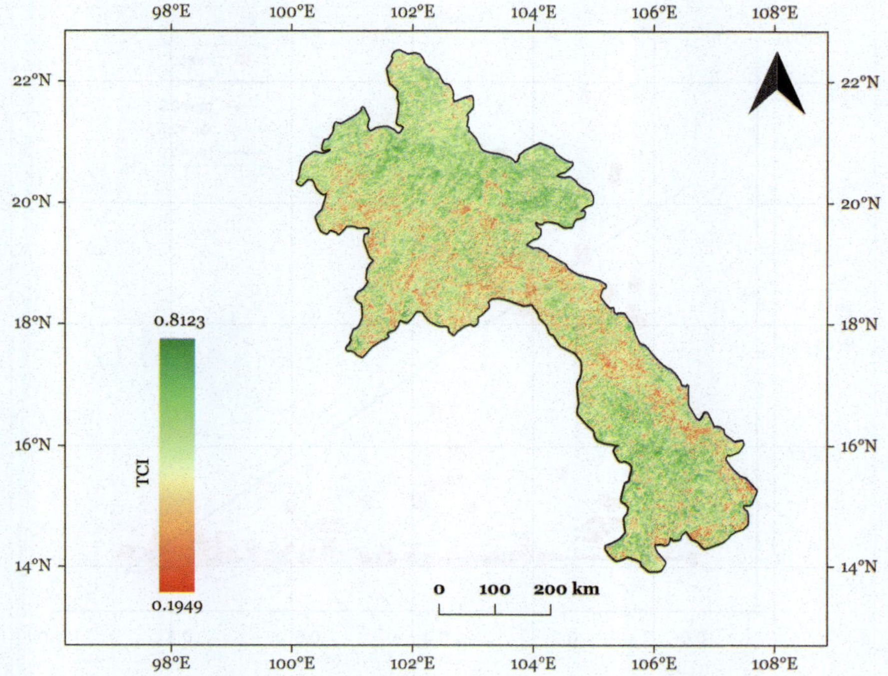

Fig. 8 Temperature condition index (TCI) map of Laos during the dry season (averaged data from March–May) 2021. MODIS land surface data have been used as inputs. The lower TCI values indicate dryness, and higher values indicate higher moisture

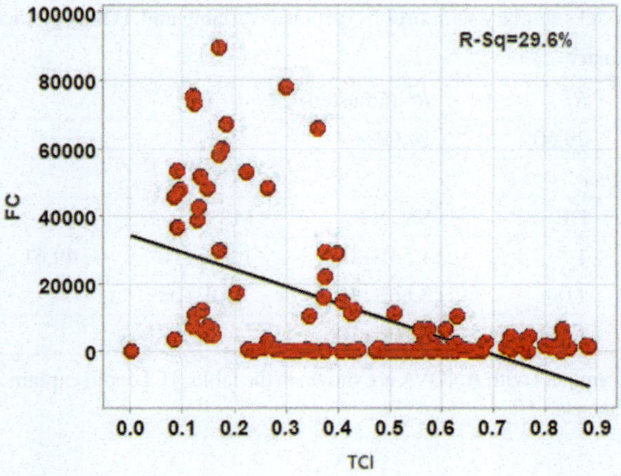

Fig. 9 OLS regression between the TCI and FC. TCI could explain 29.6% variation in the fire datasets

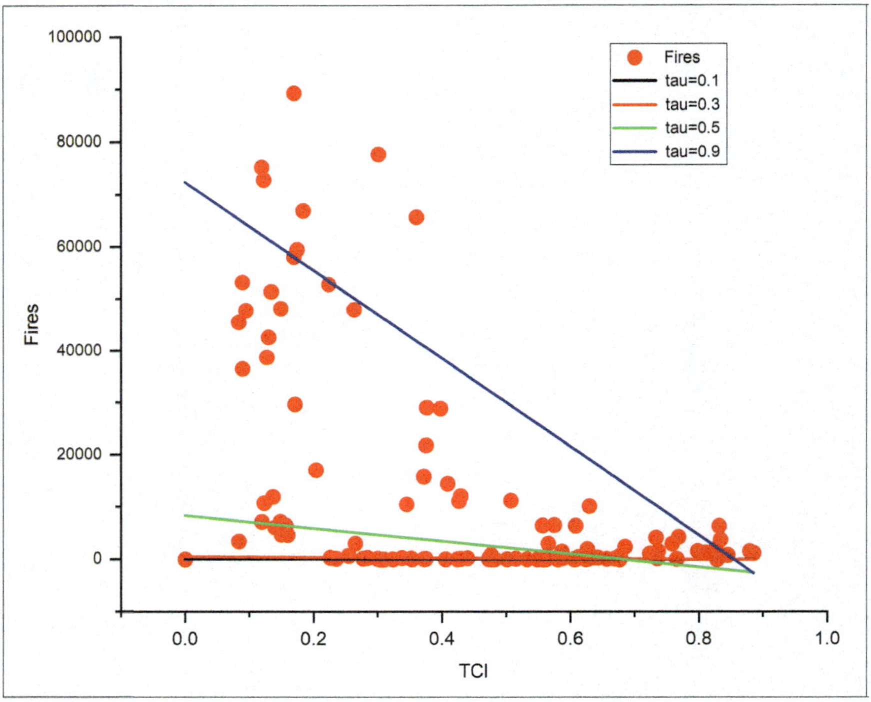

Fig. 10 Quantile regression between the TCI and FC. See text for more details

Table 2 OLS model summary with fires as response variable and TCI as predictor

OLS model summary					
S	R^2	R^2-(adjusted)			
17,607.5	29.60%	29.00%			
Analysis of variance					
Source	DF	SS	MS	*F*	*P*
Regression	1	1.54E+10	1.54E+10	49.61	0
Error	118	3.66E+10	3.10E+08		
	119	5.20E+10			

The regression parameters with ANOVA are shown in the table. TCI could explain 29.0% of variation in the Laos fire datasets

late May or early June. Various other crops are grown as intercrops, including maize, pumpkin, taro, cassava, chilies, sesame, sweet potato, eggplants, ginger, pigeon peas, etc. (Roder 2001). The fallow land is used for grazing cattle, and livestock is one of the important sources of cash income. During the past few decades, a literature review suggests that increasing population pressure and land use regulations have resulted in the expansion of shifting cultivation areas and shortening the fallow period

Table 3 Quantile regression results

Quantile regression	tau = 0.1	tau = 0.3	tau = 0.5	tau = 0.9
Number of points	120	120	120	120
Degrees of freedom	118	118	118	118
Raw sum of deviations	137,794.1	411,978.3	678,928.5	633,451.9
Min sum of deviations	137,789.49477	411,406.61938	655,723.95	411,728.83
Pseudo-*R*-squared	3.34211E−5	0.00139	0.034	0.35

(Douangsavanh et al. 2006; Inoue et al. 2010). However, such an expansion is not consistent. For example, recent studies suggest that the shifting cultivation varies differently in the provinces. For example, Fox et al. (2009) reported decreasing shifting cultivation, especially in Luang Prabang and Oudomxay provinces, with increasing permanent crops such as rice, maize, and sesame, including tree crops, mainly due to demand for these products from the neighboring countries (Alexander et al. 2010; Leek 2007). Governmental policies might have also reduced slash-and-burn practices and preference toward cash crops, especially in the Houaphan Province (Viau et al. 2009). The importance of market forces driving local agricultural practices in Bokeo and Luang provinces with a preference toward smallholder rubber plantations is documented in Ducourtieux et al. (2006), Nowak et al. (2008) and Lestrelin (2010). Similar is the case in the Xayabouly Province, with mostly maize grown to export to neighboring Thailand. Also, earlier researchers reported that poorer farmers still practice slash-and-burn agriculture as they cannot afford to invest in chemical fertilizers or pay taxes to the local government (Hurni et al. 2013) than rich farmers. With the rapidly changing economies in Southeast Asia, the demand for different agricultural products will continue. It remains to be seen how the slash-and-burn agriculture, including fires, will shape local ecosystems, including livelihoods in Laos. We infer that more ground-based studies linking satellite data with socioeconomic surveys are needed to address questions relating to sustainable natural resource management and forest-linked livelihood issues in Laos. Most importantly, for effective management and mitigation of fires, people's participation through consultation, cooperation, and collective action is much needed.

Table 4 Quantile regression parameters (UCL and LCL denote upper and lower confidence intervals, respectively, whereas Prob > |*t*| is the *p*-value for the two-tailed test)

	Value	Standard error	95% LCL	95% UCL	*t*-value	Prob > |*t*|
tau = 0.1						
Const	43.12	17.48013	8.50633	77.73716	2.4669	0.015
TCI	− 48.45	34.91584	117.59457	20.69118	− 1.38767	0.167
tau = 0.3						
Const	481.55	146.17	192.09	771.02	3.29	0.001
TCI	− 577.91	291.98	− 1156.11	0.285	− 1.97	0.050
tau = 0.5						
Const	8335.58	996.18	6362.86	10,308.29	8.37	0.00
TCI	− 12,206.24	1989.83	− 16,146.66	− 8265.83	− 6.13	0.00
tau = 0.9						
Const	72,470.39	5299.51	61,975.91	82,964.87	13.67	4.32635E−26
TCI	− 84,843.14	10,585.56	− 105,805.43	− 63,880.85	− 8.01	8.87437E−13

Acknowledgements This research is funded by the NASA Land Cover/Land Use Change Program, South/Southeast Asia Research Initiative (SARI) to the PI.

References

Albar, I., I. Jaya, B.H. Saharjo, B. Kuncahyo, and K.P. Vadrevu. 2018. Spatio-temporal analysis of land and forest fires in Indonesia using MODIS active fire dataset. In *Land-atmospheric research applications in South and Southeast Asia*, 105–127. Cham: Springer.

Alexander, K.S., J. Millar, and N. Lipscombe. 2010. Sustainable development in the uplands of Lao PDR. *Sustainable Development* 18 (1): 62–70.

Badarinath, K.V.S., and K.V. Prasad. 2011. Carbon dioxide emissions from forest biomass burning in India. *Global Environmental Research* 15: 45–52.

Badarinath, K.V.S., S.K. Kharol, K.M. Latha, T.K. Chand, V.K. Prasad, A.N. Jyothsna, and K. Samatha. 2007. Multiyear ground-based and satellite observations of aerosol properties over a tropical urban area in India. *Atmospheric Science Letters* 8 (1): 7–13.

Badarinath, K.V.S., S. Kumar Kharol, V. Krishna Prasad, A. Rani Sharma, E.U.B. Reddi, H.D. Kambezidis, and D.G. Kaskaoutis. 2008. Influence of natural and anthropogenic activities on UV Index variations—A study over tropical urban region using ground based observations and satellite data. *Journal of Atmospheric Chemistry* 59 (3): 219–236.

Badarinath, K.V.S., S.K. Kharol, A.R. Sharma, and V.K. Prasad. 2009. Analysis of aerosol and carbon monoxide characteristics over Arabian Sea during crop residue burning period in the Indo-Gangetic plains using multi-satellite remote sensing datasets. *Journal of Atmospheric and Solar-Terrestrial Physics* 71 (12): 1267–1276.

Biswas, S., K.D. Lasko, and K.P. Vadrevu. 2015a. Fire disturbance in tropical forests of Myanmar—Analysis using MODIS satellite datasets. *IEEE Journal of Selected Topics in Applied Earth Observations and Remote Sensing* 8 (5): 2273–2281.

Biswas, S., K.P. Vadrevu, Z.M. Lwin, K. Lasko, and C.O. Justice. 2015b. Factors controlling vegetation fires in protected and non-protected areas of Myanmar. *PLoS ONE* 10 (4): e0124346.

Biswas, S., K.P. Vadrevu, M.S. Mon, and C. Justice. 2021. Contemporary forest loss in Myanmar: Effect of democratic transition and subsequent timber bans on landscape structure and composition. *Ambio* 50 (4): 914–928.

Bond, W.J., and B.W. Van Wilgen. 2012. *Fire and plants*, vol. 14. Springer Science & Business Media.

Buchhorn, M., M. Lesiv, N.-E. Tsendbazar, M. Herold, L. Bertels, and B. Smets. 2020a. Copernicus global land cover layers—Collection 2. *Remote Sensing* 12 (6): 1044. https://doi.org/10.3390/rs12061044.

Buchhorn, M., L. Bertels, B. Smets, B. De Roo, M. Lesiv, N.E. Tsendbazar, D. Masiliunas, and L. Li. 2020b. Copernicus global land service: Land cover 100 m: Version 3 globe 2015–2019: Algorithm theoretical basis document. https://doi.org/10.5281/zenodo.3938968.

Choi, Y., S.A. Vay, K.P. Vadrevu, A.J. Soja, J.H. Woo, S.R. Nolf, G.W. Sachse, G.S. Diskin, D.R. Blake, N.J. Blake, and H.B. Singh. 2008. Characteristics of the atmospheric CO_2 signal as observed over the conterminous United States during INTEX-NA. *Journal of Geophysical Research: Atmospheres, 113*(D7).

Crutzen, P.J., and M.O. Andreae. 1990. Biomass burning in the tropics: Impact on atmospheric chemistry and biogeochemical cycles. *Science* 250 (4988): 1669–1678.

Dansereau, P.R., and Y. Bergeron. 1993. Fire history in the southern boreal forest of northwestern Quebec. *Canadian Journal of Forest Research* 23 (1): 25–32.

Douangsavanh, L., A. Polthanee, and R. Katawatin. 2006. Food security of shifting cultivation systems: Case studies from Luang Prabang and Oudomxay Provinces, Lao PDR. *Journal of Mountain Science* 3 (1): 48–57.

Ducourtieux, O., P. Visonnavong, and J. Rossard. 2006. Introducing cash crops in shifting cultivation regions—The experience with cardamom in Laos. *Agroforestry Systems* 66 (1): 65–76.

Eaturu, A., and K.P. Vadrevu. 2021. Evaluation of Sentinel-3 SLSTR data for mapping fires in forests, peatlands, and croplands—A case study over Australia, Indonesia, and India. In *Biomass burning in South and Southeast Asia*, 39–59. CRC Press.

Fox, J., Y. Fujita, D. Ngidang, N. Peluso, L. Potter, N. Sakuntaladewi, J. Sturgeon, and D. Thomas. 2009. Policies, political-economy, and Swidden in Southeast Asia. *Human Ecology* 37 (3): 305–322.

Goldammer, J.G., and C. Price. 1998. Potential impacts of climate change on fire regimes in the tropics based on MAGICC and a GISS GCM-derived lightning model. *Climatic Change* 39 (2): 273–296.

Goldammer, J.G., and B. Seibert. 1990. The impact of droughts and forest fires on tropical lowland rain forest of East Kalimantan. In *Fire in the tropical biota*, 11–31. Berlin, Heidelberg: Springer.

Gupta, P.K., V.K. Prasad, C. Sharma, A.K. Sarkar, Y. Kant, K.V.S. Badarinath, and A.P. Mitra. 2001a. CH_4 emissions from biomass burning of shifting cultivation areas of tropical deciduous forests–experimental results from ground-based measurements. *Chemosphere-Global Change Science* 3 (2): 133–143.

Gupta, P.K., V.K. Prasad, Y. Kant, C. Sharma, A.B. Ghosh, M.C. Sharma, A.K. Sarkar, S.L. Jain, O.P. Tripathi, R.C. Sharma, K.V.S. Badarinath. 2001b. Study of trace gases and aerosol emissions due to biomass burning at shifting cultivation sites in East Godavari District (Andhra Pradesh) during INDOEX IFP-99. *Current Science*, 186–196.

Hao, L., D.Q. Naiman, and D.Q. Naiman. 2007. *Quantile regression*, vol. 149. Sage.

Hayasaka, H., I. Noguchi, E.I. Putra, N. Yulianti, and K. Vadrevu. 2014. Peat-fire-related air pollution in Central Kalimantan, Indonesia. *Environmental Pollution* 195: 257–266.

Hurni, K., C. Hett, A. Heinimann, P. Messerli, and U. Wiesmann. 2013. Dynamics of shifting cultivation landscapes in Northern Lao PDR between 2000 and 2009 based on an analysis of MODIS time series and Landsat images. *Human Ecology* 41 (1): 21–36.

Inoue, Y. 2018. Ecosystem carbon stock, atmosphere, and food security in slash-and-burn land use: A geospatial study in mountainous region of Laos. In *Land-Atmospheric Research Applications in South and Southeast Asia*, 641–665. Cham: Springer.

Inoue, Y., Y. Kiyono, H. Asai, Y. Ochiai, J. Qi, A. Olioso, T. Shiraiwa, T. Horie, K. Saito, and L. Dounagsavanh. 2010. Assessing land-use and carbon stock in slash-and-burn ecosystems in tropical mountain of Laos based on time-series satellite images. *International Journal of Applied Earth Observation and Geoinformation* 12 (4): 287–297.

Justice, C., G. Gutman, and K.P. Vadrevu. 2015. NASA land cover and land use change (LCLUC): An interdisciplinary research program. *Journal of Environmental Management* 148: 4–9.

Kant, Y., V.K. Prasad, and K.V.S. Badarinath. 2000a. Algorithm for detection of active fire zones using NOAA AVHRR data. *Infrared Physics & Technology* 41 (1): 29–34.

Kant, Y., A.B. Ghosh, M.C. Sharma, P.K. Gupta, V.K. Prasad, K.V.S. Badarinath, and A.P. Mitra. 2000b. Studies on aerosol optical depth in biomass burning areas using satellite and ground-based observations. *Infrared Physics & Technology* 41 (1): 21–28.

Kharol, S.K., K.V.S. Badarinath, A.R. Sharma, D.V. Mahalakshmi, D. Singh, and V.K. Prasad. 2012. Black carbon aerosol variations over Patiala city, Punjab, India—A study during agriculture crop residue burning period using ground measurements and satellite data. *Journal of Atmospheric and Solar-Terrestrial Physics* 84: 45–51.

Koenker, R., and K.F. Hallock. 2001. Quantile regression. *Journal of Economic Perspectives* 15 (4): 143–156.

Koenker, R., and J.A. Machado. 1999. Goodness of fit and related inference processes for quantile regression. *Journal of the American Statistical Association* 94 (448): 1296–1310.

Kogan, F.N. 1995. Application of vegetation index and brightness temperature for drought detection. *Advances in Space Research* 15 (11): 91–100.

Lasko, K., and K. Vadrevu. 2018. Improved rice residue burning emissions estimates: Accounting for practice-specific emission factors in air pollution assessments of Vietnam. *Environmental Pollution* 236: 795–806.

Lasko, K., K.P. Vadrevu, V.T. Tran, E. Ellicott, T.T. Nguyen, H.Q. Bui, and C. Justice. 2017. Satellites may underestimate rice residue and associated burning emissions in Vietnam. *Environmental Research Letters* 12 (8): 085006.

Lasko, K., K.P. Vadrevu, and T.T.N. Nguyen. 2018. Analysis of air pollution over Hanoi, Vietnam using multi-satellite and MERRA reanalysis datasets. *PLoS ONE* 13 (5): e0196629.

Lasko, K., K.P. Vadrevu, V. Bandaru, T.N.T. Nguyen, H.Q. Bui. 2021. $PM_{2.5}$ emissions from biomass burning in South/Southeast Asia–Uncertainties and trade-offs. In *Biomass burning in South and Southeast Asia*, 149–169. CRC Press.

Leek, K.E. 2007. *Rural livelihood strategies and natural resources in Oudomxay, Lao PDR*. M.Sc. thesis. University of Natural Resources and Applied Life Sciences, Vienna; Lincoln University, Christchurch, NZ, Natural Resources Management and Ecological Engineering.

Lestrelin, G. 2010. Land degradation in the Lao PDR: Discourses and policy. *Land Use Policy* 27 (2): 424–439.

MODIS. 2022. MODIS burnt areas. https://modis-fire.umd.edu/files/MODIS_C61_BA_User_Guide_1.1.pdf.

Myers, N. 1996. Two key challenges for biodiversity: Discontinuities and synergisms. *Biodiversity & Conservation* 5 (9): 1025–1034.

Nowak, A., B. Southavilay, P. Souliyan, R. Bounmi, I. Ponepaseuth, N. Morel, F. Jullien, and V. Singkeuth. 2008. Participative analysis of the maize supply value chain in Lao PDR. Ministry of Agriculture and Forestry (MAF), Programme of Capitalisation in Support of Rural Development Policy (PCADR), Central Unit (UC), Vientiane.

Petropoulos, G.P., K.P. Vadrevu, and C. Kalaitzidis. 2013. Spectral angle mapper and object-based classification combined with hyperspectral remote sensing imagery for obtaining land use/cover mapping in a Mediterranean region. *Geocarto International* 28 (2): 114–129.

Prasad, V.K., and K.V.S. Badarinth. 2004. Land use changes and trends in human appropriation of above ground net primary production (HANPP) in India (1961–98). *Geographical Journal* 170 (1): 51–63.

Prasad, K.V., and K.V.S. Badarinath. 2006. Soil surface nitrogen losses from agriculture in India: A regional inventory within agroecological zones (2000–2001). *The International Journal of Sustainable Development and World Ecology* 13 (3): 173–182.

Prasad, V.K., Y. Kant, P.K. Gupta, C. Sharma, A.A. Mitra, and K.V.S. Badarinath. 2001a. Biomass and combustion characteristics of secondary mixed deciduous forests in Eastern Ghats of India. *Atmospheric Environment* 35 (18): 3085–3095.

Prasad, V.K., Y. Kant, and K.V.S. Badarinath. 2001b. CENTURY ecosystem model application for quantifying vegetation dynamics in shifting cultivation areas: A case study from Rampa Forests, Eastern Ghats (India). *Ecological Research* 16 (3): 497–507.

Prasad, V.K., Y. Kant, P.K. Gupta, C. Elvidge, and K.V.S. Badarinath. 2002a. Biomass burning and related trace gas emissions from tropical dry deciduous forests of India: A study using DMSP-OLS data and ground-based measurements. *International Journal of Remote Sensing* 23 (14): 2837–2851.

Prasad, V.K., Y. Kant, and K.V.S. Badarinath. 2002b. Land use changes and modeling carbon fluxes from satellite data. *Advances in Space Research* 30 (11): 2511–2516.

Prasad, V.K., M. Lata, and K.V.S. Badarinath. 2003. Trace gas emissions from biomass burning from northeast region in India—Estimates from satellite remote sensing data and GIS. *The Environmentalist* 23 (3): 229–236.

Prasad, V.K., K.V.S. Badarinath, S. Yonemura, and H. Tsuruta. 2004. Regional inventory of soil surface nitrogen balances in Indian agriculture (2000–2001). *Journal of Environmental Management* 73 (3): 209–218.

Prasad, V.K., E. Anuradha, and K.V.S. Badarinath. 2005. Climatic controls of vegetation vigor in four contrasting forest types of India—Evaluation from National oceanic and atmospheric administration's advanced very high resolution radiometer datasets (1990–2000). *International Journal of Biometeorology* 50 (1): 6–16.

Roder, W. 2001. *Slash-and-burn rice systems in the hills of northern Lao PDR: Description, challenges, and opportunities.*

Schroeder, W., P. Oliva, L. Giglio, and I.A. Csiszar. 2014. The New VIIRS 375 m active fire detection data product: Algorithm description and initial assessment. *Remote Sensing of Environment* 143: 85–96.

Thapa, P., K.P. Vadrevu, A. Eaturu, and S. Biswas. 2021. Vegetation fire status and management in Bhutan. In *Biomass burning in South and Southeast Asia*, 135–157. CRC Press.

Turner, M.G. 1989. Landscape ecology: The effect of pattern on process. In *Annual review of ecology and systematics*, 171–197.

Vadrevu, K.P. 2008. Analysis of fire events and controlling factors in eastern India using spatial scan and multivariate statistics. *Geografiska Annaler: Series A, Physical Geography* 90 (4): 315–328.

Vadrevu, K.P. 2015. International regional science meeting of NASA-LCLUC: Land Cover/Land Use Change (LC/LUC) and environmental impacts in South Asia, Coimbatore, India, 19–23 January 2013. *Journal of Environmental Management* 148: 1–163.

Vadrevu, K.P., and K.V.S. Badarinath. 2009. Spatial pattern analysis of fire events in Central India—A case study. *Geocarto International* 24 (2): 115–131.

Vadrevu, K.P., and C.O. Justice. 2011. Vegetation fires in the Asian region: Satellite observational needs and priorities. *Global Environmental Research* 15 (1): 65–76.

Vadrevu, K., and K. Lasko. 2015. Fire regimes and potential bioenergy loss from agricultural lands in the Indo-Gangetic plains. *Journal of Environmental Management* 148: 10–20.

Vadrevu, K.P., K.V.S. Badarinath, and E. Anuradha. 2008. Spatial patterns in vegetation fires in the Indian region. *Environmental Monitoring and Assessment* 147 (1): 1–13.

Vadrevu, K.P., I. Csiszar, E. Ellicott, L. Giglio, K.V.S. Badarinath, E. Vermote, and C. Justice. 2012. Hotspot analysis of vegetation fires and intensity in the Indian region. *IEEE Journal of Selected Topics in Applied Earth Observations and Remote Sensing* 6 (1): 224–238.

Vadrevu, K.P., L. Giglio, and C. Justice. 2013. Satellite based analysis of fire–carbon monoxide relationships from forest and agricultural residue burning (2003–2011). *Atmospheric Environment* 64: 179–191.

Vadrevu, K.P., T. Ohara, and C. Justice. 2014a. Air pollution in Asia. In *Environmental pollution (Barking, Essex: 1987)*, vol. 195, 233–235.

Vadrevu, K.P., K. Lasko, L. Giglio, and C. Justice. 2014b. Analysis of Southeast Asian pollution episode during June 2013 using satellite remote sensing datasets. *Environmental Pollution* 195: 245–256.

Vadrevu, K., T. Ohara, and C. Justice. 2017. Land cover, land use changes and air pollution in Asia: A synthesis. *Environmental Research Letters* 12 (12): 120201.

Vadrevu, K., A. Heinimann, G. Gutman, and C. Justice. 2019. Remote sensing of land use/cover changes in South and Southeast Asian Countries. *International Journal of Digital Earth* 12 (10): 1099–1102.

Vadrevu, K.P., A. Eaturu, S. Biswas, K. Lasko, S. Sahu, J.K. Garg, and C. Justice. 2020. Spatial and temporal variations of air pollution over 41 cities of India during the COVID-19 lockdown period. *Scientific Reports* 10 (1): 1–15.

Vadrevu, K.P., T. Le Toan,, S.S. Ray, and C. Justice. 2022a. *Remote sensing of agriculture and Land Cover/Land Use Changes in South and Southeast Asian Countries.* Cham: Springer.

Vadrevu, K., A. Eaturu, E. Casadaban, K. Lasko, W. Schroeder, S. Biswas, L. Giglio, and C. Justice. 2022b. Spatial variations in vegetation fires and emissions in South and Southeast Asia during COVID-19 and pre-pandemic. *Scientific Reports* 12 (1): 1–21.

Vadrevu, K.P., T. Ohara, and C. Justice, eds. 2018. *Land-atmospheric research applications in South and Southeast Asia.*

Vadrevu, K.P., T. Ohara, and C. Justice, eds. 2021a. *Biomass burning in South and Southeast Asia: Impacts on the biosphere*, vol. 2. CRC Press.

Vadrevu, K.P., T. Ohara, and C. Justice, eds. 2021b. *Biomass burning in South and Southeast Asia: Mapping and monitoring*, vol. 1. CRC Press.

Vay, S.A., Y. Choi, K.P. Vadrevu, D.R. Blake, S.C. Tyler, A. Wisthaler, A. Hecobian, Y. Kondo, G.S. Diskin, G.W. Sachse, and J.H. Woo. 2011. Patterns of CO_2 and radiocarbon across high northern latitudes during international polar year 2008. *Journal of Geophysical Research: Atmospheres*, *116*(D14).

Vázquez, A., and J.M. Moreno. 2001. Spatial distribution of forest fires in Sierra de Gredos (Central Spain). *Forest Ecology and Management* 147 (1): 55–65.

Viau, J., A. Keophosay, and J.-C. Castella. 2009. Impact of maize expansion on traditional rice production systems in Northern Lao PDR. A case study in Xiengkhor district, Huaphan province. Catch-Up Programme, National Agriculture and Forestry Research Institute, Vientiane.

Wooster, M.J., G.J. Roberts, L. Giglio, D.P. Roy, P.H. Freeborn, L. Boschetti, C. Justice, C. Ichoku, W. Schroeder, D. Davies, and A.M. Smith. 2021. Satellite remote sensing of active fires: History and current status, applications and future requirements. *Remote Sensing of Environment* 267: 112694.

Yu, K., Z. Lu, and J. Stander. 2003. Quantile regression: Applications and current research areas. *Journal of the Royal Statistical Society: Series D (The Statistician)* 52 (3): 331–350.

Vegetation Fires, Fire Radiative Power, and Intermediate Fire Occurrence-Intensity (IFOI) Hypothesis Testing in Myanmar, Laos, and Cambodia

Krishna Prasad Vadrevu, Aditya Eaturu, Thav Sopheak, Chittana Phompila, and Sumalika Biswas

Abstract In this study, we used VIIRS (375m) satellite data to characterize vegetation fire counts (FC) and fire radiative power (FRP) variations in Myanmar, Laos, and Cambodia. We specifically tested the intermediate fire occurrence-intensity (IFOI) hypothesis, which states that the fire occurrence in units of time per unit of an area increases with fire intensity up until a threshold is reached above which occurrence decreases with increasing intensity, in essence, a humped relationship. We used Gaussian, Power, and Piecewise curve fitting between the FRP and fire counts, to test the statistical behavior and the IFOI hypothesis. The results from averaging VIIRS data from 2012 to 2020 suggested Myanmar with a higher mean FC per year (350,943), followed by Cambodia (183,906) and Laos (139,416). In Myanmar, the results suggested the Piecewise model with a relatively higher r^2 (0.54) than the Power ($r^2 = 0.47$) and Gaussian model ($r^2 = 0.41$) for the FRP-FC relationship. In contrast, the Power and Piecewise models in Laos had almost similar r^2 (0.50), whereas Gaussian was lower than both ($r^2 = 0.41$) models. In Cambodia, Piecewise ($r^2 = 0.83$) and Power ($r^2 = 0.82$) models had a better performance for FRP-FC relationship than the Gaussian ($r^2 = 0.74$) model. The Piecewise models suggested a relatively higher clustering of fires with lower FRP (MW) for fires in Myanmar than in Laos and Cambodia. The humped relationship between FRP and FC highlighted in earlier studies was not fully apparent in Southeast Asian countries which might be attributed to variations in topography, fuels, climate, and anthropogenic drivers.

K. P. Vadrevu (✉)
NASA Marshall Space Flight Center, Huntsville, AL, USA
e-mail: Krishna.p.vadrevu@nasa.gov

A. Eaturu
University of Alabama, Huntsville, USA

T. Sopheak
Royal University of Agriculture, Phnom Penh, Cambodia

C. Phompila
National University of Laos, Vientiane, Laos

S. Biswas
University of California, Los Angeles, USA

K. P. Vadrevu et al. (eds.), *Vegetation Fires and Pollution in Asia*,
https://doi.org/10.1007/978-3-031-29916-2_13

Keywords Vegetation fires · Fire radiative power · Intermediate fire occurrence hypothesis · Southeast Asia

1 Introduction

Fires in South/Southeast Asia (S/SEA) are common and result from both climatic and anthropogenic drivers (Badarinath et al. 2007, 2008, 2009; Vadrevu et al. 2019; Biswas et al. 2015a, b, 2021). In different S/SEAn countries, fires are commonly used to convert forests to agriculture, such as through slash-and-burn agriculture (Brady 1996; Albar et al. 2018). Also, in most countries, fire is used to burn agricultural residues after harvest to clear the fields and plant the next crop impacting agroecosystem characteristics (Lasko et al. 2017, 2018; Vadrevu et al. 2018). Additional reasons for fires in the region include lightning, accidental burning, promoting grass growth for livestock consumption, hunting, etc. (Vadrevu et al. 2021a, b). As a result of these practices, fire regimes vary widely in the region, and some landscapes are more resilient to changes in fire regimes (Murdiyarso and Lebel 2007; Prasad et al. 2001a, b, 2002a, b, 2003a, b, 2004; Prasad and Badarinath 2004; Vadrevu 2008; Hayasaka et al. 2014). The impacts of fires on ecosystems and the environment vary widely, such as disruption of nutrient cycles, soil erosion, and vegetation loss, including the release of aerosols and other gases such as carbon dioxide, carbon monoxide (CO), oxides of nitrogen (NO_x), and other reactive trace gases, impacting local and regional air quality (Choi et al. 2008; Kant et al. 2000; Prasad et al. 2005; Prasad and Badarinth 2006; Badarinath and Prasad 2011; Huang et al. 2013; Lata et al. 2001; Lasko and Vadrevu 2018; Vadrevu et al. 2022a, b; Vay et al. 2011). Fires can also cause property damage, and the smoke released can have severe health implications. Furthermore, the smoke plumes can travel long distances and affect regional air quality. Specifically, air pollution caused by aerosol particles can reduce visibility, including health impacts (Mauderly and Chow 2008; Kharol et al. 2012; Vadrevu and Lasko 2015; Vadrevu 2015). In addition, the small aerosol particles can penetrate the respiratory system and cause lung infections and cancers (Karthikeyan et al. 2006). Thus, it is important to map and monitor fires, including addressing the impacts.

Satellite remote sensing has been proven effective for mapping and monitoring fires due to its multi-spectral, multitemporal, synoptic, and repetitive coverage capabilities (Giglio et al. 2003; Petropoulos et al. 2013; Vadrevu and Badarinath 2009; Vadrevu and Justice 2011; Justice et al. 2015; Wooster et al. 2021). Besides providing robust estimates on the location of fires, satellite data can be effectively used to capture the size of burned areas, the type of vegetation burnt, and the degree of burn severity, including fire intensity, using fire radiative power (FRP), and danger (Giglio et al. 2003; Freeborn et al. 2014; Vadrevu 2008; Vadrevu et al. 2008a, b, 2013, 2014, 2017,

2018, 2019, 2021a, b; Vadrevu 2021). Several active fire products have been developed in recent years utilizing satellite remote sensing. A review of the different algorithms can be found in Roberts and Wooster (2014). Some of the active fire products integrate data from satellites such as Tropical Rainfall Measuring Mission (TRMM; Giglio et al. 2003), the Along Track Scanning Radiometer (ATSR; Arino and Rosaz 1999), the Moderate Resolution Imaging Spectroradiometer (MODIS) (Giglio et al. 2003; Kaufman et al. 1998), and Visible Infrared Imaging Radiometer Suite (VIIRS) (Schroeder et al. 2014). These products, over a period of time, are refined to offer improved sensitivity to a variety of fires. However, still, it is challenging to map and monitor small fires, such as in agricultural residue burning, where they are short-lived and can occur in small field sizes. However, the active fire products also include instantaneous estimates of the power released by a fire (fire radiative power, FRP) integrated into the active fire detection products, which are quite helpful in assessing fire impacts in different ways (Wooster et al. 2005). For example, Kaufman et al. (1996) used the FRP measures as metric relating relative rates of vegetation fire smoke emission for the SCAR-B and SCAR-C airborne campaigns. Wooster et al. (2005) and Kremens et al. (2012) showed FRP to be directly related to combustion rates in laboratory-scale experiments. Boschetti and Roy (2009) integrated FRP into burned area estimation algorithms. Traditionally, Sieler and Crutzen's (1980) approach to emissions estimation includes the fire-affected area (units: m^2), the pre-burn fuel load (kg m^{-2}), combustion completeness (unitless: 0–1), and emission factors (g/kg). Compared to this approach, Kaufman et al. (1996) proposed that the rate of emission of fire radiative energy could be used to indicate the rate of combustion. Since then, measurements of fire radiative power (FRP) from polar-orbiting sensors have been used to characterize active fire (AF) properties, including fire radiative energy (FRE) to the rates of fuel consumption and trace gas and aerosol emission in a variety of landscapes (Roberts and Wooster 2008; Ichoku and Ellison 2014; Vermote et al. 2009; Vadrevu and Lasko 2018). Many different Earth Observation (EO) instruments have been employed in such efforts, and multiple algorithms were developed for the detection of actively burning fires and determination of their FRP, e.g., (Wooster et al. 2003; Kaufman et al. 1998; Giglio et al. 2003; Elvidge et al. 2013; Peterson et al. 2013; Schroeder et al. 2014; Eaturu and Vadrevu 2021; Elvidge et al. 2021). One of the most widely used FRP retrieval approaches is the mid-infrared (MIR) radiance method, developed by Wooster et al. (2003), now used for FRP derivation from a variety of satellites such as MODIS, SEVERI, VIIRS, GOES imagers, and Sentinel SLSTR (Xu et al. 2010; Wooster et al. 201220132013; Giglio et al. 2016). Integrating FRP in time provides an estimate of the total energy released (fire radiative energy, FRE), which can be converted into burned biomass estimates useful for estimating the emissions (Boschetti and Roy 2009).

In this study, we characterize the FRP variations in different SEAn countries. Studies from Pausas and Ribeiro (2013) and Luo et al. (2017) have shown that fire occurrence, defined as the number of remotely detected active fires in units of time per unit of area, increases with fire intensity up until a threshold is reached above which occurrence decreases with increasing intensity, in essence, a humped relationship. They named this intermediate fire occurrence-intensity (IFOI) hypothesis. We tested this hypothesis in this study for the South/Southeast Asian countries individually using the VIIRS fire counts (FC) and FRP data. The results identify the hotspots of FC and FRP, including the FRP-FC threshold results in different countries, useful for fire management and mitigation purposes.

2 Study Area

We focused our study on three different countries in S/SEA which included Myanmar, Laos, and Cambodia. These countries cover a longitudinal gradient in the SEA region. Myanmar, formerly known as Burma, lies between latitudes 9° and 29° N and longitudes 92° and 102° E. It is the largest country in mainland Southeast Asia. Laos is the only landlocked country in Southeast Asia. It is at the heart of the Indochinese Peninsula and bordered by Myanmar and China to the northwest, Vietnam to the east, Cambodia to the southeast, and Thailand to the west and southwest. Cambodia is a country located in the southern portion of the Indochinese Peninsula in Southeast Asia. In these countries, recurrent fires are most common; however, they show significant spatial and temporal variations driven by topography, climate, vegetation, and anthropogenic drivers.

3 Data

The first VIIRS sensor is on board the Suomi National Polar-orbiting Partnership (Suomi NPP) satellite that was launched in 2011. Both Suomi NPP and Aqua satellites cross the equator at approximately 1:30 a.m. (descending orbit) and 1:30 p.m. (ascending orbit) local times. To address the FC and FRP variations in different countries, we used the 375m active fire product derived from the VIIRS instruments onboard the Suomi National Polar-orbiting Partnership (S-NPP) and NOAA-20 satellites. In contrast to other coarser resolution satellite fire detection products such as MODIS ($\geq$ 1 km), the improved 375m data provide increased detection of smaller fires and enhanced mapping of large fire perimeters. The VIIRS 375m fire product builds on the earlier MODIS fire product heritage, using a multi-spectral contextual algorithm to identify sub-pixel fire activity and other thermal anomalies in the Level 1 (swath) input data. The algorithm uses all five 375m VIIRS channels to detect fires and separate land, water, and cloud pixels in the image (Schroeder et al. 2014). In the VIIRS data, for the daytime data, cloud pixels are classified using brightness temperature (BT) tests in channel 5 (< 265 K) or reflectance in I-band channel $\rho_1 + \rho_2 >$ 0.9 and $BT_5 < 295$ K or $\rho_1 + \rho_2 > 0.7$ and $BT_5 < 285$ K where ρ_i is the reflectance in I-band channel I and BT_i is the brightness temperature in I-band channel (i). For nighttime data, cloud pixels are classified based on the brightness temperature of channels I4 and I5 as $BT_5 < 265$ K and $BT_4 < 295$ K. Using these tests, the fire algorithm skips all day and nighttime pixels classified as cloud-covered, and their data are excluded from the calculation of fire pixel background conditions. Also, the VIIRS active fire algorithm derives the FRP based on the method proposed by Wooster et al. (2005).

4 Methods

To address the spatial variations in FRP, we first aggregated the VIIRS FRP data at 5-min intervals for Myanmar, Laos, and Cambodia. We then analyzed FC and FRP data using various statistical metrics. Apart from using descriptive statistics, we also tested the normality of the FC and FRP datasets using the Anderson–Darling (AD) test. The two hypotheses for the AD test for the normal distribution are H0: The data follow the normal distribution and H1: The data do not follow the normal distribution. The *p*-value is the probability of getting a result that is more extreme if the null hypothesis is true. Thus, if the *p*-value is low (e.g., ≤ 0.05), we can conclude that the data do not follow the normal distribution and vice versa. In addition, we used various curve-fitting models such as Gaussian, Power, and Piecewise 2-segment linear to explore the relationship between the FRP and FC data at 5-min intervals to test the IFOI hypothesis. The Gaussian equation is an exponentially decaying curve centered around the mean of the distribution scaled by a factor. The graph of the Gaussian distribution depends on two factors—the mean and the standard deviation. The mean of the distribution determines the location of the center of the graph, and the standard deviation determines the height and width of the graph. The height is determined by the scaling factor, and the width is determined by the factor in the power of the exponential. When the standard deviation is large, the curve is short and wide; when the standard deviation is small, the curve is tall and narrow. All Gaussian distributions look like symmetric, bell-shaped curves. A power law relationship between two variables denotes that a change in one variable can lead to a large change in the other, regardless of the initial quantities; such a relationship for fire behavior has been shown by several researchers (Wooster et al. 2004; Kumar et al. 2011; Laurent et al. 2019). In contrast, the basic idea of Piecewise linear regression is that if the data follow different linear trends over different regions of the data, then modeling the regression function in "pieces" can yield a robust regression estimate. The Piecewise regression model is formulated into two separate linear functions connected at some value called the "knot value or breakpoint value". The regression results and the parameter values for different curve-fitted models were presented for FRP-FC relationships for Myanmar, Laos, and Cambodia and discussed in the context of the IFOI hypothesis, including implications for fire management.

5 Results and Discussion

Spatial variations in FC and FRP for Myanmar, Laos, and Cambodia are given in Figs. 1a–c, 2a–c and 3a–c). In Myanmar, the fires were dominant in Shan, Sagaing, and Kayin and stretching from Magway, Bago, and Ayeyarwady delta (Fig. 1b). The higher FRP can be seen in Southern Shan, Southern Sagaing, Magway, Bago, and Southern Ayeyarwady delta (Fig. 1c). In Laos, fires are primarily concentrated in the northern regions, especially Houaphan, Louangphrabang, Oudomxai, Loung Namtha, Bokeo, Xiagnabouri, and south-central regions of Saravan, Champasak, and Attapu provinces (Fig. 2b). The FRP variations followed the fire patterns in the country (Fig. 2c). In Cambodia, relatively higher FC can be seen in northern provinces such as Preh Vihear and Stoeng Treng and in eastern provinces of Rotanokiri, Mondoil Kiri, and Kracheh (Fig. 3b). Higher FRP can be found for fires in Rotanokiri, Stoeng Treng, and Northern Kracheh provinces (Fig. 3c).

A comparison of total fire counts for different years in Myanmar, Laos, and Cambodia is shown in Fig. 4. Of the different countries, Myanmar had higher mean FC (Myanmar higher mean FC (350,943), followed by Cambodia (183,906) and Laos (139,416). The summary statistics of fire counts (FC) and FRP (MW) from the VIIRS data aggregated at 5-min intervals for the year 2020 is shown in Figs. 5a, b, 6a, b and 7a, b for Myanmar, Laos, and Cambodia, respectively. For all countries, the *p*-value for both the FC and FRP was lower than 0.05; thus, we rejected the null hypothesis that the data are from a normal distribution. The total number of fires was relatively higher for Myanmar, Laos, and Cambodia. However, the mean FC, median, and maximum including standard deviation, variance, and skewness were relatively higher for Cambodia, followed by Laos and Myanmar. The kurtosis, which indicates the degree to which scores cluster in the tails or the peak of a frequency distribution, is relatively higher for Laos, followed by Cambodia and Myanmar. In the case of the FRP, fires in Laos had a higher mean, standard deviation, and variance followed by Cambodia and Myanmar. The skewness and kurtosis in FRP were higher for Myanmar, followed by Laos and Cambodia. The maximum FRP was for Laos, followed by Myanmar and Cambodia. These results indicated that the fires in Laos burn more intensively than in Cambodia or Myanmar.

The results from the curve fitting of FRP versus FC using Gaussian 3-parameter peak, Power 2-parameter, and Piecewise 2-segment linear ones were shown in Figs. 8a–c, 9a–c, and 10a–c and Tables 1, 2 and 3 for Myanmar, Laos, and Cambodia, respectively. In the case of Myanmar, the Piecewise 2-segment linear fitting had a relatively higher r^2 (0.54) followed by the Power ($r^2 = 0.47$) and Gaussian fitting ($r^2 = 0.41$). In contrast, for Laos, the Power fitting and Piecewise 2-segment linear fitting had almost the same r^2 (0.50), whereas Gaussian was lower than both ($r^2 = 0.41$). In Cambodia, FRP-FC relationship was closer for Piecewise 2-segment linear ($r^2 = 0.83$) and Power ($r^2 = 0.82$) than the Gaussian ($r^2 = 0.74$) fitting. Tables 1, 2 and 3 presents the parameters for these models.

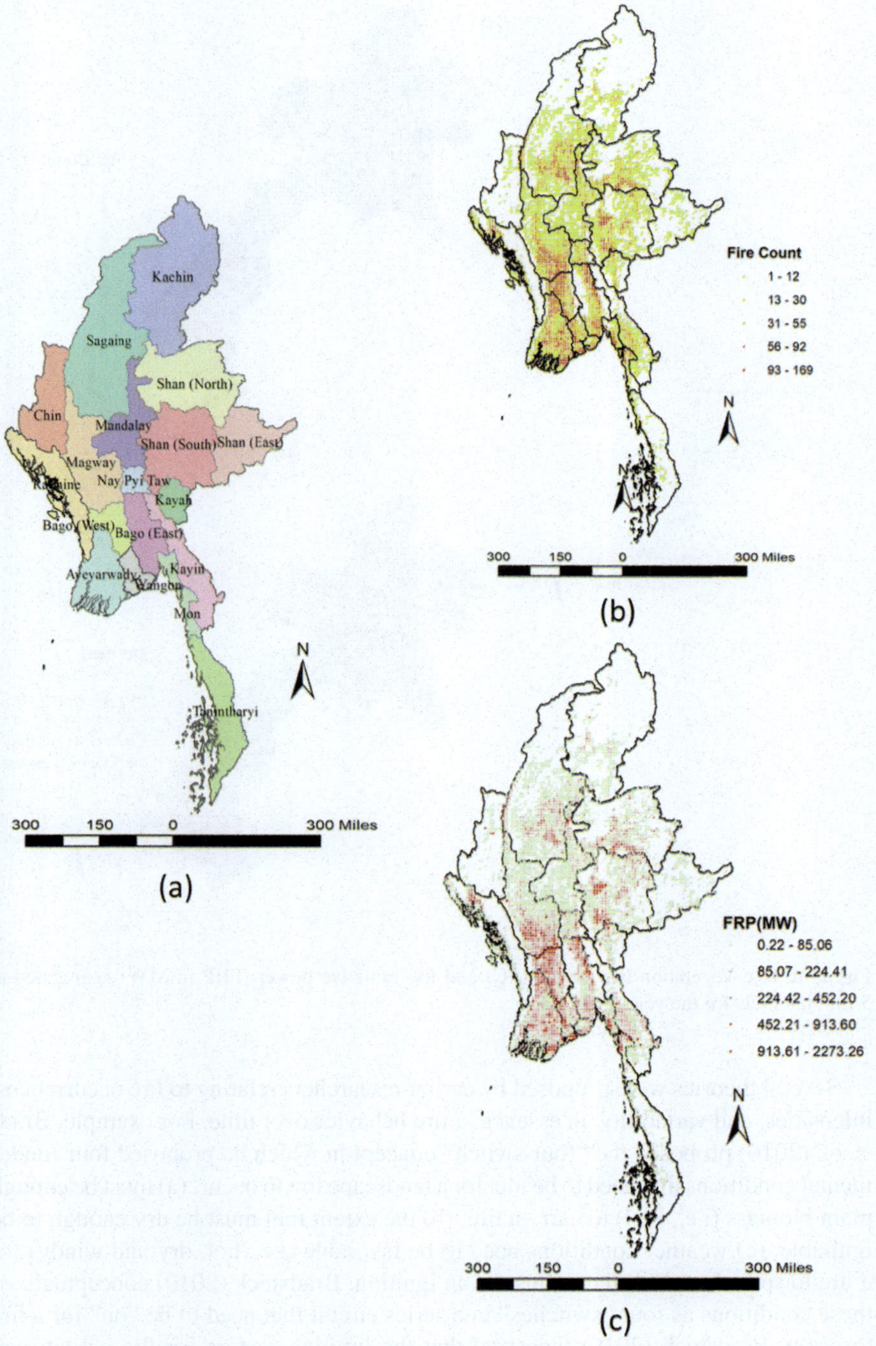

Fig. 1 a, b, c Vegetation fire counts (FC) and fire radiative power (FRP in MW) aggregated at 5-min intervals for the year 2020, Myanmar

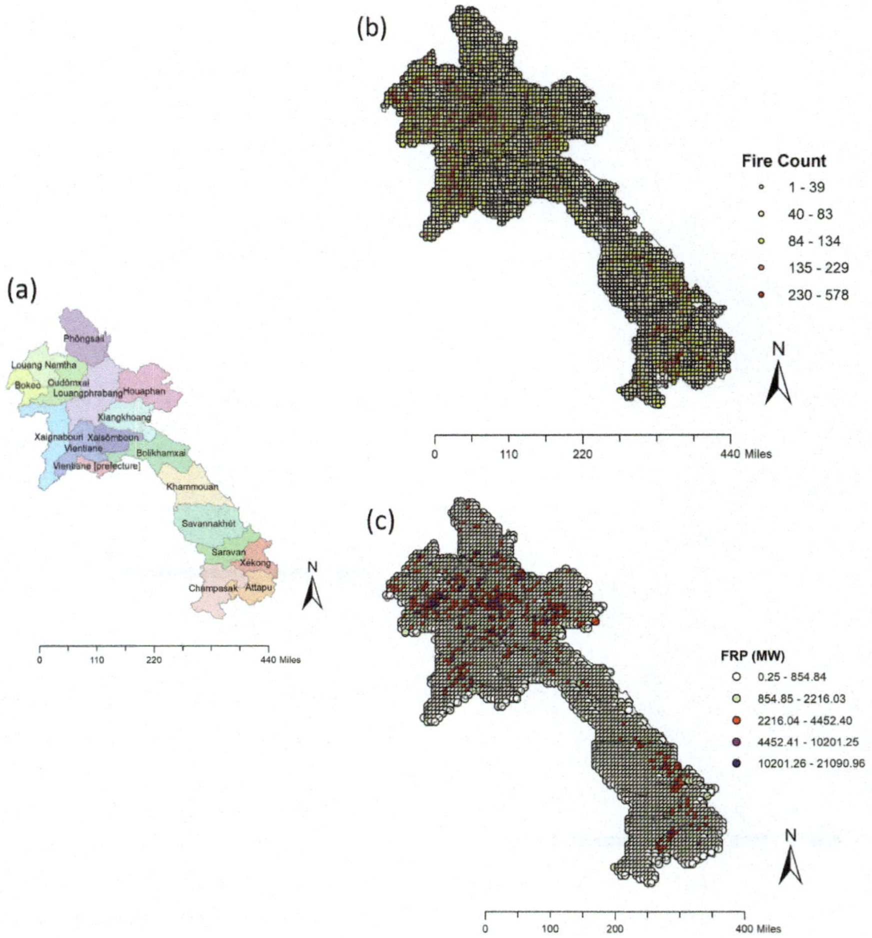

Fig. 2 **a**, **b**, **c** Vegetation fire counts (FC) and fire radiative power (FRP in MW) aggregated at 5-min intervals for the year 2020, Laos

Several theories were proposed by earlier researchers relating to fire occurrences, intensities, and variability, in essence, a fire behavior over time. For example, Bradstock (2010) proposed the "four-switch" concept in which he proposed four fundamental conditions that need to be met for a landscape fire to occur: (a) must be enough plant biomass (i.e., fuel) to carry a fire, (b) the extant fuel must be dry enough to be ignitable, (c) weather conditions need to be favorable (i.e., hot, dry and windy) for a fire to spread, and (d) there must be an ignition. Bradstock (2010) conceptualized these conditions as four "switches" in a series circuit that need to be "on" for a fire to occur. Bradstock (2010) theorized that the limiting factors for fire activity are like switches; fire activity will increase when those switches turn on, thus, switching theory for fire occurrence. With the enhancement of fire intensity, certain primary

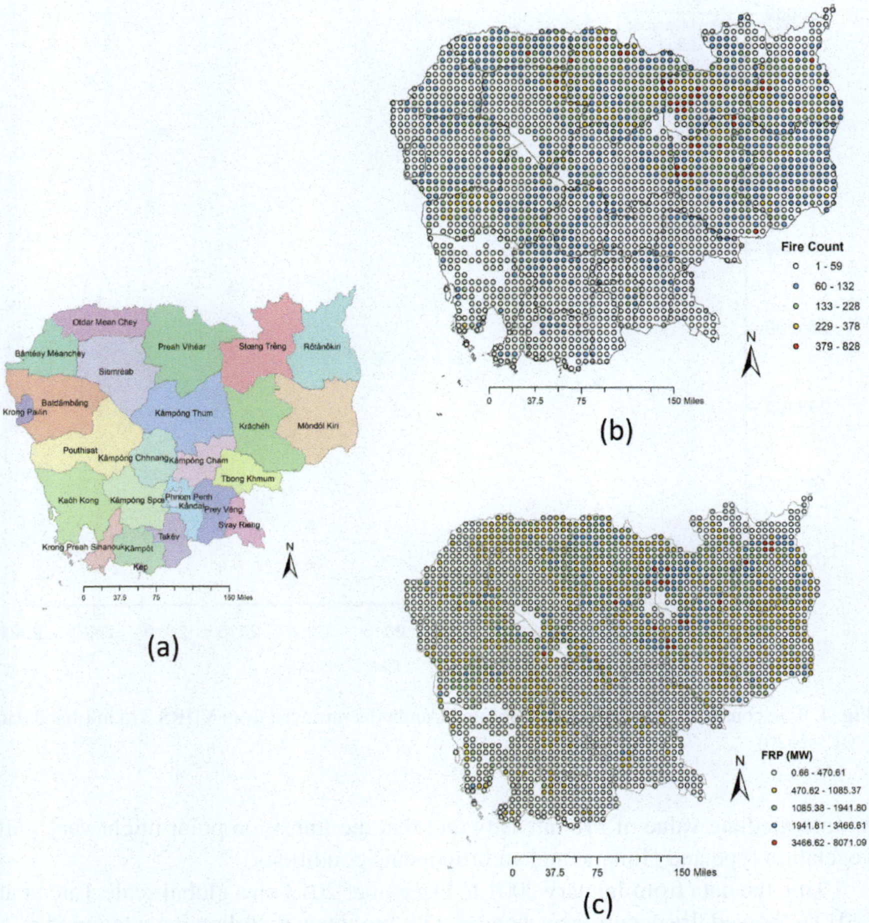

Fig. 3 **a**, **b**, **c** Vegetation fire counts (FC) (**b**) and fire radiative power (FRP in MW) (**c**) aggregated at 5-min intervals for the year 2020, Cambodia

constraints might be relieved, and potentially, the environmental conditions would benefit fire occurrence. Pausas and Ribeiro (2013) proposed the burning resources limit theory which states that fire occurrence will be limited if there is not enough time for fuel accumulation and a negative correlation between both of them. In contrast to these studies, fire occurrence is also reported to be positively related to fire intensity locally (Oliveira et al. 2015). Compared to these theories, the IFOI theory is an intermediate version of other theories. The IFOI hypothesis assumes that the relationship between fire occurrence and fire intensity is neither positive nor negative monotonously; instead, they have the same trend at the beginning, but after a transition point, their trends turn to a different direction or even reverse. Further, Luo et al. (2017) stated that the transition point corresponds to the peak of fire occurrence, but

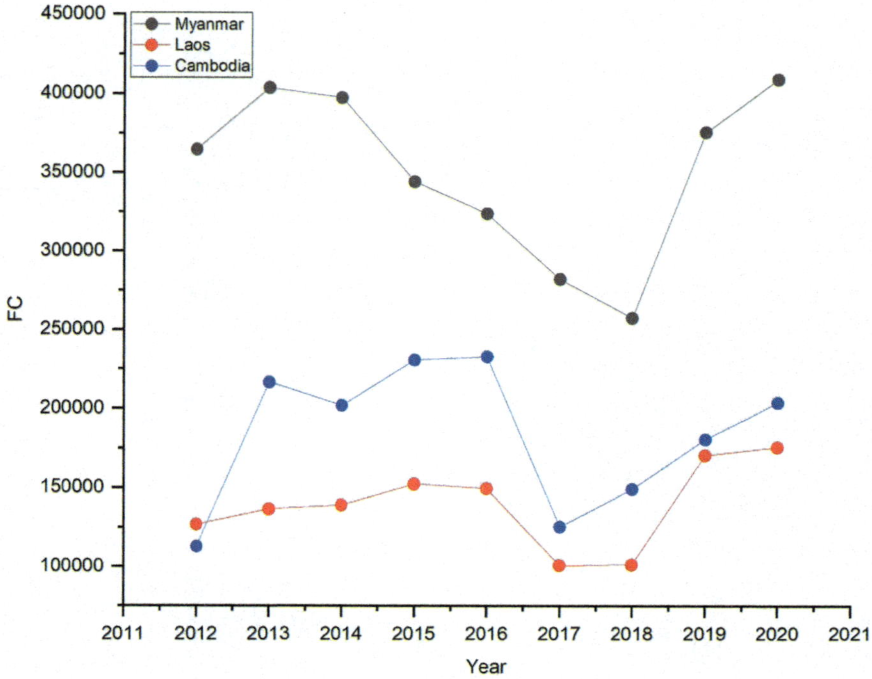

Fig. 4 Fire counts (FC) in Myanmar, Laos and Cambodia retrieved from VIIRS 375 m I-band data (2012–2020)

an intermediate value of fire intensity and that the transition point might vary with vegetation type and climatic and anthropogenic conditions.

Using the data from January 2001 to December 2013 on a global scale, Luo et al. (2017) showed fire occurrence changes with fire intensity following a humped relationship. Our examination of FRP versus FC using different curve-fitting models suggested Power and Piecewise 2-segment regression models with the best performance compared to the Gaussian models in the studied countries. The Power models were also shown to be better than linear, exponential, logarithm, or polynomial models by Luo et al. (2017); however, the humped relationship found at a global scale in their study was not apparent in our case for all the three countries at 5-min grid intervals. In the Piecewise regression model, the *y* in Tables 1, 2 and 3) are coefficients before and after the breakpoint time, and *T* represents the region where the threshold of the regression line changes the slope as determined by the Marquardt–Levenberg algorithm. Thus, the threshold FRP (MW) dividing the FRP-FC relationship into $T1$ versus $T2$ regions in the case of Myanmar was 532.3 (MW), Laos (1170.46 MW), and Cambodia (1577.5 MW). These results suggest a relatively higher clustering of fires with lower FRP (MW) in the case of Myanmar than in Laos and Cambodia,

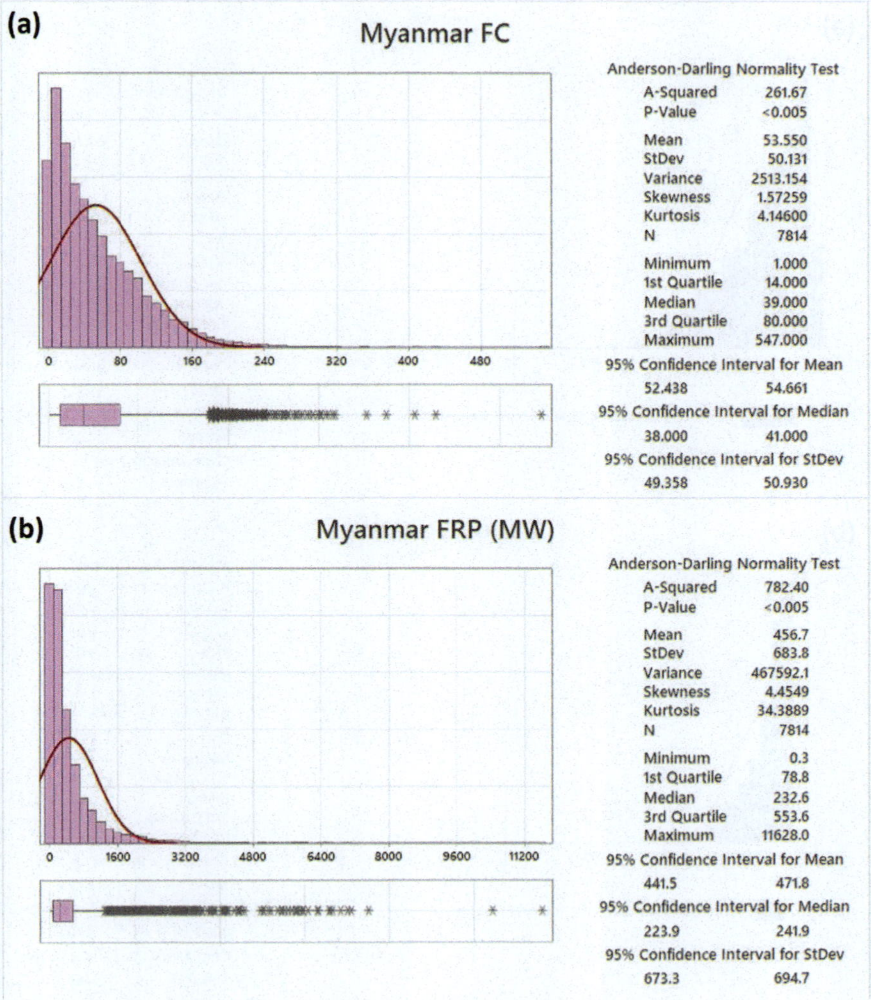

Fig. 5 **a, b** Descriptive statistics for fire counts (FC) and fire radiative power (FRP in MW) aggregated at 5-min intervals for the year 2020, Myanmar. The *N* denotes total number of cells for the 5-min intervals

which might be attributed to a combination of forest and agricultural fires occurring at small spatial scales. The relatively higher FRP-FC found in Cambodia than the Laos and Myanmar also suggests relatively higher amounts of biomass consumed and resulting emissions at smaller spatial scales. The role of vegetation as one of the driving factors in fire behavior, specifically FRP versus fire patch size, was evaluated by Laurent et al. (2019) using the Power models and shown that in most fire regions of the world, such relationship saturates for a threshold of intermediate-intensity

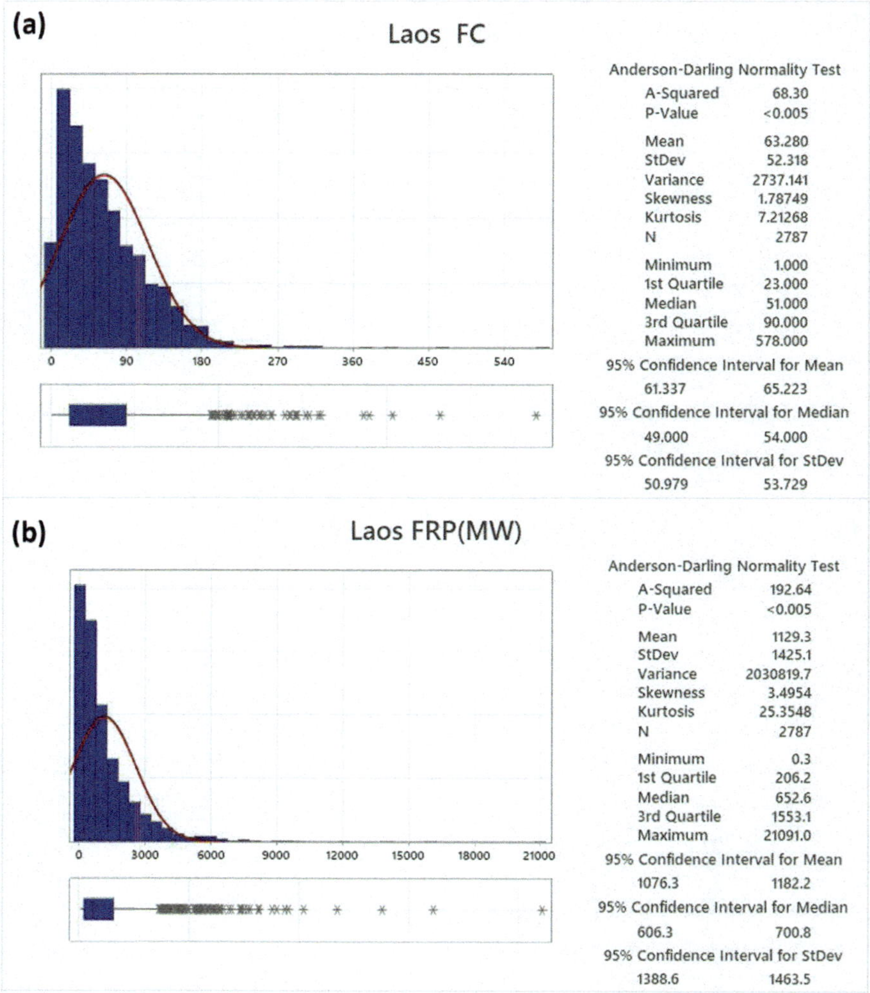

Fig. 6 **a**, **b** Descriptive statistics for fire counts (FC) and fire radiative power (FRP in MW) aggregated at 5-min intervals for the year 2020, Laos. The *N* denotes total number of cells for the 5-min intervals

fires and that threshold differs from one region to another and depends on vegetation type. Luo et al. (2017) inferred similar conclusions in their study, suggesting that fire occurrence is constrained beyond a certain fire intensity threshold, probably due to the limited available fuels. They also inferred that the highest fire occurrence was relevant to intermediate fire intensity (FRP). The exact intermediate values of fire intensity differed in different vegetation types and were substantially affected by other factors, like human activities and fire types. Our results are in line with these studies. Although we captured FRP-FC differences and thresholds in different countries, the drivers of these variations were not addressed, such as the influence of

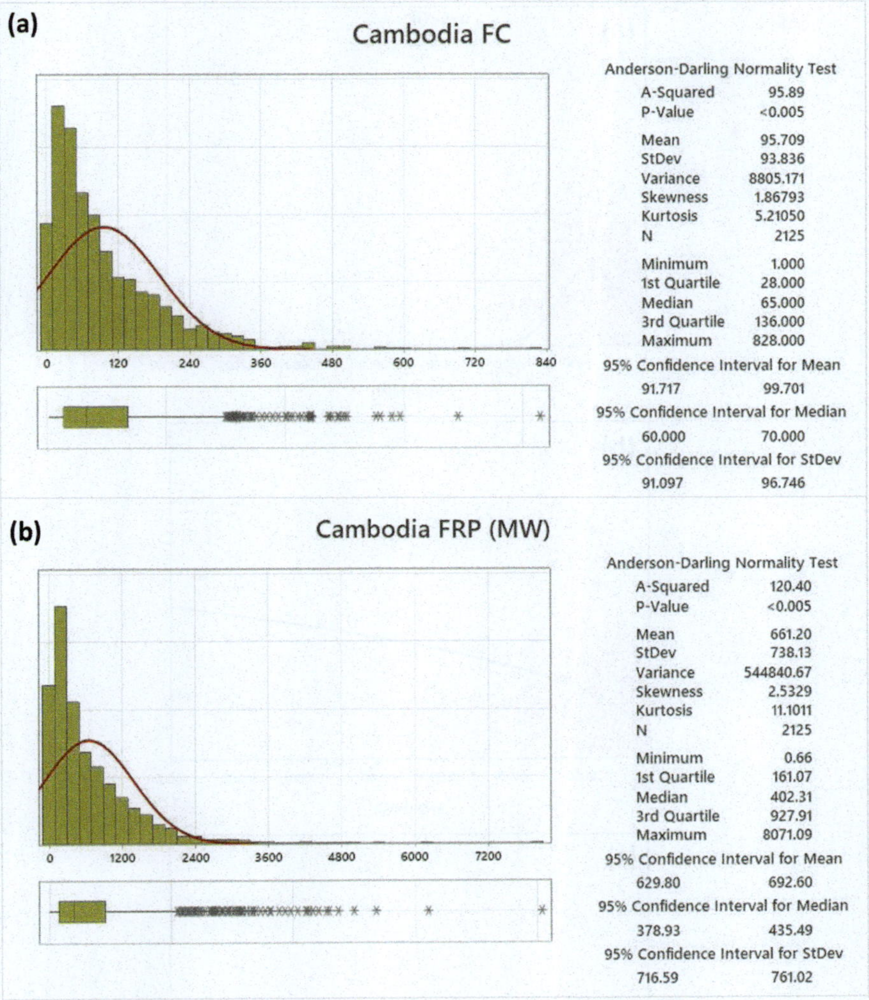

Fig. 7 **a**, **b** Descriptive statistics for fire counts (FC) and fire radiative power (FRP in MW) aggregated at 5-min intervals for the year 2020, Cambodia. The *N* denotes total number of cells for the 5-min intervals

topography, fuel amount, climate, and other human-related parameters which might govern these variations. Nevertheless, our results highlight that FRP-FC variations in different countries in SEA did not have a clear humped relationship as in Luo et al. (2017) and thus need a more thorough evaluation integrating ground-based information with the remotely sensed data, which will be the focus of our research in the coming years.

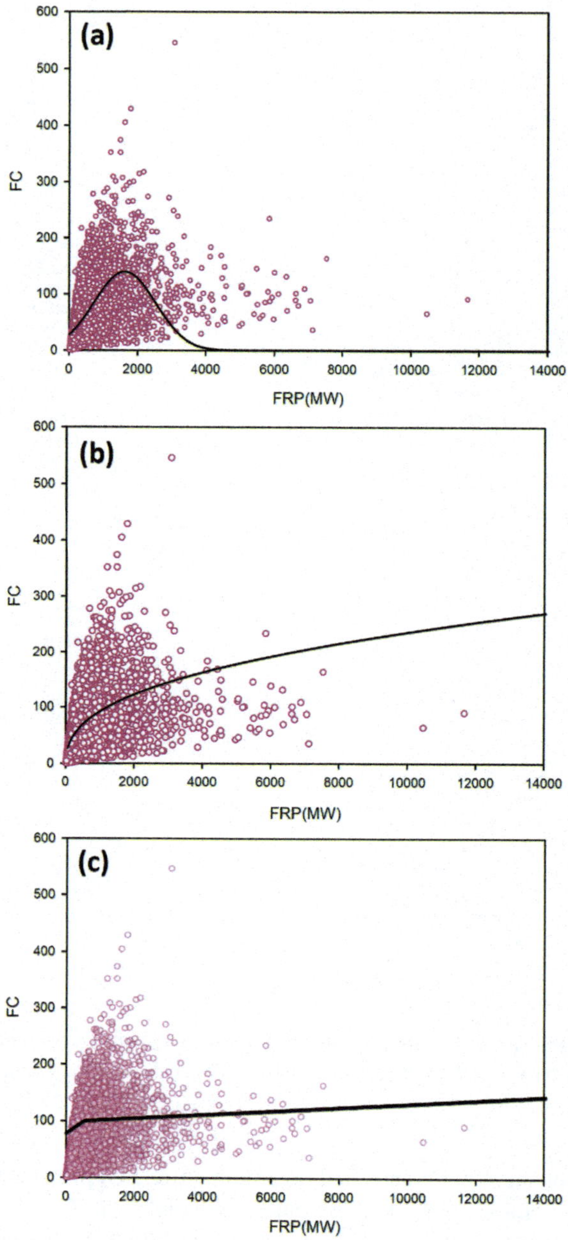

Fig. 8 **a**, **b**, **c** Gaussian, Power, and Piecewise 2-segment regression fitting for FRP and FC data, 2020, Myanmar. See details in text for distribution fitting parameters

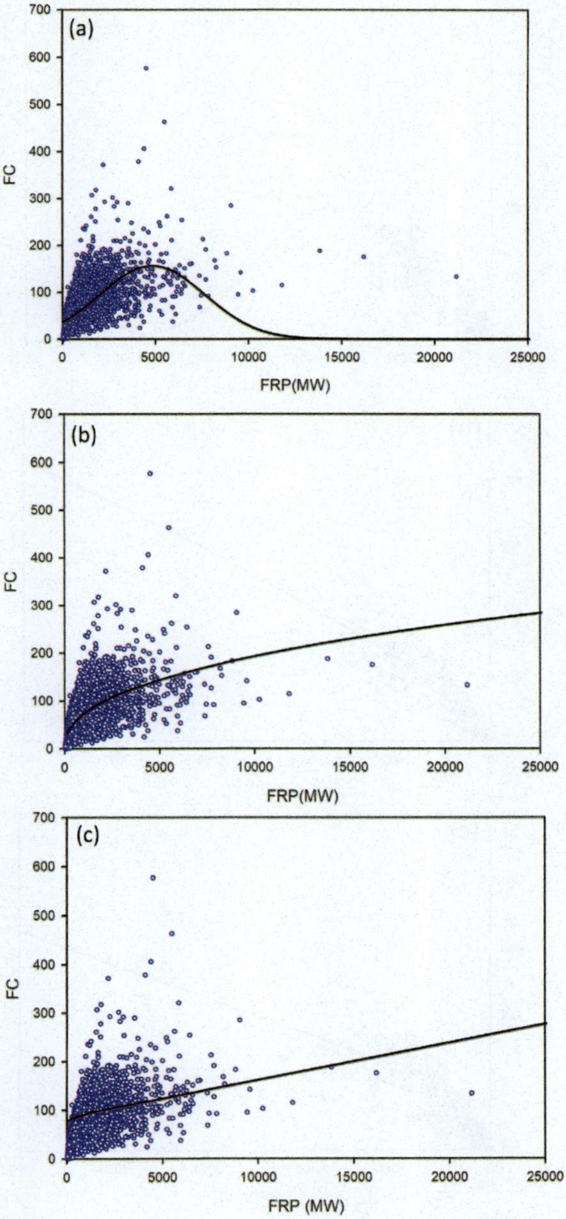

Fig. 9 **a**, **b**, **c** Gaussian, Power, and Piecewise 2-segment regression fitting for FRP and FC data, 2020, Laos. See details in text for distribution fitting parameters

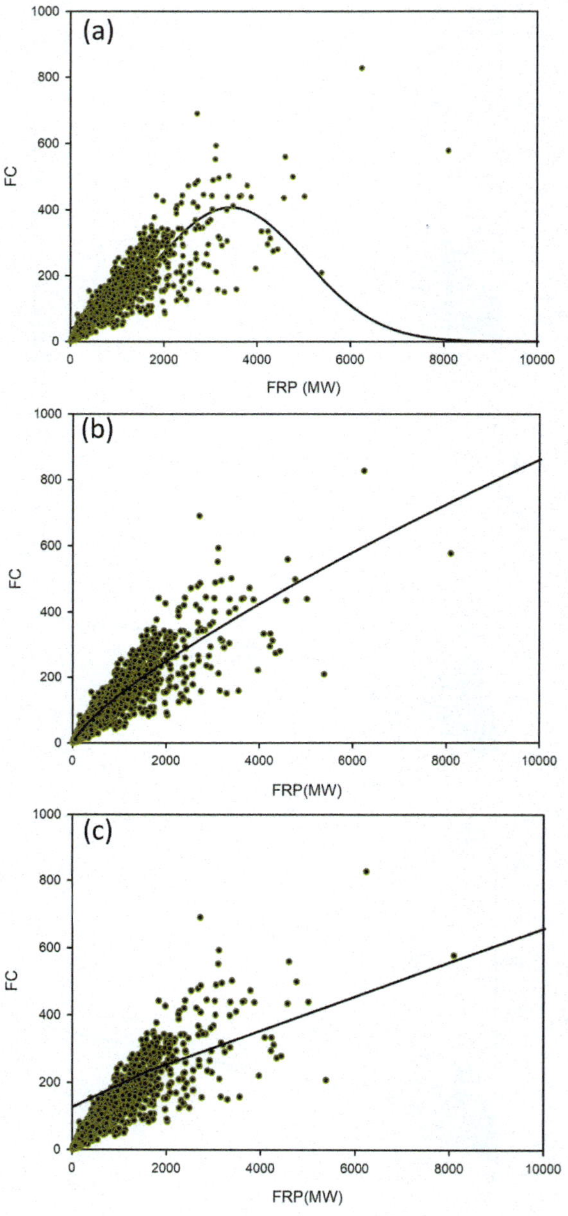

Fig. 10 **a**, **b**, **c** Gaussian, Power, and Piecewise 2-segment regression fitting for FRP and FC data, 2020, Cambodia. See details in text for distribution fitting parameters

Table 1 FRP-FC statistical models at 5-min intervals in Myanmar

(a) Gaussion 3-parameter peak fitting				
$f = a * \exp(-0.5 * ((x - x0)/b)^2)$				
R	R^2	Adj R^2	**Standard error of estimate**	
0.6449	0.4159	0.4158	38.3181	
Parameter	**Coefficient**	**Std. error**	t	P
a	139.6828	1.4868	93.9471	< 0.0001
b	890.6003	11.0244	80.7847	< 0.0001
$x0$	1605.9617	15.5996	102.9491	< 0.0001
(b) Power 2-parameter fitting				
$f = a * x^b$				
R	R^2	Adj R^2	**Standard error of estimate**	
0.6875	0.4726	0.4726	36.4077	
Parameter	**Coefficient**	**Std. error**	t	P
a	5.5694	0.2185	25.4915	< 0.0001
b	0.4067	0.0059	69.3137	< 0.0001
(c) Piecewise 2-segment linear fitting				
$t1 = \min(t)$				
$t2 = \max(t)$				
$\text{region1}(t) = (y1 * (T1 - t) + y2 * (t - t1))/(T1 - t1)$				
$\text{region2}(t) = (y2 * (t2 - t) + y3 * (t - T1))/(t2 - T1)$				
$f = \text{if}(t \leq T1, \text{region1}(t), \text{region2}(t))$				
Global goodness of fit				
R	R^2	Adj R^2	**Standard error of estimate**	
0.7367	0.5427	0.5425	33.9080	
Parameter estimates				
Parameter	**Coefficient**	**Std. error**	t	P
$y1$	4.2310	8.5225	62.4761	< 0.0001
$y2$	99.9177	0.7022	6.0251	< 0.0001
$y3$	143.9766	0.9723	102.7649	< 0.0001
$T1$	532.4502	8.3241	17.2964	< 0.0001

Table 2 FRP-FC statistical models at 5-min intervals in Laos

(a) Gaussion 3-parameter peak fitting				
$f = a * \exp(-0.5 * ((x - x0)/b)^2)$				
R	R^2	Adj R^2	**Standard error of estimate**	
0.6436	0.4143	0.4139	40.0546	
Parameter	**Coefficient**	**Std. error**	t	P

(continued)

Table 2 (continued)

a	155.7224	2.9688	52.4535	< 0.0001
b	2800.4586	68.6649	40.7844	< 0.0001
$x0$	4815.5373	105.7538	45.5354	< 0.0001
(b) Power 2-parameter fitting				
$f = a * x^b$				
R	R^2	Adj R^2	**Standard error of estimate**	
0.7104	0.5047	0.5045	36.8282	
Parameter	**Coefficient**	**Std. error**	t	P
a	3.7773	0.2777	13.5998	< 0.0001
b	0.4274	0.0097	43.8358	< 0.0001
(c) Piecewise 2-segment linear fitting				
$t1 = \min(t)$				
$t2 = \max(t)$				
$\text{region1}(t) = (y1 * (T1 - t) + y2 * (t - t1))/(T1 - t1)$				
$\text{region2}(t) = (y2 * (t2 - t) + y3 * (t - T1))/(t2 - T1)$				
$f = \text{if}(t \leq T1, \text{region1}(t), \text{region2}(t))$				
R	R^2	Adj R^2	**Standard error of estimate**	
0.7108	0.5052	0.5047	36.8205	
Parameter	**Coefficient**	**Std. error**	t	P
$y1$	13.4879	1.3411	10.0575	< 0.0001
$y2$	94.0300	1.8751	50.1468	< 0.0001
$y3$	277.2811	13.6277	20.3469	< 0.0001
$T1$	1170.4636	44.2630	26.4434	< 0.0001

Table 3 FRP-FC statistical models at 5-min intervals in Cambodia

(a) Gaussion 3-parameter peak fitting				
$f = a * \exp(-0.5 * ((x - x0)/b)^2)$				
R	R^2	Adj R^2	**Standard error of estimate**	
0.8640	0.7465	0.7462	47.2705	
Parameter	**Coefficient**	**Std. error**	t	P
a	407.3447	6.3379	64.2714	< 0.0001
b	1586.7429	23.4679	67.6132	< 0.0001
$x0$	3421.7138	44.3322	77.1836	< 0.0001

(continued)

Table 3 (continued)

(b) Power 2-parameter fitting				
$f = a * x^b$				
R	***R*²**	**Adj *R*²**	**Standard error of estimate**	
0.9107	0.8293	0.8292	38.7796	
Parameter	**Coefficient**	**Std. error**	***t***	***P***
a	0.7260	0.0467	15.5317	< 0.0001
b	0.7688	0.0087	88.0472	< 0.0001

(c) Piecewise 2-segment linear fitting				
$t1 = \min(t)$				
$t2 = \max(t)$				
$\text{region1}(t) = (y1 * (T1 - t) + y2 * (t - t1))/(T1 - t1)$				
$\text{region2}(t) = (y2 * (t2 - t) + y3 * (t - T1))/(t2 - T1)$				
$f = \text{if}(t \leq T1, \text{region1}(t), \text{region2}(t))$				
R	***R*²**	**Adj *R*²**	**Standard error of estimate**	
0.9144	0.8361	0.8359	38.0141	
Parameter	**Coefficient**	**Std. error**	***t***	***P***
*y*1	6.9855	1.3277	5.2615	< 0.0001
*y*2	232.9090	6.7065	34.7291	< 0.0001
*y*3	660.1136	16.9790	38.8783	< 0.0001
*T*1	1577.5100	54.8307	28.7705	< 0.0001

Acknowledgements The first author is grateful to the NASA Land Cover/Land Use Change Program for funding the South/Southeast Asia Research Initiative under which the study has been undertaken. Authors thank the VIIRS product developers for freely sharing the fire data.

References

Albar, I., I. Jaya, B.H. Saharjo, B. Kuncahyo, and K.P. Vadrevu. 2018. Spatio-temporal analysis of land and forest fires in Indonesia using MODIS active fire dataset. In *Land-atmospheric research applications in South and Southeast Asia*, 105–127. Cham: Springer.

Arino, O., and J.M. Rosaz. 1999. June 1997 and 1998 world ATSR fire atlas using ERS-2 ATSR-2 data. In *Proceedings from the joint fire science conference*, 177–182. Boise, ID.

Badarinath, K.V.S., and K.V. Prasad. 2011. Carbon dioxide emissions from forest biomass burning in India. *Global Environmental Research* 15: 45–52.

Badarinath, K.V.S., S.K. Kharol, K.M. Latha, T.K. Chand, V.K. Prasad, A.N. Jyothsna, and K. Samatha. 2007. Multiyear ground-based and satellite observations of aerosol properties over a tropical urban area in India. *Atmospheric Science Letters* 8 (1): 7–13.

Badarinath, K.V.S., S. Kumar Kharol, V. Krishna Prasad, A. Rani Sharma, E.U.B. Reddi, H.D. Kambezidis, and D.G. Kaskaoutis. 2008. Influence of natural and anthropogenic activities on UV Index variations—A study over tropical urban region using ground based observations and satellite data. *Journal of Atmospheric Chemistry* 59 (3): 219–236.

Badarinath, K.V.S., S.K. Kharol, A.R. Sharma, and V.K. Prasad. 2009. Analysis of aerosol and carbon monoxide characteristics over Arabian Sea during crop residue burning period in the Indo-Gangetic Plains using multi-satellite remote sensing datasets. *Journal of Atmospheric and Solar-Terrestrial Physics* 71 (12): 1267–1276.

Biswas, S., K.D. Lasko, and K.P. Vadrevu. 2015a. Fire disturbance in tropical forests of Myanmar—Analysis using MODIS satellite datasets. *IEEE Journal of Selected Topics in Applied Earth Observations and Remote Sensing* 8 (5): 2273–2281.

Biswas, S., K.P. Vadrevu, Z.M. Lwin, K. Lasko, and C.O. Justice. 2015b. Factors controlling vegetation fires in protected and non-protected areas of Myanmar. *PLoS ONE* 10 (4): e0124346.

Biswas, S., K.P. Vadrevu, M.S. Mon, and C. Justice. 2021. Contemporary forest loss in Myanmar: Effect of democratic transition and subsequent timber bans on landscape structure and composition. *Ambio* 50 (4): 914–928.

Boschetti, L., and D.P. Roy. 2009. Strategies for the fusion of satellite fire radiative power with burned area data for fire radiative energy derivation. *Journal of Geophysical Research: Atmospheres* 114 (D20).

Bradstock, R.A. 2010. A biogeographic model of fire regimes in Australia: Current and future implications. *Global Ecology and Biogeography* 19: 145–158.

Brady, N.C. 1996. Alternatives to slash-and-burn: A global imperative. *Agriculture, Ecosystems & Environment* 58 (1): 3–11.

Choi, Y., S.A. Vay, K.P. Vadrevu, A.J. Soja, J.H. Woo, S.R. Nolf, G.W. Sachse, G.S. Diskin, D.R. Blake, N.J. Blake, and H.B. Singh. 2008. Characteristics of the atmospheric CO_2 signal as observed over the conterminous United States during INTEX-NA. *Journal of Geophysical Research: Atmospheres* 113 (D7).

Eaturu, A., and K.P. Vadrevu. 2021. Evaluation of Sentinel-3 SLSTR data for mapping fires in forests, Peatlands, and Croplands—A case study over Australia, Indonesia, and India. In *Biomass burning in South and Southeast Asia*, 39–59. CRC Press.

Elvidge, C.D., M. Zhizhin, F.-C. Hsu, and K.E. Baugh. 2013. VIIRS nightfire: Satellite pyrometry at night. *Remote Sensing* 5: 4423–4449.

Elvidge, C.D., M. Zhizhin, K. Baugh, and F.C. Hsu. 2021. Identification of smoldering Peatland fires in Indonesia via triple-phase temperature analysis of VIIRS nighttime data. In *Biomass burning in South and Southeast Asia*, 25–38. CRC Press.

Freeborn, P.H., M.J. Wooster, D.P. Roy, and M.A. Cochrane. 2014. Quantification of MODIS fire radiative power (FRP) measurement uncertainty for use in satellite-based active fire characterization and biomass burning estimation. *Geophysical Research Letters* 41 (6): 1988–1994.

Giglio, L., J.D. Kendall, and R. Mack. 2003. A multi-year active fire dataset for the tropics derived from the TRMM VIRS. *International Journal of Remote Sensing* 24 (22): 4505–4525.

Giglio, L., W. Schroeder, and C.O. Justice. 2016. The collection 6 MODIS active fire detection algorithm and fire products. *Remote Sensing of Environment* 178: 31–41.

Hayasaka, H., I. Noguchi, E.I. Putra, N. Yulianti, and K. Vadrevu. 2014. Peat-fire-related air pollution in Central Kalimantan, Indonesia. *Environmental Pollution* 195: 257–266.

Huang, K., J.S. Fu, N.C. Hsu, Y. Gao, X. Dong, S.C. Tsay, and Y.F. Lam. 2013. Impact assessment of biomass burning on air quality in Southeast and East Asia during BASE-ASIA. *Atmospheric Environment* 78: 291–302.

Ichoku, C., and L. Ellison. 2014. Global top-down smoke-aerosol emissions estimation using satellite fire radiative power measurements. *Atmospheric Chemistry and Physics* 14: 6643–6667.

Justice, C., G. Gutman, and K.P. Vadrevu. 2015. NASA land cover and land use change (LCLUC): An interdisciplinary research program. *Journal of Environmental Management* 148: 4–9.

Kant, Y., V.K. Prasad, and K.V.S. Badarinath. 2000. Algorithm for detection of active fire zones using NOAA AVHRR data. *Infrared Physics & Technology* 41 (1): 29–34.

Karthikeyan, S., R. Balasubramanian, and K. Iouri. 2006. Particulate air pollution from bushfires: Human exposure and possible health effects. *Journal of Toxicology and Environmental Health, Part A* 69 (21): 1895–1908.

Kaufman, Y.J., L. Remer, R. Ottmar, D. Ward, R.R. Li, R. Kleidman, R.S. Fraser, L. Flynn, D. McDougal, and G. Shelton.1996. Relationship between remotely sensed fire intensity and rate of emission of smoke: SCAR-C experiment. In *Global biomass burning*, 685–696. Cambridge, MA, USA: MIT Press.

Kaufman, Y.J., C.O. Justice, L.P. Flynn, J.D. Kendall, E.M. Prins, L. Giglio, D.E. Ward, W.P. Menzel, and A.W. Setzer. 1998. Potential global fire monitoring from EOS-MODIS. *Journal of Geophysical Research: Atmospheres* 103 (D24): 32215–32238.

Kharol, S.K., K.V.S. Badarinath, A.R. Sharma, D.V. Mahalakshmi, D. Singh, and V.K. Prasad. 2012. Black carbon aerosol variations over Patiala city, Punjab, India—A study during agriculture crop residue burning period using ground measurements and satellite data. *Journal of Atmospheric and Solar-Terrestrial Physics* 84: 45–51.

Kremens, R.L., M.B. Dickinson, and A.S. Bova. 2012. Radiant flux density, energy density and fuel consumption in mixed-oak forest surface fires. *International Journal of Wildland Fire* 21(6): 722–730

Kumar, S.S., D.P. Roy, L. Boschetti, and R. Kremens. 2011. Exploiting the power law distribution properties of satellite fire radiative power retrievals: A method to estimate fire radiative energy and biomass burned from sparse satellite observations. *Journal of Geophysical Research: Atmospheres* 116 (D19).

Lasko, K., and K. Vadrevu. 2018. Improved rice residue burning emissions estimates: Accounting for practice-specific emission factors in air pollution assessments of Vietnam. *Environmental Pollution* 236: 795–806.

Lasko, K., K.P. Vadrevu, V.T. Tran, E. Ellicott, T.T. Nguyen, H.Q. Bui, and C. Justice. 2017. Satellites may underestimate rice residue and associated burning emissions in Vietnam. *Environmental Research Letters* 12 (8): 085006.

Lasko, K., K.P. Vadrevu, and T.T.N. Nguyen. 2018. Analysis of air pollution over Hanoi, Vietnam using multi-satellite and MERRA reanalysis datasets. PLoS ONE 13 (5): e0196629.

Lata, K.M., C.H. Sankar Rao, V. Krishna Prasad, K.V.S. Badrinath, and V. Raghavaswamy, 2001. Measuring urban sprawl: A case study of Hyderabad. GIS Development 5 (12): 22–25.

Laurent, P., F. Mouillot, M.V. Moreno, C. Yue, and P. Ciais. 2019. Varying relationships between fire radiative power and fire size at a global scale. *Biogeosciences* 16 (2): 275–288.

Luo, R., D. Hui, N. Miao, C. Liang, and N. Wells. 2017. Global relationship of fire occurrence and fire intensity: A test of intermediate fire occurrence-intensity hypothesis. *Journal of Geophysical Research: Biogeosciences* 122 (5): 1123–1136.

Mauderly, J.L., and J.C. Chow. 2008. Health effects of organic aerosols. *Inhalation Toxicology* 20 (3): 257–288.

Murdiyarso, D., and L. Lebel. 2007. Local to global perspectives on forest and land fires in Southeast Asia. *Mitigation and Adaptation Strategies for Global Change* 12 (1): 3–11.

Oliveira, S.L.J., S.W. Maier, J.M.C. Pereira, and J. Russell-Smith. 2015. Seasonal differences in fire activity and intensity in tropical savannas of northern Australia using satellite measurements of fire radiative power. *International Journal of Wildland Fire* 24 (2): 249–260.

Pausas, J.G., and E. Ribeiro. 2013. The global fire–productivity relationship. *Global Ecology and Biogeography* 22(6): 728–736

Peterson, D., J. Wang, C. Ichoku, E. Hyer, and V. Ambrosia. 2013. A sub-pixel-based calculation of fire radiative power from MODIS observations: 1: Algorithm development and initial assessment. *Remote Sensing of Environment* 129: 262–279.

Petropoulos, G.P., K.P. Vadrevu, and C. Kalaitzidis. 2013. Spectral angle mapper and object-based classification combined with hyperspectral remote sensing imagery for obtaining land use/cover mapping in a Mediterranean region. *Geocarto International* 28 (2): 114–129.

Prasad, V.K., Y. Kant, P.K. Gupta, C. Sharma, A.A. Mitra, and K.V.S. Badarinath. 2001a. Biomass and combustion characteristics of secondary mixed deciduous forests in Eastern Ghats of India. *Atmospheric Environment* 35 (18): 3085–3095.

Prasad, V.K., Y. Kant, and K.V.S. Badarinath. 2001b. CENTURY ecosystem model application for quantifying vegetation dynamics in shifting cultivation areas: A case study from Rampa Forests, Eastern Ghats (India). *Ecological Research* 16 (3): 497–507.

Prasad, V.K., Y. Kant, P.K. Gupta, C. Elvidge, and K.V.S. Badarinath. 2002a. Biomass burning and related trace gas emissions from tropical dry deciduous forests of India: A study using DMSP-OLS data and ground-based measurements. *International Journal of Remote Sensing* 23 (14): 2837–2851.

Prasad, V.K., Y. Kant, and K.V.S. Badarinath. 2002b. Land use changes and modeling carbon fluxes from satellite data. *Advances in Space Research* 30 (11): 2511–2516.

Prasad, V.K., M. Lata, and K.V.S. Badarinath. 2003a. Trace gas emissions from biomass burning from northeast region in India—Estimates from satellite remote sensing data and GIS. *The Environmentalist* 23 (3): 229–236.

Prasad, V.K., K.V.S. Badarinath, H. Tsuruta, S. Sudo, S. Yonemura, J. Cardina, B. Stinner, R. Moore, D. Stinner, and C. Hoy. 2003b. Implications of land use changes on carbon dynamics and sequestration—Evaluation from forestry datasets, India. *The Environmentalist* 23 (2): 175–187.

Prasad, V.K., and K.V.S. Badarinth. 2004. Land use changes and trends in human appropriation of above ground net primary production (HANPP) in India (1961–1998). *Geographical Journal* 170 (1): 51–63.

Prasad, K.V., and K.V.S. Badarinath. 2006. Soil surface nitrogen losses from agriculture in India: A regional inventory within agroecological zones (2000–2001). *The International Journal of Sustainable Development and World Ecology* 13 (3): 173–182.

Prasad, V.K., K.V.S. Badarinath, S. Yonemura, and H. Tsuruta. 2004. Regional inventory of soil surface nitrogen balances in Indian agriculture (2000–2001). *Journal of Environmental Management* 73 (3): 209–218.

Prasad, V.K., E. Anuradha, and K.V.S. Badarinath. 2005. Climatic controls of vegetation vigor in four contrasting forest types of India—Evaluation from National oceanic and atmospheric administration's advanced very high resolution radiometer datasets (1990–2000). *International Journal of Biometeorology* 50 (1): 6–16.

Roberts, G.J., and M.J. Wooster. 2008. Fire detection and fire characterization over Africa using Meteosat SEVIRI. *IEEE Transactions on Geoscience and Remote Sensing* 46: 1200–1218.

Roberts, G., and M.J. Wooster. 2014. Development of a multi-temporal Kalman filter approach to geostationary active fire detection & fire radiative power (FRP) estimation. *Remote Sensing of Environment* 152: 392–412.

Schroeder, W., P. Oliva, L. Giglio, and I.A. Csiszar. 2014. The New VIIRS 375 m active fire detection data product: Algorithm description and initial assessment. *Remote Sensing of Environment* 143: 85–96.

Seiler, W., and P.J. Crutzen. 1980. Estimates of gross and net fluxes of carbon between the biosphere and the atmosphere from biomass burning. *Climatic Change* 2 (3): 207–247.

Vadrevu, K.P. 2008. Analysis of fire events and controlling factors in eastern India using spatial scan and multivariate statistics. *Geografiska Annaler: Series A, Physical Geography* 90 (4): 315–328.

Vadrevu, K.P. 2015. International regional science meeting of NASA-LCLUC: Land cover/land use change (LC/LUC) and environmental impacts in South Asia, Coimbatore, India, 19–23 January 2013. *Journal of Environmental Management.* 148: 1–163.

Vadrevu, K.P. 2021. *Fire danger indices and methods: An appraisal. in biomass burning in South and Southeast Asia*, 201–232. CRC Press.

Vadrevu, K.P., and K.V.S. Badarinath. 2009. Spatial pattern analysis of fire events in Central India—A case study. *Geocarto International* 24 (2): 115–131.

Vadrevu, K.P., and C.O. Justice. 2011. Vegetation fires in the Asian region: Satellite observational needs and priorities. *Global Environmental Research* 15 (1): 65–76.

Vadrevu, K., and K. Lasko. 2015. Fire regimes and potential bioenergy loss from agricultural lands in the Indo-Gangetic Plains. *Journal of Environmental Management* 148: 10–20.

Vadrevu, K., and K. Lasko. 2018. Intercomparison of MODIS AQUA and VIIRS I-Band fires and emissions in an agricultural landscape—Implications for air pollution research. *Remote Sensing* 10 (7): 978.

Vadrevu, K.P., K.V.S. Badarinath, and E. Anuradha. 2008a. Spatial patterns in vegetation fires in the Indian region. *Environmental Monitoring and Assessment* 147 (1): 1–13.

Vadrevu, K.P., J. Cardina, F. Hitzhusen, I. Bayoh, R. Moore, J. Parker, B. Stinner, D. Stinner, and C. Hoy. 2008b. Case study of an integrated framework for quantifying agroecosystem health. *Ecosystems* 11 (2): 283–306.

Vadrevu, K.P., L. Giglio, and C. Justice. 2013. Satellite based analysis of fire–carbon monoxide relationships from forest and agricultural residue burning (2003–2011). *Atmospheric Environment.* 64: 179–191.

Vadrevu, K.P., T. Ohara, and C. Justice. 2014. Air pollution in Asia. In *Environmental pollution (Barking, Essex: 1987)*, vol. 195, 233–235.

Vadrevu, K., T. Ohara, C. Justice. 2017. Land cover, land use changes and air pollution in Asia: A synthesis. *Environmental Research Letters* 12(12)120201.

Vadrevu, K.P., T. Ohara, and C. Justice, eds. 2018. *Land-atmospheric research applications in South and Southeast Asia.* Cham: Springer.

Vadrevu, K., A. Heinimann, G. Gutman, and C. Justice. 2019. Remote sensing of land use/cover changes in South and Southeast Asian Countries. *International Journal of Digital Earth.* 12 (10): 1099–1102.

Vadrevu, K.P., T. Ohara, and C. Justice, eds. 2021a. *Biomass burning in South and Southeast Asia: Impacts on the biosphere*, vol. 2. CRC Press.

Vadrevu, K.P., T. Ohara, and C. Justice, eds. 2021b. *Biomass burning in South and Southeast Asia: Mapping and monitoring*, vol. 1. CRC Press.

Vadrevu, K.P., T. Le Toan, S.S. Ray, and C. Justice. 2022a. *Remote sensing of agriculture and land cover/land use changes in South and Southeast Asian Countries*. Cham: Springer.

Vadrevu, K., A. Eaturu, E. Casadaban, K. Lasko, W. Schroeder, S. Biswas, L. Giglio, and C. Justice. 2022b. Spatial variations in vegetation fires and emissions in South and Southeast Asia during COVID-19 and pre-pandemic. *Scientific Reports* 12 (1): 1–21.

Vay, S.A., Y. Choi, K.P. Vadrevu, D.R. Blake, S.C. Tyler, A. Wisthaler, A. Hecobian, Y. Kondo, G.S. Diskin, G.W. Sachse, and J.H. Woo. 2011. Patterns of CO_2 and radiocarbon across high northern latitudes during international polar year 2008. *Journal of Geophysical Research: Atmospheres* 116 (D14).

Vermote, E., E. Ellicott, O. Dubovik, T. Lapyonok, M. Chin, L. Giglio, and G.J. Roberts. 2009. An approach to estimate global biomass burning emissions of organic and black carbon from MODIS fire radiative power. *Journal of Geophysical Research: Atmospheres* 114 (D18).

Wooster, M.J., and Y.H. Zhang. 2004. Boreal forest fires burn less intensely in Russia than in North America. *Geophysical Research Letters* 31: L20505. https://doi.org/10.1029/2004GL020805.

Wooster, M.J., B. Zhukov, and D. Oertel. 2003. Fire radiative energy for quantitative study of biomass burning: Derivation from the BIRD experimental satellite and comparison to MODIS fire products. *Remote Sensing of Environment* 86: 83–107.

Wooster, M.J., G. Roberts, G.L.W. Perry, and Y.J. Kaufman. 2005. Retrieval of biomass combustion rates and totals from fire radiative power observations: FRP derivation and calibration relationships between biomass consumption and fire radiative energy release. *Journal of Geophysical Research: Atmospheres* 110 (D24).

Wooster, M.J., W. Xu, and T. Nightingale. 2012. Sentinel-3 SLSTR active fire detection and FRP product: Pre-launch algorithm development and performance evaluation using MODIS and ASTER datasets. *Remote Sensing of Environment* 120: 236–254.

Wooster, M.J., G.J. Roberts, L. Giglio, D.P. Roy, P.H. Freeborn, L. Boschetti, C. Justice, C. Ichoku, W. Schroeder, D. Davies, and A.M. Smith. 2021. Satellite remote sensing of active fires: History and current status, applications and future requirements. *Remote Sensing of Environment.* 267: 112694.

Xu, W., M.J. Wooster, G. Roberts, and P. Freeborn. 2010. New GOES imager algorithms for cloud and active fire detection and fire radiative power assessment across North, South and Central America. *Remote Sensing of Environment* 114: 1876–1895.

Analyzing Fire Behavior and Calibrating a Fire Growth Model in a Seasonally Dry Tropical Forest Area

Gernot Ruecker, Veerachai Tanpipat, and Kobsak Wanthongchai

Abstract Dry broad-leaved seasonal forests are widespread in Southeast Asia. They are characterized by drought deciduous tree species, which are adapted to a severe dry season that lasts several months each year. Forest fires are frequent in this vegetation type. To further understanding of fire behavior and fire impact, a series of fire field experiments implemented in the Huay Kha Khaeng (HKK) Wildlife Sanctuary (Uthai Thani Province, Thailand) between 2008 and 2016 was analyzed. A fire behavior model based on the Canadian Fire Behavior Prediction System (Prometheus) was calibrated using the experimental data for the deciduous dipterocarp forest fuel type. The model was then tested on a remotely observed large wildfire in Thailand. Our results confirm the slow fire spread and low to moderate fire intensities observed for this forest type in earlier studies. The fire spread model performs well compared to satellite observations but tends to overestimate area burned and fuel consumption and, consequently, fire emissions when used in air pollution models. Our results indicate that widely used global databases may substantially overestimate fuel consumption and hence fire emissions for this forest type.

Keywords Fire behavior models · Fire growth · Dry tropical forests · Thailand

G. Ruecker (✉)
ZEBRIS Geo-IT GmbH, Munich, Germany
e-mail: gruecker@zebris.com

V. Tanpipat · K. Wanthongchai
Upper ASEAN Wildland Fire Special Research Unit, Faculty of Forestry, Forestry Research Center, Kasetsart University, Bangkok, Thailand
e-mail: fforvrc@ku.ac.th

K. Wanthongchai
e-mail: fforvrc@ku.ac.th

K. P. Vadrevu et al. (eds.), *Vegetation Fires and Pollution in Asia*,
https://doi.org/10.1007/978-3-031-29916-2_14

1 Introduction

1.1 Fire in Dry and Deciduous Dipterocarp Forests

Seasonally dry tropical and subtropical forests are one of the global biomes identified by Olsen et al. (2001). These forests are characterized by a severe dry season that lasts several months each year. Dry broad-leaved seasonal forests are often characterized by drought deciduous tree species (Bullock et al. 1995). The remaining cover of this forest type has been estimated to be roughly 1 million km^2 globally (Miles et al. 2006). In Southeast Asia, dry deciduous dipterocarp (DDF) and mixed (deciduous and non-deciduous) forests (MDF) are key forest types. They are restricted to areas with a total annual rainfall of 1000–1500 mm and a pronounced dry season and are characterized by significant seasonal changes in tree phenology (Rundel and Boonpragob 1995). DDF often has an open canopy allowing abundant grass cover.

Although the leaves of the dominant species are quite thick and large, virtually all dominant species in DDF shed their leaves during the dry season. Fires are frequent (with return intervals from one to three years) and feed on leaf litter and grasses. The dominant dipterocarp tree species show morphological adaptations to fire, such as thick barks, and germination and seed dispersal are also adapted to fire (Baker and Bunyavejchewin 2006). Consequently, a study in the aftermath of the El Nino–Southern Oscillation (ENSO) event of 1997/1998 that had devastating consequences on logged lowland dipterocarp forests in Indonesia (Siegert et al. 2001) did not find substantial impacts in mixed and dry deciduous dipterocarp forests in our study area (Baker et al. 2009; Baker and Bunyavejchewin 2009). However, high fire frequencies or intensities may be critical as slow-growing tree seedlings are vulnerable to fire (Wanthongchai et al. 2014), and a fire-mediated tree-recruitment bottleneck has been postulated for DDF making them very similar to savannas (Nguyen et al. 2019).

Fire experiments have been carried out to characterize fires in MDF and DDF in Northern and Central Thailand. Junpen et al (2013) report on fire experiments in Chiang Mai Province in Northern Thailand. Fuel loads were, on average, 3.9 t/ha; fires had low rates of the spread between 0.51 and 2.55 m/min; and fire intensities between 40 and close to 400 kW/m. In fire experiments conducted in DDF in our study area, the Huai Kha Khaeng Wildlife Sanctuary, Wiriya and Kaitpraneet (2009) found fuel loads between 0.7 and 4.6 t/ha. The head fire rate of spread was, on average, 1.9 m/min (0.57–3.94 m/min), and the fire intensity was 190 kW/m (36–372 kW/m) with average flame lengths of 0.9 m (0.41–1.44 m). In summary, fires lit in typical conditions (high relative humidity, high temperatures, low wind speed) in mixed and dry deciduous dipterocarp forests spread slowly and burn at very low to low intensity (Wiriya and Kaitpraneet 2009; Wanthongchai et al. 2011).

1.2 Fire Behavior and the Canadian Forest Fire Behavior Prediction System

Fire behavior modeling provides a means to predict different parameters of fire behavior (e.g., rate of spread, intensity, the direction of spread). The Canadian Fire Weather Index (CFWI) is a set of weather indices used to characterize fire danger and to predict fire behavior for a given fuel type (Stocks et al. 1989), which is possible in conjunction with the Canadian Fire Behavior Prediction System (CFBP) (Forestry Canada Fire Danger Group 1992). CFBP can be used to estimate fire behavior for fuel types that have been calibrated against the CFWI and its sub-indices.

Although originally designed and calibrated for Canadian fuel types, the CFWI is widely used outside Canada to predict fire danger since the indices are found helpful in predicting the probability of fire occurrence (Bianchi and Defosse 2013; Jong et al. 2016; Steenkamp et al. 2012; Dimitrakopoulos et al. 2011). It is also calculated on a global scale (Field et al. 2014; Field 2020; Vitolo et al. 2019). CFWI is used across Southeast Asia in an adapted version to assess fire weather in the region (Groot et al. 2006; Manomaiphiboon et al. 2017). However, to use the CFBP to predict fire behavior for fuel types outside Canada, these fuel types need to be identified and calibrated (Dymond et al. 2004; Fogarty et al. 1998; Pearce et al. 2008).

To provide spatially explicit estimates, the Canadian Fire Behavior Prediction System has been coupled with a fire propagator, which enables the modeling of fire spread for single fire events. This fire spread model is publicly available under the name of Prometheus (Tymstra et al. 2010). In Prometheus, fires are propagated from points making up the vertices of a perimeter polygon (or an ignition point). Using the Prometheus fire growth model, it is possible to predict the position of a fire front at a given time under given weather, fuel, and terrain conditions, as well as fire intensity and rate of spread. Here, we describe an approach using remotely sensed and field data to calibrate the CFBP model for a dry and mixed dipterocarp forest type and then test it using the Prometheus fire growth simulator.

2 Study Area

The Huay Kha Khaeng (HKK) Wildlife Sanctuary is located in Uthai Thani Province, Thailand. It has been a UNESCO World Heritage site since December 1991. There are four types of forests in the HKK Wildlife Sanctuary: dry evergreen forest, mixed deciduous forest, dry deciduous dipterocarp forest, and savanna forest. Forest fires in this area are typically surface fires. The HKK Wildlife Sanctuary regularly witnesses a long period of continuous burning, sometimes lasting up to two months (Baker et al. 2009). Fires occur from mid-December to late April, with the peak fire season in March. Air temperatures averaged from 1979 to 2020 (Hersbach et al. 2020) show an annual mean maximum temperature of 32.3 °C with the highest temperature (35.4 °C) in April, and the annual mean temperature is 24.0 °C. The average precipitation is 1663 mm. Mean relative air humidity varies from 79% in April to 93% in September.

3 Data and Methods

3.1 Fire Experiments

Field data were collected in a total of 71 annual fire experiments (Fig. 1) implemented between 2008 and 2016 in DDF in the HKK area. Experiments were carried out during the height of the fire season from February through April of each year and on different sites within the study area. Experiments comprised four different slope classes (0°–10°, 10°–20°, 0°–30°, and 30°–40°) to assess the effect of slope on fire propagation.

Fuel consumption was estimated by sampling fuels before and after the burn. Sampling was done by collecting all material (litter and live plants) in 50 by 50-cm sampling frames. Fuel was weighed in the field and then oven-dried at the HKK research station to determine fine fuel moisture content.

The methodology of the experimental burns was similar to other works in the area (Junpen et al. 2013; Wiriya and Kaitpraneet 2009). The experimental plots were circular, having a 30 m radius. Fires were ignited using a point ignition in the center of the circle. The fire rate of spread was assessed by measuring fire arrival times at evenly spaced measuring rods in a star-shaped pattern oriented in the eight main compass directions.

From these data, head, flanking, and tail fire rates of spread were determined. Flame height was estimated from the height of the flames at the poles when the fire

Fig. 1 Fire experiments in the Huay Kha Khaeng (HKK) Wildlife Sanctuary. Left: fuel sampling. Right: Fire experiment with measuring rod for ROS and flame height

front was passing. Fire intensity was calculated according to Byram by multiplying the head fire rate of spread with the fuel consumption and the low heat of combustion (Alexander 1982), and the low heat of combustion was assumed to be 18,330 kJ/kg (Junpen et al. 2013).

3.2 *Active Fire, Fuel Map, and Fire Weather Data*

We used Landsat 8-derived fire fronts as ignitions for the fire spread model and Suomi National Polar-orbiting Partnership (Suomi-NPP) Visible Infrared Imaging Radiometer Suite (VIIRS) data to assess its spread. Landsat 8 is flying in a sun-synchronous orbit and has a local equatorial overpass at about 10 am. The Landsat Operational Land Imager (OLI) sensor features short-wave infrared bands at 30 m resolution, which are suited to detect active fires with a flaming component (Schroeder et al. 2015; Kumar and Roy 2018; Sofan et al. 2020). Landsat flies about three and a half hours ahead of S-NPP VIIRS, which has a local equatorial overpass time of about 1:30 am and 1:30 pm. Schroeder et al. (2014, 2017) developed an algorithm that blends the features of the two bands VIIRS mid-wave infrared (MWIR) bands into a single fire detection product that also provides fire radiative power (FRP), which is related to fuel consumption rates (Wooster et al. 2005). This relationship can be used to estimate fuel consumption and fire smoke emissions from space; it has been used in a large number of studies (Ichoku and Kaufman, 2005; Vermote et al. 2009; Ellicott et al. 2009; Roberts et al. 2011; Mota and Wooster 2018) and is the theoretical basis of the European Copernicus Atmospheric Monitoring Service (Kaiser et al. 2012). Fire fronts were extracted from Landsat using the short-wave infrared bands employing the algorithm by (Schroeder et al. 2015). We used the publicly available 30 m DEM produced by the shuttle radar topography mission (SRTM) (Farr et al. 2007) to derive slope, aspect, and height above sea level data for our study area. We used the official Thai land cover data (Land Development Department 2015) as the basis of our fuel map. The deciduous forest class of the land cover map was assigned the new fuel model "Deciduous Dipterocarp Forest."

Fire weather data were retrieved from the Global Fire Weather database (Field et al. 2014; Field 2020) (GFWD) for the dates and grid cell of the fire experiments for the years 2008–2016. The data represent conditions at local noon. For the field data, recordings of the 2 m air temperature and relative humidity were available for the time the burnings were implemented, but no wind measurements. The exact hour of burning was not recorded. Therefore, we used the GFWD data and assigned the (noon) CFWI data of the corresponding grid cell to each experiment.

Meteorological data for the 2016 fire season (January to April) were retrieved from the ERA Global reanalysis database (Hersbach et al. 2020). The data were fed into the R-Version of the Canadian Fire Weather Index calculation algorithm (Wang et al. 2017) to calculate daily and hourly values of the CFWI weather indices, which were then used to drive the Prometheus fire simulator.

3.3 Experimental Calibration for a Deciduous Dipterocarp Forest Fuel Type

The relationship between the rate of spread (ROS) and the initial spread index (ISI) of the CFWI exerts the primary control for fire behavior prediction in CFBP. ISI is derived from the fine fuel moisture code (FFMC), reflecting fine fuels' curing and wind speed. In CFBP, the relationship between ISI and ROS is derived from experimental data by fitting a double-exponential function of the form:

$$\mathrm{RSI} = a \times \left[1 - \mathrm{e}^{(-b\times \mathrm{ISI})}\right]^{c} \tag{1}$$

to the data (Forestry Canada Fire Danger Group 1992). In Eq. (1), RSI is the equilibrium rate of spread, and *a*, *b*, and *c* are fuel-type-specific constants that have been determined for Canadian fuels using regression models from experiments. The double-exponential function results for most fuels in an S-shaped curve of variable slope steepness. The steeper the slope before leveling off for very high ISI values, the stronger the increase in the rate of spread with increasing ISI. To derive the *a*, *b*, and *c* parameters, a nonlinear least squares (NLS) regression is used to fit the double-exponential model to the data.

3.4 Burned Area

We initialize the Prometheus fire spread model using Landsat 8-derived fire fronts as ignitions. Since the Landsat satellite overpass happens any time after the real ignition(s) of the fire, the area already burned has to be considered to prevent the model from simulating fire growth in areas already burned, especially to prevent the fire from burning backward. Already burned areas where therefore derived from the Landsat data. The Landsat Collection 1 data were downloaded from the Landsat Data archive and converted to top of the atmosphere reflectance. An important requirement for the burned area map was that the methods applied should be as robust as possible regarding pixels contaminated by smoke, especially if the short-wave infrared bands of Landsat are used. Two spectral indices were calculated, the widely used normalized burn ratio (NBR) (García and Caselles 1991) and the mid-infrared burn index (MIRBI) (Smith et al. 2007; Trigg and Flasse 2001). To avoid misclassifications of cloud shadows and mask clouds, a cloud mask was applied using the FMask algorithm (Zhu et al. 2015).

Mapping burned areas was done by a) detecting changes in the two indices between two acquisitions and b) using a single-scene threshold for MIRBI. The two scenes used for change detection were co-registered, and detection was based on threshold values 0.3 for the difference MIRBI (t2–t1) and 0 for the difference NBR (t2–t1). Pixels above the threshold for MIRBI difference and below the NBR difference were classified as burned. The single-scene threshold for MIRBI was 1.4. This threshold was necessary since clouds and smoke may obscure scenes, and obtaining two observations of the same spot for change detection is impossible. Since this algorithm had to be more conservative to avoid false alarms, the error of omission is higher in areas where it was applied. The change detection results were retained only when no clouds or cloud shadows were detected in both scenes, whereas the single-step results were always retained.

3.5 *Model Setup*

To perform a test of the usability of the calibrated fuel model, we assessed a fire front system in HKK using the Prometheus fire growth model.

Figure 2 shows a flowchart describing the modeling process. Burned areas were inserted into the fuel maps and classified as non-fuel. Evergreen forests—found in the higher areas—were also classified as non-fuel as it was observed that fires rarely spread into this forest type. To successfully run the model in Prometheus and avoid memory limitations, the ignition points derived from Landsat were clustered based on their spatial proximity. The model was run on each of these clusters. Landsat 8-derived fire fronts observed during two Landsat 8 overpasses on 02.03.2016 at 03:43 UTC (09:43 local time) and on 18th March 2016 at 03:43 UTC (09:43 local time) were used as ignitions for the model. Each cluster was assigned the weather data of the nearest weather grid point. Hourly CFWI data were directly fed into Prometheus. The model was then run for 15 hourly time steps starting from the time of the Landsat overpass until shortly after the VIIRS nighttime overpass at approximately 1:30 local time.

4 Results and Discussion

4.1 *Field Experiments*

Figure 3 shows the main results of the fire experiments in boxplot charts. A total of 71 experiments were implemented. Weather conditions were dry during all burns, and FFMC was above 90 in all cases. Wind speed was slow, with speeds between 6.6 and a maximum of 13.3 km/h, resulting in a low ISI between 8.6 and 14.9. ROS was 1.06 ± 0.61 m/min across all slope classes. Faster spread on steeper slopes was only observed in the experiments implemented in the steepest slope class (above 40°), where the mean ROS was 1.6 m/min (± 08.8 m/min). The mean ROS for all other slope classes was below 1 m/min. These low spread rates confirm observations in HKK and Northern Thailand (Wiriya and Kaitpraneet 2009) and (Junpen et al. 2013) described above.

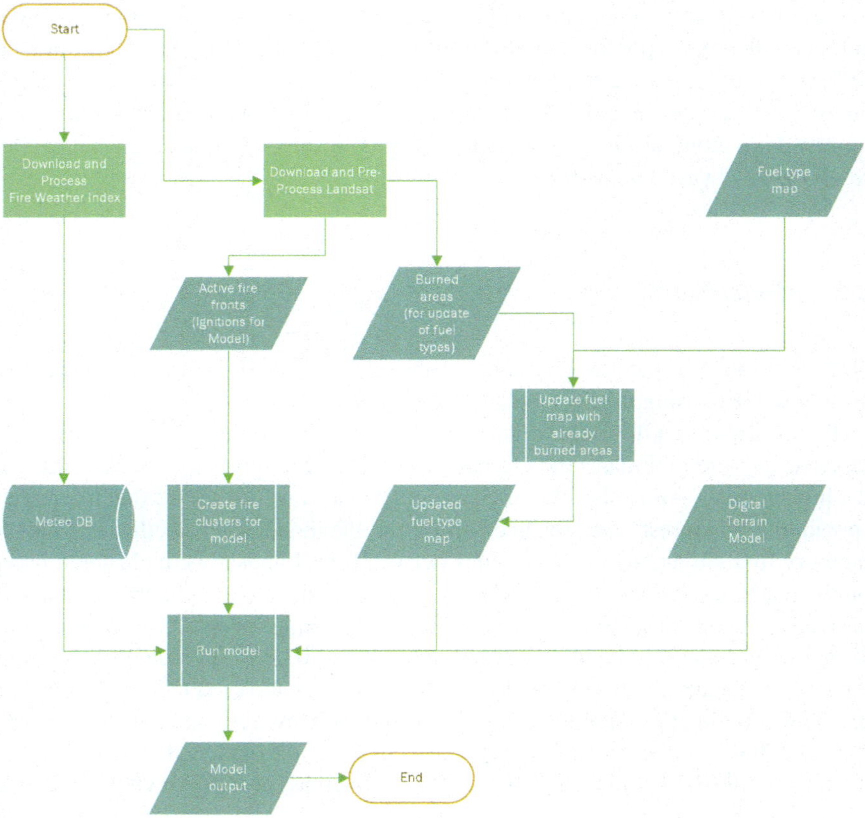

Fig. 2 Fire behavior model data processing workflow

FC was very similar across all slope classes with a mean of 5.4 t/ha (± 1.26 t/ha), a maximum of 8.9 t/ha, and a minimum of 2.66 t/ha. This is higher than the FC of 3.1 t/ha reported by Junpen et al. (2013) in a drier area of Northern Thailand but similar to measurements in HKK by Wanthongchai et al. (2011), who reported FC range from 4.3 to 8.1 t/ha, depending on the frequency of fire at the investigated sites. Also, in HKK, (Wiriya and Kaitpraneet 2009) reported mean fuel loads of 4.64 t/ha (no FC is given). These values are an order of magnitude lower than those given for dry tropical forests in the Global Fuel Consumption Database (van Leeuwen et al. 2014), lacking examples from Southeast Asia. In this compilation, examples from dry tropical forests in Brazil and Mexico are cited with an FC of 61 and 91 t/ha, respectively. Both of these examples, however, were from studies investigating conversion from forests to pasture (Kauffman et al. 1998, 2003), while in Southeast Asia, frequent fires with low FC occur without conversion. To compare the experimental FC values to another global database, the 0.25° × 0.25° resolution

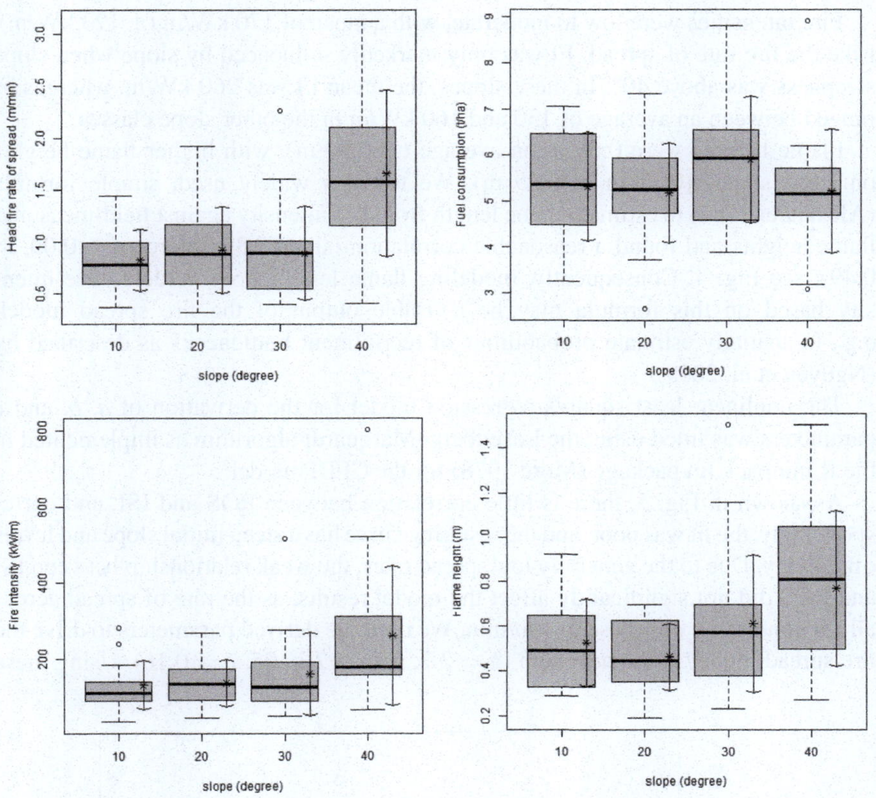

Fig. 3 Boxplots of main fire characteristics of the experimental fires. Top left: head fire rate of spread, top right: fuel consumption, bottom left: fire intensity, and bottom right: flame height. Arrows to the right of boxplot indicate mean (*) and standard deviation (arrow)

Global Fire Emissions database (GFED4s) (van der Werf et al. 2017), we downloaded the 2016 data. We calculated FC from burned area fraction, grid cell size, and the dry matter (DM) emissions provided in the dataset. FC values in GFED over the HKK in February and March 2016 were about 150 t/ha, i.e., about 30 times higher than the FC values obtained from the experiments. A possible reason for this overestimation is that GFED classifies fires over the tropical forest as "deforestation fires" if they show repeated (persistent) fire detections over a more extended time period (van der Werf et al. 2010). Indeed, the five 0.25° grid cells covering HKK showed a deforestation fire fraction between 73 and 99%, whereas no deforestation occurred in this protected area. Persistent fire detections of HKK—and most probably over other mixed and dry deciduous dipterocarp forests—may result from the specific fire behavior with slow-moving fires that consume relatively little fuel while causing persistent fire detections. Hence, this classification method and the subsequent modeling of deforestation fires over this forest type may lead to a systematic bias in GFED.

Fire intensities were low to moderate, with a mean of 170 kW/m (± 123 kW/m). Like the fire rate of spread, FI was only markedly influenced by slope when slope steepness was above 40°. In these slopes, the mean FI was 260 kW/m, whereas FI ranged between an average of 130 and 160 kW/m in the other slope classes.

Flame heights were 0.65 m on average (± 0.59 m), with higher flame heights on steep slopes (0.77 m ± 0.35 m). We tested a widely used, simple formula (Alexander 1982) to estimate flame length from fire intensity against field-measured flame heights and found a reasonable correlation (slope 0.89, intercept − 0.09, r^2 0.49); see Fig. 4. Consequently, modeling flame length from modeled fire intensity based on this formula may be a usable output of the fire spread model, e.g., to spatially estimate probabilities of recruitment bottlenecks as described by (Nguyen et al. 2019).

The nonlinear least squares regression model for the derivation of *a*, *b*, and *c* parameters was fitted using the Levenberg–Marquardt algorithm as implemented in the R minpack.lm package (Moré 1978) for the CFBP model.

As shown in Fig. 5, there is little correlation between ROS and ISI, and correspondingly, the fit was poor, and the resulting curve has a steep initial slope and levels off quickly. Due to the generally low spread rates, the weak relationship between ISI and ROS did not significantly affect the model results, as the rate of spread across all ISI conditions was close to 1 m/min. We used the derived parameters to drive the fire spread model in the next step ($a = 3.2$, $b = -$ 7E−05, $c = 0.16$). Using these

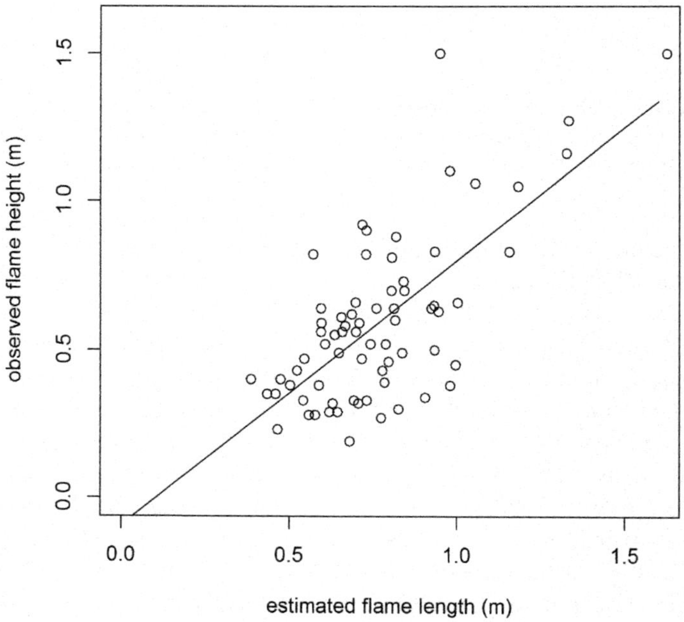

Fig. 4 Flame length estimated after Byram versus field observed flame height. The line indicates the regression line

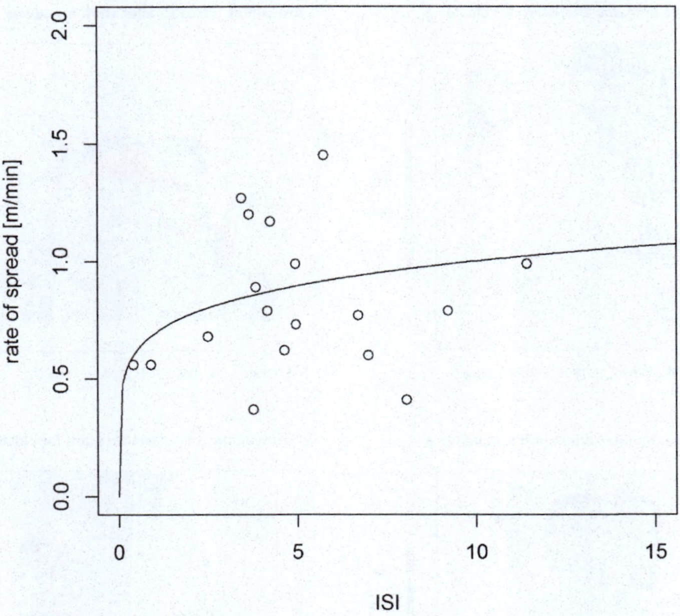

Fig. 5 ISI-ROS function derived from the experimental fires in the HKK. The line indicates the ROS (ISI) function obtained through NLS fitting

parameters, both conditions with high wind speeds and steep slopes may result in an underprediction of the rate of spread due to the early flattening of the ISI-ROS relationship. Inaccurate modeling of steep slopes may arise from the way CFBPS handles fire spread on slopes which is done by increasing the ISI value used for ROS prediction with increasing slope steepness.

4.2 Model Results

Modeling outputs for the fire clusters for March 2 and March 18 are presented in Fig. 6, showing the initial Landsat fire detections at about 10:45 am, the final fire perimeters at about 1:30 am local time on the next day, and the VIIRS and MODIS fire detections. To assess model performance, we assess the spatial proximity of the VIIRS nighttime detections to the modeled fire perimeters at the time of the VIIRS overpass. We only used the 375 m resolution detections associated with the last time step, since, due to the slow spread rate, detections, e.g., at noon, would often fall so close to the original ignition that a meaningful evaluation would not be possible. The position of the fire fronts derived from modeling was generally close to the coincident VIIRS fire detections and often within the positional accuracy of the VIIRS data. Most of the VIIRS fire detections (119 out of 125 on March 2, 2016)

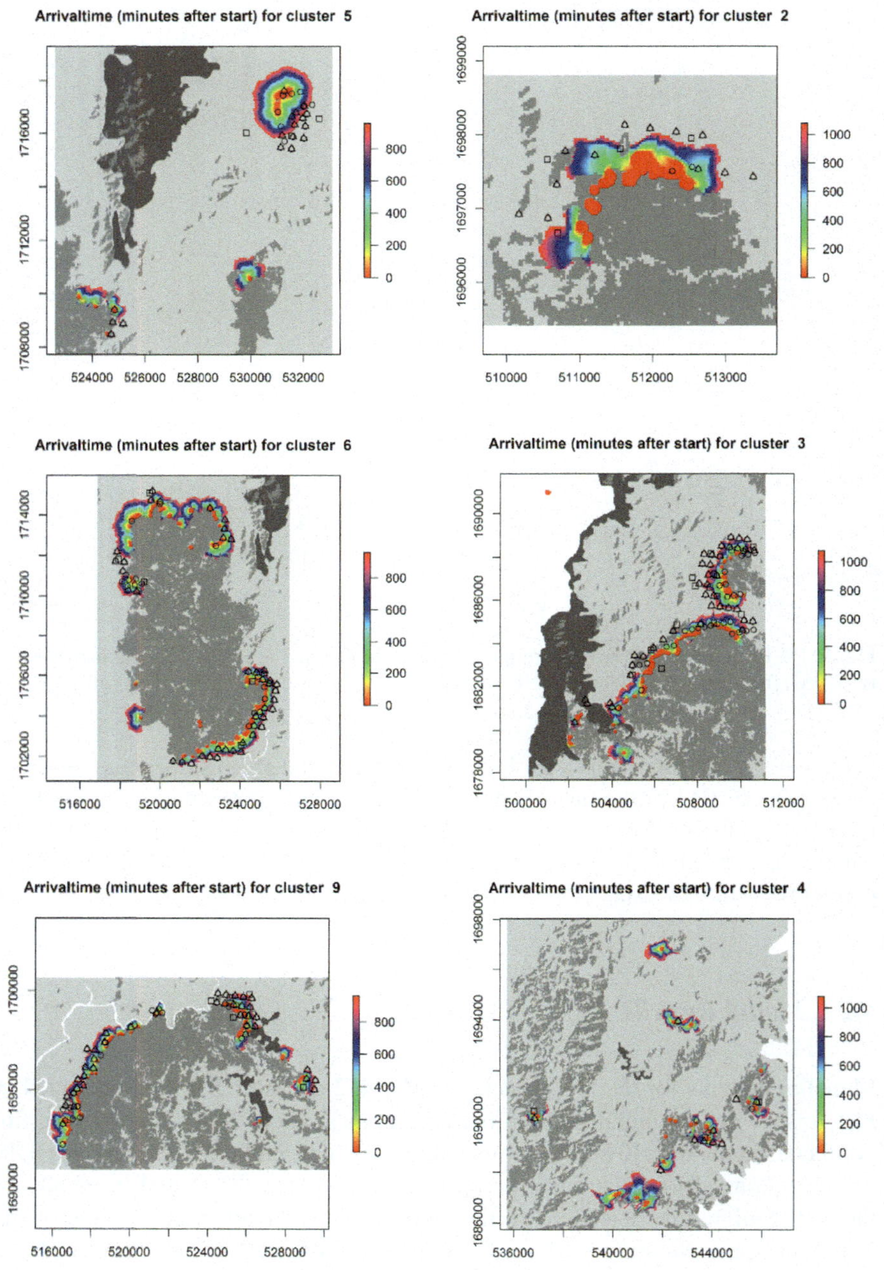

Fig. 6 Model output and satellite fire detections: left column: March 02, 2016, right column: March 18, 2016. Background colors: light gray: deciduous dipterocarp forest, medium gray: already burned area/non-fuel, dark gray: evergreen dipterocarp forest, and white: water/no data. Square markers: MODIS fire detections, circle markers: VIIRS fire detections, and triangle markers: VIIRS fire detections at the end of the simulation run. Axes coordinates are UTM zone 47 N coordinates

over the HKK area were within 750 m (i.e., two-pixel sizes) of the modeled final perimeters. Detections outside the modeled perimeters seemed to be associated with new ignitions of that day (i.e., these ignitions could not be modeled). On March 2, 44% or 95 of the detected pixel's center coordinates were within 250 m of the modeled fire perimeter, and 71% or 155 VIIRS fire pixels centers were within 500 m of the modeled final perimeter. Similarly, on March 18, 47% or 209 pixels of the VIIRS detections were within 250 m, and 73% or 326 detections were within 500 m of the modeled perimeters for that day.

While active fires were detected over all clusters, not all frontal systems within the individual clusters had detections. A substantial number of the smaller fires detected by Landsat did not evolve as modeled. This is not surprising since, due to the coarseness of the input fuel map, minor barriers were not considered in the model. At the low intensities observed, even small barriers, fuel discontinuities, or moist areas can stop fire spread as the fire cannot jump these obstacles.

One potential application of automated fire behavior modeling is to improve air quality forecasts. To achieve such a forecast, fire emissions need to be modeled based on fuel consumption rates using the fire propagation model.

To test the suitability of our model for such a task, the fire radiative power detected by infrared remote sensors can be used as a proxy to fuel consumption rate. The conversion between total fuel consumption and simulated FRE (i.e., the FRP summed over the time interval, unit MJ) can be done using a scaling factor of 0.368 MJ/kg (Wooster et al. 2005). Fuel consumption for hourly intervals can thus be estimated from the total fuel consumption and the fire arrival time. Modeled fire radiative energy (MJ) for each hourly interval was estimated by multiplying the modeled burned area for the interval by the modeled fuel consumption and applying the above-mentioned scaling factor to convert to FRE. The average FRP (unit MW) for the hourly time interval was then calculated by dividing hourly FRE by 3600.

Satellite-derived fire emissions estimates based on FRP are often systematically lower than those obtained by fuel consumption modeling based on the burned area. This effect is stronger under canopy cover than in open landscapes (Roberts et al. 2018). To explore the magnitude of this effect, a scaling factor derived from a comparison of satellite data to FC data in the GFCD by (Andela et al. 2016) was applied to simulate the FRP detected by the satellite sensor. To constrain FRP estimates, the uncertainty of the field-derived fuel consumption estimates was applied to the modeled FRP (95% confidence interval).

Both fire detections by the MODIS and VIIRS sensors were used to compare against the model output. Figure 7 shows the results for the same clusters as depicted in Fig. 6. Not all clusters had fire detections for all possible satellite overpasses. Observed FRP was lower than modeled FRP in most cases. Especially at night, modeled FRP was substantially higher than the observed FRP. Most likely, the most important source for this discrepancy is the overestimation of the rate of spread and fuel consumption at nighttime for significant parts of the fire fronts and the fact that many of the smaller fire fronts detected by Landsat in the morning were not active

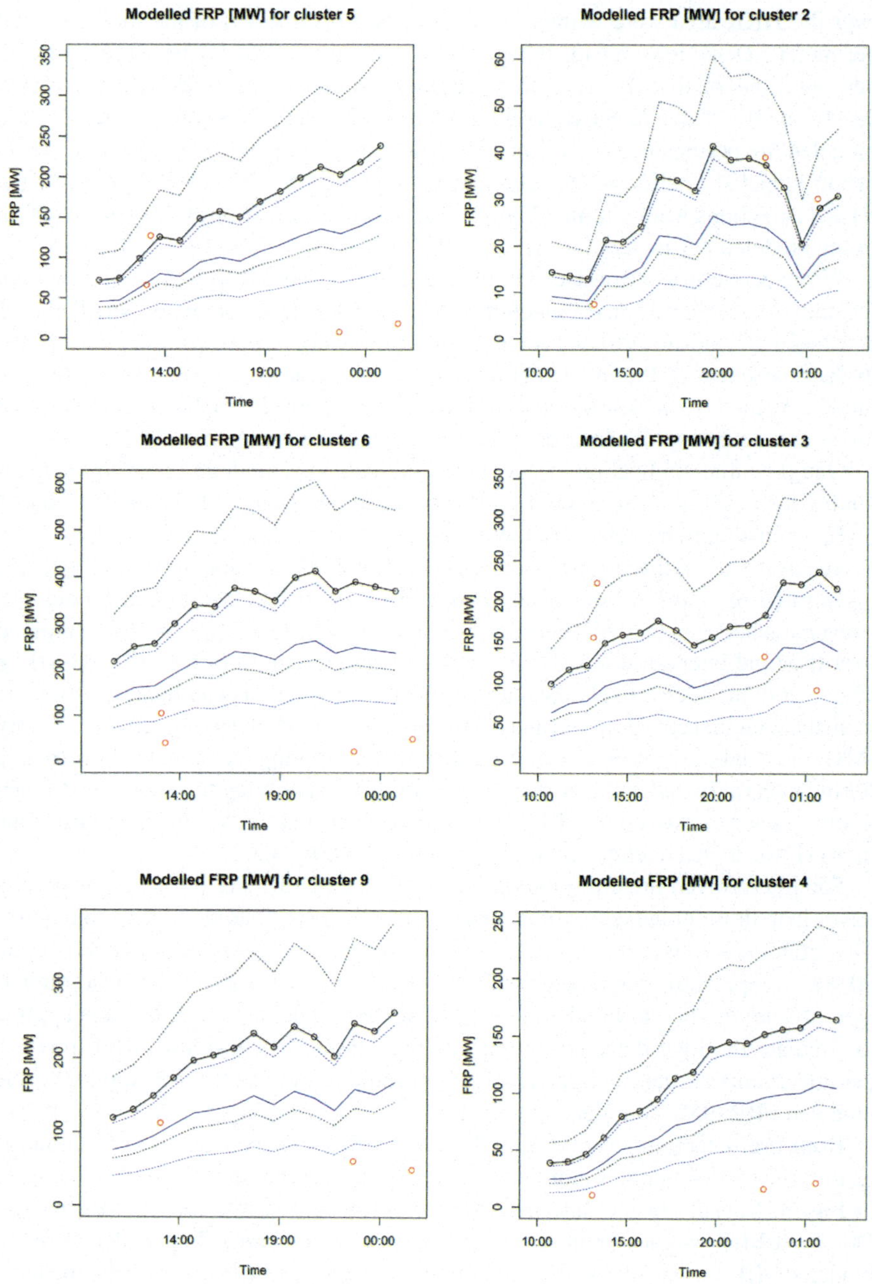

Fig. 7 Model-derived FRP versus observed FRP for March 03, 2016 (left column), and March 18, 2016 (right column). Black line with circle dots: FRP estimated following (Wooster et al. 2005). Black dotted lines 95% confidence interval for the estimate of fuel consumption. Blue solid line: FRP estimated applying the correction factor of Andela et al. (2016). Blue dotted lines: bounds of this FRP estimate defined by 95% confidence interval of field-derived FC

anymore in the evening, whereas in the model, these ignitions were sustained and kept growing. Thus, in clusters 2, 3, and 9 of Fig. 7, the observed FRP was within the (wide) bounds of modeled FRP.

5 Conclusions and Outlook

We tested the possibility of calibrating the CFBP for deciduous dipterocarp forests using globally available weather data, remote sensing data, and data from field experiments.

Our results indicate that even during drought, the fire rate of spread for fires in dry dipterocarp forests is slow, and fire intensity is generally low to moderate. Fuel consumption is similar to that found in other studies in the area and substantially lower than given for this (broad) forest type in an international reference database, which is based on deforestation fires (van Leeuwen et al. 2014).

The poor fit of the ISI-ROS NLS regression may be attributed to the inaccurate weather data and the small range of observed values. This relatively small variability of weather conditions may also cause a relatively small range of fire behavior in deciduous dipterocarp forests. To establish fire behavior further and to better calibrate the fire growth models for practical purposes, a more extensive study on a wider range of samples and fire situations will be necessary.

Despite this limitation, the comparison of modeled fire growth with remote sensing observations of actual wildfires in the same area was reasonably accurate for those that continued spreading during the modeled time span. However, a number of the (mostly smaller) fire fronts detected in the morning did not continue to burn in the evening at the end of the model run. This led to an overestimation of the area burned and of total fuel consumption, especially at night, and would equally lead to an overestimation of emissions if the model was applied for smoke forecasting. Due to their low intensity, many fires are easily stopped at natural barriers that are not in the fuel map data. Hence, improvements in the accuracy and resolution of the fuel map could improve outputs substantially. Our results also indicate that the fuel consumption and emission estimates for the particular fire observed in our study area in the Global Fire Emissions database (van der Werf et al. 2017) are an order of magnitude higher than the FC found in our field data, and that this overestimate in GFED may be systematic and is possibly caused by the repeat detection of the slow-spreading, low-intensity fires which are classified as deforestation fires in GFED.

Acknowledgements G. R. acknowledges financial support by the ZIM program of the German Ministry of Economy, Grant Number 16KN052420.

Author Contributions Conceptualization, G. R., V. T., K. W.; methodology, G. R., V. T., K. W.; formal analysis, G. R., V. T., K. W.; data curation, G. R., V. T., K. W.; writing, G. R., V. T.; and funding acquisition, G. R. All authors have read and agreed to the published version of the manuscript.

References

Alexander, M.E. 1982. Calculating and interpreting forest fire intensities. *Canadian Journal of Botany* 60: 349–357.

Andela, N., G.R. van der Werf, J.W. Kaiser, T.T. van Leeuwen, M.J. Wooster, and C.E.R. Lehmann. 2016. Biomass burning fuel consumption dynamics in the tropics and subtropics assessed from satellite. *Biogeosciences* 13 (12): 3717–3734.

Baker, P.J., and S. Bunyavejchewin. 2006. Bark thickness and the influence of forest fire on tree population structures in a seasonal evergreen tropical forest. *Natural History Bulletin of the Siam Society* 54 (2): 215–225.

Baker, P.J., and S. Bunyavejchewin. 2009. Fire behavior and fire effects across the forest landscape of continental Southeast Asia. In *Tropical fire ecology*, 311–334. Springer.

Baker, P.J., S. Bunyavejchewin, and A.P. Robinson. 2009. The impacts of large-scale, low-intensity fires on the forests of continental South-East Asia. *International Journal of Wildland Fire* 17 (6): 782–792.

Bianchi, L.O., and G.E. Defosse. 2014. Ignition probability of fine dead surface fuels in native Patagoniaforests of Argentina. *Forest Systems* 23 (1): 129–138.

Bullock, S.H., H.A. Mooney, and E. Medina. 1995. *Seasonally dry tropical forests*. Cambridge University Press.

Dimitrakopoulos, A.P., A.M. Bemmerzouk, and I.D. Mitsopoulos. 2011. Evaluation of the Canadian fire weather index system in an eastern Mediterranean environment. *Meteorological Applications* 18 (1): 83–93.

Dymond, C.C., O. Roswintiarti, and M. Brady. 2004. Characterizing and mapping fuels for Malaysia and western Indonesia. *International Journal of Wildland Fire* 13: 323–334.

Ellicott, E., E. Vermote, L. Giglio, and G. Roberts. 2009. Estimating biomass consumed from fire using MODIS FRE. *Geophysical Research Letters* 36 (13): L13401.

Farr, T.G., P.A. Rosen,, E. Caro, R. Crippen, R. Duren, S. Hensley, M. Kobrick, M. Paller, E. Rodriguez, and L. Roth. 2007. The shuttle radar topography mission. *Reviews of Geophysics* 45(2).

Field, R.D. 2020. Evaluation of global fire weather database reanalysis and short-term forecast products. *Natural Hazards and Earth System Sciences* 20 (4): 1123–1147.

Field, R.D., A.C. Spessa, N.A. Aziz, A. Camia, A. Cantin, R. Carr, W.J. de Groot, A.J. Dowdy, M.D. Flannigan, K. Manomaiphiboon, F. Pappenberger, V. Tanpipat, and X. Wang. 2015. Development of a global fire weather database for 1980–2012. *Natural Hazards and Earth System Sciences* 15: 1407–1423.

Fogarty, L.G., H.G. Pearce, W.R. Catchpole, and M.E. Alexander. 1998. Adoption vs. adaptation: Lessons from applying the Canadian forest fire danger rating system in New Zealand. In *Proceedings, 3rd International Conference on Forest Fire Research and 14th Fire and Forest Meteorology Conference, Luso, Coimbra, Portugal*, 1011–1028

Forestry Canada Fire Danger Group. 1992. *Development and structure of the Canadian forest fire behaviour prediction system*. Ottawa: Forestry Canada—Science and Sustainable Development Directorate.

García, M.J.L., and V. Caselles. 1991. Mapping burns and natural reforestation using thematic mapper data. *Geocarto International* 6 (1): 31–37.

Groot, W.J.D., R.D. Field, M.A. Brady, O. Roswintiarti, and M. Mohamad. 2006. Development of the Indonesian and Malaysian fire danger rating systems. *Mitigation and Adaptation Strategies for Global Change* 12(1), 165–180.

Hersbach, H., B. Bell, P. Berrisford, S. Hirahara, A. Horányi, J. Muñoz-Sabater, J. Nicolas, C. Peubey, R. Radu, and D. Schepers. 2020. The ERA5 global reanalysis. *Quarterly Journal of the Royal Meteorological Society* 146 (730): 1999–2049.

Ichoku, C., and Y.J. Kaufman. 2005. A method to derive smoke emission rates from MODIS fire radiative energy measurements. *IEEE Transactions on Geoscience and Remote Sensing* 43 (11): 2636–2649.

Jong, M.C.D., M.J. Wooster, K. Kitchen, C. Manley, R. Gazzard, and F.F. McCall. 2016. Calibration and evaluation of the Canadian forest fire weather index (FWI) system for improved wildland fire danger rating in the United Kingdom. *Natural Hazards and Earth System Sciences* 16(5): 1217–1237.

Junpen, A., S. Garivait, S. Bonnet, and A. Pongpullponsak. 2013. Fire spread prediction for deciduous forest fires in Northern Thailand. *Science Asia* 39: 535–545.

Kaiser, J.W., A. Heil, M.O. Andreae, A. Benedetti, N. Chubarova, L. Jones, J.J. Morcrette, M. Razinger, M.G. Schultz, M. Suttie, and G.R. van der Werf. 2012. Biomass burning emissions estimated with a global fire assimilation system based on observed fire radiative power. *Biogeosciences* 9 (1): 527–554.

Kauffman, J.B., D.L. Cummings, and D.E. Ward. 1998. Fire in the Brazilian Amazon 2. Biomass, nutrient pools and losses in cattle pastures. *Oecologia* 113 (3): 415–427.

Kauffman, J.B., M.D. Steele, D.L. Cummings, and V.J. Jaramillo. 2003. Biomass dynamics associated with deforestation, fire, and conversion to cattle pasture in a Mexican tropical dry forest. *Forest Ecology and Management* 176 (1–3): 1–12.

Kumar, S.S., and D.P. Roy. 2018. Global operational land imager Landsat-8 reflectance-based active fire detection algorithm. *International Journal of Digital Earth* 11 (2): 154–178.

Land Development Department. 2015. *Land cover map of Thailand*.

Manomaiphiboon, K., V. Tanpipat,, P. Nhuchaiya, N. Jaroonrattanapak, and C. Buaniam. 2017. An operational fire danger forecast system for lower mekong river region: Technical concepts and current implementation. *Regional Haze and Climate Change Management*. Chiang Mai: Haze Free Thailand Project.

Miles, L., A.C. Newton, R.S. DeFries, C. Ravilious, I. May, S. Blyth, V. Kapos, and J.E. Gordon. 2006. A global overview of the conservation status of tropical dry forests. *Journal of Biogeography* 33 (3): 491–505.

Moré, J.J. 1978. The Levenberg-Marquardt algorithm: Implementation and theory. In *Numerical analysis*, 105–116. Springer.

Mota, B., and M.J. Wooster. 2018. A new top-down approach for directly estimating biomass burning emissions and fuel consumption rates and totals from geostationary satellite fire radiative power (FRP). *Remote Sensing of Environment* 206: 45–62.

Nguyen, T.T., B.P. Murphy, and P.J. Baker. 2019. The existence of a fire-mediated tree-recruitment bottleneck in an Asian savanna. *Journal of Biogeography* 46 (4): 745–756.

Olson, D.M., E. Dinerstein, E.D. Wikramanayake, N.D. Burgess, G.V.N. Powell, E.C. Underwood, J.A. D'Amico, I. Itoua, H.E. Strand, and J.C. Morrison. 2001. Terrestrial ecoregions of the world: A new map of life on Earth: A new global map of terrestrial ecoregions provides an innovative tool for conserving biodiversity. *BioScience* 51 (11): 933–938.

Pearce, H.G., S.A.J. Anderson, V.R. Clifford. 2008. *A manual for predicting fire behaviour in New Zealand fuels*. Scion Rural Fire Research Group.

Roberts, G., M.J. Wooster, P.H. Freeborn, and W. Xu. 2011. Integration of geostationary FRP and polar-orbiter burned area datasets for an enhanced biomass burning inventory. *Remote Sensing of Environment*.

Roberts, G., M.J. Wooster, N. Lauret, J.P. Gastellu-Etchegorry, T. Lynham, and D. McRae. 2018. Investigating the impact of overlying vegetation canopy structures on fire radiative power (FRP) retrieval through simulation and measurement. *Remote Sensing of Environment* 217: 158–171.

Rundel, P.W., and K. Boonpragob. 1995. Dry forest ecosystems of Thailand. In *Seasonally dry tropical forests*, 93–123.

Schroeder, W. 2017. Visible infrared imaging radiometer suite (VIIRS) 375 m & 750 m active fire detection data sets based on NASA VIIRS land science investigator processing system (SIPS) reprocessed data—Version 1. Product User's Guide Version 3. NASA.

Schroeder, W., P. Oliva, L. Giglio, and I. Csiszar. 2014. The new VIIRS 375 m active fire detection data product: Algorithm description and initial assessment. *Remote Sensing of Environment* 143: 85–96.

Schroeder, W., Oliva, P., Giglio, L., Quayle, B., Lorenz, E., and F. Morelli. 2015. Active fire detection using Landsat-8/OLI data. *Remote Sensing of Environment.*

Siegert, F., G. Ruecker, A. Hinrichs, and A.A. Hoffmann. 2001. Increased damage from fires in logged forests during droughts caused by El Nino. *Nature* 414: 437–440.

Smith, A., N. Drake, M. Wooster, A. Hudak, Z.A. Holden, and C. Gibbons. 2007. Production of Landsat ETM+ reference imagery of burned areas within Southern African savannahs: Comparison of methods and application to MODIS. *International Journal of Remote Sensing* 28 (12): 2753–2775.

Sofan, P., D. Bruce, E. Jones, M.R. Khomarudin, and O. Roswintiarti. 2020. Applying the tropical Peatland combustion algorithm to Landsat-8 operational land imager (OLI) and Sentinel-2 multi spectral instrument (MSI) imagery. *Remote Sensing* 12 (23): 3958.

Steenkamp, K.C., K.J. Wessels, F. van den Bergh, G.A. McFerren, P.E. Frost, C.W. Lai, and D. Swanepoel. 2012. Quantitative comparison of fire danger index performance using fire activity, *IEEE International Geoscience and Remote Sensing Symposium*, IEEE: 3375-3378

Stocks, B.J., T.J. Lynham, B.D. Lawson, M.E. Alexander, C.E.V. Wagner, R.S. McAlpine, and D.E. Dube. 1989. Canadian forest fire danger rating system: An overview. *The Forestry Chronicle* 65 (4): 258–265.

Trigg, S., and S. Flasse. 2001. An evaluation of different bi-spectral spaces for discriminating burned shrub-savannah. *International Journal of Remote Sensing* 22 (13): 2641–2647.

Tymstra, C., R.W. Bryce, B.M. Wotton, S.W. Taylor, and O.B. Armitage. 2010. *Development and structure of prometheus: The Canadian wildland fire growth simulation model.* Natural Resources Canada, Canadian Forest Service, Northern Forestry Centre, Information Report NOR-X-417. Edmonton, AB.

van der Werf, G.R., J.T. Randerson, L. Giglio, G.J. Collatz, M. Mu, P.S. Kasibhatla, D.C. Morton, R.S. DeFries, Y. Jin, and T.T. van Leeuwen. 2010. Global fire emissions and the contribution of deforestation, savanna, forest, agricultural, and peat fires (1997–2009). *Atmospheric Chemistry and Physics* 10 (23): 11707–11735.

van der Werf, G.R., J.T. Randerson, L. Giglio, T.T. van Leeuwen, Y. Chen, B.M. Rogers, M. Mu, M.J.E. van Marle, D.C. Morton, G.J. Collatz, R.J. Yokelson, and P.S. Kasibhatla. 2017. Global fire emissions estimates during 1997–2016. *Earth System Science Data* 9 (2): 697–720.

van Leeuwen, T.T., G.R. van der Werf, A.A. Hoffmann, R.G. Detmers, G. Rücker, N.H.F. French, S. Archibald, J.A. Carvalho Jr., G.D. Cook, W.J. de Groot, C. Hély, E.S. Kasischke, S. Kloster, J.L. McCarty, M.L. Pettinari, P. Savadogo, E.C. Alvarado, L. Boschetti, S. Manuri, C.P. Meyer, F. Siegert, L.A. Trollope, and W.S.W. Trollope. 2014. Biomass burning fuel consumption rates: A field measurement database. *Biogeosciences* 11 (24): 7305–7329.

Vermote, E., E. Ellicott, O. Dubovik, T. Lapyonok, M. Chin, L. Giglio, and G.J. Roberts. 2009. An approach to estimate global biomass burning emissions of organic and black carbon from MODIS fire radiative power. *Journal of Geophysical Research* 114 (D18): D18205.

Vitolo, C., F. Di Giuseppe, B. Krzeminski, and J. San-Miguel-Ayanz. 2019. A 1980–2018 global fire danger re-analysis dataset for the Canadian fire weather indices. *Scientific Data* 6(1), 1–10.

Wang, X., B.M. Wotton, A.S. Cantin, M.-A. Parisien, K. Anderson, B. Moore, and M.D. Flannigan. 2017. cffdrs: An R package for the Canadian forest fire danger rating system. *Ecological Processes* 6 (1): 1–11.

Wanthongchai, K., J.G. Goldammer, and J. Bauhus. 2011. Effects of fire frequency on prescribed fire behaviour and soil temperatures in dry dipterocarp forests. *International Journal of Wildland Fire* 20 (1): 35–45.

Wanthongchai, K., J. Bauhus, and J.G. Goldammer. 2014. Effects of past burning frequency on woody plant structure and composition in dry dipterocarp forest. *Thai Journal of Forestry* 33 (3): 109–130.

Wiriya, K., and S. Kaitpraneet. 2009. Fire behavior in dry deciduous dipterocarp forest at Huai Kha Khaeng Wildlife Sanctuary, Uthai Thani province. *Agriculture and Natural Resources* 43 (5): 41–47.

Wooster, M.J., G. Roberts, G.L.W. Perry, Y.J. Kaufman. 2005. Retrieval of biomass combustion rates and totals from fire radiative power observations: FRP derivation and calibration relationships between biomass consumption and fire radiative energy release. *Journal of Geophysical Research* 110(D24).

Zhu, Z., S. Wang, and C.E. Woodcock. 2015. Improvement and expansion of the Fmask algorithm: Cloud, cloud shadow, and snow detection for Landsats 4–7, 8, and Sentinel 2 images. *Remote Sensing of Environment* 159: 269–277.

Greenhouse Gas Emissions and Air Pollution

Spatiotemporally Resolved Pollutant Emissions from Biomass Burning in Asia

Rui Xiong, Yatai Men, and Guofeng Shen

Abstract Asia is a densely populated continent where several people use biomass as energy or non-energy source. As a result, several regions in Asia experience significant air pollutant emissions that degrade air quality and human health. In this study, emissions of primary air pollutants, including $PM_{2.5}$, OC, BC, CO, and SO_2, from incomplete burning of biomass covering natural fires, residential stoves, open burning of agricultural wastes, power plants, and industry sources from Asia were discussed. In addition, temporal and spatial distributions of targeted air pollutants from biomass burning are analyzed. The results suggest that the total biomass consumption accounts for about one-third of the global total, with a high amount of biomass burning as open fires, including natural fires (27%), open burning of agricultural wastes (3%), and energy sources in residential stoves (48%). Estimated emissions of $PM_{2.5}$, OC, BC, CO, and SO_2 in Asia from biomass burning were 14.6 (11.1–18.7) Tg, 7.4 (5.8–11.2) Tg, 1.9 (1.4–2.6) Tg, 177.8 (146.9–215.0) Tg, and 1.4 (1.3–1.5) Tg, respectively, that were 20%, 26%, 20%, 19%, and 1% of the global total. Nearly 70% of the biomass burning emissions were from South and Southeast Asia, followed by East Asia (~ 25%). Open burning of biomass contributed to 20–50% of the total pollutant emissions, varying in different pollutant types. From 1960 to 2014, anthropogenic emissions increased and then started to decline since the early 1990s; however, there were substantial differences in the sectoral contributions and temporal trends among different countries. Our results contribute to a better understanding of biomass burning pollutants on the regional air quality and climate change in Asia.

Keywords Biomass burning · Consumption · Air pollution · Emissions · Temporal trend · Spatial distributions · Asia

R. Xiong · Y. Men · G. Shen (✉)
College of Urban and Environmental Sciences, Peking University, Beijing, China
e-mail: gfshen12@pku.edu.cn

K. P. Vadrevu et al. (eds.), *Vegetation Fires and Pollution in Asia*,
https://doi.org/10.1007/978-3-031-29916-2_15

1 Introduction

Asia is a densely populated continent, and many Asian countries suffer from severe air pollution problems (Adam et al. 2021; Yadav et al. 2017). Earlier researchers mapped high emissions of major air pollutants like CO, primary $PM_{2.5}$ (particulate matter with an aerodynamic diameter less than 2.5 μm), and black carbon (BC) in this region. It had been estimated that the emissions of primary $PM_{2.5}$ from Asia in 2014 accounted for nearly 40% of the total global emissions, with high contributions from the industry and residential sectors (Huang et al. 2014). The energy sector in most Asian countries still relies on traditional solid fuels like coal and biomass. According to the energy statistics from the International Energy Agency (IEA) and the PKU inventory study (http://inventory.pku.edu.cn/), about 20% of the energy was from biomass burning globally, and in 2019, approximately 26% of the biomass was consumed in Asia (IEA 2020; Tao et al. 2018). Some studies pointed out that the IEA estimation of biomass use in the residential sector, a significant source of biomass consumption in many developing Asian countries, was overestimated. It is because of a significant transition to cleaner modern energies in the residential sector in countries like China (Tao et al. 2018; Shen et al. 2022a). However, there is less doubt that Asian anthropogenic biomass consumption contributes mainly to global consumption. Besides anthropogenic biomass-related emissions, fires from the natural forests and crop residue burning also contribute significantly to the local air quality, human health, and climate change, which is much prominent in some regions and can affect the whole earth system through long-range transboundary pollution (Gustafsson et al. 2009; Xu et al. 2018; Yang et al. 2021). Therefore, air pollutant emissions and associated impacts on humans and the ecosystems from biomass burning in Asia are significant and should be quantified not only at the regional scale but from a global perspective (Streets et al. 2003; Badarinath et al. 2007, 2009, 2011; Hayasaka et al. 2014; Kant et al. 2000; Kharol et al. 2012; Vadrevu et al. 2017, 2018, 2021a, b). In this study, we discussed emissions of important air pollutants including $PM_{2.5}$, OC, BC, CO, and SO_2, from incomplete burning of biomass from natural fires, residential stoves, open burning of agricultural wastes, power plants, and industry that use biomass fuels as energy sources in Asia, based primarily on the global PKU inventory which has been adopted in several regional/global air quality simulation studies and its impact assessments (Huang et al. 2014; Xu et al. 2021; Shrivastava et al. 2017; Ren et al. 2021). In addition, temporal and spatial distributions of targeted air pollutants from biomass burning in Asia are discussed. Results are expected to contribute a better understanding of biomass burning on regional air quality and climate change in Asia.

2 Biomass Consumption in Asia

Biomass fuels are burned as energy sources in sectors like residential, industrial, and power plants or unintentionally as forest fires. It was estimated that the total biomass consumption in Asia was ~ 2470 (75% CI: 2270–2680) Tg in 2014, accounting for 32% of the global total biomass consumption. South and Southeast Asia had high consumption and were responsible for 76% of the total Asian biomass consumption (Fig. 1). From 1960 to 2014, total biomass consumption in Asia increased substantially until the early 1990s and became stable. The early increase was driven by the increased anthropogenic biomass consumption in residential, power plants, and industry sectors, along with population growth and socioeconomic development. As readily available and affordable fuels, biomass fuels have been widely used as energy sources, especially in the residential sector in rural areas. From 1960 to 1990, anthropogenic biomass consumption in South and Southeast Asia was higher (~ 10%) than that in East Asia, accounting for 50%, and there was much less consumption in Western Asia and Central Asia (< 4%). Since the 1990s, anthropogenic biomass consumption in Asia has been relatively stable at 1750 Tg. While the consumption in East Asia had declined, this was offset by the increased consumption in South and Southeast Asia. Natural biomass burning was around 500 $\pm$ 200 Tg since the 1980s, except for some peaks associated with unexpected large fires like in 1997. Natural biomass burning in the South and Southeast Asian area was much more apparent than in other regions (Vadrevu et al. 2021a, b).

In most developing Asian countries, biomass consumption was relatively high in sectors like natural fires (Prasad et al. 2001, 2002a, b, 2003; Albar et al. 2018; Hayasaka et al. 2014), residential consumption, and the open burning of agricultural wastes (Lasko and Vadrevu 2018; Lasko et al. 2021), while in the developed countries, biomass fuels were mainly consumed in power plants and industry. Anthropogenic biomass consumption in 2014 was 1800 Tg, accounting for ~ 73% of the total biomass consumption in Asia, as seen in Fig. 1. Among different anthropogenic sources,

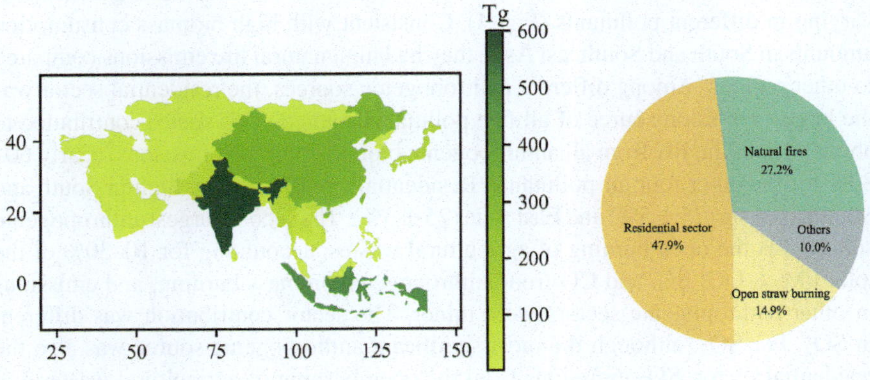

Fig. 1 Biomass consumption of different sectors in Asia in 2014

residential consumption was the largest on the regional average, although its relative contribution to the total anthropogenic biomass consumption decreased from 65% in 1960 to 48% in 2014. The second largest sector was the open burning of agricultural wastes in the field, with an increasing trend, accounting for 8% in 1960 to 15% in 2014. Biomass consumption in other sectors, including power plants, industry, and transportation, also increased, being responsible for 3% in 1960 to 10% in 2014.

3 Emissions of Typical Air Pollutants from Biomass Burning

This section discusses the emission amounts of typical air pollutants, including $PM_{2.5}$ (particles with aerodynamic diameters $\leq 2.5\ \mu m$), organic carbon (OC), black carbon (BC), CO, and SO_2 in Asia, and the spatiotemporal distributions of these emissions during 1960–2014.

3.1 *Biomass Burning Emissions in 2014*

In 2014, emissions of $PM_{2.5}$, OC, BC, CO, and SO_2 in Asia from biomass burning, including both natural fires and anthropogenic burning, were 14.6 (11.1–18.7) Tg, 7.4 (5.8–11.2) Tg, 1.9 (1.4–2.6) Tg, 177.8 (146.9–215.0) Tg, and 1.4 (1.3–1.5) Tg, respectively. These accounted for about 20%, 26%, 20%, 19%, and 1% of the total global emissions of $PM_{2.5}$, OC, BC, CO, and SO_2, respectively. Nearly 70% of the biomass burning emissions were from South and Southeast Asia, followed by East Asia (~ 25%) (Figs. 2 and 3).

As mentioned above, the biomass burning here included open fires and those used as energy sources in sectors like residential, industrial, and power plants. Generally, in Asia, natural fires contributed to 18–40% of the total biomass burning emissions, varying in different pollutants (Fig. 4). Consistent with high biomass consumption amounts in South and Southeast Asia, they had high natural fire emissions compared to other regions. Among different anthropogenic sources, the residential sector was the largest emission source of all five pollutants studied. This sector contributed to about 90% of the BC from all anthropogenic biomass burning emissions, nearly 60–80% for the other four air pollutants. Residential emissions were high in South and Southeast Asia ($75 \pm 5\%$) and East Asia ($25 \pm 3\%$). The second largest anthropogenic source was the open burning of agricultural wastes, accounting for 10–30% of the total $PM_{2.5}$, OC, BC, and CO from anthropogenic biomass burning, and emissions in other anthropogenic sectors were minor. The sector contribution was different in SO_2. For SO_2, although the most significant anthropogenic source was also the residential sector, SO_2 emissions from the open burning of agricultural wastes and other sectors, including industry and power plants, were comparable.

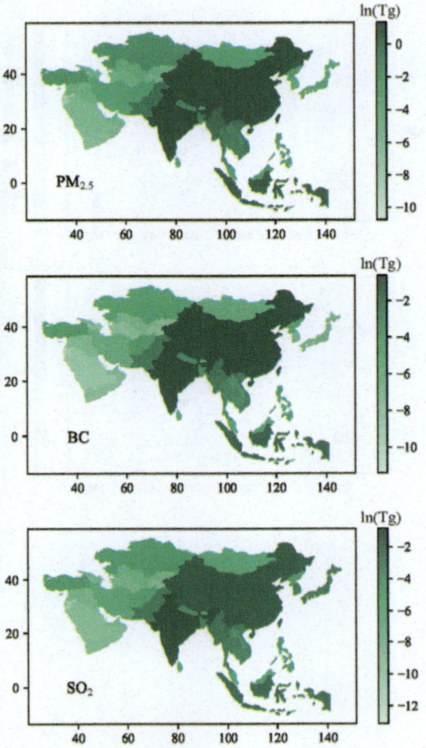

Fig. 2 Total emission amounts and emission densities of air pollutants from both natural and anthropogenic biomass burning in 2014

Sector-wise contribution variations across different countries in Asia are shown in Fig. 5. In most developing countries, residential biomass burning emissions were higher than other anthropogenic ones due to this sector's large consumption and relatively high emission factors. However, in most developed countries like Japan, the anthropogenic $PM_{2.5}$, OC, BC, and CO emissions from biomass burning were mainly from the small-scale burning of agricultural wastes.

3.2 Historical Changes in Biomass Emissions from 1960 to 2014

Though the absolute emission amounts varied, the temporal trends in $PM_{2.5}$, OC, BC, and CO were similar except in SO_2. For the former four, the Asian biomass burning emissions increased gradually till the early 1990s and then decreased slightly. In the early increasing stage, these pollutant emissions increased more than two times. Meanwhile, the relative contributions of Asian biomass emissions to the total global

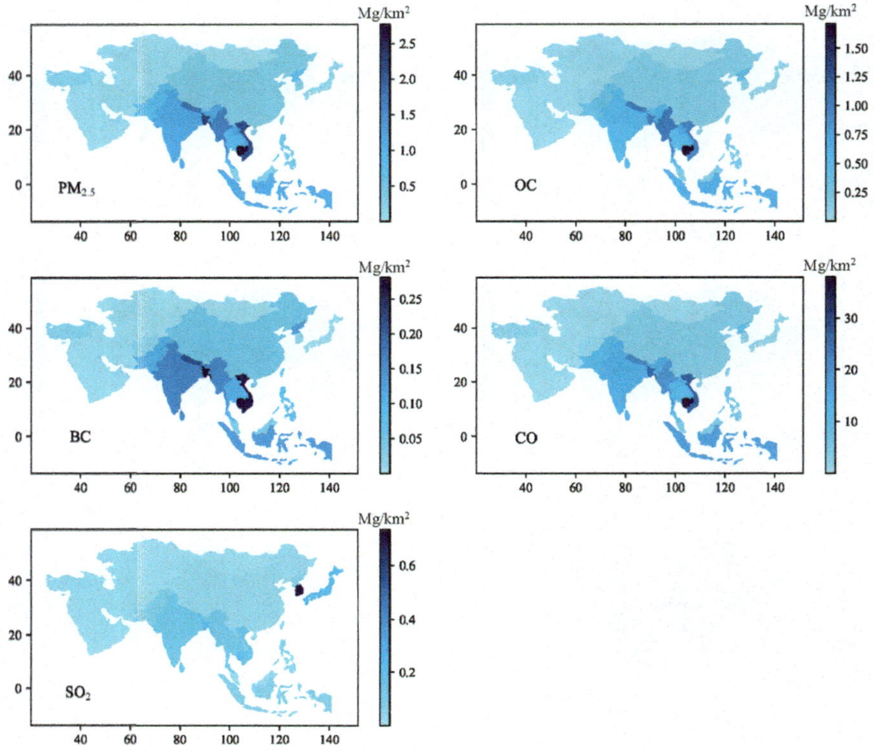

Fig. 3 Emission densities of air pollutants from both natural and anthropogenic biomass burning in 2014

emissions increased slightly. For example, $PM_{2.5}$ increased from 12 to 20%. For SO_2, Asian biomass emission contribution is gradually increasing, from 0.6 (0.5–0.7) Tg in 1960 to 1.4 (1.3–1.5) Tg in 2014, which was different in trends for the other four pollutants. However, it is important to note that Asian biomass burning SO_2 only accounted for < 1% of the total global emissions from biomass and non-biomass sources (Fig. 6).

Spatially, the biomass burning emissions of these five air pollutants in South and Southeast Asia continuously increased from 1960 to 2014. However, in East Asia, there was an upward trend before the 1990s and then started to decline. On the other hand, the emissions in Western and Central Asia did not show significant temporal trends. $PM_{2.5}$, OC, BC, and CO emissions from biomass burning in most developed countries, such as Japan, showed a slightly decreasing trend, but the SO_2 emissions kept rising. While for most developing countries except China, the emissions of five air pollutants had an upward trend. For the emissions in China, there was also an increasing trend before the early 1990s but then started to decline, mainly due to a reduction in the residential sector and lower emissions from the open burning of agricultural wastes.

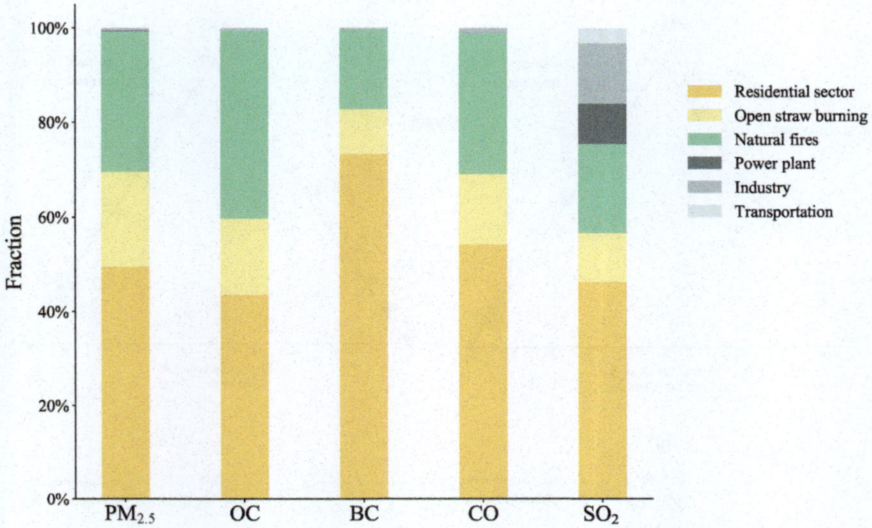

Fig. 4 Shares of different sectors to the total biomass burning emissions in Asia in 2014

Different temporal trends across different regions changed the spatial distribution of air pollutants over time. For example, before the early 1990s, the biomass burning emissions of $PM_{2.5}$, OC, BC, and CO in East Asia and South and Southeast Asia were almost equal except few years with big wildfires, accounting for 40–50% of the total Asian biomass burning emissions. However, emissions from South and Southeast Asia gradually became the high-emission region, contributing to about 70–80% in 2014.

3.3 Temporal Changes in Sector Contributions

The emissions of $PM_{2.5}$, OC, BC, CO, and SO2 from open natural forest fires in Asia fluctuated between 1.0–5.0 Tg, 1.0–4.0 Tg, 0.1–0.4 Tg, 15.0–60.0 Tg, and 0.1–0.4 Tg with few peak emissions (like in 1997), respectively. They accounted for 10–30% of Asia's total biomass burning emissions. There were no significant temporal trends in the natural fire emission contributions. However, the relative contribution of natural fire emissions was high in the South and Southeast Asian areas.

For different anthropogenic sectors, the residential emissions unsurprisingly contributed the most and dominated the changing trend of the total biomass burning emissions in Asia, i.e., the residential biomass burning emissions increased first until the early 1990s and then started to decline. In the increasing period, residential emissions in almost all Asian countries increased. These emissions contributed to not only ambient air pollution (Wolf et al. 2021; Rooney et al. 2019; Shen et al. 2021) but largely to indoor air pollution (Yun et al. 2020; Shen et al. 2020; Luo et al. 2021)

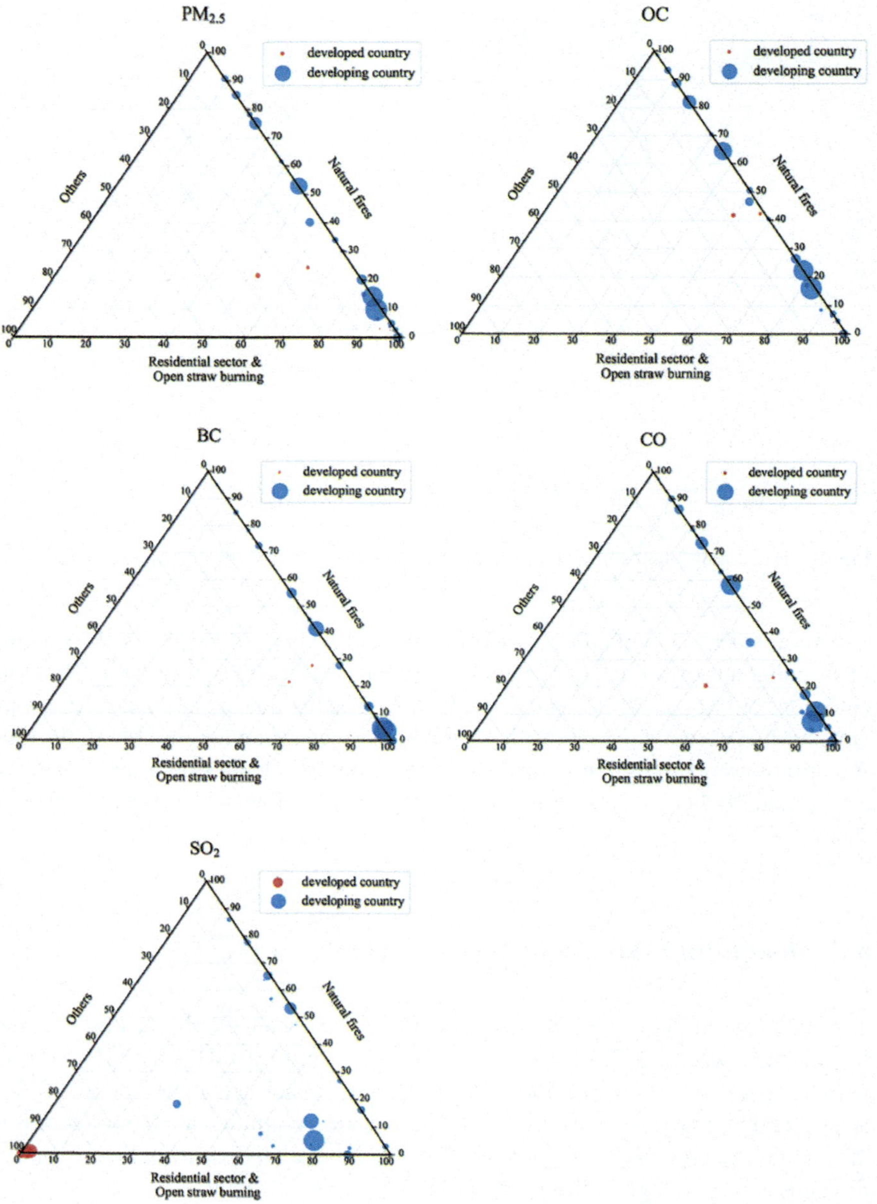

Fig. 5 Relative contribution of sectoral emissions in developed and developing countries in Asia

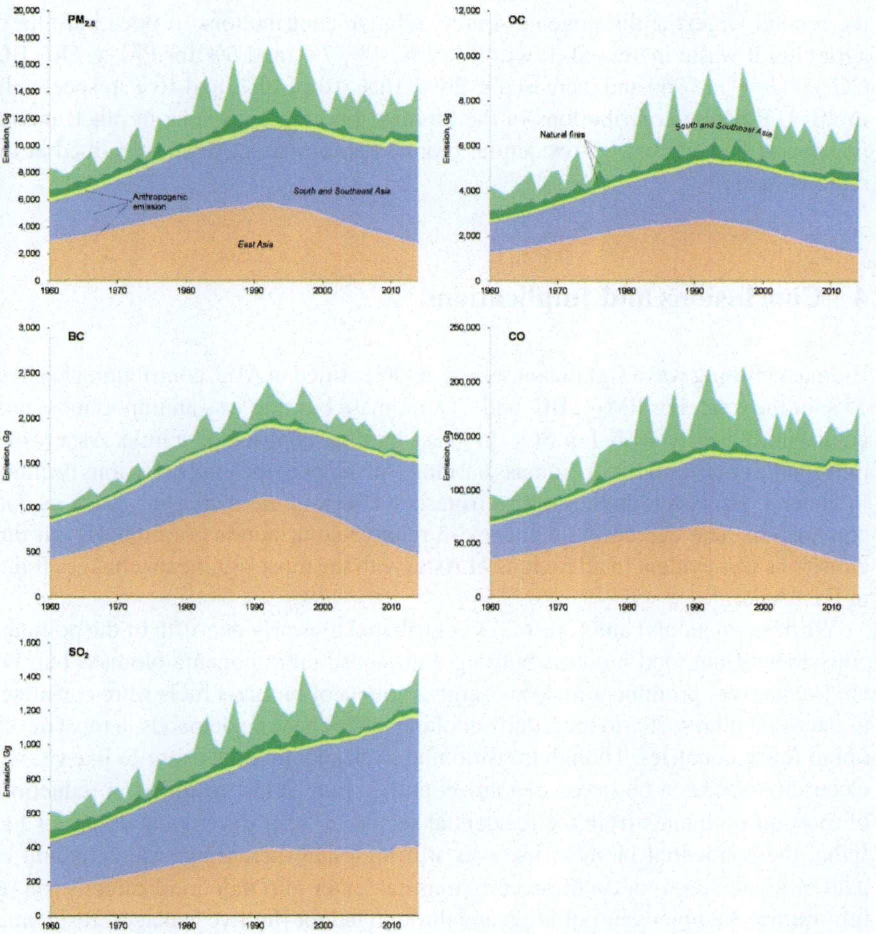

Fig. 6 Temporal changing trends of typical air pollutants emissions in Asia from 1960 to 2014

resulting in high exposure; thus, emissions from this sector are of serious concern and effectively controlled. In the past several decades, along with the promotion of governmental policies and public awareness of the importance of indoor air pollution, several countermeasures were taken in some East Asian countries that led to a decline in emissions from this sector. For instance, in China, there was an effective replacement of clean energy, extensive stove upgrading, and reprocessing of biomass fuel all around the country, especially in the rural regions where most of the biomass is consumed as fuel (Shen et al. 2022b; Ellison et al. 2020). However, it is necessary to note that residential emissions in South and Southeast Asia are still increasing. The increase was largely due to the fast growth of the population and the relatively backward economy. By 2014, about 70–80% of the residential emissions were in South and Southeast Asia, with only 20–30% in East Asia. Meanwhile, for the second largest anthropogenic source, relative contributions of open burning of agricultural waste increased. It was 9%, 7%, 4%, 7%, and 6% for $PM_{2.5}$, OC, BC, CO, and SO_2 in 1960 and increased to 20%, 16%, 10%, 15%, and 10%, respectively, in 2014. Relative contributions in the biomass burning emissions in other anthropogenic also increased, as residential biomass emissions generally declined at the regional scale.

4 Conclusions and Implications

Biomass burning was a significant source of OC emitted in Asia, contributing to about 75% of the total. For $PM_{2.5}$, BC, and CO, biomass burning was an important source contributing nearly 40%. For SO_2, biomass burning contributed a little. As a result, the relative contributions of biomass burning emissions to the total emissions declined in general. However, the emissions from other sectors are increasing, such as coal combustion. The decreasing tendency in relative contributions of biomass burning emissions was evident in all regions of Asia, with the most significant changes found in East Asia.

While open natural and forest fires contributed to nearly one-fifth of the pollutant emissions of the total biomass burning emissions, anthropogenic biomass burning emissions were prominent in Asia. Large amounts of biomass fuels were consumed in the residential sector to meet daily cooking and/or heating demands in most developing Asian countries. Though transitioning to cleaner modern energies like gas and electricity had been observed in some countries like China, resulting in reductions of most air pollutants from the residential sector, in other developing countries like India, the residential biomass use was still high and increasing, which should be paid more attention to. As these emissions can enter into indoor air directly via re-infiltration of ambient air pollution and through indoor fugitive leakages, residential biomass burning emission is one top environmental risk factor that can cause adverse human health effects. Thus, taking effective measures and countermeasures on these emissions is necessary. In addition, the open burning of agricultural wastes in many Asian countries is not effectively controlled yet, although there was a ban on policies in some regions. Therefore, examining and assessing these policies, including their implementation effectiveness, is necessary.

Emission estimates are often associated with uncertainties due to a lack of reliable localized emission factors, technology information, and energy consumption activity data. This also applies to our present estimate here. More future studies are needed to address these issues. For example, due to the stacked energy use (Zhu et al. 2019; Ruiz-Mercado et al. 2015; Shen et al. 2022c), statistics on residential energy consumption in most developing countries are often associated with high uncertainties and biases. The population of Asia is ~ 4.54 billion, accounting for 60% of the global total, but the region is also one severely polluted area on Earth. Therefore, it is imperative to improve Asian air quality by cutting down emissions from different sectors, including biomass burning. This includes control strategies and intervention actions at various levels and is not limited to residential energy transitions, stove upgrading, and open burning of agricultural wastes. In addition, there would be obvious health and climate co-benefits if transitioned to cleaner technologies.

Acknowledgements We are grateful to the scientists who developed the PKU fuel and inventory database, which is the basis of the analysis in this study. The inventory research work is partly funded by the National Natural Science Foundation of China (42077328 and 41922057).

References

Adam, M.G., P.T.M. Tran, and N. Bolan. 2021. Biomass burning-derived airborne particulate matter in Southeast Asia: A critical review. *Journal of Hazardous Materials* 407: 124760.

Albar, I., I. Jaya, B.H. Saharjo, B. Kuncahyo, and K.P. Vadrevu. 2018. Spatio-temporal analysis of land and forest fires in Indonesia using MODIS active fire dataset. In *Land-atmospheric research applications in South and Southeast Asia*, 105–127. Cham: Springer.

Badarinath, K.V.S., S.K. Kharol, K.M. Latha, T.K. Chand, V.K. Prasad, A.N. Jyothsna, and K. Samatha. 2007. Multiyear ground-based and satellite observations of aerosol properties over a tropical urban area in India. *Atmospheric Science Letters* 8 (1): 7–13.

Badarinath, K.V.S., S.K. Kharol, A.R. Sharma, and V.K. Prasad. 2009. Analysis of aerosol and carbon monoxide characteristics over Arabian Sea during crop residue burning period in the Indo-Gangetic plains using multi-satellite remote sensing datasets. *Journal of Atmospheric and Solar-Terrestrial Physics* 71 (12): 1267–1276.

Badarinath, K.V.S., and K.V. Prasad. 2011. Carbon dioxide emissions from forest biomass burning in India. *Global Environmental Research* 15: 45–52.

Ellison, C., Y. Li, and F. Yu. 2020. Household transitions to clean energy in a multiprovincial cohort study in China. *Nature Sustainability* 3: 42–50.

Gustafsson, O., M. Krusa, and Z. Zencak. 2009. Brown clouds over South Asia: Biomass or fossil fuel combustion. *Science* 323: 495–498.

Hayasaka, H., I. Noguchi, E.I. Putra, N. Yulianti, and K. Vadrevu. 2014. Peat-fire-related air pollution in Central Kalimantan, Indonesia. *Environmental Pollution* 195: 257–266.

Huang, Y., H.Z. Shen, and H. Chen. 2014. Quantification of global primary emissions of $PM_{2.5}$, PM_{10}, and TSP from combustion and industrial process sources. *Environmental Science & Technology* 48: 13834–13843.

International Energy Agency (IEA). 2020. IEA world energy statistics and balances. http://www.oecd-ilibrary.org/statistics. Accessed April 12, 2020.

Kant, Y., A.B. Ghosh, M.C. Sharma, P.K. Gupta, V.K. Prasad, K.V.S. Badarinath, and A.P. Mitra. 2000. Studies on aerosol optical depth in biomass burning areas using satellite and ground-based observations. *Infrared Physics & Technology* 41 (1): 21–28.

Kharol, S.K., K.V.S. Badarinath, A.R. Sharma, D.V. Mahalakshmi, D. Singh, and V.K. Prasad. 2012. Black carbon aerosol variations over Patiala city, Punjab, India—A study during agriculture crop residue burning period using ground measurements and satellite data. *Journal of Atmospheric and Solar-Terrestrial Physics* 84: 45–51.

Lasko, K., and K. Vadrevu. 2018. Improved rice residue burning emissions estimates: Accounting for practice-specific emission factors in air pollution assessments of Vietnam. *Environmental Pollution* 236: 795–806.

Lasko, K., K.P. Vadrevu, V. Bandaru, T.N.T. Nguyen, and H.Q. Bui. 2021. $PM_{2.5}$ emissions from biomass burning in South/Southeast Asia–Uncertainties and trade-offs. In *Biomass burning in South and Southeast Asia*, 149–169. CRC Press.

Luo, Z., L. Zhang, and G. Li. 2021. Evaluating co-emissions into indoor and outdoor air of EC, OC, and BC from in-home biomass burning. *Atmospheric Research* 248: 105247.

Prasad, V.K., Y. Kant, P.K. Gupta, C. Sharma, A.A. Mitra, and K.V.S. Badarinath. 2001. Biomass and combustion characteristics of secondary mixed deciduous forests in Eastern Ghats of India. *Atmospheric Environment* 35 (18): 3085–3095.

Prasad, V.K., Y. Kant, and K.V.S. Badarinath. 2002a. Land use changes and modeling carbon fluxes from satellite data. *Advances in Space Research* 30 (11): 2511–2516.

Prasad, V.K., Y. Kant, P.K. Gupta, C. Elvidge, and K.V.S. Badarinath. 2002b. Biomass burning and related trace gas emissions from tropical dry deciduous forests of India: A study using DMSP-OLS data and ground-based measurements. *International Journal of Remote Sensing* 23 (14): 2837–2851.

Prasad, V.K., M. Lata, and K.V.S. Badarinath. 2003. Trace gas emissions from biomass burning from northeast region in India—Estimates from satellite remote sensing data and GIS. *The Environmentalist* 23 (3): 229–236.

Ren, Y.A., G.F. Shen, and H.Z. Shen. 2021. Contribution of biomass burning to global and regional SO_2 emissions. *Atmospheric Research* 206: 105709.

Rooney, B., R. Zhao, and Y. Wang. 2019. Impacts of household sources on air pollution at village and regional scales in India. *Atmospheric Chemistry and Physics* 19: 7719–7742.

Ruiz-Mercado, I., and O. Masera. 2015. Patterns of stove use in the context of fuel-device stacking: Rationale and implications. *EcoHealth* 12: 42–56.

Streets, D.G., T.C. Bond, G.R. Carmichael, S.D. Fernandes, Q. Fu, D. He, Z. Klimont, S.M. Nelson, N.Y. Tsai, M.Q. Wang, and J.H. Woo. 2003. An inventory of gaseous and primary aerosol emissions in Asia in the year 2000. *Journal of Geophysical Research: Atmospheres, 108*(D21).

Shen, G., W. Du, and Z. Luo. 2020. Fugitive emissions of CO and $PM_{2.5}$ from indoor biomass burning in chimney stoves based on a newly developed carbon balance approach. *Environmental Science & Technology Letters* 7: 128–134.

Shen, H., Z.H. Luo, and R. Xiong. 2021. A critical review of pollutant emission factors from fuel combustion in home stoves. *Environment International* 157: 106841.

Shen, G., R. Xing, and Y. Zhou. 2022a. Revisiting the proportion of clean household energy users in rural China by accounting for energy stacking. *Sustainable Horizons* 1: 100010.

Shen, G.F., R. Xiong, and Y.L. Tian. 2022b. Substantial transition to clean household energy mix in rural households in China. *National Science Review*: nwac050.

Shen, G., R. Xiong, Y. Tian, Z. Luo, B. Jiangtulu, W. Meng, W. Du, J. Meng, Y. Chen, B. Xue, and B. Wang. 2022c. Substantial transition to clean household energy mix in rural China. *National Science Review* 9 (7): nwac050.

Shrivastava, M., S. Lou, and A. Zelenyuk. 2017. Global long-range transport and lung cancer risk from polycyclic aromatic hydrocarbons shielded by coating of organic aerosol. *PNAS* 114: 1246–1251.

Tao, S., M.Y. Ru, and W. Du. 2018. Quantifying the rural residential energy transition in China from 1992 to 2012 through a representative national survey. *Nature Energy* 3: 567–573.

Vadrevu, K., T. Ohara, and C. Justice. 2017. Land cover, land use changes and air pollution in Asia: A synthesis. *Environmental Research Letters* 12 (12): 120201.

Vadrevu, K.P., T. Ohara, and C. Justice, eds. 2018. *Land-atmospheric research applications in South and Southeast Asia.*

Vadrevu, K.P., T. Ohara, and C. Justice, eds. 2021a. *Biomass burning in South and Southeast Asia: Impacts on the biosphere*, vol. 2. CRC Press.

Vadrevu, K.P., T. Ohara, and C. Justice, eds. 2021b. *Biomass burning in South and Southeast Asia: Mapping and monitoring*, vol. 1. CRC Press.

Wolf, T., L.H. Pettersson, and I. Esau. 2021. Dispersion of particulate matter ($PM_{2.5}$) from wood combustion for residential heating: Optimization of mitigation actions based on large-eddy simulations. *Atmospheric Chemistry and Physics* 21: 12463–12477.

Xu, R.G., X.X. Tie, and G.H. Li. 2018. Effect of biomass burning on black carbon (BC) in South Asia and Tibetan plateau: The analysis of WRF-Chem modeling. *Science of the Total Environment* 645: 901–912.

Xu, H.R., Y.A. Ren, and W.X. Zhang. 2021. Updated global black carbon emissions from 1960 to 2017: Improvements, trends, and drivers. *Environmental Science & Technology* 55: 7869–7879.

Yadav, I.C., N.L. Devi, and J. Li. 2017. Biomass burning in Indo-China peninsula and its impacts on regional air quality and global climate change—a review. *Environmental Pollution* 227: 414–427.

Yang, J.H., Z.M. Ji, and S.C. Kang. 2021. Contribution of South Asian biomass burning to black carbon over the Tibetan plateau and its climatic impact. *Environmental Pollution* 207: 116195.

Yun, X. G.F. Shen, and H.Z. Shen. 2020. Residential solid fuel emissions contribute significantly to air pollution and associated health impacts in China. *Science Advances* 6: eaba7621.

Zhu, X., X. Yun, and W.J. Meng. 2019. Stacked use and transition trends of rural household energy in mainland China. *Environmental Science & Technology* 53: 521–529.

Twenty-Year (2000–2019) Variations of Aerosol Optical Depth Over Asia in Relation to Anthropogenic and Biomass Burning Emissions

Syuichi Itahashi, Junichi Kurokawa, and Toshimasa Ohara

Abstract Space-borne aerosol optical depth (AOD) was analyzed over the 13 countries that participated in the Acid Deposition Monitoring Network in East Asia (EANET) in the 20 years from 2000 to 2019. Long-term trends in AOD and available surface PM_{10} observations from EANET were generally consistent. The country-averaged annual mean AOD showed large year-to-year variation, and the anthropogenic and biomass burning emissions were analyzed with the aim of explaining the variation. Over East Asia (China, the Republic of Korea, and Japan), long-term AOD trends showed a gradual decline from 2015 to 2019, which, based on previous studies, was caused mainly by the reduction of anthropogenic emissions, especially over China. Over North and Southeast Asia, biomass burning emissions were crucial in AOD variation. Emissions from biomass burning sources showed large interannual variation in intensity and spatial expansion; accordingly, a large year-to-year variation of AOD was observed over these countries. The analyses of the difference (defined as the difference from the 20-year average value) in AOD and the difference in biomass burning emissions showed a moderate correlation (around 0.55) over Vietnam, Cambodia, Lao PDR, and Thailand, a high correlation (0.67) over the Russian Far East, and a very high correlation (0.97) over Indonesia. To decrease the AOD values and hence aerosol pollution, biomass burning emissions must be managed appropriately in North and Southeast Asian countries. In addition, because of the decline in anthropogenic emissions, the importance of biomass burning emissions is likely to emerge in the near future over East Asia. Thus, Asian biomass burning emissions should be monitored closely to understand aerosol pollution.

S. Itahashi (✉)
Sustainable System Research Laboratory (SSRL), Central Research Institute of Electric Power Industry (CRIEPI), Abiko, Chiba, Japan
e-mail: isyuichi@criepi.denken.or.jp

J. Kurokawa
Asia Center for Air Pollution Research (ACAP), Niigata, Niigata, Japan

T. Ohara
Center for Environmental Science in Saitama (CESS), Kazo, Saitama, Japan

K. P. Vadrevu et al. (eds.), *Vegetation Fires and Pollution in Asia*,
https://doi.org/10.1007/978-3-031-29916-2_16

Keywords Aerosol optical depth · Biomass burning emissions · Asia · Acid Deposition Monitoring Network in East Asia (EANET)

1 Introduction

With the recent acceleration of increases in anthropogenic emissions, Asia has experienced the most severe aerosol pollution globally (e.g., Chen et al. 2019; Itahashi et al. 2018a, b, 2020). Because exposure to air pollution is closely related to morbidity and mortality (Cohen et al. 2017), the status of aerosol pollution must be understood in the short and long term. Ground-based measurements are essential for capturing surface aerosol pollution; however, their spatial coverage is limited, especially in rural and remote areas. Most of the available ground-based measurements are over densely populated urban areas. Another recent approach is the use of space-based measurements. For example, NO_2 column density has been measured by satellites in order to capture surface NO_x emissions (Miyazaki et al. 2017; Itahashi et al. 2019). For aerosols, the space-borne aerosol optical depth (AOD), which represents the attenuation of sunlight by aerosols and is measured as the aerosol column concentration, is used to capture aerosol pollution (Kaufman et al. 2002). Combined with numerical modeling simulation, surface aerosol pollution can be derived from AOD (van Donkelaar et al. 2010, 2016). Although the instrument on board polar-orbit satellites has limited temporal coverage, this kind of space-borne measurement has excellent potential for capturing the status of aerosol pollution (Kant et al. 2000; Kharol et al. 2012).

Previous studies have reported dramatic changes in aerosol pollution status in Asia. For example, China has been the focus of much research because of its intense anthropogenic emissions (Janssens-Maenhout et al. 2015; Kurokawa and Ohara 2020); however, emissions in China are changing. Anthropogenic SO_2 emissions, which cause aerosol sulfate pollution, began to decline around 2005 (Kurokawa and Ohara 2020). Aerosol pollution over China has been decreasing since emission regulations were tightened (Ma et al. 2019; Zhang et al. 2019) and decreased over the downwind region of China, i.e., the East China Sea to the Republic of Korea and Japan (Itahashi et al. 2012, 2021; Uno et al. 2020). Over North Asia, a huge amount of episodical biomass burning emissions impacts air quality over the source areas and a much broader area (Jeong et al. 2008; Ikeda et al. 2015; Yasunari et al. 2018). Previous studies have highlighted the critical role of biomass burning in Southeast Asia (Badarinath et al. 2007, 2008, 2009; Kharol et al. 2012; Vadrevu et al. 2011, 2015, 2018, 2021a, b; Wooster et al. 2021), and intensive monitoring projects such as 7-SEAS (Tsay et al. 2013) have been conducted. Each study has revealed an important aspect of aerosol pollution in Asia; however, there needs to be a greater understanding of the long-term trends in aerosol pollution in Asia (Vadrevu et al. 2014). Because space-based AOD measurements have been accumulated over the last 20 years, it is essential to integrate this information about aerosol pollution with the latest inventories of anthropogenic and biomass burning emissions. In the present study, we clarify the 20-year variation of AOD over Asia, thereby contributing to understanding long-term aerosol pollution over Asia.

2 Dataset

To obtain the longest trends in AOD over Asia, AOD measured by the Moderate Resolution Imaging Spectroradiometer (MODIS) on board the Terra satellite was analyzed. The MODIS product in the latest Collection 6.1 (Levy et al. 2013) has been available from February 24, 2000, to the present. Level 3 of the MOD08_3D dataset gridded into a $1.0° \times 1.0°$ grid was used, and the product of AOD with the Dark Target/Dark Blue algorithms (parameter name of "AOD_550_Dark_Target_Deep_Blue_Combined_Mean") was analyzed (NASA 2021). The algorithm for determining aerosol characteristics at 550 nm has been validated with co-located surface observations from the aerosol robotic network's direct sun/sky radiometers. The expected errors are $\pm$ (0.05 + 15%) over land and + (0.04 $\pm$ 10%) to − (0.02 + 10%) over ocean (Levy et al. 2013). Using the daily MODIS dataset, annual means were calculated. Although part of the year 2000 was missing from the available data, the annual mean AOD over Asia from 2000 to 2019 was analyzed.

To explain the AOD variations, anthropogenic and biomass burning emissions were analyzed. Emissions were taken from the latest version of the Regional Emission inventory in Asia (REAS) version 3.2 for anthropogenic sources (Kurokawa and Ohara 2020) and Global Fire Emission Database (GFED) version 4 for biomass burning sources (van der Werf et al. 2017). REAS version 3.2 provides anthropogenic emissions over Asian countries during 1950–2015 with monthly variation with a $0.25° \times 0.25°$ grid resolution. Because version 3.2 did not include Russia as an emission estimation target, anthropogenic emissions over Russia were obtained from version 2.1 during 2000–2008 (Kurokawa et al. 2013). GFED provides global biomass burning emissions from 1997 to 2016 with a monthly variation on a $0.25° \times 0.25°$ grid resolution. The beta version has been updated to the latest available date; therefore, the years from 2000 to 2019 were analyzed. For REAS and GFED, annual emissions were calculated and analyzed from 2000 to 2015 and 2019, respectively. The emissions analyzed were total particulate matter (TPM), which includes carbonaceous aerosols of black and organic carbon and other particulate matter components. Although ambient aerosol also consists of secondary aerosol produced in the atmosphere (e.g., SO_2 oxidized to sulfate), TPM emissions are used as a proxy for primary emissions to consider AOD variation.

The regions for which AOD and emissions were analyzed included 13 countries in the Acid Deposition Monitoring Network in East Asia (EANET) (Fig. 1). The Russian Far East from 90° E and to 60° N covers four EANET sites (three sites around Lake Baikal and one site in Primorskaya on the eastern edge of Russia near the border with North Korea). In addition, there were 2, 3, 11, and 2 EANET sites

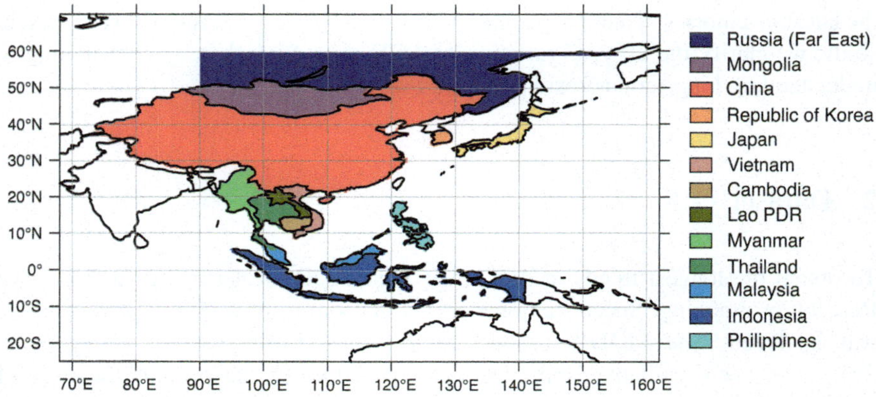

Fig. 1 Map of the 13 EANET countries

in China, the Republic of Korea, Japan, and Thailand, respectively, where surface PM_{10} was observed over ten years. These surface PM_{10} observations were used to analyze the relationship between AOD and surface aerosol pollution.

3 Results and Discussion

Spatial distributions of annual mean AOD over Asia from 2000 to 2019 are shown for each year in Fig. 2. AOD had constant high values greater than 0.6 (yellow to red shading, Fig. 2) over Eastern China and Northern India. The values were moderately high, around 0.5 (green shading, Fig. 2) over the Taklamakan Desert, which is the primary source of Asian dust, but the values showed year-to-year variation. There were moderately high values over a wide area of the Taklamakan Desert (green to yellow shading, Fig. 2) in 2001–2003, 2006, 2007, and 2010–2013, whereas this was not observed in 2000, 2004, 2005, 2009, and 2017. Moderately high values around 0.5 (green shading, Fig. 2) were seen over the Russian Far East in 2003, 2006, 2011–2014, 2016, and 2019 and were related to the episodical emissions from biomass burning sources. These emissions play an essential role in both the source and the wider area (Jeong et al. 2008; Ikeda et al. 2015; Yasunari et al. 2018). The AOD values over Indonesia and Malaysia also showed significant year-to-year variations with higher values over 0.8 (red shading, Fig. 2) in 2002, 2006, 2014, 2015, and 2019. The importance of biomass burning emissions over the Indochina Peninsula has been reported (Gautam et al. 2013; Takami et al. 2020). The year-to-year spatial distribution of AOD demonstrated that its interannual variation over Southeast Asia was also large.

The long-term trends in AOD over 13 EANET countries were analyzed, and the annual mean and average over each country is shown in Fig. 3. Overall, as indicated by the spatial distribution pattern (Fig. 2), the year-to-year variations of AOD analyzed

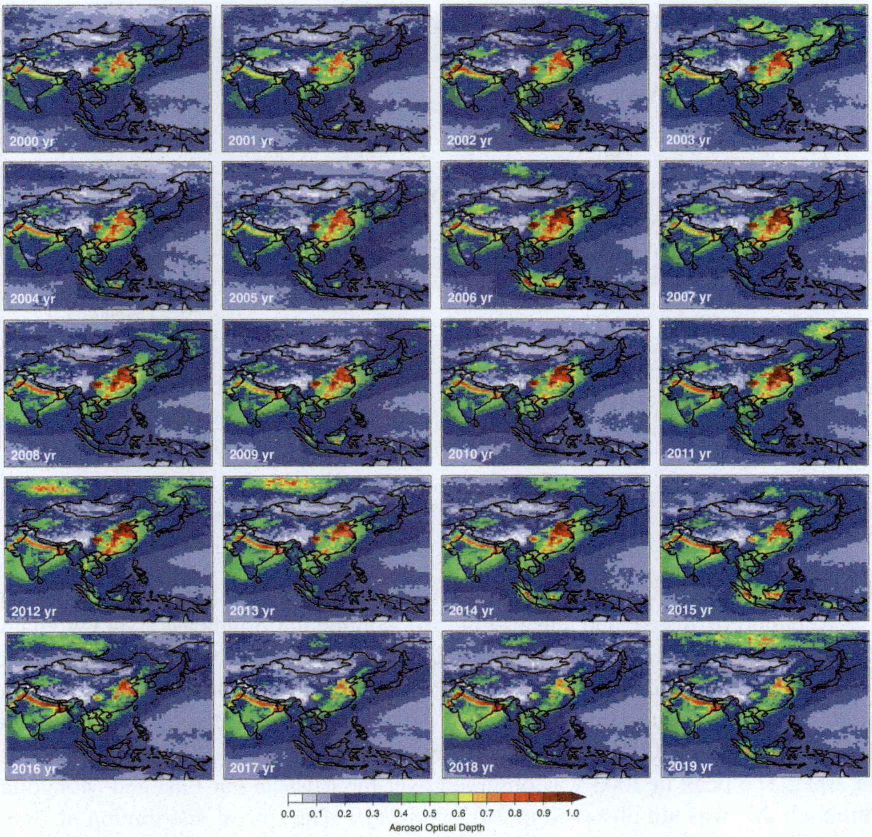

Fig. 2 Spatial distribution of annual mean AOD from 2000 to 2019

at the country level were large, except for Mongolia and the Philippines. These two countries showed almost flat trends with lower AOD values than the others. These complex trends arose because AOD contains both fine- and coarse-mode aerosols and comes from anthropogenic and natural sources. As observed in the spatial distribution (Fig. 2), the year-to-year variation of Asian dust and biomass burning events changed dynamically and influenced AOD. This variation in AOD, which is an aerosol column concentration, was compared with surface PM_{10} observations (Fig. 4). The available PM_{10} observations over 10 years were limited (gray circles on the center map, Fig. 4); however, these observations were compared with the country-averaged annual mean AOD over China, the Republic of Korea, Japan, and Thailand. The result showed a general correspondence between AOD and PM_{10} with a correlation coefficient of 0.59 for all available data. Thus, AOD is suitable as a measurement for discussing the aerosol pollution status in Asia.

To infer AOD trends in each country, the change in the annual mean AOD from its 20-year (2000–2019) average was analyzed further and separated into the following

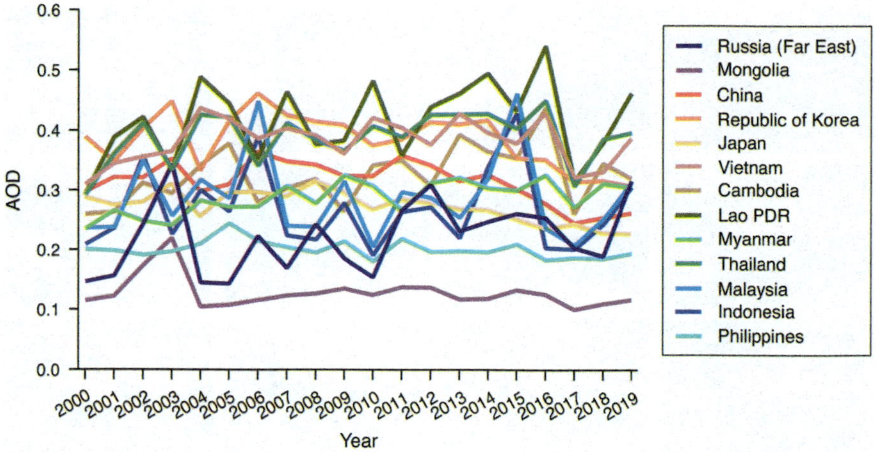

Fig. 3 Temporal variation of annual mean and country-averaged AOD from 2000 to 2019 over 13 EANET countries

four regions: North Asia (Russian Far East and Mongolia; Fig. 5a), East Asia (China, the Republic of Korea, and Japan; Fig. 5b), continental Southeast Asia (Vietnam, Cambodia, Lao PDR, Myanmar, and Thailand; Fig. 5c), and oceanic Southeast Asia (Malaysia, Indonesia, and the Philippines; Fig. 5d). Over North Asia (Fig. 5a), there was a positive difference (i.e., a high AOD compared with the 20-year average) for the Russian Far East in 2003, 2012, and 2019 with a value around + 0.1 and for Mongolia in 2003. The peak in 2003 was observed over the Russian Far East and Mongolia, although this was not observed in 2012 and 2019. The spatial distribution of AOD (Fig. 2) suggested that the location and affected areas of biomass burning in Siberia changed. In 2003, greater effects were observed from Lake Baikal eastward (Jeong et al. 2008), whereas in 2012 and 2019, the affected region was only over Eastern Russia and did not affect Mongolia. Over East Asia (Fig. 5b), the AOD values showed an inverse U-shaped curve with higher values of + 0.05 from the late 2000s to early 2010s, turning into a sharp decline after 2015, reaching around − 0.1. Owing to the reduction of anthropogenic emissions from China, a decrease in $PM_{2.5}$ concentration has been reported since 2013 (Zhang et al. 2019). This also influenced regions (Uno et al. 2020; Itahashi et al. 2021) downwind; therefore, the declining trends in AOD after 2015 were observed over all the East Asian countries (China, the Republic of Korea, and Japan). The variation over continental and oceanic Southeast Asia was generally similar (Fig. 5c, d). Over continental Southeast Asia, the highest positive difference of + 0.1 was seen in 2016 over Cambodia and Lao PDR. Over oceanic Southeast Asia, there were positive peaks in 2006 and 2015, with changes higher than + 0.1. These peaks were clear over Malaysia and Indonesia. In contrast, the variation over the Philippines was unclear, as also observed in the annual mean AOD trend in Fig. 3. In 2006 and 2015, the spatial mapping of AOD showed high values greater than 0.8 (red shading, Fig. 2) over Indonesia.

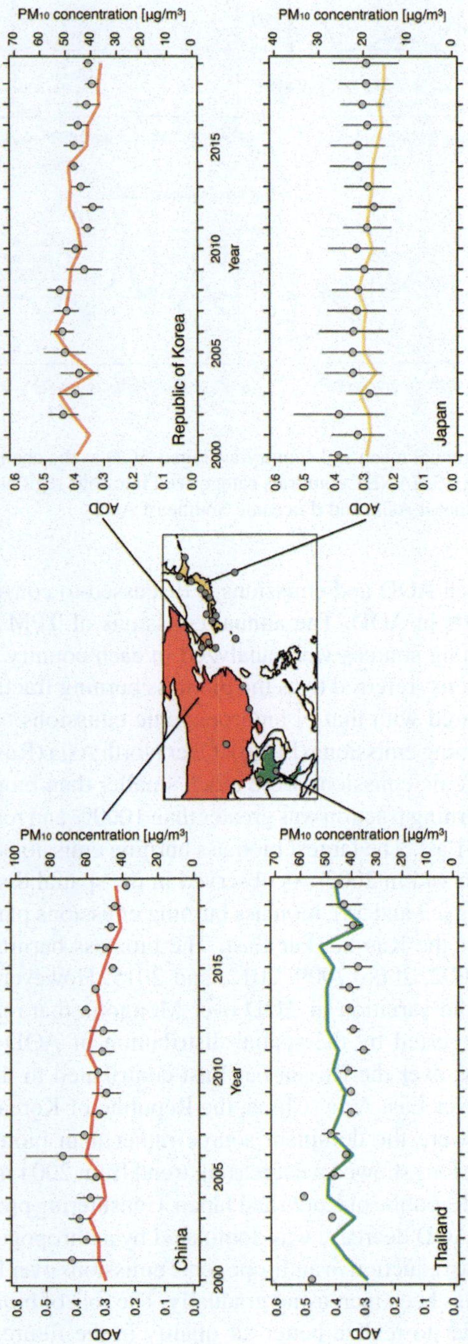

Fig. 4 Temporal variation of annual mean and country-averaged AOD (left axis) and surface PM_{10} observation (right axis) from 2000 to 2019 over China, the Republic of Korea, Japan, and Thailand

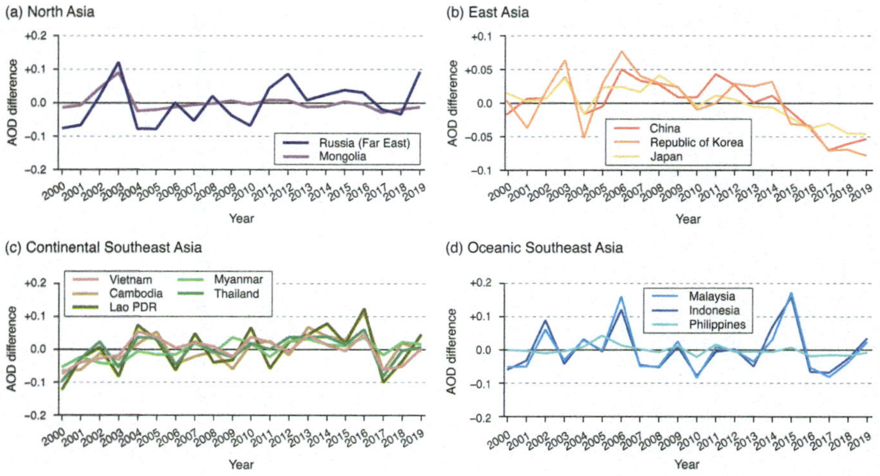

Fig. 5 Temporal variation of annual mean and country-averaged AOD as the change from the 20-year (2000–2019) average over 13 EANET countries categorized into four regions: **a** North Asia, **b** East Asia, **c** continental Southeast Asia, and **d** oceanic Southeast Asia

The relationship between AOD and emissions is discussed to consider the year-to-year variation and trends in AOD. The annual emissions of TPM from anthropogenic and biomass burning sources were analyzed in each country. The fraction of biomass burning emissions (referred to as the biomass burning fraction hereafter) was calculated and compared with that of anthropogenic emissions to identify the importance of biomass burning emissions (Fig. 6). Over North Asia (Russian Far East and Mongolia), anthropogenic emissions were much smaller than biomass burning emissions. The biomass burning fraction was greater than 1000% and reached 4000% in 2003 in the Russian Far East. The largest biomass burning emissions and biomass burning fraction were observed in 2003. As observed in the spatial distribution and trend analyses of AOD (Figs. 2 and 5a), biomass burning emissions played a critical role in the high AOD over the Russian Far East. The biomass burning fraction in Mongolia was larger in 2002, 2006–2009, 2012, and 2015. However, these peaks were not consistent with the variation in AOD over Mongolia that reached a peak in 2003 (Fig. 5a). As suggested by the spatial distribution of AOD (Fig. 2), the biomass burning emissions over the Russia Far East contributed to the high AOD over Mongolia in 2003. Over East Asia (China, the Republic of Korea, and Japan), anthropogenic emissions were the dominant source rather than biomass burning. Anthropogenic TPM emissions showed a decreasing trend from 2004 in China and a continuous decrease in the Republic of Korea and Japan. Considering previous studies over East Asia, the recent AOD decrease was dominated by anthropogenic emission sources (Fig. 5b). Due to the reduction in anthropogenic emissions over East Asia, the biomass burning fraction has been increasing gradually. The role of biomass burning must be considered in order to realize better air quality in the future. Over continental Southeast Asia (Vietnam, Cambodia, Lao PDR, Myanmar, and Thailand) and oceanic Southeast Asia (Malaysia, Indonesia, and the Philippines), biomass burning emissions are an important source of emissions (Vadrevu et al. 2021a, b). In particular, over Cambodia, Lao PDR, and Myanmar, biomass burning emissions surpassed anthropogenic emissions, and the biomass burning fraction was greater than 1000% and reached around 4000% in some cases. Over Vietnam, Thailand, and Malaysia, the amounts of anthropogenic and biomass burning emissions were comparable; thus, the biomass burning fraction was around 100%. Compared with anthropogenic emissions, the year-to-year changes in biomass burning emissions were greater, and the biomass burning fraction also showed year-to-year variation. Over Indonesia, some years showed comparable amounts of anthropogenic and biomass burning emissions, and some years showed a distinct peak in biomass burning emissions. Owing to the high biomass burning emission amounts, the biomass burning fraction reached 800%. Over the Philippines, anthropogenic emissions were greater than biomass burning emissions, as over East Asian countries, but the anthropogenic emissions remained constant due to the constant trend in AOD (Fig. 5d).

Finally, the relationship between the variation of AOD and biomass burning emissions was determined. The difference in AOD from the 20-year average value (same as the analyses in Fig. 5) and the difference in biomass burning emissions from the 20-year average value are shown as a scatter plot (Fig. 7). North and Southeast Asia were analyzed, except Mongolia and the Philippines, where AOD remained almost constant (Fig. 5). The correlation between the difference in AOD and difference in biomass burning ranged from 0.22 to 0.97. The correlation values were low (0.2–0.3; not significant according to Student's *t*-test) over Myanmar and Malaysia, whereas they were moderate (around 0.55; $p < 0.05$ or 0.01) over Vietnam, Cambodia, Lao PDR, and Thailand. The correlation value was relatively high (0.67; $p < 0.01$) over the Russian Far East, with a distinct positive–positive difference in 2003. The correlation value over Indonesia was very high (0.97; $p < 0.001$) with a positive–positive difference. Over these countries, biomass burning emissions were vital in causing AOD variations, leading to aerosol pollution.

4 Conclusion

The long-term variation in space-borne AOD was analyzed over 20 years (2000–2019). The country-averaged annual mean AOD showed large year-to-year variation, and to explain this variation, anthropogenic and biomass burning emissions were analyzed. AOD, a measurement of the aerosol column concentration, generally corresponded to the surface PM_{10} observations. Over East Asia (China, the Republic of Korea, and Japan), AOD showed a long-term gradual decline after 2015 owing to China's dramatic reduction of anthropogenic emissions. Over North Asia and Southeast Asia, biomass burning emissions were key in AOD variations; thus, AOD showed significant year-to-year variations. To improve air quality related to aerosol over North and Southeast Asia, biomass burning emissions must be managed

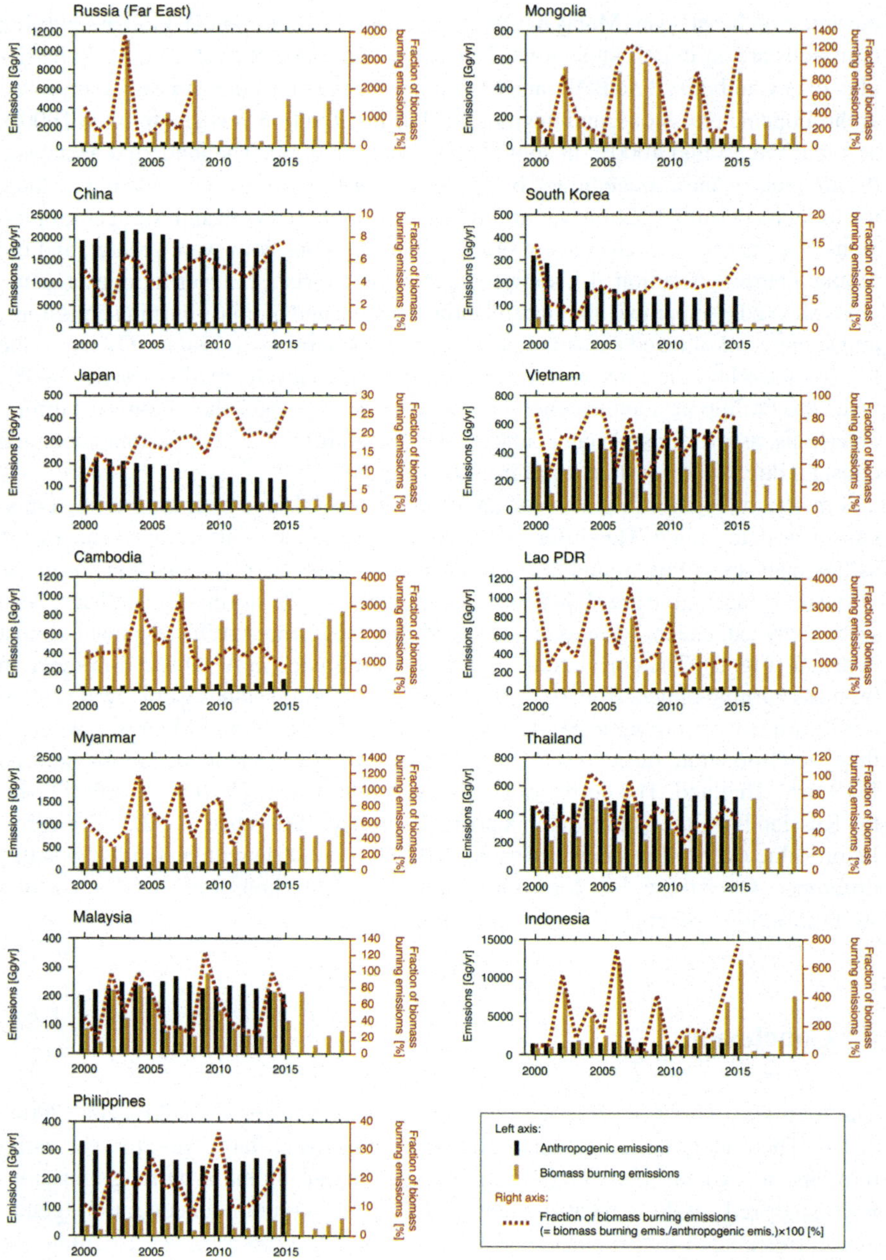

Fig. 6 Annual amount of anthropogenic and biomass burning emissions (left axis) and fraction of biomass burning emissions compared with anthropogenic emissions (right axis) over 13 EANET countries

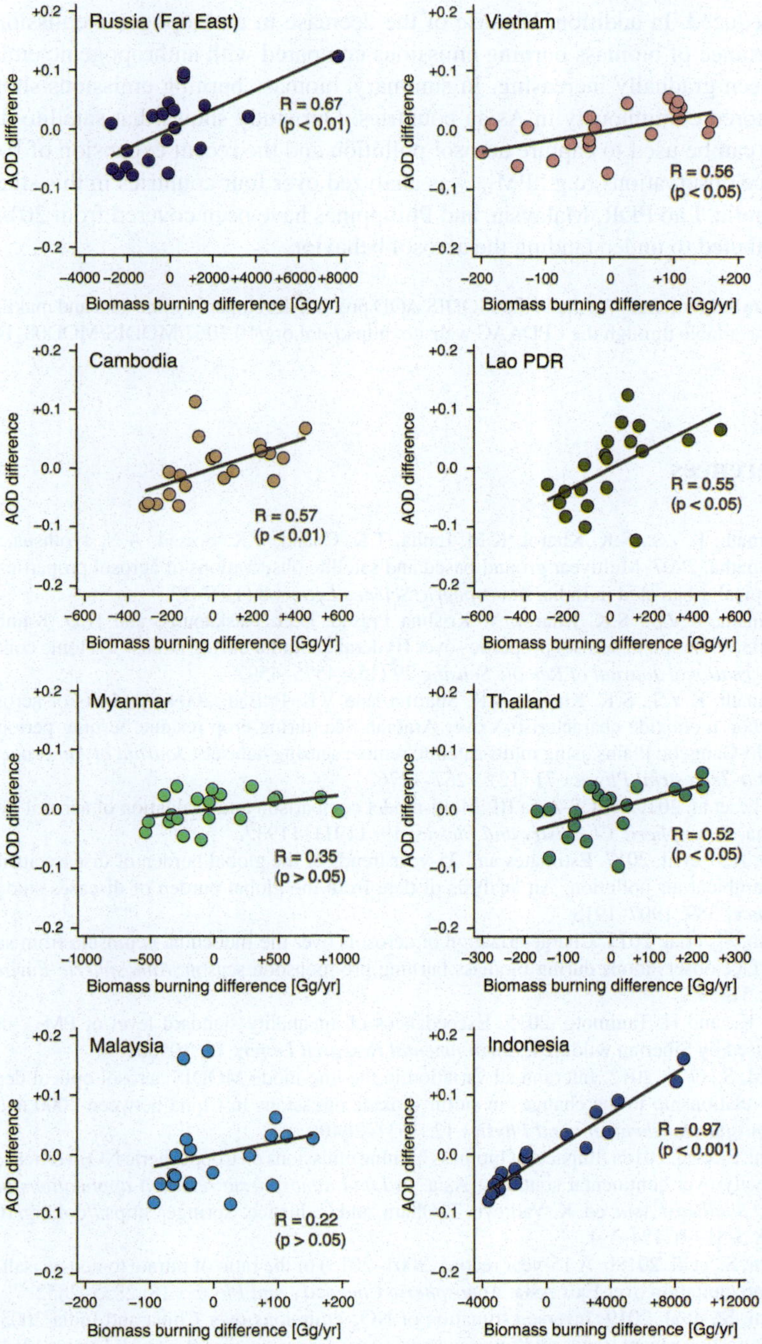

Fig. 7 Scatter plot between AOD difference (change from the 20-year average) and biomass burning emission difference over the Russian Far East, Vietnam, Cambodia, Lao PDR, Myanmar, Thailand, Malaysia, and Indonesia. The correlation value and its significance levels calculated by Student's *t*-test are included

and reduced. In addition, because of the decrease in anthropogenic emissions, the importance of biomass burning emissions compared with anthropogenic emissions has been gradually increasing. In summary, biomass burning emissions should be monitored continuously in Asian countries. Our study shows that satellite-derived AOD can be used to capture aerosol pollution and the recent expansion of EANET surface observations (e.g., PM_{10} was analyzed over four countries in this study, but Mongolia, Lao PDR, Malaysia, and Philippines have been covered from 2014) also contributed to understanding the aerosol behavior.

Acknowledgements We thank the MODIS AOD product developers for the data and making them freely available through the LPDAAC website: https://doi.org/10.5067/MODIS/MOD08_D3.061.

References

Badarinath, K.V.S., S.K. Kharol, K.M. Latha, T.K. Chand, V.K. Prasad, A.N. Jyothsna, and K. Samatha. 2007. Multiyear ground-based and satellite observations of aerosol properties over a tropical urban area in India. *Atmospheric Science Letters* 8 (1): 7–13.

Badarinath, K.V.S., S.K. Kharol, V. Krishna Prasad, D.G. Kaskaoutis, and H.D. Kambezidis. 2008. Variation in aerosol properties over Hyderabad, India during intense cyclonic conditions. *International Journal of Remote Sensing* 29 (15): 4575–4597.

Badarinath, K.V.S., S.K. Kharol, A.R. Sharma, and V.K. Prasad. 2009. Analysis of aerosol and carbon monoxide characteristics over Arabian Sea during crop residue burning period in the Indo-Gangetic Plains using multi-satellite remote sensing datasets. *Journal of Atmospheric and Solar-Terrestrial Physics* 71 (12): 1267–1276.

Chen, L., et al. 2019. MICS-Asia III: Multi-model comparison and evaluation of aerosol over East Asia. *Atmospheric Chemistry and Physics* 19: 11911–11937.

Cohen, A.J., et al. 2017. Estimates and 25-year trends of the global burden of disease attributable to ambient air pollution: An analysis of data from the global burden of diseases study 2015. *Lancet* 389: 1907–1918.

Gautam, R., et al. 2013. Characterization of aerosols over the Indochina peninsula from satellite-surface observations during biomass burning pre-monsoon season. *Atmospheric Environment* 78: 51–59.

Ikeda, K., and H. Tanimoto. 2015. Exceedances of air quality standard level of $PM_{2.5}$ in Japan caused by Siberian wildfires. *Environmental Research Letters* 10: 105001.

Itahashi, S., et al. 2012. Interannual variation in the fine-mode MODIS aerosol optical depth and its relationship to the changes in sulfur dioxide emissions in China between 2000 and 2010. *Atmospheric Chemistry and Physics* 12: 2631–2640.

Itahashi, S., et al. 2018a. Impacts of biomass burning emissions on tropospheric NO_2 vertical column density over continental southeast Asia. In *Land-atmospheric research applications in South and Southeast Asia*, ed. K. Vadrevu, T. Ohara, and C. Justice. Springer. https://doi.org/10.1007/978-3-319-67474-2_4.

Itahashi, S., et al. 2018b. A 15-year record (2001–2015) of the ratio of nitrate to non-seasalt sulfate in precipitation over East Asia. *Atmospheric Chemistry and Physics* 18: 2835–2852.

Itahashi, S., et al. 2019. Inverse estimation of NO_x emissions over China and India 2005–2016: Contrasting recent trends and future perspectives. *Environmental Research Letters* 14: 124020.

Itahashi, S., et al. 2020. MICS-Asia III: Overview of model comparison and evaluation of acid deposition over Asia. *Atmospheric Chemistry and Physics* 20: 2667–2693.

Itahashi, S., et al. 2021. Long-term trends of satellite-based fine-mode aerosol optical depth over the Seto Inland Sea, Japan, over two decades (2001–2020). *Environmental Research Letters* 16: 064062.

Janssens-Maenhout, G., et al. 2015. HTAP_v2.2: A mosaic of regional and global emission grid maps for 2008 and 2010 to study hemispheric transport of air pollution. *Atmospheric Chemistry and Physics* 15: 11411–11432.

Jeong, J.I., R.J. Park, and D. Youn. 2008. Effects of Siberian forest fires on air quality in East Asia during May 2003 and its climate implication. *Atmospheric Environment* 42: 8910–8922.

Kant, Y., A.B. Ghosh, M.C. Sharma, P.K. Gupta, V.K. Prasad, K.V.S. Badarinath, and A.P. Mitra. 2000. Studies on aerosol optical depth in biomass burning areas using satellite and ground-based observations. *Infrared Physics & Technology* 41 (1): 21–28.

Kaufman, Y.J., D. Tanre, and O. Boucher. 2002. A satellite view of aerosols in the climate system. *Nature* 419: 215–223.

Kharol, S.K., K.V.S. Badarinath, A.R. Sharma, D.V. Mahalakshmi, D. Singh, and V.K. Prasad. 2012. Black carbon aerosol variations over Patiala city, Punjab, India—A study during agriculture crop residue burning period using ground measurements and satellite data. *Journal of Atmospheric and Solar-Terrestrial Physics* 84: 45–51.

Kurokawa, J., and T. Ohara. 2020. Long-term historical trends in air pollutant emissions in Asia: Regional emission inventory in ASia (REAS) version 3. *Atmospheric Chemistry and Physics* 20: 12761–12793.

Kurokawa, J., et al. 2013. Emissions of air pollutants and greenhouse gases over Asian regions during 2000–2008: Regional Emission inventory in ASia (REAS) version 2. *Atmospheric Chemistry and Physics* 13: 11019–11058.

Levy, R.C., et al. 2013. The collection 6 MODIS aerosol products over land and ocean. *Atmospheric Measurement Techniques* 6: 2989–3034.

Ma, X., C. Wang, B. Dong, G. Gu, R. Chen, Y. Li, H. Zou, W. Zhang, and Q. Li. 2019. Carbon emissions from energy consumption in China: Its measurement and driving factors. *Science of the Total Environment* 648: 1411–1420.

Miyazaki, K., et al. 2017. Decadal changes in global surface NO_x emissions from multi-constituent satellite data assimilation. *Atmospheric Chemistry and Physics* 17: 807–837.

NASA. 2021. LAADS DAAC, MOD08_D3—MODIS/Terra aerosol cloud water vapor ozone daily L3 global 1Deg CMG, https://doi.org/10.5067/MODIS/MOD08_D3.061.

Takami, K., H. Shimadera, K. Uranishi, and A. Kondo. 2020. Impacts of biomass burning emission inventories and atmospheric reanalyses on simulated PM_{10} over Indochina. *Atmosphere* 11: 160.

Tsay, S.-C., et al. 2013. From BASE-ASIA to 7-SEAS: A satellite-surface perspective of boreal spring biomass-burning aerosols and clouds in Southeast Asia. *Atmospheric Environment* 78: 20–34.

Uno, I., et al. 2020. Paradigm shift in aerosol chemical composition over regions downwind of China. *Scientific Reports* 10: 6450.

Vadrevu, K.P., E. Ellicott, K.V.S. Badarinath, and E. Vermote. 2011. MODIS derived fire characteristics and aerosol optical depth variations during the agricultural residue burning season, north India. *Environmental Pollution* 159 (6): 1560–1569.

Vadrevu, K.P., T. Ohara, and C. Justice. 2014. Air pollution in Asia. *Environmental Pollution.* 12: 233–235.

Vadrevu, K.P., K. Lasko, L. Giglio, and C. Justice. 2015. Vegetation fires, absorbing aerosols and smoke plume characteristics in diverse biomass burning regions of Asia. *Environmental Research Letters* 10 (10): 105003.

Vadrevu, K.P., et al., eds. 2018. *Land-atmospheric research applications in South and Southeast Asia.* Springer. https://doi.org/10.1007/978-3-319-67474-2.

Vadrevu, K.P., T. Ohara, and C. Justice, eds., 2021a. *Biomass burning in South and Southeast Asia: Impacts on the biosphere*, vol. 2. CRC Press.

Vadrevu, K.P., T. Ohara, and C. Justice, eds. 2021b. *Biomass burning in South and Southeast Asia: Mapping and monitoring*, vol. 1. CRC Press.

van der Werf, G.R., et al. 2017. Global fire emissions estimates during 1997–2016. *Earth System Science Data* 9: 697–720.

van Donkelaar, A., et al. 2010. Global estimates of ambient fine particulate matter concentrations from satellite-based aerosol optical depth: Development and application. *Environmental Health Perspectives* 118 (6): 847–855.

van Donkelaar, A., et al. 2016. Global estimates of fine particulate matter using a combined geophysical-statistical method with information from satellites, models, and monitors. *Environmental Science & Technology* 50: 3762–3772.

Wooster, M.J., G.J. Roberts, L. Giglio, D.P. Roy, P.H. Freeborn, L. Boschetti, C. Justice, C. Ichoku, W. Schroeder, D. Davies, and A.M. Smith. 2021. Satellite remote sensing of active fires: History and current status, applications and future requirements. *Remote Sensing of Environment* 267: 112694.

Yasunari, T., et al. 2018. Extreme air pollution events in Hokkaido, Japan, tracked back to early snowmelt and large-scale wildfires over East Eurasia: Case studies. *Scientific Reports* 8: 6413.

Zhang, Q., et al. 2019. Drivers of improved $PM_{2.5}$ air quality in China from 2013 to 2017. *Proceedings of the National Academy of Sciences of the United States of America* 116: 24463–24469.

Light Absorption Properties of Biomass Burning Emissions in Bangladesh: Current State of Knowledge

Shahid Uz Zaman, Md Safiqul Islam, Shatabdi Roy, Farah Jeba, and Abdus Salam

Abstract Biomass burning (BB) is a significant source of pollution, both regionally and globally, with severe ramifications for climate change, air quality, and human health. Biomass combustion also contributes substantially to atmospheric pollution in South Asia, particularly Bangladesh. To estimate the environmental impacts of BB, we need to know more about its emissions, sources, transport, and transformation. This article reviews research on the light absorption properties of BB pollutants in Bangladesh. Elevated concentrations of trace elements were found from biomass burning in the rural cooking stoves. Methanol-soluble brown carbon (MeS-BrC) in the commonly used biomass demonstrated higher absorbance than water-soluble BrC (WS-BrC). MeS-BrC has a higher babs-BrC value than WS-BrC, implying that the rate of light absorption on BrC extracted in methanol was higher. Absorption emission factors at 370 nm were consistently higher than at 880 nm. The mass absorption efficiency of black carbon (BC) and brown carbon varied among the common biomass. Biomass burning associated with $PM_{2.5}$ were dominant throughout the year, particularly in the non-monsoon seasons. The study highlights the current knowledge on light absorption properties, chemical characterization, emission factors, and contribution to ambient PM in Bangladesh.

Keywords Biomass burning · Light absorption properties · Emission factor · Source contribution · Brown carbon

1 Introduction

Biomass burning (BB) has been shown to significantly impact local air quality, tropospheric chemistry, ecosystem processes, visibility, public health, the atmosphere's

S. U. Zaman · M. S. Islam · S. Roy · F. Jeba · A. Salam (✉)
Department of Chemistry, University of Dhaka, Dhaka 1000, Bangladesh
e-mail: asalam@gmail.com; aasalam@du.ac.bd

S. U. Zaman
Department of Chemistry, Faculty of Science, Bangladesh University of Engineering and Technology, Dhaka, Bangladesh

K. P. Vadrevu et al. (eds.), *Vegetation Fires and Pollution in Asia*,
https://doi.org/10.1007/978-3-031-29916-2_17

radiative balance, and climate change (Salam et al. 2013; Reddington et al. 2015). BB releases a massive quantity of hazardous compounds such as particulate matter (PM), black carbon, carbon monoxide, formaldehyde, and organic carbon (OC), the light-absorbing component of which is known as "brown carbon" (BrC). Acute respiratory infections, asthma, chronic obstructive pulmonary disease, nasopharyngeal and laryngeal cancers, tuberculosis, and vision impairments are only a few of the health concerns caused by these hazardous pollutants (Andreae and Merlet 2001). BB takes place all across the world, including North America, Asia, South America, and Africa. Therefore, it is necessary to assess the impacts of BB on local, regional, and global scales on the environment and atmosphere useful for policymaking (Vadrevu et al. 2017, 2021a, b). Understanding the impacts of biomass burning on air pollutants is crucial, as is understanding the geographical and temporal variations in emission characteristics for each kind of fire (Bouarar et al. 2017).

Bangladesh is a rapidly expanding Southeast Asian country. Many researchers attempted to study air pollution in Bangladesh, which is a big issue. Apart from different sources of pollution, BB has been shown as one of the top sources of fine airborne particulates ($PM_{2.5}$) (Salam et al. 2013, 2021; Rahman et al. 2020; Runa et al. 2021; Hasan et al. 2009; Chowdhury et al. 2012; Begum et al. 2009; Khalequzzaman et al. 2007; Ahmed et al. 2018). For example, in one of the BB studies by Hasan et al. (2009), the emission of lead, iron, cadmium, calcium, potassium, and magnesium in mixed ash was significantly higher than in rice husk coil. Choudhury et al. (2012) reported that the chemical composition of $PM_{2.5}$ generated by cooking was 59–60% organic carbon and 29–30% elemental carbon. Begum et al. (2009) observed lower PM concentrations in residences that used LPG (liquefied petroleum gas) rather than other biomass fuels. Khalequzzaman et al. (2007) found a significant association between the biomass fuel-using population and respiratory symptoms. These studies reveal the importance of BB studies in Bangladesh. In this study, we focus on the emission characteristics of BB in Bangladesh and review in detail our current understanding and contributions to fine atmospheric particulate, including the needs and priorities for future work.

2 Study Area

Bangladesh is located on the northeastern edge of the Indo-Gangetic Plain (IGP). The climate is characterized by hot and humid summers throughout most of the year, with substantial seasonal variations in precipitation. Bangladesh has four seasons: pre-monsoon (March–May), monsoon (June–September), post-monsoon (October–November), and winter (December–February) (Salam et al. 2013; Zaman et al. 2021). Middle- and upper-income households in metropolitan Bangladesh frequently utilize electricity or reasonably clean cooking fuels such as liquefied petroleum gas (LPG) or natural gas. In contrast, low-income semi-urban and rural families heavily rely on biomass fuels. These fuels encompass wood, dried leaves, animal dung, and crop residues such as rice husks, straws, bagasse (fiber derived from sugar manufacturing), jute sticks, and other similar materials (Sarkar and Islam 1998). Meteorological and economic factors may also impact the year-round use of biomass fuels. Most biomass fuels are considered low grade, with higher emissions and potential health risks (Begum et al. 2009).

3 Data and Methods

3.1 Analytical Methods

3.1.1 X-ray Fluorescence (XRF) Spectroscopy

XRF spectroscopic analysis is a valuable technique that is widely used to measure the concentrations of trace metals. Salam et al. (2013) used an X-ray fluorescence spectrometric method (serial number 1199902 and model SL 80175) to determine the amount of K, Ca, Fe, Rb, Ti, Mn, Pb, Sr, Zn, Br, Y, and Zr in smoke deposits. Begum et al. (2009) employed energy-dispersive X-ray fluorescence (EDXRF) spectrometer, which is a radioisotope-induced spectrometer (MicroMatter Co., Eastsounds, WA, USA) to determine the elemental composition of K, Ca, Fe, Cu, Cr, Ti, Mn, and Zn.

3.1.2 Atomic Absorption Spectrophotometer (AAS)

AAS is another useful analytical technique to determine the concentration of trace elements. The quantities of trace elements (Pb, Ca, Cd, Mg, K, and Fe) in black solids accumulated from BB at the cooking burners were measured by using an AAS (Shimadzu AA-680, Japan) (Hasan et al. 2009).

3.1.3 UV–visible Spectrophotometer

The concentration of ions such SO_4^{2-}, PO_3^-, and NO_3^- in smoke samples can be determined using a UV–visible spectrophotometer (Salam et al. 2013). The absorbance at 380 nm for sulfate, 358 nm for phosphate, and 410 nm for nitrate are measured using a UV–visible spectrophotometer with a reagent blank.

3.2 Light Absorption Properties

3.2.1 Absorption Coefficient (b_{abs})

The absorption coefficients for brown carbon at 365 nm can be calculated using the following formula (Runa et al. 2021; Hecobian et al. 2010).

$$b_{abs-BrC}(Mm^{-1}) = \frac{[A_{365} \times V_l \times \text{total filter area} \times \ln \ln(10)]}{V_a \times \text{extracted filter area} \times L} \quad (1)$$

Here, A_{365} indicates the absorbance attributed to water-soluble brown carbon (WS-BrC) at wavelength 365 nm, V_l denotes the amount of deionized water in m^3 used during the aerosol sample extraction, V_a represents the quantity of air filtered in m^3 during aerosol sampling through a filter-substrate, and L signifies the path length of the cell which is 1 cm. The filter's total surface area is 13.85 cm^2, of which 3.46 cm^2 is obtained in deionized water and used to assess different parameters (Runa et al. 2021).

3.2.2 Mass Absorption Efficiency (MAE)

The MAE is an important factor in converting the BrC mass concentration calculated by chemical transport models to the atmospheric absorption coefficient required by radiative transfer models (Cheng et al. 2016). The MAE for BrC can be assessed by the following equation:

$$MAE(m^2g^{-1}) = \frac{b_{abs-BrC}(Mm^{-1})}{WS - BrC} \quad (2)$$

WS-BrC (μgm^{-3}) denotes the mass concentration of BrC that is extracted water for each filter sample, and $b_{abs\text{-}BrC}$ (Mm^{-1}) represents the absorption coefficient BrC.

3.2.3 Absorption Angstrom Exponent (AAE)

The AAE value describes the wavelength dependency of aerosol optical thickness or aerosol extinction coefficient. The smaller the particle, the larger the exponent. The spectrum dependence of light-absorbing aerosol characteristics is based on the power law as Eq. 3 (Hecobian et al. 2010).

$$b_{abs} = B * \gamma^{-AAE} \quad (3)$$

Here, AAE is the absorption angstrom exponent, B is a constant applied to the mass concentration of aerosols, and γ represents the wavelength of light. The B value can be obtained by subtracting the empty filter's weight from the aerosol-loaded filter's weight due to biomass combustion and then dividing it by the volume of air utilized to load the sample. It is worth noting that each biomass has a specific B value.

3.3 *Absorption Emission Factor (AEF) Estimation*

The absorption cross section emitted to the atmosphere per MJ produced or per kilogram of fuel burned is referred to as the absorption emission factor (AEF), which is computed in $m^2\ kg^{-1}$ as follows:

$$\mathrm{AEF}(\lambda) = \frac{\mathrm{FGV}}{M_{\mathrm{fuel}}} \times b_{\mathrm{abs}(\lambda)} \times 10^{-6} \tag{4}$$

where FGV is the volume of flue-gas (m^3) produced by the burning of wood fuel under normal temperature, pressure, and dry conditions and M_{fuel} denotes the mass (kg) of the fuel.

3.4 *Biomass Burning $PM_{2.5}$ Contributions*

From November through April, the IGP, notably in Northern India, is noted for crop-burning emissions (Badarinath et al. 2007, 2009; Kharol et al. 2012; Bhuvaneshwari et al. 2019). A strategy for creating an elemental detector for this particle was developed to determine the biomass contributions to Dhaka's $PM_{2.5}$ pollution (Rahman et al. 2020). Particulate potassium (K) is an element found in many agricultural and wood combustion particles, and it has been used as a tracer for biomass burning in the past (Kim et al. 2003; Li et al. 2003).

4 Results and Discussion

4.1 *Chemical Characterization of Biomass Burning Emissions*

Previous research has suggested that biomass burning may be one of Bangladesh's primary causes of trace metal emissions (Salam et al. 2013; Hasan et al. 2009; Ahmed et al. 2018). In a study conducted in Narsingdi, Bangladesh, Salam et al. (2013) employed arjun, madhabilata, bamboo, mango, coconut, mahogany, various kinds of leaves and trees, rice husk coil, and plum as fuel biomass. The X-ray fluorescence (XRF) technique was used to determine potassium, titanium, cobalt, manganese, calcium, iron, bromine, copper, zinc, rubidium, zirconium, strontium, molybdenum, yttrium, niobium, and lead. A UV–visible spectrophotometer was used to measure sulfate, nitrate, and phosphate. The trace element concentrations varied throughout these nine biomass burning samples. The results suggested that the biomass from mango has the greatest trace metal concentration, whereas bamboo has the lowest (Table 1).

Hasan et al. (2009) used the AAS method to evaluate the chemical composition of black solid materials from cooking stoves employing two distinct types of biomasses (Rice husk coils and mixed biomasses). The average amounts of lead, iron, cadmium, calcium, potassium, and magnesium in mixed biomass were substantially greater than in rich husk coils. In rice husk coils, lead, iron, cadmium, calcium, potassium, and magnesium were 95.6, 11,520, 8.33, 1635, 17.1, and 443.1 mg kg^{-1}, respectively. In mixed biomasses, lead, iron, cadmium, calcium, potassium, and magnesium were 125.2, 12,360, 12.0, 1648, 21.5, and 534.2 mg kg^{-1}, respectively. When compared to mixed biomasses, rice husk coils produced around 31% less lead, 44% less cadmium, 26% less potassium, and 21% less magnesium. For both samples, the greatest concentration was found for iron, while the lowest concentration was found for cadmium. Burning mixed biomasses (a mixture of diverse biomasses such as straw, bamboo, cow dung, and dried leaves of trees and plants) emits higher trace metal pollution than rice husk coils.

4.2 Absorption Coefficients of Brown Carbon (b_{abs})

WSOC and OC are widely used to replace water-extractable BrC (also known as water-soluble brown carbon or WS-BrC) and methanol-extractable BrC (also known as methanol-soluble brown carbon or MeS-BrC), respectively (Satish et al. 2017). WS-BrC is a core part of BrC that can absorb light directly. It can also indirectly affect the atmosphere by increasing the capability of aerosol particles to function as a cloud condensation nucleus. To evaluate the light absorption coefficients of diverse biomasses, 14 unique biomasses that are frequently used during cooking were gathered from Savar, Dhaka, and the b_{abs} of BrC at 365 nm were obtained (Runa et al. 2021). Raintree has the highest absorption coefficient (1.63 × 105 and 2.00 × 105 Mm^{-1}) in both water and methanol of all the biomasses measured, whereas jute stick has the lowest absorption value (4.73 × 10–3 and 3.34 × 10–5 Mm^{-1}) (Fig. 1). Except for palm leaves, dhaincha, and koroi tree, methanol extracts have higher b_{abs}-BrC values than water-extracted solutions in all experimental samples. An increase in b_{abs}-BrC indicates a larger quantity of BrC extracts since the absorption coefficient reflects the light absorption rate.

Table 1 The concentration of trace elements and ions in the biomass burning smoke deposits from cooking stoves in Narsingdi, Bangladesh

Elements	Arjun	Bamboo	Coconut	Rice husk coil	Madhabilata	Mahogany	Mango	Mixed ash	Plum	Average
K	12,983	38,169	25,838	15,772	20,030	11,927	7294	14,653	29,985	19,627
Ca	6664	4189	6995	5008	7000	5991	10,059	4009	7846	6418
Fe	4147	2624	3356	3898	3805	4247	47.7	4555	2621	3256
Ti	717	559	600	621	685	808	706	659	636	665
Mn	76.9	72.6	98.8	108	80.2	99.7	78.44	171	93.6	98
Co	21.4	24.4	31.5	30.9	28.7	37.3	46.3	40.0	18.0	31.0
Cu	19.1	9.94	13.2	17.8	11.2	20.4	219.0	27.6	12.3	39.0
Zn	149	77.2	107	172	217	175	6.22	72.9	159	126
Br	20.2	47.7	39.5	27.0	21.6	15.5	19.1	37.1	44.4	30.0
Rb	26.6	94.0	71.9	51.8	41.8	31.6	20.6	89.5	45.8	52.0
Sr	48.7	51.1	60.4	51.0	53.6	46.3	56.6	57.2	60.7	54.0
Y	10.7	11.5	9.9	14.3	9.7	14.2	11.6	16.0	7.8	12.0
Zr	147	273	175	166	187	214	151	245	130	187
Nb	27.6	30.6	26.1	25.5	32.4	23.8	17.6	36.6	16.6	26.0
Mo	32.5	56.1	10.3	13.5	43.0	2.10	54.2	35.9	17.1	29.0
Pb	314	50.8	210	506	177	458	217	191	231	262
SO_4^{2-}	55.3	40.3	19.9	136.3	15.3	17.3	20.3	17.8	26.8	38.0
PO_4^{3-}	0.79	0.99	0.51	0.25	0.50	0.95	0.48	0.35	0.52	0.60
NO_3^{2-}	BDL	BDL	1.00	0.33	4.73	BDL	BDL	BDL	0.47	0.73

Trace elements were determined with X-ray fluorescence (XRF), and ions were determined with UV–visible spectrophotometer. All the units are in mg kg^{-1} (Salam et al. 2013)

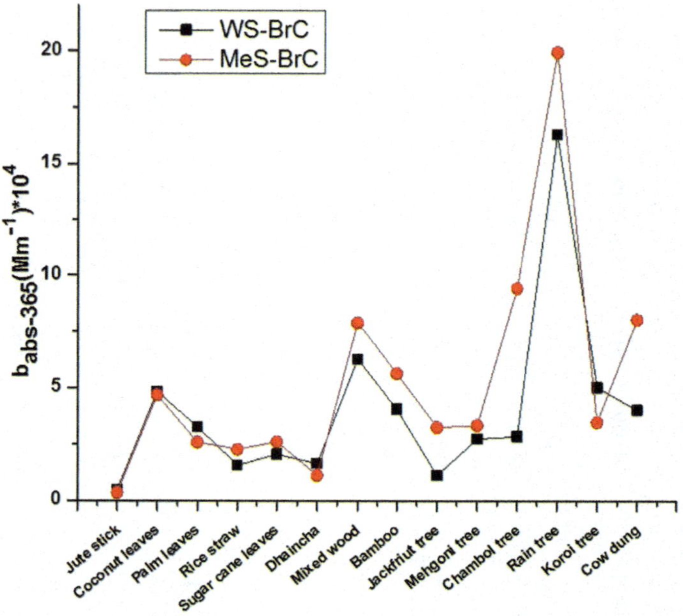

Fig. 1 Absorption coefficient ($b_{abs\text{-}BrC}$ at 365 nm) of BrC extracted in water and methanol (Runa et al. 2021)

4.3 Mass Absorption Efficiency (MAE)

MAE is the proportion of light absorption to mass concentration. It can be used to figure out how much light can be absorbed. By monitoring atmospheric radiation, these absorption characteristics are widely employed to determine the impact of aerosols on climate (Tang et al. 2020; Zhang et al. 2020). Correlation studies between the aerosol absorption coefficient and mass concentration are essential, as MAE is one of the most critical optical properties of the aerosol. It assists in comprehending the scattering effects of the sample particles. Pavel et al. (2023) measured MAEBC and MAEBrC for nine biomasses from Gazipur and Cumilla, Bangladesh, and found that MAE_{BC} values fluctuated from 1.15 to 15.06 m^2g^{-1}, with an average of 7.46 ± 4.09 m^2 g^{-1}. Cow dung combustion had the highest MAEBC, with a mean of 15.06 ± 2.35 m^2 g^{-1}, followed by bamboo (10.50 ± 2.68 m^2 g^{-1}), bagasse (7.84 ± 2.32 m^2 g^{-1}), coconut leaves (6.95 ± 1.91 m^2 g^{-1}), jackfruit leaves (5.72 ± 1.14 m^2 g^{-1}), koroi leaves (5.44 ± 1.52 m^2 g^{-1}), and paddy straw (3.70 ± 1.58 m^2 g^{-1}).

On the other hand, MAE_{BrC} values for various fuels ranged from 1.92 to 22.86 m^2 g^{-1}, with an average of 12.66 ± 7.54 m^2 g^{-1}. The MAEBrC values followed the same trends as the MAE_{BC}, with jackfruit having the highest average value of 22.86 ± 3.99 m^2 g^{-1}, followed by cow dung (21.00 ± 3.33 m^2 g^{-1}), dhaincha (18.70 ± 3.98 m^2 g^{-1}), bagasse (16.81 ± 3.27 m^2 g^{-1}), coconut leaves (12.03 ± 2.54 m^2 g^{-1}), and bamboo (8.40 ± 2.51 m^2 g^{-1}) (Fig. 2).

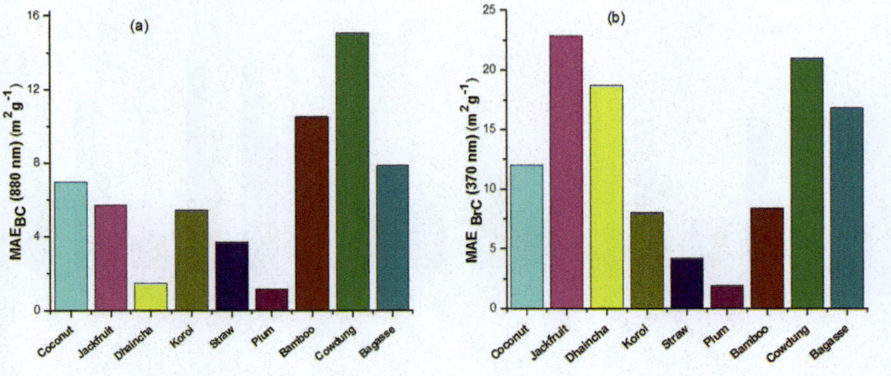

Fig. 2 MAE_{BC} **a** and MAE_{BrC} **b** values for various biomass fuel combustion in the conventional cooking stoves (Pavel et al. 2023)

Runa et al. (2021) also measured MAE for both WS-BrC and MeS-BrC and reported that MeS-BrC had a higher MAE_{BrC} than WS-BrC. The jackfruit tree has the highest MAE_{BrC} content, whereas the jute stick has the lowest (Fig. 3). At 365 nm, MAE_{BrC} was consistently higher in methanol extracts than in water extracts, indicating that BrC extracted only with methanol (MeS-BrC) was more light-absorbing than water extracts (WS-BrC).

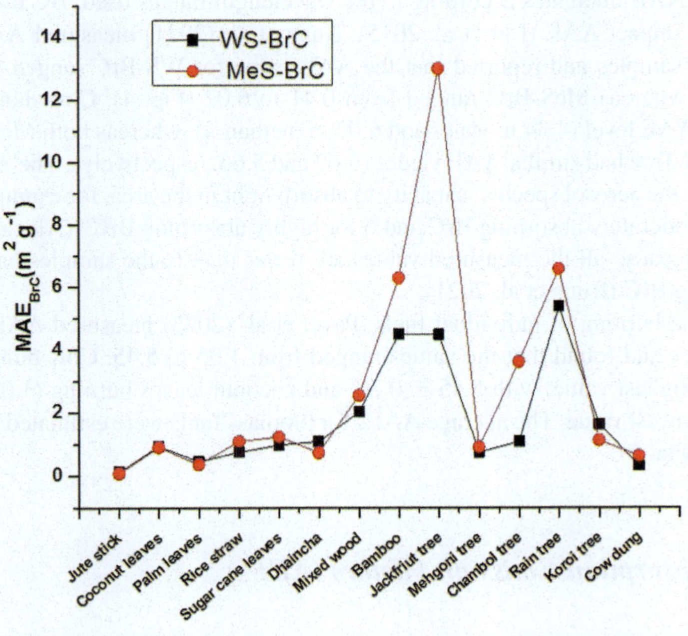

Fig. 3 Obtained MAE_{BrC} of WS-BrC and MeS-BrC for various biomasses (Runa et al. 2021)

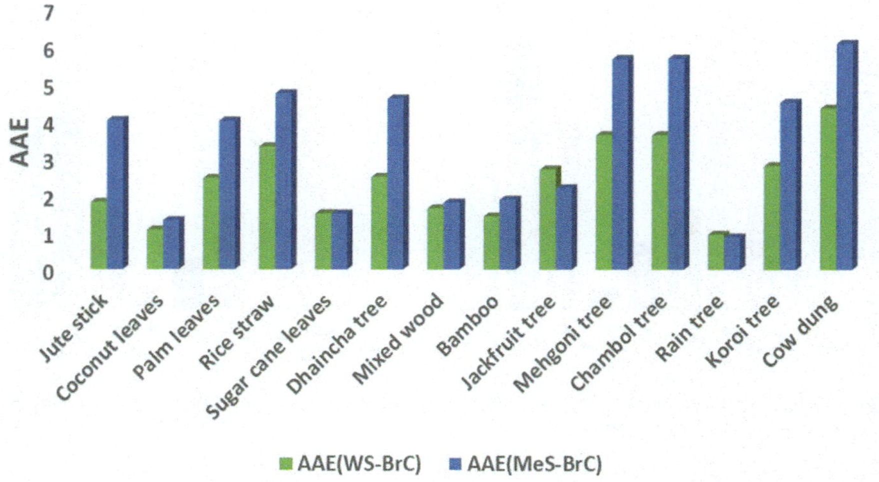

Fig. 4 Obtained AAE value of WS-BrC and MeS-BrC from different biomasses (Runa et al. 2021)

4.4 Absorption Angstrom Exponent (AAE)

The AAE shows the difference in absorption as a function of wavelength and is very dependent on the particles' size, geometry, and composition (Li et al. 2016). In practice, the AAE fluctuates according to the wavelength ranges used. BC morphology can also impact AAE (Liu et al. 2015). Runa et al. (2021) measured AAE for 14 biomass samples and reported that the AAE value for WS-BrC ranged from 0.33 to 4.34, whereas MeS-BrC ranged from 0.41 to 6.07 (Fig. 4). Cow dung had the highest AAE level (4.34 in water and 6.07 in methanol), whereas both Mehgoni and Chambal Tree had similar AAE values (3.62 and 5.66, respectively). The AAE value indicates the aerosol species' capacity to absorb light in the area, for example, "AAE 11 for moderately absorbing BrC and 6 for highly absorbing BrC" (Alexander et al. 2008). Because all the measured values are fewer than 6, the samples have a high absorbing BrC (Runa et al. 2021).

For the burning of individual fuels, Pavel et al. (2023) measured AAE for nine biomasses and found that the values ranged from 1.05 to 5.45. Cow dung burning had the highest value, with 5.45 $\pm$ 0.70, and coconut leaves burning (1.05 $\pm$ 0.19) had the lowest value. The average AAEs for biomass fuels were estimated to be 2.68 $\pm$ 0.28 (Fig. 5).

4.5 Absorption Emission Factors (AEFs)

Absorption emission factors consider the mass of the fuel consumed rather than the mass of the BrC. As a result, the AEFs are now being utilized to explain

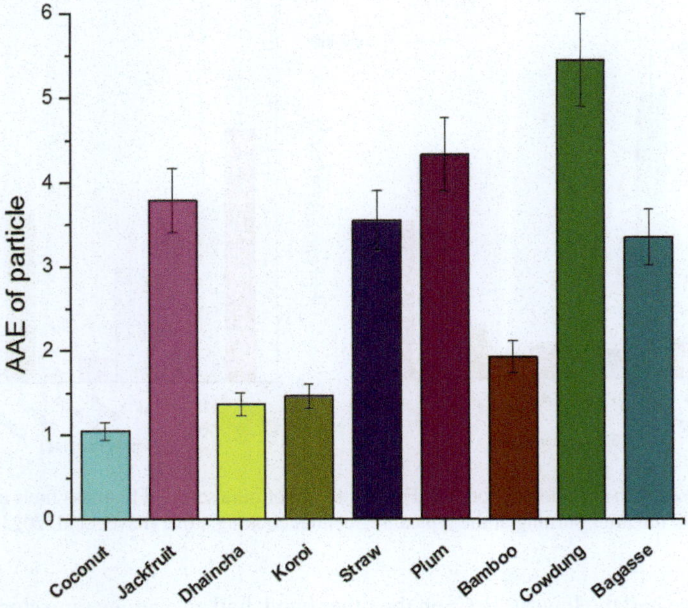

Fig. 5 Absorption angstrom efficiency values for nine biomasses (Pavel et al. 2023)

the light-absorbing characteristics of direct source particles. Residential biomass burning emission factors are critical because they measure the quantity of pollutants discharged into the atmosphere per mass of fuel burned. The AEF varied from 1.20 to 15.61 $m^2\ kg^{-1}$ at 370 nm, with a mean of $5.23 \pm 5.47\ m^2\ kg^{-1}$, whereas the AEF at 880 nm was lower, with a mean of $0.39 \pm 0.26\ m^2\ kg^{-1}$ and a span of 0.03 to 0.81 $m^2\ kg^{-1}$. Among the biomass samples, highest AEF_{370nm} values were observed for Jackfruit leaves ($15.61 \pm 2.42\ m^2\ kg^{-1}$), followed by the combustion of cow dung ($12.08 \pm 2.83\ m^2\ kg^{-1}$), bagasse ($8.63 \pm 1.45\ m^2\ kg^{-1}$), bamboo ($2.89 \pm 0.32\ m^2\ kg^{-1}$), coconut leaves ($2.02 \pm 0.25\ m^2\ kg^{-1}$), paddy straw ($1.78 \pm 0.20\ m^2\ kg^{-1}$), dhaincha ($1.58 \pm 0.27\ m^2\ kg^{-1}$), koroi ($1.30 \pm 0.24\ m^2\ kg^{-1}$), and plum ($1.20 \pm 0.18\ m^2\ kg^{-1}$) (Fig. 6).

In contrast, the highest and lowest AEF_{880nm} values were observed for coconut leaves ($0.81 \pm 0.20\ m^2\ kg^{-1}$) and plum ($0.03 \pm 0.01\ m^2\ kg^{-1}$), respectively. The results obtained in this study were lower than in previous studies. Martinsson et al. (2015) found AEF values for BC and BrC of 6.23–18.24 $m^2\ kg^{-1}$ and $-$ 0.38 to 23.58 $m^2\ kg^{-1}$ for fresh birch logs under various circumstances at = 370 nm, respectively, whereas these values were 5.90–17.67 $m^2\ kg^{-1}$ and -0.76–19.95 $m^2\ kg^{-1}$ for processed birch logs.

Ahmed et al. (2018) measured emission factors of BC and BrC for seven biomasses (dry leaves, albizia, jackfruit, rain, cow dung, mahogany, and mango). According to their study, the average emission factor for BC of all biomasses was 1.09 mgg^{-1}. The dry leaves of the mahogany tree had the greatest BC emission factor, while the

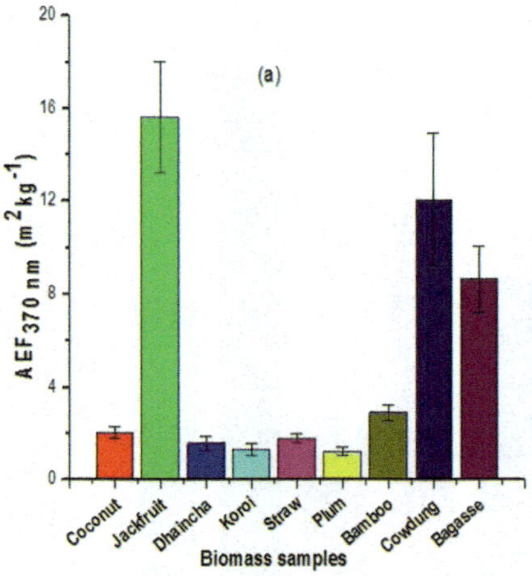

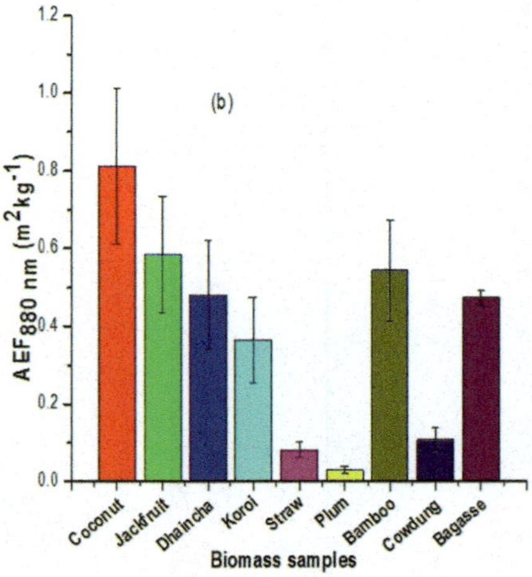

Fig. 6 Absorption emission factors (AEFs) (m^2 kg^{-1}) of nine selected biomass fuels **a** at 370 nm and **b** at 880 nm after burning at the typical residential cooking stove (Pavel et al. 2023)

mango tree had the lowest. BrC, on the other hand, had an average emission factor of 2.35 mgg^{-1}. The albizia tree had the greatest BrC emission factor, while the mango tree had the lowest.

4.6 Biomass Burning Contribution of $PM_{2.5}$

In the Asian continent, crop burning is the most common and seasonal practice. Post-harvest crop stubble burning is prevalent in northern India between November and December (rice straw burning) and April and May (wheat residue burning) (Bhuvaneshwari et al. 2019). During the winter, these emissions frequently lead to noticeable haze in this region. Prior research has shown that dust and agricultural waste burning particles may reach more than 1000 km (Uranishi et al. 2019). As a result, pollutants generated in the IGP might be carried to Dhaka by the prevailing westerly winds. Rahman et al. (2020) assessed how crop-burning pollution impacts Dhaka during the non-monsoon season. According to their study, on average, the $PM_{2.5}$ mass contribution was dominated by biomass burning-associated mass, with fossil fuel-associated sources (sulfur-related $PM_{2.5}$ mass) accounting for 21.6% (19.5 $\pm$ 11.8 μgm^{-3}), biomass burning sources (adjusted potassium-associated $PM_{2.5}$ mass) accounting for 40.2% (36.3 $\pm$ 29.9 μgm^{-3}), and other sources (other PM mass) accounting for 38.2% (34.5 $\pm$ 40.3 μgm^{-3}) (Fig. 7).

This was notably true during the pre-monsoon season, when biomass burning predominated, and the contributions were as follows: fossil fuels associated $PM_{2.5}$ 28.6% (21.2 $\pm$ 10.8 μgm^{-3}), biomass burning $PM_{2.5}$ 54.2% (40.2 $\pm$ 23.5 μgm^{-3}), and other $PM_{2.5}$ 17.2% (12.8 $\pm$ 21.4 μgm^{-3}). Similarly, they were 18.9% (22.3 $\pm$

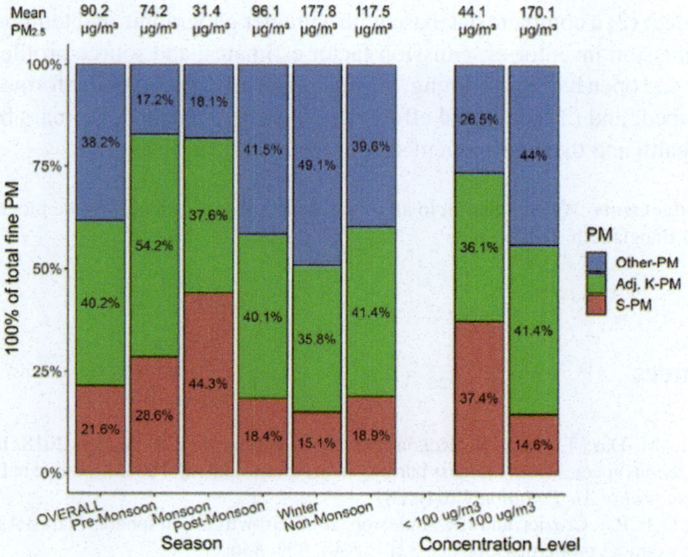

Fig. 7 Relative contribution of the sulfur-associated fossil fuel combustion $PM_{2.5}$ (SPM), adjusted potassium-associated biomass burning $PM_{2.5}$ (Adj. K-PM), and the rest of the $PM_{2.5}$ (Other-PM) on the total $PM_{2.5}$ concentrations in Dhaka by seasons (Rahman et al. 2020)

12.6 μgm^{-3}), 41.4% (48.6 ± 32.3 μgm^{-3}), and 39.6% (46.6 ± 55.3 μgm^{-3}) for the full non-monsoon season, respectively. Thus, throughout the non-monsoon season including the winter season, $PM_{2.5}$ from biomass burning such as crop burning and other sources such as dust was dominated. In contrast, during the monsoon season, when no crop residue burning was prevalent, $PM_{2.5}$ from fossil fuel combustion prevailed (Rahman et al. 2020).

5 Conclusions

This study highlights the findings of earlier researchers on the contribution of biomass burning to overall air pollution in Bangladesh. Several aspects were discussed, especially chemical characterizations of black solid materials, light absorption properties, mass absorption efficiency, emission factor determination, and biomass burning contribution to $PM_{2.5}$. The review suggests that open field burning and residential burning for heating and cooking can generate significant amounts of air pollutants, with seasonal and regional fluctuations. In addition, biomass burning aerosols can also get transported across long distances to Bangladesh. In the study, some of the shortcomings of existing research were also discussed with the following future directions: (1) The emission characteristics of biomass burning sources for different types of fuel and different types of fires need to be more extensively studied and documented; (2) a complete database with a greater geographic and temporal resolution of emission inventories, emission factor estimates, and source profiles for both domestic and open biomass burning, as well as thorough speciation information, must be developed; and (3) additional efforts to study the impacts of biomass burning on human health and the environment should be undertaken.

Acknowledgements We are grateful to all of the authors that contributed to the biomass burning research in Bangladesh.

References

Ahmed, M., M. Das, T. Afser, M. Rokonujjaman, T. Akther, and A. Salam. 2018. Emission of carbonaceous species from biomass burning in the traditional rural cooking stove in Bangladesh. *Open Journal of Air Pollution* 7 (04): 287.

Alexander, D.T., P.A. Crozier, and J.R. Anderson. 2008. Brown carbon spheres in East Asian outflow and their optical properties. *Science* 321 (5890): 833–836.

Andreae, M.O., and P. Merlet. 2001. Emission of trace gases and aerosols from biomass burning. *Global Biogeochemical Cycles* 15 (4): 955–966.

Badarinath, K.V.S., S.K. Kharol, K.M. Latha, T.K. Chand, V.K. Prasad, A.N. Jyothsna, and K. Samatha. 2007. Multiyear ground-based and satellite observations of aerosol properties over a tropical urban area in India. *Atmospheric Science Letters* 8 (1): 7–13.

Badarinath, K.V.S., S.K. Kharol, A.R. Sharma, and V.K. Prasad. 2009. Analysis of aerosol and carbon monoxide characteristics over Arabian Sea during crop residue burning period in the Indo-Gangetic Plains using multi-satellite remote sensing datasets. *Journal of Atmospheric and Solar-Terrestrial Physics* 71 (12): 1267–1276.

Begum, B.A., S.K. Paul, M.D. Hossain, S.K. Biswas, and P.K. Hopke. 2009. Indoor air pollution from particulate matter emissions in different households in rural areas of Bangladesh. *Building and Environment* 44 (5): 898–903.

Bhuvaneshwari, S., H. Hettiarachchi, and J.N. Meegoda. 2019. Crop residue burning in India: policy challenges and potential solutions. *International Journal of Environmental Research and Public Health* 16 (5): 832.

Bouarar, I., X. Wang, and G.P. Brasseur. eds. 2017. *Air pollution in Eastern Asia: an integrated perspective*, vol. 16. Basel: Springer.

Cheng, Y., G. Engling, H. Moosmüller, W.P. Arnott, L.W.A. Chen, C.E. Wold, W.M. Hao, and K.B. He. 2016. Light absorption by biomass burning source emissions. *Atmospheric Environment* 127: 347–354.

Chowdhury, Z., L.T. Le, A. Al Masud, K.C. Chang, M. Alauddin, M. Hossain, A.B.M. Zakaria, and P.K. Hopke. 2012. Quantification of indoor air pollution from using cookstoves and estimation of its health effects on adult women in northwest Bangladesh. *Aerosol and Air Quality Research* 12 (4): 463-475.

Hasan, M., A. Salam, and A.S. Alam. 2009. Identification and characterization of trace metals in black solid materials deposited from biomass burning at the cooking stoves in Bangladesh. *Biomass and Bioenergy* 33(10): 1376–1380.

Hecobian, A., X. Zhang, M. Zheng, N. Frank, E.S. Edgerton, and R.J. Weber. 2010. Water-Soluble Organic Aerosol material and the light-absorption characteristics of aqueous extracts measured over the Southeastern United States. *Atmospheric Chemistry and Physics* 10 (13): 5965–5977.

Khalequzzaman, M., M. Kamijima, K. Sakai, N.A. Chowdhury, N. Hamajima, and T. Nakajima. 2007. Indoor air pollution and its impact on children under five years old in Bangladesh. *Indoor Air* 17 (4): 297–304.

Kharol, S.K., K.V.S. Badarinath, A.R. Sharma, D.V. Mahalakshmi, D. Singh, and V.K. Prasad. 2012. Black carbon aerosol variations over Patiala city, Punjab, India—a study during agriculture crop residue burning period using ground measurements and satellite data. *Journal of Atmospheric and Solar-Terrestrial Physics* 84: 45–51.

Kim, E., P.K. Hopke, and E.S. Edgerton. 2003. Source identification of Atlanta aerosol by positive matrix factorization. *Journal of the Air & Waste Management Association* 53 (6): 731–739.

Li, J., C. Liu, Y. Yin, and K.R. Kumar. 2016. Numerical investigation on the Ångström exponent of black carbon aerosol. *Journal of Geophysical Research: Atmospheres* 121 (7): 3506–3518.

Li, J., M. Pósfai, P.V. Hobbs, and P.R. Buseck. 2003. Individual aerosol particles from biomass burning in southern Africa: 2, Compositions and aging of inorganic particles. *Journal of Geophysical Research: Atmospheres* 108 (D13).

Liu, D., J.W. Taylor, D.E. Young, M.J. Flynn, H. Coe, and J.D. Allan. 2015. The effect of complex black carbon microphysics on the determination of the optical properties of brown carbon. *Geophysical Research Letters* 42 (2): 613–619.

Martinsson, J., A.C. Eriksson, I.E. Nielsen, V.B. Malmborg, E. Ahlberg, C. Andersen, R. Lindgren, R. Nystrom, E.Z. Nordin, W.H. Brune, and B. Svenningsson. 2015. Impacts of combustion conditions and photochemical processing on the light absorption of biomass combustion aerosol. *Environmental Science & Technology* 49 (24): 14663–14671.

Pavel, M.R.S., S.U. Zaman, S. Paul, et al. 2023. Light absorption properties of black carbon and brown carbon emitted from biomass combustion at the typical rural cooking stoves in Bangladesh. *Air Quality, Atmosphere & Health.* https://doi.org/10.1007/s11869-023-01302-7.

Rahman, M.M., B.A. Begum, P.K. Hopke, K. Nahar, and G.D. Thurston. 2020. Assessing the $PM_{2.5}$ impact of biomass combustion in megacity Dhaka, Bangladesh. *Environmental Pollution* 264: 114798.

Reddington, C.L., E.W. Butt, D.A. Ridley, P. Artaxo, W.T. Morgan, H. Coe, and D.V. Spracklen. 2015. Air quality and human health improvements from reductions in deforestation-related fire in Brazil. *Nature Geoscience* 8 (10): 768–771.

Runa, F., M. Islam, F. Jeba, and A. Salam. 2021. Light absorption properties of brown carbon from biomass burning emissions. *Environmental Science and Pollution Research* 1–11.

Salam, A., A. Andersson, F. Jeba, M.I. Haque, M.D. Hossain Khan, and O. Gustafsson. 2021. Wintertime air quality in megacity Dhaka, Bangladesh strongly affected by influx of black carbon aerosols from regional biomass burning. *Environmental Science & Technology* 55 (18): 12243–12249.

Salam, A., M. Hasan, B.A. Begum, M. Begum, and S.K. Biswas. 2013. Chemical characterization of biomass burning deposits from cooking stoves in Bangladesh. *Biomass and Bioenergy* 52: 122–130.

Sarkar, M.A.R., and S.M.N. Islam. 1998. Rural energy and its utilization in Bangladesh. *Energy* 23 (9): 785–789.

Satish, R., P. Shamjad, N. Thamban, S. Tripathi, and N. Rastogi. 2017. Temporal characteristics of brown carbon over the central Indo-Gangetic Plain. *Environmental Science & Technology* 51 (12): 6765–6772.

Tang, J., J. Li, Y. Mo, M.S. Khorram, Y. Chen, J. Tang, Y. Zhang, J. Song, and G. Zhang. 2020. Light absorption and emissions inventory of humic-like substances from simulated rainforest biomass burning in Southeast Asia. *Environmental Pollution* 262: 114266.

Uranishi, K., F. Ikemori, H. Shimadera, A. Kondo, and S. Sugata. 2019. Impact of field biomass burning on local pollution and long-range transport of $PM_{2.5}$ in Northeast Asia. *Environmental Pollution* 244: 414–422.

Vadrevu, K., T. Ohara, and C. Justice. 2017. Land cover, land use changes and air pollution in Asia: a synthesis. *Environmental Research Letters* 12 (12): 120201.

Vadrevu, K.P., T. Ohara, and C. Justice. (eds.). 2021a. *Biomass burning in south and Southeast Asia: impacts on the biosphere*, vol. 2. CRC Press.

Vadrevu, K.P., T. Ohara, C. Justice (eds.). 2021b. *Biomass burning in South and Southeast Asia: mapping and monitoring*, vol. 1. CRC Press.

Zaman, S.U., M. Yesmin, M. Pavel, R. Sarkar, F. Jeba, and A. Salam. 2021. Indoor air quality indicators and toxicity potential at the hospitals' environment in Dhaka, Bangladesh. *Environmental Science and Pollution Research* 28 (28): 37727–37740.

Zhang, L., Z. Luo, W. Du, G. Li, G. Shen, H. Cheng, and S. Tao. 2020. Light absorption properties and absorption emission factors for indoor biomass burning. *Environmental Pollution* 267: 115652.

Remote Sensing of Greenhouse Gases and Aerosols from Agricultural Residue Burning Over Pakistan

Salman Tariq, Hasan Nawaz, Zia Ul-Haq, and Usman Mehmood

Abstract Biomass burning is a significant source of particulate matter and greenhouse gas emissions. It has harmful effects on human health and degrades air quality worldwide. The drivers of biomass burning vary, such as clearing the natural forest for agriculture through slash and burn, to remove shrubs, bushes, and weeds, including stubble burning to prepare the land for the next crop and other reasons. Approximately 8700 Tg of dry matter is charred annually. Several studies analyzed the physical aspects of biomass burning. However, quantifying interannual and seasonal variations in aerosols and greenhouse gas emissions is still challenging. Estimates show that ~ 66% of basmati rice, 61% of non-basmati rice, 61% of the wheat crop, 32% of maize, and 78% of the crop area of sugarcane is burnt annually in Pakistan. These numbers are increasing every year because of the need for more government policies. Thus, adding a large amount of reactive gases and particulate matter into the atmosphere substantially disturbs tropospheric oxidation capacity and energy budget.

Keywords Crop residue burning · Aerosols · Greenhouse gases · Remote Sensing

S. Tariq · H. Nawaz · Z. Ul-Haq · U. Mehmood
Remote Sensing, GIS and Climatic Research Lab (National Center of GIS and Space Applications), Centre for Remote Sensing, University of the Punjab, New-Campus, Lahore, Pakistan

S. Tariq (✉)
Department of Space Science, University of the Punjab, Lahore, Pakistan
e-mail: salmantariq_pu@yahoo.com

H. Nawaz
Centre for Atmospheric Chemistry, School of Earth, Atmospheric and Life Sciences, University of Wollongong, Wollongong, Australia

U. Mehmood
Department of Political Science, University of Management and Technology, Lahore, Pakistan

K. P. Vadrevu et al. (eds.), *Vegetation Fires and Pollution in Asia*,
https://doi.org/10.1007/978-3-031-29916-2_18

1 Introduction

Biomass burning events are a seasonal phenomenon in the developing countries of Asia including Pakistan. Important drivers of fires in Asia include slash-and-burn agriculture, clearing of forested lands for oil palm expansion, agricultural residue or waste burning for clearing of land for the next crop, etc. (Prasad et al. 2001a, b, 2002, 2003; Biswas et al. 2015; Justice et al. 2015; Hayasaka et al. 2014; Albar et al. 2018; Tariq and Ali 2015; Tariq and Ul-Haq 2018; Lasko et al. 2017, 2018; Vadrevu et al. 2018; Vadrevu and Lasko 2015). At a regional and local scales, the burning of biomass from these activities can result in the release of large amounts of radiatively active gases, aerosols, and other chemically active species that significantly alter the Earth's radiation balance and atmospheric chemistry (Andreae and Merlet 2001; Kant et al. 2000; Gupta et al. 2001; Vadrevu and Justice 2011; Justice et al. 2015). Fires can result in the loss of biodiversity and economic losses and result in adverse health effects due to the smoke released during the biomass burning. The use of remote sensing data for mapping and monitoring of different sources of biomass burning such as slash-and-burn agriculture and agricultural residue burning, including emissions estimation, has been demonstrated successfully by different researchers in Asia (Badarinath et al. 2007, 2008, 2009; Kant et al. 2000; Kharol et al. 2012; Prasad et al. 2000; Gupta et al. 2001; Hayasaka et al. 2014; 2020; Lasko and Vadrevu 2018; Tariq et al. 2014, 2017; Ul-Haq et al. 2014, 2015a, b, 2017; Vadrevu and Justice 2011; Vadrevu et al. 2013, 2014a, b, 2017, 2018, 2021a, 2021b; Yan et al. 2022). The emissions from the biomass burning comprise carbon dioxide (CO_2), nitrous oxide and methane (CH_4), and chemically reactive gases like carbon monoxide (CO) and nitric oxide (Koppmann et al. 2005; Andreae 2019). These biomass burning emissions affect population and Earth's radiation budget (Alam et al. 2011; Zafar et al. 2018; Bilal et al. 2021a, b).

The anthropogenic biomass burning activities has increased dramatically over the past three decades (Tariq et al. 2015). This gives rise to tropospheric ozone (O_3), photo-oxidants, and atmospheric brown cloud formation. These clouds influence the scattering and absorption of incoming solar radiation and also cause solar dimming and surface cooling (Ramanathan et al. 2005). Biomass burning also revealed a strong association with acid deposition in the tropical areas (United Nations 2018).

The rapid increase in urbanization and industrialization has resolved the problem of human settlement and employment, respectively, but has resulted in environmental, social, and econometric consequences at multiple scales (Waleed and Sajjad 2021). Moreover, in future, the amplified use of biofuels as an alternative energy source will further increase biomass burning events and thus can affect the chemical composition of the atmosphere. Pakistan, being fifth most populous country in South Asia, is susceptible to global warming and climate change (Sajjad 2021). Therefore, there is need to understand the chemical composition and resulting consequences of biomass burning emissions on the atmosphere for better management of air quality and pollution in the region.

This paper reviews the existing body of literature regarding GHGs and particulate matter emissions from biomass burning including recommendations on the mitigation aspects.

2 Study Area

Pakistan lies within the latitudes of 23.5° N–37° N and longitudes of 61° E–77° E in the Northwest of Indo-Pak subcontinent. It covers an area of ~ 796,095 square kilometers and has populace of ~ 212 million. It shares its eastern border with India, western with Afghanistan, southwestern with Iran, and northern with China. It has a long coastline (~ 1046 km) along the Arabian Sea in the South of Pakistan. The climate of Pakistan, in accordance with Köppen and Geiger classification, is semi-arid and subtropical with four main seasons, i.e., winter (December–February), spring (March–May), summer (June–August), and autumn (September–November). Pakistan receives 70% of its rainfall in monsoon season (July–September) and rest 30% during the winter season (Ul-Haq et al. 2015a, b, c, d). Figure 1 portrays the study-area map of Pakistan.

3 Datasets and Methodology

3.1 Moderate Resolution Imaging Spectroradiometer (MODIS)

The MODIS onboard Aqua/Terra satellite provides global daily measurements of optical properties of aerosol particles since 2000/2002. MODIS has 36 spectral channels ranging from 0.41 to 14.5 μm and has three distinct spatial resolutions, i.e., 250, 500, and 1000 m (Levy et al. 2013). The Aqua-MODIS Deep Blue land-only AOD (550 nm) (MYD08_M3, level 2) acquired from NASA Giovanni (https://giovanni.gsfc.nasa.gov/giovanni) and active fires data from NASA FIRMS (https://firms.modaps.eosdis.nasa.gov) during July 2018 to January 2021 have been used in the current study.

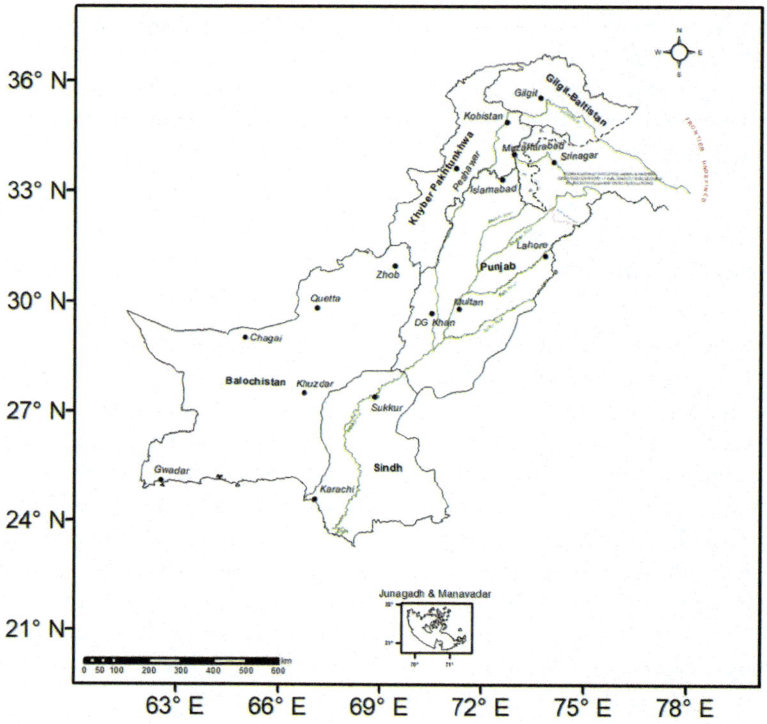

Fig. 1 Study area map of Pakistan

3.2 *Ozone Monitoring Instrument (OMI)*

The OMI abroad NASA Aura polar-orbiting satellite, launched in 2004, measures backscattered solar light using push-broom hyperspectral imaging system at a spectral resolution of 0.5 nm (Zyrichidou et al. 2013). The OMI-retrieved daily averaged tropospheric NO_2 product (30% cloud screened, OMNO2d.003, level 3) is used in this study. Tropospheric NO_2 data is retrieved within the spectral range of 405–465 nm using Differential Optical Absorption Spectroscopy assessment (Wallace and Kanaroglou 2009).

3.3 *Atmospheric InfraRed Sounder (AIRS)*

The AIRS abroad NASA Aqua satellite acquire data in 2378 channels at spectral resolution of ~ 0.5 cm^{-1} (Aumann et al. 2003) and spatial resolution of 13.5 km (Xiong et al. 2009). The AIRS tropospheric CO is retrieved within the wavelength range of 4.58–4.50 μm. Several studies validated the spatiotemporal distribution of AIRS-retrieved CO_2 data a 1% precision level (Cheng et al. 2012; Jiang et al.

2016). The monthly CO_2 column product (AIRX3C2M_005) and CO volume mixing ratio (AIRX3STM_7_0) at 925 hpa, total O_3 column (AIRS3STM v7.0), and CH_4 (AIRX3STM v7.0) at 925 hpa are used in this study.

3.4 *Modern-Era Retrospective Analysis for Research and Applications (MERRA-2)*

The MERRA-2 is global modeling GMAO product provides gridded and homogeneous datasets of the atmosphere on a regular basis since 1980. The MERRA-2 model has enhanced the presentation of cryosphere and stratosphere processes, hydrological cycle, atmospheric ozone, and joint integration of aerosols and meteorology (Gelaro et al. 2017). The model provides data at a spatial resolution of $0.5° \times 0.625°$. The monthly mean SO_2 column mass density product (M2TMNXAER_5_12_4) of MERRA-2 model from 2002 to 2021 is used in this study.

4 Results and Discussions

4.1 *Particulate's Emissions*

4.1.1 AOD and AE

Rapid population growth, urban sprawl, and industrial expansion have led to a substantial increase in particulate pollution over Pakistan. Aerosols affect energy budget, air quality, visibility, precipitation, and the hydrological cycle (Charlson et al. 1992). Ali et al. (2014) studied aerosol optical properties using AERONET data over Lahore and observed highest AOD in summer season associated with dust storm events. Alam et al. (2014) observed AOD ranging from 0.17 to 2.46 from AERONET and 0.15 to 2.45 from MODIS over Lahore. Tariq et al. (2015) studied physical and optical properties of aerosols during agricultural residue burning with instantaneous high AOD of 2.75 on October 20, 2010, indicating heavy aerosol loading over Lahore. Tariq et al. (2021) reported positive correlation between AOD and AE over northeastern Pakistan indicating presence of fine urban, industrial, and agricultural residue burning aerosol particles in the atmosphere. Bilal et al. (2021a, b) found high concentration of particulate matter ($PM_{2.5}$) with mean annual concentration of 54.7 $\mu g/m^3$ over Pakistan. Tariq et al. (2022) found higher AOD values over eastern Pakistan as compared to western Pakistan because of high population density and biomass combustion. The MODIS-retrieved active fire points; spatiotemporal distribution of AOD and AE during autumn season in Pakistan is shown in Fig. 2a–c. Figure 2a–c reveals high AOD, AE, and active fire points, respectively, in the Northeastern Pakistan during agricultural residue burning season. High AE (0.98–1.22) and low

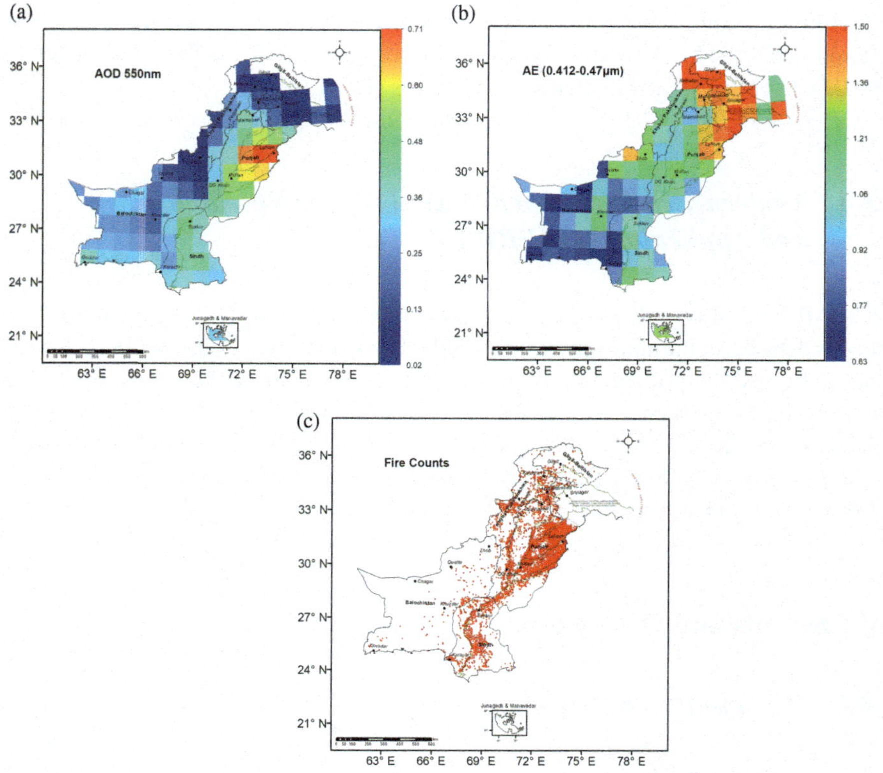

Fig. 2 Spatiotemporal distribution map of MODIS-retrieved **a** AOD, **b** AE, and **c** Active fire counts during autumn season over Pakistan from 2002 to 2021

AOD (0.25–0.36) during autumn and winter months show dominance of fine aerosol particles in the atmosphere as shown in Fig. 8.

4.2 Gaseous Emissions

4.2.1 Carbon Monoxide (CO) and Carbon Dioxide (CO_2)

Biomass burning largely contributes to the emissions of primary fine carbonaceous particles and trace gases in the global atmosphere. The CO emissions resulting from biomass combustion using satellite datasets and relating to active fires have been reported in the past (Duncan et al. 2003; Van Der Werf et al. 2004). The atmospheric assessment of CO and CO_2 concentrations over Pakistan during crop waste burning is important to improve our understandings about climate change and its impacts. Ul-Haq et al. (2017) analyzed spatiotemporal distribution of satellite-retrieved CO_2 emissions and found 147% increase in anthropogenic CO_2 emissions over Pakistan

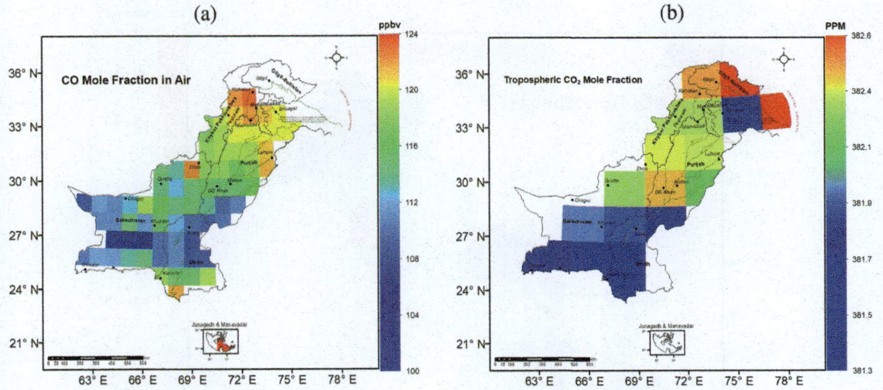

Fig. 3 Spatiotemporal distribution map of AIRS-retrieved **a** CO during 2002 to 2021 and **b** CO_2 during 2002–2012 over Pakistan during autumn season

and adjacent areas. Ul-Haq et al. (2015a) studied variation in the tropospheric CO emissions over Pakistan using remotely sensed data and found maximum CO emissions in spring season while minimum in autumn season. In another study, Lei et al. (2021) quantified CO_2 emissions over Lahore using OCO-2 soundings data and estimated ~ 6.7% increase in posterior CO_2 emissions. Figure 3a, b shows that Northern and Northeastern Pakistan are characterized by high values of CO in the ranges of 116–124 ppbv while Southern and Southwestern regions exhibit low values of CO except Karachi during autumn season. On the other hand, high concentration of CO_2 is observed in the Northern regions associated with high precipitation. The Southern Pakistan shows significantly lower values of CO_2 due to less agricultural residue burning activities, low population density, high humidity, and air circulation. Figure 8 shows the highest monthly mean concentration of CO_2 (385 PPM) during May while maximum value of CO is found to be 141 ppbv during the month of February.

4.2.2 Nitrogen Dioxide (NO_2)

Apart from CO and CO_2, biomass burning is a significant source of chemically reactive nitrogenous compounds (Delmas et al. 1995). NO_2 has adverse effects on human health as well as plants, and it also contributes to acid rain, $PM_{2.5}$, and tropospheric O_3 formation. Ul-Haq et al. (2015d) studied trends in tropospheric NO_2 over South Asia and reported ~ 42% increase in NO_2 concentration over northwestern Indo-Gangtic Basin associated with large-scale agricultural waste burning events in the autumn season. Ul-Haq et al. (2014) also found increasing trends in NO_2 concentration (3.29% per annum) with the mean annual value of $1.102 \pm 0.081 \times 10^{15}$ molecules per cm^2. Ahmad and Aziz (2013) investigated variation in ground-level concentration of NO_2 in Islamabad and Rawalpindi and found high values of NO_2

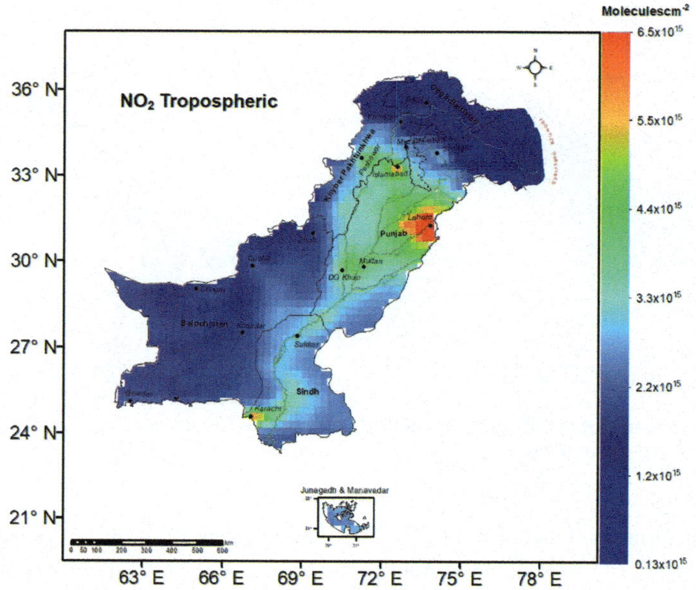

Fig. 4 Spatial and temporal distribution map of OMI-retrieved tropospheric NO_2 column during 2002 to 2021 over Pakistan during autumn season

in the areas of heavy traffic flow with an average value of 44 ± 6 molecules per cm^2. Figure 4 shows that the maximum NO_2 concentration of 6.5×10^{15} molecules per cm^2 is observed over Lahore. Islamabad, Multan, Faisalabad, Peshawar, and Karachi are characterized by NO_2 concentration of ~ 4.4×10^{15} molecules per cm^2. Figure 8 depicts the maximum NO_2 concentration of 1.396×10^{15} molecules per cm^2 in December followed 1.375×10^{15} molecules per cm^2 in November.

4.2.3 Methane (CH_4)

CH_4 is the most abundant, moderately reactive, and 2nd most important GHG present in the atmosphere (Anastasia and Moiseenko, 2013). It has harmful effects on human health, stratospheric ozone, and water vapor budget and plays a significant role in oxidizing capacity and altering chemistry of the biomass burning plumes. The main sources of CH_4 in Pakistan include enteric fermentation (63%), gas and oil sectors (184%), manure lagoons (66%), waste-water processing (70%), and municipal solid waste (149%). Ul-Haq et al. (2015c) examined spatiotemporal variations in CH_4 total column over Pakistan and its adjoining regions using satellite datasets. They found an increase in mean annual methane total column by 3.7% with average value of 1787

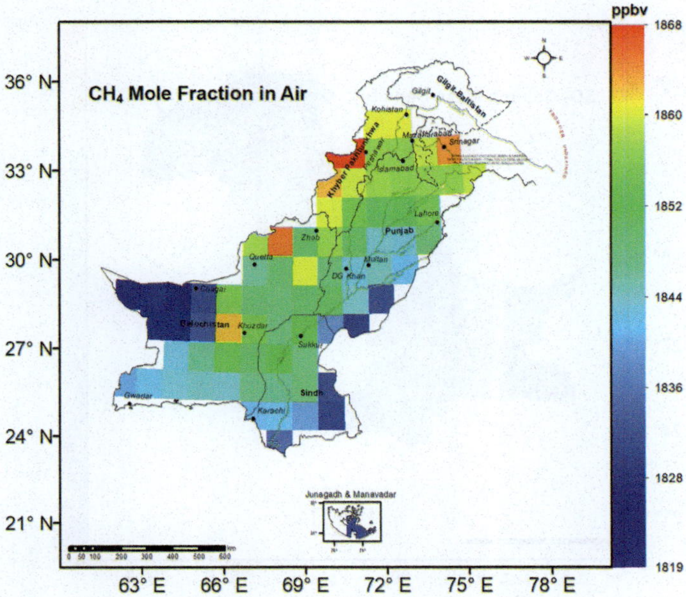

Fig. 5 Spatial and temporal distribution map of AIRS-retrieved CH_4 during 2002–2016 over Pakistan during autumn season

$\pm$ 22 ppb during January 2003 to April 2012. Mahmood et al. (2016a, b) studied CH_4 variability over Pakistan using AIRS data and found considerable increase in CH_4 concentration. Figure 5 shows spatiotemporal distribution of CH_4 over Pakistan during autumn season. High values of CH_4 are observed in Khyber Pakhtunkhwa, Zhob, Khuzdar, Islamabad, Rawalpindi, Lahore, Muzaffarabad, and Srinagar in the range of 1852–1868 ppbv. A maximum of CH_4 concentration is found to be 1837 ppbv in July while lowest concentration of 1785 ppbv is observed in January as shown in Fig. 8.

4.2.4 Sulfur Dioxide (SO_2)

SO_2 is regarded as a criteria pollutant present within the troposphere because of both natural (sea spray, biogenic and volcanic emissions) and anthropogenic activities (power generation from coal, oil and gas sectors, metals smelting, crop waste burning and wild fires) (Clarisse et al. 2012). Khattak et al. (2014) examined temporal trends in SO_2 column densities over Pakistan after Nabro volcanic eruption using satellite observations. They revealed 8.7% annual increase in SO_2 over Pakistan during 2004–2012. Mahmood et al. (2016a, b) examined SO_2 plume and concentration after Mount Nabro volcanic eruption and found significant effects of SO_2 plume on the atmospheric perturbations and air quality of Pakistan. Ul-Haq et al. (2016) observed

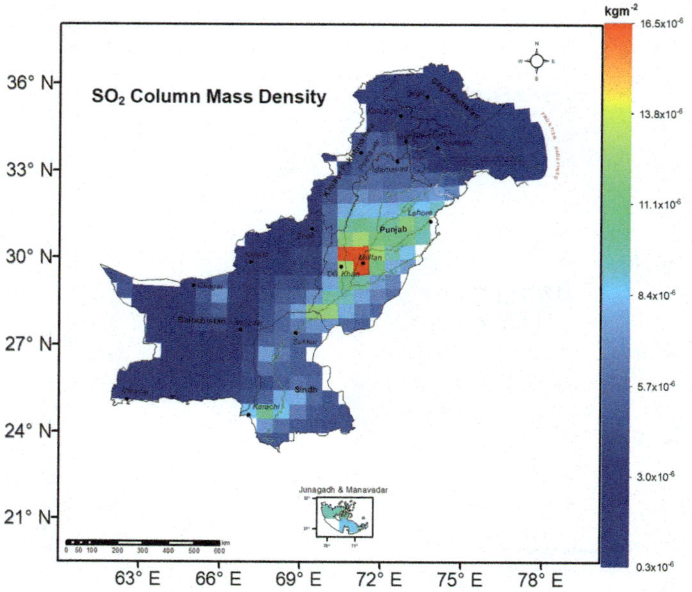

Fig. 6 Spatial and temporal distribution map of SO_2 from MERRA-2 model during 2002–2021 over Pakistan during autumn season

2.4% annual increase in SO_2 column over Pakistan associated with local meteorological patterns of human-induced emissions, agricultural waste burning, and vegetative cover. Jabeen and Khokhar (2019) observed 78% of temporal increase in SO_2 levels from anthropogenic activities across Pakistan due to industrialization and increased energy demands in the country. Figure 6 shows high SO_2 column mass densities in the ranges of 11–16 μkgm^{-2} over Lahore, Faisalabad, Multan, and Dera Ismail Khan. Figure 8 shows the maximum SO_2 value of 4.40×10^{-6} kgm^{-2} is observed in November followed 4.27×10^{-6} kgm^{-2} in December.

4.2.5 Ozone (O_3)

Tropospheric O_3 is an active GHG present in the atmosphere with 23 days of life span (Young et al. 2013). It shields surface of the Earth from harmful ultraviolet radiations in the stratosphere. While it acts as a pollutant when present in the troposphere and causes warming of the atmosphere. The presence of O_3 in the troposphere along with aerosols lowers crop yield (Burney and Ramanathan 2014) up to 15% wheat production and 4% rice production (Avnery et al. 2011). Ahmad and Aziz (2013) examined variability of ground-level ozone in twin cities of Pakistan. They found average O_3 concentration of 18.2 ± 1.24 ppb in twin cities of Pakistan. Moreover, they also revealed high concentration of O_3 in Rawalpindi as compared to Islamabad. Noreen et al. (2018) studied OMI-retrieved tropospheric O_3 and found increase in

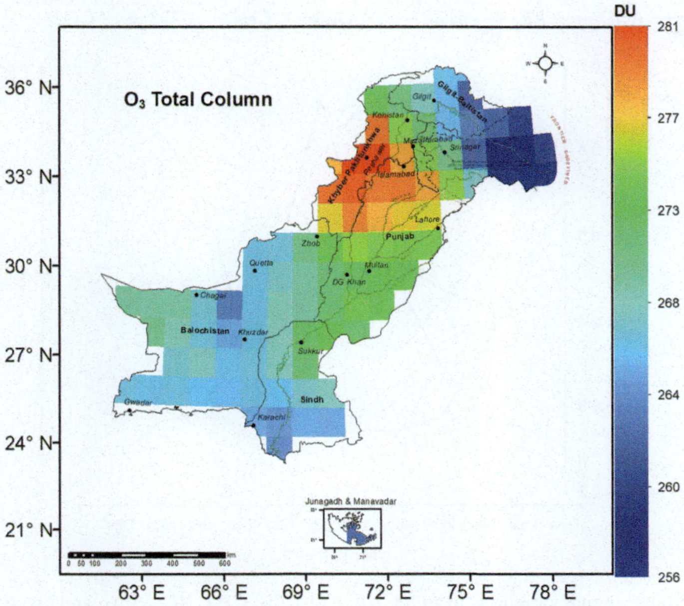

Fig. 7 Spatial and temporal distribution map of AIRS-retrieved total O_3 column during 2002 to 2016 over Pakistan during autumn season

O_3 concentration by 3.2 ± 1.1 DU over Pakistan due to urban sprawl, increase in population, and agricultural fire activities. Rafiq et al. (2017) found highest and lowest O_3 concentrations in summer and winter, respectively, while mixed trends in autumn and spring seasons. Ul-Haq et al. (2015b) examined annual spatiotemporal distribution of total O_3 column over Pakistan and its neighboring regions. They found 1.3% decadal rise in O_3 column with and average value of 278 ± 2 DU. They also observed elevated O_3 column in the regions where there is high concentration of CO and NO_2 is present. Figure 7 shows high total O_3 column in central Pakistan in the ranges of 273–281 DU, and 264–273 DU in the southern Pakistan and in Gilgit and Srinagar during autumn season. In this study, monthly mean total O_3 column reveals highest value of 297 PPM in May while the lowest value of 263 PPM is found in November as shown in Fig. 8.

5 Summary and Future Recommendations

During the last three decades, biomass burning became a major source of trace gases in the Earth's atmosphere. Insight of atmospheric dynamics and chemistry as well as their interdependencies require sufficient understanding of emissions from biomass burning and their spatiotemporal distribution. Moreover, the in-depth knowledge of composition and quantification trace gases emitted from fires helps us to combat global warming and climate change. The sources and sinks of trace gases as well

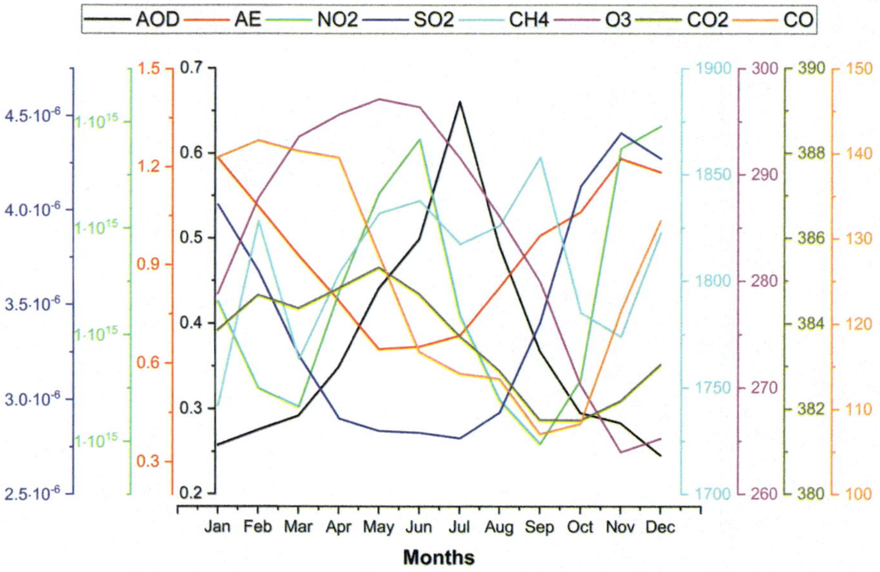

Fig. 8 The inter-annual variations in AOD, AE, CO, CO_2, CH_4, NO_2, SO_2, and O_3 over Pakistan

as aerosols are largely influenced by large-scale open fires particularly in the tropical regions where biomass burning emissions are increasing. Hence, characterizing biomass burning events integrating ground-based measurements, satellite data, and advanced algorithms is needed to evaluate the impact of biomass burning emissions on both the local and regional climate in Asia and elsewhere.

Acknowledgements The authors acknowledged NASA Giovanni for providing MODIS, OMI, AIRS, and MERRA-2 datasets.

References

Ahmad, S.S., and N. Aziz. 2013. Spatial and temporal analysis of ground level ozone and nitrogen dioxide concentration across the twin cities of Pakistan. *Environmental Monitoring and Assessment* 185 (4): 3133–3147. https://doi.org/10.1007/S10661-012-2778-7.

Alam, K., S. Qureshi, and T. Blaschke. 2011. Monitoring spatio-temporal aerosol patterns over Pakistan based on MODIS, TOMS and MISR satellite data and a HYSPLIT model. *Atmospheric Environment* 45 (27): 4641–4651. https://doi.org/10.1016/j.atmosenv.2011.05.055.

Alam, K., N. us Sahar, and Y. Iqbal. 2014. Aerosol characteristics and radiative forcing during pre-monsoon and post-monsoon seasons in an urban environment. *Aerosol and Air Quality Research* 14 (1): 99–107.https://doi.org/10.4209/AAQR.2013.05.0154.

Albar, I., I. Jaya, B.H. Saharjo, B. Kuncahyo, and K.P. Vadrevu. 2018. Spatio-temporal analysis of land and forest fires in Indonesia using MODIS active fire dataset. In *Land-atmospheric research applications in South and Southeast Asia* 105–127. Cham: Springer.

Ali, M., et al. 2014. A Study of aerosol properties over Lahore (Pakistan) by using AERONET data. *Journal of Atmospheric Science* 50 (2): 153–162. https://doi.org/10.1007/s13143-014-0004-y.

Anastasia, V, and K. Moiseenko. 2013. Methane emissions from 2000 to 2011 wildfires in Northeast Eurasia estimated with MODIS burned area data. *Atmospheric Environment* 71: 115–121. https://doi.org/10.1016/j.atmosenv.2013.02.001.

Andreae, M.O. 2019. Emission of trace gases and aerosols from biomass burning—an updated assessment. *Atmospheric Chemistry and Physics* 19 (13): 8523–8546. https://doi.org/10.5194/ACP-19-8523-2019.

Andreae, M.O., and P. Merlet. 2001. Emission of trace gases and aerosols from biomass burning. *Global Biogeochemical Cycles* 15 (4): 955–966.

Aumann, H.H. et al. 2003. AIRS/AMSU/HSB on the aqua mission: design, science objectives, data products, and processing systems. *IEEE Transactions on Geoscience and Remote Sensing* 41 (2 PART 1): 253–263. https://doi.org/10.1109/TGRS.2002.808356.

Avnery, S., et al. 2011. Global crop yield reductions due to surface ozone exposure: 1. Year 2000 crop production losses and economic damage. *Atmospheric Environment* 45 (13): 2284–2296. https://doi.org/10.1016/J.ATMOSENV.2010.11.045.

Badarinath, K.V.S., S.K. Kharol, K.M. Latha, T.K. Chand, V.K. Prasad, A.N. Jyothsna, and K. Samatha. 2007. Multiyear ground-based and satellite observations of aerosol properties over a tropical urban area in India. *Atmospheric Science Letters* 8 (1): 7–13.

Badarinath, K.V.S., S.K. Kharol, A.R. Sharma, and V.K. Prasad. 2009. Analysis of aerosol and carbon monoxide characteristics over Arabian Sea during crop residue burning period in the Indo-Gangetic Plains using multi-satellite remote sensing datasets. *Journal of Atmospheric and Solar-Terrestrial Physics* 71 (12): 1267–1276.

Badarinath, K.V.S., S. Kumar Kharol, V. Krishna Prasad, A. Rani Sharma, E.U.B. Reddi, H.D. Kambezidis, and D.G. Kaskaoutis. 2008. Influence of natural and anthropogenic activities on UV index variations—a study over tropical urban region using ground based observations and satellite data. *Journal of Atmospheric Chemistry* 59 (3): 219–236.

Bilal, M.,A. Mhawish, J.E. Nichol, Z. Qiu, M. Nazeer, M.A., Ali, et al. 2021a. Air pollution scenario over Pakistan: Characterization and ranking of extremely polluted cities using long-term concentrations of aerosols and trace gases. *Remote Sensing of Environment* 264 (July): 112617. https://doi.org/10.1016/j.rse.2021.112617.

Bilal, M.,A. Mhawish, J.E. Nichol, Z. Qiu, M. Nazeer, Ali, M.A. Ali, et al. 2021b. Air pollution scenario over Pakistan: characterization and ranking of extremely polluted cities using long-term concentrations of aerosols and trace gases. *Remote Sensing of Environment* 264: 112617. https://doi.org/10.1016/j.rse.2021.112617.

Biswas, S., K.D. Lasko, and K.P. Vadrevu. 2015. Fire disturbance in tropical forests of Myanmar—analysis using MODIS satellite datasets. *IEEE Journal of Selected Topics in Applied Earth Observations and Remote Sensing* 8 (5): 2273–2281.

Burney, J., and V. Ramanathan. 2014. Recent climate and air pollution impacts on Indian agriculture. *Proceedings of the National Academy of Sciences of the United States of America* 111 (46): 16319–16324. https://doi.org/10.1073/PNAS.1317275111/-/DCSUPPLEMENTAL.

Charlson, R.J., et al. 1992. Climate forcing by anthropogenic aerosols. *Science* 255 (5043): 423–430. https://doi.org/10.1126/SCIENCE.255.5043.423.

Cheng, M.M., H. Jiang, and Z. Guo. 2012. Evaluation of long-term tropospheric NO_2 columns and the effect of different ecosystem in Yangtze River Delta. *Procedia Environmental Sciences* 13: 1045–1056. https://doi.org/10.1016/j.proenv.2012.01.098.

Clarisse, L., et al. 2012. Retrieval of sulphur dioxide from the infrared atmospheric sounding interferometer (IASI). *Atmospheric Measurement Techniques* 5 (3): 581–594. https://doi.org/10.5194/AMT-5-581-2012.

Delmas, R. et al. 1995. Nitrogen compound emission from biomass burning in tropical African savanna FOS/DECAFE 1991 experiment (Lamto, Ivory Coast). *Journal of Atmospheric Chemistry* 22 (1): 175–193. https://doi.org/10.1007/BF00708188.

Duncan, B.N. et al. 2003. Interannual and seasonal variability of biomass burning emissions constrained by satellite observations. *Journal of Geophysical Research: Atmospheres* 108 (2). https://doi.org/10.1029/2002JD002378.

Gelaro, R., et al. 2017. The modern-era retrospective analysis for research and applications, version 2 (MERRA-2). *Journal of Climate* 30 (14): 5419–5454. https://doi.org/10.1175/JCLI-D-16-0758.1.

Gupta, P.K., V.K. Prasad, C. Sharma, A.K. Sarkar, Y. Kant, K.V.S. Badarinath, and A.P. Mitra. 2001. CH_4 emissions from biomass burning of shifting cultivation areas of tropical deciduous forests–experimental results from ground-based measurements. *Chemosphere-Global Change Science* 3 (2): 133–143.

Hayasaka, H., I. Noguchi, E.I. Putra, N. Yulianti, and K. Vadrevu. 2014. Peat-fire-related air pollution in Central Kalimantan, Indonesia. *Environmental Pollution* 195: 257–266.

Hayasaka, H., A. Usup, and D. Naito. 2020. New approach evaluating peatland fires in Indonesian factors. *Remote Sensing* 12 (12): 2055.

Jabeen, Z., and M.F. Khokhar. 2019. Extended database of SO_2 column densities over Pakistan by exploiting satellite observations. *Atmospheric Pollution Research* 10 (3): 997–1003. https://doi.org/10.1016/J.APR.2019.01.009.

Jiang, X., et al. 2016. CO_2 annual and semiannual cycles from multiple satellite retrievals and models. *Earth and Space Science* 3 (2): 78–87. https://doi.org/10.1002/2014EA000045.

Justice, C., G. Gutman, and K.P. Vadrevu. 2015. NASA land cover and land use change (LCLUC): an interdisciplinary research program. *Journal of Environmental Management* 148: 4–9.

Kant, Y., V.K. Prasad, and K.V.S. Badarinath. 2000. Algorithm for detection of active fire zones using NOAA AVHRR data. *Infrared Physics & Technology* 41 (1): 29–34.

Kharol, S.K., K.V.S. Badarinath, A.R. Sharma, D.V. Mahalakshmi, D. Singh, and V.K. Prasad. 2012. Black carbon aerosol variations over Patiala city, Punjab, India—a study during agriculture crop residue burning period using ground measurements and satellite data. *Journal of Atmospheric and Solar-Terrestrial Physics* 84: 45–51.

Khattak, P., M.F. Khokhar, and N. Yasmin. 2014. Spatio-temporal analyses of atmospheric sulfur dioxide column densities over Pakistan by using SCIAMACHY data. *Aerosol and Air Quality Research* 14 (7): 1883–1896. https://doi.org/10.4209/AAQR.2013.12.0357.

Koppmann, R., K. Von. Czapiewski, and J.S. Reid. 2005. A review of biomass burning emissions, part I: gaseous emissions of carbon monoxide, methane, volatile organic compounds, and nitrogen containing compounds. *Atmospheric Chemistry and Physics Discussions* 5 (5): 10455–10516. https://doi.org/10.5194/acpd-5-10455-2005.

Lasko, K., and K. Vadrevu. 2018. Improved rice residue burning emissions estimates: accounting for practice-specific emission factors in air pollution assessments of Vietnam. *Environmental Pollution* 236: 795–806.

Lasko, K., K.P. Vadrevu, and T.T.N. Nguyen. 2018. Analysis of air pollution over Hanoi, Vietnam using multi-satellite and MERRA reanalysis datasets. *PLoS ONE* 13 (5): e0196629.

Lasko, K., K.P. Vadrevu, V.T. Tran, E. Ellicott, T.T. Nguyen, H.Q. Bui, and C. Justice. 2017. Satellites may underestimate rice residue and associated burning emissions in Vietnam. *Environmental Research Letters* 12 (8): 085006.

Lei, R., et al. 2021. Fossil fuel CO_2 emissions over metropolitan areas from space: a multi-model analysis of OCO-2 data over Lahore Pakistan. *Remote Sensing of Environment* 264: 112625. https://doi.org/10.1016/J.RSE.2021.112625.

Levy, R.C., et al. 2013. The Collection 6 MODIS aerosol products over land and ocean. *Atmospheric Measurement Techniques* 6 (11): 2989–3034. https://doi.org/10.5194/AMT-6-2989-2013.

Mahmood, I., M.I. Shazad, M.F. Iqbal, K. Ullah, A. Waqas, and A.A. Kidwai. 2016a. Satellite based detection of volcanic SO_2 over Pakistan. *Global NEST Journal* 18 (3): 591–598. Available at: www.sacs.aeronomie.be. Accessed 17 Feb 2022.

Mahmood, I., et al. 2016b. Spatiotemporal Monitoring of CO_2 and CH_4 over Pakistan using atmospheric infrared sounder (AIRS). *International Letters of Natural Sciences* 58: 35–41. https://doi.org/10.18052/WWW.SCIPRESS.COM/ILNS.58.35.

Noreen, A., et al. 2018. Spatio-temporal assessment and seasonal variation of tropospheric ozone in Pakistan during the last decade. *Environmental Science and Pollution Research* 25 (9): 8441–8454. https://doi.org/10.1007/S11356-017-1010-2/TABLES/2.

Prasad, V.K., P.K. Gupta, C. Sharma, A.K. Sarkar, Y. Kant, K.V.S. Badarinath, T. Rajagopal, and A.P. Mitra. 2000. NOx emissions from biomass burning of shifting cultivation areas from tropical deciduous forests of India–estimates from ground-based measurements. *Atmospheric Environment* 34 (20): 3271–3280.

Prasad, V.K., Y. Kant, and K.V.S. Badarinath. 2001a. CENTURY ecosystem model application for quantifying vegetation dynamics in shifting cultivation areas: a case study from Rampa Forests, Eastern Ghats (India). *Ecological Research* 16 (3): 497–507.

Prasad, V.K., Y. Kant, P.K. Gupta, C. Sharma, A.A. Mitra, and K.V.S. Badarinath. 2001b. Biomass and combustion characteristics of secondary mixed deciduous forests in Eastern Ghats of India. *Atmospheric Environment* 35 (18): 3085–3095.

Prasad, V.K., Y. Kant, P.K. Gupta, C. Elvidge, and K.V.S. Badarinath. 2002. Biomass burning and related trace gas emissions from tropical dry deciduous forests of India: a study using DMSP-OLS data and ground-based measurements. *International Journal of Remote Sensing* 23 (14): 2837–2851.

Prasad, V.K., M. Lata, and K.V.S. Badarinath. 2003. Trace gas emissions from biomass burning from northeast region in India—estimates from satellite remote sensing data and GIS. *The Environmentalist* 23 (3): 229–236.

Rafiq, L., S. Tajbar, and S. Manzoor. 2017. Long term temporal trends and spatial distribution of total ozone over Pakistan. *Egyptian Journal of Remote Sensing and Space Science* 20 (2): 295–301. https://doi.org/10.1016/J.EJRS.2017.05.002.

Ramanathan, V., et al. 2005. Atmospheric brown clouds: impacts on South Asian climate and hydrological cycle. *Proceedings of the National Academy of Sciences of the United States of America* 102 (15): 5326–5333. https://doi.org/10.1073/pnas.0500656102.

Sajjad, M. 2021. Disaster resilience in Pakistan: a comprehensive multi-dimensional spatial profiling. *Applied Geography* 126: 102367. https://doi.org/10.1016/J.APGEOG.2020.102367.

Tariq, S., and M. Ali. 2015. Analysis of optical and physical properties of aerosols during crop residue burning event of October 2010 over Lahore Pakistan. *Atmospheric Pollution Research* 6 (6): 969–978.

Tariq, S., et al. 2021. Investigating the relationship of aerosols with enhanced vegetation index and meteorological parameters over Pakistan. *Atmospheric Pollution Research* 12 (6): 101080. https://doi.org/10.1016/j.apr.2021.101080.

Tariq, S. et al. 2022. Long-term spatiotemporal trends in aerosol optical depth and its relationship with enhanced vegetation index and meteorological parameters over South Asia. *Environmental Science and Pollution Research* 0123456789. https://doi.org/10.1007/s11356-021-17887-4.

Tariq, S., M. Ali, K. Mahmood, S.A. Batool, and A.D. Rana. 2014. A study of tropospheric NO_2 variability over Pakistan using OMI data. *Atmospheric Pollution Research* 5 (4): 709–720.

Tariq, S., Z. Ul-Haq, and M. Ali. 2015. Analysis of optical and physical properties of aerosols during crop residue burning event of October 2010 over Lahore, Pakistan. *Atmospheric Pollution Research* 6 (6): 969–978. https://doi.org/10.1016/J.APR.2015.05.002.

Tariq, S., Z. Ul-Haq, A. Imran, U. Mehmood, M.U. Aslam, and K. Mahmood. 2017. CO_2 emissions from Pakistan and India and their relationship with economic variables. *Applied Ecology and Environmental Research* 15 (4): 1301–1312.

Tariq, S. and Z. Ul-Haq. 2018. Satellite remote sensing of aerosols and gaseous pollution over Pakistan. In *Land-atmospheric research applications in South and Southeast Asia* 523–549. Cham: Springer.

Ul-Haq, Z. et al. 2014. A study of tropospheric NO_2 variability over Pakistan using OMI data. *Atmospheric Pollution Research* 5 (4): 709–720. https://doi.org/10.5094/APR.2014.080.

Ul-Haq, Z. et al. 2015a. Carbon monoxide (CO) emissions and its tropospheric variability over Pakistan using satellite-sensed data. *Advances in Space Research* 56 (4): 583–595.https://doi.org/10.1016/j.asr.2015.04.026.

Ul-Haq, Z. et al. 2015b. Satellite remote sensing of total ozone column (TOC) over Pakistan and neighbouring regions. 36 (4): 1038–1054. https://doi.org/10.1080/01431161.2015.1007255.

Ul-Haq, Z., S. Tariq, and M. Ali. 2015c. Atmospheric variability of methane over Pakistan, Afghanistan and adjoining areas using retrievals from SCIAMACHY/ENVISAT. *Journal of Atmospheric and Solar-Terrestrial Physics* 135: 161–173. https://doi.org/10.1016/J.JASTP.2015.11.002.

Ul-Haq, Z., S. Tariq, and M. Ali. 2015d. Tropospheric NO_2 trends over south Asia during the last decade (2004–2014) using OMI Data. *Advances in Meteorology* 2015 (2). https://doi.org/10.1155/2015/959284.

Ul-Haq, Z. et al. 2016. Sulphur dioxide loadings over megacity Lahore (Pakistan) and adjoining region of Indo-Gangetic Basin. 37 (13): 3021–3041. https://doi.org/10.1080/01431161.2016.1192701.

Ul-Haq, Z., S. Tariq, and M. Ali. 2017. Spatiotemporal assessment of CO_2 emissions and its satellite remote sensing over Pakistan and neighboring regions. *Journal of Atmospheric and Solar-Terrestrial Physics* 152–153: 11–19.https://doi.org/10.1016/J.JASTP.2016.11.001.

United Nation. 2018. Remote sensing for spatio-temporal mapping of smog in Punjab and identification of the underlying causes using gis techniques (r- smog).

Vadrevu, K., and K. Lasko. 2015. Fire regimes and potential bioenergy loss from agricultural lands in the Indo-Gangetic Plains. *Journal of Environmental Management* 148: 10–20.

Vadrevu, K., T. Ohara, and C. Justice. 2017. Land cover, land use changes and air pollution in Asia: a synthesis. *Environmental Research Letters* 12 (12): 120201.

Vadrevu, K.P., and C.O. Justice. 2011. Vegetation fires in the Asian region: satellite observational needs and priorities. *Global Environmental Research* 15 (1): 65–76.

Vadrevu, K.P., L. Giglio, and C. Justice. 2013. Satellite based analysis of fire–carbon monoxide relationships from forest and agricultural residue burning (2003–2011). *Atmospheric Environment* 64: 179–191.

Vadrevu, K.P., K. Lasko, L. Giglio, and C. Justice. 2014a. Analysis of Southeast Asian pollution episode during June 2013 using satellite remote sensing datasets. *Environmental Pollution* 195: 245–256.

Vadrevu, K.P., T. Ohara, and C. Justice. 2014b. Air pollution in Asia. *Environmental Pollution (Barking, Essex: 1987)* 195: 233–235.

Vadrevu, K.P., T. Ohara, and C. Justice (eds.). 2018. Land-atmospheric research applications in South and Southeast Asia.

Vadrevu, K.P., T. Ohara, and C. Justice (eds.). 2021a. *Biomass burning in South and Southeast Asia: impacts on the biosphere*, vol. 2. CRC Press.

Vadrevu, K.P., T. Ohara, and C. Justice (eds.). 2021b. *Biomass burning in South and Southeast Asia: mapping and monitoring*, vol. 1. CRC Press.

Van Der Werf, G.R., et al. 2004. Continental-scale partitioning of fire emissions during the 1997 to 2001 El Niño/La Niña period. *Science* 303 (5654): 73–76. https://doi.org/10.1126/SCIENCE.1090753/SUPPL_FILE/VAN_DER_WERF_SOM.PDF.

Waleed, M., and M. Sajjad. 2021. Leveraging cloud-based computing and spatial modeling approaches for land surface temperature disparities in response to land cover change: evidence from Pakistan. *Remote Sensing Applications: Society and Environment* 25: 100665. https://doi.org/10.1016/j.rsase.2021.100665.

Wallace, J., and P. Kanaroglou. 2009. The sensitivity of OMI-derived nitrogen dioxide to boundary layer temperature inversions. *Atmospheric Environment* 43 (22–23): 3596–3604. https://doi.org/10.1016/J.ATMOSENV.2009.03.049.

Xiong, X., et al. 2009. Methane plume over south Asia during the monsoon season: satellite observation and model simulation. *Atmospheric Chemistry and Physics* 9 (3): 783–794. https://doi.org/10.5194/ACP-9-783-2009.

Yan, J., H. He, L. Wang, H. Zhang, D. Liang, and J. Zhang. 2022. Inter-comparison of four models for detecting forest fire disturbance from MOD13A2 time series. *Remote Sensing* 14 (6): 1446.

Young, P.J., et al. 2013. Pre-industrial to end 21st century projections of tropospheric ozone from the atmospheric chemistry and climate model intercomparison project (ACCMIP). *Atmospheric Chemistry and Physics* 13 (4): 2063–2090. https://doi.org/10.5194/ACP-13-2063-2013.

Zafar, Q., S. Zafar, and B. Holben. 2018. Seasonal assessment and classification of aerosols transported to Lahore using AERONET and MODIS deep blue retrievals. *International Journal of Climatology* 38 (2): 1022–1040. https://doi.org/10.1002/JOC.5230.

Zyrichidou, I., et al. 2013. Evaluation of high resolution simulated and OMI retrieved tropospheric NO_2 column densities over Southeastern Europe. *Atmospheric Research* 122: 55–66. https://doi.org/10.1016/J.ATMOSRES.2012.10.028.

A Comparative Study of Energy, Emissions, and Economic Efficiency of Various Cookstoves in Nepal

Narayan P. Adhikari, Prajwal R. Shakya, Shubha Laxmi Shrestha, and Suyesh Prajapati

Abstract The study has been conducted to compare nine different categories of stoves using various available fuels in Nepal. All the stoves were tested at Renewable Energy Test Station (RETS) following ISO-IWA recommended Water Boiling Test (WBT) version 4.2.3 along with the $PM_{2.5}$ and CO emission measurement using Laboratory Emission Measurement System (LEMS) developed by Aprovecho Research Centre, USA. The high-power thermal efficiencies of tested stoves are found as induction stoves (90.63%), infrared stoves (75.58%), heating coils (48.26%), LPG (57.09%), kerosene stoves (46.56%), biogas (41.24%), pellet (42.15), biomass rocket stove (28.14%), and chimney stoves (24.75%). The maximum $PM_{2.5}$ emission values are found for biomass fuel wood-burning stoves ranging from 289.144 to 458.068 mg/MJd, and no emission for LPG and biogas stoves. The CO emission ranged from 5.81 to 6.391 g/MJd for biomass burning stoves. Similarly, for other biogas stoves (4.67 g/MJd), LPG stoves (1.367 g/MJd), and kerosene stoves (0.314 g/MJd), the cost of cooking a standard meal was calculated to be the least in the highly efficient induction stove at NRs. 13.06 and cost the most in kerosene stoves at NRs. 47.29 per meal. Considering all the stoves' performance parameters and the fuel cost, electric stoves are the best solution. However, the lack of a reliable grid electricity supply and the abundance of biomass resources in households indicates the need for further research and development to improve the efficiency of biomass and biogas stoves in Nepal until the cooking energy demand can be fully met through grid electricity.

Keywords Stove performance · Biomass stove · LPG stove · Kerosene stove · Biogas stove · Electric stove · Efficiency · Emissions · CO · $PM_{2.5}$ · Fuel consumption · Water Boiling Test

---

N. P. Adhikari (✉) · S. L. Shrestha
Alternative Energy Promotion Center, Kathmandu, Nepal
e-mail: narayan.adhikari@aepc.gov.np

P. R. Shakya
Institute of Engineering, Tribhuwan University, Kathmandu, Nepal

S. Prajapati
MinErgy Private Limited, Kathmandu, Nepal

K. P. Vadrevu et al. (eds.), *Vegetation Fires and Pollution in Asia*,
https://doi.org/10.1007/978-3-031-29916-2_19

1 Introduction

Clean cooking has been one of the pertinent issues under the Sustainable Development Goals (globalgiving.org). Recent literature has revealed that achieving the target of clean cooking to all by 2030 by prevailing policies and programs would not be possible (Lambooij 2020). Rural households in most South and Southeast Asian countries still rely heavily on traditional biofuels such as wood, animal waste, and crop waste for domestic fuel needs as they do not have access to commercial energy sources. The efficiencies of biofuel use are very low, and most of the bioenergy potential is wasted because of non-scientific conventional technologies. The biomass burning resulting from such activities is an important source of greenhouse gas emissions and aerosols (Adhikari 2017; Kant et al. 2000; Prasad et al. 2000; 2002; 2003; Sheesley et al. 2003; Vadrevu and Lasko 2015; Lasko and Vadrevu 2018). Biomass burning indoors and outdoors can emit substantial amounts of particulate matter (PM) and other pollutants, impacting the local air quality and human health (Crutzen and Andreae 1990; Gupta et al. 2001; Badarinath et al. 2007; 2008; 2009; Kharol et al. 2012; Lasko et al. 2017; 2018; Vadrevu et al. 2014; 2017; 2018; 2021a, b; 2022a, b). Despite national and international efforts, effective actions/policy intervention is necessary for 700 million people to use biomass and traditional stoves for cooking in South Asia (Cameron et al. 2016). For example, the impact of ongoing programs on the cooking sector in India is not satisfactory because of needing to address the full dimensions of societal energy needs (Khandelwal et al. 2017). Similar experiences were observed in Africa (Nhamo et al. 2020), where the issues were associated with technology transfer, required budget, and governance (Mallard et al. 2020). With several options available on the types of stoves beneficial for a clean environment, it raises pertinent issues, especially regarding their implications for the national economy. The selection of particular cooking fuel and technology in the community should be made with a holistic approach by considering environmental and health impacts, their adaptability, and economic viability (Vaccari et al. 2017). For example, while assessing barriers to biogas in Rwanda, among four key barriers of financial, technical, socio-cultural, and institutional, Mukeshimaana et al. (2021) noted economics as the most influential parameter. However, such barriers are contextual and demand the country's specific capabilities regarding energy resource availability, financial strength, and socio-cultural aspects. The initiations and recommendations given to India and some Sub-Sahara African countries to replace traditional biomass cooking with LPG stoves (Pye et al. 2020; Gould and Urpelainen 2020; Murshed et al. 2020) may not be relevant to the countries with no commercial petroleum reserves because of their additional burden on the national economy. Among available cooking options, Zhang et al. (2020) recommended promoting battery storage solar home systems for meeting cooking energy needs in six low-middle-income countries in South East Asia. "The program for energy-efficient cooking" (PEC) in Ecuador has targeted to create demand for electric cooking from hydroelectricity to replace imported subsidized LPG by more than 90% of households and thus reduce the burden of the national economy (Gould et al. 2020). While considering free availability, the focus has also been given to utilizing biomass by adopting cleaner technology in developing countries by demonstrating more efficient and economical rice husk stoves than commercial LPG (Mehetre et al. 2017). In the absence of evidence-based studies, decision-makers, in most cases, have been in a dilemma to choose the right cooking technology for their territories due to a lack of scientific evidence of competitive analysis among candidate technologies.

1.1 Nepalese Context

Many studies have analyzed the Nepali cooking sector by considering different dimensions of technical, socio-cultural, economic, environmental, and health with recommendations (Manibog 1984; Singh et al. 2012; Gawande et al. 2013; Pokharel 2004; Pokhrel et al. 2015; Acharya and Marhold 2018). Some studies explored various recommendations vis-à-vis prevailing barriers to the promotion of biogas technology in Nepal (Bhattarai and Risal 1970; Karki et al. 2005; Gautam et al. 2009; Singh and Maharjan 2003), while some focused on the promotion of electric stoves by deploying country's immense hydropower potential (Shrestha and Bhattarai 1995). In considering the country's biomass potential, the emphasis has also been given to utilizing biomass in cleaner forms by adopting efficient and modern technologies (Pradhan and Limmeechokchai 2017).

On the one hand, LPG imports have proliferated, negatively impacting the national economy. On the other hand, hydropower's electricity has been spilling in some cases because of the need for more transmission and distribution infrastructure and storage facilities. In the same way, considering biomass potential, there is a good opportunity to utilize it efficiently and cleanly. The migration of the population from rural to urban areas has led to a decline in biomass use in urban areas and underutilized in rural areas. The modeling by Pradhan et al. (2019) shows that biogas and electric stoves could significantly reduce the share of LPG and fuelwood consumption in Nepal.

As revealed, the use of biomass will increase, but its share of fuelwood consumption in total energy consumption will be reduced from 85% in 2015 to 64.4% and 40.2% in 2030 and 2050, respectively (Pradhan and Limmeechokchai 2017). Similarly, Bhandari and Pandit (2018) observed the requirement of an additional 1925 MW hydropower electricity to meet both growing cooking energy needs and complete substitution of LPG by 2035, by which the country could save 21.72 mil USD (2016) to 70.8 mil USD (2035) each year (Bhandari and Pandit 2018) Given the context, selecting the right cooking technology is critical and should be done based on their performance. Although some attempts are made to evaluate the efficiency of different stoves, Pokharel (2004) conducted an economic analysis of five different cooking stoves by conducting lab experiments to find system efficiency, useful energy requirement, and energy use cost. Furthermore, Dahal (2020) compared a cost–benefit analysis between LPG and electric induction stoves in one of the peri-urban areas and found that subsidized electricity costs are needed for two-thirds of LPG-adopted households to get economic benefits for switching to electric induction stoves (Dahal 2020).

Several studies have also been explicitly conducted on emission measurements and suggestions for better technologies based on household air pollution reduction. For example, Weyant et al. (2019a, b) measured emissions from six LPG, 16 wood, and 57 biogas cooking stoves in Nepal. They found no significant differences between the values of emissions of $PM_{2.5}$ and Elemental Carbon (EC) in LPG and biogas stoves, while wood stoves emitted more than 50 times higher $PM_{2.5}$ than biogas. The emission factors measured using carbon balance methods for biogas stove were $PM_{2.5}$ (7.4 ± 10.9 mg/MJ), CO (1.1 ± 0.5 g/MJ), EC (0.2 ± 0.3 mg/MJ); LPG stoves $PM_{2.5}$ (9.5 ± 6.8 mg/MJ), CO (0.4 ± 0.2 g/MJ), and EC (0.3 ± 0.3 mg/MJ); wood stove $PM_{2.5}$ (408 ± 160 mg/MJ), CO (5.1 ± 1.3 g/MJ), and EC (45.6 ± 24.5 mg/MJ) (Cheryl et al. 2019). Johnston et al. (2020) compared $PM_{2.5}$ emission levels inside brick workers' homes in Nepal from LPG and wood stoves. The $PM_{2.5}$ geometric mean values for both indoor stoves were 79.32 mg/m^3 and 541.14 mg/m^3, respectively, exceeding WHO's proposed 24 h limit of 25 mg/m^3. A study conducted by Adhikari et al. (2020) on cookstove smoke impact on ambient air quality found 66% of $PM_{2.5}$ and 80% of BC emissions from biomass cookstoves directly escape into the ambient air. Ambient $PM_{2.5}$ concentrations in the rural sites were 37% higher than in the nearby background location. Such higher levels can lead to approximately 82 cases of annual premature deaths among the rural population based on WHQ's AirQ + model simulation (Adhikari et al. 2020).

Some of the above studies have provided comparative information about stoves, mainly in terms of specific performance. However, a single comparative study was lacking on the performance of different stoves, mainly based on energy efficiency, environment, and economics for various types of stoves (based on solid biomass, electricity, fossil fuel, and biogas) that are being widely used in the country. This forms the basis for the current study for getting a complete sketch of the stoves, which helps adopt the best available cooking technology in the country.

2 Study Area

The study was conducted within the premises of Renewable Energy Test Station (RETS) located in Kathmandu valley of Nepal. RETS is government body under Nepal Academy of Science and Technology and mandated for testing and certification of renewable energy technologies and equipment.

3 Methodology

3.1 Selection of Stoves

Nine different types of cooking stoves that were widely used in Nepal are considered for the study. Based on the fuel used, these stoves were categorized under four groups, i.e., solid biomass, biogas, electricity, and fossil fuel. The stoves using solid biomass fuel consisted of a forced draft stove, a natural draft stove with a chimney, and a natural draft stove without a chimney. Two different types of GGC-2047 domestic biogas models were referred to under biogas fuel. Similarly, induction cookers, infrared cookers, and heating coil-based cookers were considered under electric stoves. Finally, LPG stoves and kerosene were considered under fossil fuel-based stoves.

3.2 Stove Testing Process

The stoves were tested by following ISO-IWA-recommended Water Boiling Test (WBT) version 4.2.3 and BIS-recommended IS 13152 standard testing procedure. The WBT is done to evaluate how effectively fuel can heat the water in a pot and the emissions produced during the process (Alliance 2013). The test includes three consecutive phases (i) cold start—high-power phase, (ii) hot start—high-power phase, and (iii) simmering—low power phase. During the cold start—high-power phase, the test starts at room temperature by weighing a certain quantity of fuel to boil a fixed measured amount of water in a standard pot. Once the water boils, the initial water would be replaced by the same quantity of water at room temperature in the second phase by using a pre-weighed amount of fuel.

Through these two phases, the performance of stoves at cold and hot stages will be obtained. In addition, the quantity of fuel needed to simmer a measured amount of water at just below boiling point for 45 min is assessed during the simmer phase. This is done to evaluate stove performance during cooking food, such as pulses and legumes, which require a long time to cook. By selecting fuel on WBT software, the actual fuel consumption for non-solid biomass stoves can be obtained.

Carbon monoxide (CO) and particulate matter (PM) concentrations at stove exhaust were measured during all three phases. Laboratory Emission Measurement System (LEMS) of Aprovecho Research Centre (ARC), USA, with $PM_{2.5}$ Gravimetric measurement, CO measurement with Electrochemical cell sensor, and CO_2 measurement with non-dispersive infrared (NDIR) sensor were used for emission measurement.

a. Natural draft chimney stove(FWS-C) b. Natural draft rocket stove(FWS-R) c. Force draft pellet stove(PS)

Fig. 1 Types of solid biomass stoves tested

3.2.1 Solid Biomass Fuel-Based Stoves

Two widely used natural draft stoves with chimney-burning fuel wood, one natural draft stove without chimney-burning fuel wood, and one force draft stove burning biomass pellet were tested.

The fuel wood-burning stove with chimney is a fixed-type stove which is a locally built stove with two potholes enabling the user to cook two items simultaneously. The stove is specially designed to improve the indoor air quality of rural households driving the smoke out of the indoor cooking space. The firepower of the stove is considerably higher, ranging from 3 to 6 kW during different test phases. Only fugitive emission values were measured for this type of stove during this study.

The fuel wood-burning stove without a chimney is a portable metallic stove built on the rocket principle. During different operation phases, the stove's firepower ranges from 2.5 to 5 kW. Total emission values were measured for this type of stove during this study.

Force draft stove falls under the most efficient form of biomass burning stove with lower emissions. The stove can be operated with chopped fuelwood as well as biomass pellets. The stove consists of a speed-adjustable fan for controlled thermal power and burns with biomass fuel gasification. During different test phases, the stove's firepower ranges from 1.5 to 3.5 kW. Total emission values were measured for this type of stove during this study (Fig. 1).

3.2.2 Biogas Stoves

Biogas stoves of two widely adopted burners (smaller and larger) were tested. The rated thermal power of a small burner is 1.5 KW, whereas it is 5 KW for a larger burner. The biogas flow meter of specification SC300 G2.5 was used to measure

a. Biogas Stove (BS1) b. Biogas Stove (BS2)

Fig. 2 Types of biogas stoves tested

biogas consumption in cubic meters during the process, which was converted to weight by relating to density (Fig. 2).

3.2.3 Fossil Fuel-Based Stoves (LPG and Kerosene Stoves)

The widely used LPG and kerosene stoves were tested under this category. The weights of the consumed kerosene and LPG before and after different test phases were measured by using an electronic scale. The selected kerosene and LPG stove firepower ranged from 2 to 3.5 kW and 1.5 to 2 kW, respectively (Fig. 3).

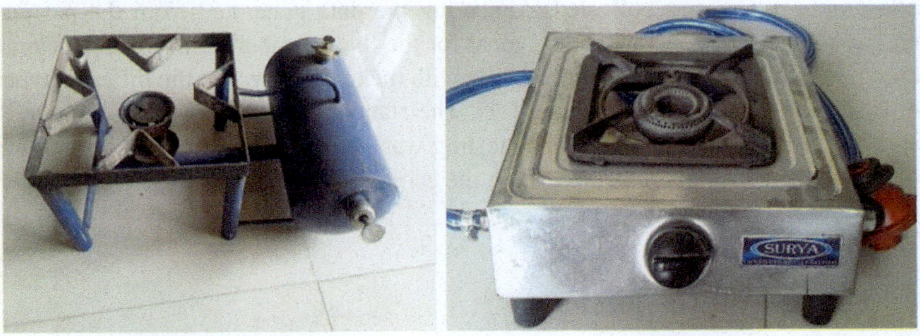

a. Kerosene stove (KS) b. LPG Stove

Fig. 3 Types of fossil fuel-based stoves

a. Infrared cooker (IR) b. Induction cooker (IC) c. Heating coil (HC)

Fig. 4 Types of electric stoves

3.2.4 Electric Stoves

Locally available, three different types of electric stoves, induction (IC), infrared (IR), and heating coil (HC), with a rated power of 2000 W, 2000 W, and 1000 W, respectively, were considered. Fluke 345 power clamp meter with data logger was used to measure the input electrical energy of these stoves. The data was analyzed using Power Log Software version 4.0 (Fig. 4).

3.3 Field Testing

For the per capita energy consumption assessment, a series of cooking activities were conducted in the five-family member household using IC, LPG Stove, and firewood-based stove. The same dish was cooked during the field-based testing, and energy consumption data was collected. Analyzing the data generated from the per day energy required to cook a standard meal for the five-member family, the per capita energy requirement was calculated along with the energy cost for the different types of cooking technology with other fuels. The standard meal consists of rice, lentil, and curry, prepared twice daily. The testing was done to compare the consumption of fuel at the household level by which the economic analysis was carried out.

4 Energy Conversion Factor

A major objective of the study was to determine the equivalent fuel consumption of different stoves. For this, the energy equivalent matrix was developed in reference to the test results. The matrix defines the energy equivalent relation between two stoves.

Under similar ambient conditions, the heat absorbed by 5 L of water to reach the boiling point is the same for all stoves. The total energy required for boiling water was determined by

$$E\_b = \left(\Delta E_{\mathrm{H_2O,Heat}} + \Delta E_{\mathrm{H_2O,evp}}\right)$$

The input energy varies significantly depending on the stove technology and design, type of fuel and condition, combustion efficiency, and the efficacy of technology to transfer the input energy to the pot. Therefore, the input energy varies with the variation depending on the technology under consideration. Losses can be expected due to absorption by the stove, heat loss via radiation, conduction, and evaporation, due to time of water heating, and direct heat loss.

A direct relation has been used to equate the energy input, providing a general equivalent relation;

$$E_{\mathrm{absorb,tech1}} = \alpha \times E_{\mathrm{absorb,tech2}} \tag{1}$$

where α is the energy equivalent conversion factor for the stoves. It defines the energy losses during the cooking along with the energy losses due to time.

$$\eta = \frac{E_{\mathrm{absorbbywater}}}{E_{\mathrm{Input,technology}}} = \frac{\Delta E_{\mathrm{H_2O,Heat}} + \Delta E_{\mathrm{H_2O,evp}}}{E_{\mathrm{input,technology}}} \tag{2}$$

Thermal efficiency of the stove can be deduced using the relation below;

$$\eta = \frac{4.186 \times \sum_{j=1}^{4}(Pj_{ci} - Pj)\left(Tj_{cf} - Tj_{ci}\right) + 2260 \times w_{cv}}{f_{cd} \times \mathrm{LHV}} \tag{3}$$

where,

f_{cd}	Equivalent dry wood consumed (grams)
LHV	Lower heating value
T_b	Local boiling point of water (°C)
j	Numbers of pots
Tj_{cf}	Final temperature of pot
Tj_{ci}	Initial temperature of pot
Pj_{ci}	Initial weight of pot with water

Water Vaporized (grams)

$$w_{cv} = \sum_{j=1}^{4}\left(Pj_{ci} - Pj_{cf}\right) \tag{4}$$

The effective mass of water boiled (Boiled water remaining at end of test)

$$w_{cr} = \sum_{j=1}^{4}\left((Pj_{ci} - Pj_{cf})\left(\frac{Tj_{cf} - Tj_{ci}}{T_b - Tj_{ci}}\right)\right) \tag{5}$$

where,

Pj_{cf} Final weight of pot with water

The factor $\frac{Tj_{cf}-Tj_{ci}}{T_b-Tj_{ci}}$ is used to "discount" the water heated in additional pots that does not come to a full boil.

$$E_{\text{Input,tech1}} = \alpha \times E_{\text{Input,tech2}} \tag{6}$$

$$\alpha = \frac{E_{\text{Input,tech1}}}{E_{\text{Input,tech2}}} \tag{7}$$

The energy equivalent conversion factor (a) is the non-unit value that compares an energy input between the two technologies for the same output. The factor gives the ratio of input energy in compared technology to referenced technology, which can also be expressed in percentage against the reference technology. Using the above conversion factors, all the energy consumption of different stoves in different forms was converted to an equivalent electrical unit that helps to compare the two stoves directly. The main purpose of using this conversion factor is to generalize the energy consumption of stoves by converting different energy conversion factors into a single unit.

5 Results and Discussion

1.1 *The Stove performance results from the WBT test for all stove categories are shown in* Table 1.

The thermal efficiency for all the tested stoves was calculated for the high-power phase, and the average energy consumption for boiling 5 L of water was calculated for this phase. Although for electric stoves, the test was conducted by burning 2.5 L of water due to constraints on using larger capacity pots, it was later converted to 5 L assuming two 2.5 L tests equivalence.

For electric stoves, the energy required to boil 5 L of water is obtained in terms of kWh, which is converted into kJ by using a relationship (1 Wh = 3.6 kJ) as recommended by RETS.

To compare different fuel-burning stoves, the average fuel consumed during the test was converted into a single energy unit using standard heating values of the fuel based on the calorific value test carried out for each fuel at RETS (Table 2).

Table 1 Performance of different stoves

Stove type	Thermal efficiency (%) (High power)	Average energy consumption to boil 5L of water (kJ)	Average time to boil 5L of water (min)
Natural draft chimney stove.(FWS-C)	24.75% [SD = 1.16%]	9418.05 [SD = 594.00]	38.48 [SD = 9.41]
Natural draft rocket stove.(FWS-R)	28.14% [SD = 0.45%]	8162.48 [SD = 459.19]	26.74 [SD = 3.68]
Force draft pellet stove.(PS)	42.10% [SD = 2.33%]	5270.21 [SD = 663.60]	30.13 [SD = 4.60]
Biogas stove (BS 1,2)	41.24% [SD = 4.11%]	6426.05 [SD = 977.99]	36.84 [SD = 5.49]
Kerosene stove (KS)	46.56% [SD = 7.62%]	4827.95 [SD = 888.20]	28.72 [SD = 5.29]
LPG stove	57.09% [SD = 8.13%]	4260.00 [SD = 339.35]	31.67 [SD = 0.58]
Heating coil (HC)	48.26% [SD = 1.45%]	7169.50 [SD = 1399.93]	151.64 [SD = 30.10]
Infrared cooker (IR)	75.58% [SD = 8.84%]	2994.30 [SD = 323.40]	43.64 [SD = 2.95]
Induction cooker (IC)	90.63% [SD = 3.72%]	1988.50 [SD = 75.26]	27.78 [SD = 5.91]

Table 2 Heating value of various fuels

S. No.	Fuel	Heating value	Remarks/source
1	LPG	48,000 kJ/Kg	RETS test report
2	Kerosene	43,300 kJ/Kg	RETS test report
3	Pellet	16,666 kJ/Kg	RETS test report
4	Biomass (Firewood)	19,724 kJ/Kg	RETS test report
5	Biogas	21,739 kJ/Kg	RETS test report

Table 3 also shows the energy requirement for each of the stoves in kWh and the energy conversion factor for each stove with respect to the induction cookstove.

The results show that the induction cookstove has highly efficient with the least energy consumption among all the stoves. The energy equivalent conversion factor calculated above gives the ratio of equivalent energy of the stove to the referenced stove, which is an induction cooker in the above case. It indicates how often energy is required with respect to the energy consumed by the reference induction cookstove to carry out the same cooking activity.

The test result showed that fuel wood-based chimney stoves consumed the highest energy, i.e., 4.74 times the energy consumed by induction cookstoves for conducting

Table 3 Energy requirement and energy conversion factor of stoves

Stove type	IC	IR	HC	LPG	KS	BS	PS	FWS-R	FWS-C
Energy requirement to boil 5 L of water (kWh)	0.552	0.832	1.992	1.183	1.341	1.785	1.464	2.267	2.616
Energy equivalent Conversion factor wrt IC ($E_{\text{input tech-XX}}/E_{\text{input tech-IC}}$)	1.00	1.51	3.61	2.14	2.43	3.23	2.65	4.10	4.74

Table 4 Equivalent fuel consumption matrix of stoves

Stove types	IC	IR	HC	LPG	KS	BS	PS	FWS-R	FWS-C
IC	1.00	1.51	3.61	2.14	2.43	3.23	2.65	4.10	4.74
IR	0.66	1.00	2.39	1.42	1.61	2.15	1.76	2.73	3.15
HC	0.28	0.42	1.00	0.59	0.67	0.90	0.74	1.14	1.31
LPG	0.47	0.70	1.68	1.00	1.13	1.51	1.24	1.92	2.21
KS	0.41	0.62	1.48	0.88	1.00	1.33	1.09	1.69	1.95
BS	0.31	0.47	1.12	0.66	0.75	1.00	0.82	1.27	1.47
PS	0.38	0.57	1.36	0.81	0.92	1.22	1.00	1.55	1.79
FWS-R	0.24	0.37	0.88	0.52	0.59	0.79	0.65	1.00	1.15
FWS-C	0.21	0.32	0.76	0.45	0.51	0.68	0.56	0.87	1.00

the same task. On the other hand, the biogas stove consumed more energy than the pellet stove, i.e., 3.23 and 2.65 times, respectively, as compared to the induction cookstove.

Equivalent Fuel Consumption Matrix

The equivalent fuel consumption matrix below illustrates the direct conversion factor of energy required to boil water on any stove about another stove (Table 4).

Firewood-based cookstoves are energy intensive compared to others. The firewood stove with a chimney requires 2.616 kWh energy equivalent, while the natural draft firewood stove without a chimney required 2.267 kWh and the pellet stove requires 1.464 kWh energy. The energy required for the natural draft biomass stove and chimney stoves using firewood as fuel is 55% and 79% higher than the high-quality pellet stove. Likewise, a kerosene stove needs 1.341 kWh energy, which is 13% more than LPG stove but 25% lesser than a biogas stove. In the case of electric cookstoves, IC dominates all of the stoves. The energy required for IC to boil 5L water is 0.552 kWh which is 33% less than the IR cooker and 72% less than the heating coil type. Overall, IC seems to be the best alternative in terms of energy consumption. In the case of biomass stoves, the pellet stove is better for household cooking, followed by the natural draft rocket stove.

The energy delivery to the pot in the induction cooker and LPG stove is controllable with minimum losses, while controlling power input in biomass stoves is challenging. Also, significant loss in biomass quantities is observed, such as degradation and loss due to weathering in storage and transportation.

Emissions

The results from the emission tests for measurement of particulate matter ($PM_{2.5}$) and carbon monoxide (CO) emitted during the conduction of the water boiling test from each stove type are presented in (Table 5). The emission test for electric stoves was not carried out as it does not emit any emissions.

Table 5 Results of emission test for different stoves

Stoves	High power $PM_{2.5}$ (mg/MJ_d)	Indoor Emission rate $PM_{2.5}$ (mg/min)	High Power CO (g/MJ_d)	Indoor Emission rate of CO (g/min)
FWS-C	24.85 [SD = 19.27][a]	2.34 [SD = 1.55]	0.30 [SD = 0.33][a]	0.03 [SD = 0.03]
FWS-R	458.068 [SD = 29.26]	35.322 [SD = 1.59]	6.391 [SD = 0.71]	0.680 [SD = 0.09]
PS	289.144 [SD = 33.78]	27.784 [SD = 13.68]	5.81 [SD = 0.64]	0.489 [SD = 0.07]
BS	0.00	0.00	4.64 [SD = 1.84]	NA
KS	114.166 [SD = 4.54]	8.160 [SD = 1.04]	0.314 [SD = 0.03]	0.024 [SD = 0.00]
LPG	0.00	0.00	1.367 [SD = 0.46]	NA

A symbolizes the fugitive emissions for chimney stoves

The emission values recorded for the chimney stoves are only the fugitive emissions due to leakages from the stove opening. The total emissions due to fuel burning were vented out of the laboratory through the chimney and were not captured during the measurement. However, for all other fuel-burning stoves, total emission was measured in terms of emission concentration per energy delivered to the pot for boiling the water and the emission rate.

Total $PM_{2.5}$ and CO emissions were found to be comparatively high on solid biomass-based stoves, indicating incomplete combustion and, thus, less efficiency due to inherited fuel properties and the inability to maintain and control the firepower as per the requirement.

Emissions from LPG, kerosene, and biogas stoves were significantly lower compared to biomass stoves. Kerosene stove released traces of CO (0.314 g/MJd), but a considerable amount of $PM_{2.5}$ (114.16 mg/MJd). Biogas did not release $PM_{2.5}$, but a significant amount of CO (4.64 g/MJd) emission was detected, which could be because of the moisture and other gases interfering with methane combustion. In the case of LPG, no $PM_{2.5}$ emission was detected, but traces of CO (1.367 g/MJd) was measured. The result shows that LPG is cleaner than biomass, kerosene, and biogas stoves.

Overall, the emission of $PM_{2.5}$ and CO depends on the efficacy of the stoves. The compiled information on the stove performance measurement shows that the higher the stove's efficiency, the lower the fuel consumption and emission. This signifies that the efficient technology burns out cleaner releasing more energy and CO_2.

WHO global air-quality guideline 2021 has recommended the new short-term (24-h) AQG level for $PM_{2.5}$ as 15 μg/with an interim target 1, 2, 3, and 4 being75 μg/m^3, 50 μg/m^3, 37.5 μg/m^3, and25 μg/m^3, respectively. Similarly, the short-term (24-h) AQG level for carbon monoxide is 4 mg/m^3, with an interim target 1 as 7 mg/m^3 (WHO 2021).

Meeting AQG interim target 1 for $PM_{2.5}$ would require the stove emission rate to meet 0.23 mg/min for unvented conditions and 0.80 mg/min for vented conditions. Similarly, meeting the AQG level for carbon monoxide would require the stove's CO emission rate to meet 0.16 g/min for unvented conditions and 0.59 g/min for vented conditions (WHO 2010).

Comparing the obtained emission level for the stoves with the WHO Indoor AQG standard, the $PM_{2.5}$ released from all three solid biomass fuel-based stoves could not meet the standard. However, indoor CO emission for chimney stoves and pellet stoves was found within the WHO AQG Standard. Also, the CO released in the kerosene stove is within the limit, while the $PM_{2.5}$ release is significantly high.

Therefore, in terms of emissions released, electricity is the best option, followed by gaseous fuel and biomass. In the case of biomass stoves, force draft stoves are superior to natural draft stoves (with a chimney or chimneyless). However, the deduced information shows that the technology and fuel used for household cooking, especially for biomass-based fuel, must be improved so that the overall emission of $PM_{2.5}$ and CO lie within the tolerance level.

Economy

The results obtained from field testing to prepare a standard meal of a day are presented in (Table 6). Apart from biogas, the cost for all fuel was directly monetized by obtaining an actual cost from the household. The feedstock used for the production of biogas is dung and water. Hence, no standard was developed to monetize these in commercial value, and the equivalent cost for biogas was derived in reference to the cost of electricity.

In the firewood stove, a significant amount of the fuelwood is consumed, about 1.7 kg in the chimney stove and 1.27 kg in the natural draft rocket stove. The cost associated with the firewood stoves ranges between 25 and 34 per day. While only 0.95 kg of the pellet is consumed in the pellet stove per day, the cost accounted for NRs. 26 per day. Likewise, the energy cost per day for cooking in biogas is NRs. 32 for consuming 3.2 kWh equivalent electric energy. Hence, using the pellet for daily household energy is cheaper than other solid biomass, including biogas.

Similarly, for the LPG and kerosene users, the fuel consumption is 0.23 kg and 0.28 kg, respectively, and the energy cost associated with it is NRs. 25.51 and NRs.47.27, respectively.

The cost associated with the modern electric stoves is the cheapest option compared to any other fuel-burning stove with a per day energy cost of NRs. 13.06 and 15.82 for induction and infrared cookstove, respectively.

6 Conclusion

This study has tested various cookstoves available in Nepal using different fuel types and conducted a comparative analysis in terms of energy efficiency, emissions, and economic performance.

Table 6 Results from field testing for preparation of standard meal

Parameters	IC	IR	HC	BS	LPG	KS	PS	FWS-R	FWS-C
Energy consumption (kWh/Day)	1.306	1.582	2.152	3.255	–	–	–	–	–
Gas/Kerosene (kg/day)	–	–	–	–	0.230	0.280	–	–	–
Biomass (kg/day)	–	–	–	–	–	–	0.949	1.265	1.695
Unit cost (NRs.)	10.00	10.00	10.00	10.00	110.92	168.91	27.00	20.00	20.00
Per day energy cost (NRs.)	13.06	15.82	21.52	32.55	25.51	47.29	25.62	25.30	33.90
Per capita daily energy cost (In NRs.)	2.61	3.16	4.30	6.51	5.10	9.46	5.12	5.06	6.78
Per capita annual energy cost (In NRs.)	953.38	1154.86	1570.90	2376.38	1862.35	3452.52	1870.04	1846.95	2474.70

From the energy prospect, electric stoves like Induction and Infrared cookstoves were highly efficient, with 90.63% and 75.58% thermal efficiency, respectively. On the other hand, the efficiency of LPG, heating coil, kerosene stove, pellet stove, biogas, natural draft biomass stoves without chimneys, and natural draft biomass stoves with chimneys were 57.09%, 48.26%, 46.56%, 42.10%, 41.24%, 28.14%, and 24.75%, respectively.

Likewise, the energy required to boil 5L of water in the induction, infrared, heating coil, LPG, kerosene, biogas, pellet stove, and firewood stoves without and with chimney were, respectively, 0.522 kWh, 0.832 kWh, 1.992 kWh, 1.183 kWh, 1.341 kWh, 1.785 kWh, 1.464 kWh, 2.267 kWh, and 2.616 kWh. The results showed that the induction cooker requires about half the energy than LPG, one-third of biogas, less than half compared to the pellet stove, and almost one-fourth less energy than the chimneyless biomass stoves under similar conditions. An energy conversion factor was developed based on efficiency and energy consumption, and the equivalent energy consumption table was generated. Through this conversion factor, the energy consumption from one stove can be directly converted to another at a comparable level. This factor also acts as the stove's performance indicator and uses its value, and one can now compare the energy consumption.

From the $PM_{2.5}$ and CO emission measurement tests conducted for all the fuel-burning stoves, LPG and Biogas stoves did not emit $PM_{2.5}$, but they released traces of CO, 1.367 g/MJd and 4.64 g/MJd, respectively. On the other hand, among the biomass-based stoves, chimney stoves were found to be cleaner for indoor environments as all pollutants are emitted outside through the chimney.

The biomass-based and kerosene stoves could not perform to achieve the $PM_{2.5}$ emission rate fulfilling WHO AQG Standard ($PM_{2.5}$, 15 $\mu g/m^3$). However, the chimney stove, pellet stove, and kerosene stove fulfilled the CO emission requirement (CO, 4 mg/m^3).

The energy cost for cooking a standard meal for a family of five members showed electricity-based cookstoves to be the cheapest compared to other fuel-based stoves. The per capita cooking energy cost per year for stove users in ascending order are induction (NRs. 953.38), infrared (NRs. 1154.86), heating coil (NRs. 1570.90), biomass rocket stove (NRs. 1846.95), LPG (NRs. 1862.35), pellet (NRs. 1870.04), biogas (NRs. 2376.38), chimney stoves (NRs. 2474.70), and kerosene (NRs. 3452.52). The calculated cost for cooking on firewood-based stoves applies to the users buying fuelwood for cooking; however, if the fuelwood availability is free, especially in rural areas, only the indirect cost for the collection of fuel will be involved. Similarly, if the feedstock is freely available at their promises, no energy cost will need to be considered for biogas users. With the escalating rate of petroleum-based fuel, the cooking energy cost associated with kerosene stoves and LPG stoves is expected to increase at different timeframes.

Considering all the performance parameters of the stove and the cost of fuel, electric stoves are the ultimate solution. However, for rural areas without proper grid supply and quality electricity feed, biogas and biomass-based stoves could be the interim solution based on local biomass resources and feedstock availability. Although LPG and kerosene stoves perform better than biomass stoves, the need for importing fossil fuels and the escalating price in the international market demands a reduction in dependency on these fuels. Similarly, the study also highlights the need to conduct further research and development to improve the performance of biomass and biogas stoves to fulfill Nepal's economic and cleaner cooking requirements.

References

Acharya, B., and K. Marhold. 2018. Determinants of household energy use and fuel switching behavior in Nepal. *Energy* 2019 (169): 1132–1138. https://doi.org/10.1016/j.energy.2018.12.109.

Adhikari, N.P. 2017. *Spatial variation of biomass energy supply and demand in rural Nepal*. Doctoral dissertation, Universitäts-und Landesbibliothek Bonn.

Adhikari, S., P.S. Mahapatra, C.P. Pokheral, and S.P. Puppala. 2020. Cookstove smoke impact on ambient air quality and probable consequences for human health in rural locations of southern Nepal. *International Journal of Environmental Research and Public Health* 17 (2): 550.

Alliance, C.C. 2013. The water boiling test, version 4.2. 3: cookstove emissions and efficiency in a controlled laboratory setting. Global Alliances for Clear Cookstoves, vol. 2, 52.

Badarinath, K.V.S., S.K. Kharol, K.M. Latha, T.K. Chand, V.K. Prasad, A.N. Jyothsna, and K. Samatha. 2007. Multiyear ground-based and satellite observations of aerosol properties over a tropical urban area in India. *Atmospheric Science Letters* 8 (1): 7–13.

Badarinath, K.V.S., S.K. Kharol, A.R. Sharma, and V.K. Prasad. 2009. Analysis of aerosol and carbon monoxide characteristics over Arabian Sea during crop residue burning period in the Indo-Gangetic Plains using multi-satellite remote sensing datasets. *Journal of Atmospheric and Solar-Terrestrial Physics* 71 (12): 1267–1276.

Badarinath, K.V.S., S. Kumar Kharol, V. Krishna Prasad, A. Rani Sharma, E.U.B. Reddi, H.D. Kambezidis, and D.G. Kaskaoutis. 2008. Influence of natural and anthropogenic activities on UV Index variations–a study over tropical urban region using ground based observations and satellite data. *Journal of Atmospheric Chemistry* 59 (3): 219–236.

Bhandari, R., and S. Pandit. 2018. Electricity as a cooking means in Nepal—a modelling tool approach. *Sustainability* 10: 1–17. https://doi.org/10.3390/su10082841.

Bhattarai, N., and S. Risal. 1970. Barrier for implementation of improved cook stove program in Nepal. *Journal of the Institute of Engineering* 7: 116–120. https://doi.org/10.3126/jie.v7i1.2069.

Cameron, C., S. Pachauri, N.D. Rao, D. McCollum, J. Rogelj, and K. Riahi. 2016. Policy trade-offs between climate mitigation and clean cook-stove access in South Asia. *Nature Energy* 2016 (1): 1–5. https://doi.org/10.1038/nenergy.2015.10.

Crutzen, P.J. and M.O. Andreae. 1990. Biomass burning in the tropics: impact on atmospheric chemistry and biogeochemical cycles. Science 250 (4988): 1669–1678.

Dahal, G.R. 2020. Cost-benefit analysis of replacing LPG stoves with induction stoves in rural households of Kavre district Nepal. Humboldt State University.

Gautam, R., S. Baral, and S. Herat. 2009. Biogas as a sustainable energy source in Nepal: present status and future challenges. *Renewable and Sustainable Energy Reviews* 13: 248–252. https://doi.org/10.1016/j.rser.2007.07.006.

Gawande, D., B. Shrestha, and A. Gawande. 2013. Can improved cooking stoves work? The Nepal Chulo experience. Himalayan Research Papers Archive.

Gould, C.F., S.B. Schlesinger, E. Molina, M.L. Bejarano, A. Valarezo, and D.W. Jack. 2020. Household fuel mixes in peri-urban and rural Ecuador: explaining the context of LPG, patterns of continued firewood use, and the challenges of induction cooking. *Energy Policy* 136: 111053. https://doi.org/10.1016/j.enpol.2019.111053.

Gould, C.F., and J. Urpelainen. 2020. The gendered nature of liquefied petroleum gas stove adoption and use in rural India. *Journal of Development Studies* 56: 1309–1329. https://doi.org/10.1080/00220388.2019.1657571.

Gupta, P.K., V.K. Prasad, C. Sharma, A.K. Sarkar, Y. Kant, K.V.S. Badarinath, and A.P. Mitra. 2001. CH4 emissions from biomass burning of shifting cultivation areas of tropical deciduous forests–experimental results from ground-based measurements. *Chemosphere-Global Change Science* 3 (2): 133–143.

Johnston, J.D., M.E. Hawks, H.B. Johnston, L.A. Johnson, and J.D. Beard. 2020. Comparison of liquefied petroleum gas cookstoves and wood cooking fires on $PM_{2.5}$ trends in brick workers' homes in Nepal. *International Journal of Environmental Research and Public Health* 17 (16): 5681.

Kant, Y., V.K. Prasad, and K.V.S. Badarinath. 2000. Algorithm for detection of active fire zones using NOAA AVHRR data. *Infrared Physics & Technology* 41 (1): 29–34.

Karki, A.B., J.N. Shrestha, and S. Bajgain. 2005. *Biogas as renewable source of energy in Nepal: theory and development*. Kathmandu, Nepal: BSP-Nepal.

Khandelwal, M., M.E. Hill, P. Greenough, J. Anthony, M. Quill, M. Linderman, et al. 2017. Why have improved cook-stove initiatives in India failed? *World Development* 92: 13–27. https://doi.org/10.1016/j.worlddev.2016.11.006.

Kharol, S.K., K.V.S. Badarinath, A.R. Sharma, D.V. Mahalakshmi, D. Singh, and V.K. Prasad. 2012. Black carbon aerosol variations over Patiala city, Punjab, India—a study during agriculture crop residue burning period using ground measurements and satellite data. *Journal of Atmospheric and Solar-Terrestrial Physics* 84: 45–51.

Lambooij, E.H.A. 2020. Achieving universal access to electricity and cleaner cooking fuels in Sub-Saharan Africa: a stakeholder influence analysis of energy sector development in Rwanda. Master's thesis. https://studenttheses.uu.nl/handle/20.500.12932/37620.

Lasko, K., and K. Vadrevu. 2018. Improved rice residue burning emissions estimates: accounting for practice-specific emission factors in air pollution assessments of Vietnam. *Environmental Pollution* 236: 795–806.

Lasko, K., K.P. Vadrevu, and T.T.N. Nguyen. 2018. Analysis of air pollution over Hanoi, Vietnam using multi-satellite and MERRA reanalysis datasets. *PLoS ONE* 13 (5): e0196629.

Lasko, K., K.P. Vadrevu, V.T. Tran, E. Ellicott, T.T. Nguyen, H.Q. Bui, and C. Justice. 2017. Satellites may underestimate rice residue and associated burning emissions in Vietnam. *Environmental Research Letters* 12 (8): 085006.

Mallard, K., L. Garbuio, and V. Debusschere. 2020. Towards sustainable business model and sustainable design of a hydro generator system dedicated to isolated communities. *Procedia CIRP* 90: 251–255. https://doi.org/10.1016/j.procir.2020.02.004.

Manibog, F.R. 1984. Improved cooking stoves in developing countries: problems and opportunities. *Annual Review of Energy* 9: 199–227.

Mehetre, S.A., N.L. Panwar, D. Sharma, and H. Kumar. 2017. Improved biomass cookstoves for sustainable development: a review. *Renewable and Sustainable Energy Reviews* 73: 672–687. https://doi.org/10.1016/j.rser.2017.01.150.

Mukeshimana, M.C., Z.Y. Zhao, M. Ahmad, and M. Irfan. 2021. Analysis on barriers to biogas dissemination in Rwanda: AHP approach. *Renewable Energy* 163: 1127–1137.

Murshed, M., S.R. Ali, S. Banerjee. 2020. Consumption of liquefied petroleum gas and the EKC hypothesis in South Asia: evidence from cross-sectionally dependent heterogeneous panel data with structural breaks. *Energy, Ecology and Environment*. https://doi.org/10.1007/s40974-020-00185-z.

Nhamo, G., G.O.A., Odularu, V. Mjimba. 2020. Scaling up SDGs implementation. In *Sustainable development goals series*. Charm: Springer.

Pokharel, S. 2004. Energy economics of cooking in households in Nepal. *Energy* 29 (4): 547–559.

Pokhrel, A.K., M.N. Bates, J. Acharya, P. Valentiner-Branth, R.K. Chandyo, P.S. Shrestha, et al. 2015. $PM_{2.5}$ in household kitchens of Bhaktapur, Nepal, using four different cooking fuels. *Atmospheric Environment* 113: 159–68. https://doi.org/10.1016/j.atmosenv.2015.04.060.

Pradhan, B.B., B. Limmeechokchai, and R.M. Shrestha. 2019. Implications of biogas and electric cooking technologies in residential sector in Nepal—a long term perspective using AIM/Enduse model. *Renewable Energy* 143: 377–389. https://doi.org/10.1016/j.renene.2019.05.026.

Pradhan, B.B., and B. Limmeechokchai. 2017. Electric and biogas stoves as options for cooking in Nepal and Thailand. *Energy Procedia* 138: 470–475. https://doi.org/10.1016/j.egypro.2017.10.227.

Prasad, V.K., P.K. Gupta, C. Sharma, A.K. Sarkar, Y. Kant, K.V.S. Badarinath, T. Rajagopal, and A.P. Mitra. 2000. NOx emissions from biomass burning of shifting cultivation areas from tropical deciduous forests of India–estimates from ground-based measurements. *Atmospheric Environment* 34 (20): 3271–3280.

Prasad, V.K., Y. Kant, P.K. Gupta, C. Elvidge, and K.V.S. Badarinath. 2002. Biomass burning and related trace gas emissions from tropical dry deciduous forests of India: a study using DMSP-OLS data and ground-based measurements. *International Journal of Remote Sensing.* 23 (14): 2837–2851.

Prasad, V.K., M. Lata, and K.V.S. Badarinath. 2003. Trace gas emissions from biomass burning from northeast region in India—estimates from satellite remote sensing data and GIS. *The Environmentalist* 23 (3): 229–236.

Pye, A., S. Ronzi, B.H.M. Ngahane, E. Puzzolo, A.H. Ashu, and D. Pope. 2020. Drivers of the adoption and exclusive use of clean fuel for cooking in Sub-Saharan Africa: learnings and policy considerations from Cameroon. *International Journal of Environmental Research and Public Health* 17: 1–24. https://doi.org/10.3390/ijerph17165874.

Sheesley, R.J., J.J. Schauer, Z. Chowdhury, G.R. Cass, and B.R. Simoneit. 2003. Characterization of organic aerosols emitted from the combustion of biomass indigenous to South Asia. *Journal of Geophysical Research: Atmospheres* 108 (D9).

Shrestha, R.M., and G.B. Bhattarai. 1995. Utility planning implications of efficient electric cooking in a developing country: case of Nepal. *Energy* 20: 195–203. https://doi.org/10.1016/0360-5442(94)00077-G.

Singh, A., B. Tuladhar, K. Bajracharya, and A. Pillarisetti. 2012. Assessment of effectiveness of improved cook stoves in reducing indoor air pollution and improving health in Nepal. *Energy for Sustainable Development* 16: 406–414. https://doi.org/10.1016/j.esd.2012.09.004.

Singh, M., K.L. Maharjan. 2003. Contribution of biogas technology in well-being of rural hill areas of Nepal: a comparative study between biogas users and non-users. *Journal of International Development and Cooperation* 9: 43–63. https://doi.org/10.15027/14401.

Vaccari, M., F. Vitali, and T. Tudor. 2017. Multi-criteria assessment of the appropriateness of a cooking technology: a case study of the Logone Valley. *Energy Policy* 109: 66–75. https://doi.org/10.1016/j.enpol.2017.06.052.

Vadrevu, K., and K. Lasko. 2015. Fire regimes and potential bioenergy loss from agricultural lands in the Indo-Gangetic Plains. *Journal of Environmental Management* 148: 10–20.

Vadrevu, K., T. Ohara, and C. Justice. 2017. Land cover, land use changes and air pollution in Asia: a synthesis. *Environmental Research Letters* 12 (12): 120201.

Vadrevu, K.P., A. Eaturu, E. Casadaban, and S. Biswas. 2022a. Agricultural fires in South Asian countries and implications. In *Remote sensing of agriculture and land cover/land use changes in South and Southeast Asian countries* 501–516. Cham: Springer.

Vadrevu, K.P., Toan Le, S.S. Ray, and C. Justice. 2022b. *Remote sensing of agriculture and land cover/land use changes in South and Southeast Asian countries*. Cham: Springer. https://doi.org/10.1007/978-3-030-92365-5

Vadrevu, K.P., T. Ohara, and C. Justice (eds.). 2018. *Land-atmospheric research applications in South and Southeast Asia*. Cham: Springer.

Vadrevu, K.P., T. Ohara, and C. Justice (eds.). 2021a. *Biomass burning in South and Southeast Asia: Impacts on the biosphere*, vol. 2. CRC Press.

Vadrevu, K.P., T. Ohara, and C. Justice (eds.). 2021b. *Biomass burning in South and Southeast Asia: mapping and monitoring*, vol. 1. CRC Press.

Vadrevu, K.P., T. Ohara, and C. Justice. 2014. Air pollution in Asia. *Environmental Pollution (Barking, Essex: 1987)* 195: 233–235.

Weyant, Cheryl L., Ryan Thompson, Nicholas L. Lam, et al. 2019a. In-Field emission measurements from biogas and liquified petroleum gas (LPG) stoves. *Atmosphere* 10: 729. https://doi.org/10.3390/atmos10120729.

Weyant, C.L., R. Thompson, N.L. Lam, B. Upadhyay, P. Shrestha, S. Maharjan, K. Rai, C. Adhikari, M.C. Fox, and A.K. Pokhrel. 2019b. In-field emission measurements from biogas and liquified petroleum gas (LPG) stoves. *Atmosphere* 10 (12): 729.

WHO Gobal Air Quality Guidelines. 2021. https://apps.who.int/iris/handle/10665/345329

WHO Guidelines for Indoor Air Quality. 2010. https://www.euro.who.int/_data/assets/pdf_file/0009/128169/e94535.pdf

Zhang, J., R. Raufer, and L. Liu. 2020. Solar home systems for clean cooking: a cost-health benefit analysis of lower-middle-income countries in Southeast Asia. *Sustainability* 12: 1–14. https://doi.org/10.3390/su12093909.

Estimation of Ultrafine Particulate Matter Emissions from Biomass Burning Using Satellite Imaging and Burn Severity

Perapong Tekasakul, Narissara Nuthammachot, Rachane Malinee, John Morris, and Racha Dejchanchaiwong

Abstract Fine and ultrafine particles emitted from open biomass burning seriously affect air quality in many countries in Southeast Asia during haze episodes. In this study, $PM_{0.1}$ and $PM_{2.5}$ emissions were estimated for Thailand using agricultural residue burning. Data from Landsat-8 and Sentinel-2 satellites was obtained from 2016 to 2019. Burn severity was evaluated using NIR and SWIR bands. Results suggested agricultural residue burning as a significant contributor accounting for 82–90% of the total burned area, and the forest fires contributed the remainder. Further, the burned areas from agricultural waste during El Niño years were up to 2 times as high as those during La Niña years. Rice residue burning was the most significant contributor, followed by sugarcane and maize. Major burned areas of rice and sugarcane residue were in the northeast. Annual average $PM_{2.5}$ and $PM_{0.1}$ emissions from crop residue burning in Thailand were about 43.3 and 4.1 Gg/year, respectively. $PM_{0.1}$ emissions during El Niño years were also larger at 4.2–5.2 Gg/year compared to 3.7–3.4 Gg/year during La Niña. $PM_{0.1}$ emissions from crop residue

P. Tekasakul · R. Malinee · R. Dejchanchaiwong (✉)
Air Pollution and Health Effect Research Center, Prince of Songkla University, Songkhla, Thailand
e-mail: racha.d@psu.ac.th

P. Tekasakul
Department of Mechanical and Mechatronics Engineering, Faculty of Engineering, Prince of Songkla University, Songkhla, Thailand

N. Nuthammachot
Faculty of Environmental Management, Prince of Songkla University, Songkhla, Thailand

R. Malinee
Energy Technology Program, Department of Specialized Engineering, Faculty of Engineering, Prince of Songkla University, Songkhla, Thailand

J. Morris
School of Industrial Education and Technology, King Mongkut's Institute of Technology Ladkrabang, Bangkok, Thailand

R. Dejchanchaiwong
Department of Chemical Engineering, Faculty of Engineering, Prince of Songkla University, Songkhla, Thailand

K. P. Vadrevu et al. (eds.), *Vegetation Fires and Pollution in Asia*,
https://doi.org/10.1007/978-3-031-29916-2_20

burning were about 10% of $PM_{2.5}$. These emissions affect local air quality in Thailand and neighboring countries due to long-range transport and vice versa. Annual mean $PM_{0.1}$ concentrations during the open biomass burning period ranged from 6.6 to 18.9 $\mu g/m^3$. $PM_{0.1}$-bound PAH concentrations were found to be 3–8 times higher than the background. The study highlights the estimation of fine and ultrafine particle emissions from open biomass burning useful for other researchers.

Keywords Agricultural residue burning · $PM_{2.5}$ and $PM_{0.1}$ · Burn severity · Southeast Asia

1 Introduction

Southeast Asia (SEA), including Brunei, Cambodia, Indonesia, Laos, Malaysia, Myanmar, Philippines, Singapore, Thailand, and Vietnam, is growing rapidly, leading to many unavoidable problems, including air pollution (Adam et al. 2021). Most of it is attributed to biomass burning (Prasad et al. 2001a, b; Biswas et al. 2015; Albar et al. 2018; Lasko and Vadrevu 2018; Vadrevu et al. 2021a, b). Since SEA countries rely on conventional agriculture and agro-industry businesses, pre-and post-harvest removal of undesirable biomass is necessary. Burning is the easiest and least expensive path for clearing and preparing the land for the next crop (Kim Oanh et al. 2011), but this results in large-scale air pollution as gas and fine ($PM_{2.5}$) and ultrafine ($PM_{0.1}$) particles. Forest fires result from land clearing, resource collection, accidents, land tenure conflicts, logging, etc. Thailand has also been heavily hit by haze from biomass burning from forest fires and agricultural waste burning (Phairuang, 2021). Most air pollutants can travel long distances and cause transboundary haze. The region observes haze pollution almost annually (Adam et al. 2021; Mahasakpan et al. 2023). Dense haze impacts air quality and climate and harms human health (Apte et al. 2018).

Heavy haze pollution was recorded in 1997, 2005, 2006, 2009, 2013, 2015–2016, and 2019 (Adam et al. 2021). The El Niño Southern Oscillation (ENSO) enhanced their severity (Hayasaka et al. 2014; Adam et al. 2021). Haze in this region is a transboundary problem; pollution in one country can travel to many countries. The haze problem in the upper SEA, including Myanmar, Thailand, Laos, Cambodia, and Vietnam, usually occurs in the first few months of each year, January to April, as a result of open burning from forest fires and agricultural waste (Thuy et al. 2018; Dejchanchaiwong et al. 2020; Sresawasd et al. 2021). The lower SEA has been constantly hit by burning of peatland fires in Sumatra, Indonesia, in the second half of the year, with severe detriments to air quality (Chomanee et al. 2020; Amin et al. 2021; Mahasakpan et al. 2023). Fine and ultrafine particles are generally the primary cause of haze (Badarinath et al. 2009; Kharol et al. 2012). Kumar et al. (2014) report that forest fires produce particles with distributions peaking at ~ 120 nm and higher concentrations than vehicle exhaust. Recently, nanoparticles or $PM_{0.1}$ have been studied more intensely, opening the understanding of ultrafine particles and suggesting ways to reduce health risks. SEA showed a general 50% increase in annual mean $PM_{0.1}$ during open biomass burning periods over 2014–2019 (Thuy et al. 2018; Boongla et al. 2021; Sresawasd et al. 2021; Hamid et al. 2022; Mahasakpan et al. 2023). Local biomass burning and long-range transport from other countries contributed to $PM_{0.1}$ concentrations during the dry season (Dejchanchaiwong et al. 2020; Boongla et al. 2021). The transport of pollutants, especially particulates, is strongly influenced by a carrying wind, whereas precipitation has the opposite effect by washing out and depositing the particulates (Ouyang et al. 2015). Once the PM is transported to any area, local meteorological conditions, e.g., temperature, relative humidity, and wind, in combination with geography, decide how long the pollution persists. The key factors are thermal or temperature inversions, planetary boundary layer heights, and ventilation rates (Feng et al. 2020).

Because open burning of biomass is a serious issue, assessment of the burning scale and location is essential to understand its impact before implementing management strategies to reduce the effects (Vadrevu et al. 2021a, b). Burn severity is a key factor in post-fire assessment, traditionally restricted to fieldwork, but remote sensing (Sentinel-2 and Landsat-8 satellites) now allows monitoring and more accurately assessing the severity—over large areas and rapidly changing land use, including agriculture and wildfire events (Roy et al. 2019). In addition, high spatial resolution burned area maps can improve environmental management, post-fire assessment, and remediation (Hantson et al. 2015).

Research on PM emission from open biomass burning in this region has previously focused on $PM_{2.5}$, but a few studies on the emission of ultrafine particles or $PM_{0.1}$ from the burning of agricultural residues in Thailand have now become available (Samae et al. 2021, 2022). Limited emission inventory (EI) data of forest fire and agricultural residue burning is a gap here, and there is an urgent need to estimate $PM_{0.1}$ emissions from these sources to set achievable goals for $PM_{0.1}$ reduction and strategies to limit $PM_{0.1}$ release.

Since $PM_{0.1}$ emission factors (EFs) for the burning of agricultural residues were obtained only in Thailand, total $PM_{0.1}$ emission will be calculated based on the statistical and satellite data (Hao and Liu 1994; Boonman et al. 2014; Kim Oanh et al. 2018). Therefore, the $PM_{0.1}$ emissions from agricultural burning in Thailand will be developed and updated to the recent years from 2016 to 2019. Moreover, the potential influence of El Niño on $PM_{0.1}$ emission in Thailand was investigated across the whole of Thailand using data from two El Niño (2016 and 2019) and two La Niña (2017 and 2018) events (Wang et al. 2019).

2 Study Area

Thailand is a dynamic country in Southeast Asia, with a fast-growing economy and population: the total area was 513,140 km^2 (United Nations 2021). Thailand and neighboring countries rely primarily on agricultural economies. The southwest monsoon influences weather in Thailand from May to October and the northeast

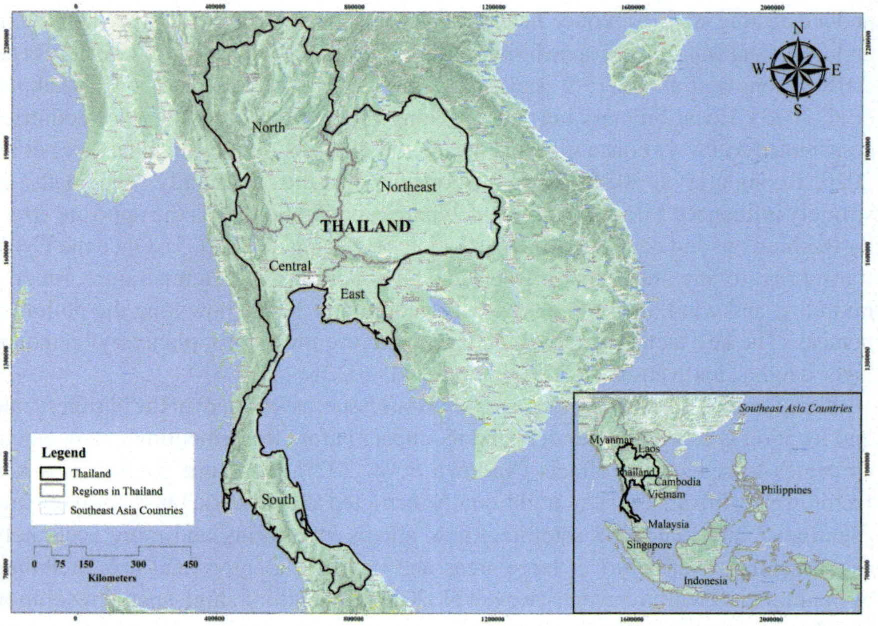

Fig. 1 Map of Thailand and its five regions

monsoon from November to January (Mahasakpan et al. 2023). The total agricultural area was 24 M ha or 46.5%. The northeastern region is the largest agricultural area (~ 10 M ha), followed by the northern and central regions (Office of Agricultural Economics 2022). Typically, agricultural waste is burned openly for clearing and preparing for the next crop (Kim Oanh et al. 2011). January to April is the biomass-intensive burning period, not only in Thailand but also in other countries in the Mekong Subregion (Chantara et al. 2019). Air pollution in Thailand is influenced by diverse sources, including vehicles, industry, long-range transport, and secondary aerosol formation. Overall, the open burning of biomass from forest fires and agricultural waste burning has a tremendous effect (Chomanee et al. 2020; Dejchanchaiwong et al. 2020; Sresawasd et al. 2021). The location of the study area is shown in Fig. 1.

3 Data and Methods

This section describes the method for burned area mapping in Thailand using images from two satellites and the burn severity estimation to evaluate emissions of $PM_{0.1}$ from different agricultural residue burning.

Table 1 Details of Landsat-8 and Sentinel-2 satellite bands used for burned area mapping

Year	Satellites	Bands	Wavelength (nm)	Resolution (m)	Number of pre-fire images	Number of post-fire images
2016	Landsat-8	B5 (NIR)	865	30	284	253
2017		B7 (SWIR)	2200	30	296	233
2018					275	339
2019					263	260
2016	Sentinel-2	B8 (NIR)	842	10	315	1405
2017		B12 (SWIR)	2190	20	2816	979
2018					2668	2627
2019					3426	3455

3.1 Satellite Data

Satellite images of the burned area from 2016 to 2019 were acquired by the Landsat-8 Operational Land Imager (OLI) and Sentinel-2 MultiSpectral Instrument (MSI), Level-1C. Near-infrared (NIR) and shortwave-infrared (SWIR) wavelengths were used in the present study (Roy et al. 2019). Both datasets were obtained from high-resolution optical sensors suited for burned area mapping (Roy et al. 2019). However, there were some limitations from adverse atmospheric conditions and areas with extensive high cloud cover and hyper-arid regions and thus a challenging issue for tropical locations. The Landsat-8 and Sentinel-2 satellites are in circular orbits with 16-day and 5-day repeat cycles with 9 (435–2200 nm) and 13 (443–2190 nm) reflective wavelength bands (Roy et al. 2019). The satellite bands used for burned area mapping are in Table 1.

3.2 Burn Severity

We used a combined Landsat-8 OLI and Sentinel-2 MSI burned area mapping algorithm. Both satellites fused data based on resolution merge and a mosaic dataset technique (Roy et al. 2019). Land use data for rice, sugarcane, and maize (Land Development Department 2022) was used to assign burned areas of crops.

Normalized burn ratio (NBR) was used to identify burned areas and estimate burn severity using the two satellites' NIR and SWIR reflective wavelength bands (Saulino et al. 2020). Healthy vegetation areas, before a fire, had very high NIR reflectance and a low SWIR response, whereas recently burned areas responded the opposite way (Saulino et al. 2020). NBR was defined from a ratio of NIR and SWIR values as,

$$\text{NBR} = \frac{\text{NIR} - \text{SWIR}}{\text{NIR} + \text{SWIR}}$$

The difference between NBR pre-fire (NBR_{pre}) and NBR post-fire (NBR_{post}), dNBR, was used to identify the optimal burn severity measures. Images of pre- and post-fire areas were used to estimate the burn severity using dNBR, and we used an iso-cluster unsupervised classification to classify burned areas. Images were acquired for August to November (pre-fire) and March to May (post-fire), following the biomass burning period (December to March). They were pre-processed in the Google Earth Engine (GEE) platform and the burn severity was analyzed using QGIS software (Roy et al. 2019; Saulino et al. 2020). Images were reassigned to three layers of severity: unburned, moderate, and high severity, following the United States Geological Survey (Roy et al. 2019) as,

$$\text{dNBR} = \text{NBR}_{\text{pre}} - \text{NBR}_{\text{post}}$$

3.3 *Estimation of Ultrafine Particles ($PM_{0.1}$) Emission*

The key contributors to open biomass burning in SEA were crop residues from rice straw (60–80%), followed by sugarcane (6–12%) and maize (6–7%) (Kim Oanh et al. 2018); thus, we focused on these three crops. $PM_{0.1}$ emissions from residue burning from 2016 to 2019 were estimated following Hao and Liu (1994) as,

$$Em_{ij} = \sum_{j} M_j \times EF_{ij}$$

where Em_{ij} is the emission of particles i from land cover type j (Gg/year) and EF_{ij} is the emission factor of particles i from land cover type j (g/kg of dry matter) and M_j is the mass of burned biomass on land cover type j (ton/year). The quantity of burned biomass from forest fires and agricultural residues can be expressed as,

$$M_j = A_{ba} \times \rho_j \times \eta_j$$

where A_{ba} is the actual burned area (ha/year), ρ_j. is the dry matter density (ton/ha), i.e., rice: 5.81 ton/ha, sugarcane: 8.71 ton/ha, and maize: 4 ton/ha and η_j is the burning efficiency: rice: 0.89, sugarcane: 0.68, and maize: 0.92 (Streets et al. 2003; Sornpoon et al. 2014; Arunrat, Pumijumnong and Sereenonchai, 2018; Junpen et al. 2018; Kim Oanh et al. 2018). The actual burned area was evaluated from the dNBR images of Sentinel-2 and Landsat-8, see Sect. 3.2 for detail.

4 Results and Discussion

This section assessed burn severity resulting from forest and agricultural burning in Thailand during El Niño and La Niña events from 2016 to 2019. Burned areas from the three major crops—rice, sugarcane, and maize—were calculated in different regions. PM emission factors and chemical components were reviewed and used to estimate emission sources. Finally, impacts on atmospheric $PM_{0.1}$ concentration are reviewed and discussed.

4.1 Burn Severity in Thailand During the El Niño and La Niña Events

Burn severity during El Niño and La Niña events is mapped in Fig. 2. 2016 and 2019 had strong El Niños, whereas 2017 and 2018 were La Nina years (Wang et al. 2019). From 2016 to 2019, the total burned area was ~ 179,752 km^2, with the largest area in 2016, ~ 62,000 km^2 or 12% of total land. Agricultural waste burning was predominant, accounting for 82–90% of the total burned area, whereas forest fires covered 10–18%, mainly in the north. Burned areas from agricultural waste during El Niño events were up to 2 times higher than during La Niña events. Generally, El Niño led to more severe climate changes and fire risks (Wang et al. 2019). Burned area, estimated from calculated dNBRs, agreed well with active fires from MODIS with $R^2 = 0.90$.

4.2 Agriculture Residue Burning Areas in Thailand

Burned areas from agricultural residue burning—rice, sugarcane, and maize—in Thailand during the peak burning season (December–March) from 2016 to 2019 are shown in Table 2. The burned area was mainly in the northeastern (41%) followed by northern (34%) and central regions (19%), whereas eastern and southern regions had only 6% of the total. This was little changed from the study of the same areas in 2010 (Boonman et al. 2014). The contribution of the burned area to the total area was rice (49–61%), sugarcane (31–40%), and maize (8–11%). Kim Oanh et al. (2011) reported that rice production dominated crop production in Thailand. The burned areas from rice and sugarcane residue burning mainly were in the northeastern region accounting for ~ 46% and ~ 39%, respectively, while the burned area from maize was mainly in the northern region, about ~ 66%. Overall, burned areas from agricultural residue during El Niño years were roughly twice those in La Niña.

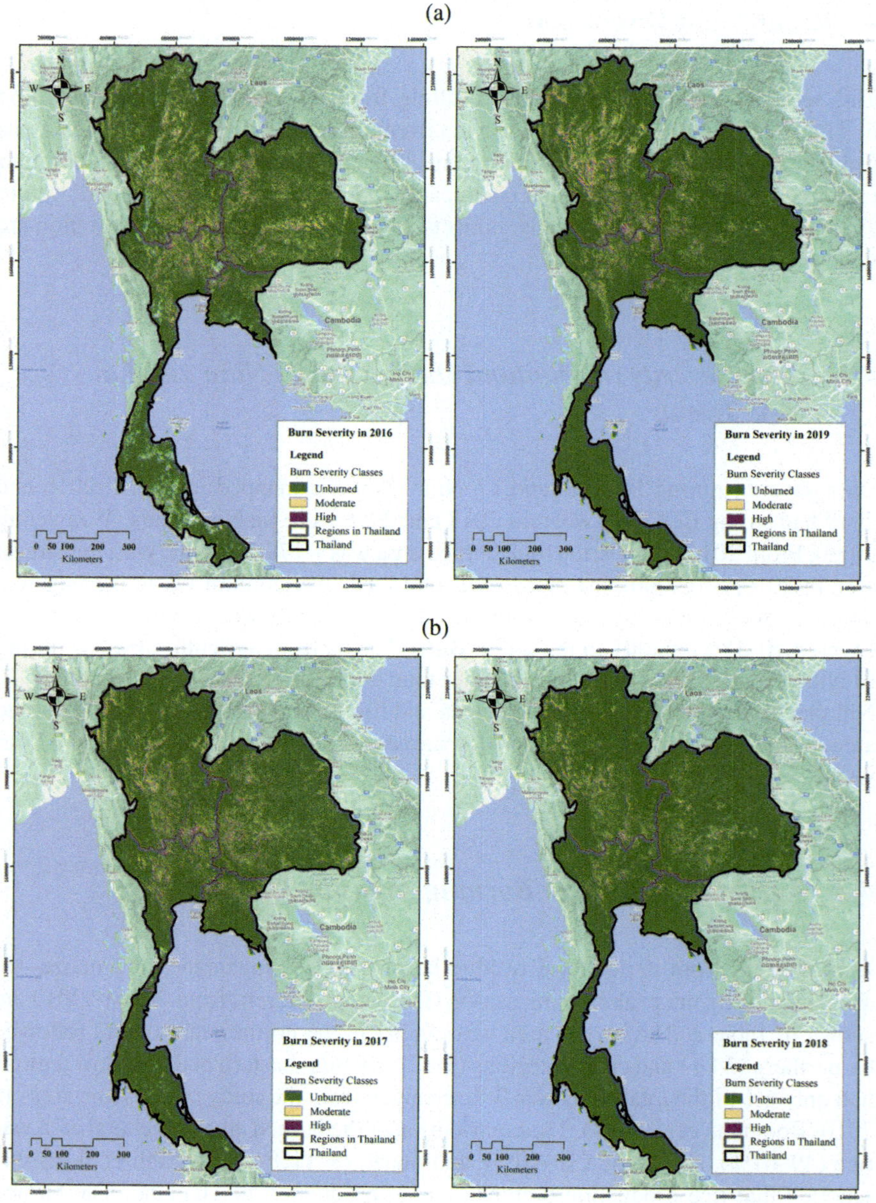

Fig. 2 Burn severity in Thailand during **a** El Niño—2016 and 2019 and **b** La Niña—2017 and 2018

Table 2 Burned areas from major crops—rice, sugarcane (SC), and maize—from Sentinel-2 and Landsat-8 over Thailand—2016–2019 (Unit: km^2)

Years	North			Central			Northeast			East			South		
	Rice	SC	Maize	Rice	SC	Maize	Rice	SC	Maize	Rice	SC	Maize	Rice	SC	Maize
El Niño															
2016	7860	3971	2564	6235	3470	367	11,531	5883	668	1563	675	36	579	–	–
2019	6425	4859	3152	3130	3696	262	7477	5643	504	949	530	32	324	–	–
La Niña															
2017	4655	3561	1736	2701	3258	240	9350	4410	484	1299	618	35	67	–	–
2018	4992	3398	2082	2081	2867	169	8735	3929	308	999	391	28	202	–	–

4.3 *Emission of Fine and Ultrafine Particles from Agricultural Residue Burning and Forest Fires*

4.3.1 Emission Factor of $PM_{2.5}$ and $PM_{0.1}$ from Agricultural Burning and Forest Fires in SEA

Open burning from agricultural waste and forest fires in SEA produced high haze levels, especially in the dry season. Biomass emissions contributed significant amounts of fine and ultrafine particles, toxic gases, and inorganic and organic compounds, i.e., water-soluble ions, heavy metals, and organic carbon, PAHs, etc., to ambient air (Kim Oanh et al. 2011; 2015; Kanokkanjana and Garivait, 2013; Chantara et al. 2019; Pham et al. 2021). Table 3 summarizes chemical and PM emission factors for different biomass and fire types in SEA.

Previous SEA studies tended to focus on $PM_{2.5}$ emissions—see Table 3. However, some recent $PM_{0.1}$ studies are also shown there. $PM_{2.5}$. EFs from laboratory studies were lower than those from field experiments, which are affected by many factors, including biomass moisture content and composition, surface soil moisture, and weather conditions (McMeeking et al. 2009). $PM_{0.1}$ EFs from rice, sugarcane, and maize were similar, 0.14–0.22 g/kg, but PAHs from maize residue were 1.5 to 4 times higher than from rice and sugarcane.

For forest vegetations, EFs in upper SEA, data from dry dipterocarp forest (DDF), and mixed deciduous forest (MDF) were collected in Thailand only ($PM_{2.5}$ EF: ~ 40 g/kg) (Chantara et al. 2019). In lower SEA, EFs from peatlands from the tropical forest were mostly collected in Indonesia and Malaysia (Stockwell et al. 2016; Jayarathne et al. 2018; Wooster et al. 2018; Watson et al. 2019). Forest $PM_{2.5}$ EFs from peat burning in Indonesia ranged from 17 to 23 g/kg (mean 19.3 g/kg). Peat properties, including moisture and carbon content, chemical composition, and burn conditions, i.e., flaming or smoldering combustion (Wooster et al. 2018), affected these EFs: with OC EFs ranging from 12 to 18 and EC EFs from 0.0005 to 0.28 g/kg. OC emissions contributed 72–80% of $PM_{2.5}$ emissions in laboratory studies. Stockwell et al. (2016) noted that high OC EFs and low EC EFs were consistent with purely smoldering combustion. No study of $PM_{0.1}$ emission from SEA forest vegetation has been reported.

4.3.2 Emission of $PM_{2.5}$ and $PM_{0.1}$ from Agricultural Residue Burning in Thailand

Table 4 summarizes annual $PM_{2.5}$ and $PM_{0.1}$ emissions from crop residue burning in Thailand from 2016 to 2019, estimated as described in Sect. 3, ranged from 36.0 to 53.8 Gg/year (mean: 43.3 Gg/year) for $PM_{2.5}$ and 3.4 to 5.2 Gg/year (mean: 4.1 Gg/year) for $PM_{0.1}$. Kim Oanh et al. (2018) reported that $PM_{2.5}$ emissions from 8 crop residues during 2010 were 84 Gg/year. Both $PM_{2.5}$ and $PM_{0.1}$ emissions were significantly higher (1.2 to 1.5 times) for El Niño than La Niña years, even

Table 3 PM and chemical emission factors for different crop and vegetation types

Types	Country	Efs of $PM_{2.5}$				Efs of $PM_{0.1}$		References
		$PM_{2.5}$ (g/kg)	PAHs (mg/kg)	WSI (g/kg)	OC and EC (g/kg)	$PM_{0.1}$ (g/kg)	PAHs (mg/kg)	
Crop residue burning								
Rice	Thailand (Laboratory)	2.13 ± 0.09				0.18 ± 0.01	16.7	Samae et al. (2021)
	Thailand (Spread burning)	8.3 ± 2.7	34 ± 35					Kim Oanh et al. (2011)
	Thailand (Laboratory)	5.3 ± 3.6	0.83 ± 0.14					Kim Oanh et al. (2015)
	Thailand (Laboratory)	3.88 ± 2.1		0.59 ± 0.39 (Cation) 0.58 ± 0.35 (Anion)				Chantara et al. (2019)
	Thailand (Piles in field)	19.21 ± 15.76						Kanokkanjana and Garivait (2013)
	Vietnam (Piles in field)	34.0 ± 17.6	10.5 ± 6.5					Pham et al. (2013)
	Vietnam (Laboratory)	10.2 ± 8.5	0.6 ± 0.63					
Sugarcane	Thailand	2.04 ± 0.12				0.22 ± 0.01	7.02	Samae et al. (2021)
Maize residue	Thailand	4.71 ± 0.18				0.14 ± 0.01	28.08	Samae et al. (2021)
	Thailand	6.00						Arunrat et al. (2018)

(continued)

Table 3 (continued)

Types	Country	Efs of $PM_{2.5}$				Efs of $PM_{0.1}$		References
		$PM_{2.5}$ (g/kg)	PAHs (mg/kg)	WSI (g/kg)	OC and EC (g/kg)	$PM_{0.1}$ (g/kg)	PAHs (mg/kg)	
	Thailand	2.2 ± 0.9		0.31 ± 0.30 (Cation) 0.51 ± 0.28 (Anion)				Chantara et al. (2019)
Forest fires								
Peatlands from tropical forest	Indonesia	21.5			16.0 and 0.005			Stockwell et al. (2016)
	Indonesia	17 ± 6			12.4 ± 5.4 and 0.24 ± 0.10			Jayarathne et al. (2018)
	Central Kalimantan	18 ± 7						Wooster et al. (2018)
	Borneo, Malaysia	23 ± 3			18 ± 2 and 0.28 ± 0.11			Watson et al. (2019)
Leaf litter	Thailand (DDF)	33.3 ± 1.0		0.18 ± 0.03 (Cation) 0.16 ± 0.02 (Anion)				Chantara et al. (2019)
	Thailand (MDF)	44 ± 0.3		0.19 ± 0.1 (Cation) 0.18 ± 0.08 (Anion)				

though crop production did not differ significantly (FAO 2022). The cultivated area affected PM emission, with the highest contribution to total $PM_{0.1}$ emission found in the northeastern (42%) followed by northern (32%) and central (20%) regions, with low emission shares from eastern and southern regions (~ 6%). Rice residue burning dominated total $PM_{0.1}$ emissions (55%), followed by sugarcane (41%) and maize (4%). The northeastern region, where rice paddies and sugarcane plantations are concentrated, contributed 46% of rice and 39% of sugarcane to $PM_{0.1}$ emissions. Maize, mostly in northern Thailand, contributed 80% of total $PM_{0.1}$ emissions. Although $PM_{0.1}$ emissions from crop residue burning constituted ~ 10% of the mass of $PM_{2.5}$, it should be of concern because $PM_{0.1}$ has worse effects on human health and the atmospheric environment (Apte et al. 2018).

Table 4 $PM_{2.5}$ and $PM_{0.1}$ emissions from agricultural residue burning in Thailand—2016–2019 (Unit: Gg/year)

Regions	Crop types	$PM_{2.5}$ (Gg/year)				$PM_{0.1}$ (Gg/year)			
		El Niño		La Niña		El Niño		La Niña	
		2016	2019	2017	2018	2016	2019	2017	2018
Northern	Rice	8.7	7.1	5.1	5.5	0.9	0.7	0.5	0.6
	Sugarcane	4.8	5.9	4.3	4.1	0.5	0.6	0.5	0.4
	Maize	4.4	5.5	3.0	3.6	0.1	0.2	0.1	0.1
Central	Rice	6.9	3.4	3.0	2.3	0.7	0.4	0.3	0.2
	Sugarcane	4.2	4.5	3.9	3.5	0.5	0.5	0.4	0.4
	Maize	0.6	0.5	0.4	0.3	0.02	0.01	0.01	0.01
Northeastern	Rice	12.7	8.2	10.3	9.6	1.3	0.9	1.1	1.0
	Sugarcane	7.1	6.8	5.3	4.7	0.8	0.7	0.6	0.5
	Maize	1.2	0.9	0.8	0.5	0.03	0.03	0.02	0.02
East	Rice	1.7	1.0	1.4	1.1	0.2	0.1	0.1	0.1
	Sugarcane	0.8	0.6	0.7	0.5	0.1	0.1	0.1	0.1
	Maize	0.1	0.1	0.1	0.05	0.002	0.002	0.002	0.001
Southern	Rice	0.6	0.4	0.1	0.2	0.1	0.04	0.01	0.02
Thailand	**Rice**	**30.6**	**20.2**	**19.9**	**18.7**	**3.2**	**2.1**	**2.1**	**1.9**
	Sugarcane	**16.9**	**17.8**	**14.3**	**12.8**	**1.8**	**1.9**	**1.5**	**1.4**
	Maize	**6.3**	**6.8**	**4.3**	**4.5**	**0.2**	**0.2**	**0.1**	**0.1**
	Total	53.8	44.8	38.5	36.0	5.2	4.2	3.7	3.4

4.4 Effect of Biomass Burning on $PM_{0.1}$ Concentration in SEA

The open burning of biomass—from both forest fires and crop residue burning—releases significant amounts of particulates into the atmosphere and causes local air pollution and long-range aerosol transport, which affects many neighboring countries. $PM_{0.1}$ and $PM_{2.5}$ concentrations and their chemical components (PAHs) are in Table 5. In upper SEA, annual mean $PM_{0.1}$ concentrations during the open biomass burning period ranged from 6.6 to 18.9 $\mu g/m^3$. The large Thai cities were also higher than the background, 2–4 times for Bangkok and 4–5 times for Chiang Mai. A high concentration observed from January to April is attributed to local open burning, especially crop residue in Thailand, Vietnam, and Cambodia. Long-range transport from neighboring countries also contributed during the dry season in Bangkok and Chiang Mai (Dejchanchaiwong et al. 2020; Sresawasd et al. 2021). In lower SEA, PM levels were significantly enhanced by transboundary haze from open peatland fires in Sumatra and Kalimantan (Chomanee et al. 2020; Amin et al. 2021; Jamhari et al. 2018; Mahasakpan et al. 2023). Average $PM_{0.1}$ concentrations ranged from 2.0 to 16.8 $\mu g/m^3$. Southern Thailand was more severely affected by this haze in 2015 than in 2019—El Niño enhanced its severity (Adam et al. 2021). These peatland fires caused $PM_{0.1}$ to be ~ 34% of total emissions (Mahasakpan et al. 2023). In the 2015 and 2019 severe haze episodes, peak PAH levels were also 3–8 times higher than the background (Chomanee et al. 2020). $PM_{0.1}$ measurements showed that open biomass burning significantly enhanced them, especially during El Niño years (Chomanee et al. 2020; Dejchanchaiwong et al. 2020; Sresawasd et al. 2021).

5 Conclusion

The largest burned area in Thailand was ~ 62,000 km^2 in 2016—about 12% of the total land. Agricultural waste burning contributed to ~ 82–90% of that area, whereas the remainder was from forest fires, mostly in the north. Burned areas during El Niño years were up to 2 times higher than those during La Niña. Contributions to the burned area from the three major crop residues were rice 49–61%, sugarcane 31–40%, and maize. Burned rice and sugarcane residues were mostly found in the northeast, whereas maize residue was mostly burned in the north.

Annual particulate emissions from crop residue burning in Thailand were estimated from emission factors and burned areas to be about $PM_{2.5}$ 43.3 Gg/year and $PM_{0.1}$ 4.1 Gg/year. $PM_{0.1}$ emissions during El Niño years were 4.2–5.2 Gg/year, ~ 1.2–1.5 times higher than during a La Niña (3.7–3.4 Gg/year), tracking $PM_{2.5}$ emissions. Rice residue burning dominated $PM_{0.1}$ emissions, contributing about 55%, followed by sugarcane (41%) and maize (4%). $PM_{0.1}$ emissions from crop residue burning constituted ~ 10% of $PM_{2.5}$ in Thailand.

Table 5 PM and PAHs concentration in SEA during open biomass burning: 2015–2019

Locations	Periods	$PM_{0.1}$ (μg/m^3)	$PM_{2.5}$ (μg/m^3)	PAHs in $PM_{0.1}$ (ng/m^3)	References
Upper SEA					
Chiang Mai, Thailand	Apr 2017, Mar–Apr 2018 and Mar 2019	6.6–13.0 ± 3.3	46.8–88.9 ± 21.2	–	Sresawasd et al. (2021)
Bangkok, Thailand	Dec 2018-Jan 2019	11.3 ± 0.8–11.4 ± 4.7	58.3 ± 1.6–67.8 ± 19.8	0.32 ± 0.1–0.51 ± 0.2	Dejchanchaiwong et al. (2020)
	Jan–Feb 2020	18.9 ± 4.0	73.4 ± 16.3	–	Boongla et al. (2021)
Hanoi, Vietnam	Aug 2015	5.97–6.06	46.84–58.14	–	Thuy et al. (2018)
	Oct–Dec 2015	5.36–11.9	70.37–136.87	–	
Lower SEA					
Padang, Indonesia	Mar 2018	8.17 ± 1.48–14.56 ± 3.22	26.21 ± 4.54–42.30 ± 4.70	–	Amin et al. (2021)
Medan, Indonesia	Feb–Mar 2019	13.1 ± 3.8–16.8 ± 4.0	61.7 ± 12.9–213.1 ± 37.8	–	
Hat Yai, Thailand	Sep–Oct 2015	14.2 ± 10.0	73.7 ± 49.8	2.27 ± 1.92	Chomanee et al. (2020)
Thepha, Thailand	Jun–Sep 2019	2.0 ± 1.0	12.9 ± 0.8	0.15 ± 0.01–0.28 ± 0.02	Mahasakpan et al. (2023)
Kuala Lumpur, Malaysia	Feb–Dec 2017	3.98 ± 0.51	21.6 ± 2.4–31.4	–	Jamhari et al. (2022)

Thus, open burning of forest fires and agricultural residues releases significant amounts of fine and ultrafine particles into the atmosphere. Transport of these pollutants affects local air quality as well as neighboring countries. Annual mean $PM_{0.1}$ concentrations during the open biomass burning period ranged from 6.6 to 18.9 $\mu g/m^3$. $PM_{0.1}$ concentrations in Bangkok and Chiang Mai, during haze periods, were 2–4 times and 4–5 times as high as those of the background. During 2015 and 2019 haze episodes in southern Thailand, PAH concentrations in $PM_{0.1}$ were 3–8 times higher than in the background. $PM_{0.1}$ levels in this region confirmed the significant effects of open biomass burning, especially during El Niño years.

This approach can be used to estimate emissions of fine and ultrafine particles from open biomass burning in other countries based on satellite burned area data and pollutant emission factors. With this data, proposals to reduce burning and emissions can be further proposed.

Acknowledgements We thank the National Research Council of Thailand (NRCT), Thailand Science Research and Innovation (TSRI), Biodiversity-based Economy Development Office of Thailand (BEDO) and Prince of Songkla University for research funding and support, and Ms. Nobchonnee Nim for data preparation.

References

Adam, M.G., P.T.M. Tran, N. Bolan, and R. Balasubramanian. 2021. Biomass burning-derived airborne particulate matter in Southeast Asia: a critical review. *Journal of Hazardous Materials* 407: 124760.

Albar, I., I. Jaya, B.H. Saharjo, B. Kuncahyo, and K.P. Vadrevu. 2018. Spatio-temporal analysis of land and forest fires in Indonesia using MODIS active fire dataset. In *Land-atmospheric research applications in South and Southeast Asia* 105–127. Cham: Springer.

Amin, M., R.M. Putri, F. Goembira, M. Hata, and M. Furuuchi. 2021. Carbon characterization of size-segregated particulate matters (PMs) in residential and educational areas in Padang city, Indonesia. *Journal of Geoscience and Environment Protection* 09 (08): 225–238.

Apte, J.S., M. Brauer, A.J. Cohen, M. Ezzati, and C.A. Pope. 2018. Ambient $PM_{2.5}$ reduces global and regional life expectancy. *Environmental Science & Technology Letters* 5 (9): 546–551. 560.

Arunrat, N., N. Pumijumnong, and S. Sereenonchai. 2018. Air-pollutant emissions from agricultural burning in Mae Chaem Basin, Chiang Mai Province Thailand. *Atmosphere* 9 (4): 145.

Badarinath, K.V.S., S.K. Kharol, A.R. Sharma, and V.K. Prasad. 2009. Analysis of aerosol and carbon monoxide characteristics over Arabian Sea during crop residue burning period in the Indo-Gangetic Plains using multi-satellite remote sensing datasets. *Journal of Atmospheric and Solar-Terrestrial Physics* 71 (12): 1267–1276.

Biswas, S., K.P. Vadrevu, Z.M. Lwin, K. Lasko, and C.O. Justice. 2015. Factors controlling vegetation fires in protected and non-protected areas of Myanmar. *PLoS ONE* 10 (4): e0124346.

Boongla, Y., P. Chanonmuang, M. Hata, M. Furuuchi, and W. Phairuang. 2021. The characteristics of carbonaceous particles down to the nanoparticle range in Rangsit city in the Bangkok Metropolitan Region Thailand. *Environmental Pollution* 272: 115940.

Boonman, T., S. Garivait, S. Bonnet, and A. Junpen. 2014. An inventory of air pollutant emissions from biomass open burning in Thailand using MODIS burned area product (MCD45A1). 10.

Chantara, S., D. Thepnuan, W. Wiriya, S. Prawan, and Y.I. Tsai. 2019. Emissions of pollutant gases, fine particulate matters and their significant tracers from biomass burning in an open-system combustion chamber. *Chemosphere* 224: 407–416.

Chomanee, J., K. Thongboon, S. Tekasakul, M. Furuuchi, R. Dejchanchaiwong, and P. Tekasakul. 2020. Physicochemical and toxicological characteristics of nanoparticles in aerosols in southern Thailand during recent haze episodes in lower Southeast Asia. *Journal of Environmental Sciences* 94: 72–80.

Dejchanchaiwong, R., P. Tekasakul, S. Tekasakul, W. Phairuang, N. Nim, C. Sresawasd, K. Thongboon, T. Thongyen, and P. Suwattiga. 2020. Impact of transport of fine and ultrafine particles from open biomass burning on air quality during 2019 Bangkok haze episode. *Journal of Environmental Sciences* 97: 149–161.

FAO. 2022. *Crops and livestock products statistic division*, viewed 15 Mar 2022. https://www.fao.org/faostat/en/#data/QCL.

Feng, X., S. Wei, and S. Wang. 2020. Temperature inversions in the atmospheric boundary layer and lower troposphere over the Sichuan Basin, China: climatology and impacts on air pollution. *Science of the Total Environment* 726: 138579.

Hamid, H.H., P. Tekasakul, W. Phairuang, M. Hata, M. Furuchi, and N.F. Rajab. 2022. Seasonal variation and size distribution of inorganic and carbonaceous components, source identification of size-fractioned urban air particles in Kuala Lumpur Malaysia. *Chemosphere* 287: 132309.

Hantson, S., S. Pueyo, and E. Chuvieco. 2015. Global fire size distribution is driven by human impact and climate: Spatial trends in global fire size distribution. *Global Ecology and Biogeography* 24 (1): 77–86.

Hao, W.M., and M.-H. Liu. 1994. Spatial and temporal distribution of tropical biomass burning. *Global Biogeochemical Cycles* 8 (4): 495–503.

Hayasaka, H., I. Noguchi, E.I. Putra, N. Yulianti, and K. Vadrevu. 2014. Peat-fire-related air pollution in Central Kalimantan, Indonesia. *Environmental Pollution* 195: 257–266.

Jayarathne, T., C.E. Stockwell, A.A. Gilbert, K. Daugherty, M.A. Cochrane, K.C. Ryan, E.I. Putra, B.H. Saharjo, A.D. Nurhayati, I. Albar, and R.J. Yokelson. 2018. Chemical characterization of fine particulate matter emitted by peat fires in Central Kalimantan, Indonesia, during the 2015 El Niño. *Atmospheric Chemistry and Physics* 18 (4): 2585–2600.

Jamhari, A.A., M.T. Latif, M.I.A. Wahab, H. Hassan, M. Othman, T. Abd Jayarathne, C.E. Stockwell, A.A. Gilbert, K. Daugherty, M.A. Cochrane, K.C. Ryan, E.I. Putra, B.H. Saharjo, A.D. Nurhayati, I. Albar, R.J. Yokelson, and E.A. Stone. 2018. Chemical characterization of fine particulate matter emitted by peat fires in Central Kalimantan, Indonesia, during the 2015 El Niño. *Atmospheric Chemistry and Physics* 18 (4): 2585–2600.

Junpen, A., J. Pansuk, O. Kamnoet, P. Cheewaphongphan, and S. Garivait. 2018. Emission of air pollutants from rice residue open burning in Thailand, 2018. *Atmosphere* 9 (11): 449.

Kanokkanjana, K. and S. Garivait. 2013. Alternative rice straw management pactices to reduce field open burning in Thailand. *International Journal of Environmental Science and Development* 119–123.

Kharol, S.K., K.V.S. Badarinath, A.R. Sharma, D.V. Mahalakshmi, D. Singh, and V.K. Prasad. 2012. Black carbon aerosol variations over Patiala city, Punjab, India—a study during agriculture crop residue burning period using ground measurements and satellite data. *Journal of Atmospheric and Solar-Terrestrial Physics* 84: 45–51.

Kim Oanh, N.T., B.T. Ly, D. Tipayarom, B.R. Manandhar, P. Prapat, C.D. Simpson, and L.-J. Sally Liu. 2011. Characterization of particulate matter emission from open burning of rice straw. *Atmospheric Environment* 45 (2): 493–502.

Kim Oanh, N.T., D.A. Permadi, P.K. Hopke, K.R. Smith, N.P. Dong, and A.N. Dang. 2018. Annual emissions of air toxics emitted from crop residue open burning in Southeast Asia over the period of 2010–2015. *Atmospheric Environment* 187: 163–173.

Kim Oanh, N.T., A. Tipayarom, T.L. Bich, D. Tipayarom, C.D. Simpson, D. Hardie, and L.-J. Sally Liu. 2015. Characterization of gaseous and semi-volatile organic compounds emitted from field burning of rice straw. *Atmospheric Environment* 119: 182–191.

Kumar, P., L. Morawska, W. Birmili, P. Paasonen, M. Hu, M. Kulmala, R.M. Harrison, L. Norford, and R. Britter. 2014. Ultrafine particles in cities. *Environment International* 66: 1–10.

Land Development Department. 2022. *Land use and land cover in 2019*, viewed 1 Mar 2022. https://www.ldd.go.th/home/.

Lasko, K., and K. Vadrevu. 2018. Improved rice residue burning emissions estimates: Accounting for practice-specific emission factors in air pollution assessments of Vietnam. *Environmental Pollution* 236: 795–806.

Mahasakpan, N., P. Chaisongkaew, M. Inerb, N. Nim, W. Phairuang, S. Tekasakul, M. Furuuchi, M. Hata, T. Kaosol, P. Tekasakul, and R. Dejchanchaiwong. 2023. Fine and ultrafine particle- and gas-polycyclic aromatic hydrocarbons affecting southern Thailand air quality during transboundary haze and potential health effects. *Journal of Environmental Sciences* 124: 253–267.

McMeeking, G.R., S.M. Kreidenweis, S. Baker, C.M. Carrico, J.C. Chow, J.L. Collett, W.M. Hao, A.S. Holden, T.W. Kirchstetter, W.C. Malm, H. Moosmüller, A.P. Sullivan, and C.E. Wold. 2009. Emissions of trace gases and aerosols during the open combustion of biomass in the laboratory. *Journal of Geophysical Research* 114 (D19): D19210.

Office of Agricultural Economics. 2022. *Agricultural production during the year 2016–2019*, viewed 15 Mar 2022. https://www.oae.go.th/view/1/.

Ouyang, W., B. Guo, G. Cai, Q. Li, S. Han, B. Liu, and X. Liu. 2015. The washing effect of precipitation on particulate matter and the pollution dynamics of rainwater in downtown Beijing. *Science of the Total Environment* 505: 306–314.

Pham, C.-T., B.-T. Ly, T.-D. Nghiem, T.H.-P. Pham, N.-T. Minh, N. Tang, K. Hayakawa, and A. Toriba. 2021. Emission factors of selected air pollutants from rice straw burning in Hanoi Vietnam. *Air Quality, Atmosphere and Health* 14 (11): 1757–1771.

Phairuang, W. 2021. Biomass burning and their impacts on air quality in Thailand. In Biomass Burning in South and Southeast Asia (pp. 21–38). CRC Press.

Prasad, V.K., Y. Kant, and K.V.S. Badarinath. 2001a. Century ecosystem model application for quantifying vegetation dynamics in shifting cultivation areas: a case study from Rampa Forests, Eastern Ghats (India). *Ecological Research* 16 (3): 497–507.

Prasad, V.K., Y. Kant, P.K. Gupta, C. Sharma, A.A. Mitra, and K.V.S. Badarinath. 2001b. Biomass and combustion characteristics of secondary mixed deciduous forests in Eastern Ghats of India. *Atmospheric Environment* 35 (18): 3085–3095.

Roy, D.P., H. Huang, L. Boschetti, L. Giglio, L. Yan, H.H. Zhang, and Z. Li. 2019. Landsat-8 and Sentinel-2 burned area mapping—a combined sensor multi-temporal change detection approach. *Remote Sensing of Environment* 231: 111254.

Samae, H., S. Tekasakul, P. Tekasakul, and M. Furuuchi. 2021. Emission factors of ultrafine particulate matter (PM$<$ 0.1 μm) and particle-bound polycyclic aromatic hydrocarbons from biomass combustion for source apportionment. *Chemosphere* 262: 127846.

Samae, H., S. Tekasakul, P. Tekasakul, W. Phairuang, M. Furuuchi, and S. Hongtieab. 2022. Particle-bound organic and elemental carbons for source identification of PM$<$ 0.1 μm from biomass combustion. *Journal of Environmental Sciences* 113: 385–393.

Saulino, L., A. Rita, A. Migliozzi, C. Maffei, E. Allevato, A.P. Garonna, and A. Saracino. 2020. Detecting burn severity across mediterranean forest types by coupling medium-spatial resolution satellite imagery and field Data. *Remote Sensing* 12 (4): 741.

Sornpoon, W., S. Bonnet, P. Kasemsap, P. Prasertsak, and S. Garivait. 2014. Estimation of emissions from sugarcane field burning in Thailand using bottom-up country-specific activity data. *Atmosphere* 5 (3): 669–685.

Sresawasd, C., T. Chetiyanukornkul, P. Suriyawong, S. Tekasakul, M. Furuuchi, M. Hata, R. Malinee, P. Tekasakul, and R. Dejchanchaiwong. 2021. Influence of meteorological conditions and fire hotspots on $PM_{0.1}$ in northern Thailand during strong haze episodes and carbonaceous aerosol characterization. *Aerosol and Air Quality Research* 21 (11): 210069.

Stockwell, C.E., T. Jayarathne, M.A. Cochrane, K.C. Ryan, E.I. Putra, B.H. Saharjo, A.D. Nurhayati, I. Albar, D.R. Blake, I.J. Simpson, E.A. Stone, and R.J. Yokelson. 2016. Field measurements of trace gases and aerosols emitted by peat fires in Central Kalimantan, Indonesia, during the 2015 El Niño. *Atmospheric Chemistry and Physics* 16 (18): 11711–11732.

Streets, D.G., K.F. Yarber, J.-H. Woo, and G.R. Carmichael. 2003. Biomass burning in Asia: annual and seasonal estimates and atmospheric emissions. *Global Biogeochemical Cycles* 17 (4): n/a-n/a.

Thuy, N.T.T., N.T. Dung, K. Sekiguchi, L.B. Thuy, N.T.T. Hien, and R. Yamaguchi. 2018. Mass concentrations and carbonaceous compositions of $PM_{0.1}$, $PM_{2.5}$, and PM_{10} at urban locations of Hanoi, Vietnam. *Aerosol and Air Quality Research* 18 (7): 1591–1605.

United Nations. 2021. *Demographic Yearbook (2020)*. United Nations.

Vadrevu, K.P., T. Ohara, and C. Justice. (eds.). 2021a. *Biomass burning in South and Southeast Asia: impacts on the biosphere,* vol. 2. CRC Press.

Vadrevu, K.P., T. Ohara, and C. Justice. (eds.). 2021b. *Biomass burning in South and Southeast Asia: mapping and monitoring,* vol. 1. CRC Press.

Wang, X., S. Zhong, X. Bian, and L. Yu. 2019. Impact of 2015–2016 El Niño and 2017–2018 La Niña on $PM_{2.5}$ concentrations across China. *Atmospheric Environment* 208: 61–73.

Watson, J.G., J. Cao, L.-W.A. Chen, Q. Wang, J. Tian, X. Wang, S. Gronstal, S.S.H. Ho, A.C. Watts, J.C. Chow. 2019. Gaseous, $PM_{2.5}$ mass, and speciated emission factors from laboratory chamber peat combustion. *Atmospheric Chemistry and Physics* 19 (22): 14173–14193.

Wooster, M., David Gaveau, M. Salim, T. Zhang, W. Xu, D. Green, V. Huijnen, D. Murdiyarso, D. Gunawan, N. Borchard, M. Schirrmann, B. Main, and A. Sepriando. 2018. New tropical peatland gas and particulate emissions factors indicate 2015 Indonesian fires released far more particulate matter (but less methane) than current inventories imply. *Remote Sensing* 10 (4): 495.

Characteristics of Transboundary Haze and General Aerosol Over Pulau Pinang, Malaysia

Lim Hwee San, Brent N. Holben, Ezekiel Kaura Makama, and Mohamad Farid Izzat Bin Zahari

Abstract Profound episodes of transboundary haze originating from Indonesian regions were observed in the year 2015 across the states of Peninsular Malaysia. In this study, we characterize the aerosol properties in Pulau Pinang in the following year, 2016, from a ground-based Aerosol Robotic Network (AERONET) observations. Using the AERONET 500 nm data, the aerosol optical depth (AOD) parameters, anomalies, and sudden increases or drops in daily values were identified and investigated. The highest mean AOD was found during the pre-monsoon season, with the corresponding least value during the post-monsoon due to the down washing of aerosol occasioned by constant rainfall. The spikes in the pre-monsoon season are attributed to aerosol transportation by a prevailing wind. The dominant aerosol type for the study domain was determined using α–τ scatter plots for the 500 nm AOD and 440–870 nm Angstrom exponent (AE), which compared well with many other reported thresholds. Biomass aerosol (BMA), being the dominant aerosol, is seasonally dependent in 2016, particularly during the pre-monsoon season, followed closely by post-monsoon, suggesting weather as the most influencing factor of aerosol for each season. Historical AERONET AOD, air pollution index (API), and MODIS data showed an increasing annual trend at the beginning of the northeast monsoon (NEM) season in 2016, which corresponds with the period of the transboundary haze recorded in 2015.

Keywords Transboundary haze · Aerosol · AERONET · AOD

L. H. San (✉) · M. F. I. B. Zahari
School of Physics, Universiti Sains Malaysia, 11800 Pulau Pinang, Malaysia
e-mail: hslim@usm.my

Brent N. Holben
NASA Goddard Space Flight Center, Greenbelt, MD, USA

E. K. Makama
Department of Physics, University of Jos, Jos PMB 2084, Nigeria

K. P. Vadrevu et al. (eds.), *Vegetation Fires and Pollution in Asia*,
https://doi.org/10.1007/978-3-031-29916-2_21

1 Introduction

Aerosol, being an uncertain but critical player in regional and global climate change (Dubovik et al. 2002), impacts the balance of the earth's radiation budget, either directly through the scattering and absorption of incoming solar radiation or indirectly by modifying the microphysical properties of clouds in the form of condensation nuclei (Kant et al. 2000; Kaufman et al. 2005; Badarinath et al. 2007, 2008, 2009; Kharol et al. 2012; Rosenfeld et al. 2019). Besides, several studies have posited that the increase in aerosol emissions has not only enhanced aerosol radiative effects that partly unbalance the greenhouse effect (Stocker et al. 2013) but poses a threat to public health (Pöschl 2005).

Rapid industrial and economic developments in recent years are primarily responsible for the increase in anthropogenic aerosol emissions. Malaysia, a fast-developing country in Southeast Asia, has its fair share of air pollution concerns due to increased anthropogenic aerosol emissions. The study area, Pulau Pinang in Peninsular Malaysia, is one of the densely populated cities characterized by large industrial and vehicular activities. Moreover, persistent uncontrolled forest fires in the past few years (Field et al. 2009), mainly from Indonesia, a neighboring country, have aggravated air quality across the study site (Hayasaka et al. 2014). Malaysia has a long history of transboundary haze effects, with the earliest official study conducted in 1980 (Palanissamy 2013). Based on the air pollution index (API) report for 2015, Pulau Pinang, along with other states in Malaysia, was affected by a haze incident that originated in the Indonesian region. The anthropogenically spurred haze, transported by the prevailing Southwest monsoon wind to the study site, is attributed to agricultural open-burning activities done by farmers of local plantations in the affected region (Albar et al. 2018).

These incidences are further compounded by the structure of the earth's atmosphere, which is characterized by complex wind circulations, precipitation, and temperature regimes. Such atmospheric complexities have impeded comprehensive atmospheric studies due mainly to data uncertainties from cloud contamination (Hyer et al. 2011) or its complete removal occasioned by strict screening stages (Van Donkelaar 2011). The worsening air quality due to the contamination of the atmospheric space demands immediate assessment in both the short and long-term measures by the source and receptor communities (Reid et al. 2013; Vadrevu et al. 2018; 2021a, b). Although small-scale studies of aerosol optical properties are considered accurate, temporal and spatial coverage over a larger study area are sacrificed. Previous localized studies, from both the sun and sky-scanning radiometers of AERONET, can best be used as complementary investigations in more detailed aerosol studies (Tan et al. 2015a, b). Frequent data voids in AERONET optical depth over the current study site due to prevalent cloud cover threaten the accurate evaluation of aerosol optical density (AOD). Aside from AERONET, Moderate-resolution Imaging Spectroradiometer (MODIS), a global atmospheric monitoring system, is widely used to study AOD. However, the newly launched collection of MODIS AOD observation requires that its suitability in aerosol research and in tandem with AERONET AOD in this part of Southeast Asia be evaluated.

Though the immediate effect of the transboundary haze on Pulau Pinang has been characterized, its long-term effect has yet to be confirmed. Therefore, this study focuses mainly on investigating the aerosol properties based on the effect of the 2015 transboundary haze on AERONET AOD data in 2016, as well as evaluating the suitability of MODIS AOD data over Pulau Pinang.

2 Data and Methods

The study area, Pulau Pinang, an island in the northwestern part of Peninsular Malaysia connected to its mainland by the Penang bridge, is shown in Fig. 1. The area has 1.746 million inhabitants within an area of about 1048 km^2.

The AOD dataset used in this study is from the AERONET, while the historical weather records of temperature, wind speed, and direction are from Weather Underground (WU). The Air Pollution Index (API) was obtained from the Malaysian Administrative Modernization and Management Planning Unit (MAMPU).

The AERONET site, codenamed USM_Penang, has been in operation since 2011. The USM_Penang site is located at an observatory platform of the School of

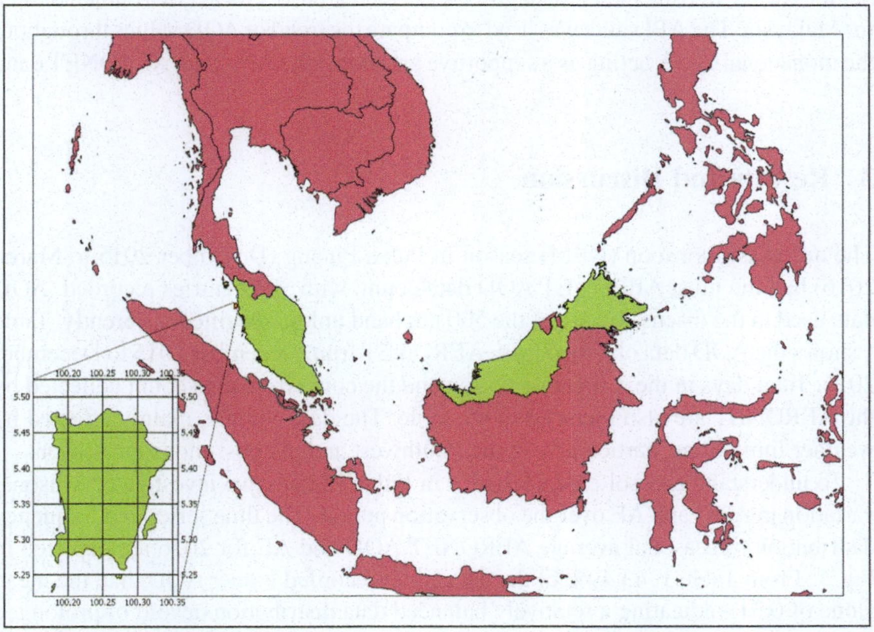

Fig. 1 Overview of the study area (enlarged in bottom left corner). The focus is the island part of Pulau Pinang

Physics (5.358 N, 100.302 E), Universiti Sains Malaysia (USM), Pulau Pinang. The AERONET suite is equipped with a CIMEL sun photometer which makes measurements of direct-sun irradiance in eight spectral channels (340, 380, 440, 500, 675, 870, 940, and 1020 nm) and sky radiance in four channels (440, 675, 870, and 1020 nm) (Holben et al. 1998). The AERONET provides instantaneous data and daily average by calculating the diurnal average of the instantaneous values (Holben et al. 1998). More descriptions of the instrument's working principle, calibration, data processing, and quality control have been provided elsewhere (Dubovik et al. 2002; Holben et al. 1998). In this study, daily AOD, Angstrom exponents (AE), and precipitable water (PW) from December 2015 to December 2016, covering the entire monsoonal cycle in the year 2016 over the region of interest, were retrieved from the AERONET website (version 3, level 2). This dataset is high-quality data that is cloud-screened with instrument anomalies removed (Smirnov et al. 2000; Giles et al. 2019). The daily average data has been used to calculate the monthly and seasonal mean of the aerosol parameters of our interest.

Weather data such as wind speed (m/s), wind direction, and daily average temperature (OC), required for a robust explanation of the possible outcome of AERONET data, were downloaded from Weather Underground (WU) (https://www.wunderground.com). In addition, hourly and daily averaged data from the Penang International Airport, Bayan Lepas, approximately 8.4 km to the south-southeast of the AERONET site, were used. Also, daily averaged API data for December 2015 to December 2016 in Pulau Pinang were downloaded from MAMPU—Data Terbuka Sektor Awam website (www.data.gov.my), which archives local environmental data for Malaysia. The API value was used to compare the trend of AOD values throughout the monsoonal cycle, acting as a supportive ground truth to evaluate AERONET data.

3 Results and Discussion

The northeast monsoon (NEM) season in Pulau Pinang (December 2015 to March 2016) had the most AERONET AOD data count, with 3029 entries recorded. AOD data used in the discussion are of the 500 nm band unless specified differently. Table 1 shows the AOD data obtained from AERONET from December 2015 to December 2016. Total days in the respective season and the concurrent data count collected by the AERONET are also included in the table. The data count is mainly affected by weather limitations, particularly in the Southwest and the post-monsoon season.

To understand aerosol characteristics in Pulau Pinang, we investigated seasonal variation in AOD and AE over the observation period. The time series and frequency distribution of seasonal average AERONET AOD and AE for 2016 are depicted in Fig. 2. From Table 1, 43.49% of the data count recorded higher AOD than the mean value of 0.31, indicating a relatively balanced data distribution devoid of prolonged series of extremes, notwithstanding the presence of intermittent (Fig. 2a). The result showed that daily AOD over Pulau Pinang varied in the range of 0.03 to 1.78 with the minimum and maximum values observed on 9 September and 24 April, respectively.

Table 1 Seasonal duration and data count for December 2015 to December 2016

Season	Days in total	Total data recorded
NEM (December 2015–March 2016)	122	3029
Pre-monsoon (April–May)	61	591
SWM (June–September)	122	720
Post-monsoon (October–November 2016)	61	271

This is contrary to the hypothesized maximum in the SWM period when the annual transboundary haze transport is reported to occur. These periods of maximum and minimum AOD in 2016 are discussed subsequently in this section.

The modal value for the frequency distribution of AOD 500 nm in 2016 is 0.4, as seen in Fig. 2b. From the figure, the distribution is more populated on the lower spectrum of AOD, with the frequency greatly decreasing past the modal value, with values > 1 occurring < 1% of the time. The rare occurrence of AOD > 1.0 in the study area is most likely due to anomalies such as the influx of transported aerosols or noises that may impair the reading of AERONET instruments. However, it could also be suggestive that there was no heavy and sustained pollution throughout the 2016 monsoonal cycle.

The AOD 440 nm wavelength band, a general assumption of an environment background cleanliness proposed by Toledano et al. (2007), was also analyzed. They opined that a clean background ideally should have an AOD 440 nm value less than 0.05. The average value of 0.36 for Pulau Pinang classifies the area as generally unclean, however, in some instances, lower AOD values than the specified average, with some days accounting for as low as 0.03. However, these occurred briefly and did not represent the overall average picture of AOD in Pulau Pinang for the 2016 monsoonal cycle.

The time series and frequency distribution plots of AE (400–870 nm) for Pulau Pinang in 2016 are shown, respectively, in Fig. 2a, b, with a large daily average variation being 1.30. The high AE value indicates that the prevailing atmospheric aerosols are small (or fine-sized) across the study area since AE is inversely related to the particle size distribution. This category of aerosol is of anthropogenic sources (e.g., open burning, vehicular and industrial pollutions) (Holben et al. 2001), which is particularly true for Penang, listed among the states in Malaysia with the highest vehicular population (Afroz et al. 2003). The frequency distribution indicates that most recorded AOD are associated with large AE values (bin 1.6 being the maximum), showing 33% dominance of fine-sized aerosol over Pulau Pinang in the 2016 monsoon cycle.

The time series plot and frequency distribution of PW for the whole monsoon cycle in 2016 are depicted in Fig. 3, with a seasonal average of 4.45 cm. The averages for each season showed an increasing trend, with the calculated values being 4.11, 4.27, 4.53, and 4.88 cm for NEM, pre-monsoon, SWM, and post-monsoon seasons, respectively. Though PW and rainfall have been reported to exhibit a strong relationship

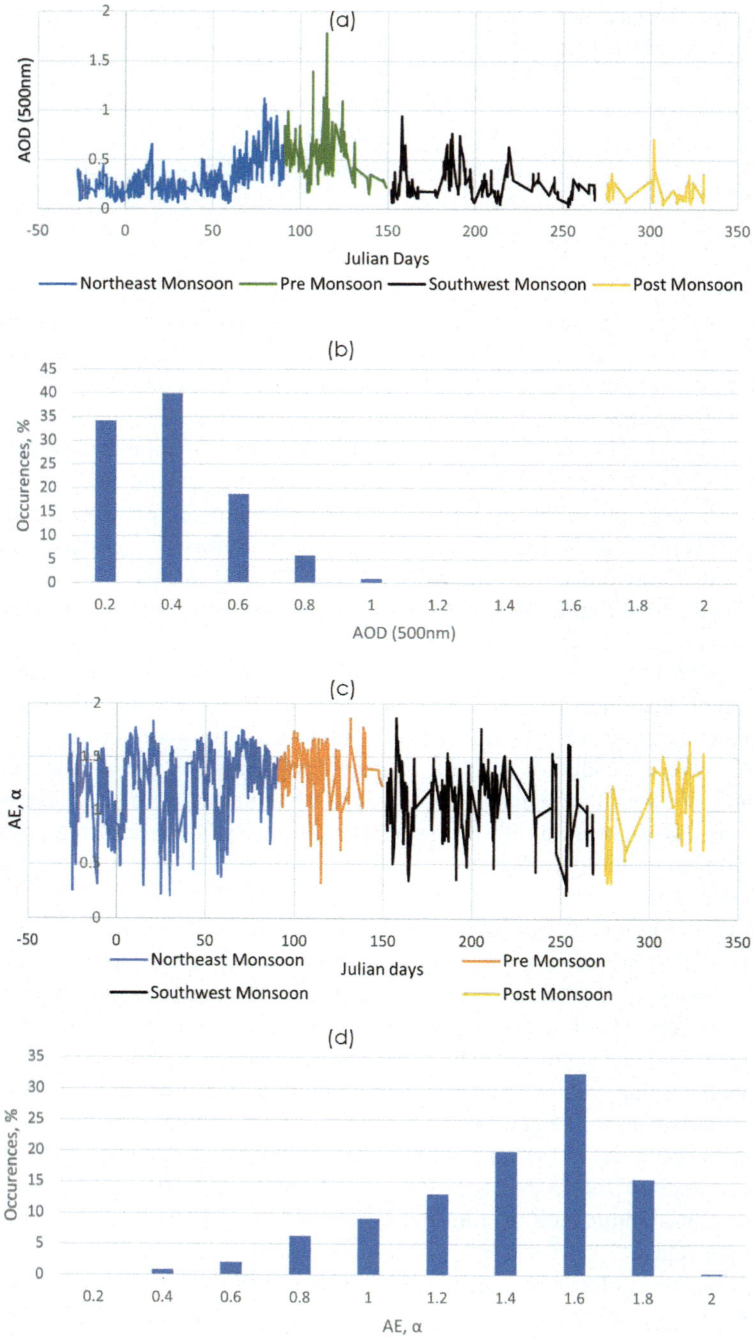

Fig. 2 Monsoonal cycle of the USM_Penang AERONET data for 2016: **a** AERONET 500 nm AOD time series, **b** AERONET 500 nm AOD frequency distribution, **c** AERONET 440–870 nm Angstrom Exponent (AE) time series, AERONET 440–870 nm, and **d** AE frequency distribution

(Trenberth and Zhang 2018), an increase in the former does not necessarily result in the latter's occurrence. Although higher water vapor content aloft could indirectly lead to the formation of clouds and subsequent precipitation (Reichardt et al. 2012), which may interfere with AERONET reading, other factors like relative humidity and temperature have also been noted to induce formation.

The cut-out section from the yearly trend for the NEM season in 2016 is shown in Fig. 4a with a mean AOD of 0.30 ± 0.13, slightly below the yearly mean of 0.31 ± 0.16. A distinct behavior from the graph depicts higher AOD, mainly toward the end of the season (March 2016). The parameters in Table 2 are also suspected to be partly, if not entirely, responsible for the increase in AOD values within this period in 2016.

A report from WU (Table 2) showed that from January to March 2016, daily average precipitation rapidly decreased from 3.38 mm in December 2015 to 0.02 mm in March 2016. Days in March were also paired with a relatively high daily mean temperature of 29.1 °C, compared to the 27.3 °C mean obtained in December 2015. Additionally, mean daily AERONET PW also showed a decreasing pattern toward the end of this season. It is evident from Table 2 that the later part of the NEM season is drier and hotter than the earlier part. It could be inferred, therefore, that the dry and hot weather, devoid of precipitation, may be responsible for the aerosol build-up in the atmosphere (Ali et al. 2014). The NEM season was also drier than other seasons, as reported in a simple, 30-year-long analysis by International Association for Medical to Travelers (IAMAT) (https://www.iamat.org/country/malaysia/climate-data).

A cursory inspection of Fig. 4a reveals a drop in AOD on day 15 (15 January 2016), between 12:00 UTC and 04:00 UTC, when weather data indicated rainy activity. This decrease in AOD, which is ~ 83% (from 0.66 to 0.11), may be attributed to the slight difference in the weather data location relative to the AERONET site, best described as a frontal passage (Song et al. 2019).

The frequency distribution of AOD for the NEM season, grouped into bins of the 0.2 range, is presented in Fig. 4b. Modal AOD is the 0.4 bin, which represents almost 46% of the entire data recorded for the season, while for the total data, the 0.2 and the 0.6 bins represent about 33% and 15%. Bins with larger values (0.8 and

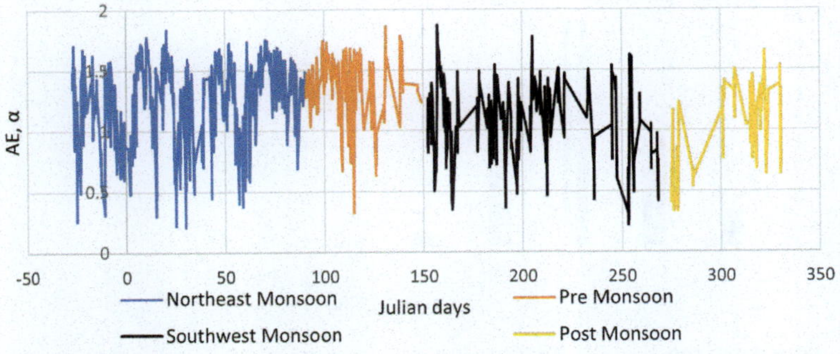

Fig. 3 Monsoonal cycle of AERONET AE for the USM_Penang site in 2016

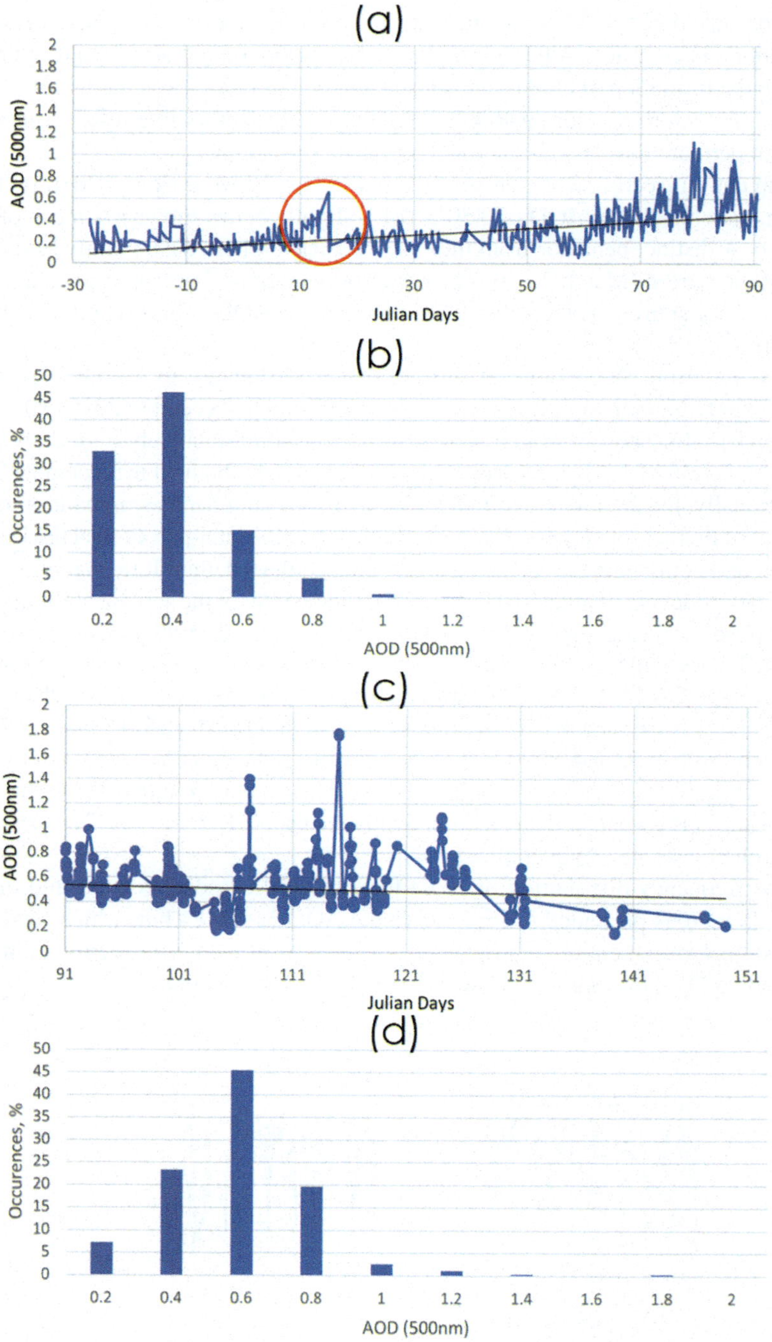

Fig. 4 AERONET 500 nm seasonal AOD and frequency distribution at the USM_Penang site in 2016 for: northeast monsoon (**a** and **b**), pre-monsoon (**c** and **d**), southwest monsoon (**e** and **f**), and post-monsoon (**g** and **h**)

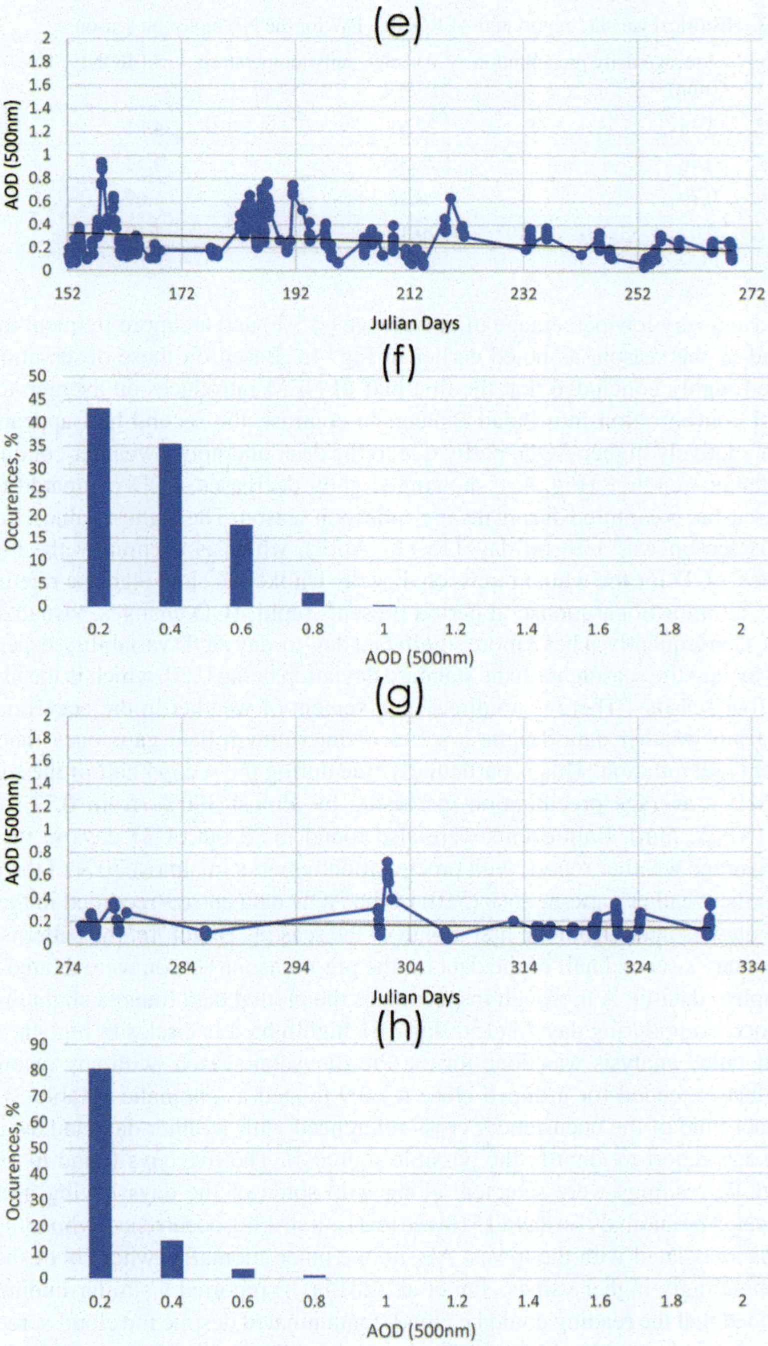

Fig. 4 (continued)

Table 2 Historical weather report and AERONET PW for the NE monsoon season

Month	Average daily precipitation (mm)	Average daily temperature (°C)	AERONET PW (cm)
Dec-15	3.38	27.31	4.61
Jan-16	0.56	28.32	4.53
Feb-16	0.99	28.28	4.09
Mar-16	0.02	29.10	4.10

> 1) have a very low percentage of occurrence (< 5%) and are more frequent toward the end of the season, as noted earlier in Fig. 4a. Based on these observations, it can be roughly concluded that the first half of NEM introduces an average-to-low aerosol concentration into Pulau Pinang. In contrast, the second half appeared to have a relatively higher AOD, partly due to the drier and hotter weather conditions.

A flatter trendline (Fig. 4c), showing slightly decreased AOD compared to the NEM season, is captured during the pre-monsoon season. The highest value recorded for this season was 1.78 on day 115 (24 April), which is eventually the highest recorded AOD for the whole monsoonal cycle. Unlike the slow increase rate in late NEM, the transitional monsoon period presents rapid AOD changes, visualized as spikes. Consequently, it has a more significant day-to-day AOD variability, especially mid-way into the season, with the standard deviation being 0.20, which is the highest in all four seasons. There is no direct involvement of weather in the occurrence of AOD from weather data despite a series of intermittent data gaps occasioned by frequent precipitation. This is particularly true during the second half of the season, with May's average precipitation increasing by almost 300% (from 0.77 mm in April to 3.22 mm). Rain events were also found in 22 out of 31 days in the WU daily average weather report, with precipitation remarks in almost 70% of the days. Due to the weather pattern, 86% (510) of the AOD data entries recorded in the pre-monsoon season are from the first month of the season. Therefore, the patterns seen in this study's second half of the data for the pre-monsoon season were treated as an incomplete dataset. A thorough inspection of the plotted data found a slight drop of AOD occurring during day 15 (15th Jan-16), highlighted in circles in Fig. 4a.

A detailed analysis was done for the Pre-monsoon season, with any records of AOD that exceeded (or grouped close to) 0.9 treated as anomalies (Table 3) with the exact time of the occurrences cross-referenced with weather data and available local news report to identify the possible source(s). The five days found to exhibit high AOD readings were selected, along with some of the days having multiple instances. The anomaly on day 115 (in red text) is a striking occurrence, whose highest AOD is associated with the lowest AE, unlike other anomalies where both showed correspondingly higher values. Tan et al. (2015a, b) reported a similar finding and concluded that the reading could be cloud-contaminated despite the cloud-screening process implemented by AERONET prior to data distribution. This is probably the case in our finding since the data averaged 70% higher AOD, vastly different from other spikes recorded.

Table 3 List of anomalies and their details (AE for 550–870 nm)

Day	Instances (UTC)		AOD value (500 nm)	AE (500–870 nm)	Wind direction	Wind speed (mph)	Remarks
107	3	03:30	1.3965	1.1816	NNE	2	Cloudy
		03:50	1.3468	1.1645	N	3	Cloudy
		04:20	1.1445	1.3496	N	3	Cloudy
113	2	05:00	1.1285	1.3956	Calm	0	Cloudy
		05:30	1.039	1.1592	Calm	0	Cloudy
115	2	12:30	1.7568	0.3961	NNE	3	Thunder
		12:45	1.7788	0.3234	NNE	3	Thunder
116	1	01:15	1.0119	1.1779	Calm	0	Cloudy
124	5	02:15	1.0733	1.0129	Calm	0	Cloudy
		02:30	0.9993	1.0443	NNW	1	Cloudy
		02:45	1.0911	1.0636	NNW	1	Cloudy
		03:00	1.0768	1.0354	NE	3	Thunder
		03:45	0.9066	1.0126	N	2	Thunder

Other instances of anomalies with several similar weather properties also existed. For instance, a wind direction that persists from mainly north (N) and sometimes north-northeast (NNE) for most of the time, blowing into the study site, is also a weather phenomenon of interest. Geographically, Gelugor and Georgetown in Pulau Pinang match the northeasterly wind direction and based on AE ($\alpha > 1.0$) less aerosol composition was prevalent in the atmosphere during this period. One inference that suits both characteristics was the fine surplus particles emitted by a source(s) in that region and transported by the wind. Figure 5, the map of Pulau Pinang, depicts the position of Georgetown relative to the AERONET site used in this study.

The relative position of the study site to the Georgetown area holds the possibility of the prevailing NNE wind transporting an influx of aerosol into the study site. It thus causes an increase in the overall reading of AOD in the destination area, up to five times the average background conditions (Pérez-Ramírez et al. 2017). We attempt to compare the general weather conditions during those instances, which are primarily cloudy ~ 70% of the time. However, cloudy conditions were common in the study area, not only on the days shown above but throughout the year. It is, therefore, difficult to say with certainty if cloudy conditions are associated with the spikes in the AOD reading. Although cloudy conditions have been fingered to enhance the overall AOD by 0.05 with cloud fraction (CF) of 0.8–0.9 in the current study area (Chand et al. 2012; Christophe et al. 2010), the degree of the possible enhancement is relatively negligible.

Furthermore, the spikes happen during midnight and early morning (i.e., from 12.00 am to 05.30 UTC). Inversion could form between those time frames, but the AOD value's erratic behavior at that time makes it seem unlikely. However, it is

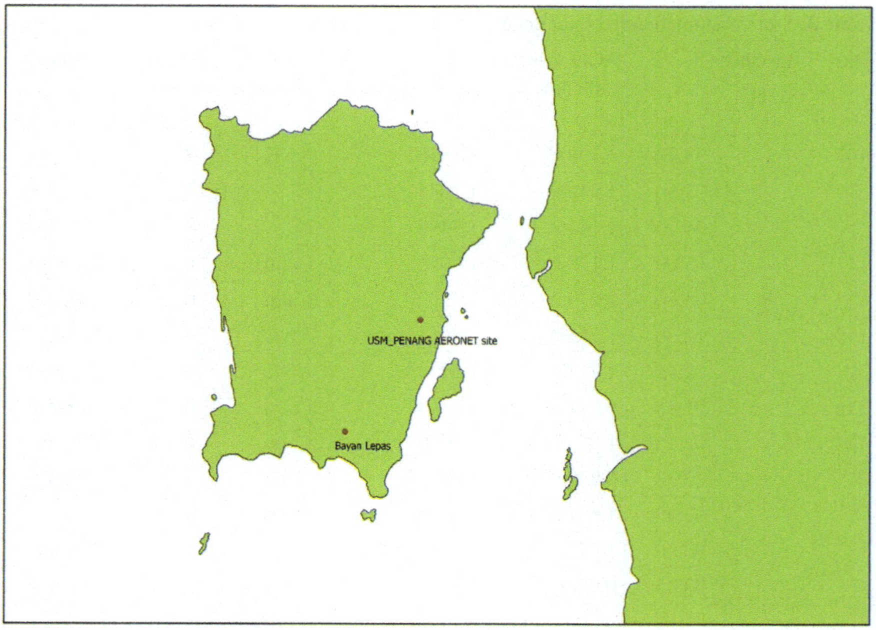

Fig. 5 Map of Pulau Pinang showing the location of the USM_Penang site and Georgetown (possible source)

outside the scope of the present study to determine whether the concerned weather properties can sustain the formation of an inversion (i.e., cold nights in urban areas). Furthermore, no linkable information was found from the local news on reported events. The time frame also does not suggest any known yearly events that could contribute to a sudden increase in aerosol loading. Therefore, the idea of an event with significant importance being the cause of the observed spikes was discarded.

Figure 4d shows the AERONET 500 nm AOD frequency distribution for the USM_Penang site for the 2016 pre-monsoon season. From the frequency distribution, the 0.6 bin presented the highest frequency for the pre-monsoon season, representing ~ 45% of the data, followed closely by the 0.4 and 0.8 bins at 23% and 19%, respectively. The high percentage in the 0.6 bin, compared to the 0.4 bin during the previous season, suggests an increase in aerosol loading overall for the pre-monsoon season. However, the need for more information on such activities makes it difficult to explain the increase in overall aerosol loading during this season.

The SWM season was initially suspected of having a series of extreme AOD values due to the transboundary haze from Indonesia that occurs around July and August annually. However, the data retrieved from the AERONET is plagued with large gaps of almost one week in certain instances throughout the season, as can be seen in Fig. 4e, which presents AOD distribution for the SWM season. The data count for this season, which is 720, is relatively low, representing a 76.2% reduction compared to the NEM season, despite having a similar day count in terms of seasonal duration of 122 days. The data also lacks the predicted extreme highs, with a calculated average mean < 0.272 and a flat-to-decreasing trendline. Acknowledging the necessary details and differences between the 2015 and 2016 SWM seasons is essential. Firstly, the forest fire in Indonesia in 2016 was not as intense as in previous years. Secondly, Penang's resulting transboundary haze (if any) during that period (June to September 2016) was not as heavy as in previous years. Finally, the data count during SWM is minimal compared to the NEM (ratio of 1:4), possibly due to the intrusive weather pattern during the season. Therefore, the average calculated is not likely a representation of aerosol characteristics during the SWM season due to sparse data.

From Fig. 4e, relatively low average values were observed for the whole season compared to the 0.27 seasonal average calculated, which is lower than the yearly average of 0.31. This may be due to constant rainfall recorded during the season, in which case suspended particles in the air are down washed [Song et al. (2019) and Holben et al. (2001)], thereby impairing the AOD reading for the whole season. The yearly minimum also occurred this season on day 253 (9 September), where AOD reached an all-time low for 2016 at 0.03. This corresponds to the heavy rain from midnight of day 252 (8 September) to the afternoon of the next day and intermittent light showers with occasional cloud cover, reported in the historical weather records.

The frequency distribution for the SWM season is shown in Fig. 4f. The low-valued bins hold most of the data distribution, indicating that 43% of the data were below 0.2, followed by the 0.4 bin and the 0.6 bin values, which are 35.5% and 17.9%, respectively. The 0.4 and 0.2 bins covered over two-third of the total frequency of occurrences, depicting the trend of a low AOD in the season. Aerosol scavenging via precipitation, suspected to be the case in the current study, has been shown to reduce columnar mass loading up to 64%, depending on the particle size distribution (Saha and Krishna Murthy 2004).

Figure 4g, h respectively depicts AERONET 500 nm AOD time series and its frequency distribution for the post-monsoon season. The season presented only 272 entries (~ 5.90%) of the 4614 data recorded in the full monsoon cycle, and as expected, the plotted data is extremely plagued with significant gaps. These data voids may again be attributed to unsuitable prevailing weather conditions, which are thought to impair AERONET from running its pre-programmed measurements. This conclusion is based on the weather observations which reported frequent rain and thunderstorms in the post-monsoon months. Average precipitation peaked at 56.6 cm, where 50 out of 61 days in October and November are marked with up to 81% of rainfall. This also explains the low mean AOD (0.1567) for this season, with a 48.89% departure from the yearly mean and the lowest across all the seasons in 2016.

The frequency distribution (Fig. 4h) indicates that most of the data (80.8%) lies within the 0.2 bin, followed by 14.7% in the 0.4 bin. The rest of the data are thinly distributed in the 0.6 and 0.8 bins (3.7% and 1.1%, respectively). Overall, the post-monsoon season is the least polluted, primarily due to the frequent precipitation within the period, showing weather impact on AOD reading. Radke et al. (1980) reported that the degree of scavenging efficiency by precipitation on atmospheric aerosol depends on the sizes of aerosols and the rain droplets. However, since raindrop size, as one of the accountable parameters, is outside the scope of this study, an in-depth explanation of the influence of precipitation on the AOD data observed in all four seasons is difficult to substantiate.

The α-τ scatter plots for each season were constructed and presented in Fig. 6. Using the threshold comparison method, the predominant aerosol in the atmosphere for each season can be identified in most cases. The data density pattern for NEM (Fig. 6a) can be identified, with the majority of the recorded AOD values (< 0.5) corresponding to a broad spectrum of AE ranging from 0.2 to 1.8. The diversity of AE in the AOD spectrum decreases as AOD increases. AOD readings larger than 0.6 can be seen to have a corresponding AE value of at least greater than 1.0. This suggests that the AOD spikes recorded at AE > 1.0 were mostly due to particles with small sizes suspected of anthropogenic origin, which is a common characteristic in urban areas (Li et al. 2015).

During the pre-monsoon season (Fig. 6b), the spread of AE data can be seen tightened to an approximate range of 1.0–1.8 with few outliers. The AOD values are concentrated in the 0.4 to 0.7 region, most of which are associated with a large AE (> 1), implying a majority of fine-sized aerosol, as suggested earlier. The spikes that occurred in the mid-season are, therefore, associated with transported aerosol from the urban area of Georgetown to the study site. Interestingly, a few readings at the far right of Fig. 6b exhibit large AOD with extremely small AE (circled in red), as previously discussed. Although cloud particles have been reported to enhance the value of AOD with minor AE (Tan et al. 2015a, b), dust and marine (salt particle) aerosols are also suspects, particularly the latter, given the geographical location of the study site.

During the SWM season (Fig. 6b), a more spread out yet similar trend as in the NEM period was observed, perhaps due to weather influence, which is responsible for a lower data yield. On the lower spectrum of AOD, AE ranges from 0.2 to 1.6, with the AE calculated to be > 0.8 as AOD increases (> 0.4). Traces of relatively high AOD readings (circled in red) with low AE values can be seen on the right side of Fig. 6c. Again, cloud contamination could be the case here, just like during the pre-monsoon season, but with significantly lower AOD than the previous occurrences. Finally, in the Post-monsoon season (Fig. 6d), a semblance of the NEM can still be seen, although with a much lower data count. Lower AOD reading has a wide range of calculated AE, everywhere from 0.2 to 1.7, while higher AOD is mainly attributed to a higher AE. This reinforces the idea that the study site's aerosol characteristic is mostly fine.

The main aerosol types in an atmosphere are best understood by analyzing the relationship between AOD and AE, a popular procedure deployed over a wide range of environments (e.g., Pokharel et al. 2019; Rupakheti et al. 2019a, b, 2020, 2021). To differentiate the aerosol types, we used the threshold values of AOD and AE similar to Tan et al. (2015a, b), based on the thresholds of (1) dust aerosol (DA), maritime aerosol (MA), urban and industrial aerosol (UIA) with free dust and sea salt, and biomass aerosol (BMA) (Salinas et al. 2009), (2) coarse DA, MA, continental/ urban/industrial aerosol, and BMA (Abd Jalal et al. 2012), and (3) DA, MA, continental/urban/industrial aerosol, and BMA (Toledano et al. 2007).

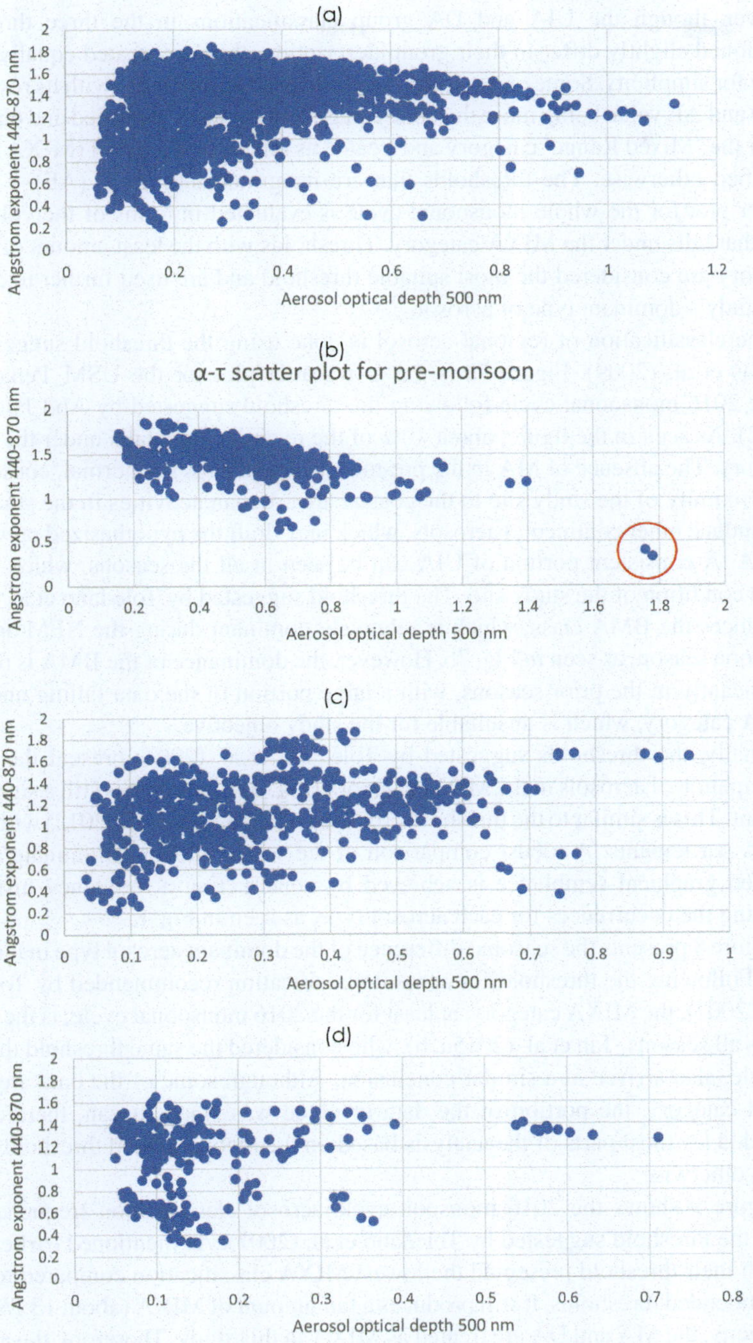

Fig. 6 Scatter plots of α–τ for AERONET data at the USM_Penang site in 2016: **a** northeast monsoon (NEM), **b** pre-monsoon, **c** southwest monsoon (SWM), and **d** post-monsoon

Even though the UIA and DA group classifications in the three thresholds mentioned slightly differ in their group description, they are treated equally in this study for simplicity. Some exceptional cases where a measurement with its respective AOD and AE value being more than one categorization were identified and classified under the 'Mixed Range' category and treated as mixed type aerosol (MIXA) unless specified otherwise. The thresholds that are integrated into the α (AE)–τ (AOD) scatter plot for the whole monsoonal cycle is evaluated in terms of the volume of data that falls under the MIXA category. Thresholds with the least amount of MIXA category are considered the most suitable threshold and are used further to analyze this study's dominant type of aerosol.

The classification of regional aerosol is done using the threshold suggested by Salinas et al. (2009). Figure 7a is the α–τ scatter plot for the USM_Penang site in the 2016 monsoonal cycle following the threshold suggested by Abd Jalal et al. (2012). As seen in the figure, about 40% of the recorded data falls under the MIXA category. The absence of MA in the pre-monsoon season is not normal, considering the proximity of the study site to the coastal area. Urban activities in the study area may induce other continental aerosols, which may draft the hypothesized prevalence of MA. A consistent portion of UIA can be seen in all the seasons, which fits the urban condition of the study site. The threshold suggested by Toledano et al. (2007) introduces the BMA class, which is relatively dominant during the NEM and Pre-monsoon season, as seen in Fig. 7b. However, the dominance of the BMA is reduced significantly in the prior seasons, with a large portion of the data falling under the MIXA category, which is unsuitable for the study objective.

Finally, the thresholds suggested by Toledano et al. (2007) present the lowest participation of aerosols in the MIXA category (Fig. 7a), though it is still a significant amount. This is similar to the threshold proposed by Abd Jalal et al. (2012) with lesser MIXA participants. A robust comparison of the scatter plots to illuminate a more detailed graphical semblance is achieved by constructing a frequency histogram detailing the occurrences for each aerosol type, as seen in Fig. 8.

Figure 8 presents the seasonal difference of the dominant aerosol type in the study area. Following the threshold for aerosol classification recommended by Toledano et al. (2007), the MIXA category, at least for the 2016 monsoonal cycle, is the lowest across all seasons. Tan et al. (2015a, b), who considered the same threshold the most suitable, also arrived at a similar conclusion. Although some of the data are in the mixed category, the portion of the data involved was small. It can, therefore, be excluded for most parts of the analysis based on the recommended threshold unless stated otherwise.

Figure 9 shows the 2016 monsoon cycle aerosol classification frequency plot using the threshold suggested by Toledano et al. (2007). As mentioned earlier, even though their threshold presented the lowest MIXA classification compared to other recommended thresholds, it still produces a fair amount of MIXA (about 13.5%). For simplicity, the MA and DA are treated as MIXA in this study. Therefore, the amount of discarded data, along with the Mixed category, was approximately 14.15%.

In the NEM season, the top three aerosol types are the BMA, UIA, and MA, calculated at 30.2%, 26.0%, and 22.8%, respectively. The geographical location of

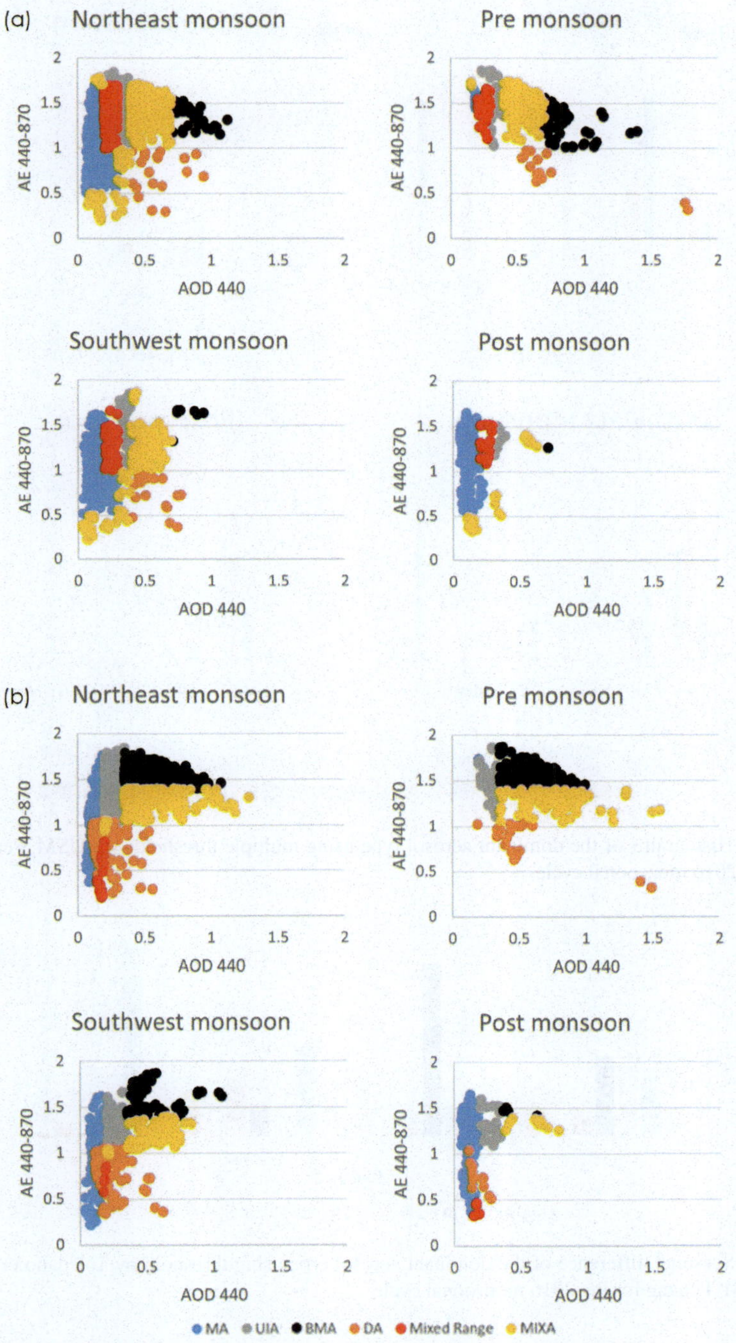

Fig. 7 The 2016 monsoonal cycle α–τ scatter plots using the threshold of: **a** Abd Jalal et al. (2012), **b** Toledano et al. (2007) for AERONET site at USM_Penang

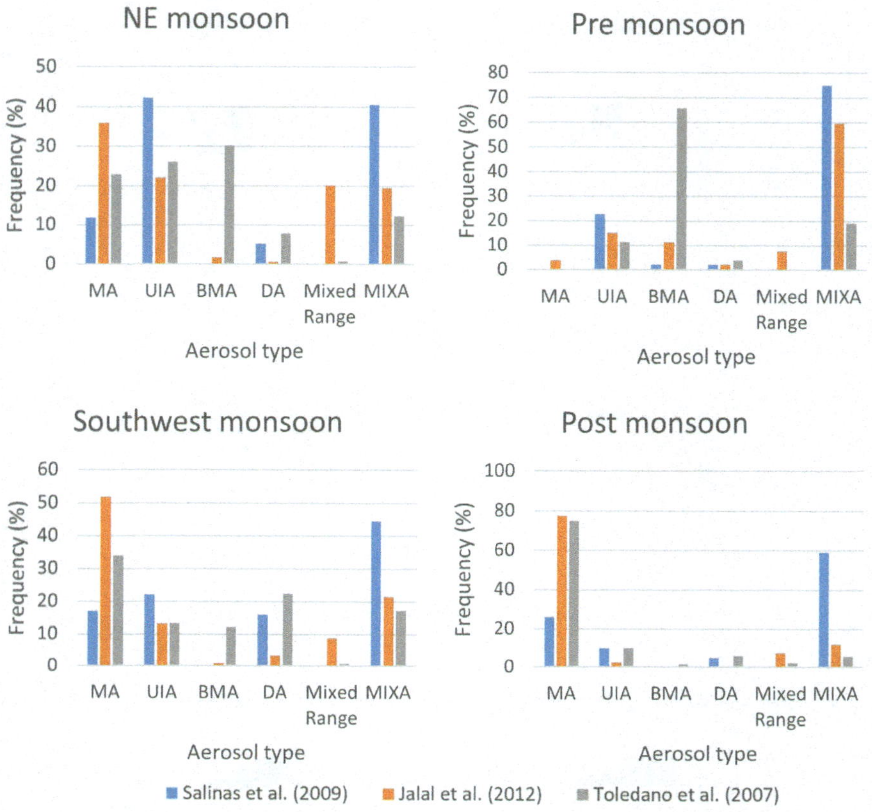

Fig. 8 Histograms of the dominant aerosol type using multiple thresholds for USM_Penang site for the 2016 monsoonal cycle

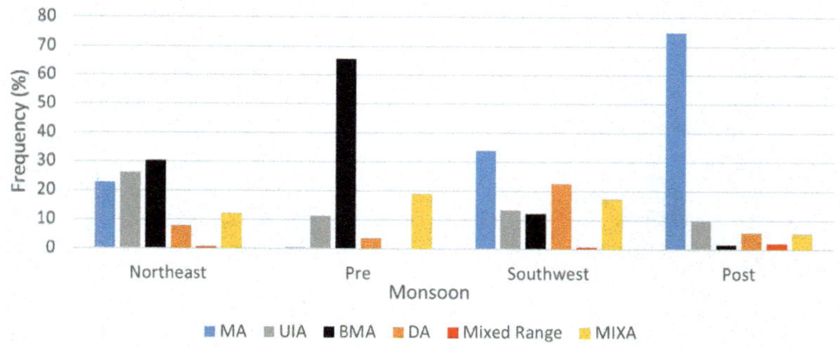

Fig. 9 Seasonal difference of the dominant aerosol type using threshold by Toledano et al. 2007 for USM_Penang for the 2016 monsoonal cycle

the study site and the heavy industrial activities surrounding the island are most likely responsible for these results. DA had the least frequency due mainly to the location and topography of the study site, which is an unlikely source. The presence of MA, however, may be due to the coastal surroundings of the study site.

BMA class is the dominant aerosol type in the pre-monsoon season, accounting for 65.6% of all recorded data. This increase can be tied to the dry condition of the month, where prolonged AOD accumulation in the atmosphere is a likelihood (Kaskaoutis et al. 2012). Recall that the weather condition during the pre-monsoon season, a dry month of April, presented multiple AOD spikes, followed by frequent precipitation during May. Although AOD data for May was incomplete, it can be inferred that at least in the preceding month alone, there was an upsurge of BMA.

In the SWM season, BMA dropped significantly to around 12.2%. At the same time, the MA was dominant, with a frequency of 33.9%, partly due to heavy precipitation during this season as well as the coastal proximity of the study area. DA unexpectedly had some significant presence during this season, with about 22.3% frequency of occurrence. Available data within the study period needs to convincingly explain the reason for the involvement of DA during the SWM season. Finally, during the post-monsoon season, a large increment of MA is conspicuously observed in the figure, with a calculated frequency of 75%. Being a coastal site with a perpetual source for MA, this again highlights the influence of local geographical properties of the study site on the collected AOD data. The presence of other aerosols like UIA and DA, with calculated frequency at 9.9% and 5.8%, respectively, are less pronounced. This also indicates that the post-monsoon season was relatively clean, with most of the aerosol present in the atmosphere as natural.

MODIS-AERONET AOD correlation test

MODIS AOD data (Aqua and Terra) for the 2016 monsoonal cycle was obtained from the LAADS DAAC webpage using the product search tools. Table 4 summarizes the total data obtained after the extraction and validation process.

Very few data points were obtained compared to the 500 nm AERONET AOD data, partly due to the scale of the study site relative to the maximum resolution of the MODIS instruments on both satellites and the constant cloud cover in the region. More so, MODIS AOD data consists of 1–2 readings per day. Regarding the limitation of the MODIS AOD data mentioned earlier, for this study, the data obtained consists of measurements done in the early morning (3–4 AM for Terra satellite and 6–7 AM for Aqua). Bearing this limitation, only the morning averaged AERONET AOD data, consisting of averaged AOD from 3:00 AM to 8:00 AM, were correlated with the MODIS AOD.

Table 4 AOD data count obtained from MODIS for the 2016 monsoonal cycle

Satellite	3 km swath data count
Aqua	68
Terra	75

Figure 10 is a plot of the MODIS-retrieved AOD against AERONET AOD data, from which the Terra satellite is observed to have a slightly better correlation ($r = 0.741$) with AERONET observation compared to the Aqua satellite, whose correlation coefficient is 0.706. The correlation between the two sources of data was observed to be better at lower AOD values, which is consistent with the findings of Green et al. (2009). The good correlation between the AERONET AOD and the MODIS values reveals the potential of the latter to complement the former, except for its significantly small data size.

The annual trend of AERONET AOD from the 5-year data is shown in Fig. 11, from which the highest AOD spike (6.13) was observed on 21 October 2015. Based on API reading, the site is associated with the transboundary haze period. Although some AOD spikes were also observed in 2013 and 2014, the magnitudes were less prominent than those obtained in 2015 (Fig. 11).

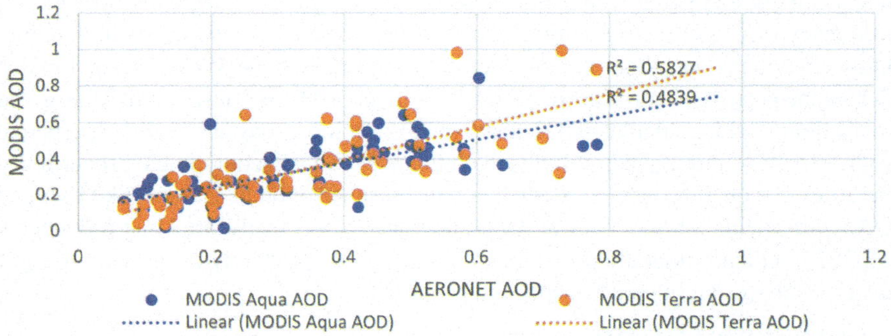

Fig. 10 Correlation between daily average AOD from MODIS and AERONET AOD sources for the 2016 monsoonal cycle

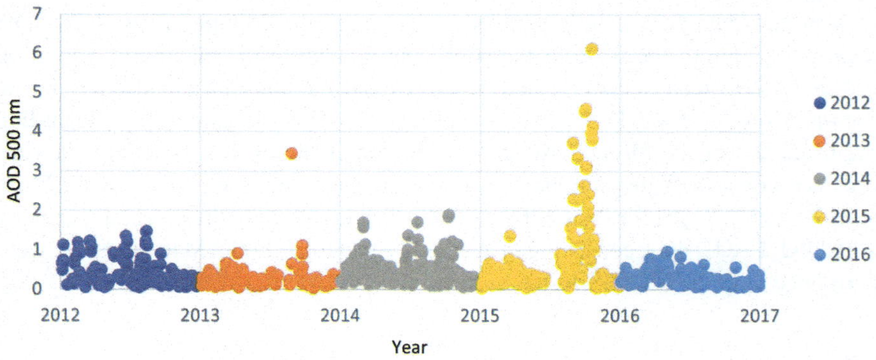

Fig. 11 5-year trend of AERONET 500 nm AOD in USM_Penang site

4 Conclusions

Patterns of aerosol distribution for the 2016 monsoonal cycle were identified and analyzed, with each season showing a distinct distribution of aerosol concentration and type. Data analyzed include the AERONET data (AOD, AE, and PW), MODIS AOD, weather data, and supplementary ground truth, albeit limited to local newspaper reports. The SWM season was initially expected to have the highest AOD for Pulau Pinang, mainly due to repeated transboundary haze, but the result for 2016 showed otherwise. This was attributed to much lower hotspots detected in the Sumatran region of Indonesia for that year because of frequent precipitation that scavenged aerosol in the season. An increasing annual trend was observed from the beginning of the NEM season, with a couple of dips that were later associated with precipitation during the event. The overall increase was, however, tied to the monsoon season's dry and hot weather, which promoted the build-up of aerosol in the atmosphere. AOD in the subsequent pre-monsoon season showed a flat trendline but with significant fluctuation, especially during days 101–130. AOD spikes (< 1.0) were listed and cross-referenced with available weather parameters and supplementary ground truth to find possible contributing factors. The cause was identified as the aerosol transport from the Georgetown area based on the island's prevailing wind direction and topography. Though the possibility of inversions during certain spikes was also discussed, it took more information to conclude. The SWM and post-monsoon periods had similar flat trendlines, plagued with gaps throughout the respective seasons. Additionally, the average AOD for these two seasons was low compared to the other seasons. This was mainly due to frequent precipitation during the 81% rainy days for the post-monsoon period. AE data for the 2016 monsoonal cycle showed that the study site was dominated by fine aerosol, in which over 80% of this data was less than 1.0. The distribution of PW data described a drier atmosphere for the first half of the cycle, with multiple days dipping below 3.0 cm. The reverse was seen in the second half, with much of the PW recorded being well above 4.0 cm. This distribution showed a well-fitted inverse relationship with the AOD behavior, implying a negative correlation between them. A–τ scatter plots, together with the threshold method, showed that the dominant aerosol type depends on the monsoonal seasons. The BMA prevailed during the SWM and post-monsoon periods in 2016 but was surpassed by MA in the NEM and pre-monsoon seasons. These changes had a weather pattern tied closely to them.

Acknowledgements The authors would like to acknowledge the Ministry of Higher Education (MOHE) for funding provided under the Fundamental Research Grant Scheme (FRGS), with reference code FRGS/1/2021/STG08/USM/02/2.

References

Abd Jalal, K., A. Asmat, and N. Ahmad. 2012. Retrievals of aerosol optical depth and ae for identification of aerosols at Kuching, Sarawak. *Advanced Materials Research* 518–523: 5734–5737. https://doi.org/10.4028/www.scientific.net/AMR.518-523.5734.

Afroz, R., M.N. Hassan, and N.A. Ibrahim. 2003. Review of air pollution and health impacts in Malaysia. *Environmental Research* 92 (2): 71–77. https://doi.org/10.1016/S0013-9351(02)00059-2.

Albar, I., I. Jaya, B.H. Saharjo, B. Kuncahyo, and K.P. Vadrevu. 2018. Spatio-temporal analysis of land and forest fires in Indonesia using MODIS active fire dataset. In *Land-atmospheric research applications in South and Southeast Asia* 105–127. Cham: Springer

Ali, M., S. Tariq, K. Mahmood, A. Daud, A. Batool, and Zia-Ul-Haq. (2014). A study of aerosol properties over Lahore (Pakistan) by using AERONET data. *Asia-Pacific Journal of Atmospheric Sciences* 50 (2): 153–162.https://doi.org/10.1007/s13143-014-0004-y.

Badarinath, K.V.S., S.K. Kharol, K.M. Latha, T.K. Chand, V.K. Prasad, A.N. Jyothsna, and K. Samatha. 2007. Multiyear ground-based and satellite observations of aerosol properties over a tropical urban area in India. *Atmospheric Science Letters* 8 (1): 7–13.

Badarinath, K.V.S., S. Kumar Kharol, V. Krishna Prasad, A. Rani Sharma, E.U.B. Reddi, H.D. Kambezidis, and D.G. Kaskaoutis. 2008. Influence of natural and anthropogenic activities on UV Index variations—a study over tropical urban region using ground based observations and satellite data. *Journal of Atmospheric Chemistry* 59 (3): 219–236.

Badarinath, K.V.S., S.K. Kharol, A.R. Sharma, and V.K. Prasad. 2009. Analysis of aerosol and carbon monoxide characteristics over Arabian Sea during crop residue burning period in the Indo-Gangetic Plains using multi-satellite remote sensing datasets. *Journal of Atmospheric and Solar-Terrestrial Physics* 71 (12): 1267–1276.

Chand, D., R. Wood, S.J. Ghan, M. Wang, M. Ovchinnikov, P.J. Rasch, S. Miller, B. Schichtel, and T. Moore. 2012. Aerosol optical depth increase in partly cloudy conditions. *Journal of Geophysical Research Atmospheres* 117 (17): 1–8. https://doi.org/10.1029/2012JD017894.

Christophe, E., A. Song Chia, T. Yin, and L. Keong Kwoh. (2010). *2009 Earthquakes in Sumatra: the use of L-band interferometry in a SAR-hostile environment* 1202–1205. https://doi.org/10.1109/IGARSS.2010.5654173.

Dubovik, O., B.N. Holben, T. Lapyonok, A. Sinyuk, M.I. Mishchenko, P. Yang, and I. Slutsker. 2002. Non-spherical aerosol retrieval method employing light scattering by spheroids. *Geophysical Research Letters* 29 (10): 54–61.

Field, R.D., G.R. van der Werf, and S.S.P. Shen. (2009). Human amplification of drought-induced biomass burning in Indonesia since 1960. *Nature Geoscience* 2: 185. Retrieved from https://doi.org/10.1038/ngeo443.

Giles, D.M., A. Sinyuk, M.G. Sorokin, J.S. Schafer, A. Smirnov, I. Slutsker, T.F. Eck, B.N. Holben, J.R. Lewis, J.R. Campbell, and E.J. Welton. 2019. Advancements in the Aerosol Robotic Network (AERONET) version 3 database–automated near-real-time quality control algorithm with improved cloud screening for Sun photometer aerosol optical depth (AOD) measurements. *Atmospheric Measurement Techniques* 12 (1): 169–209.

Green, M., S. Kondragunta, P. Ciren, and C. Xu. (2009). Comparison of GOES and MODIS Aerosol Optical Depth (AOD) to Aerosol Robotic Network (AERONET) AOD and improve PM2.5 mass at Bondville, Illinois. *Journal of the Air and Waste Management Association* 59 (9): 1082–1091. https://doi.org/10.3155/1047-3289.59.9.1082.

Hayasaka, H., I. Noguchi, E.I. Putra, N. Yulianti, and K. Vadrevu. 2014. Peat-fire-related air pollution in Central Kalimantan, Indonesia. *Environmental Pollution* 195: 257–266.

Holben, B.N., T.F. Eck, I. Slutsker, D. Tanr_e, J.P. Buis, A. Setzer, E. Vermote, J.A. Reagan, Y.J. Kaufman, T. Nakajima, F. Lavenu, I. Jankowiak, and A. Mirnov. 1998. AERONET—a federated instrument network and data archive for aerosol characterization. *Remote Sensing of Environment* 66 (98): 00031–5. https://doi.org/10.1016/S0034-4257.

Holben, B.N., D. Tanré, A. Smirnov, T.F. Eck, I. Slutsker, N. Abuhassan, W.W. Newcomb, J.S. Schafer, B. Chatenet, F. Lavenu, Y.J. Kaufman, J.V. Castle, A. Setzer, B. Markham, D. Clark, R. Frouin, R. Halthore, A. Karneli, N.T. O'Neill, C. Pietras, R.T. Pinker, K. Voss, and G. Zibordi. 2001. An emerging ground-based aerosol climatology: Aerosol optical depth from AERONET. *Journal of Geophysical Research: Atmospheres* 106 (D11): 12067–12097. https://doi.org/10.1029/2001JD900014.

Hyer, E.J., J.S. Reid, and J. Zhang. 2011. An over-land aerosol optical depth data set for data assimilation by filtering, correction, and aggregation of MODIS Collection 5 optical depth retrievals. *Atmospheric Measurement Techniques* 4 (3): 379–408. https://doi.org/10.5194/amt-4-379-2011.

Kant, Y., A.B. Ghosh, M.C. Sharma, P.K. Gupta, V.K. Prasad, K.V.S. Badarinath, and A.P. Mitra. 2000. Studies on aerosol optical depth in biomass burning areas using satellite and ground-based observations. *Infrared Physics and Technology* 41 (1): 21–28.

Kaskaoutis, D.G., R. Gautam, R.P. Singh, E.E. Houssos, D. Goto, S. Singh, A. Bartzokas, P.G. Kosmopoulos, M. Sharma, N.C. Hsu, B.N. Holben, and T. Takemura. 2012. Influence of anomalous dry conditions on aerosols over India: transport, distribution, and properties. *Journal of Geophysical Research Atmospheres* 117 (9): 1–17. https://doi.org/10.1029/2011JD017314.

Kaufman, Y.J., O. Boucher, D. Tanré, M. Chin, L.A. Remer, and T. Takemura. 2005. Aerosol anthropogenic component estimated from satellite data. *Geophysical Research Letters* 32 (17).

Kharol, S.K., K.V.S. Badarinath, A.R. Sharma, D.V. Mahalakshmi, D. Singh, and V.K. Prasad. 2012. Black carbon aerosol variations over Patiala city, Punjab, India—a study during agriculture crop residue burning period using ground measurements and satellite data. *Journal of Atmospheric and Solar-Terrestrial Physics* 84: 45–51.

Li, S., T. Wang, M. Xie, Y. Han, and B. Zhuang (2015). Observed aerosol optical depth and AE in urban area of Nanjing, China. *Atmospheric Environment* 123, 350–356. https://doi.org/10.1016/j.atmosenv.2015.02.048.

Palanissamy, A. 2013. Haze free air in Singapore and Malaysia—the spirit of the law in South East Asia. *International Journal of Education and Research* 1 (8): 1–8.

Pérez-Ramírez, D., M. Andrade-Flores, T.F. Eck, A.F. Stein, N.T. O'Neill, H. Lyamani, S. Gassó, D.N. Whiteman, I. Veselovskii, F. Velarde, and L. Alados-Arboledas. (2017). Multi-year aerosol characterization in the tropical Andes and in adjacent Amazonia using AERONET measurements. *Atmospheric Environment* 166: 412–432. https://doi.org/10.1016/j.atmosenv.2017.07.037.

Pokharel, M., J. Guang, B. Liu, S. Kang, Y. Ma, B.N. Holben, et al. 2019. Aerosol properties over Tibetan Plateau from a decade of AERONET measurements: baseline, types, and influencing factors. *Journal of Geophysical Research: Atmospheres* 124: 13357–13374. https://doi.org/10.1029/2019JD031293.

Pöschl, U. 2005. Atmospheric aerosols: composition, transformation, climate and health effects. *Angewandte Chemie International Edition* 44 (46): 7520–7540.

Radke, L.F., P.V. Hobbs, and M.W. Eltgroth. 1980. Scavenging of aerosol particles by precipitation. *Journal of Applied Meteorology and Climatology* 19 (6): 715–722.

Reichardt, J., U. Wandinger, V. Klein, I. Mattis, B. Hilber, and R. Begbie. 2012. RAMSES: German Meteorological Service autonomous Raman lidar for water vapor, temperature, aerosol, and cloud measurements. *Applied Optics* 51 (34): 8111–8131.

Reid, J.S., E.J. Hyer, R.S. Johnson, B.N. Holben, R.J. Yokelson, J. Zhang, J.R. Campbell, S.A. Christopher, L. Di Girolamo, L. Giglio, R.E. Holz, C. Kearney, J. Miettinen, E.A. Reid, F.J. Turk, J. Wang, P. Xian, G. Zhao, R. Balasubramaniam, B.N. Chew, S. Janjai, N. Langrosas, P. Lestari, N.H. Lin, M. Mahmud, A.X. Nguyen, B. Norris, N.T.K. Oanh, M. Oo, S.V. Salinas, E.J. Welton, and S.C. Liew. (2013). Observing and understanding the Southeast Asian aerosol system by remote sensing: an initial review and analysis for the Seven Southeast Asian Studies (7SEAS) program. *Atmospheric Research* 122: 403–468. https://doi.org/10.1016/j.atmosres.2012.06.005.

Rosenfeld, D., Y. Zhu, M. Wang, Y. Zheng, T. Goren, S. Yu. 2019. Aerosol-driven droplet concentrations dominate coverage and water of oceanic low-level clouds. *Science* 363 (6427): eaav0566.

Rupakheti, D., S. Kang, M. Bilal, J. Gong, X. Xia, Z. Cong. 2019a. Aerosol optical depth climatology over Central Asian countries based on Aqua-MODIS Collection 6.1 data: aerosol variations and sources. *Atmospheric Environment* 207: 205e214. https://doi.org/10.1016/j.atmosenv.2019.03.020.

Rupakheti, D., S. Kang, M. Rupakheti, Z. Cong, A.K. Panday, B.N. Holben. (2019b). Identification of absorbing aerosol types at a site in the northern edge of Indo-Gangetic Plain and a polluted valley in the foothills of the central Himalayas. *Atmospheric Research* 223 (November 2018): 15–23. https://doi.org/10.1016/j.atmosres.2019.03.003

Rupakheti, D., M. Rupakheti, S.F. Abdullaev, X. Yin, and S. Kang. (2020). Columnar aerosol properties and radiative effects over Dushanbe, Tajikistan in Central Asia. *Environmental Pollution* 265.https://doi.org/10.1016/j.envpol.2020.114872.

Rupakheti, D., M. Rupakheti, X. Yin, J. Hofer, M. Rai, Y. Hu, S.F. Abdullaev, S. Kang. (2021). Modifications in aerosol physical, optical and radiative properties during heavy aerosol events over Dushanbe, Central Asia. *Geoscience Frontiers* 12 (6). https://doi.org/10.1016/j.gsf.2021.101251

Saha, A., and K. Krishna Moorthy. 2004. Impact of precipitation on aerosol spectral optical depth and retrieved size distributions: a case study. *Journal of Applied Meteorology* 43 (6): 902–914. https://doi.org/10.1175/1520-0450(2004)043%3c0902:iopoas%3e2.0.co;2.

Salinas, S.V., B.N. Chew, and S.C. Liew. 2009. Retrievals of aerosol optical depth and Ångström exponent from ground-based Sun-photometer data of Singapore. *Applied Optics* 48 (8): 1473–1484. https://doi.org/10.1364/AO.48.001473.

Smirnov, A., B.N. Holben, T.F. Eck, O. Dubovik, and I. Slutsker. 2000. Cloud-screening and quality control algorithms for the AERONET database. *Remote Sensing of Environment* 73 (3): 337–349.

Song, Z., D. Fu, X. Zhang, X. Han, J. Song, J. Zhang, J. Wang, and X. Xia. (2019). MODIS AOD sampling rate and its effect on PM 2.5 estimation in North China. *Atmospheric Environment,* 209 (October 2018): 14–22. https://doi.org/10.1016/j.atmosenv.2019.04.020.

Stocker, T.F., D. Qin, G.K. Plattner, M. Tignor, S.K. Allen, J. Boschung, A. Nauels, Y. Xia, V. Bex, and P.M. Midgley. 2013. Climate change 2013: the physical science basis. In *Intergovernmental panel on climate change, working group I contribution to the IPCC fifth assessment report (AR5). New York.*

Tan, F., H.S. Lim, K. Abdullah, T.L. Yoon, and B. Holben. 2015a. Monsoonal variations in aerosol optical properties and estimation of aerosol optical depth using ground-based meteorological and air quality data in Peninsular Malaysia. *Atmospheric Chemistry and Physics* 15 (7): 3755–3771. https://doi.org/10.5194/acp-15-3755-2015.

Tan, Fuyi, H. San Lim, K. Abdullah, T.L. Yoon, and B. Holben. 2015b. AERONET data–based determination of aerosol types. *Atmospheric Pollution Research* 6 (4): 682–695. https://doi.org/10.5094/APR.2015.077.

Toledano, C., V.E. Cachorro, A. Berjon, A.M. De Frutos, M. Sorribas, B.A. De la Morena, and P. Goloub. 2007. Aerosol optical depth and Ångström exponent climatology at El Arenosillo AERONET site (Huelva, Spain). *Quarterly Journal of the Royal Meteorological Society: A Journal of the Atmospheric Sciences, Applied Meteorology and Physical Oceanography* 133 (624): 795–807.

Trenberth, K.E., and Y. Zhang. 2018. How often does it really rain? *Bulletin of the American Meteorological Society* 99 (2): 289–298.

Vadrevu, K.P., T. Ohara, and C. Justice. (eds.). 2018. *Land-atmospheric research applications in South and Southeast Asia.* Springer.

Vadrevu, K.P., T. Ohara, and C. Justice. (eds.). 2021a. *Biomass burning in south and Southeast Asia: impacts on the biosphere*, vol. 2. CRC Press.

Vadrevu, K.P., T. Ohara, and C. Justice (eds.). 2021b. *Biomass burning in South and Southeast Asia: mapping and monitoring*, vol. 1. CRC Press.

Van Donkelaar, A., R.V. Martin, R.C. Levy, A.M. da Silva, M. Krzyzanowski, N.E. Chubarova, E. Semutnikova, and A.J. Cohen. 2011. Satellite-based estimates of ground-level fine particulate matter during extreme events: a case study of the Moscow fires in 2010. *Atmospheric Environment* 45 (34): 6225–6232. https://doi.org/10.1016/j.atmosenv.2011.07.068.

Measurements of Atmospheric Carbon Dioxide Emissions from Fire-Prone Peatlands in Central Kalimantan, Indonesia, Using Ground-Based Instruments

Masafumi Ohashi, Windy Iriana, Osamu Kozan, Masahiro Kawasaki, and Kenichi Tonokura

Abstract To reduce the effect of fire-induced peatland disruption on the global carbon balance, monitoring atmospheric carbon dioxide (CO_2) concentrations in fire-prone peatlands is essential. In this study, we measured the column-averaged atmospheric mixing ratios, XCO_2, using a portable instrument in the tropical peatlands of Central Kalimantan, Indonesia. Combining the measured increments above the background level and the estimated increments from airport visibility records, the raw mean increment, $<\Delta XCO_2>$, was 7.8 ppm during the fire season from September to November 2014. Droughts in peatlands lower the groundwater table and induce changes in the microbial communities of the aerobic soil zone to emit CO_2 into the atmosphere. During the peat-respiration periods, July–August 2014 and April–May 2015, $<\Delta XCO_2>$ was 4.8 ppm. CO_2 emission from the ground peat soil was measured using the nocturnal temperature-inversion trap method.

Keywords Hot fire · Cold fire · Underground fire · Column density · Latent method · MRV

M. Ohashi
Graduate School of Science and Engineering, Kagoshima University, Kagoshima 890–8580, Japan

W. Iriana
Center for Environmental Studies, Bandung Institute of Technology, Bandung 40132, Indonesia

W. Iriana · K. Tonokura
Graduate School of Frontier Sciences, The University of Tokyo, Kashiwa 277–8563, Japan

O. Kozan
Center for Southeast Asian Studies, Kyoto University, Kyoto 606–8304, Japan

O. Kozan · M. Kawasaki (✉)
Research Institute for Humanity and Nature, Kyoto 603–8047, Japan
e-mail: kawasaki@moleng.kyoto-u.ac.jp

K. P. Vadrevu et al. (eds.), *Vegetation Fires and Pollution in Asia*,
https://doi.org/10.1007/978-3-031-29916-2_22

1 Introduction

Tropical peatlands play an important role in controlling the global carbon cycle and regulating atmospheric carbon dioxide (CO_2) concentration. The undisturbed peatlands have been functioning as net carbon sinks for millenniums. However, environmental and anthropogenic forcing on the carbon balance of peatland ecosystems has increased atmospheric CO_2 concentrations over the last few decades (Hooijer et al. 2012). To reduce the effect of peatland disruption on the global carbon balance, it is essential to quantify the net CO_2 exchange between peatlands and atmospheric ecosystems by monitoring the atmospheric concentration in peatlands. The CO_2 emission from tropical peatlands is caused by cold and hot fires, depending on the conditions of drainage and climate variability, as shown in Fig. 1. The rewetting of peatlands, a cost-effective method for preventing forest and peatland fires, can be supported via measurement–reporting–validation (MRV) activities for carbon emission reduction. Monitoring of the atmospheric CO_2 in peatlands is one of these activities. Cold fire is caused by the decomposition of organic materials by microorganisms under aerobic conditions of peat soil, which results in the steady release of CO_2 from the ground to the atmosphere. During drought conditions, peatlands have an increased susceptibility to fire (Albar et al. 2018; Hayasaka and Sepriando 2018). Droughts lower the groundwater table in abandoned peatlands and change the peat conditions to aerobic (Hayasaka et al. 2021). The high aerobic condition and vegetation change modify the microbial communities in the constantly aerated peat layers, causing them to decompose organic material at a higher rate, resulting in a steady release of CO_2 from the ground to the atmosphere (Minkkinen et al. 1999). Organic materials in the aerobic zone of peat soil decompose 50 times faster than those in the anaerobic zone (Clymo 1984). In Central Kalimantan, large amounts of CO_2 enter the atmosphere from the ground during the non-fire season.

Hot fires are accelerated by drought and can be classified into surface and underground fires. In Indonesia, the occurrence of forest and peatland fires are more

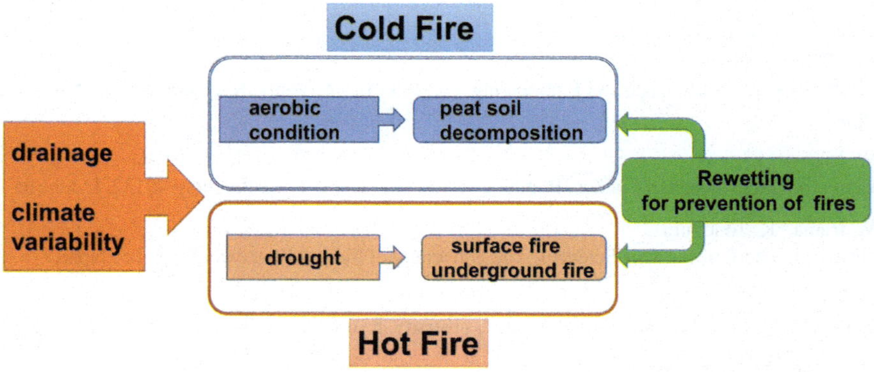

Fig. 1 Drainage and climate variability induce two types of peatland fires

pronounced in El Niño Southern Oscillation (ENSO) years (Iriana 2019; Kusumaningtyas et al. 2021). The ENSO condition is defined as the anomalous warming of ocean temperatures over a scale of years. Most peatland fires in Indonesia are anthropogenic because open burning is a cost-effective method for clearing farmlands and producing ash for fertilizer (Usup et al. 2004). Peatland fires have occurred in Central Kalimantan during the dry season every year except for La Niña years (Yulianti et al. 2020). One of the worst fire events in Indonesia occurred in 1997 during a long dry season associated with ENSO, during which 0.95–2.57 Gt of carbon was estimated to be released into the atmosphere (Page et al. 2002). From September to October 2015, the mean carbon emission rate was estimated to be 4 Mt C/day, which exceeded the CO_2 emissions of the European Union at that time (Huijnen et al. 2016). Indonesia's peatlands are located in the low altitude coastal areas of Sumatra, Kalimantan, and Papua. In Central Kalimantan, where our study was conducted, the peatlands were disturbed and abandoned after an unsuccessful drainage construction project, the Mega Rice Project. Figure 2 shows a map of the observation sites, the Tjilik Riwut Airport in Palangka Raya, and the Kalampangan site. Moderate Resolution Imaging Spectroradiometer (MODIS), a NASA satellite sensor, provides data on fire hotspots. In Palangka Raya, the fire-prone season is from August to October. Because of the yearly decline in precipitation, the groundwater table has lowered to a level that induces large-scale wildfires in the perturbed peatland.

A fire image sensor onboard the TET-1 satellite can monitor low-intensity peatland fire fronts through smoke, allowing underground fires that emit CO_2 and particulate matter (PM) to be monitored (Atwood et al. 2016). Fire front edges observed in images from TET-1 indicate that fire was slowly spreading through the deeper peat layer for two weeks in the peatland of Central Kalimantan. Although controlling underground fires is complex, combining a thermal camera and an unmanned aerial vehicle may identify invisible fires and aid in fire extinguishing (Kameoka et al. 2021).

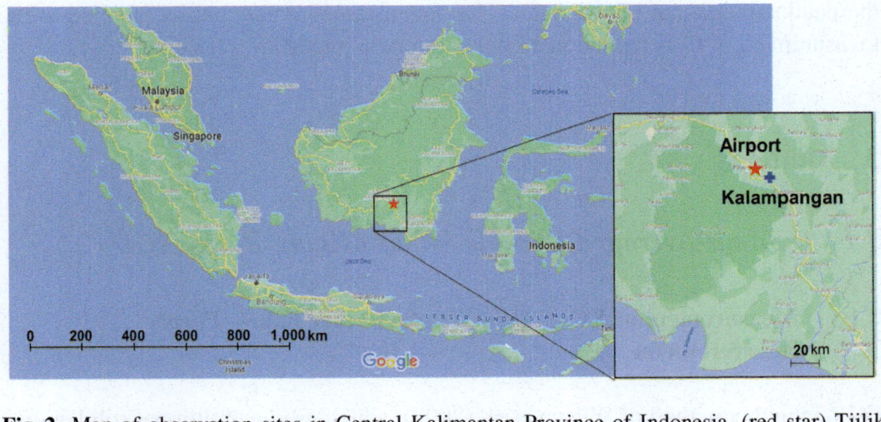

Fig. 2 Map of observation sites in Central Kalimantan Province of Indonesia. (red star) Tjilik Riwut Airport in Palangka Raya, (blue cross) Kalampangan site. Image from Google Maps

2 Location and Description of Study Areas

The inset of Fig. 2 shows the study sites in the tropical peatland area of Central Kalimantan in 2° S–3.5° S and 113.2° E–115° E. Palangka Raya is the capital city of Central Kalimantan province. It is 140 km away from the southern coastal area, and the surrounding is forests, natural conservation areas, and lowland peat swamp ecosystems. CO_2 column densities were measured using a Fiber-Etalon Solar carbon dioxide sensor (model FES-C, Meisei Electric, Japan) at the Tjilik Riwut meteorological station at Palangka Raya airport (2.224° S, 113.946° E, 10 m ASL). We also performed measurements via the nocturnal temperature-inversion trap method (NTIT) at the Kalampangan site (2.320° S, 114.04° E), 20 km southeast of the meteorological station.

The Ex-Mega Rice Project area is in the peatland area of Central Kalimantan province. The project, which constructed canals involving the extensive removal of trees and draining peat deposits in 1996, resulted in severe peatland disturbance and degradation (Page et al. 2002). After the failure of this project, the abandoned peatland has become a significant source of carbon emissions, as aerobic peat soil provides favorable conditions for microbial decomposition and increases the risk of peat fires.

The tropical climate in Central Kalimantan is associated with high annual temperatures and precipitation. The seasonal pattern of precipitation in this area is determined by the monsoon climate, which features dry and rainy seasons. The dry monsoon typically occurs from July to October when prevailing southeasterly wind blows from the Australian and Indian oceans. From December to April, a prevailing northwestern wind from Central Asia brings wet monsoon conditions. This seasonally varying wind direction allows the measurement of CO_2 flux in the targeting area via the latent flux method. By factoring the difference in the measured column-averaged CO_2 dry-air molar mixing ratio (XCO_2) at the out- and in-flow points of the targeting area with air mass flux, the total amount of CO_2 emitted from hot and cold fires in the peatland area can be estimated. The latent measurement can be applied to the measurement of CO_2 reduction amounts as one of the MRV activities.

3 Methods

3.1 Column-Averaged Dry-Air Molar Mixing Ratio, XCO_2

3.1.1 Measurement of XCO_2 Using Optical Fiber Fabry–Perot Interferometer

Following the method of Wilson et al. (2007), which used a traditional solid etalon, our study used an optical fiber Fabry–Perot interferometer (FFPI, Nippon Electric Glass Co., Japan) to resolve the overtone band of CO_2 in the near-infrared region and measure the atmospheric column density. Because this portable instrument is strong against high humidity and hot temperature, it is suitable to use under tropical weather conditions. A schematic of this single optical path spectrometer is shown in Fig. 3. The outdoor part comprised a fiber collimator installed as a telescope on a portable sun tracker, and a long bandpass filter was placed in front of the object lens of the solar telescope. The indoor part comprised an FFPI (13 mm long, 1.25 mm diameter, free spectral range = 0.324 nm, FWHM = 0.025 nm) and a solar intensity monitor, which was placed in a small temperature-controlled box. The solar signal collected by the telescope was guided to both the FFPI and solar intensity monitor via an optical fiber cable attached to a beam separator. The solar intensity fluctuation caused by occasional cloud coverage of solar light was properly compensated. The wavelength of the solar spectrum transmitted through the FFPI was controlled by changing the temperature. Thus, the transmitted light was aligned/unaligned with the CO_2 rotational lines centered at 1572 nm (Kobayashi et al. 2010). By modulating the FFPI temperature with an interval of 40 s/cycle, the intensity ratio of the incident to transmitted light can be deduced using the Beer–Lambert law. As described below, to evaluate CO_2 column densities, the optical transmission spectrum of the FFPI optics is convoluted with simulated CO_2 spectra for various XCO_2 around 400 ppm.

Evaluation of column-averaged dry-air molar mixing ratios

When solar light passes through the atmospheric layer, it interacts with trace gas molecules (CO_2, CH_4, and H_2O), lowering the rovibronic band intensities in the near-infrared through molecular absorption. Spectral data for this study were obtained from the HITRAN2008 database (Rothman et al. 2009). The spectral range used to analyze CO_2 concentrations was 1568–76 nm. As the molecular parameters for the absorption lines depend on pressure and temperature, we used the vertical profiles of meteorological parameters for the spectral simulation.

Herein, we describe how the column-averaged dry-air molar mixing ratio, XCO_2, was obtained. We assumed homogeneous vertical profiles of the mixing ratio from altitudes of 0–48 km. The profiles were divided into 28 sublayers. Afterward, we

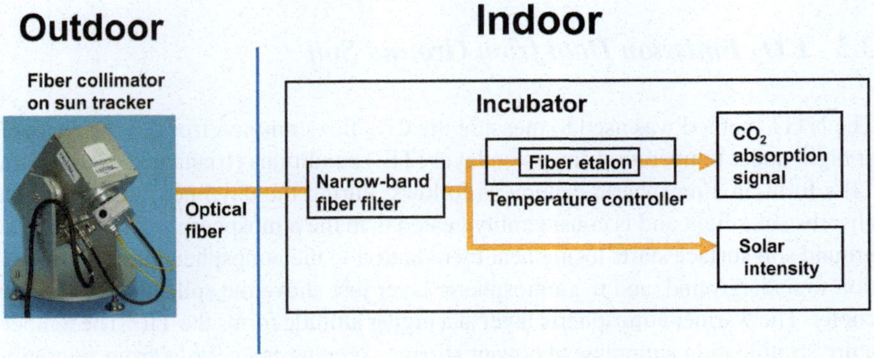

Fig. 3 Schematic of the FES-C instrument. Solar signal collection and data acquisition system

applied the Beer–Lambert law to CO_2 absorption in each layer for a certain CO_2 molar mixing ratio. The corresponding total absorbance for the CO_2 molar mixing ratio was obtained by summing the absorbance values with a line-by-line calculation. Using the obtained calibration curves, the observed optical densities were converted to XCO_2. The air masses of the layers were calculated from the solar zenith angles from 39.30° to 1.93° obtained from the FES-C instrument's GPS time and geographical data.

Validation of XCO_2 data for the commercial model FES-C

XCO_2 data from the FES-C were compared with a co-located TCCON Fourier transform spectrometer instrument at the National Institute for Environmental Studies in Tsukuba, Japan. Referring to the Fourier transform spectrometer data, the obtained scale factor for the commercial model of the FES-C was 0.996 ± 0.005 (one standard deviation; σ). To confirm vertical uniformity of the CO_2 concentration near the ground under conditions prone to fire, measurements were performed using a tethered balloon (1.5 m diameter, 6 m length) from August 21 to 22, 2011, at Palangka Raya peatland in the fire season. Figure 4 shows that the in situ CO_2 concentration was almost uniform up to 700 m altitudes. The air was homogeneously mixed vertically by a southerly wind.

3.1.2 Greenhouse Gases Observing Satellite Data (GOSAT)

The Greenhouse Gases Observing Satellite "GOSAT/IBUKI" is a spacecraft that measures CO_2 column densities from space (GOSAT Data Archive Service). The primary sensor of GOSAT is a NIR-looking Fourier transform spectrometer. XCO_2 data were available only above the Java Sea between the Kalimantan and Java Islands in 2014. The background level of XCO_2 was determined by the combination of the non-fire wet season data of FES-C and the GOSAT annual trend coefficient of 2.00 ppm/year. The measured XCO_2 above the background level corresponds to the increments of XCO_2 caused by hot and cold fires.

3.2 CO_2 Emission Data from Ground Soil

The NTIT method was used to measure the CO_2 flux emission from the ground soil at night under temperature-inversion layer (TIL) conditions (Iriana et al. 2016). The TILs form an atmospheric canopy as follows: during the daytime, the soil surface absorbs solar light and is considerably heated than the atmosphere. After sunset, the ground soil surface starts losing heat by radiation to the atmosphere and by conduction to underground, and the atmospheric layer just above the soil surface becomes cooler. The warmer atmospheric layer at a higher altitude forms the TIL. The temperature stratification suppresses buoyant stirring, keeping trace gases from migrating vertically and trapping the emitted CO_2 within a natural chamber (Fochesatto 2015). The conditions for stable TIL formation for appreciable gas accumulation are strong

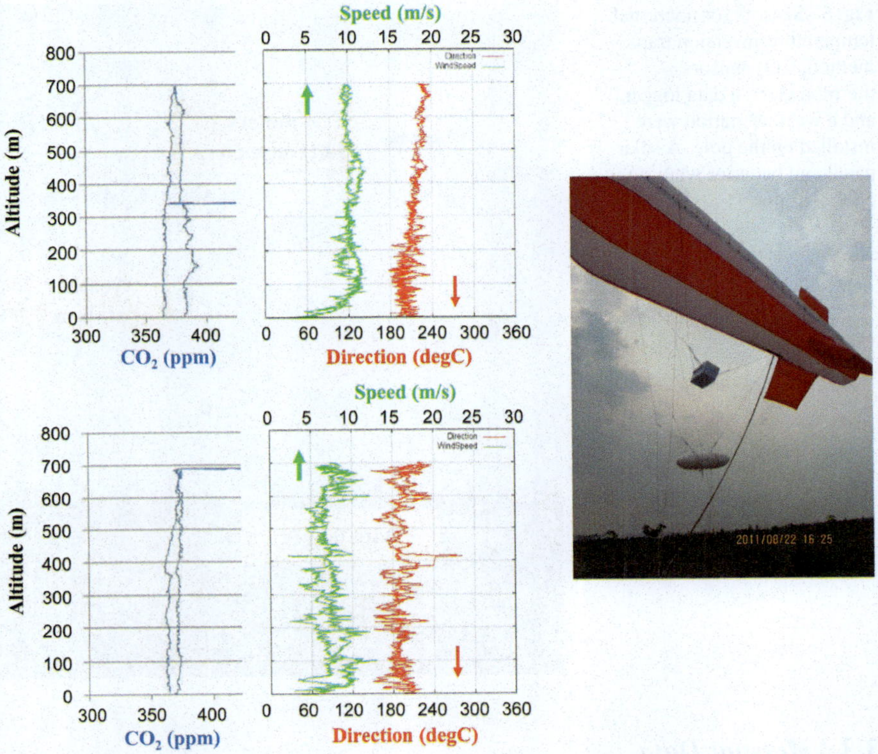

Fig. 4 Tethered balloon measurements up to 700 m. (Blue) CO_2 concentration in ppm, (green) wind speed in m, (red) wind direction. August 21 (top) and 22 (bottom), 2011, at the Palangka Raya peatland. Note that the scale for CO_2 is shifted by 10 ppm

solar radiation, clear nights, and slow surface winds (Mathieu et al. 2005). The NTIT method is inefficient under windy or cloudy/rainy conditions because these conditions restrain radiative cooling from the ground soil surface. Furthermore, unstable TIL nights are more likely to occur during the rainy season, restricting measurements. Despite these limitations, the NTIT method is a practical method for estimating CO_2 emissions from disturbed tropical peatlands in developing countries, where it is difficult to conduct measurements with sophisticated instruments. The NTIT measurements were performed using plastic pipes, small CO_2 sensors, a data logger, and a solar battery, as shown in Fig. 5. Under stable nocturnal conditions, the coverage distance is 10 km × 10 km (Chambers et al. 2011).

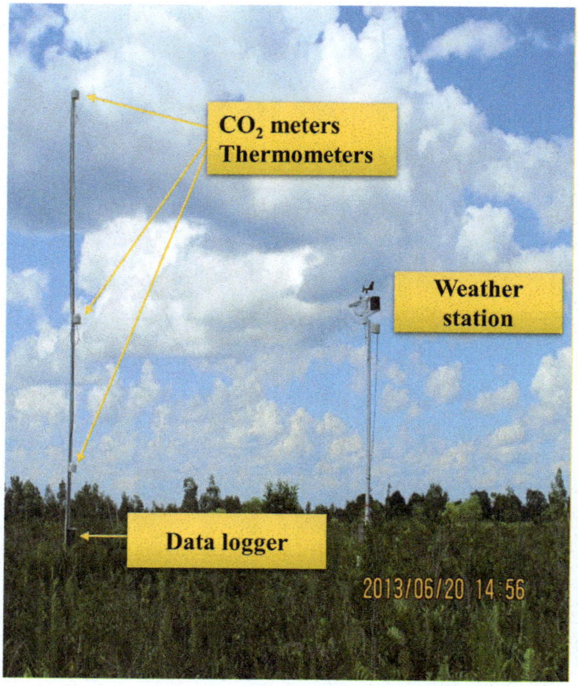

Fig. 5 Sensors for nocturnal temperature-inversion trap method. CO_2 sensors, thermometers, a data logger, and a weather station were installed on the pole. A solar panel and batteries were set on the ground

3.3 Hotspot Data

Satellite data from NASA

Daily active fire detection data from MODIS observation have been used as peatland fire indices. The daily data collection 6 for MODIS hotspots was obtained from the Fire Information for Resources Management System (NASA FIRMS 2022).

3.4 Doppler Radar Detection of Plume Trails of Large-Scale Fires

The C-band Doppler weather radar at the Badan Meteorologi, Klimatologi, dan Geofisika (BMKG) station in Palangka Raya detected echoes at the height of less than 2 km and a slant range of 100 km (Rahman et al. 2021). Weather radar monitoring of the fire smoke layer-top images in Fig. 6 revealed plume trails of large-scale fires. The formation of the top layer with refractive eddies induced Bragg scattering in the radio wave. The fire smoke layer-top trails in the radar echo images in Fig. 6 extend from the MODIS hotspot points (as marked with the red circles) toward the northern areas. The corresponding AHI/HIMAWARI-8 true-color images also show

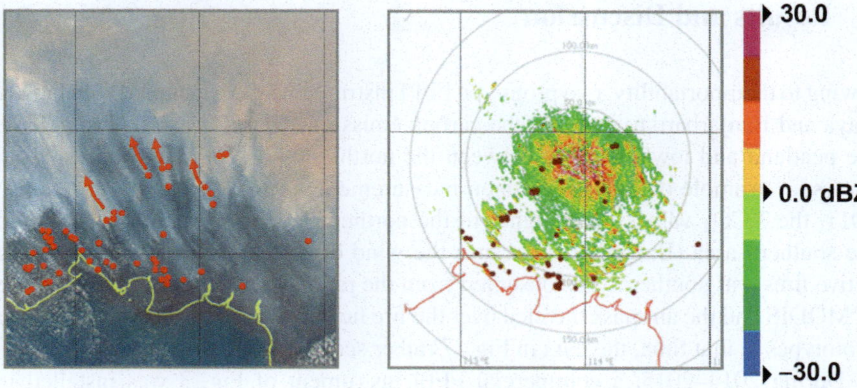

Fig. 6 Radar echo image at 03Z (10LT) on October 15, 2015 (right). Study areas are covered by smoke plumes: image from AHI/HIMAWARI-8/JMA (left). Red dots are MODIS/NASA fire hotspots. Red arrows are forward trajectories at 1000 m altitude from 00Z to 03Z: calculation with NOAA HYSPLIT MODEL/GDAS (Rahman et al. 2021)

the trails of the fire smoke layer top. To confirm the origins of the trails, we analyzed the smoke trajectories starting from the hotspots, as indicated by the red arrows. Four trajectories were projected in the study areas using the calculated forward trajectories for 00Z−03Z (07LT−10LT) at 1000 m altitude. We used the NOAA hybrid single-particle Lagrangian integrated trajectory (HYSPLIT) model/global data assimilation system (GDAS) (Stein et al. 2015; Air Resources Laboratory 2018). These trajectories matched the smoke layer-top trails in the radar echo images. These data suggest the potential application of the latent flux method to estimate CO_2 emissions from the peatlands of Kalimantan.

3.5 Remote Monitoring of Soil Temperatures During the Underground Fire

In November 2019, we used an unmanned aerial vehicle (drone model Inspire 1, DJI, China) with an attached infrared camera (Zenmuse XT, DJI, China) at 100 m altitude to obtain mapping data of soil surface temperatures for Tanjung Leban near Kalampangan. Soil surface temperatures were above 100 °C where no surface fires were observed (Kameoka et al. 2021). Usup et al. (2004) measured fire temperatures in the field below the ground surface using thermocouple sensors and found that during peat soil burning days in Kalampangan in August 2002, the soil temperature at a depth of 0.1 m was 60 °C.

4 Results and Discussion

Owing to their portability, two prototype FFPI instruments were installed at Palangka Raya and Banjarbaru to measure the carbon emissions from widespread sources in the peatland and lowland areas between the northern and southern sites. Figure 7 shows an example of carbon emission measurements in the target area. In August 2011, the XCO_2 values were higher in the northern area (Palangka Raya) than in the southern area (Banjarbaru) because the wind direction on the peatland during active fires was southernly, as observed from the plume direction in the field image of MODIS and the air mass traveled over the fire field. Because the instruments were prototypes at that time, the data in Fig. 7 rather scattered.

During 2014–2015, a commercial FFPI instrument of Fig. 3 was installed in Palangka Raya (Iriana et al. 2018). Figure 8 shows XCO_2 measured during the fire and non-fire seasons. The increments in XCO_2 above the background level are due to emissions from hot and cold fires in the peatland (Iriana et al. 2018). These periods are shown in the upper part of Fig. 7. For the cold fire during the peat-respiration period of 122 days, the raw mean increment $\langle \Delta XCO_2 \rangle$ was 4.8 ppm. During the hot fire period, the increments in XCO_2, hotspot count, and airport visibility were highly correlated. On heavy fire days, smoke covered the sky over Palangka Raya, resulting in cloudy weather without solar signals. On these days, we estimated missing data from airport visibility data, as described below.

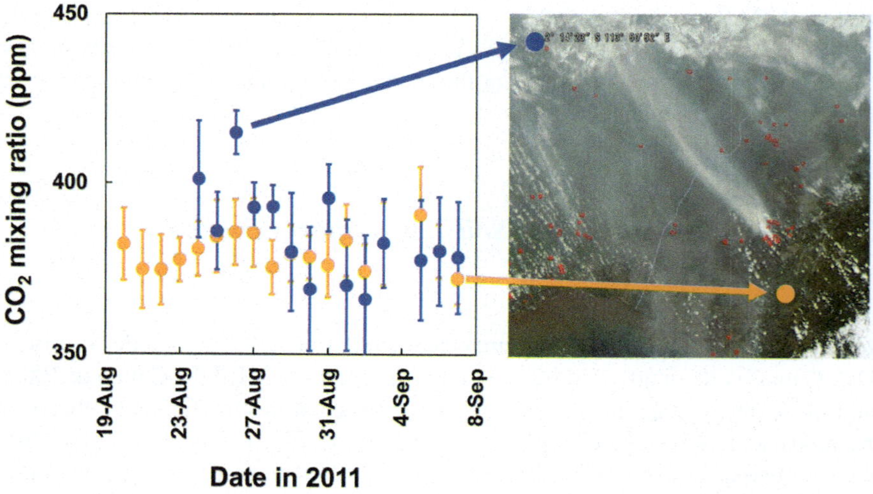

Fig. 7 Example of the latent flux method for estimating the total amount of CO_2 emitted from the peatland between Palangka Raya and Banjarbaru apart by 95 km where peat fires occurred on August 25, 2011. By factoring the difference in XCO_2 at out- and in-flow points with air mass fluxes, the total amount of CO_2 emitted in the area may be estimated. (red dot) Fire hot spots from MODIS/TERRA/NASA

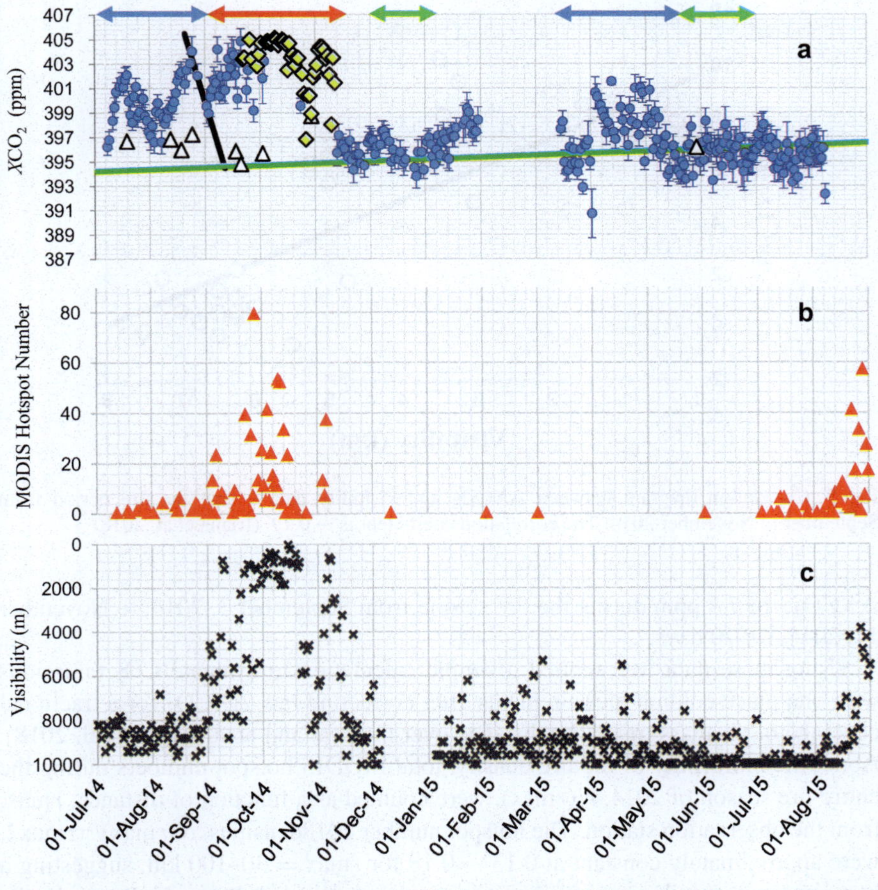

Fig. 8 **a** Time series of XCO_2; (blue circle) FES-C average for UTC = 3–7. The error bar shows one-σ, (Δ) GOSAT data on sea, (green solid line) background level, (yellow $\Diamond$) ΔXCO_2 estimated from the airport visibility record, (blue $\leftrightarrow$) peat soil respiration or cold fire period, (red $\leftrightarrow$) hot fire period, (green $\leftrightarrow$) background period. The black solid line for eye clarity indicates the contribution of the peat soil respiration. **b** MODIS hotspot count per day in a 50 km circle centered at the observation station. **c** Visibility data presented inversely (Iriana et al. 2018)

Visibility is a measure of the transparency of the atmosphere. It is defined as the greatest distance at which a target object can be recognized against the horizon sky at night and day. Field et al. (2009) reported a high correlation between the extinction coefficient and total PM emissions from fires, using the visibility records from the BMKG stations in Kalimantan. Because the air masses from peatland fires contain both CO_2 and PM, we can estimate XCO_2 using an experimental calibration curve. Figure 9 shows a linear correlation between the daily visibility data at Tjilik Riwut Airport in Palangka Raya and the measured XCO_2. Using this calibration curve, we estimated XCO_2 for very heavy smoke cases, as marked by the yellow diamonds in Fig. 8a. Combining the measured and estimated values, we obtained a raw mean <

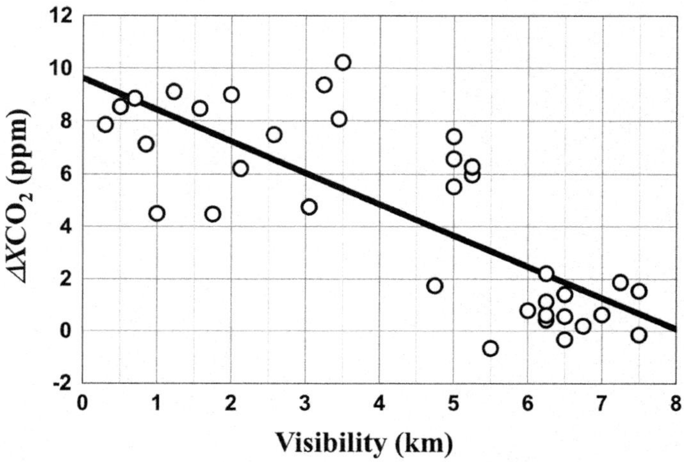

Fig. 9 Correlation diagram between ΔXCO_2 and visibility data during the fire period from September to November 2014. The correlation coefficient is − 0.79 (Iriana et al. 2018)

ΔXCO_2> of 7.8 ppm during the fire season from September 3, 2014, to November 11, 2014, for 70 days.

Using these data, we applied a simple calculation to evaluate CO_2 emissions with one-place column measurements after correcting raw ΔXCO_2 for a fractional coverage of air mass trajectory distance over the peatland area (Iriana et al. 2018). About the uniformity of the fire density, total MODIS hotspot numbers during the entire fire season of 2014, $N(r\text{max})$, were counted as a function of distance, $r$max, from the observation station. The hotspot number 2018densities, $N(r\text{max})/\pi r\text{max}2$, were approximately constant at 0.13 − 0.18 for $r\text{max} = 30$–100 km, suggesting a homogeneous distribution of hotspots over the peatland of Palangka Raya. As the average wind speed was 1.9 m/s, corresponding to 160 km/day, the air mass over the peatland should be replaced with fresh air from the sea within a day. The calculated total sum of daily ΔXCO_2 (corrected) was 889 ppm for the fire season of 70 days and 720 ppm for the non-fire season of 122 days. The CO_2 emissions during both the fire and non-fire seasons were almost the same during the periods shown in Fig. 8. Using these data, the total amount of CO_2 emitted from the peatland in the fire season was estimated to be 0.1 Gt C/y, which was considerably smaller than the emissions in ENSO years, e.g., 4 Mt C/day from September to October 2015 (Huijnen et al. 2016). The present observation period, from July 2014 to May 2015, was just before the start of the 2015 ENSO period.

For cold fire, the total carbon emission was 3000 ± 140 gC/m^2 for the two non-fire periods in Fig. 8 from July 2014 to May 2015. The annual net ecosystem CO_2 exchange, predominantly controlled by oxidative peat decomposition, was 1500 ± 70 gC/m^2year averaged for the two non-fire seasons from 2014 to 2015. This estimate might be the maximum estimate because the measurement was conducted only during the dry season. Hirano et al. (2012) measured the CO_2 emission using the eddy

Table 1 Raw data of accumulated ΔXCO_2 for various temperature-difference criterions

ΔT (K)	Accumulated CO_2 concentration (ppm m/12 h)	Number of data	1σ (ppm m/12 h)
> 3.5	2500	34	990
> 3.0	2500	52	940
> 2.5	2400	77	2300
> 2.0	2100	118	3700

covariance technique at two locations: a drained swamp forest (2.35° S, 114.14° E) and a drained burned swamp forest (2.34° S, 114.04° E) located within the Mega Rice Project area of Palangka Raya. Their observed values exhibited clear seasonal variations with functions in the groundwater level, ranging from 105 to 532 ppm or 427 to 571 gC/m^{2}year for 2004–2008. There was a large discrepancy between our estimated values and the reported values of Hirano et al. (2012). This is partly due to the differences in the hydrological environments of the observation sites. Additionally, the average temperature during our observation period was ~ 1.5 °C higher than that from 2004 to 2008.

Furthermore, the NTIT measurements in Kalampangan were conducted during the dry season of 2013, from the end of June to early November (135 days). The classification criteria examined were $\Delta T > 1.5$ °C to $\Delta T > 3.5$ °C with 0.5 °C steps, where ΔT stands for the difference between the temperatures at the tower top and bottom. The time series of the CO_2 concentration for ΔT peaked at approximately 05:00 LT, corresponding to the amount of CO_2 accumulated under the natural canopy for 12 h. The accumulated CO_2 concentrations until 05:00 LT for each ΔT are listed in Table 1. The larger ΔT results in higher accumulation efficiency because a more stable natural canopy is formed. In contrast, a weak inversion layer or small ΔT results in highly scattered data, as shown in Table 1. Considering that there is no appreciable change in the accumulated CO_2 data under $\Delta T > 3.5$ °C and > 3.0 °C, ΔT > 3.0 °C was adopted for data analysis because of the smaller standard deviation. After the accumulated CO_2 data were plotted as a function of wind speed, extrapolating to a wind speed of 0 m/s, the intrinsic accumulated CO_2 was 3700 ppm m/12 h or 1300 gC/m^{2}year, consistent with the value obtained from the column density measurements.

Acknowledgements The authors thank BMKG (Indonesia) for providing the airport visibility data. We acknowledge the products provided by GOSAT, WDCGG, HITRAN, TCCON, and the Goddard Earth Sciences Data and Information Services Center of NASA. This work was supported by a KAKENHI Grant (18KK0294) from the Japan Society for Promotion of Sciences and a project (No. 14200117) from the Research Institute for Humanity and Nature.

References

Air Resources Laboratory. 2018. NOAA's Office of Atmospheric Research, National Oceanic and Atmospheric Administration, U.S.A. https://ready.arl.noaa.gov/hypub-bin/trajtype.pl

Albar, I., I. Jaya, B.H. Saharjo, B. Kuncahyo, and K.P. Vadrevu. 2018. Spatio-temporal analysis of land and forest fires in Indonesia using MODIS active fire dataset. In *Land-atmospheric research applications in South and Southeast Asia*, 105–127. Cham: Springer.

Atwood, E.C., S. Englhart, E. Lorenz, W. Halle, W. Wiedemann, and F. Siegert. 2016. Detection and characterization of low temperature peat fires during the 2015 fire catastrophe in Indonesia using a new high-sensitivity fire monitoring satellite sensor (FireBird). *PLoS ONE* 11: e0159410. https://doi.org/10.1371/journal.pone.0159410.

Chambers, S., A.G. Williams, W. Zahorowski, A. Griffiths, and J. Crawford. 2011. Separating remote fetch a local mixing influence on vertical radon measurements in the lower atmosphere. *Tellus* 63B: 843–859. https://doi.org/10.1111/j.1600-0889.2011.00565.x.

Clymo, R.S. 1984. The limits to peat bog growth. *Philosophical Transactions of the Royal Society of London. B, Biological Science* 303: 605–654. https://doi.org/10.1098/rstb.1984.0002.

Field, R.D., G.R. van der Werf, and S.P. Shen. 2009. Human amplification of drought-induced biomass burning in Indonesia since 1960. *Nature Geoscience* 2: 185–188. https://doi.org/10.1038/ngeo443.

Fochesatto, G.J. 2015. Methodology for determining multilayered temperature inversions. *Atmospheric Measurement Techniques* 8: 2051–2060. https://doi.org/10.5194/amt-8-2051-2015.

GOSAT Data Archive Service. GOSAT Data Archive Service (nies.go.jp).

Hayasaka, H., and A. Sepriando. 2018. Severe air pollution due to peat fires during 2015 super El Nino in Central Kalimantan, Indonesia. In *Land-atmospheric research applications in South and Southeast Asia*, 127–142. Cham: Springer. https://doi.org/10.1007/978-3-319-67474-2_7

Hayasaka, H., A. Sepriando, A. Usup, and N. Daisuke. 2021. Air pollution conditions near peat fire-prone areas during El Niño in Central Kalimantan, Indonesia. In *Biomass burning in South and Southeast Asia*, 233–247. CRC Press.

Hirano, T., H. Segah, K. Kusin, S. Limin, H. Takahashi, and M. Osaki. 2012. Effects of disturbances on the carbon balance of tropical peat swamp forests. *Global Change Biology* 18: 3410–3422. https://doi.org/10.1111/j.1365-2486.2012.02793.x.

Hooijer, A., S. Page, J. Jauhiainen, W.A. Lee, X.X. Lu, A. Idris, and G. Anshari. 2012. Subsidence and carbon loss in drained tropical peatlands. *Biogeosciences* 9: 1053–1071. https://doi.org/10.5194/bg-9-1053-2012.

Huijnen, V., M.J. Wooster, J.W. Kaiser, D.L.A. Gaveau, J. Flemming, M. Parrington, A. Inness, D. Murdiyarso, B. Main, and M. van Weele. 2016. Fire carbon emissions over maritime Southeast Asia in 2015 largest since 1997. *Scientific Reports* 6: 26886. https://doi.org/10.1038/srep26886.

Iriana, W., K. Tonokura, M. Kawasaki, G. Inoue, K. Kusin, and S.H. Limin. 2016. Measurement of carbon dioxide flux from tropical peatland in Indonesia by nocturnal temperature-inversion trap method. *Environmental Research Letters* 11: 095011. https://doi.org/10.1088/1748-9326/11/9/095011.

Iriana, W. 2019. Development of monitoring methods for carbon dioxide emissions from disturbed tropical peatland. Doctral Dissertation, The University of Tokyo. Japan.

Iriana, W., K. Tonokura, G. Inoue, M. Kawasaki, O. Kozan, K. Fujimoto, M. Ohashi, M.A. Rahman, and D. Gunawan. 2018. Ground-based measurements of column-averaged carbon dioxide molar mixing ratios in a peatland fire-prone area of Central Kalimantan Indonesia. *Scientific Reports* 8: 8437. https://doi.org/10.1038/s41598-018-26477-3.

Kameoka, T., O. Kozan, S. Hadi, Asnawi. and Hasrullah. 2021. Monitoring the groundwater level in tropical peatland through UAV mapping of soil surface temperature: a pilot study in Tanjung Leban, Indonesia. *Remote Sensing Letters* 12: 542–552.https://doi.org/10.1080/2150704X.2021.1906974

Kobayashi, N., G. Inoue, M. Kawasaki, H. Yoshioka, M. Minomura, I. Murata, T. Nagahama, Y. Matsumi, T. Tanaka, I. Morino, and T. Ibuki. 2010. Remotely operable compact instruments for measuring atmospheric CO_2 and CH_4 column densities at surface monitoring sites. *Atmospheric Measurement Techniques* 3: 1103–1112. https://doi.org/10.5194/amt-3-1103-2010.

Kusumaningtyas, S.D.A., E. Aldrian, S. Sunaryo, and R. Kurniawan. 2021. Biomass burning and haze in Indonesia, long-term climate perspective, and impact on regional air quality. In *Biomass burning in South and Southeast Asia* 249–268. CRC Press. https://doi.org/10.1201/978042902 2258

NASA FIRMS. 2022. Fire information for resource management system. https://earthdata.nasa.gov/earth-observation-data/near-real-time/firms/active-fire-data

Mathieu, N., I.B. Strachan, M.Y. Leclerc, A. Karipot, and E. Pattey. 2005. Role of lowlevel jets and boundary-layer properties on the NBL budget technique. *Agricultural and Forest Meteorology* 135: 35–43. https://doi.org/10.1016/j.agrformet.2005.10.001.

Minkkinen, K., H. Vasander, S. Jauhiainen, M. Karsisto, and J. Laine. 1999. Post-drainage changes in vegetation composition and carbon balance in Lakkasuo mire, Central Finland. *Plant and Soil* 207: 107–120. https://doi.org/10.1023/A:1004466330076.

Page, S.E., F. Siegert, J.O. Rieley, H.D.V. Boehm, A. Jaya, and S. Limin. 2002. The amount of carbon released from peat and forest fires in Indonesia during 1997. *Nature* 420: 61–65. https://doi.org/10.1038/nature01131.

Rahman, Mh.A., D.S. Nugroho, M.D. Yamanaka, M. Kawasaki, O. Kozan, M. Ohashi, H. Hashiguchi, and S. Mori. 2021. Weather radar detection of planetary boundary layer and smoke layer top of peatland fire in Central Kalimantan, Indonesia. *Scientific Reports* 11: 367. https://doi.org/10.1038/s41598-020-79486-6

Rothman, L.S., et al. 2009. The HITRAN 2008 molecular spectroscopic database. *Journal of Quantitative Spectroscopy and Radiative Transfer* 110: 533–572. https://doi.org/10.1016/j.jqsrt.2009.02.013.

Stein, A.F., R.R. Draxler, G.D. Rolph, B.J.B. Stunder, and M.D. Cohen. 2015. NOAA's HYSPLIT atmospheric transport and dispersion modeling system. *American Meteorological Society*. https://doi.org/10.1175/BAMS-D-14-00110.1.

Usup, A., Y. Hashimoto, H. Takahashi, and H. Hayasaka. 2004. Combustion and thermal characteristics of peat fire in tropical peatland in Central Kalimantan. *Tropics*, 14 (1): 1–19. https://www.jstage.jst.go.jp/article/tropics/14/1/14_1_1/_article.

Wilson, E.L., E.M. Georgieva, and W.S. Heaps. 2007. Development of a Fabry-Perot interferometer for ultra-precise measurements of column CO_2. *Measurement Science and Technology* 18: 1495–1502. https://doi.org/10.1088/0957-0233/18/5/040.

Yulianti, N., K. Kusin, D. Naito, M. Kawasaki, O. Kozan, and K.E. Susatyo. 2020. The linkage of EL NIÑO-induced peat fires and its relation to current haze condition in central Kalimantan. *Journal of Wetlands Environmental Management* 8: 100–116. https://doi.org/10.20527/jwem.v8i2.221.

Air Pollution Caused by Peatland Fires in Central Kalimantan

Hiroshi Hayasaka and Aswin Usup

Abstract This study summarizes air pollution due to peatland fires near peat fire-prone areas of ex-Mega Rice Project (MRP) in Central Kalimantan. Background air pollution measured in 2002 is discussed based on some knowledge and findings from laboratory analysis of peat, field research of peatland fires, weather and fire conditions, satellite images, and recent scientific papers. Air pollutants such as SO_2, CO, O_3, NO_2, and particulate matter (PM10) were measured from 2000 to 2010. The worst air pollution occurred in 2002 due to active deep peatland fire under low groundwater level (GWL) condition. Maximum peak concentrations of PM10, SO_2, CO, and O_3 reached 1905, 85.8, 38.3, and 1003 $\times 10^{-6}$ g m^{-3}, respectively, on October 14, 2002, when the GWL was less than − 1000 mm. The O_3 peak may suggest the serious formation of photochemical smog under the high-NO_2 (= 42.5 $\times 10^{-6}$ g m^{-3}) derived from both peatland fires and engines. Low visibility less than 2 km lasted about 70 days from the middle of August to the end of October 2002. The daily changes in each air pollutant mentioned above during the 2002 fire season showed a unique feature of smoldering peat combustion. Two special types of peat fires are overhanging and peat layer fires. The occurrence of these underground peat fires can be seen from the increase in CO and PM10. In 2002, these underground peat fires occurred when the GWL dropped below about − 500 mm.

Keywords Air pollution · Haze · Peatland fire · Visibility · Smoldering · Overhanging · PM10 · SO_2 · CO · O_3 · NO_2 · Photochemical smog

H. Hayasaka (✉)
Hokkaido University, Sapporo, Japan
e-mail: hhaya@eng.hokudai.ac.jp

A. Usup
Palangka Raya University, Palangka Raya, Indonesia

K. P. Vadrevu et al. (eds.), *Vegetation Fires and Pollution in Asia*,
https://doi.org/10.1007/978-3-031-29916-2_23

1 Introduction

Biomass burning is the most common source of pollution in several countries including South/Southeast Asia (Streets et al. 2003). Several researchers in Asia reported biomass burning as an important source of greenhouse gas emissions and aerosols (Albar et al. 2018; Badarinath et al. 2007, 2008, 2009; Badarinath and Prasad 2011; Prasad et al. 2002a, b, 2003, 2008; Putra et al. 2008; Putra and Hayasaka 2011; Gupta et al. 2001; Kant et al. 2000; Eaturu and Vadrevu 2021; Hayasaka and Sepriando 2018; Hayasaka et al. 2021; Hayasaka 2022; Justice et al. 2015; Kharol et al. 2012; Lasko and Vadrevu 2018; Lasko et al. 2017, 2018, 2021). The smoke particles released during the biomass burning can impact Earth's radiation budget (Hsu et al. 2003) by their light scattering effects and impacting cloud microphysical processes (Lin et al. 2014). Also, biomass burning has also been attributed to an increase in growth rates of CO, CO_2, and CH_4 during the ENSO events (Hayasaka et al. 2014). In addition, biomass burning has been shown to influence a variety of land–atmospheric interactions at different scales, such as biogeochemical cycles, emissions, vegetation transpiration, soil erosion, albedo (Crutzen and Andreae 1990; Choi et al. 2008; Prasad et al. 2001a, b, 2003, 2004, 2005; Prasad and Badarinth 2004; Prasad and Badarinath 2006). Smoke-borne aerosols from fires disrupt normal hydrological processes and reduce rainfall, potentially contributing to regional drought. In addition to these effects on Earth's radiation, atmosphere, climate, and ecosystems, the pollutants released from the biomass burning (Vadrevu 2008; Vadrevu and Badarinath 2009; Vadrevu and Justice 2011; Vadrevu et al. 2008, 2013, 2014a, b, 2018, 2019, 2020) can have adverse health effects such as asthma, acute respiratory illness, eye irritation, cardiovascular mortality, thrombosis, and in severe cases, mortality (Yin 2023).

In Asia, mostly, biomass burning is driven by anthropogenic activities. For example, fire is used to clear the forests for agriculture through slash and burn (Biswas et al. 2015a, b) agricultural residues after crop harvest (Lasko et al. 2017, 2018; Vadrevu and Lasko 2015), to clear the land for the next crop, to clear the forested lands for plantations (Hayasaka et al. 2014; Albar et al. 2018), promoting the growth of grass in pasture lands for cattle (Stott et al. 1990), including reducing of weeds prior to planting of crops (Simorangkir 2007). In addition, the ignition source can also be intentional or accidental human activities. While most of these fires are anthropogenic, the drivers of fires can also be natural such as lightning and extreme and prolonged drought conditions. Thus, it is important to address biomass burning emissions from different sources, regions, and varied spatial scales (Vadrevu 2015, 2021; Vadrevu et al. 2021a, b, 2022a, b; Wooster et al. 2021).

Of the several countries in Asia, Indonesia has some of the world's highest rates of deforestation and forest degradation, the principal drivers of which are agricultural expansion and wood extraction in combination with an increased incidence of fires (Page et al. 2013). An average annual clearance rate of 10×10^3 km^2 during the 1980s has increased to an average of 20×10^3 km^2 per year since 1996. The largest tree-cover loss since 2001 was 24.2×10^3 km^2 in 2016 (Weisse and Goldman 2021).

Due to the thick peatland in Indonesia, peat fires become more active when the groundwater level drops. The risk of fire on peatland is increased greatly by drainage, which lowers the water table, exposing a greater volume of dry peat to combustion. This effect is demonstrated by a study of the fire regime in the former Mega Rice Project (MRP) area on peatland in southern Central Kalimantan (Hoscilo et al. 2008a, b). From 1973 to 1996, fires affected 24% of a 4.5×10^3 km^2 block C area in MRP. Following multiple fires in peat swamp forest, the number of tree species and individual trees, saplings and seedlings within the secondary vegetation are greatly reduced, and at the highest levels of degradation, succession back to forest is diverted to a retrogressive succession to communities dominated by ferns with very few or no trees (Hoscilo et al. 2008a, b; Page et al. 2009). Non-forested areas in Kalimantan had shorter average fire return intervals (FRI) than Sumatra (13 years vs. 40 years), with ferns/low shrub areas burning most frequently (Vetrita and Cochrane 2021).

Peatland fires make a major contribution to emissions of greenhouse gases, fine particulate matter, and aerosols, thus contributing to climate change as well as presenting a problem for human health (Page et al. 2013). The devastating 1997–1998 Indonesian fires were among the largest peak emission events in the recorded history of fires in equatorial Southeast Asia (Schultz et al. 2008; Van der Werf et al. 2006). According to a recent report on intense fires in 2015, peat fires in Indonesia during July–October 2015 released about 2 Tg of carbon into the atmosphere, 81% of which was in the form of carbon dioxide (CO_2), 16% carbon monoxide (CO), and 2.3% methane (CH_4) (Setyawati and Suwarsono 2018). Fine particulate matter (PM2.5) emissions from fires across Sumatra and Kalimantan (Borneo) during September–October 2015 were estimated at 7.33 Tg (Kiely et al. 2019). The level of major air pollution gases and particulate matter (PM10) at Palangka Raya near the Mega Rice Project was already reported (Hayasaka et al. 2014).

Peat can be initiated by flaming fires and embers of surface vegetation. The probability of ignition depends on the moisture content, inert content, and other chemical–physical properties (Huang and Rein 2014; Huang et al. 2015). Two mechanisms control the spread of smoldering combustion: oxygen supply and heat losses (Rein 2015; Ohlemiller 1985). Studies on smoldering peat fires introduced the overhanging combustion of peat (Huang et al. 2016). Their experiment and analysis results help a physical understanding of the spread and overhang phenomenon in peat wildfires and explain the role of moisture and oxygen supply. The spread of smoldering peat fire can be explained by the formation and collapse of overhanging. In addition to overhanging, we add "peat layer fires (underground peat fires)" which spread under topsoil without large openings. We often see "peat layer fires" in agricultural lands and under paved roads with mineral soil above a deep peat layer.

In this study, we summarize the worst air pollution measured in 2002. Our knowledge and findings from past studies including our own research results from laboratory analysis of peat, field research of peatland fires, weather and fire conditions, and satellite images are used to discuss background air pollution in MRP+ area in Central Kalimantan.

2 Study Area

Figure 1 shows the peatland distribution and study area called "MRP+" in Kalimantan (Borneo) and air pollution measuring sites in Palangka Raya (capital of Central Kalimantan). Peatlands in Kalimantan cover about 57,600 km^2, which is equivalent to peatlands of Sumatra. Central Kalimantan alone has 30,100 km^2 of peatlands. Deforested peatlands, abandoned after the MRP, became major air pollution sources because of the large accumulation of peat and their combustion after human disturbances.

MRP+ covers the MRP area and its vicinity, including the Sebangau National Park (1.75–3.5° S, 113.5–115° E). MRP is divided into five blocks, block A to block E, as shown in Fig. 1. Their actual boundaries are defined mainly by rivers. In this chapter, the approximate boundaries are defined by latitude and longitude lines. MRP+ was chosen as the study area simply because it is dominated by peatlands in Central Kalimantan and is one of the highest fire hotspot density areas in Indonesia (Yulianti et al. 2012).

Palangka Raya is located in Central Kalimantan, as shown by a circle (2.207° S, 113.917° E) in Fig. 1. A (red color) and U (blue color) near Palangka Raya in Fig. 1 stand for Tjilik Riwut Meteorology Station (TRMS) (2.23° S, 113.95° E) near

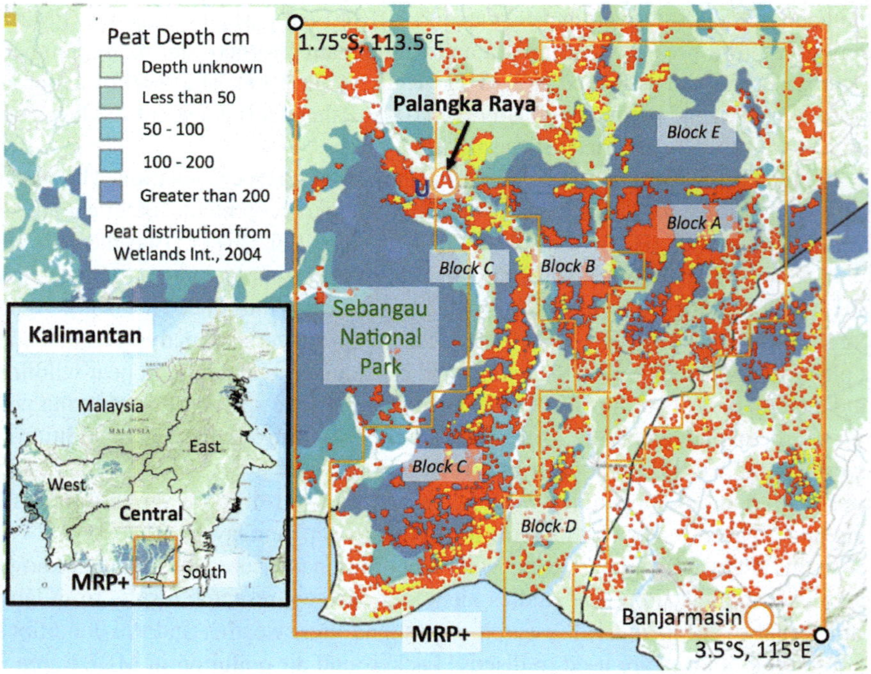

Fig. 1 Map of Kalimantan, MRP+, blocks A–E, and peatland. MODIS hotspots are shown by red (detected from DN = 170 to 330) and yellow dots (highest hotspot 1191, detected on DN = 285)

Airport and measuring site of groundwater level (GWL) measured site (2.32° S, 113.9° E), respectively. The nearest coastline distance is about 100 km, and the average altitude is only around 10 m. The MRP was built on tropical swamp forest areas on the eastern and southern sides of Palangka Raya. Before the disturbance, the tropical swamp forest could hold enough water to stay wet even in the dry season. However, the constructed 4000-km-long MRP canal built for irrigation facilitated illegal logging and loss of water through drainage from most of the peatlands in the MRP area. These disturbances are the main reasons for severe fire occurrences in the MRP area.

3 Data and Methods

3.1 *Air Pollution and Weather Data*

Air pollution in Palangka Raya had been monitored by the Air Quality Management System and regional Air Quality Center in Palangka Raya from 2001 until around 2010. The locations of the three measuring stations are Tjilik Riwut, Tilung, and Murjani (Hayasaka et al. 2014). Each station measured PM10, SO_2, CO, O_3, and NO_2 by using automated continuous analyzers. The Air Quality Center processed all air pollution data automatically every 30 min and recorded their values. Unfortunately, most of their gas analyzers were broken in 2011, and air pollution data were unavailable now (Hayasaka et al. 2014).

The weather data were measured at the Palangka Raya weather station of the Indonesian Agency for Meteorology, Climatology, and Geophysics (BMKG) at Palangka Raya International Airport—shown by "A" in Fig. 1. Underground level (GWL) was measured at "U" in Fig. 1. Daily rainfall data were used to evaluate underground level (GWL) using GWL simulation model (MODEL-0) (Hayasaka et al. 2020).

3.2 *Hotspot (Fire) Data and Satellite Imagery*

Hotspot (HS, temperature anomaly) data detected by moderate-resolution imaging spectroradiometer (MODIS) on the Terra and Aqua satellites are used to evaluate fires in boreal forests. The MODIS data are available from November 2000 (for Terra) and from July 2002 (for Aqua) to the present. MODIS HS data in 2002 are obtained from NASA FIRMS (Fire Information for Resource Management System, https://firms2.modaps.eosdis.nasa.gov/download/ (Accessed 14 April 2021)). We use only the spatial and temporal (latitude, longitude, and acquisition date and time) hotspot data. The number of daily hotspots is used to identify fire activities in 2002.

The Terra/MODIS Corrected Reflectance (True Color) imagery is obtained through the website of EOSDIS Worldview (https://worldview.earthdata.nasa.gov (Accessed 14 April 2021)) and used to grasp haze situation and estimate origin of air pollutants source.

4 Results and Discussion

4.1 Source of Air Pollutants on Deep Peatland

Figure 2a shows typical peatland fire conditions in the deep peatland. Deep peatland fires emit grayish smoke from underground peat and brownish smoke from surface fire. A mixture of these smokes constitutes the haze. Smoldering of peat emits various air pollutants (including transient gases) due to the low-temperature smolder (peat pyrolysis temperature: about 450–720 K) compared to surface flaming fire (> about 1000 K).

In the deep peatland (peat depth more than about 0.5 m), active underground peat fires can occur with cave and overhanging as shown in Fig. 2b. This type of underground peat fire is unique and occurs only in deep peatland areas such as MRP+ region in Central Kalimantan. Once deep peatland fire starts, smoldering peat fires spread horizontally along the free surface and vertically in-depth (Rein 2013; Huang and Rein 2014). Horizontally, fires often burn peat soil below the surface leaving a cave and overhang as Fig. 2b. Although the overhanging and cave will collapse easily under the weight of a person, the lateral spread of the fire continues, and some fires resurface to promote new surface peat fires. Thus, underground smoldering peat fires can be continued by keeping the heat from peat pyrolysis with little heat loss. From

Fig. 2 Haze (air pollutants) and peatland fires. **a** Fire and smoke situation. **b** Overhanging and cave in deep peatland fire

our field observations, it can be said that underground peat fires are the dominant type of fires rather than the cave and overhanging in the third fire stage. The spread of smoldering combustion is controlled by oxygen supply and heat losses (Tacconi 2003; Jakarta Post 2019).

4.2 *Fire Stages and GWL*

Deep peatland fires can be classified into three fire stages based on groundwater level (GWL) as noted from the field observations (Hayasaka et al. 2016; Usup et al. 2004). We classify the three fire stages as follows: (1) surface fire, (2) shallow peatland fire, and (3) deep peatland fire. Their GWL boundary values were about − 300 mm and about − 500 mm (Hayasaka et al. 2020). This classification is necessary to identify the air pollution status at each fire stage.

2002 fires are divided into three fire stages as shown in Fig. 3. From Fig. 3, surface fire started at the end of June (Day Number (ND) around 170). Shallow peatland fire started from around DN = 178 (July 6) with 27 hotspots (HS) day^{-1}. Deep peatland fires started from around DN = 220 (August 8) with 180 hotspots (HS) day^{-1} and lasted until the end of October. Classification of these three fire stages is necessary to assess air pollution from peatland fires. At the surface fire stage, most vegetation burns with flames (temperature higher than smoldering combustion). Smoldering peat fire in shallow peatland begins with the help of surface vegetation fires. Finally, when the thickness of flammable peat (moisture content less than around 100%) reached about − 300 mm measured on August 2 (Usup et al. 2004), the third fire stage (deep peatland fire) begins. The gradient of the straight line drawn on the accumulated hotspot (HS) curve in Fig. 3 shows the fire activity in the three fire stages. 180 HS day^{-1} for "(3) Deep peatland fire" was larger than about 160 HS day^{-1} (average of last six fire years (2002, 2004, 2006, 2009, 2014, and 2015)) (Hayasaka et al. 2020). This active fire trend during "(3) Deep peatland fire" can occur only in deep peatland.

4.3 *Air Pollution Situation in Deep Peatland*

Daily change of various air pollutants, SO_2, CO, O_3, NO_2, and particulate matter (PM10) is shown in Fig. 4. The above air pollutants can be emitted from peat and surface combustion except SO_2. As SO_2 from surface vegetation fires is usually very low, SO_2 will be one of the important signs for occurrence of smoldering peat fires. High values of CO and PM10 in the third fire stage (DN = 220–300) are also important signs for occurrence of the deep peatland fire (lower temperature smoldering peat combustion). NO_2 has two different sources, derived from vegetation fires and combustion in vehicle engines. This is why the NO_2 values are also high from the first and second fire stages (DN = 170–220).

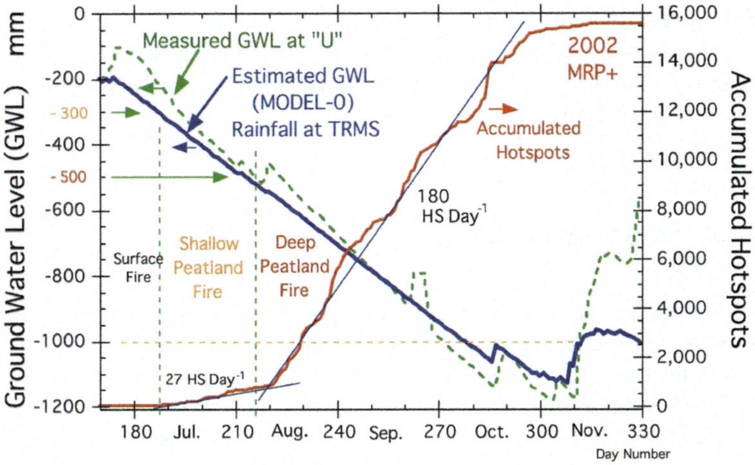

Fig. 3 Three stages of peat fire and groundwater level

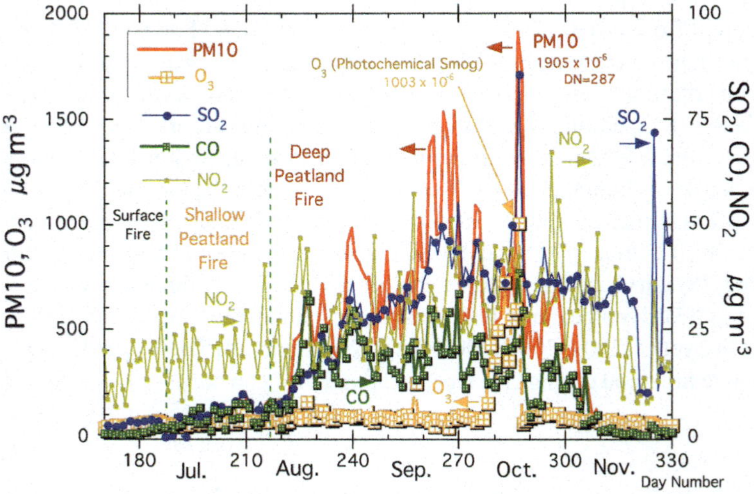

Fig. 4 PM10, O_3, SO_2, CO, and NO_2

The worst air pollution occurred on October 14 in 2002 due to active deep peatland fires in the end of the third fire stage, "(3) Deep peatland fire". GWL was lowest about – 1000 mm (see Fig. 3). Maximum peak concentrations of PM10, SO_2, CO, and O_3 reached 1905, 85.8, 38.3, and 1003 $\times 10^{-6}$ g m^{-3}, respectively. The O_3 peak may suggest the serious formation of photochemical smog under the high-NO_2 (= 42.5 $\times 10^{-6}$ g m^{-3}). This may be the world's first record of photochemical smog caused by deep peatland fires.

4.4 *Haze, Visibility, and Hotspot*

Figure 5 shows daily change of PM10, visibility, and hotspot (HS). Accumulated HS curve shows the fire activity in each fire stage. Dense haze from deep peatland fires can make unfavorable meteorological conditions such as calm winds and low air temperature. Under the dense haze condition in 2002, sunlight could not reach the ground easily, suggesting that daytime convection flow by the sunlight barely occurred on these days and the haze was confined to the surface due to low air temperature. High PM10 values and low visibilities in the third fire stage of "(3) Deep peat layer fire" are due to high fire rate of 180 HS day^{-1}.

PM10 gradually increased during two periods of surface fire (DN = 170–187) and shallow peatland fire (DN = 187–213) as shown in Fig. 5. A sharp increase in PM10 from about 100 to 400 × 10^{-6} g m^{-3} broke out in the third fire stage (around August 11 (DN = 223)). Very high PM10 (> 500 × 10^{-6} g m^{-3}) lasted until late October. The daily changes in PM10 resemble a zigzag pattern of hotspots (HS) as shown in Fig. 5. PM10 reached its maximum of 1905 × 10^{-6} g m^{-3} on October 14 (ND = 287) two days after the HS peak (HSs = 1191) on October 12.

Visibility changed in the opposite tendency to PM10. As PM10 increased, visibility decreased (Fig. 5). Visibility is gradually decreased during two periods of surface fire and shallow peatland fire. A sharp decrease in visibility from about 7 to 4 km broke out in the third fire stage (around August 11 (DN = 223)). The lowest visibility was nearly 0.16 km on October 13 (ND = 286). Low visibility less than 2 km lasted about 70 days from the middle of August to the end of October in 2002.

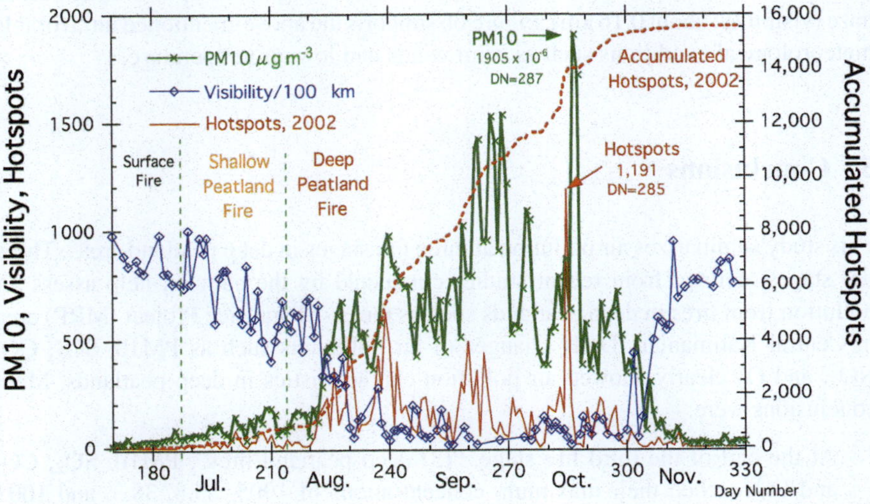

Fig. 5 PM10, visibility, and hotspots

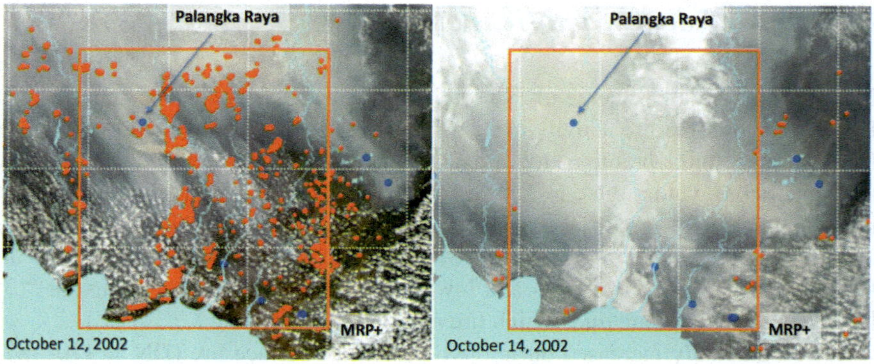

Fig. 6 Fire and haze situation. **a** Hotspot peak day (DN = 285, October 12, 2002) PM10 = 867 $\times$ 10^{-6} g m^{-3}, HSs = 1191. **b** PM10 peak day (DN = 287, October 14, 2002), PM10 = 1905 $\times$ 10^{-6} g m^{-3}, HSs = 13

About 16 days from DN = 260 in late September, visibility was less than 0.5 km and wind speed was also very slow (mean wind speed less than 2 ms^{-1}) (Hayasaka et al. 2014).

Two satellite images in Fig. 6a, b show the fire and haze situation of HS peak day (ND = 285) and PM10 maximum day (ND = 287). Figure 5a shows active fire distribution and smoke from fires. Most smoke appears to flow from the southeast to the northwest which is the prevailing wind direction during the dry (fire) season in Kalimantan. Figure 5b shows that dense smoke (haze) covers the Central Kalimantan. Figure 5b suggests that MODIS could not detect hotspots through dense haze (visibility; about 0.16 km). Figure 6b supports the above-mentioned unfavorable meteorological conditions such as calm winds and low air temperature.

5 Conclusions

This study summarizes air pollution at three fire stages in deep peatland areas. Three fire stages, defined from recent studies conducted by the authors, help assess air pollution from fires in deep peatlands such as the ex-Mega Rice Project (MRP) area in Central Kalimantan. Daily changes of air pollutants such as PM10, SO_2, CO, NO_2, and O_3 clearly showed air pollution characteristics in deep peatlands. Main conclusions were:

1. At the end of the third fire stage, "(3) deep peatland fires", PM10, SO_2, CO, and O_3 reached their maximum concentrations of 1905, 85.8, 38.3, and 1003 $\times$ 10^{-6} g m^{-3}, respectively, on October 14, 2002. On this day in 2002, the groundwater level (GWL) was the lowest, about − 1000 mm.
2. The sharp increase in PM10, SO_2, and CO at the beginning of the third fire stage was a sign of deep peatland fires (the overhanging and peat layer fires).

3. The O_3 peak may suggest the serious formation of photochemical smog under the high-NO_2 ($= 42.5 \times 10^{-6}$ g m^{-3}) derived from both peatland fires and engines.
4. Low visibility of less than 2 km persisted most of the third fire stage (about 70 days from the middle of August to the end of October 2002).
5. Major air pollutants are emitted from unique peat fire types of the overhanging and peat layer fires underground in the third fire stage.

Acknowledgements This research is partially funded by the Ministry of Agriculture, Forestry and Fisheries of Japan grant to CIFOR. Project is named "Enhancing climate-resilient livelihoods in boreal and tropical high carbon forests and peatlands". We are grateful for their support.

References

Albar, I., I. Jaya, B.H. Saharjo, B. Kuncahyo, and K.P. Vadrevu. 2018. Spatio-temporal analysis of land and forest fires in Indonesia using MODIS active fire dataset. In *Land-atmospheric research applications in South and Southeast Asia*, 105–127. Cham: Springer.

Badarinath, K.V.S., and K.V. Prasad. 2011. Carbon dioxide emissions from forest biomass burning in India. *Global Environmental Research* 15: 45–52.

Badarinath, K.V.S., S.K. Kharol, K.M. Latha, T.K. Chand, V.K. Prasad, A.N. Jyothsna, and K. Samatha. 2007. Multiyear ground-based and satellite observations of aerosol properties over a tropical urban area in India. *Atmospheric Science Letters* 8 (1): 7–13.

Badarinath, K.V.S., S. Kumar Kharol, V. Krishna Prasad, A. Rani Sharma, E.U.B. Reddi, H.D. Kambezidis, and D.G. Kaskaoutis. 2008. Influence of natural and anthropogenic activities on UV Index variations—A study over tropical urban region using ground based observations and satellite data. *Journal of Atmospheric Chemistry* 59 (3): 219–236.

Badarinath, K.V.S., S.K. Kharol, A.R. Sharma, and V.K. Prasad. 2009. Analysis of aerosol and carbon monoxide characteristics over Arabian Sea during crop residue burning period in the Indo-Gangetic Plains using multi-satellite remote sensing datasets. *Journal of Atmospheric and Solar-Terrestrial Physics* 71 (12): 1267–1276.

Biswas, S., K.D. Lasko, and K.P. Vadrevu. 2015a. Fire disturbance in tropical forests of Myanmar—Analysis using MODIS satellite datasets. *IEEE Journal of Selected Topics in Applied Earth Observations and Remote Sensing* 8 (5): 2273–2281.

Biswas, S., K.P. Vadrevu, Z.M. Lwin, K. Lasko, and C.O. Justice. 2015b. Factors controlling vegetation fires in protected and non-protected areas of Myanmar. *PLoS ONE* 10 (4): e0124346.

Choi, Y., S.A. Vay, K.P. Vadrevu, A.J. Soja, J.H. Woo, S.R. Nolf, G.W. Sachse, G.S. Diskin, D.R. Blake, N.J. Blake, and H.B. Singh. 2008. Characteristics of the atmospheric CO_2 signal as observed over the conterminous United States during INTEX-NA. *Journal of Geophysical Research: Atmospheres* 113 (D7).

Crutzen, P.J., and M.O. Andreae. 1990. Biomass burning in the tropics: Impact on atmospheric chemistry and biogeochemical cycles. *Science* 250 (4988): 1669–1678.

Eaturu, A., and K.P. Vadrevu. 2021. Evaluation of Sentinel-3 SLSTR data for mapping fires in forests, peatlands, and croplands—A case study over Australia, Indonesia, and India. In *Biomass burning in South and Southeast Asia*, 39–59. CRC Press.

Gupta, P.K., V.K. Prasad, C. Sharma, A.K. Sarkar, Y. Kant, K.V.S. Badarinath, and A.P. Mitra. 2001. CH_4 emissions from biomass burning of shifting cultivation areas of tropical deciduous forests—Experimental results from ground-based measurements. *Chemosphere-Global Change Science* 3 (2): 133–143.

Hayasaka, H. 2022. Fire weather conditions in Boreal and polar regions in 2002–2021. *Atmosphere* 13 (7): 1117.

Hayasaka, H., and A. Sepriando. 2018. Severe air pollution due to peat fires during 2015 super El Niño in Central Kalimantan, Indonesia. In *Land-atmospheric research applications in South and Southeast Asia*, 129–142. Cham: Springer.

Hayasaka, H., I. Noguchi, E.I. Putra, N. Yulianti, and K. Vadrevu. 2014. Peat fire-related air pollution in Central Kalimantan. *Environmental Pollution* 195: 257–266.

Hayasaka, H., H. Takahashi, S.H. Limin, N. Yulianti, and A. Usup. 2016. Peat fire occurrence. In *Tropical peatland ecosystems*, chap. 25, 337–395. Springer.

Hayasaka, H., A. Usup, and D. Naito. 2020. New approach evaluating peatland fires in Indonesian factors. *Remote Sensing* 12 (12): 2055. https://doi.org/10.3390/rs12122055.

Hayasaka, H., A. Sepriando, A. Usup, and N. Daisuke. 2021. Air pollution conditions near peat fire-prone areas during El Niño in Central Kalimantan, Indonesia. In *Biomass burning in South and Southeast Asia*, 233–247. CRC Press.

Hoscilo, A., S.E. Page, and K.J. Tansey. 2008a. Repeated and extensive fire as the main driver of land cover change in Block C of the former Mega Rice Project. In *Proceedings of the 13th International Peat Congress: After Wise Use—The Future of Peatlands*, 206–208. Ireland: International Peat Society.

Hoscilo, A., S.E. Page, H. Wosten, J. Jauhiainen, M. Silvius, J. Rieley, H. Ritzema, K. Tansey, L. Graham, and H. Vasander. 2008b. Restoration ecology of lowland tropical peatlands in Southeast Asia: Current knowledge and future research directions. *Ecosystems* 12: 888–905. https://doi.org/10.1016/j.envpol.2014.06.031.

Hsu, N.C., J.R. Herman, and S.C. Tsay. 2003. Radiative impacts from biomass burning in the presence of clouds during boreal spring in Southeast Asia. *Geophysical Research Letters* 30 (5).

Huang, X., and G. Rein. 2014. Smouldering combustion of peat in wildfires: Inverse modelling of the drying and the thermal and oxidative decomposition kinetics. *Combustion and Flame* 161 (6): 1633–1644. https://doi.org/10.1016/j.combustflame.2013.12.013.

Huang, X., G. Rein, and H. Chen. 2015. Computational smoldering combustion: Predicting the roles of moisture and inert contents in peat wildfires. *Proceedings of the Combustion Institute* 35 (3): 2673–2681. https://doi.org/10.1016/j.proci.2014.05.048.

Huang, X., F. Restuccia, M. Gramola, and G. Rein. 2016. Experimental study of the formation and collapse of an overhang in the lateral spread of smouldering peat fires. *Combustion and Flame*. https://doi.org/10.1016/j.combustflame.2016.01.017.

Jakarta Post. 2019. Fires in Indonesia burn 1.6 m ha of land, mostly former forests: Satellite data. Available online: https://www.thejakartapost.com/life/2019/12/02/fires-in-indonesia-burn-1-6m-ha-of-land-mostlyformer-forests-satellite-data.html. Accessed 18 January 2020.

Justice, C., G. Gutman, and K.P. Vadrevu. 2015. NASA land cover and land use change (LCLUC): An interdisciplinary research program. *Journal of Environmental Management* 148: 4–9.

Kant, Y., V.K. Prasad, and K.V.S. Badarinath. 2000. Algorithm for detection of active fire zones using NOAA AVHRR data. *Infrared Physics & Technology* 41 (1): 29–34.

Kharol, S.K., K.V.S. Badarinath, A.R. Sharma, D.V. Mahalakshmi, D. Singh, and V.K. Prasad. 2012. Black carbon aerosol variations over Patiala city, Punjab, India—A study during agriculture crop residue burning period using ground measurements and satellite data. *Journal of Atmospheric and Solar-Terrestrial Physics* 84: 45–51.

Kiely, L., D.V. Spracklen, C. Wiedinmyer, L. Conibear, C.L. Reddington, S. Archer-Nicholls, D. Lowe, S.R. Arnold, C. Knote, and M.F. Khan. 2019. New estimate of particulate emissions from Indonesian peat fires in 2015. *Atmospheric Chemistry and Physics* 19: 11105–11121.

Lasko, K., and K. Vadrevu. 2018. Improved rice residue burning emissions estimates: Accounting for practice-specific emission factors in air pollution assessments of Vietnam. *Environmental Pollution* 236: 795–806.

Lasko, K., K.P. Vadrevu, V.T. Tran, E. Ellicott, T.T. Nguyen, H.Q. Bui, and C. Justice. 2017. Satellites may underestimate rice residue and associated burning emissions in Vietnam. *Environmental Research Letters* 12 (8): 085006.

Lasko, K., K.P. Vadrevu, and T.T.N. Nguyen. 2018. Analysis of air pollution over Hanoi, Vietnam using multi-satellite and MERRA reanalysis datasets. *PLoS ONE* 13 (5): e0196629.

Lasko, K., K.P. Vadrevu, V. Bandaru, T.N.T. Nguyen, and H.Q. Bui. 2021. PM2.5 emissions from biomass burning in South/Southeast Asia—Uncertainties and trade-offs. In *Biomass burning in South and Southeast Asia*, 149–169. CRC Press.

Lin, N.H., A.M. Sayer, S.H. Wang, A.M. Loftus, T.C. Hsiao, G.R. Sheu, N.C. Hsu, S.C. Tsay, and S. Chantara. 2014. Interactions between biomass-burning aerosols and clouds over Southeast Asia: Current status, challenges, and perspectives. *Environmental Pollution* 195: 292–307.

Ohlemiller, T. 1985. Modeling of smoldering combustion propagation. *Progress in Energy and Combustion Science* 11 (4): 277–310. https://doi.org/10.1016/0360-1285(85)90004-8.

Page, S.E., A. Hoscilo, A. Langner, K.J. Tansey, F. Siegert, S. Limin, and J.O. Rieley. 2009. Tropical peatland fires in Southeast Asia. In *Tropical fire ecology, climate change, land use and ecosystem dynamics*. Berlin: Springer.

Page, S., J. Rieley, A. Hoscilo, A. Spessa, and U. Weber. 2013. Current fire regimes, impacts and the likely changes—IV, Tropical Southeast Asia. In V*egetation fires and global change—Challenges for concerted international action. A white paper directed to the United Nations and International Organizations*, 91–96. Amsterdam: Publishing House.

Prasad, K.V., and K.V.S. Badarinath. 2006. Soil surface nitrogen losses from agriculture in India: A regional inventory within agroecological zones (2000–2001). *The International Journal of Sustainable Development and World Ecology* 13 (3): 173–182.

Prasad, V.K., and K.V.S. Badarinth. 2004. Land use changes and trends in human appropriation of above ground net primary production (HANPP) in India (1961–98). *Geographical Journal* 170 (1): 51–63.

Prasad, V.K., Y. Kant, P.K. Gupta, C. Sharma, A.A. Mitra, and K.V.S. Badarinath. 2001a. Biomass and combustion characteristics of secondary mixed deciduous forests in Eastern Ghats of India. *Atmospheric Environment* 35 (18): 3085–3095.

Prasad, V.K., Y. Kant, and K.V.S. Badarinath. 2001b. CENTURY ecosystem model application for quantifying vegetation dynamics in shifting cultivation areas: A case study from Rampa Forests, Eastern Ghats (India). *Ecological Research* 16 (3): 497–507.

Prasad, V.K., Y. Kant, P.K. Gupta, C. Elvidge, and K.V.S. Badarinath. 2002a. Biomass burning and related trace gas emissions from tropical dry deciduous forests of India: A study using DMSP-OLS data and ground-based measurements. *International Journal of Remote Sensing* 23 (14): 2837–2851.

Prasad, V.K., Y. Kant, and K.V.S. Badarinath. 2002b. Land use changes and modeling carbon fluxes from satellite data. *Advances in Space Research* 30 (11): 2511–2516.

Prasad, V.K., M. Lata, and K.V.S. Badarinath. 2003. Trace gas emissions from biomass burning from northeast region in India—Estimates from satellite remote sensing data and GIS. *The Environmentalist* 23 (3): 229–236.

Prasad, V.K., K.V.S. Badarinath, S. Yonemura, and H. Tsuruta. 2004. Regional inventory of soil surface nitrogen balances in Indian agriculture (2000–2001). *Journal of Environmental Management* 73 (3): 209–218.

Prasad, V.K., E. Anuradha, and K.V.S. Badarinath. 2005. Climatic controls of vegetation vigor in four contrasting forest types of India—Evaluation from National Oceanic and Atmospheric Administration's Advanced Very High Resolution Radiometer datasets (1990–2000). *International Journal of Biometeorology* 50 (1): 6–16.

Prasad, A.S. 2008. Zinc in human health: Effect of zinc on immune cells. *Molecular Medicine* 14: 353–357. https://doi.org/10.2119/2008-00033.Prasad.

Putra, E.I., and H. Hayasaka. 2011. The effect of the precipitation pattern of the dry season on peat fire occurrence in the Mega Rice Project area, Central Kalimantan, Indonesia. *TROPICS* 19 (4): 145–156.

Putra, E.I., H. Hayasaka, H. Takahashi, and A. Usup. 2008. Recent peat fire activity in the mega rice project area, Central Kalimantan, Indonesia. *Journal of Disaster Research* 3 (5): 334–341.

Rein, G. 2013. Smouldering fires and natural fuels. In *Fire phenomena and the earth system*, 15–33. Wiley.
Rein, G. 2015. Smoldering combustion. In *SFPE handbook of fire protection engineering*, 581–603. Springer.
Schultz, M.G., A. Heil, J.J. Hoelzemann, A. Spessa, K. Thonicke, J.G. Goldammer, A.C. Held, J.M.C. Pereira, and M. van het Bolscher. 2008. Global wildland fire emissions from 1960 to 2000. *Global Biogeochemical Cycles* 22: 1–17.
Setyawati, W., and Suwarsono. 2018. Carbon emission from peat fire in 2015. *IOP Conference Series: Earth and Environmental Science* 166: 1–7.
Simorangkir, D. 2007. Fire use: Is it really the cheaper land preparation method for large-scale plantations? *Mitigation and Adaptation Strategies for Global Change* 12 (1): 147–164.
Stott, P.A., J.G. Goldammer, and W.L. Werner. 1990. The role of fire in the tropical lowland deciduous forests of Asia. In *Fire in the tropical biota*, 32–44. Berlin, Heidelberg: Springer.
Streets, D.G., T.C. Bond, G.R. Carmichael, S.D. Fernandes, Q. Fu, D. He, Z. Klimont, S.M. Nelson, N.Y. Tsai, M.Q. Wang, and J.H. Woo. 2003. An inventory of gaseous and primary aerosol emissions in Asia in the year 2000. *Journal of Geophysical Research: Atmospheres* 108 (D21).
Tacconi, L. 2003. *Fires in Indonesia: Causes, costs and policy implications*. CIFOR Occasional Paper No. 38. Bogor: CIFOR.
Usup, A., Y. Hashimoto, H. Takahashi, and H. Hayasaka. 2004. Combustion and thermal characteristics of peat fire in tropical peatland in Central Kalimantan, Indonesia. *TROPICS* 14 (1): 1–19.
Vadrevu, K.P. 2008. Analysis of fire events and controlling factors in eastern India using spatial scan and multivariate statistics. *Geografiska Annaler: Series A, Physical Geography* 90 (4): 315–328.
Vadrevu, K. 2015. International regional science meeting of NASA-LCLUC: Land cover/land use change (LC/LUC) and environmental impacts in South Asia, Coimbatore, India, 19–23 January 2013. *Journal of Environmental Management* 148: 1–163.
Vadrevu, K.P. 2021. Fire danger indices and methods: An appraisal. In *Biomass burning in South and Southeast Asia*, 201–232. CRC Press.
Vadrevu, K.P., and K.V.S. Badarinath. 2009. Spatial pattern analysis of fire events in Central India—A case study. *Geocarto International* 24 (2): 115–131.
Vadrevu, K.P., and C.O. Justice. 2011. Vegetation fires in the Asian region: Satellite observational needs and priorities. *Global Environmental Research* 15 (1): 65–76.
Vadrevu, K., and K. Lasko. 2015. Fire regimes and potential bioenergy loss from agricultural lands in the Indo-Gangetic Plains. *Journal of Environmental Management* 148: 10–20.
Vadrevu, K.P., K.V.S. Badarinath, and E. Anuradha. 2008. Spatial patterns in vegetation fires in the Indian region. *Environmental Monitoring and Assessment* 147 (1): 1–13.
Vadrevu, K.P., L. Giglio, and C. Justice. 2013. Satellite based analysis of fire–carbon monoxide relationships from forest and agricultural residue burning (2003–2011). *Atmospheric Environment* 64: 179–191.
Vadrevu, K.P., T. Ohara, and C. Justice. 2014a. Air pollution in Asia. *Environmental Pollution (Barking, Essex: 1987)* 195: 233–235.
Vadrevu, K.P., K. Lasko, L. Giglio, and C. Justice. 2014b. Analysis of Southeast Asian pollution episode during June 2013 using satellite remote sensing datasets. *Environmental Pollution* 195: 245–256.
Vadrevu, K.P., T. Ohara, and C. Justice (eds.). 2018. *Land-atmospheric research applications in South and Southeast Asia.*
Vadrevu, K., A. Heinimann, G. Gutman, and C. Justice. 2019. Remote sensing of land use/cover changes in South and Southeast Asian countries. *International Journal of Digital Earth* 12 (10): 1099–1102.
Vadrevu, K.P., A. Eaturu, S. Biswas, K. Lasko, S. Sahu, J.K. Garg, and C. Justice. 2020. Spatial and temporal variations of air pollution over 41 cities of India during the COVID-19 lockdown period. *Scientific Reports* 10 (1): 1–15.

Vadrevu, K.P., T. Ohara, and C. Justice (eds.). 2021a. *Biomass burning in South and Southeast Asia: Impacts on the biosphere*, vol. 2. CRC Press.

Vadrevu, K.P., T. Ohara, and C. Justice (eds.). 2021b. *Biomass burning in South and Southeast Asia: Mapping and monitoring*, vol. 1. CRC Press.

Vadrevu, K.P., T. Le Toan, S.S. Ray, and C. Justice. 2022a. *Remote sensing of agriculture and land cover/land use changes in South and Southeast Asian countries*. Cham: Springer.

Vadrevu, K., A. Eaturu, E. Casadaban, K. Lasko, W. Schroeder, S. Biswas, L. Giglio, and C. Justice. 2022b. Spatial variations in vegetation fires and emissions in South and Southeast Asia during COVID-19 and pre-pandemic. *Scientific Reports* 12 (1): 1–21.

Van der Werf, G.R., J.T. Randerson, L. Giglio, G.J. Collatz, and P.S. Kasibhatla. 2006. Interannual variability in global biomass burning emission from 1997 to 2004. *Atmospheric Chemistry and Physics* 6: 3423–3441.

Vetrita, Y., and M.A. Cochrane. 2021. *Land use and cover maps from Landsat, Mawas, Central Kalimantan, Indonesia, 1994–2019*. ORNL DAAC.

Weisse, M., and E.D. Goldman. 2021. *Just 7 commodities replaced an area of forest twice the size of Germany between 2001 and 2015*.

Wooster, M.J., G.J. Roberts, L. Giglio, D.P. Roy, P.H. Freeborn, L. Boschetti, C. Justice, C. Ichoku, W. Schroeder, D. Davies, and A.M. Smith. 2021. Satellite remote sensing of active fires: History and current status, applications and future requirements. *Remote Sensing of Environment* 267: 112694.

Yin, S. 2023. Decadal changes in premature mortality associated with exposure to outdoor PM2.5 in mainland Southeast Asia and the impacts of biomass burning and anthropogenic emissions. *Science of The Total Environment* 854: 158775.

Yulianti, N., H. Hayasaka, and A. Usup. 2012. Recent forest and peat fire trends in Indonesia the latest decade by MODIS hotspot data. *Global Environmental Research* 16 (1): 105–116.

Chemical Speciation of PM_{10} Emissions from Peat Burning Emission in Central Kalimantan, Indonesia

Puji Lestari, Isna Utami, Febri Juwita, Rajasekhar Balasubramanian, and Jeffrey S. Reid

Abstract Biomass burning is a significant source of particulate matter (PM_{10}) in Central Kalimantan, Indonesia. However, only limited data exist on the emission characteristics from this source. Therefore, an intensive field study was conducted in Central Kalimantan, Indonesia, during a peat fire to investigate the physical and chemical characteristics of particulate emissions in peat smoke. In this study, OC, EC, metals, as well as major inorganic ions were measured to determine the chemical composition of PM_{10}. PM_{10} samples were collected using Mini Volume Portable Air Samplers (Air metrics) and a quartz fiber filter. Nine samples of PM_{10} were collected near the sources (located at Pulau Pisau) to represent the emission from the source (it is called a burning site), and two samples were collected at the ambient air far from the source to provide an overview of the urban background site at a residential area in Palangkaraya city. The samples were then analyzed to determine concentrations of PM_{10} and its chemical composition, e.g., black carbon (BC) (carbonaceous materials which consist of organic carbon (OC) and elemental carbon (EC), water-soluble ions, and metals). The results showed a high concentration of PM_{10} near the source, and the average concentrations for PM_{10}, OC, and EC were 4141.5 $\mu g\ m^{-3}$, 2133.2 $\mu g\ m^{-3}$, and 50.9 $\mu g\ m^{-3}$, respectively, while average concentrations at the urban background site were 81.2 $\mu g\ m^{-3}$, 34.2 $\mu g\ m^{-3}$ and 1.9 $\mu g\ m^{-3}$ for PM_{10}, OC, and EC, respectively. The results also indicate that the dominant chemical component of PM_{10} from peat land burning was organic carbon (OC), ions (Cl^-, SO_4^{2-}, K^+), and metals such as Al and Zn. The average contribution of OC, EC, ions, and metals to the total PM_{10} at the burning site was ~ 60%, ~ 1%, ~ 9%, and ~ 0.004%, respectively, while at the urban residential site was ~ 43%, ~ 2%, ~ 33%, and ~ 0.3%, respectively.

P. Lestari (✉) · I. Utami · F. Juwita
Faculty of Civil and Environmental Engineering, Bandung Institute of Technology, Ganesha 10, Bandung 40132, Indonesia
e-mail: pujilest@indo.net.id

R. Balasubramanian
Department of Civil and Environmental Engineering, Faculty of Engineering, National University of Singapore, 1 Engineering Drive 2, 1A-07-03, Singapore 117576, Singapore

J. S. Reid
Marine Meteorology Division, Naval Research Laboratory, 7 Grace Hopper Ave, Stop 2, Monterey, CA 93943-5502, USA

K. P. Vadrevu et al. (eds.), *Vegetation Fires and Pollution in Asia*,
https://doi.org/10.1007/978-3-031-29916-2_24

Keywords Peat fires · Particulate matter (PM_{10}) · Carbonaceous materials · Ions · Metals

1 Introduction

In Southeast Asia (SEA), major causes of transboundary smoke haze pollution include forest, bush, and peat fires originating from Kalimantan and Sumatra, Indonesia (e.g., Gutman et al. 2000; Koe et al. 2001; Balasubramanian et al. 2003; Mahmud 2009; Hyer and Chew 2010; Reid et al. 2011; Albar et al. 2018; Vadrevu et al. 2018, 2019, 2021a, b). Not only important from an air quality point of view (Kunii et al. 2002), such emissions significantly perturb the regional and perhaps global carbon budgets (Page et al. 2002, 2011), as well as the regional radiative balance including air quality (Davidson et al. 2005; Rajeev et al. 2008). Nearly thirty percent of South/Southeast Asia had recurrent fires with the most in Laos, Cambodia, Thailand, and Myanmar, which is a matter of concern (Vadrevu and Justice 2011). Smoke effects caused due to the 1997–1998 Indonesian wildfires affected many countries in SEA, primarily Malaysia, Singapore, and Thailand (Langmann and Heil 2004; Levine 1999; Arvelyna et al. 2021; Lasko and Vadrevu 2018; Lasko et al. 2018). With the world's largest area of tropical peat lands with 27 million hectares (Wulandari 2002), tropical peat fires are the main source of smoke production in Indonesia during fire disasters (Langmann and Heil 2004). Peat fires are a problem as during drought periods, they can become exposed and be easily ignited but almost impossible to extinguish (Hayasaka et al. 2014). They can smolder deep underground indefinitely and flare up again during the next dry period, with devastating impacts on the environment and, most importantly, on human health (Kunii et al. 2002; Page et al. 2002).

There are two stages of the combustion process in biomass fires. The first stage is flaming, and combustion of pyrolysis products (gases and vapors) occurs at this stage. This occurred when pyrolysis results heated until the ignition point, at the temperature range between 325 and 350 °C. Thermal at the flaming stage accelerates the rate of the pyrolysis process and causes a lot of flammable gases to be produced. The gas is also easily oxidized and leads to a greater flame (flaming process enlarged). The second stage is smoldering. In this stage, a very thick smoke is formed after the flaming process. Based on Reid et al. (2005), smoldering combustion begins when most of the volatile material has been burned. In other words, the smoldering combustion phase starts when the fuel's volatile content is low. Smoldering generally occurs in solid fuel with oxygen-limited conditions. Despite the importance of the peat biomass burning system, there are few direct measurements of peat smoke properties, such as size, thermodynamics, or even basic chemistry. Most measurements are made, typically away from source regions (e.g., Maenhaut et al. 2002; Balasubramanian et al. 2003; He et al. 2010). Given smoke's propensity for rapid chemical and physical evolution after emission (Reid et al. 2005), there is currently a considerable knowledge gap on differences between smoke properties at the source compared to downwind receptor sites in different regions of the world (Badarinath et al. 2007, 2008, 2009; Kharol et al. 2012). To address this problem, in 2009 while contributing to the 7-Southeast Asian Studies program, an opportunity arose to measure smoke chemical properties in Southern Kalimantan, Indonesia. That year hosted a significant El Nino event which led to large-scale drought, hydrological peat land draining, and subsequent significant peat burning (Reid et al. 2011). Filter samples were collected in September and October during the peak of the peat burning episode in both source regions and nearby residential areas to serve as regional receptors. Filter samples were analyzed for mass concentration, organic carbon mass fractions, and major ions. This is, to our knowledge, the few comprehensive measurement study done for peat burning in Kalimantan. The key objectives of this study were to measure the mass concentration of airborne particles (PM_{10}) and their primary chemical contents. A specific chemical component of interest is carbonaceous materials that are total carbon (TC), which consists of organic carbon (OC) and elemental carbon (EC), water-soluble ions, and metals, some of which are of concern because of their adverse health effects, especially to individuals being inadvertently exposed.

2 Methodology

2.1 Description of Sampling Location

PM_{10} samples were collected in Central Kalimantan during a major peat fire episode in September–October 2009. The samples were collected near the source of peat burning, directly adjacent to the plume, and at a nearby regional residential area. Figure 1 shows the sampling location at the burning site, village Taruna Jaya, Pulau Pisau, in Central Kalimantan (Borneo Island), Indonesia. This site is about 30–40 km from the city of Palangkaraya (the capital city of central Kalimantan) with coordinates (S 02 20′ 08, 7″ and E 114 04′ 38, 2″). Air sampling was conducted in the open field within the peat fires. Most of the terrain was covered with open forests of *Cycas rumphii*, *Ceratopteris thalictroides*, and *Gonystylus bancanus*. The sampling equipment was placed near the peat smoke as shown in Fig. 2. To serve as a regional receptor, sampling was also conducted in a residential area of Palangkaraya city (about 40 km northwest of Pulau Pisau) which is the capital city of Central Kalimantan province (S 02° 13′ 13, 93″ and E 113° 54′ 2, 56″). It is recognized that both local emissions from industries and vehicular traffic and emissions from peat fires during the sampling period influenced the air quality of this site.

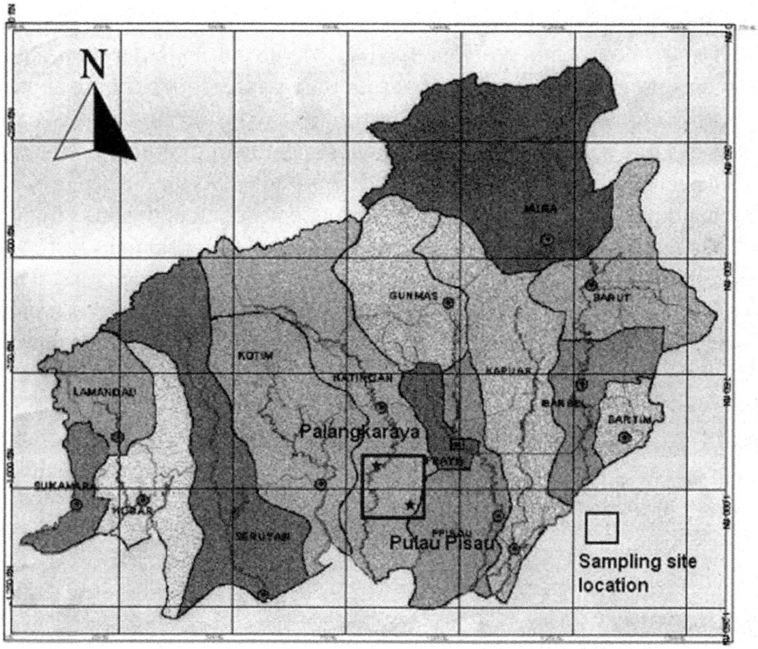

Fig. 1 Map of the sampling site in Palangkaraya

Fig. 2 Burning site condition

2.2 PM_{10} Sampling Collection

During a peat fire episode in Pulau Pisau (September 19–October 12, 2009), the particulate sampling was conducted at the locations over three weeks. Two pre-calibrated Mini-Vol portable air samplers (from Air metrics) were used simultaneously at the sampling locations to collect PM_{10} samples and run at the flow rate of 5 L min^{-1}. Two different types of filters, 47 mm pre-fired quartz fiber and Teflon membrane (Polytetrafluoroethylene, PTFE), were used for PM_{10} sampling and subsequent chemical analysis. The sampling time at the Pulau Pisau site ranged from 6 to 8 h to obtain sufficient PM mass for chemical analysis. The PTFE and quartz filters were dried in a desiccator at a constant temperature of 25 °C and a relative humidity of 35% for at least 24 h before and after the particulate sampling. The filters were then weighed with an MC5 microbalance (Sartorius AG) accurate to 1 μg. After sampling, the filters were weighed and stored in Petri dishes wrapped in aluminum foil and at 4 °C until the analysis.

2.3 Chemical Analysis

After field sampling, filters containing PM_{10} samples were placed in protected Petri dishes and returned to the laboratory for weighing and chemical analysis. Particulate mass concentration was measured using the gravimetric method by weighing samples before and after sampling, and the measurements were blank-corrected. The samples were then analyzed at different laboratories.

(a) Optical Black Carbon Measurement: Was made at ITB Bandung by reflectance measurements using an EEL type of Smoke Stain Reflectometer. Secondary standards of known black carbon concentrations are used to calibrate the reflectometer.

(b) Ions and Metals: They were analyzed at the National University of Singapore (NUS) Department of Environmental Engineering; filter samples were extracted with suprapure® HNO_3 65% and 5 ml aquadest and Microwave digested for 30 min and then rinsed with aquadest up to 25 ml volume and analyzed for 13 metals, namely, Al, Ti, V, Cr, Mn, Fe, Co, Ni, Cu, Zn, As, Cd, and Pb, on an inductively coupled plasma mass spectrometer (ICP-MS) (Perkin-Elmer Inc.) ultrapure water for ion analysis, which was then ultrasonicated at 60 °C for 1 h. The extract was subsequently filtered through a 110-mm membrane Whatman 32 filter. Then, the extract was rinsed with ultrapure water up to 25 ml volume. After the extraction, samples were analyzed for ion concentrations using Ion Chromatograph. Ions which include F^-, Cl^-, NO^{3-}, SO^{2-} and NH_4^+, K^+, Ca^+, Na^+ were determined. The ion concentrations were blank-corrected. Next to the metals analysis, standard samples were prepared by diluting multi-element standard solutions (OC517986, OC285604, and Arsenic standard solution) for the calibration of ICP-MS (Perkin-Elmer Inc., USA). The limits of detection

(LODs) for all target analyses were calculated based on three times the standard deviation of the blank. The LODs ranged from 0.01 (Pb) to 1.09 (Fe) μg/L. The precision and accuracy of the extraction procedure were evaluated using standard reference materials (SRMs) (NIST SRM 1648, urban particulate matter). The SRM samples in triplicate were processed following the same procedure as the samples and analyzed using ICP-MS. The concentrations in each fraction were added, and the total concentration was compared with the certified values. The recoveries of total elements were satisfactory and ranged between 84% (Fe) and 109% (Ni).

(c) Water-Soluble Ions' Analysis: Filter samples were extracted with ultrapure water using an ultrasonicator at 60 °C for 1 h. The extract was subsequently filtered through a 110-mm membrane Whatman 42 filter and then rinsed with ultrapure water up to 25 ml volume. After the extraction, the water samples were analyzed for ion concentrations using an Ion Chromatograph (IC). Ions which include F^-, Cl^-, NO^{3-}, SO^{2-} and NH^+, K^+, Ca^+, Na^+ were determined. The ion concentrations were blank-corrected.

(d) Measurement of Carbonaceous Materials: A total carbon (TC), elemental carbon (EC), and organic carbon (OC) analysis was carried out at the Desert Research Institute's Laboratory (DRI) using a thermo-optical technique where particulate filter samples will be heated sequentially without and with oxygen, which will produce OC and EC, respectively (Chow et al. 1993, 2001). The filter sample was placed in the EC-OC oven analyzer to monitor the reflectance or transmittance of red light (wavelength 633 nm) passing through the sample. Based on the response of the flame ionization detector (FID) and the data from laser transmission, the amount of OC and EC of corresponding samples can be determined. The sum of OC and EC is defined as the samples' total carbon (TC) (Schauer 2004). EC is operationally defined as the material that does not volatilize in the first step of the analysis. OC is the difference between TC and EC (TC = EC + OC).

3 Results and Discussion

A total of nine PM_{10} samples were collected during a recent peat fire episode at the burning sites in Central Kalimantan, Indonesia. In comparison, two samples were collected in the residential area in the city of Palangkaraya to provide the urban background information on the air quality in the vicinity area. Table 1 summarizes the PM_{10} concentrations and meteorological and sampling conditions at the location during sampling periods. Sample numbers 1–9 indicate samples collected from the burning site, while sampling numbers 10 and 11 indicate samples collected from a background residential area in Palangkaraya.

Table 1 PM_{10} concentration, sampling, and meteorological conditions

Sample number	Sampling date	Mass concentration ($\mu g\ m^{-3}$)	Remark	Meteorological condition			
				Wind direction	Wind speed (m/s)	Pressure (mmHg)	Temp (°C)
Region source							
1	19/09/2009	1721	Flaming	South East	3.3	30	33.5
2	21/09/2009	4322	Flaming	South East	2.1	30	30.5
3	22/09/2009	2151	Flaming	South	2.7	30.1	31.8
4	25/09/2009	692	Smouldering	South East	1.6	30.1	32.2
5	28/09/2009	570	Smouldering	North West and South East	1.9	30.1	31.9
6	29/09/2009	1749	Smouldering	South East	1.1	30	29.8
7	30/09/2009	15,705	Smouldering	South West and South East	1.9	30	32.6
8	1/10/2009	774	Smouldering	North East and South East	2.6	30	29.5
9	2/10/2009	9591	Smouldering	North East and South East	1.7	30.2	29
Urban residential							
10	11/10/2009	114	Residential area	South West, CALM	0.5	30	29.3
11	29/09/2009	49	Residential area	North East and North West	0.7	30.1	29

3.1 Mass Concentrations

The mass concentrations of PM_{10} at the two sampling sites were found to vary per their respective distances from the peat fires. For example, at Pulau Pisau (the burning site), the PM_{10} concentration ranged from 570 to 15,704 $\mu g\ m^{-3}$, and the average PM_{10} concentration was 4142 $\mu g\ m^{-3}$. As there were no major sources of air pollution at the burning sites, their high concentration was only due to peat fire emissions. It was also influenced by meteorological conditions at the time of sampling.

Two burning conditions (flaming and smoldering) were observed during the sampling periods. In the smoldering condition, the average concentration of PM_{10} was much higher than in the flaming condition, i.e., 4847 $\mu g\ m^{-3}$ in smoldering and 2731 $\mu g\ m^{-3}$ in the flaming conditions. These findings agreed with the previous study on wildfire mixed-evergreen forest in Portugal (Alves et al. 2010); it was observed that the particulate matter significantly enhanced under smoldering fire conditions. The variations in the mass changes influenced concentrations in the intensity of peat fires and those in the meteorological conditions (e.g., wind direction and speed) at the sampling time, while at the residential site, the average concentration of PM_{10} was 81 $\mu g\ m^{-3}$. The PM_{10} at the residential site showed low mass concentration, mainly influenced by the local urban emissions, as the site was far away from the burning of peat lands. The other sources of PM_{10}, besides local urban emissions, most likely included soil dust (wind-blown dust), waste combustion activities, and traffic emission in the vicinity of the measurement. However, during the burning seasons and certain meteorological conditions, the air quality in Palangkaraya may also be affected by the forest fire (peat burning). In this study, the concentrations of PM_{10} in the first and second measurements (samples #10 and #11) were 114 $\mu g\ m^{-3}$ and 50 $\mu g\ m^{-3}$, respectively. The two measurements were observed at different wind directions, which may cause different sources to contribute to the air quality in the residential area. The high concentration (114 $\mu g\ m^{-3}$) was observed when the wind blew from the south and southwest where the peat land burning occurred. This may indicate the possible contribution of burning peat lands to the air quality in Palangkaraya (residential site).

3.2 *Chemical Speciation*

Table 2 presents the average PM_{10}, EC, OC, ions, and metals concentrations, with their corresponding standard deviations measured at the two sampling sites. The speciated chemical components of PM_{10} were much higher at Pulau Pisau, where the peat fires took place. This trend is consistent with what was observed for the mass concentration of PM_{10}. The average concentrations of PM_{10} and chemical components varied significantly during the sampling period, which is reflected by the high standard deviation values. These variations could be caused by the prevailing meteorological conditions (wind directions and wind speeds), which frequently changed during the sampling period, and the frequent changes in peat fire intensity. The individual percentage contributions of EC, OC, metals, cations, and anions to PM_{10} for each sample are presented in Fig. 3. It can be seen that the major component of PM_{10} was OC, followed by water-soluble ions. The fraction of OC in the PM_{10} in the residential site (Palangkaraya city) was lower than in the burning site at Pulau Pisau. The average relative contributions of various chemical species to the total mass concentration of PM_{10} are illustrated in Fig. 4. In this figure, "others or residual" refers to the fraction of PM_{10} which was not identified by any of the chemical analyses used in this study, where 238 residual $= PM_{10} - (EC + OC + ions + metals)$ (1) 241. The air quality in Palangkaraya city was presumably influenced by local industries and vehicular traffic. Thus, the chemical makeup of PM_{10} was different from that in Pulau Pisau, which was influenced by emissions from peat fires. The major components ofand their possible sources are discussed below in detail.

Table 2 Mass concentrations of PM_{10} and its chemical components

	Pulau Pisau	Palangkaraya
PM 10 (μg m^{-3})		
PM_{10}	4142 ± 5184	81 ± 46
Carbonaceous particles (μg m^{-3})		
BC	22 ± 16	3 ± 0.8
EC	51 ± 70	2 ± 2.7
OC	2133 ± 2238	34 ± 18
Water-soluble ions (μg m^{-3})		
F^-	4.5 ± 3.0	0.5 ± 0.0
Cl^-	26.2 ± 28.5	2.4 ± 0.1
NO^{3-}	5.3 ± 4.1	1.0 ± 0.1
$SO_4{}^{2-}$	42.0 ± 32.3	3.0 ± 2.1
Na^+	36.7 ± 38.8	6.0 ± 1.3
$NH_4{}^+$	10.8 ± 10	0.1 ± 0.1
K^+	68.3 ± 42.9	9.1 ± 1.3
$Ca_2{}^+$	11.8 ± 10.9	1.5 ± 0.3
Metals (ng m^{-3})		
Al	379.6 ± 366.3	101.0 ± 38.5
Ti	12.5 ± 8.2	5.6 ± 2.0
V	38.9 ± 24.9	3.7 ± 0.7
Cr	15.2 ± 10.0	3.3 ± 0.6
Mn	11.3 ± 16.6	0.5 ± 0.5
Fe	27.3 ± 29.9	2.5 ± 3.1
Co	3.4 ± 5.5	0.2 ± 0.2
Ni	2.5 ± 4.5	1.4 ± 1.9
Cu	5.0 ± 14.8	
Zn	301.8 ± 460.7	46.1 ± 14.2
As	14.5 ± 9.6	1.3 ± 0.5
Cd	1.4 ± 2.5	0.1 ± 0.0
Pb	22.8 ± 48.6	1.9 ± 0.3

3.3 *Carbonaceous Particles*

The overall chemical composition analysis indicates that the major component of PM_{10} measured in this study is carbonaceous particles. At the source area or burning site (Pulau Pisau), OC and EC accounted for 2133 ± 2238 μg m^{-3} (~ 60% of PM_{10}) and 51 ± 70 μg m^{-3} (~ 1% of PM_{10}), respectively, while those at the urban residential area (Palangkaraya city) were 34 ± 18 μg m^{-3} (43% of PM_{10}) and 2 ± 3 μg m^{-3} (2% of PM_{10}) for OC and EC, respectively. Compared to the research on peat fire carbon

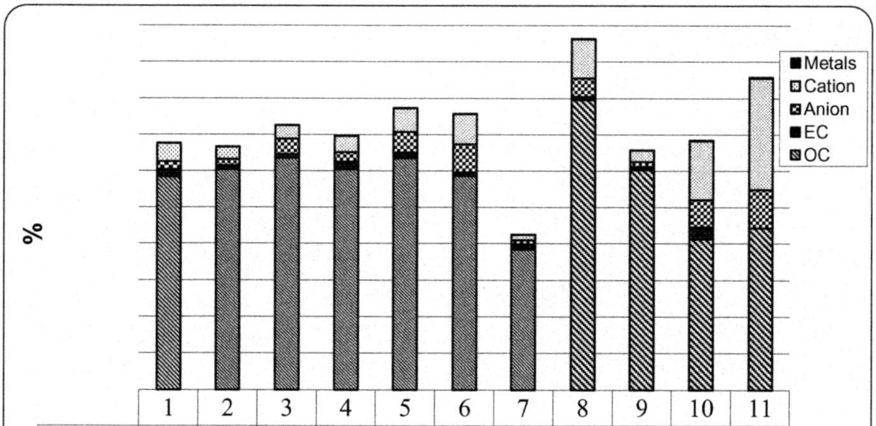

	1	2	3	4	5	6	7	8	9	10	11
■ Metals	0.01	0.01	0.03	0.01	0.07	0.09	0.01	0.09	0.02	0.19	0.25
▣ Cation	5.04	3.44	3.68	4.43	6.51	8.31	1.48	10.77	3.17	16.21	30.55
▩ Anion	2.22	1.77	4.29	2.67	5.68	7.80	1.18	5.00	1.42	7.56	10.67
■ EC	1.97	1.08	1.06	2.10	1.54	1.00	1.45	0.84	0.83	3.33	0.00
▧ OC	58.40	60.36	63.53	60.44	63.56	58.56	38.31	79.62	60.29	41.19	44.23

Fig. 3 Contribution of OC, EC, ions, and metals to total PM_{10} concentration in individual samples

emissions of PM2.5 in Sumatra (See et al. 2007), this study found a higher portion of OC in the aerosol particles than in Sumatra. However, a portion of EC in the PM_{10} was much lower than that of the Sumatra measurement. The study in Sumatra showed that OC and EC at the source region contributed about 750 ± 500 $\mu g\ m^{-3}$ (46.9% of PM2.5) and 310 ± 200 $\mu g\ m^{-3}$ (19.4% of PM2.5), respectively, while in the urban area (Pekanbaru), OC and EC concentrations and contribution were 30 ± 20 $\mu g\ m^{-3}$ (21.4% of PM2.5) and 21 ± 12 $\mu g\ m^{-3}$ (15.0% of PM2.5), respectively. The difference in concentrations of OC and EC, as well as their contribution to the total aerosol particles, may be caused by the different peatland characteristics in those two regions (Sumatra and Kalimantan). Further studies are needed to understand the differences at these sites. Lee et al. (2005) quantified the emission of carbonaceous particles from forest fires in Georgia. The study in Georgia showed that OC and EC contributed about 60.3 ± 18.5% and 3.9 ± 1.1% to the total PM2.5, respectively. The study in Portugal (Alves et al. 2010) on wildfire mixed-evergreen forests also observed that the particulate matter significantly enhanced under smoldering fire conditions. In line with PM_{10}, the average concentration of OC at the source region (burning site) was also much higher during smoldering conditions than during flaming conditions, i.e., 2370 $\mu g\ m^{-3}$ (smoldering) and 1660 $\mu g\ m^{-3}$ (flaming). The high percentage of carbonaceous particulates, especially OC, at those locations closer to the peat fires was consistent with the fact that peat lands are carbon sinks and rich in organic matter (Sabiham 1988). During peat fires, carbon is released into the air (Page et al. 2002) and more than half of the carbon is in organic form (Reid et al. 2005).

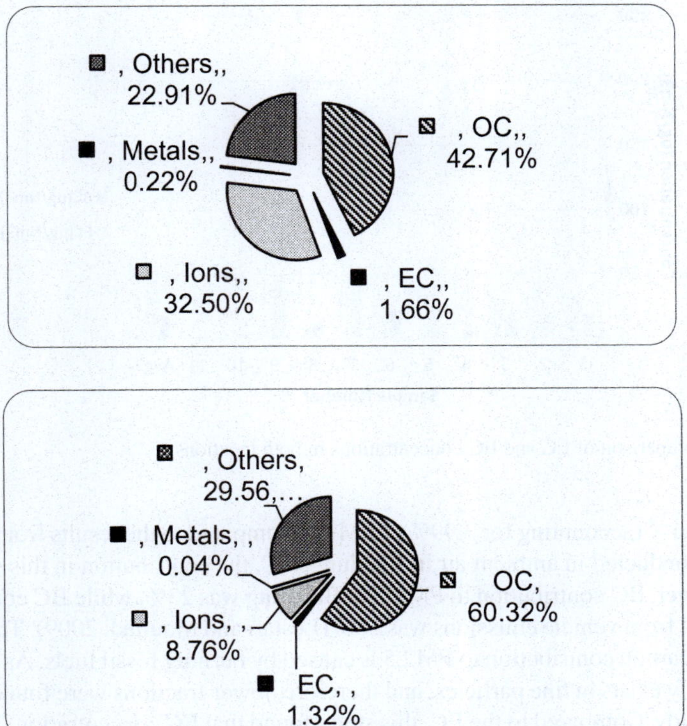

Fig. 4 Average chemical composition of PM_{10} sampled at Palangkaraya site and Pulau Pisau (Burning site)

This is also agreed with a previous study that the Indonesian peat content 54.7% carbon (C) (Akagi et al. 2011). EC constitutes a smaller fraction of the carbonaceous aerosol. EC is operationally defined as a material that is not volatile and highly absorbs carbon species with a graphitic-like structure.

3.4 Black Carbon

In addition to the EC measurement, black carbon (BC) was also measured to evaluate both results by applying different measurement methods. EC and BC were measured using thermal and optical methods, respectively. Black carbon (BC) is one of the particulate's components. BC can strongly absorb solar radiation (Martins et al. 1998), and biomass burning is responsible for about 45% of BC emissions on a global scale (Andreae et al. 1996). BC concentration in the burning site varied from 4.13 to 40.14 $\mu g\ m^{-3}$, and the average was 21.52 $\mu g\ m^{-3}$ (accounting for 0.87% of PM_{10}). In contrast, in the residential site, Palangkaraya city, the average BC concentration was

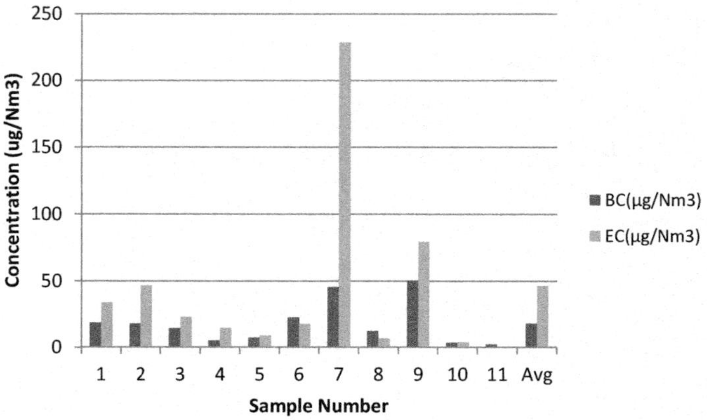

Fig. 5 Comparison of EC and BC concentrations in both locations

2.95 $\mu g\ m^{-3}$ (accounting for 3.99% of PM_{10}). Compared to the results from previous studies conducted in ambient air in Bandung, BC, the contribution in this study was much lower. BC contribution to PM2.5 in Bandung was 23%, while BC contribution to the PM from vehicle emissions was 16% (Lestari and Mauliadi 2009). The amount of black carbon contribution to PM2.5 is caused by burning fossil fuels. As we know, BC mainly exists in fine particles, and therefore, lower fractions were found in PM_{10} in this study. Compared to the EC, this study found that EC concentration (measured with the thermal method) was much higher than BC concentration (measured with the optical method), as shown in Fig. 5.

3.5 Ions

The total ions accounted for 211 $\mu g\ m^{-3}$ (~ 9% of PM_{10}) at Pulau Pisau and 24 $\mu g\ m^{-3}$ (~ 33% of PM_{10}) at Palangkaraya. The analysis of ions is divided into anions and cations. SO_4^{2-} and Cl^- are the most abundant of all anions, with an average concentration measured at Pulau Pisau being 42 $\mu g\ m^{-3}$ and 26 $\mu g\ m^{-3}$, respectively, while at Palangkaraya, the concentrations were 3 $\mu g\ m^{-3}$ and 2 $\mu g\ m^{-3}$ for SO_4^2 and Cl^-, respectively. SO_4^{2-} probably came from sulfurous materials stored in peat that were converted to SO_4^{2-} through the gas-to-particle conversion processes during combustion (See et al. 2007). Seinfeld (1986) indicated that almost all fossil fuels used by humans, such as petroleum, coal, gas, wood, organic matter, and others fuels containing sulfur, contribute about 80% to the total amount of sulfate in the atmosphere. In the Palangkaraya site, vehicles that used fossil fuels also contributed to the presence of SO_4^{2-} in PM_{10}. The dominant cation in the particulate forest and land fires was the potassium ion (K^+). The average concentration of ions measured at Pulau Pisau and Palangkaraya was 68 $\mu g\ m^{-3}$ and 9 $\mu g\ m^{-3}$, respectively. Anion and cation concentrations in this study are similar to those in Gwangju's agricultural field (Ryu et al. 2004). Ryu et al. (2004) reported that the primary anion and cations resulting from the events of grassland fire are Cl^-, SO_4^{2-}, K^+, and NH_4^+, respectively. The NH_4^+ value measured in this study is not very high and represents a minor contributor to the total concentration of ions. The difference in the NH_4^+ concentration between the two studies could be caused by differences in the types of biomass or vegetation combusted and/or the amount of moisture present before a fire occurs in plants.

3.6 *Concentrations of Metals*

Higher mass concentrations of the 13 metals were observed nearer the peat fires, but their relative proportions in PM_{10} were lower compared to TC and ions. The 13 metals together accounted for 836 ng m^{-3} (0.04% of PM_{10}) at Pulau Pisau and 168 ng m^{-3} (0.3% of PM_{10}) at Palangkaraya. Al and Zn were the most abundant trace metals at all the sampling sites, making up more than 85% of the total metals. Since the same metals are also the top two most abundant metals in the Earth's crust, it is postulated that these metals were released into the air from the local soil upon high-temperature combustion of the peat. Furthermore, due to biogeochemical processes, peat soils accumulate all kinds of metals due to atmospheric deposition (Gorham and Janssens 2005), which can be subsequently discharged into the ambient air, made airborne, and transported by the prevailing winds upon high-temperature combustion of peat soil. Therefore, it is not surprising to see an unusually high concentration of metals at the affected sites. Al and Zn are the dominant elements found in rice straw combustion in Gwangju (Ryu et al. 2004). According to Manahan (1994), Al is one element that belongs to the Earth other than Fe, Si, Ti, Na, Ca, V, and Mn. Al, as a dominant element in PM, was also found in Sumatra's research on forest and land fires (See et al. 2007).

4 Conclusions

The results of this study indicated that the dominant chemical components of PM_{10} from peat fires in Central Kalimantan (Borneo) were organic carbon and ions. Around 70% of the PM_{10} composition was identified and explained in this study, with OC contributing around 60% to the total mass concentration of PM_{10}. As for ions, the major ions identified in this study were K^+, Cl^-, and SO_4^{2-}, while Al and Zn were the significant metals in the PM_{10} samples collected from the site where peat fires took place.

Acknowledgements This research was conducted in cooperation with the National University of Singapore (NUS), Naval Research Laboratory (NRL-USA), and the Institute of Technology Bandung. The technical assistance provided by all students involved in this project is gratefully acknowledged.

References

Akagi, S.K., R.J. Yokelson, C. Wiedinmyer, M.J. Alvarado, J.S. Reid, T. Karl, J.D. Crounse, and P.O. Wennberg. 2011. Emission factors for open and domestic biomass burning for use in atmospheric models. *Atmospheric Chemistry and Physics* 11: 4039–4072.

Albar, I., I. Jaya, B.H. Saharjo, B. Kuncahyo, and K.P. Vadrevu. 2018. Spatio-temporal analysis of land and forest fires in Indonesia using MODIS active fire dataset. In *Land-atmospheric research applications in South and Southeast Asia*, 105–127. Cham: Springer.

Alves, A.C., A. Vicente, C. Monteiro, C. Goncalves, M. Evtyugina, and C. Pio. 2010. Emission of trace gases and organic components in smoke particles from a wildfire in a mixed-evergreen forest in Portugal. *Science of The Total Environment* 409: 1466–1475.

Andreae, M.O., E. Atlas, H. Cachier, W.R. Cofer, G.W. Harris, G. Helas, R. Koppmann, J.P. Lacaux, and D.E. Ward. 1996. Trace gas and aerosol emissions from savanna fires. In *Biomass burning and global change, vol. 1: Remote sensing, modeling and inventory development, and biomass burning in Africa*, ed. J.S. Levine, 278–295. MIT, MA.

Arvelyna, Y., H. Takahashi, L. Indrayanti, H. Hayasaka, K.P. Vadrevu, R. Maryani, M. Osaki, and H. Kazuyo. 2021. Peatland surface loss due to fires in Central Kalimantan, Indonesia—A case study using differential interferometry SAR (DInSAR). In *Biomass burning in South and Southeast Asia*, 77–90. CRC Press.

Badarinath, K.V.S., S.K. Kharol, K.M. Latha, T.K. Chand, V.K. Prasad, A.N. Jyothsna, and K. Samatha. 2007. Multiyear ground-based and satellite observations of aerosol properties over a tropical urban area in India. *Atmospheric Science Letters* 8 (1): 7–13.

Badarinath, K.V.S., S. Kumar Kharol, V. Krishna Prasad, A. Rani Sharma, E.U.B. Reddi, H.D. Kambezidis, and D.G. Kaskaoutis. 2008. Influence of natural and anthropogenic activities on UV Index variations—A study over tropical urban region using ground based observations and satellite data. *Journal of Atmospheric Chemistry* 59 (3): 219–236.

Badarinath, K.V.S., S.K. Kharol, A.R. Sharma, and V.K. Prasad. 2009. Analysis of aerosol and carbon monoxide characteristics over Arabian Sea during crop residue burning period in the Indo-Gangetic Plains using multi-satellite remote sensing datasets. *Journal of Atmospheric and Solar-Terrestrial Physics* 71 (12): 1267–1276.

Balasubramanian, R., W.B. Qian, S. Decesari, M.C. Facchini, and S. Fuzzi. 2003. Comprehensive characterization of PM2.5 aerosol in Singapore. *Journal of Geophysical Research* 108: 4523. https://doi.org/10.1029/2002JD002517.

Chow, J.C., J.G. Watson, L.C. Pritchett, W.R. Pierson, C.A. Frazier, and R.G. Purcell. 1993. The DRI thermal/optical analysis system: Description, evaluation and applications in U.S. air quality studies. *Atmospheric Environment. Part A. General Topics* 27A: 1185–1201.

Chow, J.C., J.G. Watson, S. Crow, D.H. Lowenthal, and T. Merrifield. 2001. Comparison of IMPROVE and NIOSH carbon measurements. *Aerosol Science and Technology* 34: 23–34.

Davidson, C.I., R.F. Phalen, and P.A. Solomon. 2005. Airborne particulate matter and human health: A review. *Aerosol Science and Technology* 39 (8): 737–749.

Gorham, E., and J.A. Janssens. 2005. The distribution and accumulation of chemical elements in five peat cores from the mid-continent to the eastern coast of North America. *Wetlands* 25: 259–278.

Gutman, G., I. Csiszar, and P. Romanov. 2000. Using NOAA/AVHRR products to monitor El Nino impacts: Focus on Indonesia in 1997–98. *Bulletin of the American Meteorological Society* 81: 1189–1205.

Hayasaka, H., I. Noguchi, E.I. Putra, N. Yulianti, and K. Vadrevu. 2014. Peat-fire-related air pollution in Central Kalimantan, Indonesia. *Environmental Pollution* 195: 257–266.

He, J., B. Zielinska, and R. Balasubramanian. 2010. Composition of semi-volatile organic compounds in the urban atmosphere of Singapore: Influence of biomass burning. *Atmospheric Chemistry and Physics* 10: 11401–11413.

Hyer, E.J., and B.N. Chew. 2010. Aerosol transport model evaluation of an extreme smoke episode in Southeast Asia. *Atmospheric Environment* 44: 1422–1427.

Kharol, S.K., K.V.S. Badarinath, A.R. Sharma, D.V. Mahalakshmi, D. Singh, and V.K. Prasad. 2012. Black carbon aerosol variations over Patiala city, Punjab, India—A study during agriculture crop residue burning period using ground measurements and satellite data. *Journal of Atmospheric and Solar-Terrestrial Physics* 84: 45–51.

Koe, L.C.C., A.F. Arellano, and J.L. McGregor. 2001. Investigating the haze transport from the 1997 biomass burning in Southeast Asia: Its impacts on Singapore. *Atmospheric Environment* 35: 2723–2734.

Kunii, O., S. Kanagawa, and I. Yajima. 2002. The 1997 haze disaster in Indonesia: Its air quality and health effects. *Archives of Environmental Health* 57: 16–22.

Langmann, B., and A. Heil. 2004. Release and dispersion of vegetation and peat fire emissions in the atmosphere over Indonesia 1997/1998. *Atmospheric Chemistry and Physics* 4: 2145–2160.

Lasko, K., and K. Vadrevu. 2018. Improved rice residue burning emissions estimates: Accounting for practice-specific emission factors in air pollution assessments of Vietnam. *Environmental Pollution* 236: 795–806.

Lasko, K., K.P. Vadrevu, and T.T.N. Nguyen. 2018. Analysis of air pollution over Hanoi, Vietnam using multi-satellite and MERRA reanalysis datasets. *PLoS ONE* 13 (5): e0196629.

Lee, S., K. Baumann, J.J. Schauer, R.J. Sheesley, L.P. Naeher, S. Meinardi, D.R. Blake, E.S. Edgerton, A.G. Russell, and M. Clements. 2005. Gaseous and particulate emissions from prescribed burning in Georgia. *Environmental Science & Technology* 39 (23): 9049–9056.

Lestari, Puji, and Yandi Dwi Mauliadi. 2009. Source apportionment of particulate matter at urban mixed site in Indonesia using PMF. *Atmospheric Environment* 43: 1760–1770.

Levine, J.S. 1999. The 1997 fires in Kalimantan and Sumatra, Indonesia: Gaseous and particulate emissions. *Geophysical Research Letters* 26 (7): 815–818.

Maenhaut, W., D.J.A. De Ridder, M.T. Fernandez-Jimenez, M.A. Hooper, B. Hooper, and M. Nurhayati. 2002. Long-term observations of regional aerosol composition at two sites in Indonesia. *Nuclear Instruments and Methods in Physics Research Section B: Beam Interactions with Materials and Atoms* 189: 259–265.

Mahmud, M. 2009. Mesoscale model simulation of low level equatorial winds over Borneo during the haze episode of September 1997. *Journal of Earth System Science* 118: 295–307.

Manahan, S.E. 1994. *Environmental chemistry*, 3rd ed. Boca Raton: Lewis Publishers.

Martins, V.J., P. Artaxo, C. Liousse, J.S. Reid, P.V. Hobbs, and Y.J. Kaufman. 1998. Effects of black carbon content, particle size, and mixing on light absorption by aerosols from biomass burning in Brazil. *Journal of Geophysical Research* 103 (D4): 32041–32050. https://doi.org/10.1029/98JD02593.

Page, S.E., F. Siegert, J.O. Rieley, H.D.V. Boehm, A. Jaya, and S. Limin. 2002. The amount of carbon released from peat and forest fires in Indonesia during 1997. *Nature* 420: 61–65.

Page, S.E., J.O. Rieley, and C.J. Banks. 2011. Global and regional importance of the tropical peatland carbon pool. *Global Change Biology* 17: 798–818.

Rajeev, K., K. Parameswaran, S.K. Nair, and S. Meenu. 2008. Observational evidence for the radiative impact of Indonesian smoke in modulating the sea surface temperature of the equatorial Indian Ocean. *Journal of Geophysical Research* 113: D17201. https://doi.org/10.1029/2007JD009611.

Reid, J.S., R. Koppmann, T. Eck, and D. Eleuterio. 2005. A review of biomass burning emissions part II: Intensive physical properties of biomass burning particles. *Atmospheric Chemistry and Physics* 5: 799–825. SRef-ID: 1680-7324/acp/2005-5-799. http://www.atmos-chem-phys.org/acp/5/799/.

Reid, J.S., P. Xian, E.J. Hyer, M.K. Flatau, E.M. Ramirez, F.J. Turk, C.R. Sampson, C. Zhang, E.M. Fukada, and E.M. Maloney. 2011. *Atmospheric Chemistry and Physics Discussions* 11: 1–80. www.atmos-chem-phys-discuss.net/11/1/2011/. https://doi.org/10.5194/acpd-11-1-2011.

Ryu, S.Y., J.E. Kim, H. Zhuanshi, Y.J. Kim, and G.U. Kang. 2004. Chemical composition of post-harvest biomass burning aerosols in Gwangju, Korea. *Journal of the Air & Waste Management Association* 54 (9): 1124–1137.

Sabiham, S. 1988. Studies on peat in the coastal plains of Sumatra and Borneo: I. Physiography and geomorphology of the coastal plains. *Southeast Asian Studies* 26 (3): 308–335.
Schauer, J.J. 2004. Evaluation of elemental carbon as a marker for diesel particulate matter. *Journal of Exposure Analysis and Environmental Epidemiology* 13: 443–453.
See, S.W., R. Balasubramanian, E. Rianawati, S. Karthikeyan, and D.G. dan Streets. 2007. Characterization and source apportionment of PM $\leq 2.5\ \mu$m in Sumatra, Indonesia during a recent peat fire episode. *Environmental Science & Technology* 41 (10): 3488–3494.
Seinfeld, J.H. 1986. *Atmospheric chemistry and physics of air pollution.* Wiley.
Vadrevu, K.P., and C.O. Justice. 2011. Vegetation fires in the Asian region: Satellite observational needs and priorities. *Global Environmental Research* 15 (1): 65–76.
Vadrevu, K.P., T. Ohara, and C. Justice (eds.). 2018. *Land-atmospheric research applications in South and Southeast Asia.*
Vadrevu, K.P., K. Lasko, L. Giglio, W. Schroeder, S. Biswas, and C. Justice. 2019. Trends in vegetation fires in South and Southeast Asian countries. *Scientific Reports* 9 (1): 1–13.
Vadrevu, K.P., T. Ohara, and C. Justice (eds.). 2021a. *Biomass burning in South and Southeast Asia: Impacts on the biosphere*, vol. 2. CRC Press.
Vadrevu, K.P., T. Ohara, and C. Justice (eds.). 2021b. *Biomass burning in South and Southeast Asia: Mapping and monitoring*, vol. 1. CRC Press.
Wulandari, F. 2002. Reckless farming in peatland forests major cause of fires. *The Jakarta Post*, 25 August 2002.

GHG Emissions' Estimation from Peatland Fires in Indonesia—Review and Importance of Combustion Factor

Bambang Hero Saharjo

Abstract The IPCC-based methodology for estimating the GHG emissions from the peatland fires requires several parameters such as the annual area burnt (ha), the mass of fuel available for combustion (tons ha^{-1}), the combustion factor (dimensionless), and emission factor for each gas (g kg^{-1}). The Indonesian National Carbon Accounting System (INCAS) broadly follows the IPCC approach integrating local parameters. Although the approach followed by INCAS is good, the parameters used for computing peatland emissions could be more accurate, especially the combustion factor. For example, INCAS recommends using a combustion factor of one for peatland burning, which suggests complete combustion. However, several studies indicate that complete combustion in peatlands may never reach one because peat is usually covered with water or has high moisture content. Similar is the case with the emission factors used in the INCAS system for computing greenhouse gas emissions using broad emission factors which are not site-specific. The current study points out some of the limitations of the INCAS approach and highlights the need for more robust ground-based measurements on combustion factors, emission factors, and site conditions helpful in estimating emissions from peatlands accurately.

Keywords GHG · Peat fires · Combustion factor · IPCC · INCAS

1 Introduction

Vegetation fires are common in South/Southeast Asia. The impact of fires on landscape structure and composition in varied landscapes of the world are well documented (Goldammer and Seibert 1990). Particularly, the tropical ecosystems are highly fire-prone; thus, with the increasing anthropogenic pressure on these systems, the vegetation is continuously degraded (Goldammer and Seibert 1990; Saharjo 1997, 2007; Saharjo and Munoz 2005; Hayasaka et al. 2014, Biswas et al. 2015a, b; Vadrevu

B. H. Saharjo (✉)
Forest Fire Laboratory, Division of Forest Protection, Department of Silviculture, Faculty of Forestry and Environment, IPB University, Bogor, Indonesia
e-mail: saharjobambangh@gmail.com

K. P. Vadrevu et al. (eds.), *Vegetation Fires and Pollution in Asia*,
https://doi.org/10.1007/978-3-031-29916-2_25

et al. 2019). The drivers of fires can include both climate as well as anthropogenic factors (Albar et al. 2018; Badarinath et al. 2007, 2008, 2009; Badarinath and Prasad 2011; Biswas et al. 2015a, b, 2021; Prasad and Badarinth 2004; Prasad et al. 2001a, b, 2002a, b, 2003, 2004, 2005; Prasad and Badarinath 2006; Vadrevu and Badarinath 2009; Vadrevu and Justice 2011; Vadrevu and Lasko 2015; Saharjo and Novita 2021). In the literature, both the positive and negative effects of fires have been highlighted by different researchers. Specific to the adverse effects, fires result in the loss of forests and their attendant ecosystem services, such as timber, recreation, nutrients, and water retention, including land degradation and bioenergy loss (Justice et al. 2015; Eaturu and Vadrevu 2021; Kant et al. 2000; Vadrevu and Lasko 2015). Repeated burning also modifies the soil's nutrient balance, primarily through pyrodenitrification (Crutzen and Andreae 1990). Specifically, in Asia, fires are used as a management tool for forest clearing, such as through slash and burn, and clearing of agricultural residues after harvest (Lasko et al. 2017, 2018; Vadrevu et al. 2008, 2013, 2014a, b, 2017; Vadrevu 2015). The biomass burning resulting from such activities is an important source of greenhouse gas emissions and aerosols (Goldammer and Seibert 1990; Kharol et al. 2012; Lasko and Vadrevu 2018; Lasko et al. 2017, 2018, 2021; Vadrevu et al. 2018, 2019). Outdoor fires, such as wildfires and slash-and-burn agriculture, can emit substantial amounts of particulate matter (PM) and other pollutants into the atmosphere. These emissions may significantly impact air quality on both local and regional scales (Crutzen and Andreae 1990; Gupta et al. 2001; Vadrevu 2008; Vadrevu et al. 2020, 2021a, b, 2022a, b) (Fig. 1).

In Indonesia, peatland fires are the most common and are the most significant contributor to fire emissions in the region (Saharjo and Nurhayati 2005, 2006; Carlson et al. 2013; Page et al. 2002). As peatland is composed of organic matter, the peat decomposition produces a significant amount of greenhouse gases (GHGs) which

Fig. 1 Smoke blankets the city of Palembang, South Sumatra, Indonesia, during the peak burning season

include carbon dioxide (CO_2) produced aerobically, methane (CH_4) anaerobically, and nitrous oxide (N_2O) both by aerobic and anaerobic process (Hatano 2019). When peat is drained, it dries out and becomes more susceptible to fires. Peatland fires can burn into these underground organic layers and smolder for weeks after the surface fire subsides (Roulston et al. 2018), resulting in substantially greater emissions than surface vegetation fires (Heil et al. 2016). Biomass burning of peatlands can result in extreme pollution with the smoke release which can get transported long distances causing human health issues including economic losses (World Bank 2016) and international tension throughout the region (Wooster et al. 2012). It is estimated that in 2015, 53% of fires in Indonesia occurred on peatlands (Miettinen et al. 2017). Peat fires are estimated to contribute 3.7% of global fire carbon emissions (van der Werf et al. 2017). For the fires in 2015, Wooster et al. (2018) found that 95% of the particulate matter (PM2.5) emissions came from peatland fires, and Wiggins et al. (2018) estimated that 85% of smoke plumes detected in Singapore originated from peat fires. In addition, peatland fires are responsible for forest habitat loss and the degradation of flora and fauna (Posa et al. 2011; Yule 2010). It is estimated that Indonesia lost US $20.1 billion during the 1997/98 fire season alone (Varma 2003). Both national and international policies were designed to reduce fires in Indonesia prior to the 2015 fire season (e.g., ASEAN Agreement on Transboundary Haze Pollution, Singapore's Transboundary Haze Pollution Act, and Indonesia's national law (Act No 41/1999) banning corporations from using fire to clear land for palm oil plantations), but with limited success (Cattaua et al. 2016). Given the variety and severity of the consequences of tropical peatland fires, it is of global interest to understand the changing disturbance regimes, reduce fire emissions, and the resulting impacts such as on emissions (Harrison et al. 2009).

In this study, the focus is on the combustion factor used to calculate biomass burning emissions. The study highlights the need for improved methodologies for estimating GHGs from biomass burning of peatlands, particularly the combustion factor. Without the same, the emissions can be significantly over or underestimated. In this study, how the peatland fire emissions are computed in the Indonesian National Carbon Accounting System (INCAS) is debated and discussed, with a strong recommendation for revising the combustion factor and other parameters for accurately quantifying the emissions (Figs. 2 and 3).

2 Indonesian National Carbon Accounting System (INCAS)

The Indonesian National Carbon Accounting System (INCAS) document developed by the Ministry of Environment and Forestry Research, Development and Innovation Agency, Indonesia, highlights different methods for carbon accounting. In particular, the methodology for estimating greenhouse gas emissions from Peatlands in Indonesia is given in the chapter "Standard Methods for Estimating Greenhouse Gas

Fig. 2 Burned peat in the plantations, Indonesia

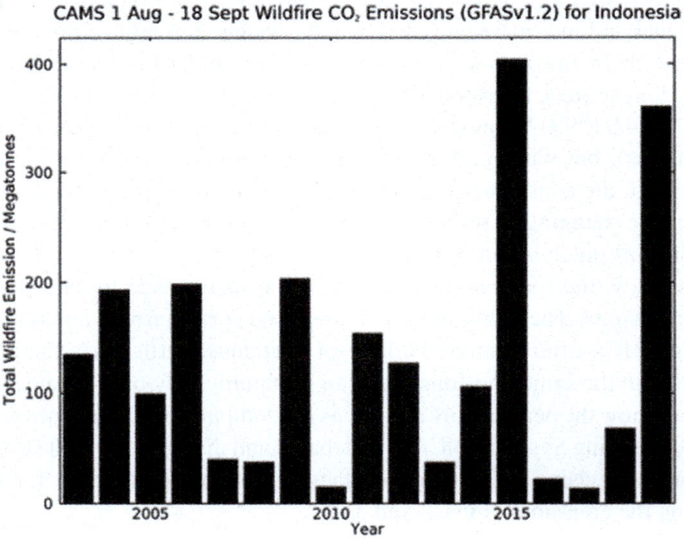

Fig. 3 Total estimated CO_2 equivalent emissions calculated for Indonesia between August 1 and September 18 for all years between 2003 and 2019 from CAMS

Emissions from Forests and Peatlands in Indonesia (version 2)" (Krisnawati et al. 2015). This document highlights data collation, analysis, quality control, quality assurance, modeling, and reporting details. This standard method defines peatland as land with organic soils (Krisnawati et al. 2015). This represents areas with an accumulation of partly decomposed organic matter, with ash content equal to or less than 35%, peat depth equal to or more than 50 cm, and organic carbon content (by weight) of at least 12% (Wahyunto et al. 2004; Agus et al. 2011). Peatland GHG emissions are estimated annually for the following sources and gases (Krisnawati et al. 2015): biological oxidation of drained peat: CO_2–C, CO_2e, peat fire: CO_2–C, CO_2, CO, CH_4, direct emissions from drained organic soils: N_2O, CH_4. In the methodology, the peat biological oxidation emission factors are taken from the IPCC (2013) Wetlands' Supplement which provides different emission factors for CO_2, DOC, and CH_4 (Krisnawati et al. 2015). Emission factors for peat fires were developed by the KFCP project in Central Kalimantan (Krisnawati et al. 2015). INCAS has adopted the data underpinning the fire emission factors for the KFCP project site from Page et al. (2002) but adapted the emission factors to meet international reporting requirements so that GHG emission estimates from organic soil fire were expressed in tons of each GHG emitted. The method used for determining country-specific emission factors for Indonesia follows the approach described in IPCC (2013) (Krisnawati et al. 2015). Hooijer et al. (2014) considered the fire emission factors resulting from the KFCP work to be more representative of normal fire conditions in Indonesia than the emission factors presented in IPCC (2013), which they consider overestimated fire GHG emissions (due to the reliance on small number of studies that were influenced by extreme conditions in 1997/98). The authors acknowledge the alternative emission factors developed from research in Central Kalimantan and the need for consensus on using the emission factors that best represent the emissions profiles in Indonesia. The authors also point out the need to review the emission factors more thoroughly to be integrated into the INCAS (Krisnawati et al. 2015).

To calculate annual CO_2–C and non-CO_2 emissions from organic soil fire, the following equations are used for the INCAS (Krisnawati et al. 2015):

$$L_{\text{fire}} = A \times \text{MB} \times C_{\text{f}} \times G_{\text{ef}} \times 10^{-3},$$

where L_{fire} = amount of CO_2 or non-CO_2 emissions (e.g., CH_4 from fire (tons)); A = total area burned annually (ha); MB = mass of fuel available for combustion (t ha^{-1}); C_{f} = combustion factor (dimensionless); G_{ef} = emission factor for each gas (g kg^{-1} dry matter burned). Further, the mass of fuel available for combustion (MB) is calculated as burned area (m^2) × burn depth (m) × bulk density (t m^{-3}). More details about the methodology can be found in Krisnawati et al. (2015).

3 Discussion

The importance of peat fire contribution to the total carbon emissions in Indonesia is highlighted in the National Greenhouse Gas Emissions Inventory done for 2016. For example, using Tier 1 and Tier 2 of the 2006 IPCC Reporting Guidelines and the IPCC Good Practice Guidelines for land use, land-use change, and forestry (LULUCF) suggests that in 2016, the total GHG emissions for the three leading greenhouse gases (CO_2, CH_4, and N_2O) excluding forestry and other land uses (FOLU) and peat fire amounted to 822,326 Gg CO_2e. Including FOLU and peat fires, the total

GHG emissions from Indonesia are ~ 1,457,774 Gg CO_2e. The main contributing sectors were AFOLU, including peat fires (51.59%), followed by energy (36.91%), waste (7.71%), and Industrial Processes and Product Use (IPPU) (3.79%). The GHG emissions (in CO_2 equivalent) were distributed unevenly between the three gases at 82.46%, 13.29%, and 4.26% for CO_2, CH_4, and N_2O, respectively (Sugardiman 2018). In addition to these estimates, the importance of peatland fire contribution to the total GHG emissions for the 2019 fires is also highlighted in the Copernicus Atmosphere Monitoring Service (CAMS) that tracks the extent and pollution from forest fires across Indonesia (CAMS 2019). The press release made by CAMS suggests that the Indonesian fires, which started in August, pumped out at least 708 megatons of CO_2 until the end of November 2019. The burning of carbon-rich peatlands and drier-than-average conditions mainly caused the fires. CAMS also highlighted that the daily total fire intensity was higher than the average of the last 16 years. In 2019, thousands of acres of ecologically significant land were burned, causing a toxic haze and threatening the health of the local population and the natural forests and wildlife (Fig. 4).

When dealing with fires at peatlands, there will be two types of fires that will occur. The first type is when the fires that occur at the surface are called surface fires on peat. In this type, the fires will consume the fuel at the surface, like fires on mineral soils. This means that the fuel at the surface can have a moisture content of 0% during the dry seasons, such as Imperata grassland. However, the peatlands are mostly covered with water beneath; thus, the fuel at the surface will never reach 0% moisture. The second type is when the fires occur on the peat itself because the groundwater level of peat is far away from the peat surface, and it then causes the peat to become dry and easy to burn, even though it will not dry with 0% moisture. The combustion factor (CF) is the fraction of biomass consumed in a fire as determined by fuel type and moisture

Fig. 4 Burning in peatland area in Ogan Komering Ilir, South Sumatra, Indonesia

conditions (Pribadi and Kurata 2017). It varies significantly among the different biomasses of different fuel types. Various methods exist to estimate the combustion factor. The most common method is to measure the biomass left on site before and after the fire to arrive at the combustion factor. It is recommended that moisture levels can be considered while estimating the emissions, as the combustion factor varies based on the fuel's moisture content, including the peat soil layers at different depths. For example, Pribadi and Kurata (2017) assumed the depth of burned peat soil based on the peat depth. In their method, the first category (peat depth < 50 cm) took the middle value between 0 and 50 cm, i.e., 25 cm, as the burned peat depth, while in the other category (peat depth > 50 cm), they assumed 51 cm as the burned peat depth (Pribadi and Kurata 2017). The peat dry bulk density was assumed to be 100 kg/m^3. These assumptions were based on previous field studies in Indonesia that reported the burned peat depth between 20–150 cm (Boehm et al. 2001) and 25–85 cm (Page et al. 2002). Such measurements are critical to emissions' estimation. Also, previous results of the research done in Central Kalimantan in the year 2015, which was published in 2016, clearly showed that burnt peat was not dry; thus, the CF is not one but less than 1. Thus, it is inferred that the calculation for INCAS that uses CF = 1 needs to be corrected and revised. The INCAS scientists argued that the using CF = 1 was based on IPCC Wetlands Supplement Eq. 2.8 from the IPCC WS 2014 sourced paper from Yokelson et al. (1997). However, my communication with Yokelson suggests otherwise. He inferred that the CF assumption of 1 is incorrect and should be less than one as "peat lands" include some wet peat, some mineral soil mixed in, some vegetation that does not burn, etc.

Another important aspect of the INCAS methodology is regarding the data inputs on the CO_2 emissions and using the IPCC formulas for calculating emissions, especially for the peat fires. The parameters taken are not from the same place in Indonesia but from different places. Other recent studies on emissions from peat fires, too, need site specificity as there are no measurements of the emission factors. For example, Setyawati and Suwarsono (2018) calculated the emissions from the 2015 peat fires, the largest after 1997. They followed the INCAS methodology to calculate the carbon emission from the peat fire in Sumatra, Kalimantan, and Papua. They focused on the CO_2, CO, and CH_4 emissions as they were the largest gaseous carbon compounds emitted by the peat fires and contribute more than 95% of total carbon emitted (Christian et al. 2003; Stockwell et al. 2014, 2016). The other trace gaseous and particulate carbons were neglected (Setyawati and Suwarsono 2018). Further, Setyawati and Suwarsono (2018) used emission factors for Kalimantan peat fires as the average of emission factors from three previous studies (Stockwell et al. 2014, 2016; Setyawati et al. 2017). Because there were no previous studies for Papua and West Papua peat fires, the authors used an emission factor for CO_2 of 1111 g/kg by extrapolating peat carbon mass fraction of 0.3053 for hemic peat (Wahyunto et al. 2006) to the linear regression equation of emission factor for smoldering peat fires (Setyawati et al. 2017). Although this method is acceptable, more ground-based studies on emission factors from the peatlands are needed. One good aspect of this study is the use of CF of less than one and not following INCAS methodology. Specifically, Setyawati and Suwarsono (2018) mention that although INCAS recommends using one as the

Fig. 5 GHG emissions during peat fires

combustion factor, which means that a peat fire is a complete combustion, they used 0.8 for Sumatra and 0.7 for Kalimantan and Papua peat fires (Christian et al. 2003; Stockwell et al. 2016). It was because peat fires were mostly dominated by smoldering combustion (Setyawati and Suwarsono 2018), a type of combustion when no flame is visually observed but apparent thin or thick smoke (Setyawati et al. 2017). However, I infer that using CF 0.8 for Sumatra and 0.7 for Kalimantan and Papua peat fires is also not true from the papers that were referred to (Christian et al. 2003; Stockwell et al. 2016). This is because the papers do not mention CF but use the term modified combustion efficiency (MCE). MCE is defined as the ratio $1CO_2/(1CO_2 + 1CO)$ and is mathematically equivalent to $1/(1 + 1CO/1CO_2)$ (Yokelson et al. 1996). These inferences suggest caution in estimating emissions from peatland fires. Overall, it is inferred that the emissions' estimation from the peatlands is complex. More ground-based measurements are needed to quantify the combustion factors and emission factors, including the drivers of biogeochemical cycling, accurately for emissions estimation in various peatlands of Indonesia (Fig. 5).

4 Conclusions

Calculating the GHG emissions using the IPCC methodology is a straightforward approach. The results from the IPCC approach can be used for a broader emissions' contribution from different sources. However, emissions' estimation from specific sources such as peatlands needs more attention. The biogeochemical processes and the resulting emissions from fires can be nonlinear as the peatlands are covered with water beneath and will never be dry like grasslands. Thus, the combustion factor used for calculating emissions from the peatlands cannot be 1, i.e., wholly combusted or burnt. In addition, the emission factors too can be different at different sites governed by varied site conditions such as soil type, moisture, peatland depth; thus, broad emission factors may not reflect the actual emissions. Therefore, it is inferred that more ground-based measurements on the combustion factors, emission factors, and site characteristics are needed for quantifying emissions accurately in Indonesia.

Acknowledgements Thanks to the Ministry of Environment and Forest Indonesia, IPB/NASA Project and Yayasan Konservasi Alam Nusantara (The Nature Conservancy) for the chance to discuss and use the data for paper.

References

Agus, F., K. Hairiah, and A. Mulyani. 2011. *Measuring carbon stock in peat soil: Practical guidelines*. Bogor: World Agroforestry Centre (ICRAF) and Indonesian Soil Research Institute. Available from http://balittanah.litbang.deptan.go.id.

Albar, I., I. Jaya, B.H. Saharjo, B. Kuncahyo, and K.P. Vadrevu. 2018. Spatio-temporal analysis of land and forest fires in Indonesia using MODIS active fire dataset. In *Land-atmospheric research applications in South and Southeast Asia*, 105–127. Cham: Springer.

Badarinath, K.V.S., and K.V. Prasad. 2011. Carbon dioxide emissions from forest biomass burning in India. *Global Environmental Research* 15: 45–52.

Badarinath, K.V.S., S.K. Kharol, K.M. Latha, T.K. Chand, V.K. Prasad, A.N. Jyothsna, and K. Samatha. 2007. Multiyear ground-based and satellite observations of aerosol properties over a tropical urban area in India. *Atmospheric Science Letters* 8 (1): 7–13.

Badarinath, K.V.S., S. Kumar Kharol, V. Krishna Prasad, A. Rani Sharma, E.U.B. Reddi, H.D. Kambezidis, and D.G. Kaskaoutis. 2008. Influence of natural and anthropogenic activities on UV Index variations—A study over tropical urban region using ground based observations and satellite data. *Journal of Atmospheric Chemistry* 59 (3): 219–236.

Badarinath, K.V.S., S.K. Kharol, A.R. Sharma, and V.K. Prasad. 2009. Analysis of aerosol and carbon monoxide characteristics over Arabian Sea during crop residue burning period in the Indo-Gangetic Plains using multi-satellite remote sensing datasets. *Journal of Atmospheric and Solar-Terrestrial Physics* 71 (12): 1267–1276.

Biswas, S., K.D. Lasko, and K.P. Vadrevu. 2015a. Fire disturbance in tropical forests of Myanmar—Analysis using MODIS satellite datasets. *IEEE Journal of Selected Topics in Applied Earth Observations and Remote Sensing* 8 (5): 2273–2281.

Biswas, S., K.P. Vadrevu, Z.M. Lwin, K. Lasko, and C.O. Justice. 2015b. Factors controlling vegetation fires in protected and non-protected areas of Myanmar. *PLoS ONE* 10 (4): e0124346.

Biswas, S., K.P. Vadrevu, M.S. Mon, and C. Justice. 2021. Contemporary forest loss in Myanmar: Effect of democratic transition and subsequent timber bans on landscape structure and composition. *Ambio* 50 (4): 914–928.

Boehm, H.-D.V., F. Siegert, J.O. Rieley, S.E. Page, J. Jauhiainen, H. Vasander, and A. Jaya. 2001. Fire impacts and carbon release on tropical peatlands in Central Kalimantan, Indonesia. In *22nd Asian conference on remote sensing*, 5–9.

CAMS. 2019. The Copernicus Atmosphere Monitoring Service (CAMS) tracks extent and pollution from forest fires across Indonesia, 20 September 2019. https://atmosphere.copernicus.eu/copernicus-atmosphere-monitoring-service-tracks-extent-and-pollution-fires-across-indonesia

Carlson, K.M., L.M. Curran, G.P. Asner, A.M.D. Pittman, S.N. Trigg, and J. Marion Adeney. 2013. Carbon emissions from forest conversion by Kalimantan oil palm plantations. *Nature Climate Change* 3 (3): 283–287. https://doi.org/10.1038/nclimate1702.

Cattaua, M.E., E.H. Mark, S. Iwan, T. Sady, U. María, and D. Ruth. 2016. Sources of anthropogenic fire ignitions on the peat-swamp landscape in Kalimantan, Indonesia. *Global Environmental Change*. https://doi.org/10.1016/j.gloenvcha.2016.05.005.

Christian, T.J., B. Kleiss, R.J. Yokelson, R. Holzinger, P.J. Crutzen, W.M. Hao, B.H. Saharjo, and D.E. Ward. 2003. Comprehensive laboratory measurements of biomass-burning emissions: 1. Emissions from Indonesian, African, and other fuels. *Journal of Geophysical Research* 108 (D23): 4719. https://doi.org/10.1029/2003JD003704.

Crutzen, P.J., and M.O. Andreae. 1990. Biomass burning in the tropics: Impact on atmospheric chemistry and biogeochemical cycles. *Science* 250 (4988): 1669–1678.

Eaturu, A., and K.P. Vadrevu. 2021. Evaluation of Sentinel-3 SLSTR data for mapping fires in forests, peatlands, and croplands—A case study over Australia, Indonesia, and India. In *Biomass burning in South and Southeast Asia*, 39–59. CRC Press.

Goldammer, J.G., and B. Seibert. 1990. The impact of droughts and forest fires on tropical lowland rain forest of East Kalimantan. In *Fire in the tropical biota*, 11–31. Berlin, Heidelberg: Springer.

Gupta, P.K., V.K. Prasad, C. Sharma, A.K. Sarkar, Y. Kant, K.V.S. Badarinath, and A.P. Mitra. 2001. CH_4 emissions from biomass burning of shifting cultivation areas of tropical deciduous forests—Experimental results from ground-based measurements. *Chemosphere-Global Change Science* 3 (2): 133–143.

Harrison, M.E., S.E. Page, and S.H. Limin. 2009. The global impact of Indonesian forest fires. *Biologist* 56: 156–183.

Hatano, R. 2019. Impact of land use change on greenhouse gases emissions in peatland: A review. *International Agrophysics* 33 (2): 167–173.

Hayasaka, H., I. Noguchi, E.I. Putra, N. Yulianti, and K. Vadrevu. 2014. Peat-fire-related air pollution in Central Kalimantan, Indonesia. *Environmental Pollution*. https://doi.org/10.1016/j.envpol.2014.06.031.

Heil, A., B. Langmann, and E. Aldrian. 2016. Indonesian peat and vegetation fire emissions: Study on factors influencing large-scale smoke haze pollution using a regional atmospheric chemistry model. *Mitigation and Adaptation Strategies for Global Change* 12 (1): 113–133.

Hooijer, A., S. Page, P. Navratil, R. Vernimmen, M. Van der Vat, K. Tansey, K. Konecny, F. Siegert, U. Ballhorn, and N. Mawdsley. 2014. *Carbon emissions from drained and degraded peatland in Indonesia and emission factors for measurement, reporting and verification (MRV) of peatland greenhouse gas emissions—A summary of KFCP research results for practitioners*. Jakarta, Indonesia: IAFCP.

IPCC. 2013. Supplement to the 2006 IPPC guidelines for national greenhouse gas inventories: Wetlands. In *Methodological guidance on lands with wet and drained soils and constructed wetlands for wastewater treatment*, ed. S. Gyldenkaerne and E. Lin. Japan: Institute for Global Environmental Strategies [IGES].

Justice, C., G. Gutman, and K.P. Vadrevu. 2015. NASA land cover and land use change (LCLUC): An interdisciplinary research program. *Journal of Environmental Management* 148: 4–9.

Kant, Y., V.K. Prasad, and K.V.S. Badarinath. 2000. Algorithm for detection of active fire zones using NOAA AVHRR data. *Infrared Physics & Technology* 41 (1): 29–34.

Kharol, S.K., K.V.S. Badarinath, A.R. Sharma, D.V. Mahalakshmi, D. Singh, and V.K. Prasad. 2012. Black carbon aerosol variations over Patiala city, Punjab, India—A study during agriculture crop residue burning period using ground measurements and satellite data. *Journal of Atmospheric and Solar-Terrestrial Physics* 84: 45–51.

Krisnawati, H., R. Imanuddin, and W.C. Adinugroho. 2015. *Standard method to estimate greenhouse gases emission from forest and peatland in Indonesia, version 2*. Bogor: The Agency for Research, Development and Innovation, The Ministry of Environment and Forestry.

Lasko, K., and K. Vadrevu. 2018. Improved rice residue burning emissions estimates: Accounting for practice-specific emission factors in air pollution assessments of Vietnam. *Environmental Pollution* 236: 795–806.

Lasko, K., K.P. Vadrevu, V.T. Tran, E. Ellicott, T.T. Nguyen, H.Q. Bui, and C. Justice. 2017. Satellites may underestimate rice residue and associated burning emissions in Vietnam. *Environmental Research Letters* 12 (8): 085006.

Lasko, K., K.P. Vadrevu, and T.T.N. Nguyen. 2018. Analysis of air pollution over Hanoi, Vietnam using multi-satellite and MERRA reanalysis datasets. *PLoS ONE* 13 (5): e0196629.

Lasko, K., K.P. Vadrevu, V. Bandaru, T.N.T. Nguyen, and H.Q. Bui. 2021. PM2.5 emissions from biomass burning in South/Southeast Asia—Uncertainties and trade-offs. In *Biomass burning in South and Southeast Asia*, 149–169. CRC Press.

Miettinen, J., A. Hooijer, R. Vernimmen, S.C. Liew, and S.E. Page. 2017. From carbon sink to carbon source: Extensive peat oxidation in insular Southeast Asia since 1990. *Environmental Research Letters* 12: 024014. https://doi.org/10.1088/1748-9326/aa5b6f.

Page, S.E., F. Siegert, J.O. Rieley, H.D. Boehm, A. Jaya, and S. Limin. 2002. The amount of carbon released from peat and forest fires in Indonesia during 1997. *Nature* 420: 61–65.

Posa, M.R.C., L.S. Wijedasa, and R.T. Corlett. 2011. Biodiversity and conservation of tropical peat swamp forests. *BioScience* 61: 49–57.

Prasad, K.V., and K.V.S. Badarinath. 2006. Soil surface nitrogen losses from agriculture in India: A regional inventory within agroecological zones (2000–2001). *The International Journal of Sustainable Development and World Ecology* 13 (3): 173–182.

Prasad, V.K., and K.V.S. Badarinth. 2004. Land use changes and trends in human appropriation of above ground net primary production (HANPP) in India (1961–98). *Geographical Journal* 170 (1): 51–63.

Prasad, V.K., Y. Kant, P.K. Gupta, C. Sharma, A.A. Mitra, and K.V.S. Badarinath. 2001a. Biomass and combustion characteristics of secondary mixed deciduous forests in Eastern Ghats of India. *Atmospheric Environment* 35 (18): 3085–3095.

Prasad, V.K., Y. Kant, and K.V.S. Badarinath. 2001b. CENTURY ecosystem model application for quantifying vegetation dynamics in shifting cultivation areas: A case study from Rampa Forests, Eastern Ghats (India). *Ecological Research* 16 (3): 497–507.

Prasad, V.K., Y. Kant, P.K. Gupta, C. Elvidge, and K.V.S. Badarinath. 2002a. Biomass burning and related trace gas emissions from tropical dry deciduous forests of India: A study using DMSP-OLS data and ground-based measurements. *International Journal of Remote Sensing* 23 (14): 2837–2851.

Prasad, V.K., Y. Kant, and K.V.S. Badarinath. 2002b. Land use changes and modeling carbon fluxes from satellite data. *Advances in Space Research* 30 (11): 2511–2516.

Prasad, V.K., M. Lata, and K.V.S. Badarinath. 2003. Trace gas emissions from biomass burning from northeast region in India—Estimates from satellite remote sensing data and GIS. *The Environmentalist* 23 (3): 229–236.

Prasad, V.K., K.V.S. Badarinath, S. Yonemura, and H. Tsuruta. 2004. Regional inventory of soil surface nitrogen balances in Indian agriculture (2000–2001). *Journal of Environmental Management* 73 (3): 209–218.

Prasad, V.K., E. Anuradha, and K.V.S. Badarinath. 2005. Climatic controls of vegetation vigor in four contrasting forest types of India—Evaluation from National Oceanic and Atmospheric Administration's Advanced Very High Resolution Radiometer datasets (1990–2000). *International Journal of Biometeorology* 50 (1): 6–16.

Pribadi, A., and G. Kurata. 2017. Greenhouse gas and air pollutant emissions from land and forest fire in Indonesia during 2015 based on satellite data. *IOP Conference Series: Earth and Environmental Science* 54: 012060.

Roulston, C., C. Paton-Walsh, T.E.L. Smith, A. Guérette, S. Evers, C.M. Yule, G. Rein, and G.R. Van der Werf. 2018. Fine particle emissions from tropical peat fires decrease rapidly with time since ignition. *Journal of Geophysical Research: Atmospheres* 123: 5607–5617. https://doi.org/10.1029/2017JD027827.

Saharjo, B.H. 1997. Fire protection and industrial plantation management in the tropics. *The Commonwealth Forestry Review* 203–206.

Saharjo, B.H. 2007. Shifting cultivation in peatlands. *Mitigation and Adaptation Strategies for Global Change* 12 (1): 135–146.

Saharjo, B.H., and C.P. Munoz. 2005. Controlled burning in peat lands owned by small farmers: A case study in land preparation. *Wetlands Ecology and Management* 13 (1): 105–110.

Saharjo, B.H., and N. Novita. 2021. The high potential of peatland fires management for greenhouse gas emissions reduction in Indonesia. *Journal of Tropical Silviculture* 2086: 8277.

Saharjo, B.H., and A.D. Nurhayati. 2005. Changes in chemical and physical properties of hemic peat under fire-based shifting cultivation. *Tropics* 14 (3): 263–269.

Saharjo, B.H., and A.D. Nurhayati. 2006. Domination and composition structure change at hemic peat natural regeneration following burning; A case study in Pelalawan, Riau Province. *Biodiversitas Journal of Biological Diversity* 7 (2).

Setyawati, W., and Suwarsono. 2018. Carbon emission from peat fire in 2015. *IOP Conference Series: Earth and Environmental Science* 166 (2018): 012041.

Setyawati, W., E. Damanhuri, P. Lestari, and K. Dewi. 2017. Emission factor from small scale tropical peat combustion. *IOP Conference Series: Materials Science and Engineering* 180: 012113. https://doi.org/10.1088/1757-899X/180/1/012113.

Stockwell, C.E., R.J. Yokelson, S.M. Kreindeweis, A.L. Robinson, P.J. DeMott, R.C. Sullivan, J. Reardon, K.C. Ryan, D.W.T. Griffith, and L. Steven. 2014. Trace gas emission from combustion of peat, crop residue, domestic biofuel, grasses and other fuels: Configuration and Fourier Transform Infrared (FTIR) component of the fourth fire lab at Missoula Experiment (FLAME-4). *Atmospheric Chemistry and Physics* 14: 9727–9754.

Stockwell, C.E., T. Jayarathne, M.A. Cochrane, K.C. Ryan, E.I. Putra, B.H. Saharjo, A.D. Nurhayati, I. Albar, D.R. Black, I.J. Simpson, E.A. Stone, and R.J. Yokelson. 2016. Field measurements of trace gases and aerosol emitted by peat fires in Central Kalimantan, Indonesia during the 2015 El Nino. *Atmospheric Chemistry and Physics Discussion*. https://doi.org/10.5194/acp-2016-411.

Sugardiman, R.A. 2018. *Indonesia second Biennial Update Report (second BUR)*. United Nations Framework Convention on Climate Change (UNFCCC). Jakarta: Directorate General of Climate Change, Ministry of Environment and Forestry MoEF.

Vadrevu, K.P. 2008. Analysis of fire events and controlling factors in eastern India using spatial scan and multivariate statistics. *Geografiska Annaler: Series A, Physical Geography* 90 (4): 315–328.

Vadrevu, K. 2015. International regional science meeting of NASA-LCLUC: Land cover/land use change (LC/LUC) and environmental impacts in South Asia, Coimbatore, India, 19–23 January 2013. *Journal of Environmental Management* 148: 1–163.

Vadrevu, K.P., and K.V.S. Badarinath. 2009. Spatial pattern analysis of fire events in Central India—A case study. *Geocarto International* 24 (2): 115–131.

Vadrevu, K.P., and C.O. Justice. 2011. Vegetation fires in the Asian region: Satellite observational needs and priorities. *Global Environmental Research* 15 (1): 65–76.

Vadrevu, K., and K. Lasko. 2015. Fire regimes and potential bioenergy loss from agricultural lands in the Indo-Gangetic Plains. *Journal of Environmental Management* 148: 10–20.

Vadrevu, K.P., K.V.S. Badarinath, and E. Anuradha. 2008. Spatial patterns in vegetation fires in the Indian region. *Environmental Monitoring and Assessment* 147 (1): 1–13.

Vadrevu, K.P., L. Giglio, and C. Justice. 2013. Satellite based analysis of fire–carbon monoxide relationships from forest and agricultural residue burning (2003–2011). *Atmospheric Environment* 64: 179–191.

Vadrevu, K.P., T. Ohara, and C. Justice. 2014a. Air pollution in Asia. *Environmental Pollution (Barking, Essex: 1987)* 195: 233–235.

Vadrevu, K.P., K. Lasko, L. Giglio, and C. Justice. 2014b. Analysis of Southeast Asian pollution episode during June 2013 using satellite remote sensing datasets. *Environmental Pollution* 195: 245–256.

Vadrevu, K., T. Ohara, and C. Justice. 2017. Land cover, land use changes and air pollution in Asia: A synthesis. *Environmental Research Letters* 12 (12): 120201.

Vadrevu, K.P., T. Ohara, and C. Justice (eds.). 2018. *Land-atmospheric research applications in South and Southeast Asia.*

Vadrevu, K., A. Heinimann, G. Gutman, and C. Justice. 2019. Remote sensing of land use/cover changes in South and Southeast Asian countries. *International Journal of Digital Earth* 12 (10): 1099–1102.

Vadrevu, K.P., A. Eaturu, S. Biswas, K. Lasko, S. Sahu, J.K. Garg, and C. Justice. 2020. Spatial and temporal variations of air pollution over 41 cities of India during the COVID-19 lockdown period. *Scientific Reports* 10 (1): 1–15.

Vadrevu, K.P., T. Ohara, and C. Justice (eds.). 2021a. *Biomass burning in South and Southeast Asia: Impacts on the biosphere*, vol. 2. CRC Press.

Vadrevu, K.P., T. Ohara, and C. Justice (eds.). 2021b. *Biomass burning in South and Southeast Asia: Mapping and monitoring*, vol. 1. CRC Press.

Vadrevu, K., T. Le Toan, S.S. Ray, and C. Justice. 2022a. *Remote sensing of agriculture and land cover/land use changes in South and Southeast Asian countries.* Cham: Springer.

Vadrevu, K., A. Eaturu, E. Casadaban, K. Lasko, W. Schroeder, S. Biswas, L. Giglio, and C. Justice. 2022b. Spatial variations in vegetation fires and emissions in South and Southeast Asia during COVID-19 and pre-pandemic. *Scientific Reports* 12 (1): 1–21.

Van Der Werf, G.R., J.T. Randerson, L. Giglio, T.T. Van Leeuwen, Y. Chen, B.M. Rogers, M. Mu, M.J. Van Marle, D.C. Morton, G.J. Collatz, and R.J. Yokelson. 2017. Global fire emissions estimates during 1997–2016. *Earth System Science Data* 9 (2): 697–720.

Varma, A. 2003. The economics of slash and burn: A case study of the 1997/1998 Indonesian forest fires. *Ecological Economics* 46: 159–171.

Wahyunto, S. Ritung, and H. Subagjo. 2004. *Peta Sebaran Lahan Gambut, Luas dan Kandungan Karbon di Kalimantan* [Map of peatland distribution area and carbon content in Kalimantan]. 2000–2002 Wetlands International—Indonesia Programme & Wildlife Habitat Canada (WHC).

Wahyunto, H.B., H. Bekti, and F. Widiastuti. 2006. *Peta-Peta Sebaran Lahan Gambut, Luas dan Kandungan Karbon di Papua* [Maps of peatland distribution, area and carbon content in Papua]. 2000–2001 Wetlands International—Indonesia Programme & Wildlife Habitat Canada (WHC).

Wiggins, E.B., Czimczik, C.I., Santos, G.M., Chen, Y., Xu, X., Holden, S.R., Randerson, J.T., Harvey, C.F., Kai, F.M. and Yu, L.E., 2018. Smoke radiocarbon measurements from Indonesian fires provide evidence for burning of millennia-aged peat. *Proceedings of the National Academy of Sciences*, 115(49): 12419–12424.

Wooster, M.J., G.L.W. Perry, and A. Zoumas. 2012. Fire, drought and El Niño relationships on Borneo (Southeast Asia) in the pre-MODIS era (1980–2000). *Biogeosciences* 9: 317–340. https://doi.org/10.5194/bg-9-317-2012.

Wooster, M., D. Gaveau, M. Salim, T. Zhang, W. Xu, D. Green, V. Huijnen, D. Murdiyarso, D. Gunawan, N. Borchard, M. Schirrmann, B. Main, and A. Sepriando. 2018. New tropical peatland gas and particulate emissions factors indicate 2015 Indonesian fires released far more particulate matter (but less methane) than current inventories imply. *Remote Sensing* 10: 495. https://doi.org/10.3390/rs10040495.

World Bank. 2016. *The cost of fire: An economic analysis of Indonesia's 2015 fire crisis* (English). Indonesia Sustainable Landscapes Knowledge. Note No. 1. Washington, DC: World Bank Group.

Yokelson, R.J., D.W.T. Griffith, and D.E. Ward. 1996. Open path Fourier transform infrared studies of large-scale laboratory biomass fires. *Journal of Geophysical Research* 101: 21067–21080.

Yokelson, R.J., D.E. Ward, R.A. Susott, J. Reardon, and D.W.T. Griffith. 1997. Emissions from smoldering combustion of biomass measured by open-path Fourier transform infrared spectroscopy. *Journal of Geophysical Research* 102: 18865–18877.

Yule, C. 2010. Loss of biodiversity and ecosystem functioning in Indo-Malayan peat swamp forests. *Biodiversity and Conservation* 19: 393–409.

Forest Fire Emissions in Equatorial Asia and Their Recent Delay Anomaly in the Dry Season

Ida Bagus Mandhara Brasika

Abstract In Equatorial Asia, forest fires are a significant contributor to greenhouse gas emissions. This study explains the conditions of forest fire emissions in Equatorial Asia and how they might connect to precipitation in the region. Integrated satellite and model data were used to relate to forest fire emissions. Several methods were applied to understand the spatial and temporal nature of emissions, namely time-series, time-averaged, correlation, and Hovmoller diagram. The results suggested that from 1980 to 2021, the emissions were generated from three regions: east of Sumatra Island, south and east of Borneo Island, and the capital city of Jakarta on Java Island. During this period, several massive emission events occurred in 1982, 1998, 2002, and 2006. South Borneo was the highest emitter for 41 years. Emissions in Sumatra and Borneo were mainly due to forest fires, while those in Jakarta were due to industries and transportation. The two main emitter islands, Borneo and Sumatra, have shown different strengths of correlation with precipitation. It is considered weak in East Sumatra, around − 0.2, while South Borneo is stronger at about − 0.5. However, from 2009 onward, Borneo had abnormal conditions, where carbon emissions remained low even in high forest fires. This is contrary to the previous decades. Apart from that, carbon emissions were also delayed after the rainy season. Borneo's forest fire emissions were due to peat organic matter burning. From 2009 to 2021, there is a possibility that the peat had regained its ability to hold water, so carbon emissions became lower and delayed. This clearly shows that forest fires have been the main contributor to carbon emissions in Equatorial Asia for several decades, and precipitation plays a crucial role in it. The emissions can be reduced when the peat soil regains its capacity as water storage in long-term precipitation periods.

Keywords Carbon emissions · Peat combustion · Borneo · Equatorial Asia · Long-term precipitation

---

I. B. M. Brasika (✉)
Department of Marine Science, Udayana University, Bali, Indonesia
e-mail: mandharabrasika@unud.ac.id

Department of Mathematics and Statistics, The University of Exeter, Exeter, UK

K. P. Vadrevu et al. (eds.), *Vegetation Fires and Pollution in Asia*,
https://doi.org/10.1007/978-3-031-29916-2_26

1 Introduction

Anthropogenic emissions have been a significant contributor to climate change, especially carbon emissions, which can increase global temperatures and stay longer in the atmosphere (Callendar 1938; Cao and Caldeira 2010). The concentration of carbon dioxide (CO_2) in the atmosphere has been increasing rapidly since the industrial era. For example, it was only 277 parts per million (ppm) in 1750 and then became 412.5 ppm in 2020 (Joos and Spahni 2008; Lindsey 2020). Carbon is emitted into the atmosphere during a combustion process. This global increase in carbon concentration is due primarily to forest fires, which can be traced back to the pre-industrial era, and fossil fuels, which have become dominant since the industrial era (Ciais et al. 2013; Friedlingstein et al. 2020). Most developed countries such as USA and China have been major contributors to global emissions from their extensive industries (Ritchie and Roser 2020), while other massive forested regions such as Central Africa, South America, and Equatorial Asia contribute to global emissions from forest fires (van Wees et al. 2021). For example, in 2015, very high El Nino tropical regions in the three continents had different dominant processes of carbon flux anomaly; they are gross primary production in South America, fires in Asia, and respiration in Africa (Liu et al. 2017).

Fires in tropical regions are not caused by natural factors only. As tropical regions have considerably higher water content in their atmosphere, they can maintain low temperatures and high humidity on the ground, even in the dry season. This condition is not suitable for fire ignition (Bonan 2008; Baker and Spracklen 2019; Uhl et al. 1988). Although many fires appear during the dry season, they burn degraded and deforestation areas, not the primary intact forest area (Brasika et al. 2021). Therefore, the primary forest maintains non-suitable conditions for a fire regime. Moreover, tropical rainforests have high precipitation (Smith 2019). However, the fact that all the central regions of tropical forest fires (Africa, South America, and Equatorial Asia) experience forest fire regimes means that climate condition is not the only main driver of these fires. Other factors, such as human interference, need to be considered.

For example, fire is used to clear the forests for agriculture through slash and burn (Albar et al. 2018; Badarinath et al. 2007, 2008, 2009; Badarinath and Prasad 2011; Biswas et al. 2015a, b, 2021; Eaturu and Vadrevu 2021; Justice et al. 2015; Prasad et al. 2001a, b, 2002a, b, 2003, 2004; Prasad and Badarinth 2004) agricultural residues after crop harvest (Lasko and Vadrevu 2018; Lasko et al. 2017, 2018, 2021), to clear the land for the next crop, to clear the forested lands for plantations (Albar et al. 2018), promoting the growth of grass in pasture lands for cattle, including reducing of weeds prior to planting of crops (Simorangkir 2007). Regardless of the ignition source, wildfires in forested areas can spread rapidly and become uncontrollable due to the local meteorological and environmental conditions. As a result, fires can threaten human lives and cause severe economic damage. In addition, several studies report vegetation fires as a critical source of greenhouse gas emissions and aerosols. In addition, vegetation fires can influence a variety of land–atmospheric interactions at different scales, such as transpiration, soil erosion, albedo, including the biogeochemical cycles (Crutzen and Andreae 1990; Kant et al. 2000; Kharol et al. 2012; Prasad et al. 2005; Prasad and Badarinath 2006; Vadrevu 2008, 2015; Vadrevu and Badarinath 2009; Vadrevu and Justice 2011; Vadrevu and Lasko 2015; Vadrevu et al. 2008, 2013, 2014a, b, 2017). In addition to these effects on Earth's radiation, atmosphere, climate, and ecosystems, the pollutants released from the biomass burning can have adverse health effects such as asthma, acute respiratory illness, eye irritation, cardiovascular mortality, thrombosis, and in severe cases, mortality (Phung et al. 2022; Vadrevu et al. 2018, 2019, 2020, 2022a, b; Vadrevu 2021; Wooster et al. 2021).

Among tropical forests, the Equatorial Asian forests have shown distinct characteristics. Although Equatorial Asia only accounts for 17% of the global tropical land area, its forest fires contributed to 41% of the global CO_2 emissions during 1990–1999 and 2000–2005. Tropical Asia is probably a strong net carbon source (Malhi 2010). Another distinct feature of the Equatorial Asian forests is their location on islands, which means that they are surrounded by or close to the ocean, which might affect the weather and climate, particularly the rainfall, of the areas where the forests are situated. Thus, tropical precipitation may critically determine the strength of the climate-carbon cycle during the twenty-first century (van der Werf et al. 2008).

Moreover, tropical forests in Equatorial Asia are also distributed, with a high concentration of peat soil with an estimated carbon content of around 70–80 Pg C, mostly on Borneo and Sumatra (Malhi 2010). Peat soil has a distinct attribute: it works like a sponge (Jaenicke et al. 2010; Turner 1757). It is highly porous, with a saturated water content of 86–94% of its volume (Hobbs 1986; Plyusnin 1964). However, it has a greater capacity for water retention than a sponge (Bacon et al. 2017), so it responds well to long-term precipitation.

Unfortunately, it also becomes a source of organic content that fuels fires when it is drained. Burned peat is a significant contributor to forest fire emissions. Peatlands in many Equatorial Asian forests are vulnerable to oxidation and fire due to human drainage systems (Page et al. 2002; Hooijer et al. 2006; Hayasaka et al. 2014), then affecting regional air quality and global concentrations of greenhouse gasses (van der Werf et al. 2008). Peatland fires emitted 0.12 ± 0.06 Pg C year^{-1}, while peatland drainage and oxidation emitted 0.17 ± 0.07 Pg C year^{-1} over the period 1997–2006 (Hooijer et al. 2010; van der Werf et al. 2008). These figures have the potential to increase as climate projections suggest future drying and warming (Li et al. 2007).

Forest fire and weather/climate are in a looping system where one affects the other. While long dry weather conditions might trigger fire ignition, emissions from forest fires contribute to massive aerosols and carbon content in the atmosphere that affect the weather/climate (Pio et al. 2008; Nakajima et al. 1999; Vadrevu et al. 2021a, b). In Equatorial Asia, the dry season (El Nino) tends to increase temperatures and atmospheric water vapor deficits and causes substantial declines in precipitation (Malhi and Wright 2004). In this study, the main features of forest fire emissions in Equatorial Asia and how long-term precipitation affects them through peat soil were evaluated.

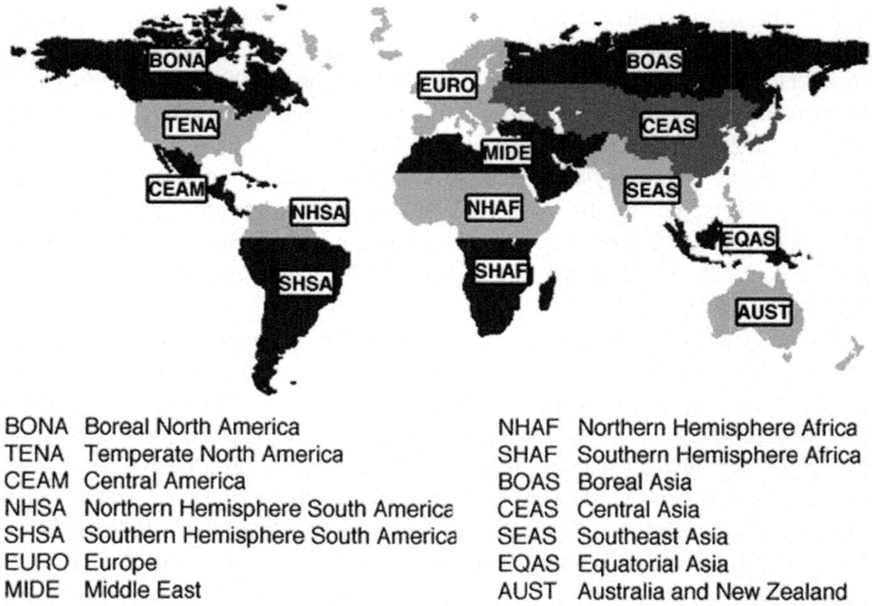

Fig. 1 Illustration of the Equatorial Asia (EQAS) region. *Source* van der Werf et al. (2006)

2 Study Area

The study area is located in Equatorial Asia, around 97–155E; 8N–15S. This area covers many countries, including Indonesia, Papua New Guinea, Malaysia, Singapore, Brunei Darussalam, and Timor-Leste. Many of them share territories on the same island. Equatorial Asia has rich tropical forests, mostly located in its three biggest islands, Borneo, Sumatra, and Papua. As island nations, their weather conditions are highly varied to the local conditions. Thus, each island might have different characteristics and responses to changes, including changes in forest fire emissions (Fig. 1).

3 Data and Methods

In this research, many different types of data were used, from the reanalysis modeled data to satellite data, as each type has advantages and disadvantages in understanding forest fire emissions.

3.1 Spatial and Temporal Patterns of Equatorial Asia's Emissions

This variable is a combination of forest fire emissions, industrial emissions, and others. By this, the dominant emissions in the area (from forest fires or industries) can be seen. The data were obtained from the MERRA-2 model (M2TMNXCHM v5.12), a reanalysis model covering 1980–2021 (41 years). The dominant spatial pattern in Equatorial Asia can be seen from the time-averaged map of CO emissions (ensemble). Meanwhile, the temporal pattern is determined by the time series of CO emissions from 1980 to 2021.

3.2 Connection Between CO Emissions and Precipitation

This study looks at the connection between emissions and precipitation. The statistics can be seen from the correlation between CO emissions and precipitation. The data on CO emissions were obtained from the MERRA-2 model. The precipitation dataset was derived from the Tropical Rainfall Measuring Mission (TRMM) data from 2000 to 2019 (GES-DISC 2011).

The variable is the monthly precipitation rates from TRMM_3B43 version 7 with a 0.25° spatial distribution. This product is claimed as the "best" precipitation estimation in a latitude band covering 50° N to 50° S as it was created using TRMM-adjusted merge microwave-infrared precipitation rate and root-mean-square precipitation-error estimates (Huffman et al. 2010).

3.3 Emission Anomaly and the Impact of Precipitation

The amount of emissions from the forest area should be connected to the number of fire counts. To understand this, the total yearly emission and fire counts were plotted together to see the potential anomaly. The CO emission data were obtained from the MERRA-2 model, while the active fire data were obtained from the Moderate Resolution Imaging Spectroradiometer (MODIS) satellite. MODIS Collection 6 (standard processing) MCD14ML with a spatial resolution of 1 km was utilized. Although the MODIS dataset has several limitations, for example, it cannot illustrate small-scale/local fires, this dataset is sufficient to show the regional-scale active fire distribution. Both datasets were collected from 2001 to 2020. To avoid potential false alarms, selective measurement was performed in filtering the datasets. This research only used high-confidence fires, with more than 80% confidence level. Then, the dynamic connection between CO emissions and precipitation was analyzed.

Hovmoller diagram was used to illustrate the spatiotemporal patterns of CO emissions and precipitation, with the CO emission data from MERRA-2 and precipitation data from TRMM.

4 Results and Discussion

4.1 Main Regions and Dominant Years of Emissions in Equatorial Asia

Time-averaged CO emissions in Equatorial Asia for 41 years from 1980 to 2021 were plotted. Most of the areas in Equatorial Asia have low average CO emissions. However, several regions have contributed to the main global carbon emissions from Equatorial Asia. They are distributed across three main islands: Borneo, Sumatra, and Java. Among the three islands, Java has the smallest area with CO emissions, concentrated around the capital city of Jakarta, northwest of the island. Emissions in Jakarta are mainly caused by industries and transportation (Lestari et al. 2022; Sodri & Garniwa, 2016), as this area is Indonesia's governmental and economic center. As for the other two islands, the CO emissions are distributed across vast areas. Compared to Jakarta, Borneo and Sumatra have higher CO emissions, and they are distributed in larger areas. The carbon source on these two islands is different from that in Jakarta, as there are no megacities in Sumatra or Borneo. Averaged CO emissions appear along the southeast shoreline of Sumatra Island, from South Sumatra to Riau Province. In Borneo, CO emissions appear in the Indonesian part of the island, known as Kalimantan. This is dominantly in the south and east parts of Borneo Island. While the east Borneo region shows some concentration of CO emissions, the south Borneo region releases extremely high CO emissions into the atmosphere. Compared to another CO emission hotspot, the southern area of Borneo has the highest average CO emissions in Equatorial Asia. Thus, the main contributor to CO emissions in Equatorial Asia is forest fires/land fires and land used change, particularly in Sumatra and Borneo (Fig. 2).

4.2 The Connection Between Emissions and Precipitation

Carbon emissions have a strong relation to the weather and climate. Despite their impact on climate change by increasing the amount of carbon as greenhouse gases, carbon emissions also directly connect to the weather, particularly to precipitation. For example, low precipitation might trigger forest fires (Fanin and Van Der Werf 2017). Carbon emissions and precipitation have a negative correlation. It means that precipitation decreases while carbon emissions increase. However, the two emitter islands in Equatorial Asia, Sumatra and Borneo, show different patterns. In Sumatra,

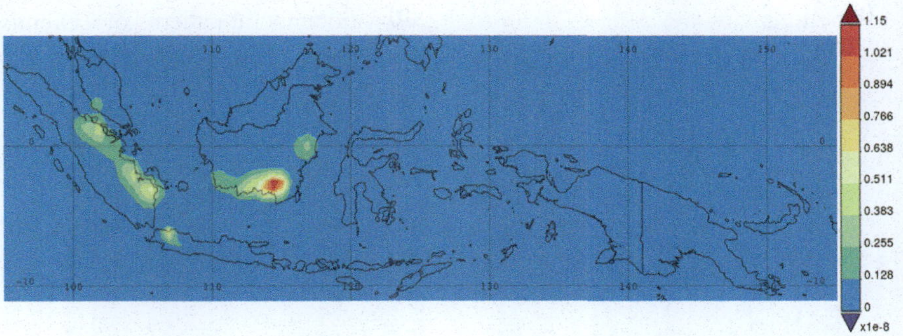

Fig. 2 Time-averaged CO emissions (kg m^{-2} s^{-1}) in Equatorial Asia from 1980 to 2021

the correlation is considerably low in the carbon emission region, which is between 0 and − 0.2 in the east part of the island, while in Borneo, the south region can reach − 0.4 and higher (Fig. 3).

4.3 The Anomaly of Forest Fire Emissions Since 2009

As Borneo has become the major contributor of carbon release and has a strong correlation with precipitation, this region was analyzed further, particularly concerning the monthly time series of CO emissions in Borneo Island from January 1980 to December 2021. Extremely high carbon releases occurred periodically from 1980 onward, during which there were some high carbon emissions almost every 3–4 years. The highest was in 1998, also known as the worst forest fire in terms of emissions in Indonesia (Langmann and Hell 2004) (Fig. 4).

Each occurrence of high CO emissions usually is similar to the increase of fire occurrence in Borneo island as the primary contributor of forest fire emissions. However, there was a changing behavior from 2009 onward. CO emissions did not follow an increase in fire counts at the same rate as in previous decades. It can be seen clearly in 2002 and 2006 when high fire counts and emissions occurred; however, fire counts were high in 2009, 2015, and 2019, whereas the CO emissions were considerably low (Fig. 5).

The Hovmoller diagram for carbon emissions and precipitation in the south Borneo region (109–117E; 0.9–3.6S) was plotted to understand this connection further. Six years with high fire counts were chosen, namely 2019, 2015, 2009, 2006, 2004, and 2002. Based on carbon emissions, they can be divided into two categories. The years with extremely high carbon emissions are 2002, 2004, and 2006. The rest are from 2009, 2015, and 2019. In all the years, carbon emissions occurred during the long period of low precipitation (the dry season), usually from July to October, sometimes extending to November. However, there were distinct features in the years of extremely high carbon emissions. It is shown that carbon emissions would occur

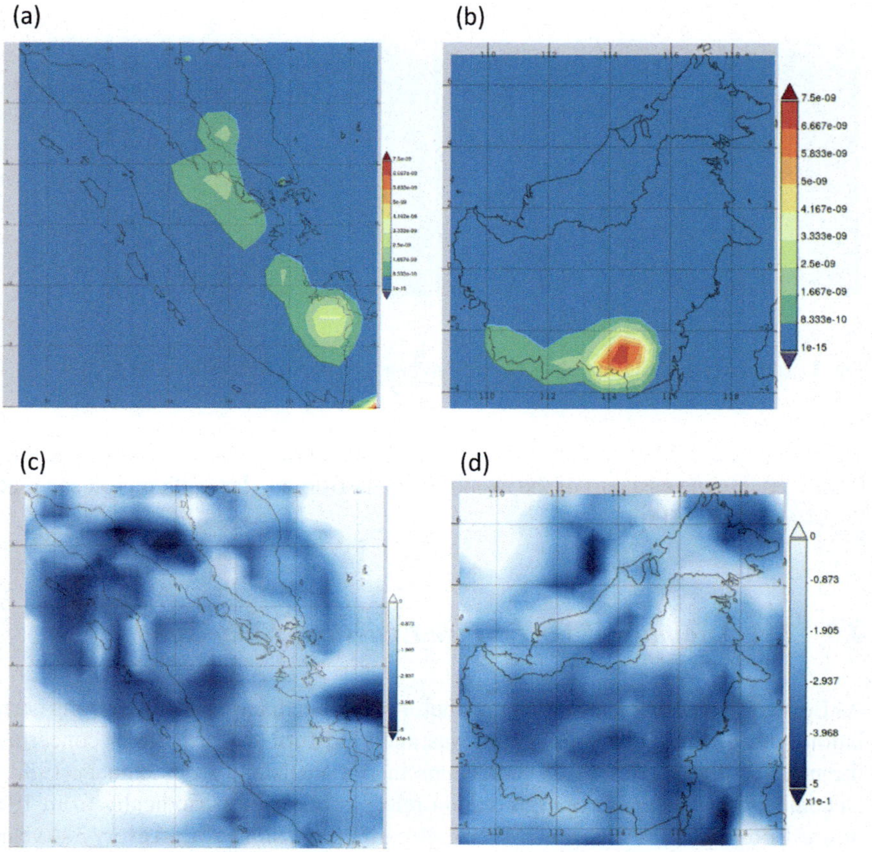

Fig. 3 Time-averaged CO emissions (kg m^{-2} s^{-1}) from 1980 to 2021 of **a** Sumatra and **b** Borneo; correlation between CO emissions and precipitation from 1980 to 2021 of **c** Sumatra and **d** Borneo

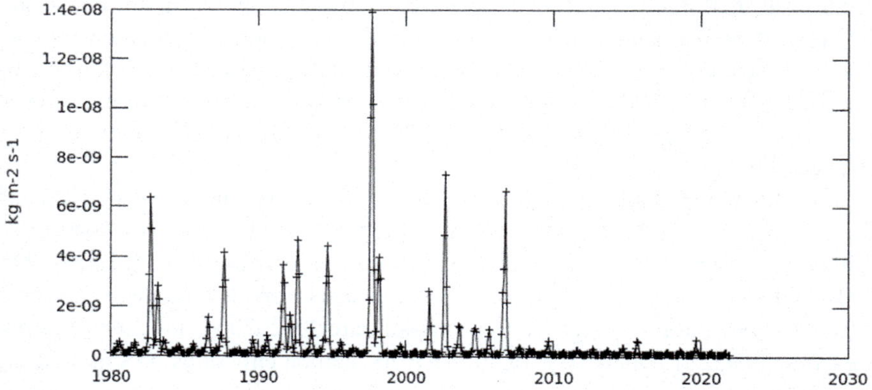

Fig. 4 Monthly time series of CO emissions (kg m^{-2} s^{-1}) in Borneo from January 1980 to December 2021

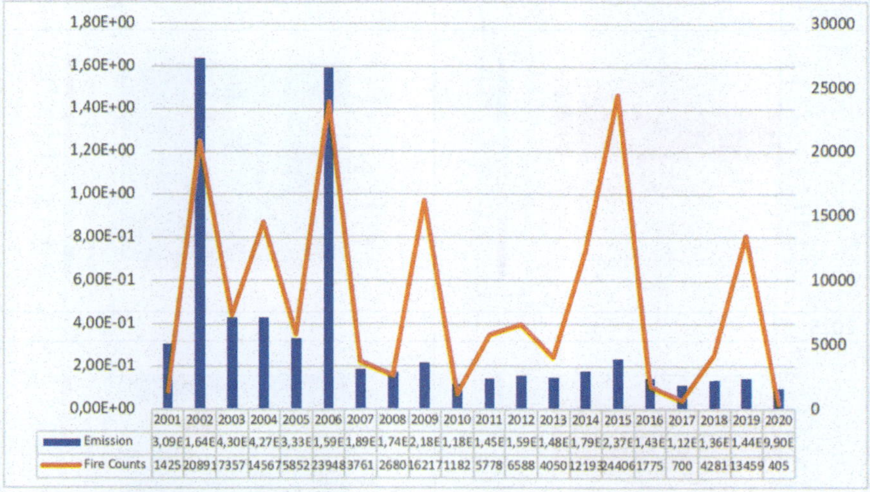

Fig. 5 Yearly time series of CO emissions (kg m^{-2} s^{-1}) and fire counts from 2001 to 2020

almost simultaneously/in the same month as low precipitation, as in 2002 and 2004. However, in the other years, from 2009 to 2019, there were delays in the onset of emissions of around 1–3 months after low precipitation. The period of emissions was also highly different. Before 2009, emissions would last for 5–6 months and cover almost all the region. After 2009, however, carbon emissions lasted for 1–3 months.

The decrease in peat soils may have caused these differences since 1990 (Miettinen and Liew 2010). Peat can store water in massive amounts. Thus, it is more persistent in drought conditions and delays potential forest fires for 1–3 months. However, in 2002 and 2006, the peat soils had dried before the drought season (Aldhous 2004), so the fires/emissions occurred in the same months as low precipitation. As a result, the peat soil affected the longer and stronger carbon emissions in 2002 and 2006 (Hooijer et al. 2010). In addition, the fire burns the area below the ground on the peatland, making it difficult to extinguish it (Lin et al. 2020). As a result, peat continues burning for a longer period. Moreover, peat soil also contains vast amounts of organic materials that release carbon emissions when combusted. As peat soil has burned for many decades, the availability of peat soil currently is much lower (Miettinen et al. 2016) (Fig. 6).

5 Conclusion

Borneo and Sumatra are two islands that contribute to the largest carbon emissions in the Equatorial Asia region caused by forest fires. These two islands had extremely higher CO emissions for 41 years (1980–2021) compared to other areas in Equatorial Asia. However, they have different correlations with precipitation. Precipitation and

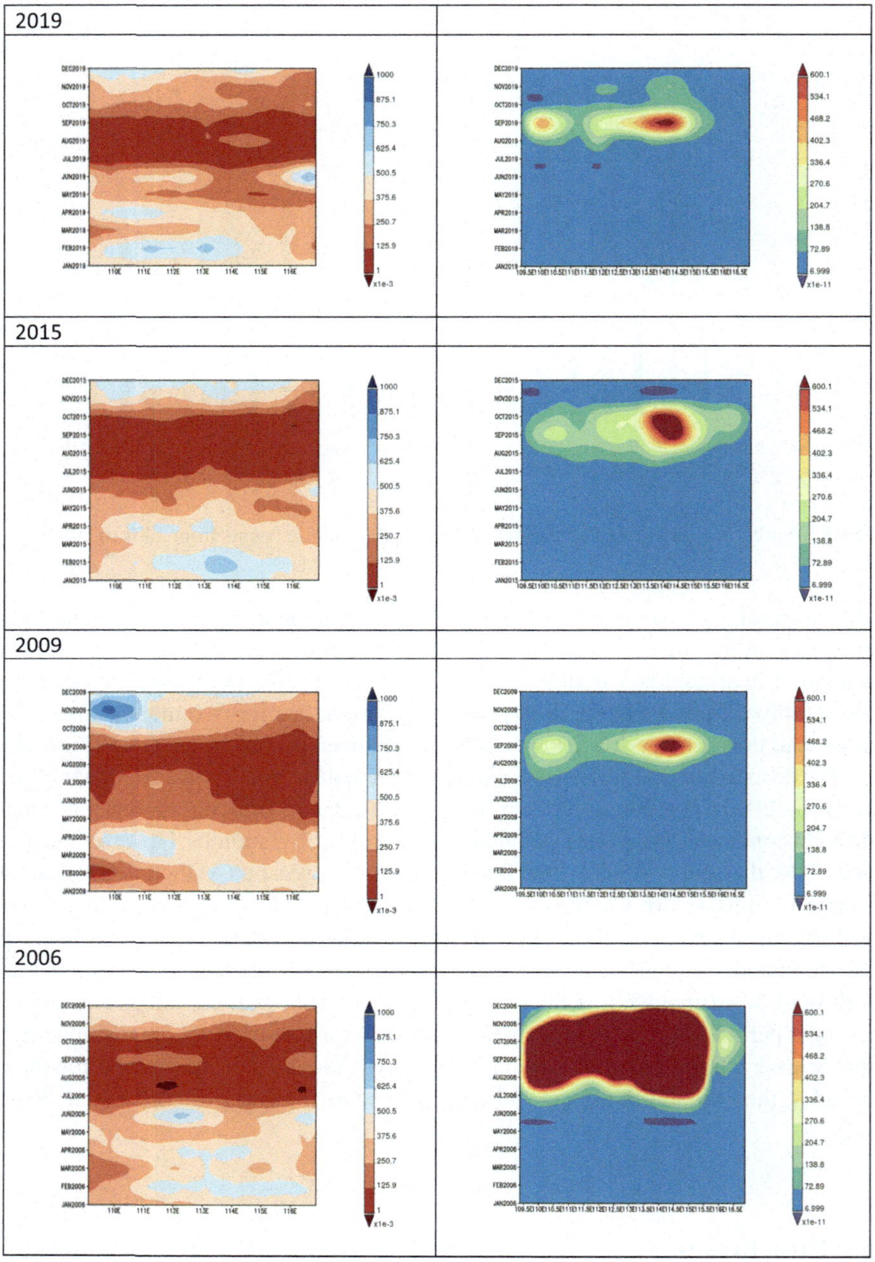

Fig. 6 Hovmoller comparison between precipitation (mm/day) in the left and CO emissions (kg m^{-2} s^{-1}) in the right in 2002, 2004, 2006, 2009, 2015, and 2019

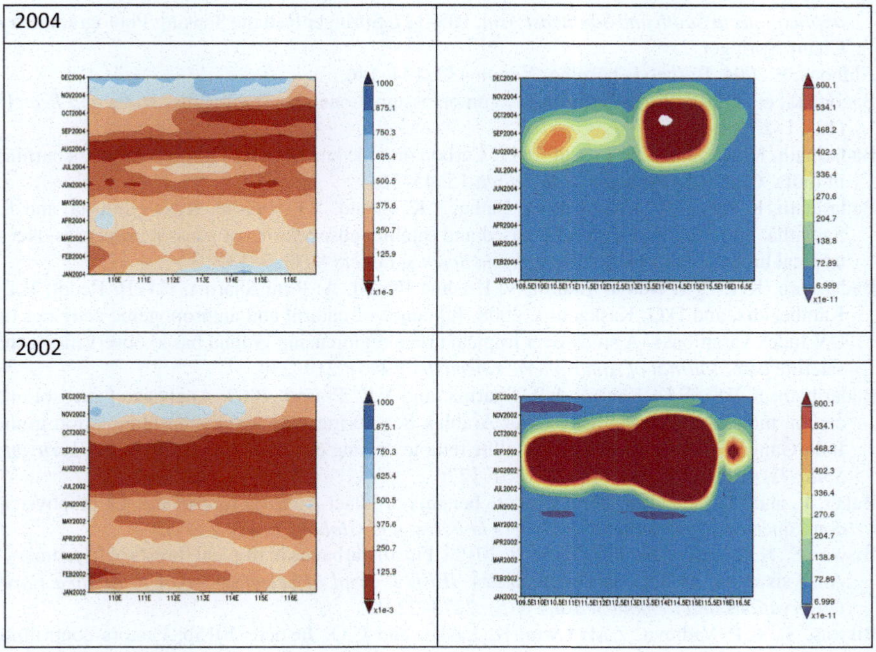

Fig. 6 (continued)

CO emissions have a weak connection as CO emissions are only – 0.2 in Sumatra. Meanwhile, CO emissions in Borneo are much more potent at almost – 0.5 around the extremely high emission area (South Borneo). High carbon emissions have consistently occurred following periodic fire events for an extended period since 1980. However, from 2009 onward, there were odd patterns. The increase in carbon emissions was considerably low compared to the number of fire counts. Moreover, carbon emissions have been delayed and occurred for a shorter period. This might have been caused by peat soil's contribution to CO emissions.

Acknowledgements The author acknowledges the use of data and/or imagery from NASA's Fire Information for Resource Management System (FIRMS) (https://earthdata.nasa.gov/firms), part of the NASA Earth Observing System Data and Information System (EOSDIS). Some of the analyses and visualizations used in this study were produced with the Giovanni online data system, developed and maintained by the NASA GES-DISC.

References

Albar, I., I. Jaya, B.H. Saharjo, B. Kuncahyo, and K.P. Vadrevu. 2018. Spatio-temporal analysis of land and forest fires in Indonesia using MODIS active fire dataset. In *Land-atmospheric research*

applications in South and Southeast Asia, 105–127. Springer Remote Sensing/Photogrammetry, Cham: Springer.
Aldhous, P. 2004. Borneo is burning. *Nature* 432: 144–146.
Bacon, K., et al. 2017. Questioning ten common assumptions about peatlands. *Mires and Peat* 19 (12): 1–23.
Badarinath, K.V.S., and K.V. Prasad. 2011. Carbon dioxide emissions from forest biomass burning in India. *Global Environmental Research* 15: 45–52.
Badarinath, K.V.S., S.K. Kharol, K.M. Latha, T.K. Chand, V.K. Prasad, A.N. Jyothsna, and K. Samatha. 2007. Multiyear ground-based and satellite observations of aerosol properties over a tropical urban area in India. *Atmospheric Science Letters* 8 (1): 7–13.
Badarinath, K.V.S., S. Kumar Kharol, V. Krishna Prasad, A. Rani Sharma, E.U.B. Reddi, H.D. Kambezidis, and D.G. Kaskaoutis. 2008. Influence of natural and anthropogenic activities on UV Index variations—A study over tropical urban region using ground based observations and satellite data. *Journal of Atmospheric Chemistry* 59 (3): 219–236.
Badarinath, K.V.S., S.K. Kharol, A.R. Sharma, and V.K. Prasad. 2009. Analysis of aerosol and carbon monoxide characteristics over Arabian Sea during crop residue burning period in the Indo-Gangetic Plains using multi-satellite remote sensing datasets. *Journal of Atmospheric and Solar-Terrestrial Physics* 71 (12): 1267–1276.
Baker, J., and D. Spracklen. 2019. Climate benefits of intact Amazon forests and the biophysical consequences of disturbance. *Frontiers in Forest and Global Change* 2 (47).
Biswas, S., K.D. Lasko, and K.P. Vadrevu. 2015a. Fire disturbance in tropical forests of Myanmar—Analysis using MODIS satellite datasets. *IEEE Journal of Selected Topics in Applied Earth Observations and Remote Sensing* 8 (5): 2273–2281.
Biswas, S., K.P. Vadrevu, Z.M. Lwin, K. Lasko, and C.O. Justice. 2015b. Factors controlling vegetation fires in protected and non-protected areas of Myanmar. *PLoS ONE* 10 (4): e0124346.
Biswas, S., K.P. Vadrevu, M.S. Mon, and C. Justice. 2021. Contemporary forest loss in Myanmar: Effect of democratic transition and subsequent timber bans on landscape structure and composition. *Ambio* 50 (4): 914–928.
Bonan, G.B. 2008. Forests and climate change: Forcings, feedbacks and the climate benefits of forests. *Science* 320 (5882): 1444–1449.
Brasika, I.B.M., I.M.O.G. Antara, and I.W.G.A. Karang. 2021. Investigating El Nino Southern Oscillation as the main driver of forest fire in Kalimantan. *Geografia* 17 (14): 305–320.
Callendar, G. 1938. The artificial production of carbon dioxide and its influence on temperature. *Quarterly Journal of the Royal Meteorological Society*.
Cao, L., and K. Caldeira. 2010. Atmospheric carbon dioxide removal: Long-term consequences and commitment. *Environmental Research Letters* 5 (2).
Ciais, P., et al. 2013. *Carbon and other biogechemical cycles*. Cambridge: Cambridge University Press.
Crutzen, P.J., M.O. Andreae. 1990. Biomass burning in the tropics: Impact on atmospheric chemistry and biogeochemical cycles. *Science* 250(4988): 1669–1678. https://doi.org/10.1007/s11027-006-9049-2
Eaturu, A., and K.P. Vadrevu. 2021. Evaluation of Sentinel-3 SLSTR data for mapping fires in forests, peatlands, and croplands—A case study over Australia, Indonesia, and India. In *Biomass burning in South and Southeast Asia*, 39–59. CRC Press.
Fanin, T., and G.R. Van Der Werf. 2017. Precipitation–fire linkages in Indonesia (1997–2015). *Biogeosciences* 14 (18): 3995–4008.
Friedlingstein, P., et al. 2020. Global carbon budget 2020. *Earth System Science Data* 12 (4): 3269–3340.
GES-DISC. 2011. Tropical rainfall measuring mission (TRMM), TRMM (TMPA/3B43) rainfall estimate L3 1 month 0.25 degree × 0.25 degree v7. [Online]. Available at: https://disc.gsfc.nasa.gov/datasets/TRMM_3B43_7/summary.
Hayasaka, H., I. Noguchi, E.I. Putra, N. Yulianti, and K. Vadrevu. 2014. Peat-fire-related air pollution in Central Kalimantan, Indonesia. *Environmental Pollution* 195: 257–266.

Hobbs, N. 1986. Mire morphology and the properties and behaviour of some British and foreign peats. *Quarterly Journal of Engineering Geology* 19: 7–80.

Hooijer, A., M. Silvius, H. Woesten, and S. Page. 2006. *PEAT-CO_2 assessment of CO_2 emission from drained peatlands in SE Asia.* Delft, The Netherlands: Delft Hydraulics.

Hooijer, A., et al. 2010. Current and future CO_2 emissions from drained peatlands in Southeast Asia. *Biogesciences* 7: 1505–1514.

Huffman, G.J., R.F. Adler, D.T. Bolvin, and E.J. Nelkin. 2010. The TRMM multi-satellite precipitation analysis (TMPA). In *Satellite applications for surface hydrology*, ed. M. Gebremichael and F. Hossain. Dordrecht: Springer.

Jaenicke, J., H. Wösten, A. Budiman, and F. Siegert. 2010. Planning hydrological restoration of peatlands in Indonesia to mitigate carbon dioxide emissions. *Mitigation and Adaptation Strategies for Global Change* 15: 223–239.

Joos, F., and R. Spahni. 2008. Rates of change in natural and anthropogenic radiative forcing over the past 20,000 years. *PNAS* 105 (5).

Justice, C., G. Gutman, and K.P. Vadrevu. 2015. NASA land cover and land use change (LCLUC): An interdisciplinary research program. *Journal of Environmental Management* 148: 4–9.

Kant, Y., V.K. Prasad, and K.V.S. Badarinath. 2000. Algorithm for detection of active fire zones using NOAA AVHRR data. *Infrared Physics & Technology* 41 (1): 29–34.

Kharol, S.K., K.V.S. Badarinath, A.R. Sharma, D.V. Mahalakshmi, D. Singh, and V.K. Prasad. 2012. Black carbon aerosol variations over Patiala city, Punjab, India—A study during agriculture crop residue burning period using ground measurements and satellite data. *Journal of Atmospheric and Solar-Terrestrial Physics* 84: 45–51.

Langmann, B., and A. Hell. 2004. Release and dispersion of vegetation and peat fire emissions in the atmosphere over Indonesia 1997/1998. *Atmospheric Chemistry and Physics* 4 (8): 2145–2160.

Lasko, K., and K. Vadrevu. 2018. Improved rice residue burning emissions estimates: Accounting for practice-specific emission factors in air pollution assessments of Vietnam. *Environmental Pollution* 236: 795–806.

Lasko, K., K.P. Vadrevu, V.T. Tran, E. Ellicott, T.T. Nguyen, H.Q. Bui, and C. Justice. 2017. Satellites may underestimate rice residue and associated burning emissions in Vietnam. *Environmental Research Letters* 12 (8): 085006.

Lasko, K., K.P. Vadrevu, and T.T.N. Nguyen. 2018. Analysis of air pollution over Hanoi, Vietnam using multi-satellite and MERRA reanalysis datasets. *PLoS ONE* 13 (5): e0196629.

Lasko, K., K.P. Vadrevu, V. Bandaru, T.N.T. Nguyen, and H.Q. Bui. 2021. PM2.5 emissions from biomass burning in South/Southeast Asia—Uncertainties and trade-offs. In *Biomass burning in South and Southeast Asia*, 149–169. CRC Press.

Lestari, P., M.K. Arrohman, S. Damayanti, and Z. Klimont. 2022. Emissions and spatial distribution of air pollutants from anthropogenic sources in Jakarta. *Atmospheric Pollution Research* 13 (9): 101521.

Li, W., et al. 2007. Future precipitation changes and their implications for tropical peatlands. *Geophysical Research Letters* 34 (01): 1–6.

Lin, S., Y.K. Cheung, Y. Xiao, and X. Huang. 2020. Can rain suppress smoldering peat fire? *Science of The Total Environment* 727: 138468. https://doi.org/10.1016/j.scitotenv.2020.138468

Lindsey, R. 2020. Climate change: Atmospheric carbon dioxide. [Online]. https://www.climate.gov/news-features/understanding-climate/climate-change-atmospheric-carbon-dioxide.

Liu, J., et al. 2017. Contrasting carbon cycle responses of the tropical continents to the 2015–2016 El Nino. *Science* 358 (6360): 1–7.

Malhi, Y. 2010. The carbon balance of tropical forest regions 1990–2005. *Current Opinion in Environmental Sustainability* 2 (4): 237–244.

Malhi, Y., and J. Wright. 2004. Spatial patterns and recent trends in the climate of tropical rainforest regions. *Philosophical Transactions of the Royal Society of London. Series B: Biological Sciences* 359 (1443): 311–329.

Miettinen, J., and S. Liew. 2010. Degradation and development of peatlands in Peninsular Malaysia and in the islands of Sumatra and Borneo since 1990. *Land Degradation & Development* 21 (3): 285–296.

Miettinen, J., C. Shi, and S.C. Liew. 2016. Land cover distribution in the peatlands of Peninsular Malaysia, Sumatra and Borneo in 2015 with changes since 1990. *Global Ecology and Conservation* 6: 67–78.

Nakajima, T., A. Higurashi, N. Takeuchi, and J.R. Herman. 1999. Satellite and ground-based study of optical properties of 1997 Indonesian Forest Fire aerosols. *Geophysical Research Letters* 26 (16): 2421–2424.

Page, S.E., et al. 2002. The amount of carbon released from peat and forest fires in Indonesia during 1997. *Nature* 420: 61–65.

Phung, V.L.H., A. Uttajug, K. Ueda, N. Yulianti, M.T. Latif, and D. Naito. 2022. A scoping review on the health effects of smoke haze from vegetation and peatland fires in Southeast Asia: Issues with study approaches and interpretation. *PLoS ONE* 17 (9): e0274433.

Pio, C., et al. 2008. Chemical composition of atmospheric aerosols during the 2003 summer intense forest fire period. *Atmospheric Environment* 42 (32): 7530–7543.

Plyusnin, I. 1964. *Reclamative soil science*, 412. Moscow: Foreign Language Press.

Prasad, K.V., and K.V.S. Badarinath. 2006. Soil surface nitrogen losses from agriculture in India: A regional inventory within agroecological zones (2000–2001). *The International Journal of Sustainable Development and World Ecology* 13 (3): 173–182.

Prasad, V.K., and K.V.S. Badarinth. 2004. Land use changes and trends in human appropriation of above ground net primary production (HANPP) in India (1961–98). *Geographical Journal* 170 (1): 51–63.

Prasad, V.K., Y. Kant, P.K. Gupta, C. Sharma, A.A. Mitra, and K.V.S. Badarinath. 2001a. Biomass and combustion characteristics of secondary mixed deciduous forests in Eastern Ghats of India. *Atmospheric Environment* 35 (18): 3085–3095.

Prasad, V.K., Y. Kant, and K.V.S. Badarinath. 2001b. CENTURY ecosystem model application for quantifying vegetation dynamics in shifting cultivation areas: A case study from Rampa Forests, Eastern Ghats (India). *Ecological Research* 16 (3): 497–507.

Prasad, V.K., Y. Kant, P.K. Gupta, C. Elvidge, and K.V.S. Badarinath. 2002a. Biomass burning and related trace gas emissions from tropical dry deciduous forests of India: A study using DMSP-OLS data and ground-based measurements. *International Journal of Remote Sensing* 23 (14): 2837–2851.

Prasad, V.K., Y. Kant, and K.V.S. Badarinath. 2002b. Land use changes and modeling carbon fluxes from satellite data. *Advances in Space Research* 30 (11): 2511–2516.

Prasad, V.K., M. Lata, and K.V.S. Badarinath. 2003. Trace gas emissions from biomass burning from northeast region in India—Estimates from satellite remote sensing data and GIS. *The Environmentalist* 23 (3): 229–236.

Prasad, V.K., K.V.S. Badarinath, S. Yonemura, and H. Tsuruta. 2004. Regional inventory of soil surface nitrogen balances in Indian agriculture (2000–2001). *Journal of Environmental Management* 73 (3): 209–218.

Prasad, V.K., E. Anuradha, and K.V.S. Badarinath. 2005. Climatic controls of vegetation vigor in four contrasting forest types of India—Evaluation from National Oceanic and Atmospheric Administration's Advanced Very High Resolution Radiometer datasets (1990–2000). *International Journal of Biometeorology* 50 (1): 6–16.

Ritchie, H., and M. Roser. 2020. CO_2 and greenhouse gas emissions. [Online]. https://ourworldindata.org/co2-emissions#citation.

Smith, J. 2019. Tropical rainforest. [Online]. https://www.britannica.com/science/tropical-rainforest.

Sodri, A., and I. Garniwa. 2016. The effect of urbanization on road energy consumption and CO_2 emissions in emerging megacity of Jakarta, Indonesia. *Procedia—Social and Behavioral Sciences* 227: 728–737.

Simorangkir, D. 2006. Fire use: Is it really the cheaper land preparation method for large-scale plantations? *Mitigation and Adaptation Strategies for Global Change* 12(1): 147–164. https://doi.org/10.1007/s11027-006-9049-2

Turner, N. 1757. *An essay on draining and improving peat bogs in which their nature and properties are fully considered*. London: Baldwin and Pew.

Uhl, C., B. Kauffman, and D.L. Cummings. 1988. Fire in the Venezuelan Amazon 2: Environmental conditions necessary for forest fires in the evergreen rainforest of Venezuela. *JSTOR* 53 (2): 176–184.

Vadrevu, K.P. 2008. Analysis of fire events and controlling factors in eastern India using spatial scan and multivariate statistics. *Geografiska Annaler: Series A, Physical Geography* 90 (4): 315–328.

Vadrevu, K. 2015. International regional science meeting of NASA-LCLUC: Land cover/land use change (LC/LUC) and environmental impacts in South Asia, Coimbatore, India, 19–23 January 2013. *Journal of Environmental Management* 148: 1–163.

Vadrevu, K.P. 2021. Fire danger indices and methods: An appraisal. In *Biomass burning in South and Southeast Asia*, 201–232. CRC Press.

Vadrevu, K.P., and K.V.S. Badarinath. 2009. Spatial pattern analysis of fire events in Central India—A case study. *Geocarto International* 24 (2): 115–131.

Vadrevu, K.P., and C.O. Justice. 2011. Vegetation fires in the Asian region: Satellite observational needs and priorities. *Global Environmental Research* 15 (1): 65–76.

Vadrevu, K., and K. Lasko. 2015. Fire regimes and potential bioenergy loss from agricultural lands in the Indo-Gangetic Plains. *Journal of Environmental Management* 148: 10–20.

Vadrevu, K.P., K.V.S. Badarinath, and E. Anuradha. 2008. Spatial patterns in vegetation fires in the Indian region. *Environmental Monitoring and Assessment* 147 (1): 1–13.

Vadrevu, K.P., L. Giglio, and C. Justice. 2013. Satellite based analysis of fire–carbon monoxide relationships from forest and agricultural residue burning (2003–2011). *Atmospheric Environment* 64: 179–191.

Vadrevu, K.P., T. Ohara, and C. Justice. 2014a. Air pollution in Asia. *Environmental Pollution (Barking, Essex: 1987)* 195: 233–235.

Vadrevu, K.P., K. Lasko, L. Giglio, and C. Justice. 2014b. Analysis of Southeast Asian pollution episode during June 2013 using satellite remote sensing datasets. *Environmental Pollution* 195: 245–256.

Vadrevu, K., T. Ohara, and C. Justice. 2017. Land cover, land use changes and air pollution in Asia: A synthesis. *Environmental Research Letters* 12 (12): 120201.

Vadrevu, K.P., T. Ohara, and C. Justice (eds.). 2018. *Land-atmospheric research applications in South and Southeast Asia.*

Vadrevu, K., A. Heinimann, G. Gutman, and C. Justice. 2019. Remote sensing of land use/cover changes in South and Southeast Asian countries. *International Journal of Digital Earth* 12 (10): 1099–1102.

Vadrevu, K.P., A. Eaturu, S. Biswas, K. Lasko, S. Sahu, J.K. Garg, and C. Justice. 2020. Spatial and temporal variations of air pollution over 41 cities of India during the COVID-19 lockdown period. *Scientific Reports* 10 (1): 1–15.

Vadrevu, K.P., T. Ohara, and C. Justice (eds.). 2021a. *Biomass burning in South and Southeast Asia: Impacts on the biosphere*, vol. 2. CRC Press.

Vadrevu, K.P., T. Ohara, and C. Justice (eds.). 2021b. *Biomass burning in South and Southeast Asia: Mapping and monitoring*, vol. 1. CRC Press.

Vadrevu, K., T. Le Toan, S.S. Ray, and C. Justice. 2022a. *Remote sensing of agriculture and land cover/land use changes in South and Southeast Asian countries*. Cham: Springer.

Vadrevu, K., A. Eaturu, E. Casadaban, K. Lasko, W. Schroeder, S. Biswas, L. Giglio, and C. Justice. 2022b. Spatial variations in vegetation fires and emissions in South and Southeast Asia during COVID-19 and pre-pandemic. *Scientific Reports* 12 (1): 1–21.

van der Werf, G., et al. 2008. Climate regulation of fire emissions and deforestation in equatorial Asia. *PNAS* 105 (51): 20350–20355.

van der Werf, G. R., J. T. Randerson, L. Giglio, G. J. Collatz, P. S. Kasibhatla, and A. F. Arellano Jr. (2006) Interannual variability in global biomass burning emissions from 1997 to 2004. *Atmospheric Chemistry and Physics* 6(11) 3423–3441. https://doi.org/10.5194/acp-6-3423-2006

van Wees, D., et al. 2021. The role of fire in global forest loss dynamics. *Global Change Biology* 27 (11): 2377–2391.

Wooster, M.J., G.J. Roberts, L. Giglio, D.P. Roy, P.H. Freeborn, L. Boschetti, C. Justice, C. Ichoku, W. Schroeder, D. Davies, and A.M. Smith. 2021. Satellite remote sensing of active fires: History and current status, applications and future requirements. *Remote Sensing of Environment* 267: 112694.

Air Pollution Modeling and Decision Support Systems

Impact of Vegetation Fires on Regional Aerosol Black Carbon Over South and East Asia

Yogesh Kant, Aryan Natwariya, Debashis Mitra, and Prakash Chauhan

Abstract Biomass burning (BB) aerosols can influence regional and global climate through radiation, clouds, and precipitation interactions. In this study, we investigated spatiotemporal variations in fire occurrences, aerosol optical depth (AOD), and black carbon (BC) over South, East, and Southeast Asia over two decades. We observed an average increase in fire anomalies by 27.3%, 11.1%, and 12.6%, and AOD by 24.5%, 4.7%, and 7.1% in South, East, and East Asia, respectively. Further, we observed a reduction in BC of about 12.3%, 6.9%, and 11.3% in South, East, and Southeast Asia, respectively, during the 2014–2020 period. The reduction in BC is attributed to several initiatives and policies by the local governments. Fire events are on a marginal decrease by 7%, 5%, and 19% during the 2014–2020 period, as observed over South Asia, East Asia, and Southeast Asia (over Indonesia), respectively. In one decade, AOD and BC increased by 6–8% over South Asia; however, rate of increase in the last five years reduced by 2–3.5%. Meanwhile, in East Asia, AOD and BC decreased by 8–10%, and the reduction was higher by 11–12% in Southeast Asia. Despite the rate of decrease in aerosol particulate matter, the magnitude of AOD and BC is still higher over East Asia (annual average 0.56 ± 0.06, $2.59 \pm 0.33\ \mu gm^{-3}$) compared to South Asia (annual average 0.51 ± 0.05, $1.52 \pm 0.47\ \mu gm^{-3}$) while Southeast Asia region had low concentration (annual average 0.43 ± 0.05, $0.72 \pm 0.24\ \mu gm^{-3}$). Our study also revealed Indo Gangetic Plain (IGP) plains with heavy mineral dust influx combined with high population-driven emissions, the East China region dominated by anthropogenic emissions, and Southeast Asia by vegetation fires. Synoptic meteorology also plays a vital role in the dispersion and transport of aerosols throughout the year. The source apportionment using Potential Source Contribution Function (PSCF) and Concentrated Weighted Trajectories (CWT) revealed both the local and external sources of pollution in the IGP and Eastern China region.

Keywords Biomass burning · Black carbon · AOD · Asia · Source apportionment · PSCF · CWT

Y. Kant (✉) · A. Natwariya · D. Mitra · P. Chauhan
Indian Institute of Remote Sensing (IIRS), ISRO Department of Space, Government of India, 4-Kalidas Road, Dehradun, India
e-mail: ykanty@gmail.com

K. P. Vadrevu et al. (eds.), *Vegetation Fires and Pollution in Asia*,
https://doi.org/10.1007/978-3-031-29916-2_27

1 Introduction

Biomass burning (BB) releases particulate matter (PM), such as black carbon (BC) aerosols have a significant influence on Earth's radiation budget, radiative forcing, human health, visibility, cloud formation, global warming, including climate change at a global scale (Kant et al. 2000a, b; Gupta et al. 2001a, b; IPCC 2007; Badarinath et al. 2007, 2008, 2009; Field et al. 2016; Vadrevu et al. 2018, 2021a, b). It is known to be the leading source of carbonaceous aerosols and reactive trace gases in the global atmosphere, especially in the form of open burning of agricultural residues, grassland, forest fires, and combustion of biofuels (Yokelson et al. 2013). Exposure to PM2.5 or PM10 is usually associated with BC, adversely affecting human health. BB is a predominant practice in Asia, where China contributes 25% of the total, followed by India (18%), Indonesia (13%), and Myanmar (8%) (Streets et al. 2003; Taylor 2010). About one-third to one-half of worldwide carbon monoxide (CO) and 20% of nitrogen oxide (NO_x) emissions are caused by uncontrollable open vegetation fires (Prasad et al. 1999, 2000a, b, c; Olivier et al. 2005; Wiedinmyer et al. 2011). Seasonal spikes in pollution levels are commonly observed due to unsustainable agricultural practices like stubble burning (SB) (Kant et al. 2022). Deforestation activities such as slash and burn agriculture, crop residue burning, peat land burning, including domestic biofuel use for cooking and heating, and combustion of fossil fuels can also aggravate the situation (Prasad et al. 2001a, b, 2002a, b, c; Kant et al. 2000a, b, 2020, Hayasaka et al. 2014; Albar et al. 2018; Venkataraman et al. 2018). While industries and power plants are significant emitters of carbonaceous aerosols (Yao et al. 2016), and in urban spaces, emission due to traffic is also one of the most prominent drivers (Titos et al. 2017). These factors add up to the emission of particulate BC concentration to larger scales polluting the environment. Air pollution causes around 5 million global deaths, of which 3 million deaths are only from PM2.5 and more than 50% of deaths are in China and India (State of Global Air 2019).

Due to the inhomogeneous nature of BB and aerosol particulates, monitoring at surface levels covering large areas is not practical and feasible; hence, remote sensing technology can play an essential role in characterizing the aerosols at large spatial and temporal scales. There have been several isolated regional studies on aerosols, BB, BC, and their impacts on air quality over parts of Asia (Field et al. 2016; Shaik et al. 2019; Roozitalab et al. 2021; Kant et al. 2022). For example, in North India, high PM2.5 concentration is due to residential biomass combustion, agriculture, and biomass burning (Venkataraman et al. 2018). In China, annual PM2.5 emissions from open straw burning are 7.8% of the total anthropogenic emissions of PM. In comparison, in Eastern China, it contributes to 56% of total emissions in summer (He et al. 2020), while Indonesian wildfire pollutes Singapore and adjoining region air and increases adverse human health impacts (Sheldon and Sankaran 2017). The study aims to quantify the long-term fire occurrences, aerosol optical depth, and BC over South, East, and Southeast Asia, highlighting the hotspots in these regions and source apportionments.

2 Study Area

The focus of the study is on Asia broadly divided into three regions; (i) South Asia (7 countries in green), East Asia (4 countries in tan), and Southeast Asia (9 countries in blue) (Fig. 1). Together, they account for about 27% of the total global landmass. The region nurtures about 56% of the global population with metropolitan cities like Shanghai, Beijing, Tokyo, Delhi, Mumbai, Jakarta, Bangkok, Kuala Lumpur, Karachi, Dhaka, etc. In addition, it consists of some of the major economies and industrial regions like China, Japan, India, the Republic of Korea, Singapore, Malaysia, Thailand, and Indonesia. South and East Asia contribute about one-third of the world's Gross Domestic Product (GDP). This region also contributes to natural and anthropogenic aerosols, including BC emissions. Natural sea salt aerosols dominate in the coastal areas due to the Indian Ocean in the south and the Pacific Ocean to the east. The primary source of BC emissions is forest fires and agricultural residue burning.

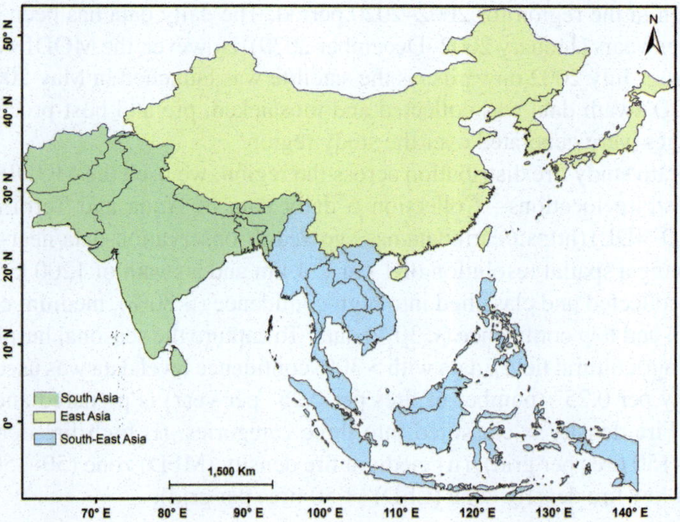

Fig. 1 Study region

3 Datasets and Methods

3.1 MODIS

MODIS provides daily information on atmospheric aerosols globally. It utilizes a radiometric resolution of 12 bits to monitor reflected sunlight from the surface, atmosphere, and emitted thermal radiation at 36 distinct wavelengths ranging from 400 m to 1440 nm (Levy et al. 2013). MODIS makes 14–15 orbits daily to cover the entire globe with a swath approximately 2330 km wide and uses two different algorithms to retrieve the aerosols over the land and ocean (Levy et al. 2007; Remer et al. 2008), and a detailed algorithm can be found at MODIS science team publications (Remer et al. 2005; Levy et al. 2013). The MODIS-AOD product is validated globally and widely used in aerosol studies (Mhawish et al. 2017). The expected error is $\pm$ 0.05 (Levy et al. 2010). In the present study, daily Level 2, Collection 6 data (https://ladsweb.modaps.eosdis.nasa.gov/) MODIS/Aqua (MOD04_3k) and MODIS/Terra (MYD04_3k) AOD with a spatial resolution of 3 km was used to study the aerosol distribution in the region for 2002–2020 period. The daily data has been processed for nineteen years (January 2002–December 2020). However, the MODIS/Aqua data is taken from July 2002 onwards, as the satellite was launched in May 2002. In this study, AOD swath data was collected and mosaicked, pre and post-processed, and daily images were generated over the study region.

Further, to study fire distribution across the region, we used the MODIS Thermal Anomalies/Fire locations—Collection 6 daily data of Aqua and Terra combined data (MCD14DL) (https://earthdata.nasa.gov/earth-observation-data/near-real-time/firms) having a spatial resolution of 1 km $\times$ 1 km and a swath of 1200 km. The fire data was collected and classified into high confidence (> 80%), medium confidence (30–80%), and low confidence (< 30%) data. To capture the seasonal burning events (< 1 ha of agricultural field), data with > 30% confidence level data was used. Further, fire density per 0.25° (number of fires per 0.25° per year) is prepared and used for analysis. Fire density is classified into three categories, (i) high fire density zone (HFD) (> 150 fires per grid); (ii) medium fire density (MFD) zone (50–150 fires per grid); and low fire density zone (LFD) (< 50 fires per grid).

3.2 AERONET

The AErosol RObotic NETwork (AERONET) is a group of well-calibrated ground-based sun-photometers developed by the National Aeronautics and Space Administration (NASA), USA (Holben et al. 1998). It maintains long-term aerosol data that is used to validate the satellite algorithms (Holben et al. 1998). The working principle, retrieval algorithms, calibration, and instrumentation of CIML can be found in studies (Holben et al. 2001). Under cloud-free conditions, the uncertainties in CIML retrieval are < $\pm$0.01. For this study, clear sky, cloud-free, and quality-assured

Level 2.0 AOD product are taken from 23 stations in 17 countries and is validated against MODIS-AOD 3 km product. The validation cities are Kolkata, New Delhi, Kanpur, Coimbatore, Pune (India), Kandahar (Afghanistan), Karachi (Pakistan), Dhaka (Bangladesh), Pokhra (Nepal), Mandalay (Myanmar), Jambi (Indonesia), Vientiane (Laos), Kuching (Malaysia), Manila (Philippines), Omkoi (Thailand), Hanoi (Vietnam), Singapore City (Singapore), Lanzhou, Beijing, Hongkong (China), Chiba (Japan), Seoul (South Korea), and Taipei (Taiwan).

3.3 MERRA-2 BC

We also used the global atmospheric reanalysis product Modern-Era Retrospective analysis for Research and Applications, version 2 (MERRA-2) with a spatial resolution of $0.5° \times 0.625°$ and 72 hybrid sigma/pressure levels are simulated using the GEOS-5, DAS (Goddard Earth Observing System, Data Assimilation System, Version 5.12.4). It is produced by NASA Global Modeling and Assimilation Office (GMAO) (Bali et al. 2017). It was replaced with the original MERRA dataset because of the assimilation system developments that further assimilate the microwave observations and modern hyperspectral radiance and GPS-Radio Occultation datasets. The MERRA-2 dataset has better accuracy than the observed satellite data, but the values are higher (Pfenninger and Staffell 2016; Kuo 2017). More information and details on this data can be found in Pfenninger and Staffell (2016), Bali et al. (2017), and Kuo (2017). The MERRA-2 datasets used in this study are surface, black carbon (BC) (https://disc.gsfc.nasa.gov/). The surface BC monthly data was processed, and annual spatial maps were prepared extensively for the 19 years. The MERRA-2 data is validated against nine cities, namely; New Delhi, Pune, Kanpur, Kolkata, Dhaka, Karachi, Manila, Beijing, and Hong Kong.

3.4 CAMS Reanalysis

We used the Copernicus Atmosphere Monitoring Service (CAMS) reanalysis is the state-of-the-art atmospheric composition dataset that consists of 3D time consistent atmospheric composition fields (aerosols, chemical species, and greenhouse gases). The reanalysis data is produced vertically at 60 hybrid sigma/pressure levels using 4Dvar data assimilation in CY42R1 of ECMWF's Integrated Forecast System (*CAMS: Reanalysis data documentation*, no date). For the winds, 10 m u and v component is taken at $0.25° \times 0.25°$ grid. The data can be obtained freely from European Centre for Medium-Range Weather Forecasts (ECMWF) portal (https://www.ecmwf.int/en/research/climate-reanalysis/cams-reanalysis). Further, the average seasonal data is produced to get the seasonal average wind speed and magnitude.

3.5 HYSPLIT-CWT/PSCF and Forward Trajectory

HYbrid Single Particle Lagrangian Integrated Trajectory (HYSPLIT) model (HYSPLIT-V5) was used to estimate the position of air mass using a time-based 3D (latitude, longitude, and altitude) Lagrangian air mass velocity algorithm and is developed by NOAA ARL. We modeled five days backward air mass trajectories from stubble burning (SB) hotspot region at 500 m above ground level using NCEP/NCAR reanalysis archived data (ftp://arlftp.arlhq.noaa.gov/pub/archives/reanalysis) during 2011–2020. The BC-emitted source regions are identified using Concentration Weighted Trajectory (CWT) analysis by incorporating satellite-measured PM2.5 data into the trajectories. The air mass trajectories were computed for clusters to investigate the sources. Further, to understand the dispersion and advection of the SB emissions to influence regions, we used the Potential Source Contribution Function (PSCF), which is a receptor model that incorporates meteorological information in its analysis scheme to produce a probability field that can be used to determine areas of the potential source contribution. The CWT analysis over the observational sites (receptor site) is calculated with the satellite-observed BC concentrations projected with 5-day back trajectories during the study period. The CWT values observed and analyzed are annual averaged BC concentrations. Further, the probability (PSCF > 0.6 corresponds to high potential BC sources, and PSCF < 0.6 represents the sources with a low probability of reaching the receptor site) of the emitted sources is estimated using PSCF over selected hotspot sites over the study region. For analysis, the 19 years are divided into three time periods T_1 (2002–2007), T_2 (2008–2013), and T_3 (2014–2020).

4 Results

4.1 Meteorology

Aerosol production, dispersion, secondary aerosol generation, and dry/wet deposition are all affected by the meteorology of the region (Joshi et al. 2016; Shaik et al. 2017). Climate and meteorology, which vary seasonally, greatly impact the spatiotemporal pattern of optical and physical characteristics of particulates in the atmosphere. CAMS reanalysis data are used to depict seasonal wind directions and magnitudes (Fig. 2). The seasonal wind patterns show similar situations and patterns of winds during autumn and winter. In summer and spring, these patterns are almost identical. Strong north-westerly winds (> 4 ms^{-1}) over South Asia during summers carry mineral dust and other pollutants to northern India and IGP region. However, reverse meteorology (change in wind directions) can be seen during the autumn and winter season with low speed (< 2 ms^{-1}). The winds over IGP in India are slow and steady in all seasons (2–3 ms^{-1}) except summer; a low boundary layer (< 1 km), favorable wind direction, and stable atmosphere help to prevent the vertical mixing of pollutants and trap the emitted BC and aerosol particulates over the region.

The north-westerly winds dominate in the East Asia region, particularly over central China, during all the seasons. The winds are strong (> 4 ms^{-1}) and carry mineral dust and other pollutants to eastern regions. The Eastern China region experiences slow, steady winds (mostly < 1 ms^{-1}) throughout. During autumn and winter, the winds from Mongolia (speed > 3 ms^{-1}) dominate the East China region and can carry mineral dust from the Gobi desert. The winds in Southeast Asia, on the other hand, remain almost constant. This region is dominated by slow and steady south-easterly winds (< 2 ms^{-1}) over land and strong winds (> 4 ms^{-1}) over the ocean during the autumn and winter season. However, in the summer season, the meteorology is reversed, and north-easterly solid winds (> 3 ms^{-1}) blow over the land, and over the ocean, the wind speed is > 5 ms^{-1}. This meteorology affects the dispersion and transport of aerosols in downward directions.

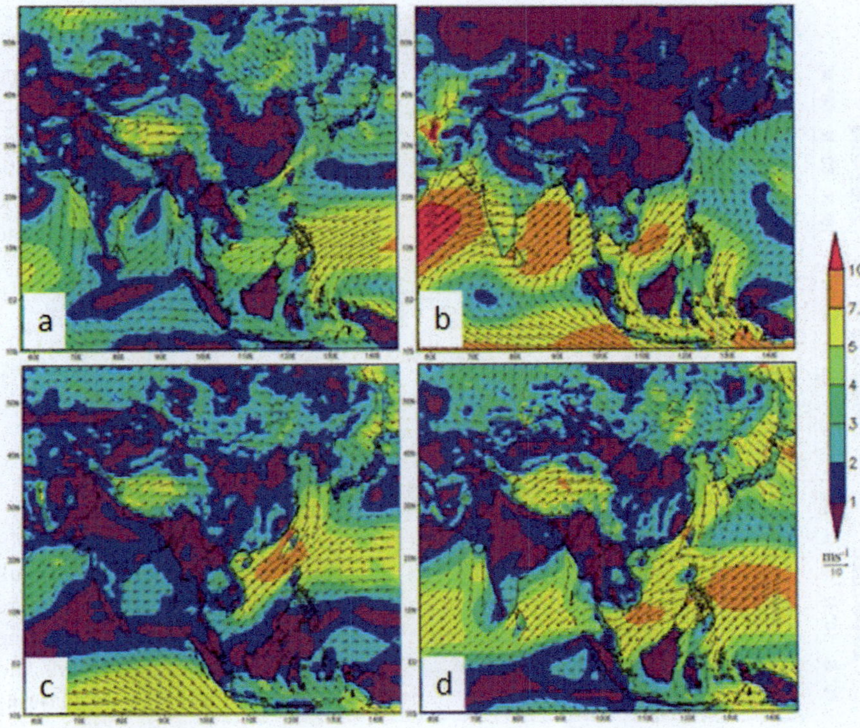

Fig. 2 Seasonal meteorology during **a** spring, **b** summer, **c** autumn, and **d** winter over the study area

4.2 Regional Variation of AOD, BC, and Fire

South Asia, East Asia, and Southeast Asia are highly populated, which have high aerosol loading, and contribute more than half of global BC emissions. The high aerosol-polluted regions in South Asia are Indo Gangetic Plain (IGP) in India, Bangladesh, and Indus plains in Pakistan. In East Asia, Eastern China (Beijing, Wuhan, and Chongqing region) and Southeast Asia—Malaysia, Indonesia, and Myanmar contribute significantly to aerosol pollution. The high spatiotemporal variability of aerosols is found at small and large spatial scales (Dey et al. 2004; Jethva et al. 2005; Gautam et al. 2011). These regions mostly have high aerosol, BC, and PM2.5 concentrations due to many anthropogenic and naturally induced aerosols. Synoptic meteorology and other local and regional geographic conditions play a vital role in particulate aerosol distribution. The satellite-based measurements from MODIS (Aqua and Terra) and MERRA-2 are used to characterize the spatiotemporal variability of fire counts, aerosols, and BC over the region (Fig. 3).

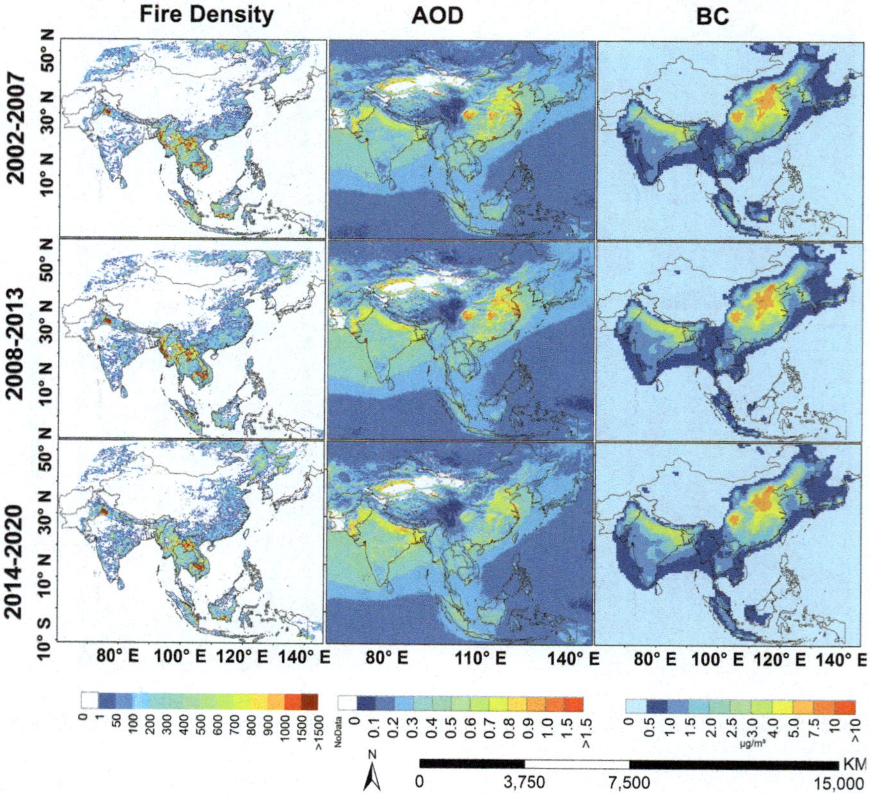

Fig. 3 Variation of fire density, AOD, and BC over the study region for time period T_1, T_2, and T_3 (density refers to number of fire points per 0.25° grid)

Table 1 Annual average number of fires, AOD and BC in South, East, and Southeast Asia

	2002–2007	2008–2013	2014–2020
	Fire—average no. of fire occurrences per year		
South Asia	79,398	97,578	101,084
East Asia	89,994	109,079	112,109
Southeast Asia	251,039	224,968	219,934
	Average AOD per year		
South Asia	0.45 ± 0.06	0.49 ± 0.04	0.51 ± 0.05
East Asia	0.53 ± 0.04	0.59 ± 0.07	0.56 ± 0.06
Southeast Asia	0.46 ± 0.04	0.44 ± 0.03	0.43 + 0.05
	Average BC conc. per year (μgm^{-3})		
South Asia	1.35 ± 0.29	1.47 ± 0.37	1.52 ± 0.47
East Asia	2.42 ± 0.43	2.53 ± 0.53	2.59 ± 0.33
Southeast Asia	0.81 ± 0.09	0.74 ± 0.13	0.72 ± 0.24

The long-term quantification of the fires, aerosol optical depth, and BC over the region is summarized in Table 1. It is inferred that from T_1 to T_3, there has been an increase in fire occurrences, AOD, and BC concentration by 27.31%, 11.19%, and 12.65% in South Asia; 24.57%, 4.72%, and 7.14% in East Asia; and a reduction of 12.39%, 6.92%, and 11.31% in Southeast Asia, respectively. This variation in fires, aerosol loading, and BC concentration in South and East Asia is attributed to increasing population, urbanization, industrialization, and biomass burning (anthropogenic and natural); however, the decrease in Southeast Asia is primarily due to the reduction in the rate of deforestation and burning of palm oil plantations.

The detailed analysis shows that in the last decade (2011–2020), there has been an annual increase in fire events, AOD, and BC in South Asia by 0.91%, 0.73%, and 0.67% yr^{-1} (particularly over IGP where fires, AOD, and BC increase is 1.14% yr^{-1} (i.e., 887 fires yr^{-1}), 0.91% yr^{-1}, 0.64% yr^{-1}). However, in the last five years (2016–2020), over India, due to government regulations in stubble burning and air pollution policies, there has been a reduction in annual increase of fire occurrences, AOD, and BC by 0.73%, 0.69%, and 0.46% yr^{-1}, respectively. East Asia, on the other hand, witnessed a 0.58% increase in fires over a decade while a 1.12% and 1.29% yr^{-1} decrease in AOD and BC concentrations. The increase in fire events is due to open straw fires in the East and southeast region of China. Further, the improvement in air quality is a result of strict local government policies (Clean Air Policy, 2013, which aimed to reduce PM2.5 levels by 25% till 2017), the shift toward green fuels in vehicles, and less dependence on thermal power plants that helped in lowering the local anthropogenic emissions. However, over the Southeast Asia region, the regions in fires in the HFD zone have shifted to the LFD zone, thereby reduction in aerosol loading and BC concentration in the region (particularly over Indonesia and the surrounding region that shows a 9% reduction in fires). Over the Indonesia region, the fires, AOD, and BC have reduced by 0.86%, 0.73%, and 1.14% yr^{-1} in the last decade due to a reduction in the rate of deforestation and the burning of palm oil plantations. This is also evident that the literature on the deforestation rate in Indonesia has been decreasing from 1.09 in 2011 to 0.15 in 2020 (Nicholas 2021). A report by the Agriculture Ministry of Indonesia states that the annual forest clearing for palm oil plantations has reduced by 4.7 million ha, down from 7.8 million ha per annum in the 1990s (FAO and UNEP 2020). On the other hand, fires, AOD, and BC in Myanmar have decreased by 0.9% yr^{-1}, 0.2% yr^{-1}, and 0.3% yr^{-1} during the 2011–20 period. Here, the meteorological conditions and topography disperse the pollutants to nearby regions (Thailand, Laos).

4.3 Regional Hotspots and Source Apportionment

Analysis of long-term fire density (Fig. 3a–c), AOD, and BC data (Fig. 4) over the region reveals that there are regional hotspots that constantly have high loading over the seasons with absorbing aerosols, either due to anthropogenic or natural phenomenon. The identified regional hotspot regions are—IGP plains in South Asia (annual average BC: 3.0–7.5 μgm^{-3}; annual average AOD: 0.6–1.0; and MFD zone); Eastern China in East Asia (annual average BC: 5–10 μgm^{-3}; annual average AOD: 0.7–1.5; and MFD zone); and Myanmar, Indonesia, and nearby region (annual average BC: 2.5–3 μgm^{-3}; annual average AOD 0.6–0.9, and MFD zone).

Pollutants from biomass burning, volcanic eruptions, and anthropogenic activities are lifted into the atmosphere and carried to other regions by wind (Shaik et al. 2019). This advection of air is studied by 5-day backward air mass trajectories. The long-range transport of BC aerosols is studied using CWT and PSCF analysis on an annual basis at major BC emitting hotspot regions, i.e., at New Delhi and Kolkata in

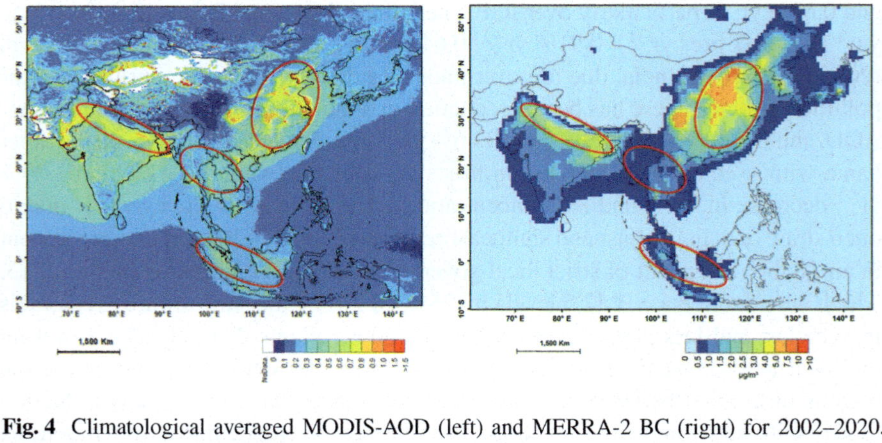

Fig. 4 Climatological averaged MODIS-AOD (left) and MERRA-2 BC (right) for 2002–2020. Red ovals represent regional hotspots

South Asia region; at Beijing over East Asia and at Mandalay (Myanmar) and Jambi (Indonesia) in Southeast Asia region.

The high PSCF values (> 0.6), i.e., the maximum probability sources to the receptor site New Delhi (Fig. 5b), show a strong probability of BC sources being located in the South-west, West, Northwest, and IGP region. Further, the high annual average BC CWT values of 2.5–4.5 μgm^{-3} are observed over western IGP and Nepal, parts of Pakistan, followed by the Arabian region (annual average 1.5–2.5 μgm^{-3}), which contributes high influx of BC concentration over Delhi region (Fig. 5a). Kolkata region, on the other hand, shows high PSCF values (> 0.6) over eastern IGP in India and Bangladesh. Also, a high contribution of high BC flux is observed (strong probabilities PSCF > 0.7) from the eastern region of India and Myanmar, indicating the high probability sources of BC (> 4.5 μgm^{-3}) influencing this hotspot region (Fig. 5c, d). The bottom line is that both local and external sources contribute to the high BC observed over these hotspot regions in South Asia. Beijing (a hotspot over East Asia) shows very high probabilities (PSCF > 0.9) over the nearby region, and high probabilities (> 0.6) are observed from Mongolia and Western China (Fig. 6b). The corresponding CWT values reveal a high amount of BC aerosols (> 7.5 μgm^{-3}) local and nearby sources (Fig. 6a). The mineral dust from the Gobi Desert in Mongolia and Taklamakan desert in China are responsible for BC (annual average 4.5–6.5 μgm^{-3}) contribution to this receptor site.

Over Southeast Asia, the high PSCF (> 0.6) values are observed over Thailand, Laos, and Vietnam in the east and over Bangladesh and Northeast India in the west of Mandalay. The corresponding CWT values are, however, low as an annual average of 0.5–1.5 μgm^{-3}. Hence, the BC contribution from these regions is small at the receptor site Mandalay (Fig. 7a, b). On the other hand, the high PSCF values in the Indonesia region are seen in the entire region with high CWT concentration (annual average 3.5–4.5 μgm^{-3}) from nearby local sources with no source contribution from external sources (Fig. 7c, d).

5 Conclusion

The present research investigates the spatial and temporal variation and their characteristics over different regions of South, East, and Southeast Asia. Long-term MODIS-AOD, fire anomalies, and MERRA-2 BC data helped us understand fire and BC aerosols' annual and seasonal variations over these regions. The key findings are as follows:

- Analysis of two decadal data reveals a drastic reduction in the rate of increase by 65–70% in fire occurrence in South Asia and East Asia. At the same time, Southeast Asia has continuously decreased over these years. Similarly, the rate of increase for AOD and BC has also been reduced by 30–50% over these three regions.

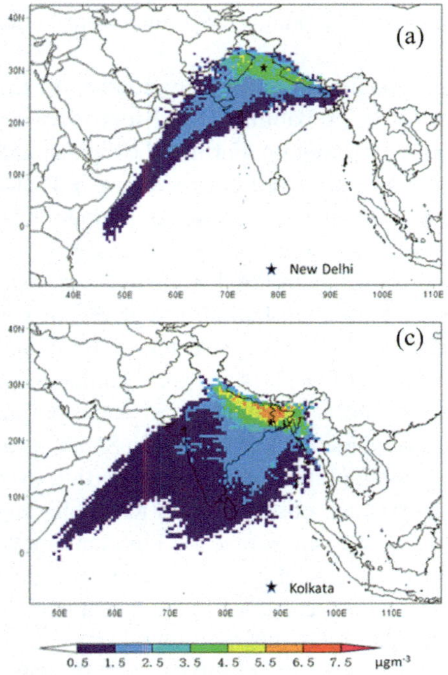

Fig. 5 CWT (**a**, **c**) and PSCF (**b**, **d**) analysis for hotspot regions in South Asia. The selected sites for 5 days backward trajectories are represented by star symbol; (i) New Delhi (top); (ii) Kolkata (bottom)

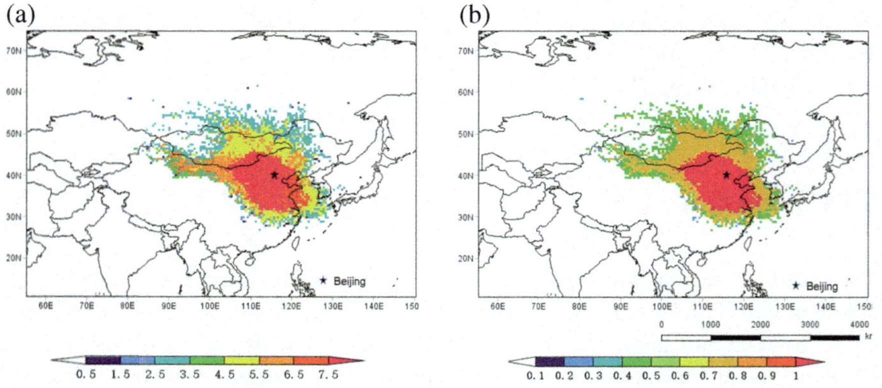

Fig. 6 CWT (**a**) and PSCF (**b**) analysis for hotspot regions in East Asia. The selected site Beijing for 5 days backward trajectories are represented by star symbol

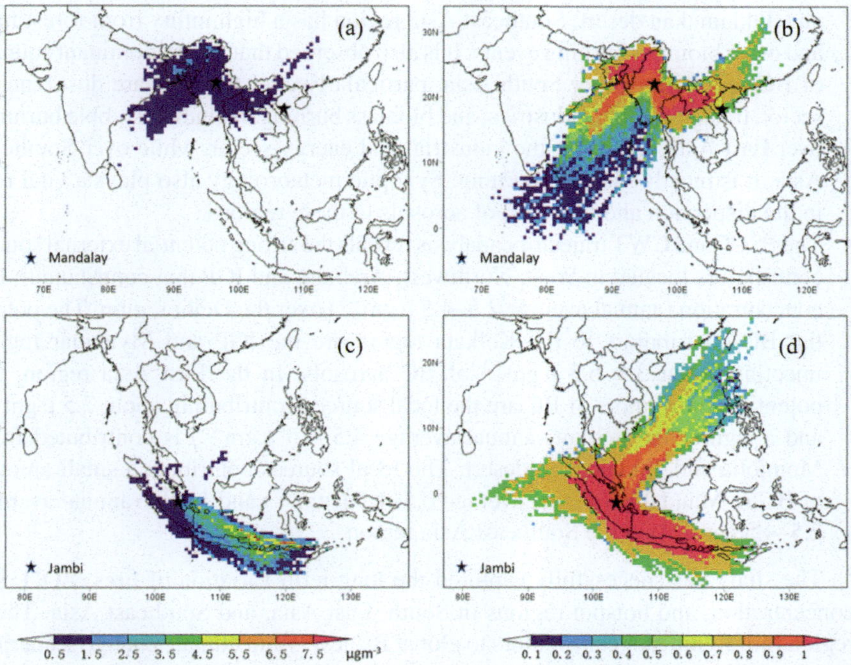

Fig. 7 CWT (**a**, **c**) and PSCF (**b**, **d**) analysis for hotspot regions in Southeast Asia. The selected sites for 5 days backward trajectories are represented by star symbol; (i) Mandalay (top); (ii) Jambi (bottom)

- In one decade, AOD and BC increased by 6–8% over South Asia; however, rate of increase in the last five years reduced by 2–3.5%. Meanwhile, in East Asia, AOD and BC decreased by 8–10% and reduced to a great extent by 11–12% over Southeast Asia. These reduced emissions are due to the measures by the government, reduced anthropogenic BB events, technological advancements, and a shift toward green energy. Despite the rate of decrease in aerosol particulate matter, the magnitude of AOD and BC is still higher in East Asia compared to South Asia.
- MODIS-AOD is validated in 23 cities in 17 countries over the entire region. The continuously available ground data ranged from 3 to 15 years spread over all seasons, and a good agreement ($R^2 = 0.72–0.83$) is observed at most stations. Further, the MERRA-2 BC is validated against literature cited values over nine regional cities ($R^2 = 0.63–0.72$). However, literature shows a high concentration of BC as these are recorded near to surface and at traffic sites.
- Regional hotspots of aerosol pollution, namely, IGP plains in South Asia, Eastern China in East Asia, Myanmar, Indonesia, and the nearby region were identified. The influx of high particulates loading over IGP is mainly from mineral dust from the Saharan and Arabian regions and regional anthropogenic activities. East China has an influx mainly from regional activities and mineral dust flow from Mongolia

and Taklamakan desert. Southeast Asia region has a high influx from forest fires and other biomass-burning events. It is also observed that the predominant sources of BC particulates over South Asia, particularly over the IGP, are dust, energy sector, transportation, industries, and biomass burning, including stubble burning; over East Asia, it is from the industrial and energy sector, while over Southeast Asia, it is mostly biomass burning. Synoptic meteorology also plays a vital role in the dispersion and transport of aerosols to these regions.

- The PSCF and CWT trajectory analysis reveals the strong potential external source contributors located in West, Northwest, Arabian, and IGP that contribute to BC concentration (annual average 2.5–4.5 μgm^{-3}) over the Delhi region. The potential BC contributors to the Kolkata region are the IGP and Myanmar region injecting about 3.5–6.5 μgm^{-3} of BC aerosols. In the East Asia region, the potential contributors of BC are the local sources contributing about 7.5 μgm^{-3}, and a significant amount (annual average 4.5–6.5 μgm^{-3}) is contributed from Mongolia and Taklamakan desert. The local sources contribute a small amount of BC to Mandalay (annual average 0.5–1.5 μgm^{-3}) and Jambi (annual average 2.5–4.5 μgm^{-3}) in the Southeast Asia region.

The study has successfully captured the long-term variation of fires, AOD, BC concentration, and hotspot regions in South, East Asia, and Southeast Asia. These regions are significant contributors to global BC levels and climate change scenarios. However, a more detailed analysis of optical and physical properties and vertical profiling is needed to understand the behavior of the BC aerosols.

Acknowledgements The authors thank the MODIS NASA team and NOAA Air Resources Laboratory (ARL) for the availability of Fire, AOD, MERRA-2, and HYSPILT model, ECMWF for CAMS reanalysis data, and AERONET.

References

Albar, I., I. Jaya, B.H. Saharjo, B. Kuncahyo, and K.P. Vadrevu. 2018. Spatio-temporal analysis of land and forest fires in Indonesia using MODIS active fire dataset. In *Land-atmospheric research applications in South and Southeast Asia*, 105–127. Cham: Springer.

Badarinath, K.V.S., S.K. Kharol, K.M. Latha, T.K. Chand, V.K. Prasad, A.N. Jyothsna, and K. Samatha. 2007. Multiyear ground-based and satellite observations of aerosol properties over a tropical urban area in India. *Atmospheric Science Letters* 8 (1): 7–13.

Badarinath, K.V.S., S. Kumar Kharol, V. Krishna Prasad, A. Rani Sharma, E.U.B. Reddi, H.D. Kambezidis, and D.G. Kaskaoutis. 2008. Influence of natural and anthropogenic activities on UV Index variations—A study over tropical urban region using ground based observations and satellite data. *Journal of Atmospheric Chemistry* 59 (3): 219–236.

Badarinath, K.V.S., S.K. Kharol, A.R. Sharma, and V.K. Prasad. 2009. Analysis of aerosol and carbon monoxide characteristics over Arabian Sea during crop residue burning period in the Indo-Gangetic Plains using multi-satellite remote sensing datasets. *Journal of Atmospheric and Solar-Terrestrial Physics* 71 (12): 1267–1276.

Bali, K., A.K. Mishra, and S. Singh. 2017. Impact of anomalous forest fire on aerosol radiative forcing and snow cover over Himalayan region. *Atmospheric Environment* 150: 264–275.

CAMS. no date. *Reanalysis data documentation*. Available at: https://confluence.ecmwf.int/display/CKB/CAMS%3A+Reanalysis+data+documentation. Accessed 9 March 2021.

Dey, S., et al. 2004. Influence of dust storms on the aerosol optical properties over the Indo Gangetic basin. *Journal of Geophysical Research: Atmospheres* 109 (D20).

FAO and UNEP. 2020. *The state of the world's forests 2020, forests, biodiversity and people*. Rome. https://doi.org/10.1515/9783035608632-002.

Field, R.D., et al. 2016. Indonesian fire activity and smoke pollution in 2015 show persistent nonlinear sensitivity to El-Nino-induced drought. *Proceedings of the National Academy of Sciences* 113 (33): 9204–9209.

Gautam, R., et al. 2011. Accumulation of aerosols over the Indo-Gangetic plains and southern slopes of the Himalayas: Distribution, properties and radiative effects during the 2009 pre-monsoon season. *Atmospheric Chemistry and Physics* 11 (24): 12841–12863.

Gupta, P.K., V.K. Prasad, Y. Kant, C. Sharma, A.B. Ghosh, M.C. Sharma, A.K. Sarkar, S.L. Jain, O.P. Tripathi, R.C. Sharma, and K.V.S. Badarinath. 2001a. Study of trace gases and aerosol emissions due to biomass burning at shifting cultivation sites in East Godavari District (Andhra Pradesh) during INDOEX IFP-99. *Current Science* 186–196.

Gupta, P.K., V.K. Prasad, C. Sharma, A.K. Sarkar, Y. Kant, K.V.S. Badarinath, and A.P. Mitra. 2001b. CH_4 emissions from biomass burning of shifting cultivation areas of tropical deciduous forests—Experimental results from ground-based measurements. *Chemosphere-Global Change Science* 3 (2): 133–143.

Hayasaka, H., I. Noguchi, E.I. Putra, N. Yulianti, and K. Vadrevu. 2014. Peat-fire-related air pollution in Central Kalimantan, Indonesia. *Environmental Pollution* 195: 257–266.

He, G., T. Liu, and M. Zhou. 2020. Straw burning, PM2.5 and death: Evidence from China. *Journal of Development Economics* 45: 102468.

Holben, B.N., et al. 1998. AERONET—A federated instrument network and data archive for aerosol characterization. *Remote Sensing of Environment* 66 (1): 1–16.

Holben, B.N., et al. 2001. An emerging ground based aerosol climatology: Aerosol optical depth from AERONET. *Journal of Geophysical Research Atmosphere* 106 (D11): 12067–12097.

IPCC. 2007. *The physical science basis*. Cambridge University Press. ISBN: 978052188009-1.

Jethva, H., S.K. Satheesh, and J. Srinivasan. 2005. Seasonal variability of aerosols over the Indo Gangetic basin. *Journal of Geophysical Research: Atmospheres* 110 (D21).

Joshi, H., et al. 2016. Investigations of aerosol black carbon from a semi-urban site in the Indo-Gangetic Plain region. *Atmospheric Environment* 125: 346–359.

Kant, Y., V.K. Prasad, and K.V.S. Badarinath. 2000a. Algorithm for detection of active fire zones using NOAA AVHRR data. *Infrared Physics & Technology* 41 (1): 29–34.

Kant, Y., A.B. Ghosh, M.C. Sharma, P.K. Gupta, V.K. Prasad, K.V.S. Badarinath, and A.P. Mitra. 2000b. Studies on aerosol optical depth in biomass burning areas using satellite and ground-based observations. *Infrared Physics & Technology* 41 (1): 21–28.

Kant, Y., et al. 2020. Black carbon aerosol quantification over north-west Himalayas: Seasonal heterogeneity, source apportionment and radiative forcing. *Environmental Pollution* 257: 113446. https://doi.org/10.1016/j.envpol.2019.113446.

Kant, Y., et al. 2022. Long term influence of groundwater preservation policy on stubble burning and air pollution over North-West India. *Scientific Reports* 12: 2090. https://doi.org/10.1038/s41598-022-06043-8.

Kuo, C.L. 2017. Assessments of Ali, Dome A, and Summit Camp for mm-wave observations using MERRA-2 reanalysis. *The Astrophysical Journal* 848 (1): 64.

Levy, R.C., L.A. Remer, and O. Dubovik. 2007. Global aerosol optical properties and application to Moderate Resolution Imaging Spectroradiometer aerosol retrieval over land. *Journal of Geophysical Research: Atmospheres* 112 (D13). https://doi.org/10.1029/2006JD007811.

Levy, R.C., et al. 2010. Global evaluation of the Collection 5 MODIS dark-target aerosol products over land. *Atmospheric Chemistry and Physics* 10 (21): 10399–10420.

Levy, R., et al. 2013. The Collection 6 MODIS aerosol products over land and ocean. *Atmospheric Measurement Techniques* 6 (11): 2989–3034.

Mhawish, A., et al. 2017. Evaluation of MODIS Collection 6 aerosol retrieval algorithms over Indo-Gangetic Plain: Implications of aerosols types and mass loading. *Remote Sensing of Environment* 201: 297–313.

Nicholas, H. 2021. *Deforestation in Indonesia hits record low*. https://news.mongabay.com/2021/03/2021-deforestation-in-indonesia-hits-record-low-but-experts-fear-a-rebound/. Accessed 1 April 2022.

Olivier, J.G.J., et al. 2005. Recent trends in global greenhouse gas emissions: Regional trends 1970–2000 and spatial distribution of key sources in 2000. *Environmental Sciences* 2 (2–3): 81–99. https://doi.org/10.1080/15693430500400345.

Pfenninger, S., and I. Staffell. 2016. Long-term patterns of European PV output using 30 years of validated hourly reanalysis and satellite data. *Energy* 114: 1251–1265.

Prasad, V.K., Y. Kant, and K.V.S. Badarinath. 1999. Vegetation discrimination using IRS-P3 WiFS temporal dataset—A case study from Rampa Forests, Eastern Ghats, A.P. *Journal of the Indian Society of Remote Sensing* 27 (3): 149–154.

Prasad, V.K., P.K. Gupta, C. Sharma, A.K. Sarkar, Y. Kant, K.V.S. Badarinath, T. Rajagopal, and A.P. Mitra. 2000a. NOx emissions from biomass burning of shifting cultivation areas from tropical deciduous forests of India—Estimates from ground-based measurements. *Atmospheric Environment* 34 (20): 3271–3280.

Prasad, V.K., T. Rajagopal, Y. Kant, and K.V.S. Badarinath. 2000b. Quantification of carbon fluxes in tropical deciduous forests using satellite data. *Advances in Space Research* 26 (7): 1101–1104.

Prasad, V.K., Y. Kant, and K.V.S. Badarinath. 2000c. Quantifying short-term carbon dynamics from land use changes using satellite data—A case study from Rampa Forests (Eastern Ghats) India. *Geocarto International* 15 (2): 71–78.

Prasad, V.K., Y. Kant, P.K. Gupta, C. Sharma, A.A. Mitra, and K.V.S. Badarinath. 2001a. Biomass and combustion characteristics of secondary mixed deciduous forests in Eastern Ghats of India. *Atmospheric Environment* 35 (18): 3085–3095.

Prasad, V.K., Y. Kant, and K.V.S. Badarinath. 2001b. CENTURY ecosystem model application for quantifying vegetation dynamics in shifting cultivation areas: A case study from Rampa Forests, Eastern Ghats (India). *Ecological Research* 16 (3): 497–507.

Prasad, V.K., Y. Kant, and K.V.S. Badarinath. 2002a. Land use changes and modeling carbon fluxes from satellite data. *Advances in Space Research* 30 (11): 2511–2516.

Prasad, V.K., Y. Kant, and K.V.S. Badarinath. 2002b. Estimation of potential GHG emissions from net primary productivity of forests—A satellite based approach. *Advances in Space Research* 29 (11): 1793–1798.

Prasad, V.K., Y. Kant, P.K. Gupta, C. Elvidge, and K.V.S. Badarinath. 2002c. Biomass burning and related trace gas emissions from tropical dry deciduous forests of India: A study using DMSP-OLS data and ground-based measurements. *International Journal of Remote Sensing* 23 (14): 2837–2851.

Remer, L.A., et al. 2005. The MODIS aerosol algorithm, products, and validation. *Journal of the Atmospheric Sciences, American Meteorological Society* 62 (4): 947–973. https://doi.org/10.1175/JAS3385.1.

Remer, L.A., et al. 2008. Global aerosol climatology from the MODIS satellite sensors. *Journal of Geophysical Research: Atmospheres* 113 (D14). https://doi.org/10.1029/2007JD009661.

Roozitalab, B., G.R. Carmichael, and S.K. Guttikunda. 2021. Improving regional air quality predictions in the Indo-Gangetic Plain—Case study of an intensive pollution episode in November 2017. *Atmospheric Chemistry and Physics* 21 (4): 2837–2860.

Shaik, D.S., et al. 2017. Assessment of aerosol characteristics and radiative forcing over northwest Himalayan region. *IEEE Journal of Selected Topics in Applied Earth Observations and Remote Sensing* 10 (12): 5314–5321.

Shaik, D.S., et al. 2019. Impact of biomass burning on regional aerosol optical properties: A case study over northern India. *Journal of Environmental Management* 244: 328–343. https://doi.org/10.1016/j.jenvman.2019.04.025.

Sheldon, T., and C. Sankaran. 2017. The impact of Indonesian forest fire on Singaporean pollution and health. *American Health Review* 107 (5): 526–529.

State of Global Air. 2019. *A special report on global exposure to air pollution to air pollution and its disease burden*. Special Report. Boston, MA: Health Effects Institute. ISSN 2578-6873.

Streets, D.G., et al. 2003. Biomass burning in Asia: Annual and seasonal estimates and atmospheric emissions. *Global Biogeochemical Cycles* 17 (4). https://doi.org/10.1029/2003gb002040.

Taylor, D. 2010. Biomass burning, humans and climate change in Southeast Asia. *Biodiversity and Conservation* 19 (4): 1025–1042.

Titos, G., et al. 2017. Spatial and temporal variability of carbonaceous aerosols: Assessing the impact of biomass burning in the urban environment. *Science of the Total Environment* 578: 613–625.

Vadrevu, K.P., T. Ohara, and C. Justice (eds.). 2018. *Land-atmospheric research applications in South and Southeast Asia*.

Vadrevu, K.P., T. Ohara, and C. Justice (eds.). 2021a. *Biomass burning in South and Southeast Asia: Impacts on the biosphere*, vol. 2. CRC Press.

Vadrevu, K.P., T. Ohara, and C. Justice (eds.). 2021b. *Biomass burning in South and Southeast Asia: Mapping and monitoring*, vol. 1. CRC Press.

Venkataraman, C., et al. 2018. Source influence on emission pathways and ambient PM2.5 pollution over India (2015–2050). *Atmospheric Chemistry and Physics* 18: 8017–8039.

Wiedinmyer, C., et al. 2011. The Fire INventory from NCAR (FINN): A high resolution global model to estimate the emissions from open burning. *Geoscientific Model Development* 4 (3): 625–641. https://doi.org/10.5194/gmd-4-625-2011.

Yao, L., et al. 2016. Characteristics of carbonaceous aerosols: Impact of biomass burning and secondary formation in summertime in a rural area of the North China Plain. *Science of the Total Environment* 557: 520–530.

Yokelson, R.J., et al. 2013. Coupling field and laboratory measurements to estimate the emission factors of identified and unidentified trace gases for prescribed fires. *Atmospheric Chemistry and Physics* 13 (1): 89–116.

Detection and Modeling of South Asian Biomass Burning Aerosols from Both Macro- and Micro-perspectives

Shuaiyi Shi, Tianhai Cheng, Yu Wu, Xingfa Gu, Xiaoyang Li, Siheng Wang, and Yuyang Wang

Abstract In this research, we introduce our work, stretching from the micro- to macro-perspective, of detecting the aerosols emitted by biomass burning in South Asia and attempting to model it. Both satellite (macro-perspective) and laboratory/simulation (micro-perspective) data are used in this research. Satellite data show that the biomass burning aerosol originating from South Asia could transport to and influence the northern part of the Indian Ocean (including the Bay of Bengal and the Arabian Sea), the northern part of the Indo-China Peninsula, South China, and even far the Pacific Ocean (including part of the East China Sea and South China Sea). The chemical, physical, and optical characteristics of biomass burning aerosols over land and ocean show different features and evolution patterns. Such difference is caused by the different ambient environment and mixed aerosol during transport (urban/industrial aerosol over land and sea salt over the ocean). During the 48-h aging process, the volume fraction of black carbon and Angstrom Exponent decreased. Meanwhile, the aerosol sphere fraction and SSA increased. The biomass burning aerosol over land shows a more obvious evolution trend than that over the ocean. The biomass burning aerosol over the ocean generally has higher SSA and lower volume fraction of black carbon, aerosol sphere fraction, and Angstrom Exponent. Laboratory/simulation data show that the BC absorption enhancement (*Eabs*) produced by flaming combustion may be up to two times more than those produced by smoldering combustion, suggesting that different combustion states could dramatically influence the absorption of carbonaceous aerosols freshly emitted from burning biomass, leading to varied estimates of *Eabs* across a wide range of flaming-dominated and smoldering-dominated combustion states. Besides, the variation in *Eabs* caused by the combustion states was also investigated according to their different types and

S. Shi · T. Cheng (✉) · Y. Wu · X. Gu · X. Li
State Key Laboratory of Remote Sensing Science, Aerospace Information Research Institute, Chinese Academy of Sciences, Beijing 100094, China
e-mail: chength@aircas.ac.cn

X. Li
University of Chinese Academy of Sciences, Beijing 100049, China

S. Wang · Y. Wang
Institute of Remote Sensing Satellite, China Academy of Space Technology, Beijing 100094, China

K. P. Vadrevu et al. (eds.), *Vegetation Fires and Pollution in Asia*,
https://doi.org/10.1007/978-3-031-29916-2_28

the humidity levels of the biomass sources. The variation in *Eabs* of wet wheat straw and dry rapeseed plants is larger than that of dry wheat straw. Our study concluded that ascertaining the aging time and considering the humidity levels of the biomass sources, combustion states, and ambient environment would help reduce the uncertainty of the biomass burning aerosol radiative forcing assessments in South Asia.

Keywords Biomass burning aerosol (BBA) · Radiative forcing · South Asia · Aerosol aging · Black carbon (BC) · BC absorption enhancement (*Eabs*) · Burning type · Humidity · Combustion states · Ambient environment

1 Introduction

Biomass burning aerosols significantly affect the Earth's radiative budget (Bond et al. 2013; Stocker et al. 2013). The biomass burning aerosol in South Asia has a complicated composition with distinct chemical, physical, optical, and radiative forcing properties (Reid et al. 2005; Sheesley et al. 2003; Shaik et al. 2019; Nirmalkar et al. 2019; Sharma et al. 2017; Badarinath et al. 2007; 2008; 2009; Kharol et al. 2012). In addition, after emission, the aging process of biomass burning aerosols (Calvo et al. 2010; Capes et al. 2008; Nikonovas et al. 2015) and the mixing process with other types of aerosols during the transportation (Gawhane et al. 2019; Jain et al. 2018; Sudheer et al. 2014; Verma et al. 2013; Reddy and Venkataraman 2000) should be considered, which increases the complexity and uncertainty of aerosols (Myhre et al. 2013; Markowicz et al. 2017). The important sources of biomass burning are due to vegetation fires resulting from the slash and burn agriculture, agricultural crop residue burning, clearing of vegetation for promoting grass growth for cattle, lemon grass, oil palm and rubber cultivation for commercial purposes, and accidental fires including lightning (Krishna Prasad et al. 2000, 2001a, b, 2002a, b; Biswas et al. 2015a, b; Thapa et al. 2021; Eaturu and Vadrevu 2021; Lasko et al. 2021; Vadrevu et al. 2011; 2018; 2021a, b; 2022). The aerosols from these sources can get mixed with the dust resulting in complex aerosol behavior. Thus, an in-depth understanding of the dynamic characteristics of aerosols originating from South Asia biomass burning is necessary.

Satellite remote sensing can observe the Earth at multiple spatial and temporal resolutions (Kant et al. 2000; Kosmopoulos et al. 2008; Wooster et al. 2021). Moreover, satellite remote sensing makes it feasible to obtain much global and repeatable aerosol information even in remote ocean regions (Junghenn Noyes et al. 2020; Shaik et al. 2019; Xue et al. 2014). Furthermore, the satellite data with the pollution transportation models makes it suitable to track the smoke plume and detect their aging characteristics in a real ambient environment (Shi et al. 2020, 2021).

Laboratory observation and model simulations allow us to study biomass burning aerosols in a controlled way, making it suitable to dig into the influence of the types and the humidity levels of the biomass sources as well as the combustion states on the characteristic of biomass burning aerosol (Wu et al. 2020). Previous theoretical modeling and laboratory measurements of BC aerosols have shown a significant enhancement up to a factor of ~ 3.5 (Jacobson 2001). However, this enhancement is considered negligible (at ~ 1.06) by some field observations, thereby causing confusion regarding the parameterizations of BC absorption and thus leading to large uncertainties regarding the aerosol radiative forcing (Cappa et al. 2013).

In this research, we detect and model the aerosols emitted by biomass burning in South Asia from macro- and micro-perspectives. From a macro-perspective, satellite data are used to detect the influence region and its long time aging characteristics in a real ambient environment. From micro-perspective, laboratory observation and model simulation data are used to study the influence of the types and the humidity levels of the biomass sources as well as the combustion states on the characteristic of biomass burning aerosol.

The paper is organized as follows. In Sect. 2, satellite and laboratory data used, as well as the techniques and models, are introduced. The results are given in Sect. 3. Finally, a discussion and conclusions are provided in Sect. 4.

2 Data and Methods

2.1 Satellite Dataset

The POLDER/PARASOL product data from 2005 to 2013 are used in this study. The official POLDER/PARASOL product can provide a series of aerosol and surface parameters. The parameters obtained from the POLDER/GRASP high-precision archive dataset and used in this research include Single Scattering Albedo (SSA), Angstrom Exponent, Sphere Fraction, Complex Refractive Index, and Surface Directional Hemispherical Reflectance, etc. Besides, the volume fractions of black carbon in the aerosol mixture can be retrieved from the complex refractive index using the Maxwell-Garnett effective medium approximation (Bohren and Huffman 2008).

The infrared band of the satellite sensor can capture open biomass burning phenomena. The MODIS Thermal Anomalies/Fire Dataset (Justice et al. 2002; Vadrevu and Justice 2011; Oliveira 2015) adopted from the Fire Information for Resource Management System (FIRMS) is used in this research to provide the possible source information of biomass burning aerosols. The fire dataset can be combined with the crop type data to determine the biomass type or with the aerosol property dataset to extract the biomass burning aerosol characteristics. The low-confidence fire data were eliminated to reduce the misjudgment of biomass burning events. We used only the high-confidence and nominal-confidence fire data.

2.2 Biomass Burning Aerosols and HYSPLIT

The HYSPLIT backward trajectory model (HYbrid Single-Particle Lagrangian Integrated Trajectory), in conjunction with the MODIS Thermal Anomalies/Fire dataset, is used in this research to select the biomass burning aerosols and determine the aging time of biomass burning aerosol after emission. HYSPLIT was developed by the NOAA Air Resources Laboratory (ARL) and has been maintained by ARL for more than 30 years (Stein et al. 2015; Rolph et al. 2017; Draxler et al. 1999). In addition, the reanalysis data from the National Center for Atmospheric Research and National Center for Environmental Prediction (NCAR/NCEP) was used in this research to calculate the backward trajectories.

HYSPLIT backward trajectory and the MODIS thermal anomalies/fire dataset are used to identify the biomass burning aerosol and estimate the aging time of the aerosol. The geolocation, time, and aerosol height information from the potential POLDER/GRASP aerosol dataset are entered into the HYSPLIT model. Then, the backward trajectory can be calculated. We trace back the MODIS fire points within the distance of 50 km and the time difference of 15 min from the backward trajectory. If and only if the MODIS fire points in South Asia are traced back successfully, the potential POLDER/GRASP aerosol dataset is identified as the biomass burning aerosol, and the aging time is determined from the traced back fire points with max FRP.

2.3 Laboratory Data and Simulation

The biomass burning experiments were performed in a combustion chamber in a laboratory environment and were conducted using dry wheat straw, wet wheat straw, and dry rapeseed plants, which are all typical crops in South Asia. Eighteen samples were directly burned in the chamber (referred to as "dry"), and four samples (referred to as "wet") were placed in humid conditions (RH > 99%) for 30 min. A 50 cm long, 1/4-inch flexible conductive silicone tube was used for aerosol sampling, and a polytetrafluoroethylene (PTFE) tube was used for gas sampling. The residence time was short (~ 6 s) to minimize the aging of the aerosols in the tube (Pan et al. 2017). The mixing ratio of CO_2 and CO was measured using a Li-7000 CO_2 analyzer (Li-COR Inc.) and an ultrafast CO analyzer (model AL5002, Aero-Laser GmbH). Optical simulations were performed using the aggregate model parameterized by the complex particle morphology of BC at different aging scales.

3 Results

3.1 *Influence Region and Aging Time of Biomass Burning Aerosols Originated from South Asia*

First of all, from a macro-perspective, we statistically analyzed the aging time of biomass burning aerosols from the satellite data and generated the map of the average age distribution that originated from South Asia. As seen in Fig. 1, over land, apart from South Asia, the biomass burning aerosols from South Asia can be transported to and influence the northern part of the Indo-China Peninsula and South China. Over the ocean, the biomass burning aerosols that originated from South Asia can be transported to and influence the northern part of Indian Ocean (including Bay of Bengal and Arabian Sea), even reaching far to the Pacific Ocean (including part of East China Sea and South China Sea). However, due to the capacity of the aerosol satellite remote sensing inversion algorithm and our screening process used in this research to ensure the data accuracy and reliability, some regions that could be affected by the biomass burning aerosols originating from South Asia cannot be reflected in this map, such as the Tibetan Plateau (Li et al. 2017; Cao et al. 2010).

To show the influence region of South Asia biomass burning aerosols, the map shown in Fig. 1 does not exclude the potential distribution of biomass burning aerosols outside South Asia. Therefore, in the following study, the biomass burning aerosols associated with the MODIS fire points in South Asia are only analyzed for more accurate analysis.

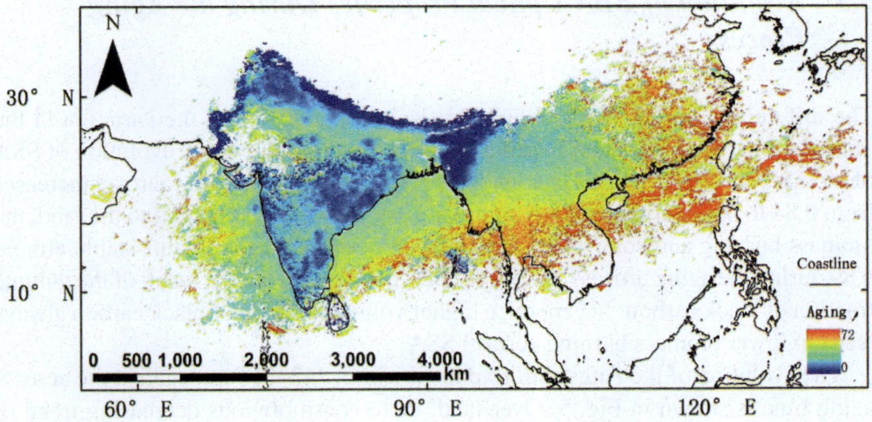

Fig. 1 Average age (in hours) of biomass burning aerosol originated from South Asia

3.2 Evolution of BBA Chemical and Physical Properties During the Aging Process

The black carbon in aerosol and the sphere fraction of aerosol are two important aerosol chemical and physical properties having a non-negligible influence on the aerosol optical properties and aerosol radiative forcing (Srivastava et al. 2017). We first investigated the evolution of the volume fraction of black carbon against the aerosol aging time, shown in Fig. 2. The analyses of biomass burning aerosol characteristics are conducted separately over land and ocean. Over land, the volume fraction of black carbon in biomass burning aerosol decreases from 4.2 to 3.5% during the 48-h aging process. Compared to the land region, the biomass burning aerosol over the ocean has a lower volume fraction of black carbon and remains steady during the aging process, around 3.1%. Sea salt mixing is considered as the possible reason causing the lower volume fraction of black carbon over the ocean.

The evolution of biomass burning aerosol sphere fraction is also investigated. Figure 3 shows that, over land, the aerosol sphere fraction increases from 35.5 to 42.5% during the 48-h aging process. Compared to the land, the biomass burning aerosols over the ocean has a lower aerosol sphere fraction, and a pronounced growth from 14.9 to 23.5% is also observed. Mixing sea salt decreases the aerosol sphere fraction of biomass burning aerosol over the ocean. The increase of aerosol sphere fraction during the aging process reflects the compaction process of aggregated small carbon spherules and the coating process by other aerosols.

3.3 Evolution of BBA Optical Properties During the Aging Process

The differences in the chemical and physical properties led to the variation in the optical properties of biomass burning aerosols. Figure 4 shows the evolution of SSA of biomass burning aerosol. Over land, the SSA of biomass burning aerosol increases from 0.84 to 0.87 during the 48-h aging process. However, compared to the land, the biomass burning aerosol over the ocean has higher SSA and remains stable around 0.89 during the aging process. Such a result is consistent with the result of the volume fraction of black carbon. In general, a higher volume fraction of black carbon always leads to lower biomass burning aerosol SSA.

The evolution of the Angstrom Exponent within 443–865 nm against the aerosol aging time is shown in Fig. 5. Over land, there is asn obvious decreasing trend of Angstrom Exponent from 1.59 to 1.49 during the 48-h aging process. Compared to the land, the biomass burning aerosol over the ocean has a lower Angstrom Exponent, and a slight decline from 1.50 to 1.47 has been observed. The decreasing trend of the Angstrom Exponent reflects the particle size growth of biomass burning aerosol during the 48-h aging process. The relatively low Angstrom Exponent over the ocean is caused by the mixing of sea salt, which consists of coarse particles.

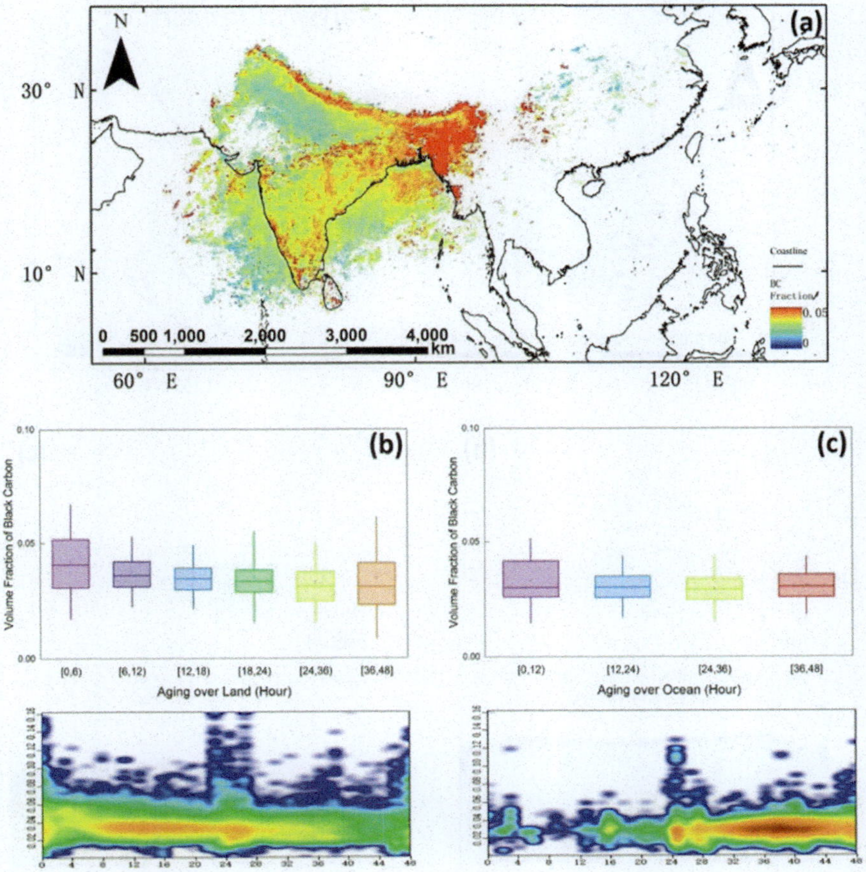

Fig. 2 Evolution of aerosol volume fraction of black carbon. **a** The geographical distribution. **b** The aging process over land. **c** The aging process over ocean

3.4 The Influence of Combustion States and Biomass Sources on Mass Absorption Cross Sections

After analyzing the influence region and its long time aging characteristics from macro-perspective, we then used the laboratory observation data and model simulation to study the influence of the types and the humidity levels of the biomass sources as well as the combustion states on the mass absorption cross sections (MAC) of biomass burning aerosol.

MAC of the BC-containing aerosols at 532 nm was estimated in Fig. 6 using the aggregate model. Simulated MAC values of freshly emitted BC particles range between 7 and 8.5 at 532 nm depending on the burned biomass's modified combustion efficiency (MCE). The MAC of the freshly emitted BC aerosols from the

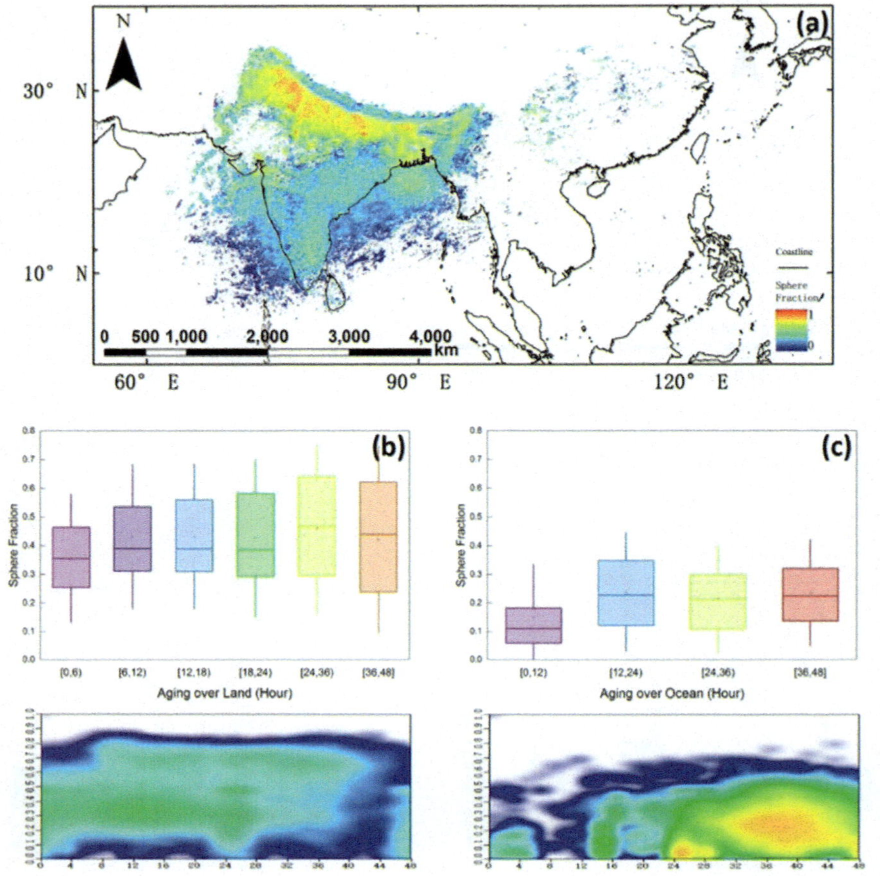

Fig. 3 Evolution of aerosol sphere fraction. **a** The geographical distribution. **b** The aging process over land. **c** The aging process over ocean

smoldering-dominated combustion states are ~ 1.09 times larger than those from flaming-dominated combustion at 532 nm, suggesting that smoldering combustion tends to produce more thickly coated BC particles than flaming combustion, thus intensifying the lens effect.

The effect of the biomass source on the MCE-dependent MAC of the BC-containing aerosols was further investigated in Fig. 7. The results show that, except for the biomass types, the humidity of biomass is also an important influence factor. The wet biomass usually has a higher MAC in smoldering-dominated states and lower MAC in flaming-dominated burning states than the dry biomass. A possible reason for this result is that the wet biomass was unfavorable for the production of BC particles in the flaming-dominated burning states and led to a smaller MAC, and the hygroscopic behavior may generate a slightly thicker non-BC coating in

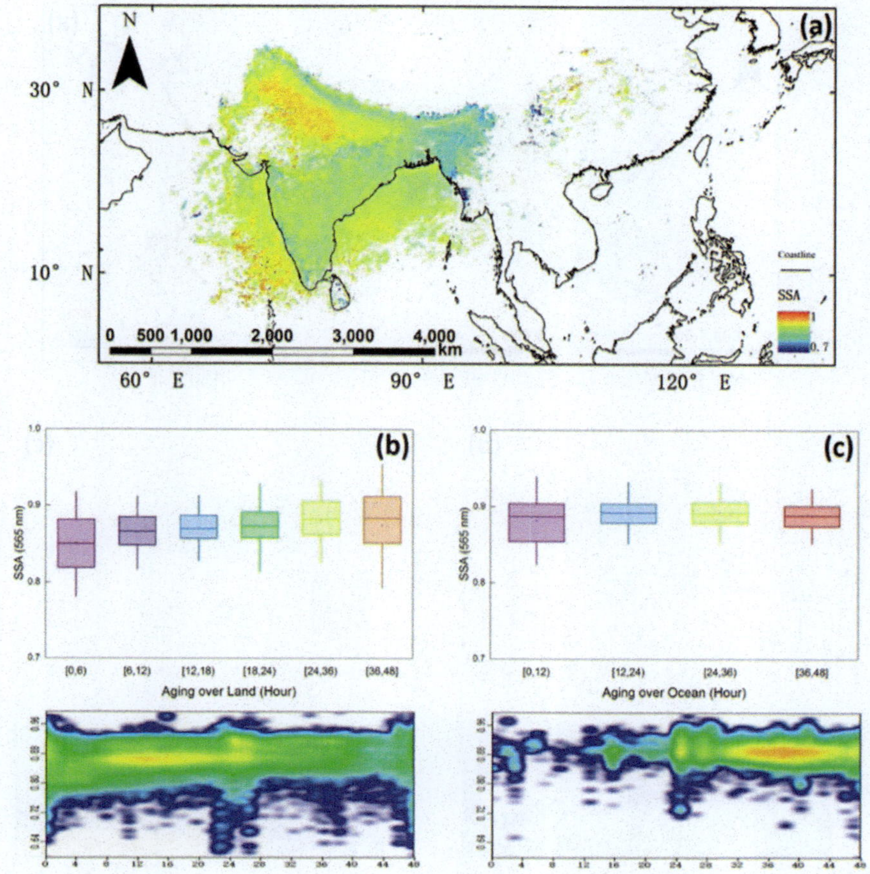

Fig. 4 Evolution of aerosol single scattering albedo at 565 nm. **a** The geographical distribution. **b** The aging process over land. **c** The aging process over ocean

the smoldering-dominated states, thus leading to a lager MAC. These discrepancies suggest that regional planting patterns and humidity conditions may influence the measured MAC of biomass burning.

3.5 *Simulated Absorption Enhancement of BC-Containing Aerosols*

Compared to the lifetime (one day to one week) of BC aerosols in the atmosphere, the atmospheric aging time of BC emissions from different biomass burning states measured in this study is sufficiently short. It can thus be considered to represent freshly emitted states for estimating *Eabs* in the ambient environment.

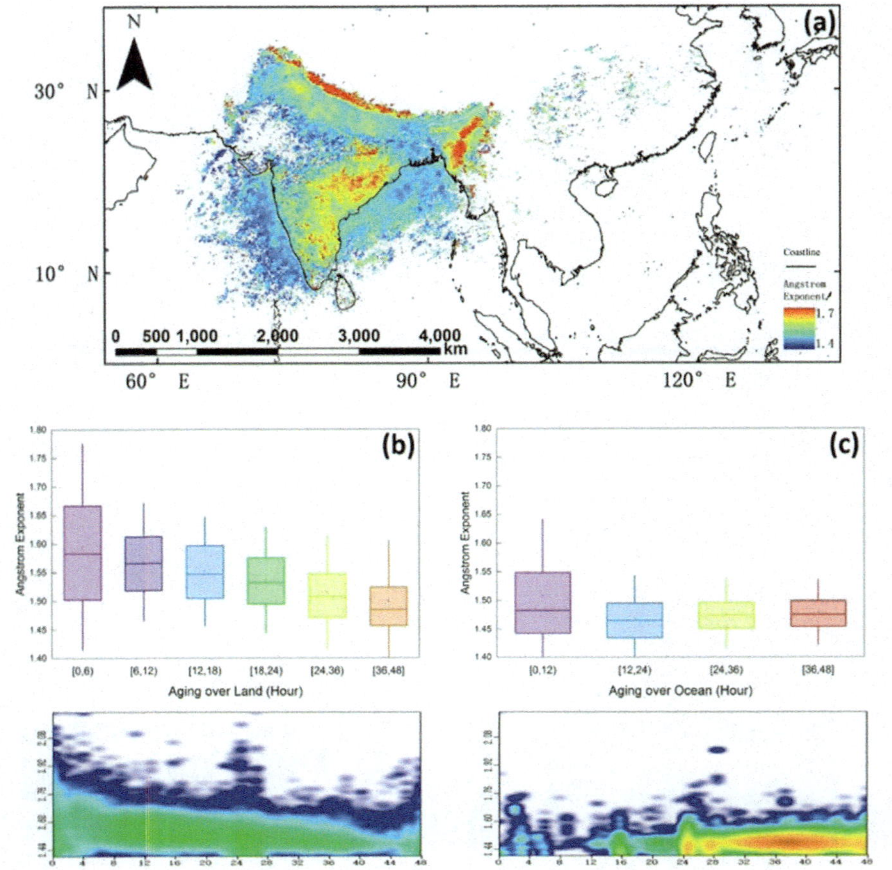

Fig. 5 Evolution of Angstrom Exponent within 443–865 nm. **a** The geographical distribution. **b** The aging process over land. **c** The aging process over ocean

Figure 8 shows that the *Eabs* between the fully aged and freshly emitted BC aerosols vary significantly with the different combustion states of biomass burning. The values of *Eabs* can reach to ~ 2.6 by assuming that the BC aerosols are freshly emitted from flaming-dominated combustion and decrease to ~ 1.8 for smoldering-dominated combustion.

The discrepancies between field observations and theoretical modeling of BC absorption enhancement can be reduced by a better understanding of carbonaceous aerosols' initial and final states. The small values of *Eabs* measured in field observations may be due to the uncertain reference to bare BC in the freshly emitted state, and it is, thus, important to provide a suitable synchronization of the field observations and theoretical modeling on the particle morphologies and to mix states of freshly emitted carbonaceous aerosols. To improve climate model predictions, it is imperative to understand better both the mass absorption cross section of the freshly

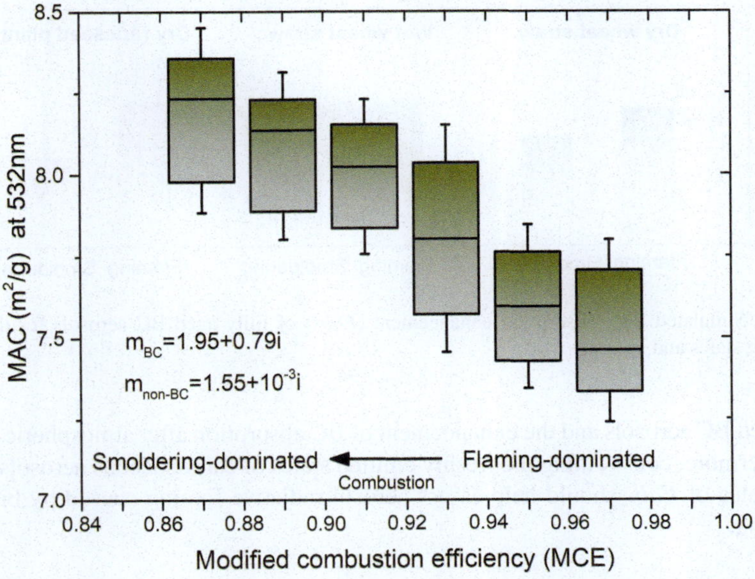

Fig. 6 The mass absorption cross sections (MAC) of BC-containing aerosols at 532 nm predicted by the aggregate model is dependent on the modified combustion efficiency (MCE) of biomass burning, which indicates different combustion states

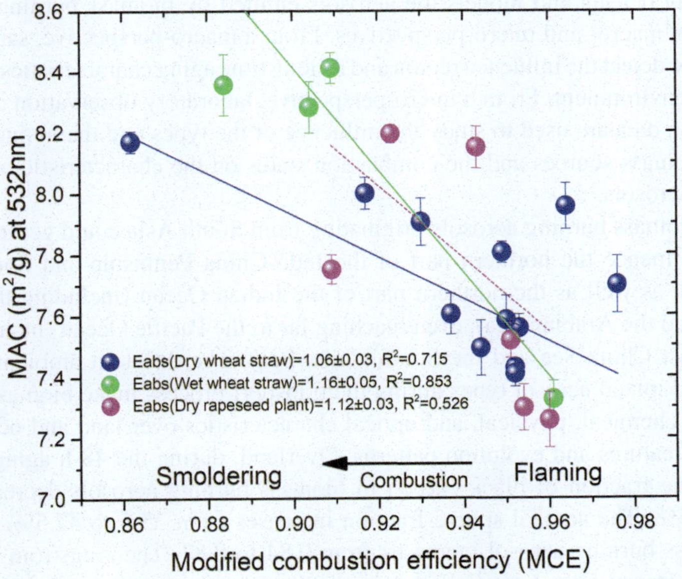

Fig. 7 Variations of the mass absorption cross sections (MAC) with modified combustion efficiency (MCE) at 532 nm for different types and humidity levels of biomass sources

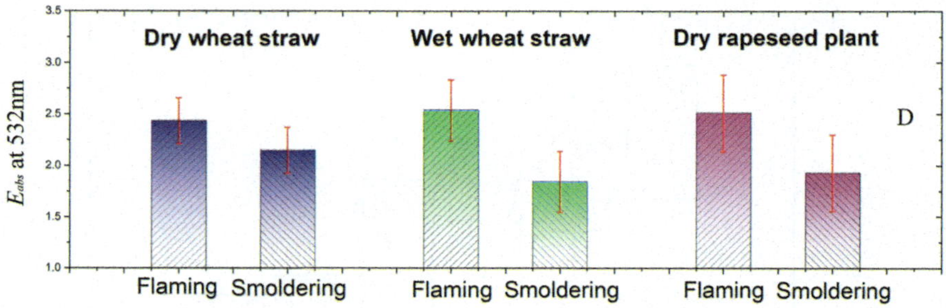

Fig. 8 Simulated light absorption enhancement (*Eabs*) of fully aged BC aerosols for different burning states and sources

emitted BC aerosols and the enhancement of BC absorption after atmospheric aging. Furthermore, constraining the freshly emitted states of carbonaceous aerosols in the estimates of *Eabs* should help assess aerosol radiative forcing caused by biomass burning.

4 Discussion and Conclusions

This study detects and models the aerosols emitted by biomass burning in South Asia from macro- and micro-perspectives. From a macro-perspective, satellite data are used to detect the influence region and its long time aging characteristics in the real ambient environment. From a micro-perspective, laboratory observation and model simulation data are used to study the influence of the types and the humidity levels of the biomass sources and the combustion states on the characteristic of biomass burning aerosols.

The biomass burning aerosols originating from South Asia could get transported to and influence the northern part of the Indo-China Peninsula and South China over land, as well as the northern part of the Indian Ocean (including the Bay of Bengal and the Arabian Sea), even reaching far to the Pacific Ocean (including part of the East China Sea and the South China Sea). The different ambient environments and mixed aerosol types during the transport process make biomass burning aerosols' chemical, physical, and optical characteristics over land and ocean show different features and evolution patterns. Overland, during the 48-h aging process, the volume fraction of black carbon in biomass burning aerosols decreases from 4.2 to 3.5%. The aerosol sphere fraction increases from 35.5 to 42.5%. The SSA of biomass burning aerosol increases from 0.84 to 0.87. The Angstrom Exponent decreases from 1.59 to 1.49. Compared to the land region, over the ocean, the volume fraction of black carbon in biomass burning aerosols is lower and remains steady during the aging process, around 3.1%. The aerosol sphere fraction is lower, and a pronounced growth from 14.9 to 23.5% is observed. The SSA is higher and remains stable around 0.89. The Angstrom Exponent is lower, and a slight decline from 1.50

to 1.47 has been observed. Besides, different combustion states and biomass sources could dramatically influence the absorption of carbonaceous aerosols emitted from burning biomass, leading to varied estimates of *Eabs* across a wide range of combustion states, biomass types, and humidity conditions. For example, the BC absorption enhancement (*Eabs*) produced by flaming combustion may be up to two times more than those produced by smoldering combustion. Besides, the variation in *Eabs* of wet wheat straw and dry rapeseed plants is larger than that of dry wheat straw.

We conclude that to reduce the uncertainty of the biomass burning aerosol radiative forcing in South Asia, it is imperative to better ascertain the aging time and the ambient environment, including the types and the humidity levels of the biomass sources and their combustion states in the assessments.

Acknowledgements This work was supported by the Natural Science Foundation of China (Grant Number: 42005104). The authors would like to acknowledge the use of POLDER data "POLDER/GRASP Level-3 data" provided initially by CNES (http://www.icare.univ-lille1.fr/) processed at AERIS/ICARE Data and Services Center with GRASP software (https://www.grasp-open.com) developed by Dubovik et al. The MODIS fire product was acquired online by courtesy of the Fire Information for Resource Management System (FIRMS) (https://firms.modaps.eosdis.nasa.gov/). The NOAA Air Resources Laboratory (ARL) is gratefully appreciated for the provision of the HYSPLIT backward trajectory (https://www.ready.noaa.gov/HYSPLIT.php). The authors appreciate Dr. Xiaole Pan and his team for the laboratory data of biomass burning in Pan et al. We thank Dr. Daniel Mackowski and Dr. Michael Mishchenko for the superposition T-Matrix code (MSTM): (http://www.eng.auburn.edu/users/dmckwski/scatcodes/). The software of the aggregate model for BC-containing particles is available on request.

References

Badarinath, K.V.S., S.K. Kharol, K.M. Latha, T.K. Chand, V.K. Prasad, A.N. Jyothsna, and K. Samatha. 2007. Multiyear ground-based and satellite observations of aerosol properties over a tropical urban area in India. *Atmospheric Science Letters* 8 (1): 7–13.

Badarinath, K.V.S., S.K. Kharol, A.R. Sharma, and V.K. Prasad. 2009. Analysis of aerosol and carbon monoxide characteristics over Arabian Sea during crop residue burning period in the Indo-Gangetic Plains using multi-satellite remote sensing datasets. *Journal of Atmospheric and Solar-Terrestrial Physics* 71 (12): 1267–1276.

Badarinath, K.V.S., S. Kumar Kharol, V. Krishna Prasad, A. Rani Sharma, E.U.B. Reddi, H.D. Kambezidis, and D.G. Kaskaoutis. 2008. Influence of natural and anthropogenic activities on UV Index variations–a study over tropical urban region using ground based observations and satellite data. *Journal of Atmospheric Chemistry* 59 (3): 219–236.

Biswas, S., K.D. Lasko, and K.P. Vadrevu. 2015a. Fire disturbance in tropical forests of Myanmar—Analysis using MODIS satellite datasets. *IEEE Journal of Selected Topics in Applied Earth Observations and Remote Sensing* 8 (5): 2273–2281.

Biswas, S., K.P. Vadrevu, Z.M. Lwin, K. Lasko, and C.O. Justice. 2015b. Factors controlling vegetation fires in protected and non-protected areas of Myanmar. *PLoS ONE* 10 (4): e0124346.

Bohren, C.F., and D.R. Huffman. 2008. *Absorption and scattering of light by small particles*. John Wiley & Sons.

Bond, T.C., S. Doherty, D.W. Fahey, P. Forster, T. Berntsen, B. Deangelo, M. Flanner, S. Ghan, B. Kärcher, D. Koch, S. Kinne, Y. Kondo, P.K. Quinn, M. Sarofim, M.G. Schultz, S. Michael, C.

Venkataraman, H. Zhang, S. Zhang and C.S. Zender. 2013. Bounding the role of black carbon in the climate system: A Scientific assessment.

Calvo, A., V. Pont, A. Castro, M. Mallet, C. Palencia, J.-C. Roger, P. Dubuisson, and R. Fraile. 2010. Radiative forcing of haze during a forest fire in Spain.

Cao, J.J., X.X. Tie, B.Q. Xu, Z.Z. Zhao, C.S. Zhu, G.H. Li, and S.X. Liu. 2010. Measuring and modeling black carbon (BC) contamination in the SE Tibetan Plateau. *Journal of Atmospheric Chemistry* 67: 45–60.

Capes, G., J. Bt, G. Mcfiggans, P. Williams, J. Haywood, and H. Coe. 2008. Aging of biomass burning aerosols over West Africa: Aircraft measurements of chemical composition, microphysical properties, and emission ratios.

Cappa, C.D., T.B. Onasch, P. Massoli, D.R. Worsnop, T.S. Bates, E.S. Cross, P. Davidovits, J. Hakala, K.L. Hayden, B.T. Jobson, K.R. Kolesar, D.A. Lack, B.M. Lerner, S.-M. Li, D. Mellon, I. Nuaaman, J.S. Olfert, T. Petäjä, P.K. Quinn, C. Song, R. Subramanian, E.J. Williams, and R.A. Zaveri. 2013. Response to comment on "radiative absorption enhancements due to the mixing state of atmospheric black carbon." *Science* 339: 393.

Draxler, R.R., B. Stunder, G. Rolph, A. Taylor. 1999. HYSPLIT 4 Users' Guide, US Department of Commerce, National Oceanic and Atmospheric Administration, Environmental Research Laboratories, Air Resources Laboratory.

Eaturu, A., & Vadrevu, K. (2021). Evaluation of Sentinel-3 SLSTR Data for Mapping Fires in Forests, Peatlands, and Croplands – A Case Study over Australia, Indonesia, and India. https://doi.org/10.1201/9780429022258-3.

Gawhane, R.D., P.S.P. Rao, K. Budhavant, D.C. Meshram, and P.D. Safai. 2019. Anthropogenic fine aerosols dominate over the Pune region, Southwest India. *Meteorology and Atmospheric Physics* 131: 1497–1508.

Jacobson, M.Z. 2001. Strong radiative heating due to the mixing state of black carbon in atmospheric aerosols. *Nature* 409: 695–697.

Jain, S., S.K. Sharma, T.K. Mandal, and M. Saxena. 2018. Source apportionment of PM10 in Delhi, India using PCA/APCS, UNMIX and PMF. *Particuology* 37: 107–118.

Junghenn Noyes, K., R. Kahn, A. Sedlacek, L. Kleinman, J. Limbacher, and Z. Li. 2020. Wildfire Smoke Particle Properties and Evolution, from Space-Based Multi-Angle Imaging. *Remote Sensing* 12: 769.

Justice, C., L. Giglio, S. Korontzi, J. Owens, J. Morisette, D. Roy, J. Descloitres, S. Alleaume, F. Petitcolin, and Y. Kaufman. 2002. The MODIS fire products. *Remote Sensing of Environment* 83: 244–262.

Kant, Y., V.K. Prasad, and K.V.S. Badarinath. 2000. Algorithm for detection of active fire zones using NOAA AVHRR data. *Infrared Physics & Technology* 41 (1): 29–34.

Kharol, S.K., K.V.S. Badarinath, A.R. Sharma, D.V. Mahalakshmi, D. Singh, and V.K. Prasad. 2012. Black carbon aerosol variations over Patiala city, Punjab, India—a study during agriculture crop residue burning period using ground measurements and satellite data. *Journal of Atmospheric and Solar-Terrestrial Physics* 84: 45–51.

Kosmopoulos, P.G., D.G. Kaskaoutis, P.T. Nastos, and H.D. Kambezidis. 2008. Seasonal variation of columnar aerosol optical properties over Athens, Greece, based on MODIS data. *Remote Sensing of Environment* 112: 2354–2366.

Krishna Prasad, V. et al. (2000). Quantifying Short-Term Carbon Dynamics from Land Use Changes Using Satellite Data-A Case Study from Rampa Forests (Eastern Ghats) India. *Geocarto International.*

Lasko, K., K.P. Vadrevu, V. Bandaru, T.N.T. Nguyen, and H.Q. Bui. 2021. PM2. 5 emissions from biomass burning in South/Southeast Asia–Uncertainties and trade-offs. In *Biomass burning in south and Southeast Asia*, 149–169. CRC Press.

Li, H.Y., Q.S. He, Q. Song, L.G. Chen, Y.J. Song, Y.H. Wang, K. Lin, Z.C. Xu, and M. Shao. 2017. Diagnosing Tibetan pollutant sources via volatile organic compound observations. *Atmospheric Environment* 166: 244–254.

Markowicz, K.M., J. Lisok, and P. Xian. 2017. Simulations of the effect of intensive biomass burning in July 2015 on Arctic radiative budget. *Atmospheric Environment* 171: 248–260.

Myhre, G., B.H. Samset, M. Schulz, Y. Balkanski, S. Bauer, T.K. Berntsen, H. Bian, N. Bellouin, M. Chin, T. Diehl, R.C. Easter, J. Feichter, S.J. Ghan, D. Hauglustaine, T. Iversen, S. Kinne, A. Kirkevåg, J.F. Lamarque, G. Lin, X. Liu, M.T. Lund, G. Luo, X. Ma, T. Van Noije, J.E. Penner, P.J. Rasch, A. Ruiz, Ø. Seland, R.B. Skeie, P. Stier, T. Takemura, K. Tsigaridis, P. Wang, Z. Wang, L. Xu, H. Yu, F. Yu, J.H. Yoon, K. Zhang, H. Zhang, and C. Zhou. 2013. Radiative forcing of the direct aerosol effect from AeroCom Phase II simulations. *Atmospheric Chemistry and Physics* 13: 1853–1877.

Nikonovas, T., P.R.J. North, and S.H. Doerr. 2015. Smoke aerosol properties and ageing effects for northern temperate and boreal regions derived from AERONET source and age attribution. *Atmospheric Chemistry and Physics* 15: 6445–6479.

Nirmalkar, J., D.K. Deshmukh, M.K. Deb, Y.I. Tsai, and S. Pervez. 2019. Characteristics of aerosol during major biomass burning events over eastern central India in winter: A tracer-based approach. *Atmospheric Pollution Research* 10: 817–826.

Oliveira, S.L.J. 2015. Seasonal differences in fire activity and intensity in tropical savannas of northern Australia using satellite measurements of fire radiative power. *The International Journal of Wildland Fire, 24*(2): 249–260

Pan, X., Y. Kanaya, F. Taketani, T. Miyakawa, S. Inomata, Y. Komazaki, H. Tanimoto, Z. Wang, I. Uno, and Z. Wang. 2017. Emission characteristics of refractory black carbon aerosols from fresh biomass burning. *A Perspective from Laboratory Experiments* 17: 13001–13016.

Prasad, V.K., Y. Kant, and K.V.S. Badarinath. 2001a. CENTURY ecosystem model application for quantifying vegetation dynamics in shifting cultivation areas: a case study from Rampa Forests, Eastern Ghats (India). *Ecological Research* 16 (3): 497–507.

Prasad, V.K., Y. Kant, and K.V.S. Badarinath. 2002a. Land use changes and modeling carbon fluxes from satellite data. *Advances in Space Research* 30 (11): 2511–2516.

Prasad, V.K., Y. Kant, P.K. Gupta, C. Elvidge, and K.V.S. Badarinath. 2002b. Biomass burning and related trace gas emissions from tropical dry deciduous forests of India: a study using DMSP-OLS data and ground-based measurements. *International Journal of Remote Sensing* 23 (14): 2837–2851.

Prasad, V.K., Y. Kant, P.K. Gupta, C. Sharma, A.A. Mitra, and K.V.S. Badarinath. 2001b. Biomass and combustion characteristics of secondary mixed deciduous forests in Eastern Ghats of India. *Atmospheric Environment* 35 (18): 3085–3095.

Reddy, M.S., and C. Venkataraman. 2000. Atmospheric optical and radiative effects of anthropogenic aerosol constituents from India. *Atmospheric Environment* 34: 4511–4523.

Reid, J.S., T.F. Eck, S.A. Christopher, R. Koppmann, O. Dubovik, D. Eleuterio, B.N. Holben, E.A. Reid, and J. Zhang. 2005. A review of biomass burning emissions part III: Intensive optical properties of biomass burning particles. *Atmospheric Chemistry and Physics* 5: 827–849.

Rolph, G., A. Stein, and B. Stunder. 2017. Real-time Environmental Applications and Display sYstem: READY. *Environmental Modelling & Software* 95: 210–228.

Shaik, D.S., Y. Kant, D. Mitra, A. Singh, H.C. Chandola, M. Sateesh, S.S. Babu, and P. Chauhan. 2019. Impact of biomass burning on regional aerosol optical properties: a case study over northern India. *Journal of Environmental Management* 244: 328–343.

Sharma, D., A.K. Srivastava, K. Ram, A. Singh, and D. Singh. 2017. Temporal variability in aerosol characteristics and its radiative properties over Patiala, northwestern part of India: Impact of agricultural biomass burning emissions. *Environmental Pollution* 231: 1030–1041.

Sheesley, R.J., J.J. Schauer, Z. Chowdhury, G.R. Cass, and B.R.T. Simoneit. 2003. Characterization of organic aerosols emitted from the combustion of biomass indigenous to South Asia. *Journal of Geophysical Research: Atmospheres, 108*.

Shi, S., T. Cheng, X. Gu, H. Guo, Y. Wu, Y. Wang, F. Bao, and X. Zuo. 2020. Probing the dynamic characteristics of aerosol originated from South Asia biomass burning using POLDER/GRASP satellite data with relevant accessory technique design. *Environment International* 145: 106097.

Shi, S., Y. Ma, F. Bao, and F. Mumtaz. 2021. A satellite data based detailed study of the aerosol emitted from open biomass burning in Northeast China. *Atmosphere* 12: 1700.

Srivastava, P., S. Dey, A.K. Srivastava, S. Singh, S.K. Mishra, and S. Tiwari. 2017. Importance of aerosol non-sphericity in estimating aerosol radiative forcing in Indo-Gangetic Basin. *Science of the Total Environment* 599: 655–662.

Stein, A.F., R.R. Draxler, G.D. Rolph, B.J.B. Stunder, M.D. Cohen, and F. Ngan. 2015. NOAA's HYSPLIT atmospheric transport and dispersion modeling system. *Bulletin of the American Meteorological Society* 96: 2059–2077.

Stocker, T.F., D. Qin, G.-K. Plattner, M. Tignor, S.K. Allen, J. Boschung, A. Nauels, Y. Xia, V. Bex, and P.M. Midgley. 2013. Climate change 2013. The physical science basis. working group I contribution to the Fifth assessment report of the Intergovernmental Panel on Climate Change. *Intergovernmental Panel on Climate Change-IPCC.*

Sudheer, A.K., R. Rengarajan, D. Deka, R. Bhushan, S.K. Singh, and M.Y. Aslam. 2014. Diurnal and seasonal characteristics of aerosol ionic constituents over an urban location in Western India: secondary aerosol formation and meteorological influence. *Aerosol and Air Quality Research* 14: 1701-U527.

Thapa, P., K.P. Vadrevu, A. Eaturu, and S. Biswas. 2021. Vegetation fire status and management in Bhutan. In *Biomass burning in South and Southeast Asia*, 135–157. CRC Press.

Vadrevu, K.P., E. Ellicott, K.V.S. Badarinath, and E. Vermote. 2011. MODIS derived fire characteristics and aerosol optical depth variations during the agricultural residue burning season, north India. *Environmental Pollution* 159 (6): 1560–1569.

Vadrevu, K.P., and C.O. Justice. 2011. Vegetation fires in the Asian region: Satellite observational needs and priorities. *Global Environmental Research* 15 (1): 65–76.

Vadrevu, K.P., A. Eaturu, E. Casadaban, and S. Biswas. 2022. Agricultural Fires in South Asian Countries and Implications. In *Remote sensing of agriculture and Land Cover/Land Use Changes in South and Southeast Asian Countries*, 501–516. Cham: Springer

Vadrevu, K.P., T. Ohara, and C. Justice, eds. 2018. Land-atmospheric research applications in South and Southeast Asia.

Vadrevu, K.P., T. Ohara, and C. Justice, eds. 2021a. *Biomass burning in south and Southeast Asia: Impacts on the biosphere*, vol. 2. CRC Press.

Vadrevu, K.P., T. Ohara, and C. Justice, eds. 2021b. *Biomass burning in South and Southeast Asia: Mapping and monitoring*, vol. 1. CRC Press.

Verma, S., S.K. Pani, and S.N. Bhanja. 2013. Sources and radiative effects of wintertime black carbon aerosols in an urban atmosphere in east India. *Chemosphere* 90: 260–269.

Wooster, M.J., Roberts, G.J., Giglio, L. et al. (2021). Satellite remote sensing of active fires: History and current status, applications and future requirements. *Remote Sensing of Environment: An Interdisciplinary Journal, 267*.

Wu, Y., T. Cheng, X. Pan, L. Zheng, S. Shi, and H. Liu. 2020. The role of biomass burning states in light absorption enhancement of carbonaceous aerosols. *Scientific Reports* 10: 12829.

Xue, Y., H. Xu, J. Guang, L. Mei, J. Guo, C. Li, R. Mikusauskas, and X. He. 2014. Observation of an agricultural biomass burning in central and east China using merged aerosol optical depth data from multiple satellite missions. *International Journal of Remote Sensing* 35: 5971–5983.

Remote Sensing of Agricultural Biomass Burning Aerosols, Gaseous Compounds, Long-Distance Transport, and Impact on Air Quality

Yonghua Wu, Yong Han, and Fred Moshary

Abstract Agricultural crop open fires (ACOF) or biomass burnings emit large amounts of gaseous and particulate pollutants, resulting in air pollution episodes. This study presents satellite remote sensing of agricultural fires, gaseous and particulate emissions, smoke vertical distribution, and optical properties using a variety of NASA satellite data (MODIS, AIRS, OMI, CALIPSO, CATS), and potential impacts on the air quality in April and June of 2015 in the Northeast and Eastern China. Aerosol, CO, HCHO, and NO_2 concentrations showed coincident hot spots in the fire-affected regions. Range-resolved observations by NASA spaceborne lidar CALIPSO/CALIOP and CATS indicated that the smoke plumes reached up to 5 km altitude, and aloft plumes contribute 60–90% to the total aerosol optical depth (AOD). The smoke mixed with dust particles is identified in spring. Ground PM_{10}, $PM_{2.5}$, and CO in the rural area became higher than those in the surrounding urban areas because of the smoke emissions nearby the fire areas. High-level ozone (O_3) is consistent with high HCHO and CO for the summer ACOF event. The NAAPS model indicated high emissions of smoke and dust particles associated with the ACOF and dust events; nonetheless, the emission constraints with satellite and ground observations are critical to improve model accuracy. In addition, the trans-Pacific transport of the plumes to the eastern US was demonstrated by the consistent high level of CO, AOD, and aerosol vertical distribution along the transport path. The aloft aerosol plumes were also observed from the CCNY-lidar and CALIPSO in the Northeast US. At the same time, the HYSPLIT backward trajectories analysis indicates that they originated from Northern China.

Keywords Agricultural biomass burning · Air quality · Remote sensing · Satellite · Lidar

Y. Wu (✉) · F. Moshary
Optical Remote Sensing Lab, The City College of New York, NY 10031, USA
e-mail: yhwu@ccny.cuny.edu

Y. Han
School of Atmospheric Sciences, Sun Yat-Sen University, Zhuhai 519082, China

K. P. Vadrevu et al. (eds.), *Vegetation Fires and Pollution in Asia*,
https://doi.org/10.1007/978-3-031-29916-2_29

1 Introduction

Satellite remote sensing techniques have been extensively applied to observe wildfire sources, burned areas, fire radiative power, emissions estimate, and smoke transport in several South/Southeast Asian countries (Prasad et al. 2003; Badarinath et al. 2007; 2008; Kharol et al. 2012; Justice et al. 2015; Biswas et al. 2015a,b; Albar et al. 2018; Thapa et al. 2021; Arvelyna et al. 2021; Lasko et al. 2018a,b; Wooster et al. 2021; Wu et al. 2015; 2017; 2018; 2020). Due to the mandatory bans or management of agricultural crop open fires (ACOF) during the post-harvest seasons (early summer and autumn) by the Chinese government, the total amount of intensive fires, frequencies, and burned areas was generally reduced (Yin et al. 2021). However, recent studies show that ACOF for farmland practices or cultivation indicates an increasing trend in spring (March–April) in Northeast (NE) China for recent years (Wang et al. 2020; Wu et al. 2020). For instance, Wang et al. (2020) analyzed the satellite-measured fire counts in NE China during 2003–2017 and indicated a dramatic increment in fire counts in March–April since 2013 over the agricultural cropland area. Wu et al. (2020) analyzed spatial and monthly variations of biomass burning during 2003–2014 in China and found small variations of the straw open fires in the North China Plain (NCP) but an increasing trend in NE China. Yin et al. (2021) analyzed the spatiotemporal variation and distribution characteristics of crop residue burning (CRB) in China from 2001 to 2018 using the remote sensing fire product and land-use product; the fire spots, fire radiative power (FRP), and average CRB indicate that spring, summer, and autumn CRB in the national wide had dropped dramatically over previous levels by 2018 due to strict regulations or bans by government, but regarding spatial distribution in the study period, spring CRB spots presented a significant increase in NE China. Thus, managing agricultural biomass burning remains a challenge in China, and strong policy measures on the alternate use of crop residues are needed.

Biomass burnings (agricultural crop fires, forest fires, grassland fires, wood and straw combustion as fuel, etc.) can emit abundant fine particles ($PM_{2.5}$ with particulate diameter of 2.5 μm or smaller), organic carbon (OC), black carbon (BC) and gaseous compounds such as carbon monoxide (CO), nitrogen oxide (NOx), volatile organic compounds (VOCs), methane (CH_4), ammonia (NH_3), carbon dioxide (CO_2) (Andreae et al. 1988, Crutzen et al. 1979; Prasad et al. 2002a,b; Badarinath et al. 2009; Lasko et al. 2017; 2021; Vadrevu et al. 2021a,b,c). Moreover, via chemical processes, VOCs can affect the formation of secondary organic aerosols (SOA) and ozone (O_3). The downwind transport of wildfire smoke can affect air quality, radiation, and human health on a regional and continental scale. The simulations by Li et al. (2018) indicate that the precursor emissions from agricultural fires play a major role in modifying O_3 photochemistry, with a maximal increase in O_3 by 20 ppb near the fire area in China. Many heavy air pollution episodes are associated with ACOP in China (Chen et al. 2017; Wu et al. 2017; Li et al. 2018; Yang et al. 2020). For instance, Wu et al. (2017) show that during ACOP, the surface PM_{10} and $PM_{2.5}$ increased up to 800 and 485 $\mu g/m^3$ in Southeast China. Wang et al. (2020)

indicate that during 2013–2017, ACOP emitted the 158.46 Gg of $PM_{2.5}$, 49.50 Gg of NO_2, and 27.0 Gg of NH_3 in NE China. More recently, on April 12–19, 2020, NE China suffered from a severe air pollution event with hourly $PM_{2.5}$ greater than 2000 μg/m_3 at some sites due to ACOP (https://www.163.com/dy/article/FAAAPJDI0514TVLI.html). Thus, ACOP emissions are one of the significant sources of air pollution episodes in China.

It is important to improve the ACOP emission inventory using satellite products (Streets et al., 2013; Penn and Holloway 2020). Top-down emission estimates (TDEs) will benefit from satellites' more frequent and high spatial resolution observations (Kim et al., 2020). With TDE estimates, an evaluation of air quality using inverse modeling can be performed to improve existing emission information on time with better accuracy. Uncertainties of current ACOP emissions in chemical transport models (CTMs) lead to inaccuracies in evaluating their impacts on haze and O_3 formations (Yang et al. 2020; Mehmood et al. 2020). Satellite retrievals provide an alternative that can be used to simultaneously quantify the emissions of AOFM and other types of open biomass burning (OBB), such as the Fire INventory from NCAR version 1.5 (FINNv1.5) (Wiedinmyer et al. 2011). Long-term satellite measurements can contribute to emission rates by detecting missing emission sources and hot spots. However, the TDE uncertainties remain relatively high because they depend on fuel sources/materials, combustion efficiency, species-specific emissions factors, meteorology, and complex chemical-physical processes (Chen et al. 2017; Yang et al. 2020; Mehmood et al. 2020; Zhou et al. 2017). There are some field campaigns to investigate wildfire emissions, chemistry, and transport in North America (https://csl.noaa.gov/projects/firex-aq/, www.eol.ucar.edu/field_projects/we-can).

This study presents two events of agricultural open fires in the spring and summer of 2015 in China, during which severe air pollution occurred. A suite of satellite and ground-based observations are analyzed to characterize the spatial distribution of smoke aerosols and gaseous compounds (CO, formaldehyde-HCHO, NO_2, and O_3), smoke particle vertical height and optical properties, and the trans-Pacific transport to US. Some model studies on the smoke transport paths and potential effects on the air quality are discussed.

2 Study Region

This study focuses on the ACOF emissions in NE and Eastern China, which is the central agricultural region with large energy consumption and high population density. The focus is on three provinces (Liaoning, Heilongjiang, and Jilin) in NE China which are important farming regions and industrial areas. The Eastern China, which includes the mid and lower reaches of the Yangtze River and Yellow River, is also a major agricultural region and accounts for one-third of China's cultivated land and almost half of the country's agricultural yields.

3 Methodology and Dataset

The Fire Information for Resource Management System (FIRMS) delivers global hot spots/fire locations measured from NASA Moderate Resolution Imaging Spectroradiometer (MODIS) onboard the Terra and Aqua satellites and the Visible Infrared Imaging Radiometer Suite (VIIRS) aboard S-NPP and NOAA 20 (formally known as JPSS-1) (https://firms.modaps.eosdis.nasa.gov/). Active fire locations correspond to the center of a 1 km pixel flagged by the MOD14/MYD14 Fire and Thermal Anomalies algorithm as containing one or more hot spots/fires within the pixel (https://firms.modaps.eosdis.nasa.gov/). The MODIS active fire products include the Fire Radiative Power product (FRP), a measure of fire intensity. Integrating the FRP over the lifetime of the fire provides an estimate of the total Fire Radiative Energy (FRE), which, for wildfires, should be proportional to the total mass of fuel biomass combusted, thus potentially permitting improved estimates of pyrogenic gaseous and aerosol emissions (Giglio et al. 2006). In addition, MODIS observations also provide aerosol optical depth (AOD) and Angstrom exponent, which are retrieved from the 'Dark Target' (DT) algorithm over land and the 'Deep Blue' (DB) algorithm over bright deserts (Levy et al. 2007). The Level-3 AOD (Collection 5.1, spatial resolution $1 \times 1°$) is used in this study. Han et al. (2015) compared MODIS-AOD with sun photometer measurements in southeast China and found a good correlation ($R > 0.8$) and a linear regression slope close to 1.0.

CO and O_3 are part of the atmospheric product of the Atmospheric Infra-Red Sounder (AIRS) instrument onboard NASA's Earth Observing System (EOS) Aqua satellite. AIRS version-5 L2 total column CO concentration is derived from the 4.55 μm region of AIRS spectra (Susskind et al. 2003). In addition, NO_2 and formaldehyde (HCHO) are measured from the Ozone Monitoring Instrument (OMI) instrument onboard the NASA Aura platform. The OMI-derived tropospheric vertical NO_2 and HCHO column densities are used to track the potential enhancement of NO_2 pollution from crop burning (Boersma et al. 2001). This study uses the Level-3 daily NO_2 and HCHO products with a spatial resolution of $0.25 \times 0.25°$ latitude-longitude.

The Cloud-Aerosol Lidar with Orthogonal Polarization (CALIOP), on board the Cloud-Aerosol Lidar and Infrared Pathfinder Satellite Observations (CALIPSO) platform, is a spaceborne polarization-sensitive two-wavelength (532 and 1064 nm) Lidar (Winker et al. 2009). It observes the aerosol/cloud vertical distribution and provides products of aerosol type classification and optical properties on a global scale. In addition, NASA's Clouds and Aerosol Transport System (CATS), a lidar remote sensing instrument, provides range-resolved profile measurements of atmospheric aerosols and clouds from the International Space Station (ISS) (Yorks et al. 2016). At present, the released CALIPSO and CATS aerosol products include: (a) Level-1 attenuated backscatter coefficient profiles or calibrated range-corrected lidar returns; (b) Level-2 aerosol and cloud layer product, vertical feature mask (VFM) including cloud aerosol discrimination (CAD) and aerosol-type classification, extinction, and backscatter coefficient profiles; (c) Level-3 aerosol globally gridded monthly profile product.

In addition, the NOAA-HYSPLIT (Hybrid Single-Particle Lagrangian Integrated Trajectory) model is used to compute air parcel trajectories and to model the dispersion and the route of dust particles (Draxler and Hess 1997). For this, input weather maps of the pressure and wind fields from the National Centers for Environmental Prediction (NCEP) reanalysis data (Kalnay et al. 1996) have been used.

4 Results and Discussion

4.1 Agricultural Crop Open Fires (ACOF) and Smoke in Spring 2015

Figure 1 shows the wildfire locations (orange symbols) and smoke (gray) on the satellite MODIS RGB image and fire radiative power from April 12–16, 2015. It can be seen that there are many fires detected in Northeast (NE) China, and the smoke plumes spread toward the Yellow sea. According to the land surface and surface record information, these fires mainly occurred over the cropland where the farmers burned for the spring planting (http://www.aimayubao.com/blog/2015/04/15/). The satellite-measured AOD, CO, HCHO, and NO_2 which are given in Fig. 2. The results indicate consistent 'hot spots' of AOD, CO, and HCO in the fire region located in NE China, and the values of AOD, CO, and HCHO are even higher than those in the urban area (Beijing, Shanghai). The NO_2 in the fire region is relatively higher but still lower than those in the urban area.

Aerosol vertical distribution, optical properties, and type classification are illustrated from the NASA CALIOP/CALIPSO observations at 17:45:56 UTC on April 14, 2015, in Eastern China (Fig. 3). The results indicate that the aerosol plumes appear at 37–53° N latitude and reach up to 5 km altitude. The dense plumes show aerosol

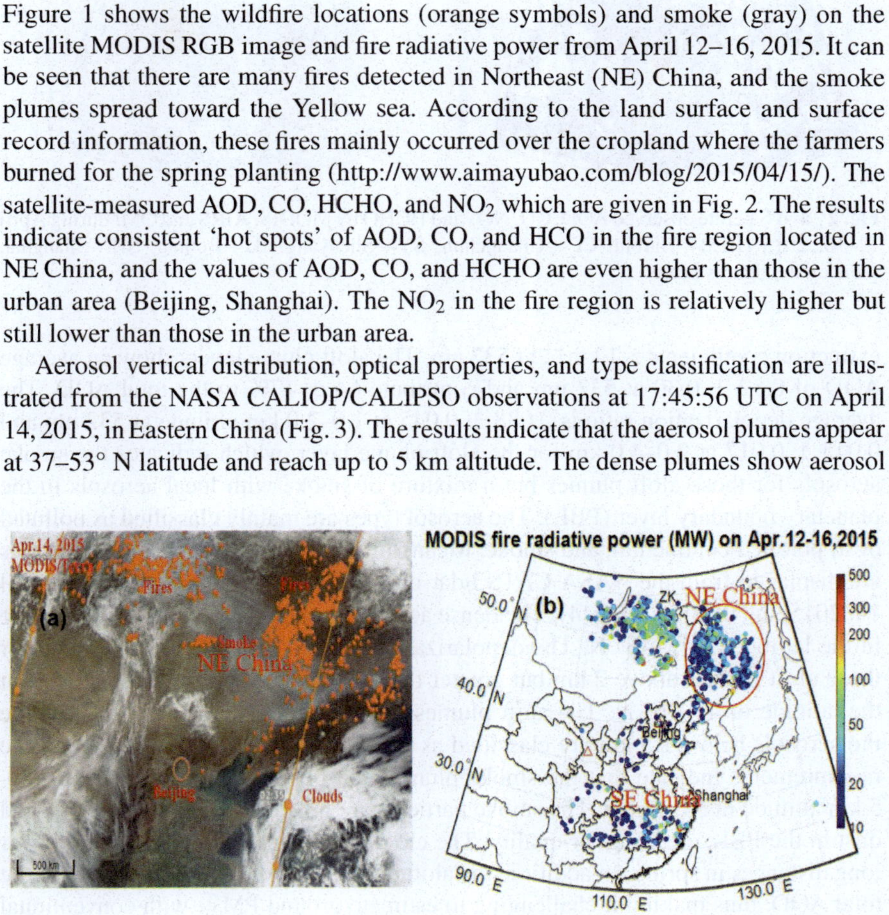

Fig. 1 **a–b** Agricultural crop residue open fires/smoke from the satellite MODIS/Terra RGB image on April 14, 2015 and fire radiative power during April 12–16, 2015

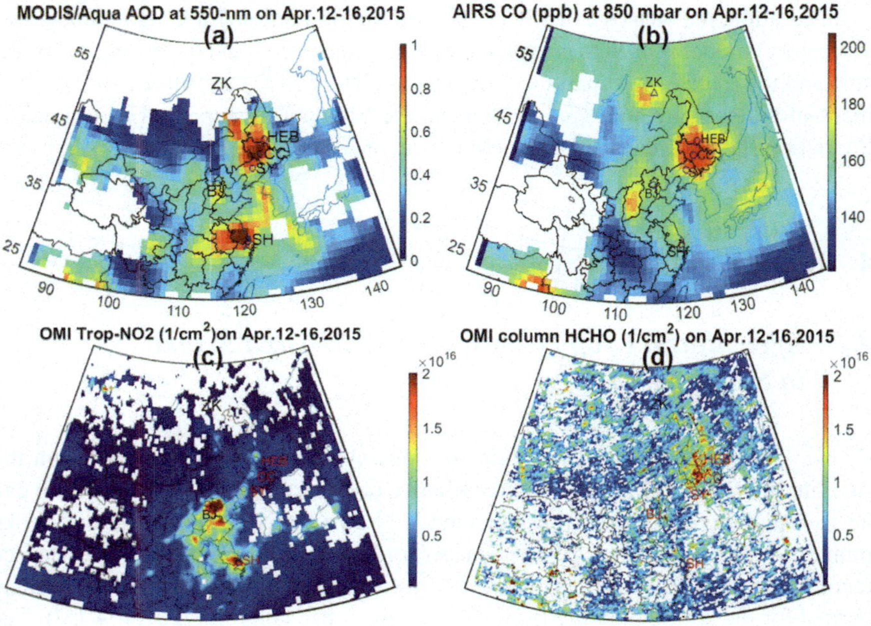

Fig. 2 **a**–**d** Satellite observed AOD, CO, NO_2 and HCHO by MODIS, AIRS and OMI during April 12–16, 2015 (HEB—Harbin, CC—Changchun, SY—Shenyang, BJ—Beijing, SH—Shanghai, same on the figures below)

extinction coefficients > 1 km^{-1} at 532 nm. The aloft plume layers show an average AOD of 0.52 ± 0.30 at 532 nm and contribute 64 ± 17% to the total AOD. The average depolarization ratio is 0.088 ± 0.045 at 1.0–2.0 km altitude at 532 nm and 0.045 ± 0.017 at 2.0–4.0 km for the aloft plume layer, which indicates the smoke aerosols for those aloft plumes but a mixture of smoke with local aerosols in the planetary boundary layer (PBL). The aerosol types are mainly classified as polluted dust, polluted continental, and smoke. Meanwhile, the aerosol vertical distributions are displayed from the NASA CATS lidar observations at 13:22:16 UTC on April 14, 2015, in NE China (Fig. 4). The dense aerosol plumes were below 4 km altitude in the latitude of 43–46° N. The depolarization ratio is generally less than 0.1 for those aloft plumes above 2 km but greater than 0.2 for the near-surface aerosols at the latitude of 44–46° N. The aloft plumes are mainly classified as smoke, while the aerosols below 1.5 km are classified as dust mixture and smoke. These profile measurements indicate that the smoke plumes can be vertically transported to 4–5 km altitude in NE China. The smoke particles are mixed with local pollutants and dust in the PBL, affecting air quality. The elevated plumes can be transported over long distances in spring. In addition, the aloft dense plumes contribute largely to the total AOD, thus making it challenging to estimate ground $PM_{2.5}$ with conventional column AOD measured by satellites.

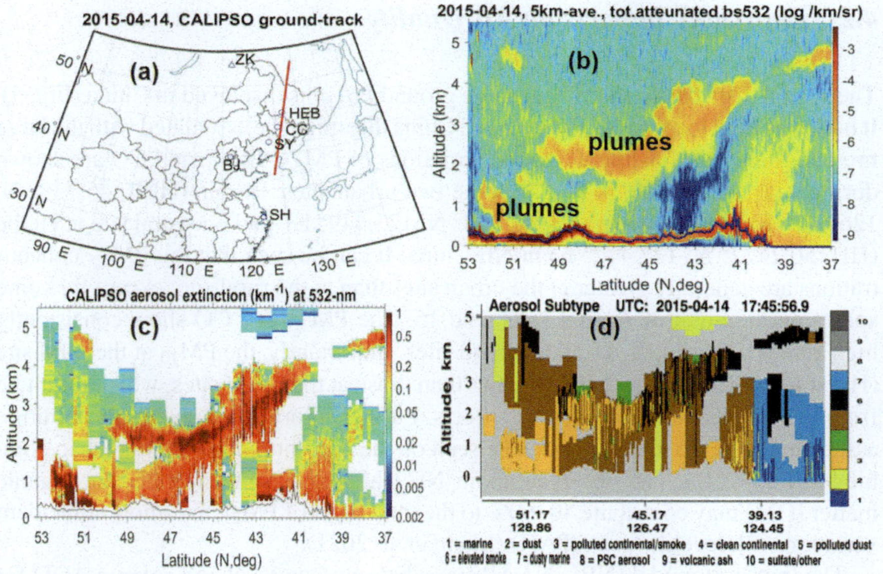

Fig. 3 **a–d** Satellite CALIPSO observed aerosol vertical distribution, extinction, and aerosol-type classification on April 14, 2015 (17:45:56 UTC)

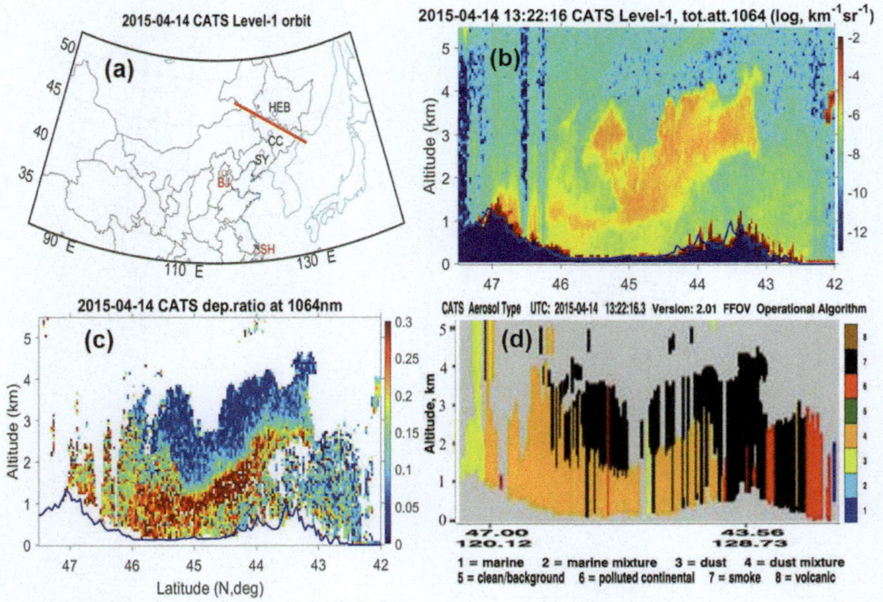

Fig. 4 **a–d** Spaceborne-lidar CATS observed aerosol vertical distribution, depolarization ratio and aerosol-type classification on Apr.14, 2015 (13:22:16 UTC)

4.2 Emissions Impacts on Air Quality

The spatial distribution of daily average ground $PM_{2.5}$ is displayed in China (Fig. 5). It indicates high loadings of $PM_{2.5}$ in NE China that might be associated with the large fire emissions. Meanwhile, temporal variations of $PM_{2.5}$, PM_{10}, and CO are shown during April 12–16, 2015 (Fig. 5) at the two urban sites, Harbin (HEB, 45.738° N, 126.648° E) and Shenyang (SY, 41.808° N, 123.449° E), and one rural site at Heihe (HH, 50.2472° N, 127.472° E) in NE China. It can be seen that the $PM_{2.5}$ concentrations are generally greater at the urban sites than at the rural site, except for some spikes on the night of April 12, 13, and 15. The $PM_{2.5}$ and CO show consistently high values on April 14–15 at the urban sites. Importantly, the PM_{10} at the rural site (HH) on April 13, 15, and 16 is greater than those at the urban sites, which might be from the smoke and dust emission effects. Further information on the air pollution associated with the fire smoke can be seen on the link (http://www.aimayubao.com/blog/2015/04/21/). The observations in NE China and NCP indicate that organic matter (OM) may contribute 40–60% to the total mass of $PM_{2.5}$ and show coincident variation with levoglucosan (PAH) (Liang et al. 2021).

The transport model (NRL-NAAPS) predicts surface-level and column AOD for the smoke, dust, and sulfate aerosols. As shown in Fig. 6, there is high-level smoke at the surface level in NE China, while high dust concentrations appear in the Gobi

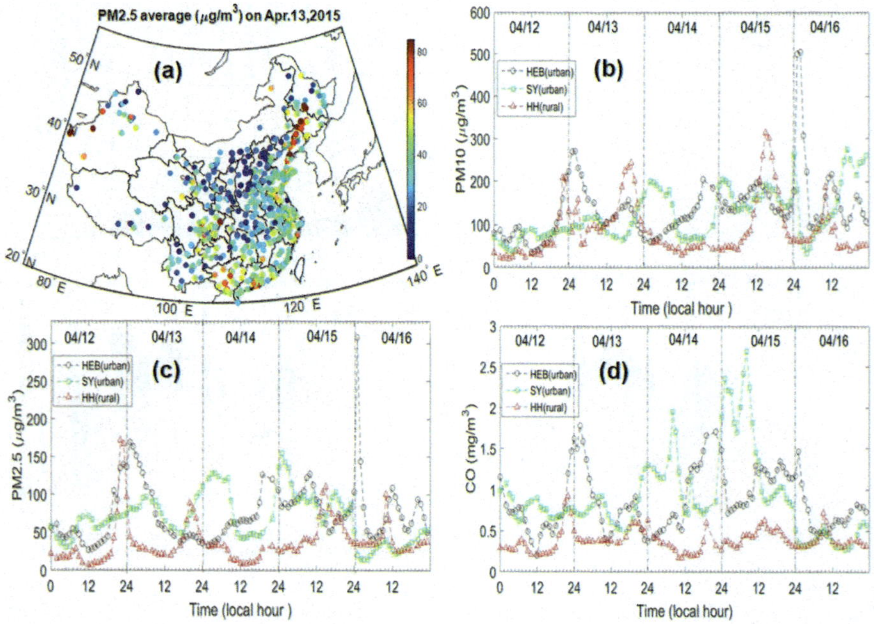

Fig. 5 **a–d** Daily average $PM_{2.5}$ in China and temporal variation of $PM_{2.5}$, PM_{10} and CO in NE China during April 12–16, 2015. HEB—Harbin (45.738° N, 126.648° E, urban), SY—Shenyang (41.808° N, 123.449° E, urban), HH—Heihe (50.2472° N, 127.472° E , rural)

deserts on the board of China and Mongolia. The sulfate aerosols mainly appear in the urban and industrial areas in Eastern China. The high concentrations of smoke near the surface are mainly from fire smoke emissions. However, validation with the ground observations is highly required to quantify these predicted concentrations of smoke and dust (Han et al. 2015; Wu et al. 2017, 2018). Wu et al. (2017) compared the $PM_{2.5}$ between the WRF-Chem model and ground observation for an ACOF smoke episode in eastern China; the results indicate that the model captured the accumulation and downwind transport of surface $PM_{2.5}$ for the phase-1 of smoke but showed a dramatic underestimate for the phase-2 when dense aerosols are present. Such a discrepancy in the model is associated with improper emission, vertical apportion of transported smoke, and atmospheric diffusion conditions when compared with the observed aerosol and wind profiles. Further, the model simulations indicate that the transported smoke can contribute 50–70% to the ground $PM_{2.5}$ in Nanjing. Using inverse modeling, Stavrakou et al. (2016) derive that satellite-based post-harvest burning fluxes are, on average, at least a factor of 2 higher than the bottom-up statistical estimates. Crop burning is attributed to an increase in surface ozone (7%) and fine aerosol concentrations (18%) in the NCP in June.

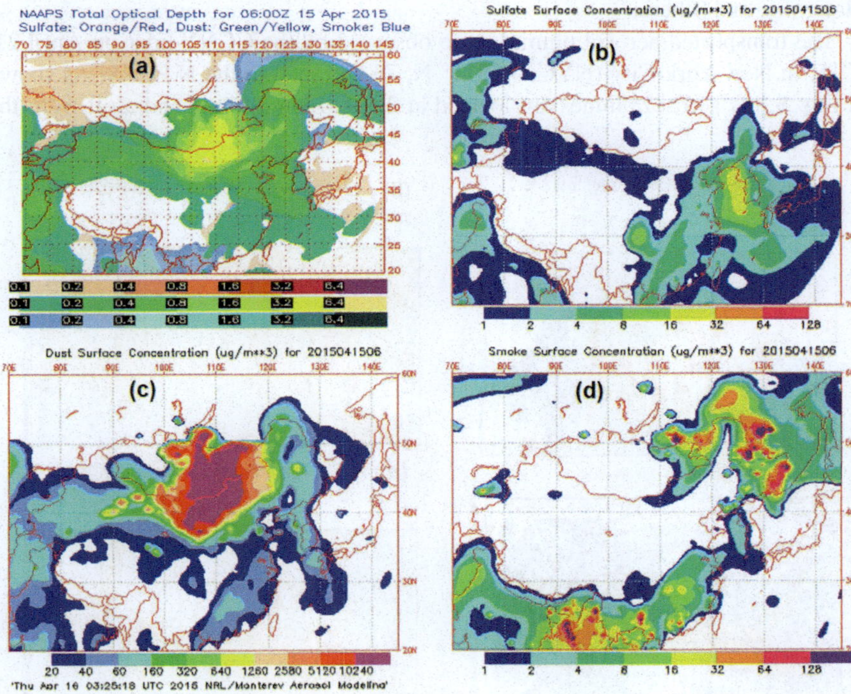

Fig. 6 a–d Aerosol species AOD and surface concentration of smoke, dust, and sulfate from the NRL-NAAPS model in China at 06:00 UTC on April 14, 2015

4.3 Trans-Pacific Transport of the Plumes to the US

The trans-Pacific transport of smoke plumes is illustrated from the spatial distribution of MODIS-AOD, AIRS-CO at 500–700 mbar, and wind patterns in Fig. 7. The average MODIS-AOD during April 12–16 shows higher values from eastern China to the west coast of the US, which indicates the transport pattern. Spatial inhomogenous AOD over the Pacific Ocean is indicated because the Asian pollutant transport is driven by the extratropical cyclone weather system that often forms clouds. The CO at 700 mbar shows similar high-level patterns with the AOD. More importantly, the CO shows a high concentration zone spreading from Canada to the eastern US at both 700 and 500 mbar, which might be linked to the fire smoke transport. Furthermore, the NCEP/NCAR reanalysis at 500 mbar indicates a westerly jet stream from eastern China to the eastern US with a wind velocity > 18 m/s. On the west coast of the US, the CALIPSO lidar observations on April 20 illustrated the smoke and polluted dust plumes at 1–5 km altitude at the 40–49° N latitude and the polluted dust layer at 3–7 km in the latitude of 30–37° N. Meanwhile, at the high latitude of 60.9–66.7° N around the Arctic, the polluted dust appeared at 4–10 km altitude. These aloft plumes are likely transported from the emissions of wildfires and dust storms in North China, Mongolia, and Siberia.

The transported aerosol plumes were observed by the CCNY-lidar on April 21, 2015, in New York City (NYC, 40.821° N, 73.949° W) in the N.E. US. As shown in Fig. 8, the aerosol plumes are located at 2.5–8.0 km altitude and aloft from the

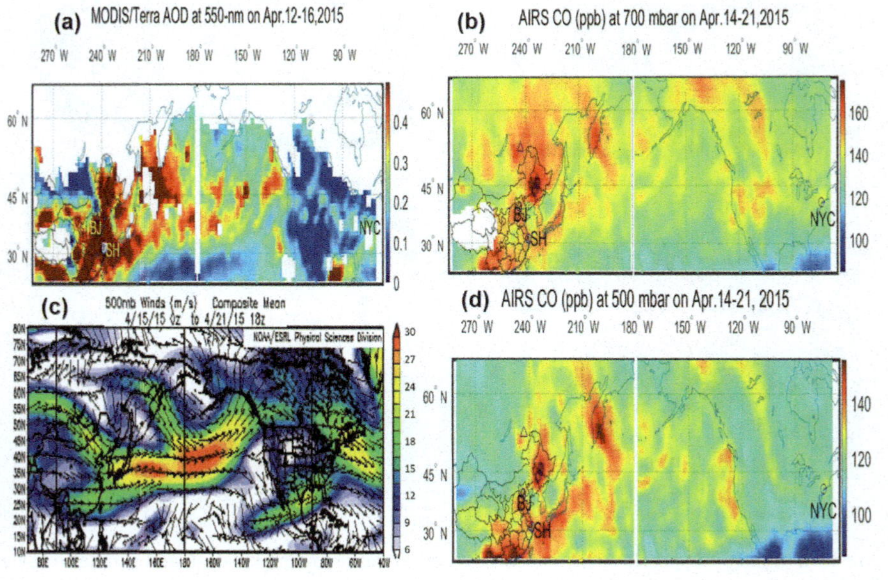

Fig. 7 a–d Aerosol optical depth, CO and NCEP/NCAR Reanalysis wind flows at 500 mbar on April 15–21, 2015 (BJ—Beijing, SH—Shanghai, NYC—New York City)

local PBL; the Angstrom exponent (A.E.) at wavelength 532–1064 is in the range of 1.0–1.5 probably indicates the mixture of smoke and dust. According to the lidar-derived aerosol extinction profile, the aloft plumes contribute about 70% to the total AOD. The NOAA-HYSPLIT backward trajectories indicate that the air at 4–6 km in the plume layer originated from North China and traveled for around one week long before arriving in NYC. In particular, the air at a 4 km altitude at CCNY originated from the ground level in North China, where there is a large amount of smoke and dust mixture. The co-located AERONET sun photometer observations show the average total AOD at 0.38 ± 0.05 at 500 nm and A.E. of 1.02 ± 0.04 at 440–87 nm. In addition, the ground $PM_{2.5}$ at the CCNY site shows an increase from 5 to 15 $\mu g/m^3$ from April 21 to April 22, 2015.

The range-resolved spatial distribution of aerosols in the eastern US was measured from the CALIPSO in the early morning of April 22, 2015. The results in Fig. 9 demonstrate that the aloft plumes are mainly located at 3–8 km altitude in the latitude of 35–41° N. The volume depolarization ratios range from 0.07–0.2 at 532 nm, while the attenuated color ratio is at 0.4–0.8 at 532–1064 nm. The aerosol plume layers are partially classified as polluted dust and dust, depending on the threshold of depolarization ratio used in the algorithm. In the PBL, the aerosols are mostly classified as polluted dust. There were no dramatic wildfire and dust events on April 19–22 in the continental US, according to the NOAA hazard mapping system (HMS) fire

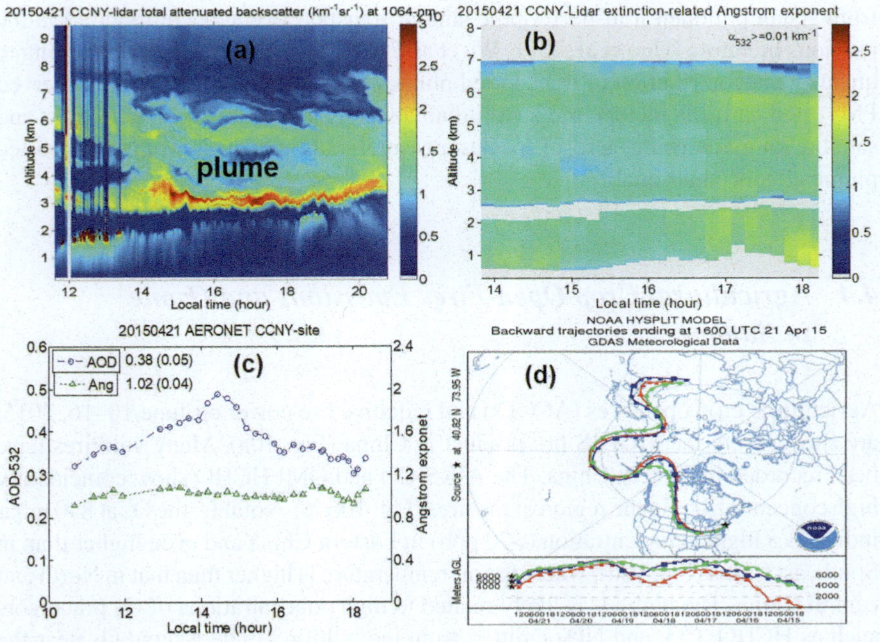

Fig. 8 a–d Aerosol plumes observed by CCNY-lidar and AERONET sun photometer at CCNY site (40.82N, 73.95W, NE US), and HYSPLIT backward trajectories ending at CCNY (4–6 km altitude) on April 21, 2015

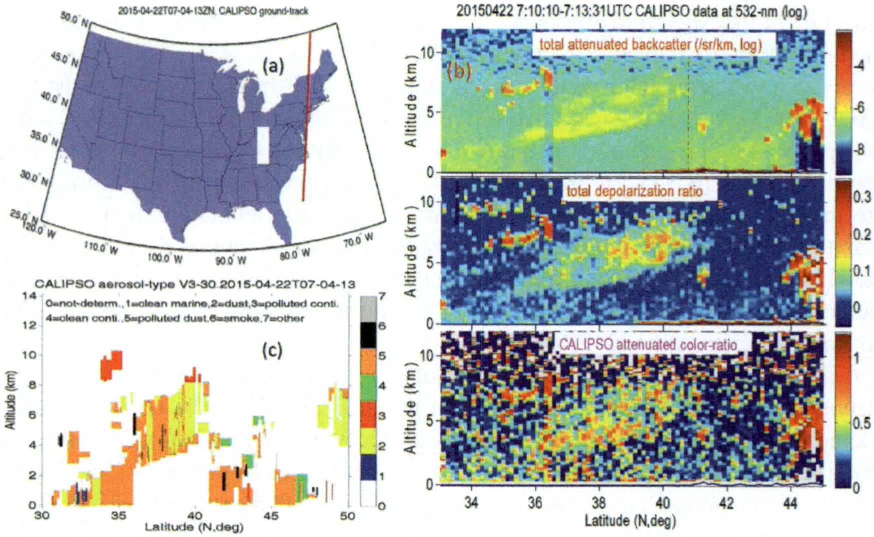

Fig. 9 **a–c** Aerosol plume vertical distribution, optical properties and aerosol-type classification from CALIPSO observation at 7:10:10 UTC on April 22, 2015 in the Eastern US

and smoke product. Thus, these plume layers in the eastern US probably originated from Asian pollutant transport. Some similar transport events were reported in the previous literature (Uno et al. 2011; Wu et al. 2015). Nonetheless, it is challenging to quantify the contribution of transported plumes to ground $PM_{2.5}$ since the observed $PM_{2.5}$ and aerosols include the local urban aerosols. In combination with lidar and satellite remote sensing, aerosol species measurements help evaluate the transported plume effects on air quality.

4.4 Agricultural Crop Open Fires Emissions and Ozone in Summer

Agricultural crop open fires (ACOFs) and radiative fire power on June 10–16, 2015, are shown from the MODIS fire product in China (Fig. 10a). Many wildfires have been recorded in Eastern China. The AIRS-CO and OMI HCHO show coincidently high concentrations in the region of the fires (Fig. 10b–c). Notably, the O_3 at 850 mbar indicates a higher concentration (>70 ppb) in Eastern China and even higher than in Southeast China (Fig. 10d), where the air temperature is higher than that in North and central China. The high O_3 is likely related to high concentrations of O_3 precursors such as HCHO, CO, and NO_2 emitted from the wildfires. The ground O_3 near the fire area at Zhengzhou exceeded the NAAQS with an hourly maximum of 200 and 230 $\mu g/m^3$ on June 9 and 10, 2015, respectively. Li et al. (2021) indicate that the fast O_3 production was driven by the formaldehyde-HCHO in the NCP in the winter-spring.

Li et al. (2018) simulated the ACOF emission effects on the O_3 using the WRF-Chem model and satellite observations in June 2016 in China. The results indicate that the precursor emissions from agricultural fires can enhance O_3 photochemistry and show a maximal O_3 increase by 20 ppb near the fire zones in China. The effects of heterogeneous uptake on soot are changing the average O_3, NO_2, OH, and HO_2 concentrations by + 0.8%, − 0.5%, − 0.7%, and + 0.8%, respectively. The study by Stavrakou et al. (2016) indicates that the top-down crop burning fluxes of VOCs in June (2005–2012) exceeded by around a factor of 2 in comparison to the other anthropogenic emissions in the North China Plain (NCP), mainly attributed to the ACOF. Crop burning is attributed to an increase in surface O_3 by 7% and fine aerosol concentrations by 18% in the NCP in June. Several field campaigns have demonstrated that there are also large emissions of VOCs from the wildfires in the western and southeast US (Liu et al. 2016, Jin et al. 2022). However, there is a lack of such field campaigns with integrated aircraft and ground observations to quantify the ACOP emissions and improve the model product accuracy in China (Fu et al. 2021).

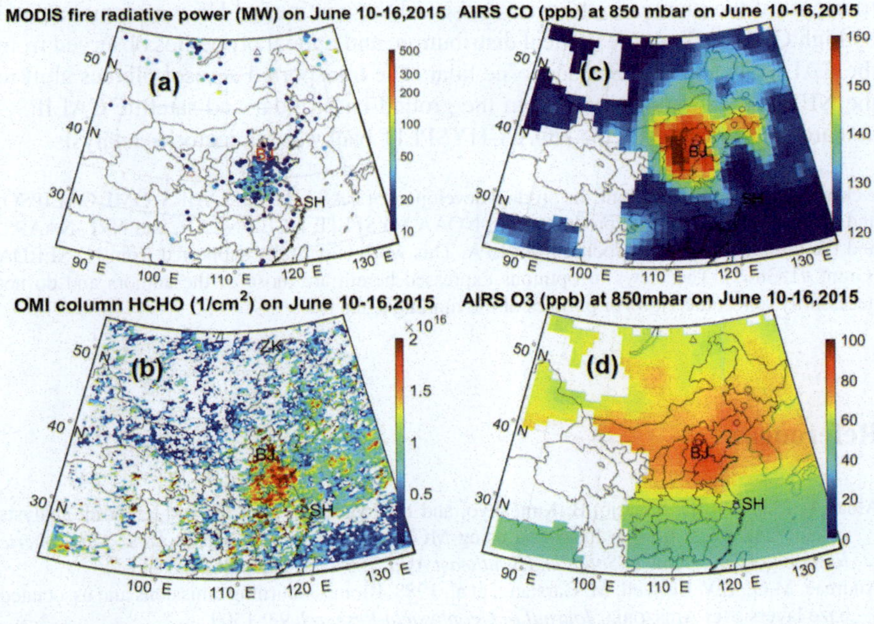

Fig. 10 **a–d** Satellite observed fire radiative power, CO, HCHO and O_3 on June 10–16, 2015 in China (BJ—Beijing, SH—Shanghai)

5 Conclusion

This study presents satellite remote sensing (MODIS, AIRS, OMI, CALISPO, and CATS) and ground observations of agricultural crop open fires, emissions of particles and gaseous compounds, smoke vertical distribution and optical properties, and potential impacts on the air quality in the NE and Eastern China in April and June of 2015, including trans-Pacific transport to the continental US. The integrated satellite observations clearly demonstrate coincident hot spots of AOD, CO, HCHO, NO_2, and O_3. The smoke plumes can be transported up to 5 km altitude and largely contribute to the total AOD. For the event in spring in NE China, the smoke plumes mix with the dust particles and can be transported across the Pacific Ocean. Much polluted dust is classified as a smoke-dust mixture. In summer, high CO, HCHO, and NO_2 may enhance the O_3 formation, thereby resulting in high O_3 concentration and haze episodes in Eastern China. Severe air pollution events are indicated by high-level PM_{10}, $PM_{2.5}$, CO, and NO_2 nearby the fire region attributed to the fire smoke emissions. The model simulations indicate high emissions of smoke particles and dramatic O_3 formation, but further constraints and/or validations with satellite and ground observations are critical to quantify fire emission effects on air quality. Driven by westerly jet streams in the middle-upper troposphere, the trans-Pacific transport of the plumes to the west coast and the continental US was demonstrated by high CO, AOD, plume vertical distribution, and optical properties observed from the CALIPSO and CATS spaceborne lidar. The transported aerosol plumes aloft to the NE US are also observed from the ground-based lidar and satellite CALIPSO and are demonstrated by the NOAA-HYSPLIT backward trajectories analysis.

Acknowledgements We thank the product developers of NASA-MODIS, AIRS, OMI, CALIPSO, and CATS, and the modeled products from NOAA-HYSPLIT, NCEP/NCAR, and NRL-NAAPS, and the ground air quality product from EPA. This work is partially supported from NYSERDA (Grant #183869). The views or opinions expressed herein are those of the authors and do not necessarily reflect the views or policies of the funding agencies

References

Albar, I., I. Jaya, B.H. Saharjo, B. Kuncahyo, and K.P. Vadrevu. 2018. Spatio-temporal analysis of land and forest fires in Indonesia using MODIS active fire dataset. In *Land-Atmospheric Research Applications in South and Southeast Asia*. 105–127. Cham: Springer.

Andreae, M.O., E.V. Browell, M. Garstang, et al. 1988. Biomass-burning emissions and associated haze layers over Amazonia. *Journal of Geophysical Research* 93: 1509.

Arvelyna, Y., H. Takahashi, L. Indrayanti, H. Hayasaka, K.P. Vadrevu, R. Maryani, M. Osaki, and H. Kazuyo. 2021. Peatland surface loss due to fires in Central Kalimantan, Indonesia–a case study using differential interferometry SAR (DInSAR). In *Biomass Burning in South and Southeast Asia*, 77–90. CRC Press.

Badarinath, K.V.S., S.K. Kharol, K.M. Latha, T.K. Chand, V.K. Prasad, A.N. Jyothsna, and K. Samatha. 2007. Multiyear ground-based and satellite observations of aerosol properties over a tropical urban area in India. *Atmospheric Science Letters* 8 (1): 7–13.

Badarinath, K.V.S., S.K. Kharol, A.R. Sharma, and V.K. Prasad. 2009. Analysis of aerosol and carbon monoxide characteristics over Arabian Sea during crop residue burning period in the Indo-Gangetic Plains using multi-satellite remote sensing datasets. *Journal of Atmospheric and Solar-Terrestrial Physics* 71 (12): 1267–1276.

Badarinath, K.V.S., S. Kumar Kharol, V. Krishna Prasad, A. Rani Sharma, E.U.B. Reddi, H.D. Kambezidis, and D.G. Kaskaoutis. 2008. Influence of natural and anthropogenic activities on UV Index variations–a study over tropical urban region using ground based observations and satellite data. *Journal of Atmospheric Chemistry* 59 (3): 219–236.

Biswas, S., K.D. Lasko, and K.P. Vadrevu. 2015a. Fire disturbance in tropical forests of Myanmar—Analysis using MODIS satellite datasets. *IEEE Journal of Selected Topics in Applied Earth Observations and Remote Sensing* 8 (5): 2273–2281.

Biswas, S., K.P. Vadrevu, Z.M. Lwin, K. Lasko, and C.O. Justice. 2015b. Factors controlling vegetation fires in protected and non-protected areas of Myanmar. *PLoS ONE* 10 (4): e0124346.

Boersma, K., E. Bucsela, E. Brinksma, and J. Gleason. 2001. NO_2, OMIEOS Algorithm. *Theoretical Basis Document: Trace gas algorithms: NO2 4*: 12–35.

Chen, J., C. Li, Z. Ristovski, A. Milic, et al. 2017. A review of biomass burning: emissions and impacts on air quality, health and climate in China. *Science of the Total Environment* 579: 1000–1034.

Crutzen, P.J., L.E. Heidt, and J.P. Krasnec. 1979. Biomass burning as a source of atmospheric gases CO, H_2, N_2O, NO, CH_3Cl, and COS. *Nature* 282 (5736): 253–356.

Draxler, R.R., and G.D. Hess. 1997. Description of the HYSPLIT_4 modeling system. *NOAA Tech. Memo. ERL ARL-224, NOAA Air Resources Laboratory*. vol 24 Silver Spring, MD.

Fu, Y., H. Gao, H. Liao, X. Tian. 2021. Spatiotemporal variations and uncertainty in crop residue burning emissions over North China Plain: Implication for atmospheric CO_2 simulation. *Remote Sensing* 13(19):3880.

Giglio, L., I. Csiszar, and C. Justice. 2006. Global distribution and seasonality of active fires as observed with Terra and Aqua Moderated Resolution Imaging Spectroradiometer (MODIS) sensors. *Journal of Geophysical Research* 111: G02016.

Han, Y., Y. Wu, T. Wang, et al. 2015. Impacts of elevated-aerosol-layer and aerosol type on the correlation of AOD and particulate matter with ground-based and satellite measurements in Nanjing, southeast China. *Science of the Total Environment* 532: 195–207.

Jin, L., W. Permar, V. Selimovic, D. Ketcherside, et al. 2022. Constraining emissions of volatile organic compounds from western US wildfires with WE-CAN and FIREX-AQ airborne observations, EGUsphere [preprint]. https://doi.org/10.5194/egusphere-2022-1107

Justice, C., G. Gutman, and K.P. Vadrevu. 2015. NASA land cover and land use change (LCLUC): An interdisciplinary research program. *Journal of Environmental Management* 148: 4–9.

Kalnay, E., et al. 1996. The NCEP/NCAR 40-year reanalysis project. *Bulletin of the American Meteorological Society* 77: 437–470.

Kharol, S.K., K.V.S. Badarinath, A.R. Sharma, D.V. Mahalakshmi, D. Singh, and V.K. Prasad. 2012. Black carbon aerosol variations over Patiala city, Punjab, India—a study during agriculture crop residue burning period using ground measurements and satellite data. *Journal of Atmospheric and Solar-Terrestrial Physics* 84: 45–51.

Kim, J., U. Jeong, M. Ahn, et al. 2020. New Era of Air Quality Monitoring from Space, Geostationary Environment Monitoring Spectrometer (GEMS). *Bulletin of the American Meteorological Society* 101 (1): E1–E22.

Lasko, K., and K. Vadrevu. 2018. Improved rice residue burning emissions estimates: Accounting for practice-specific emission factors in air pollution assessments of Vietnam. *Environmental Pollution* 236: 795–806.

Lasko, K., K.P. Vadrevu, and T.T.N. Nguyen. 2018. Analysis of air pollution over Hanoi, Vietnam using multi-satellite and MERRA reanalysis datasets. *PLoS ONE* 13(5): e0196629.

Lasko, K., K.P. Vadrevu, V. Bandaru, T.N.T. Nguyen, and H.Q. Bui. 2021. PM2. 5 emissions from biomass burning in South/Southeast Asia–Uncertainties and trade-offs. In *Biomass burning in south and Southeast Asia*. 149–169. CRC Press.

Lasko, K., K.P. Vadrevu, V.T. Tran, E. Ellicott, T.T. Nguyen, H.Q. Bui, and C. Justice. 2017. Satellites may underestimate rice residue and associated burning emissions in Vietnam. *Environmental Research Letters* 12 (8): 085006.

Levy, R.C., L.A. Remer, S. Mattoo, E.F. Vermote, and Y.J. Kaufman. 2007. Second-generation operational algorithm: retrieval of aerosol properties over land from inversion of moderate resolution imaging spectroradiometer spectral reflectance. *Journal of Geophysical Research: Atmospheres 112*(D13).

Li, K., D.J. Jacob, H. Liao, et al. 2021. Ozone pollution in the North China Plain spreading into the late-winter haze season. *PNAS* 118 (10): e2015797118.

Li, M., T. Wang, M. Xie, et al. 2018. Agricultural fire impacts on ozone photochemistry over the Yangtze River Delta region, East China. *Journal of Geophysical Research Atmospere* 123: 6605–6623.

Liang, L., G. Engling, C. Liu, et al. 2021. Measurement report: chemical characteristics of $PM_{2.5}$ during typical biomass burning season at an agricultural site of the North China Plain. *Atmospheric Chemistry and Physics* 21: 3181–3192.

Liu, X., Y. Zhang, L. G. Huey, et al. 2016. Agricultural fires in the southeastern U.S. during SEAC4RS: Emissions of trace gases and particles and evolution of ozone, reactive nitrogen, and organic aerosol. *Journal of Geophysical Research: Atmospheres* 121: 7383–7414. https://doi.org/10.1002/2016JD025040

Mehmood, K., Y. Wu, L. Wang, et al. 2020. Relative effects of open biomass burning and open crop straw burning on haze formation over central and eastern China: Modeling study driven by constrained emissions. *Atmospheric Chemistry and Physics* 20: 2419–2443.

Penn, E., and T. Holloway. 2020. Evaluating current satellite capability to observe diurnal change in nitrogen oxides in preparation for geostationary satellite missions. *Environmental Research Letters* 15: 034038.

Prasad, V.K., M. Lata, and K.V.S. Badarinath. 2003. Trace gas emissions from biomass burning from northeast region in India—estimates from satellite remote sensing data and GIS. *The Environmentalist* 23 (3): 229–236.

Prasad, V.K., Y. Kant, and K.V.S. Badarinath. 2002a. Land use changes and modeling carbon fluxes from satellite data. *Advances in Space Research* 30 (11): 2511–2516.

Prasad, V.K., Y. Kant, P.K. Gupta, C. Elvidge, and K.V.S. Badarinath. 2002b. Biomass burning and related trace gas emissions from tropical dry deciduous forests of India: a study using DMSP-OLS data and ground-based measurements. *International Journal of Remote Sensing* 23 (14): 2837–2851.

Stavrakou, T., J.F. Müller, M. Bauwens, et al. 2016. Substantial underestimation of post-harvest burning emissions in the North China plain revealed by multi-species space observations. *Science and Reports* 6: 32307.

Streets, D.G., T. Canty, Carmichael, et al. 2013. Emissions estimation from satellite retrievals: a review of current capability. *Atmospheric Environment* 77: 1011–1042.

Susskind, J., C. Barnet, and J. Blaisdell. 2003. Retrieval of atmospheric and surface parameters from AIRS/AMSU/HSB data in the presence of clouds. *IEEE Transactions on Geoscience and Remote Sensing* 2: 390–409.

Thapa, P., K.P. Vadrevu, A. Eaturu, and S. Biswas. 2021. Vegetation fire status and management in Bhutan. In *Biomass Burning in South and Southeast Asia.* 135–157. CRC Press.

Uno, I., K. Eguchi, K. Yumimoto, Z. Liu, et al. 2011. Large Asian dust layers continuously reached North America in April 2010. *Atmospheric Chemistry and Physics* 11: 7333–7341.

Vadrevu, K.P., S. Biswas, and A. Eaturu. 2021c. Spatial point patterns and scale analysis of vegetation fires in Laos and Cambodia. In *Biomass Burning in South and Southeast Asia.* 117–132. CRC Press.

Vadrevu, K.P., A. Eaturu, and S. Biswas. 2021b. Burnt area mapping in Nainital, Uttarakhand, India, using very high-resolution PlanetScope imagery. In *Biomass Burning in South and Southeast Asia.* 91–103. CRC Press.

Vadrevu, K.P., T. Ohara, and C. Justice. 2021a. Biomass burning in South/Southeast Asia–needs and priorities. In *Biomass Burning in South and Southeast Asia.* 1–21. CRC Press.

Wang, L., X. Jin, Q. Wang, et al. 2020. Spatial and temporal variability of open biomass burning in Northeast China from 2003 to 2017. *Atmospheric and Oceanic Science Letter 13*(3).

Wiedinmyer, C., S.K. Akagi, R.J. Yokelson, et al. 2011. The Fire Inventory from NCAR (FINN): a high resolution global model to estimate the emissions from open burning. *Geosci. Model Dev.* 4: 625–641.

Winker, D.M., M.A. Vaughan, A. Omar, et al. 2009. Overview of the CALIPSO Mission and CALIOP Data Processing Algorithms. *Journal of the Seismological Society of Japan* 26: 2310–2323.

Wooster, M.J., G.J. Roberts, L. Giglio, D.P. Roy, P.H. Freeborn, L. Boschetti, C. Justice, C. Ichoku, W. Schroeder, D. Davies, and A.M. Smith. 2021. Satellite remote sensing of active fires: History and current status, applications and future requirements. *Remote Sensing of Environment* 267: 112694.

Wu, J., S. Kong, F. Wu, et al. 2020. The moving of high emission for biomass burning in China: View from multi-year emission estimation and human-driven forces. *Environment International* 142: 105812.

Wu, Y., A. Arapi, J. Huang, B. Gross, and F. Moshary. 2018. Intra-continental wildfire smoke transport and impact on local air quality observed by ground-based and satellite remote sensing in New York City. *Atmospheric Environment* 187: 266–281.

Wu, Y., Y. Han, Voulgarakis, et al. 2017. An agricultural biomass burning episode in eastern China: transport, optical properties, and impacts on regional air quality. *Journal of Geophysical Research: Atmosphere* 122: 2304–2324.

Wu, Y., Z. Han, C. Nazmi, B. Gross, and F. Moshary. 2015. A trans-Pacific Asian Dust episode and its impacts to air quality in the east coast of U.S. *Atmospheric Environment* 106: 358–368.

Yang, G., H. Zhao, D.Q. Tong, et al. 2020. Impacts of post-harvest open biomass burning and burning ban policy on severe haze in the Northeastern China. *Science of the Total Environment* 716: 136517.

Yin, S., M. Guo, X. Wang, et al. 2021. Spatiotemporal variation and distribution characteristics of crop residue burning in China from 2001 to 2018. *Environmental Pollution* 268: 115849.

Yorks, J.E., M.J. McGill, S.P. Palm, et al. 2016. An overview of the CATS level 1 processing algorithms and data products. *Geophysical Research Letters* 43: 4632–4639.

Zhou, Y., X. Xing, J. Lang, et al. 2017. A comprehensive biomass burning emission inventory with high spatial and temporal resolution in China. *Atmospheric Chemistry and Physics* 17: 2839–2864.

Agricultural Fires in Northeast China: Characteristics, Impacts, and Challenges

Jiumeng Liu and Yuan Cheng

Abstract Sustainable use of crop residues remains a challenge in main agricultural regions of China, such as the Northeast China Plain. In this study, we investigated the impacts of biomass burning during a six-month-long heating season in the Harbin–Changchun (HC) metropolitan area, i.e., China's only national-level city cluster located in the severe cold climate region. Biomass burning was found to be the major contributor to $PM_{2.5}$ pollution, mainly driven by agricultural residue burning with relatively low combustion efficiencies, i.e., smoldering combustion. The smoldering nature of combustion brought substantial difficulties in model simulations, causing significant under-prediction of $PM_{2.5}$. In addition, the low combustion efficiency agricultural fires emitted a large amount of brown carbon (BrC). The synthesis of these variations overall caused an underestimation of the fraction of solar energy absorbed by BrC relative to EC. This study demonstrates that open burning is of great concern, especially in major agricultural regions such as the HC metropolitan area.

Keywords Biomass burning · Organic aerosol · Agricultural fires · Brown carbon · CMAQ

1 Introduction

Biomass burning of agricultural residues is most common in Asian countries, including China (Badarinath et al. 2007; 2008; 2009; Cheng et al. 2013; 2021a; b, 2022; Gupta et al. 2001a,b; Kharol et al. 2012; Lasko et al. 2017; 2018; Vadrevu and Lasko 2015; Lasko and Vadrevu 2018; Oanh et al. 2018; Vadrevu et al. 2014; 2017; 2018; 2022a, b). Several researchers highlighted the impacts of burning residues in the open fields, such as releasing greenhouse gas emissions and aerosols, with impacts on air quality at both local and regional scales (Vadrevu et al. 2021a,b).

J. Liu · Y. Cheng (✉)
State Key Laboratory of Urban Water Resource and Environment, School of Environment, Harbin Institute of Technology, Harbin, China
e-mail: ycheng@hit.edu.cn

K. P. Vadrevu et al. (eds.), *Vegetation Fires and Pollution in Asia*,
https://doi.org/10.1007/978-3-031-29916-2_30

Specifically, in China, in recent years, a series of clean air policies, including the "Air Pollution Prevention and Control Action Plan" and the "Three-Year Action Plan for Winning the Blue Sky Defense Battle," has been implemented to address the severe haze pollution. With the implementation of these actions, considerable decreases in $PM_{2.5}$ were identified nationwide. However, model simulations suggested that the decrease was mainly driven by reductions in industrial emissions. On the contrary, biomass burning becomes increasingly important. For example, biomass burning (BB) episodes caused by post-harvest burning of crop residues have been repeatedly observed in the North China Plain (Cheng et al. 2013), even in recent years under the toughest-ever clean air policy. Despite national and local bans on it, open burning is difficult to be eliminated without effective treatment of crop residues. Since the emissions from biomass burning could not be effectively controlled, both observational measurements and model results have suggested an increasing trend in the contribution of biomass burning in recent years, especially the influence on $PM_{2.5}$ pollution (Hu et al. 2016; Xu et al. 2019).

However, studies on $PM_{2.5}$ and $PM_{2.5}$ compositions, either measurement- or model-based, have been most frequently conducted at a limited number of air pollution hotspots in China (e.g., the North China Plain and the Yangtze River Delta). Correspondingly, other regions with different characteristics of air pollutant emissions and/or meteorological conditions are largely overlooked, prohibiting a more comprehensive understanding of the diversity of haze pollution in Chinese cities. In particular, the Harbin–Changchun (HC) metropolitan area is such a "forgotten" city cluster. Here, we present $PM_{2.5}$ composition in Harbin, the central city of the HC area, during a six-month campaign. Our results revealed the substantial impacts of BB on haze pollution in Harbin. The climate impacts and challenges for model simulation brought by BB are also discussed.

2 Study Area

The HC area is unique due to its complex air pollution sources and distinct climate. The HC city cluster is located in the severe cold climate region in Northeast China, where the daily average temperature could be below -20 °C during winter. Thus, intensive energy use (e.g., coal and biomass fuels for household space-heating in rural areas) and massive air pollutant emissions are expected during the heating season, which usually lasts as long as six months (i.e., from mid-October through mid-April). In addition, the city cluster is within the main agricultural region in China (i.e., the Northeast China Plain). Although prohibited, the open burning of crop residues has not been completely eliminated in HC and its surrounding areas. As a result, emissions from biomass burning and other sources have resulted in severe air pollution in the HC city cluster, e.g., daily averages of $PM_{2.5}$ reaching ~ 650 $\mu g/m^3$ have been reported (Li et al. 2019).

3 Data and Methods

3.1 *Data*

A total of 180 $PM_{2.5}$ quartz-fiber samples (24-h integrated) were collected in urban Harbin during a six-month period, from October 16, 2018, to April 14, 2019. The sampling was performed on the campus of Harbin Institute of Technology (HIT; 45°45′24″ N, 126°40′49″ E). The collected filters were analyzed for organic carbon (OC), elemental carbon (EC), levoglucosan (LG), mannosan (MN), water-soluble inorganic ions, and brown carbon (BrC) absorption.

The $PM_{2.5}$ concentration $(PM_{2.5})^*$ was calculated as the sum of the inorganic ions, organic matter (OM, determined as $1.6 \times$ OC), and EC. Hourly air quality data were obtained from China's National Urban Air Quality Real-Time Publishing Platform (http://106.37.208.233:20035/). Hourly meteorological were obtained from Weather Underground (https://www.wunderground.com/).

3.2 *Satellite Data*

The Visible Infrared Imaging Radiometer Suite (VIIRS) true-color images around the sampling site were created using the Fire Information for Resource Management System (FIRMS; https://firms.modaps.eosdis.nasa.gov/). The images were overlaid with the active fire detections, which were also obtained from the FIRMS system.

3.3 *Approach*

Contributions of various sources to $PM_{2.5}$ and OC were estimated by the EPA's positive matrix factorization (PMF) model (version 5.0). Two samples corresponding to firework episodes due to Lunar New Year and the Lantern Festival, and 14 samples resulting from dust episodes, were excluded from the analysis. A total of five factors were resolved (Cheng et al. 2021c).

The mass concentration and chemical composition of $PM_{2.5}$ were simulated using a revised CMAQ model (version 11; SAPRC-11). The meteorological inputs were retrieved from the Weather Research and Forecasting (WRF) model, and the emission inputs were generated by combining various inventories, including the Multi-resolution Emission Inventory for China (MEIC; http://www.meicmodel.org/) and the satellite-based Fire INventory from NCAR (FINN; Wiedinmyer et al. 2011). The simulations were performed over East Asia with a horizontal resolution of 36×36 km, and modeling results were extracted for the grid cell where the sampling site is located (for details, see Cheng et al. 2021e).

4 Results and Discussion

4.1 Biomass Burning Around Harbin: Low-Efficiency Field Combustion of Crop Residues

Figure 1 shows the temporal variations of $(PM_{2.5})^*$ and levoglucosan, with campaign-averaged chemical compositions shown by the inner pie chart. The concentrations of $(PM_{2.5})^*$ during heating season are unexpectedly high, with an average of 56.14 ± 45.44 μg/m^3, and highest value reaching ~ 300 μg/m^3. As the $(PM_{2.5})^*$ concentrations were relatively higher during the coldest months, i.e., through December to February, one may expect that the severe pollution is due to intensive energy use for heating, from coal combustion and/or biofuel burning. SO_2 did show an increasing trend when the temperature decreased, which may point to emissions from heating-induced coal combustion. However, no consistent trend was observed regarding the dependence of sulfate on temperature, making the role of coal combustion on $PM_{2.5}$ pollution vague (Cheng et al. 2021d). On the other hand, levoglucosan, as a tracer of biomass burning, exhibited similar patterns with $PM_{2.5}$, indicating the potential contribution of biomass burning.

To better illustrate the role of BB, we first investigated the influence of biomass burning on organic matter, the dominant component in $(PM_{2.5})^*$ (Fig. 1). Samples were classified into three groups (namely Cases *A*, *B* and *C*) with LG/OC ranges of < 1.5%, between 1.5 and 3.0%, and > 3.0% (on a basis of carbon mass), respectively (Fig. 2a). For the three cases, levoglucosan and OC correlated strongly ($R^2 > 0.9$) with different slopes. The strong correlations between levoglucosan and OC could not be attributed primarily to the influence of meteorological conditions (e.g., wind speed and the planetary boundary layer height), since the correlations between levoglucosan and other components, e.g., EC and sulfate, were much weaker (Cheng et al. 2021c). It could thus be inferred that Fig. 2a pointed to the importance of biomass burning

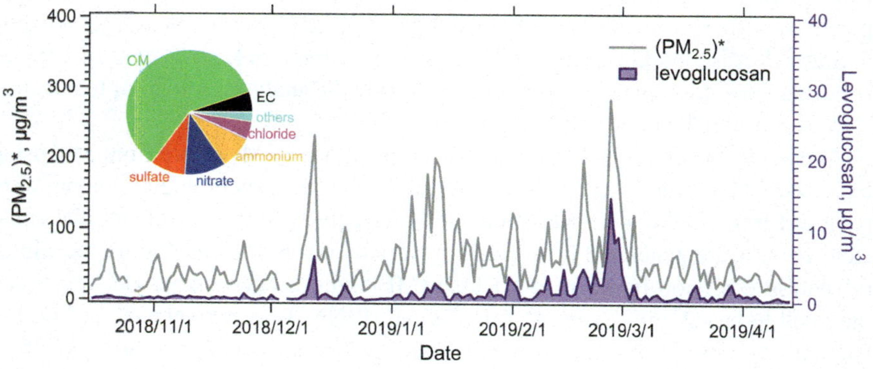

Fig. 1 Temporal variations of constructed $PM_{2.5}$ concentration, i.e., $(PM_{2.5})^*$, and levoglucosan. Campaign-averaged $(PM_{2.5})^*$ compositions are shown by the inner pie charts

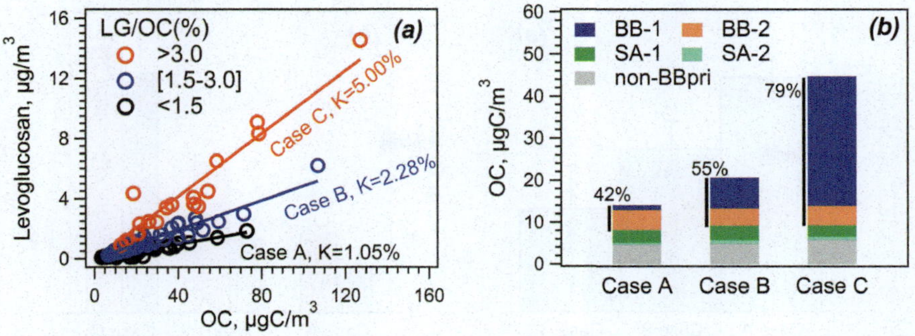

Fig. 2 **a** Classification of Cases *A*, *B*, and *C* based on the LG/OC ratios, and **b** comparison of OC source apportionment results for the three cases. In **a**, lines indicate linear regression results, with *K* denoting the slope. The LG/OC ratios and the *K* (i.e., $\Delta LG/\Delta OC$) values are presented on a basis of carbon mass

as an OC source for all the three cases. This was also supported by the source apportionment results from EPA's positive matrix factorization (PMF) model, i.e., ~ 51% of OC was attributed to biomass burning for the whole campaign while the BB contributions to OC were estimated to be ~ 42, 55 and 79% for Cases A, B, and C, respectively (Fig. 2b).

The three cases exhibited different chemical signatures, which could be related to the changes of biomass burning contribution. First, linear regression of EC on carbon monoxide (CO) showed decreasing slopes from Cases *A* through *C* (Fig. 3a). Second, levoglucosan became more abundant relative to other biomass burning tracers (i.e., mannosan and K^+) from Cases *A* through *C* (Fig. 3b). Based on a synthesis of the results from Harbin and BB source emission studies, it was inferred that the increase of biomass burning contribution from Cases *A* through *C* was mainly driven by smoldering combustion emissions. The chemical signatures also provided information on the types of the biomass fuels contributing to Harbin's BB aerosols. According to the source identification method suggested by Cheng et al. (2013), the observed $\Delta LG/\Delta K^+$ and $\Delta LG/\Delta MN$ values (Fig. 3b) were in general the characteristics of the crop residues combustions, which typically showed $EF_{LG}/EF_{K}{}^+$ and EF_{LG}/EF_{MN} values of 0.01–1 and 10–100, respectively. Although the relatively high $\Delta LG/\Delta K^+$ in Case *C* (i.e., above 1) was less commonly seen during laboratory burns of crop residues (Sullivan et al. 2008), comparable values have been observed in biomass burning episodes caused by post-harvest combustion, e.g., open burning of corn straws during autumn in the North China Plain (Liang et al. 2021). In general, the chemical signatures suggested the nature of smoldering combustion of crop residues for the agricultural fires observed in this study.

Among the two BB-related factors resolved by PMF, OC masses apportioned to BB-1 and BB-2 ($OC_{BB\text{-}1}$ and $OC_{BB\text{-}2}$) exhibited different patterns of variation, i.e., $OC_{BB\text{-}2}$ kept relatively unchanged, whereas $OC_{BB\text{-}1}$ increased by ~ 25 folds from

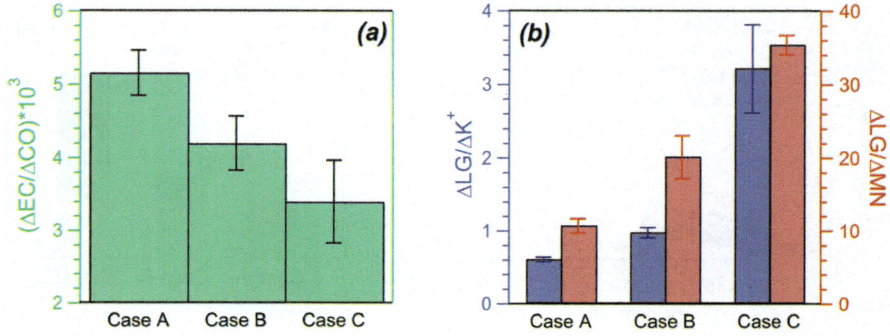

Fig. 3 Comparisons of **a** ΔEC/ΔCO, and **b** $\Delta LG/\Delta K^+$ and ΔLG/ΔMN across the three cases with different ranges of LG/OC (on a basis of carbon mass). The ratios are determined as slopes derived from linear regression

Cases *A* through *C* (Fig. 2b). Given that ambient temperature exhibited an increasing trend across the three cases (Cheng et al., 2021b) and there were typically intensive fire counts around Harbin for Cases *B* and *C* (Fig. 4), the BB-1 and BB-2 factors were inferred to be more representative of open burning and household use of biofuels for heating and cooking, respectively. As we discussed previously, the BB aerosols encountered in this campaign were mainly caused by the combustion of straw rather than wood, based on the relative abundances of levoglucosan, mannosan, and water-soluble potassium. Thus, the BB-1 and BB-2 factors could be further approximated as emissions from agricultural fires and residential burning of crop residues. In addition, PMF results showed that negligible EC was apportioned to BB-1 compared to BB-2 (Cheng et al. 2022), while a decrease in ΔEC/ΔCO occurred from Cases *A* through *C* with the ascending influence of the BB-1 factor, both pointing to the low-efficiency combustion (i.e., smoldering-dominated) characteristics of BB-1, the open burning factor. The relatively low combustion efficiencies made the BB smoke impacting Harbin different from most agricultural fires in literature (e.g., Pan et al. 2012) but to some extent similar to the tropical forest fires (Hodgson et al. 2018). Thus, the agricultural fires encountered in this study were distinct, presumably because the low ambient temperatures and the frost/snow over the land were unfavorable for flaming combustion of the crop residues.

In general, the $PM_{2.5}$ pollution in Harbin during the 2018–2019 heating season was largely impacted by the overlaying of two biomass burning-related sources, with the open burning source as the major contributor rather than household use of biomass. Furthermore, as indicated by the chemical signatures, the open burning pointed to agricultural fires of crop residues with relatively low combustion efficiencies.

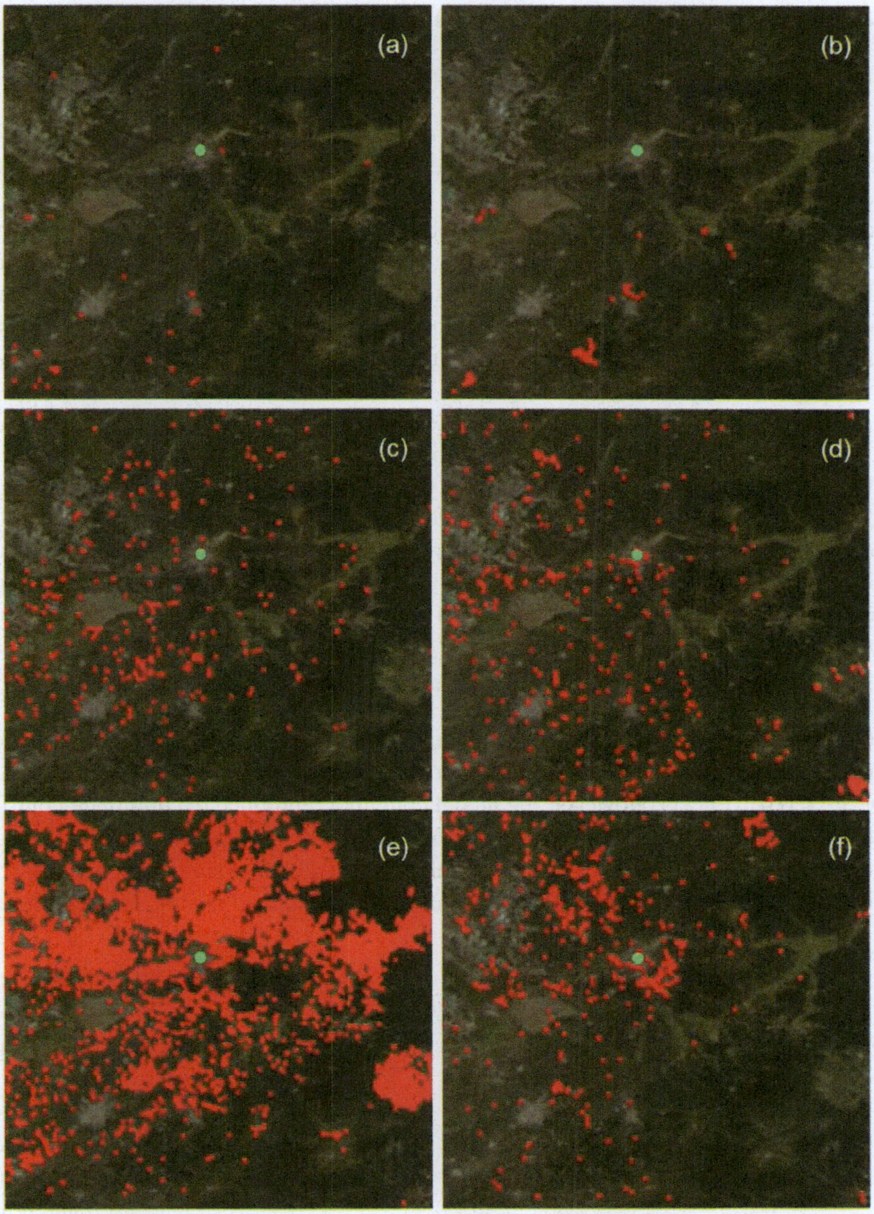

Fig. 4 Visible infrared imaging radiometer suite (VIIRS) true-color image around Harbin (highlighted by the green solid circles), overlaid with the active fire detections as red dots, during: **a** November 1–4, 2018; **b** December 5–8, 2018; **c** March 6–9, 2019; **d** April 10–13, 2019; **e** February 25–28, 2019; **f** March 28–31, 2019. The samples corresponding to **a** and **b** fell into Case *A*, the samples corresponding to **c** and **d** fell into Case *B*, and the samples corresponding to **e** and **f** fell into Case *C*, respectively. The images were created using the Fire information for resource management system (FIRMS; https://firms.modaps.eosdis.nasa.gov/)

4.2 Model versus Observation Discrepancy Associated with Open Burning

In this study, we compared not only the observed and modeled (subscripted using "obs" and "mod," respectively) $PM_{2.5}$ concentrations but also major particulate components for the entire heating season. As shown in Fig. 5a, $PM_{2.5mod}$ was significantly lower than $PM_{2.5obs}$ for a considerable fraction of samples. The highest underestimation was as much as ~ 230 $\mu g/m^3$, corresponding to the sample collected from February 26 to 27, 2019. This sample was characterized by the highest levoglucosan concentration during the campaign (14.56 $\mu g/m^3$). It was also apparent from Fig. 5a that $PM_{2.5}$ was under-predicted more significantly with increasing levoglucosan, i.e., with stronger impact of open burning.

Looking into the major components of $PM_{2.5}$, it is interesting to see that EC values were generally overestimated, opposite to bulk $PM_{2.5}$. Given that ΔEC (EC_{obs}-EC_{mod}) was independent of levoglucosan but became more negative with increasing NO_2, vehicle emissions were likely overestimated by the model. This inference also explained the negative Δnitrate at low levoglucosan. With increasing levoglucosan, however, Δnitrate increased toward zero and finally became positive, indicating the enhancement of nitrate formation by open burning. The influence of open burning on sulfate formation was apparent only at the higher end of levoglucosan concentrations (Cheng et al. 2021d).

The major component responsible for the under-prediction of bulk $PM_{2.5}$ is OC. Observational results showed that organic aerosol (OA) generally dominated the particle mass during the measurement period, with an average OA-to-$PM_{2.5}$ ratio of 0.59 ± 0.07 (Fig. 1). However, the model could not reproduce the dominant contribution of OA, as indicated by the substantially lower OA-to-$PM_{2.5}$ ratios derived from the revised CMAQ (averaging 0.40 ± 0.11). In addition, similar to the comparison for $PM_{2.5}$ concentration, the model also under-predicted the organic mass for a

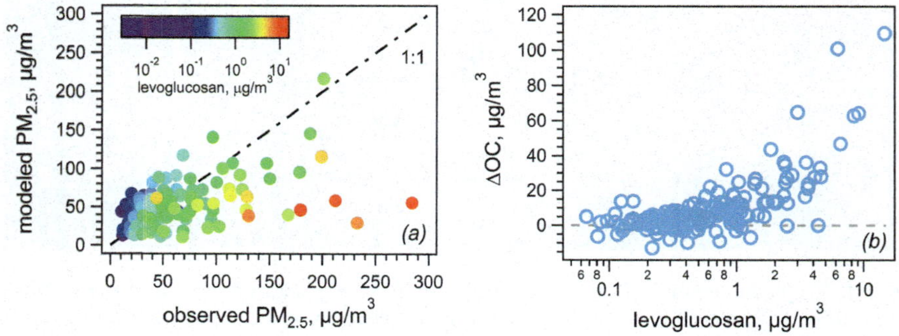

Fig. 5 **a** Comparison of observed and modeled $PM_{2.5}$ concentrations, color-coded by levoglucosan concentrations in log-scale. The dashed line indicates a one-to-one correspondence. **b** Dependence of the difference between observed and modeled OC concentrations (OC_{obs}–OC_{mod}, i.e., ΔOC) on levoglucosan. The gray dashed line indicates ΔOC value of zero

substantial fraction of samples, and the difference between observed and modeled OC (OC_{obs}-OC_{mod}, i.e., ΔOC) depended positively on levoglucosan as well (Fig. 5b). At the extremely high levoglucosan concentrations above 5 μg/m^3, ΔOC reached ~ 45–110 μgC/m^3, pointing to a dramatic underestimation of organic aerosol by the revised CMAQ when the open burning impacts are significant.

The levoglucosan-dependent under-prediction pointed to an underestimation of open burning emissions, which were obtained from the FINN inventory in the revised CMAQ. The same conclusion was reached by Uranishi et al. (2019), which suggested that for Northeast China, the FINN-based agricultural fire emissions needed to be increased by 20 times to explain the observed concentrations in the fall of 2014. On the other hand, although the under-prediction of OC by the revised CMAQ could be attributed to the underestimation of open burning emissions by the FINN inventory, this does not necessarily mean that all the ΔOC was caused by the missed primary OC (POC), i.e., missing of secondary OC (SOC) formed from agricultural fire emissions could also be partially responsible. In summary, this study demonstrates that uncertainties in open burning emissions could introduce substantial difficulties to $PM_{2.5}$ simulation for both primary components and secondary species.

4.3 Biomass Burning and Brown Carbon: Climate Impacts

Light-absorbing organic aerosol, i.e., brown carbon (BrC), has attracted much attention due to its potential contribution to global warming (Andreae and Gelencsér 2006). BB has generally been recognized as a dominant source of BrC worldwide, whereas current field observations are far from enough given that BB activities (e.g., wildfires, prescribed fires, and household burning) and the subsequent pollutant emissions and transformations are highly variable across regions (Laskin et al. 2015). This, in turn, results in substantial uncertainties in predicting and mitigating the climate impacts of BrC. In this study, we focused on BrC properties under the extremely strong BB impact.

The BrC properties, both mass concentration and light absorption, were retrieved via methanol extraction approaches. The methanol-extracted filters were also measured for EC, and the results were named $EC_{refined}$. Correspondingly, EC directly determined from untreated filters was defined as EC_{raw}. The BrC mass concentration was determined as MSOC, methanol-soluble OC, and the absorption was primarily investigated at a wavelength of 365 nm, shown as $(b_{abs})_{365}$. Persistently high $(b_{abs})_{365}$ and MSOC were observed in this campaign (Cheng et al. 2021b). $(b_{abs})_{365}$ correlated strongly with MSOC throughout the campaign, with an average mass absorption efficiency, MAE_{365}, of 1.28 ± 0.32 m^2/gC. Both $(b_{abs})_{365}$ and MSOC showed an increasing trend from Cases A through C, with ascending impacts from biomass burning. A more detailed analysis suggested that the residential burning of crop residues laid the foundation for overall lifted levels of biomass burning activities, whereas the agricultural fires, which seemed to have relatively low combustion efficiencies, were the dominant driver responsible for elevated BrC mass concentrations and increased BrC absorption (Cheng et al. 2021b).

Comparison of $EC_{refined}$ and EC_{raw}, i.e., the EC difference between methanol-extracted and untreated filters, provided insights into EC measurement uncertainties associated with the methanol-soluble OC. As shown in Fig. 6, $EC_{refined}$ was ~ 20, 30, and 40% lower than EC_{raw} for Cases *A*, *B*, and *C*, respectively. The loss of BC particles could not explain the relatively low $EC_{refined}$ during the methanol extraction process (Cheng et al. 2021b) and instead pointed to the overestimation of EC mass by EC_{raw}, i.e., EC determined following conventional procedures. Given that the BB-1 contribution increased sharply from Cases *A* through *C*, the EC overestimation was likely associated with emissions from biomass burning, especially the agricultural fires, which were inferred to have relatively low combustion efficiencies. A possible explanation was that a fraction of biomass burning OA was pyrolyzed into non-absorbing materials, which would be misclassified as EC. This artifact was more pronounced for protocols with relatively low peak inert-mode temperatures (e.g., IMPROVE-A), which might be responsible for decreasing $EC_{refined}$ to EC_{raw} ratios from Cases *A* through *C*.

In general, the agricultural fires were inferred to have relatively low combustion efficiencies, and the resulting emissions were identified as the dominant contributor to BrC. On the other hand, the agricultural fire emissions also led to an overestimation of EC mass by a factor of up to 1.6. Both could result in a substantial underestimation of the relative importance of BrC absorption compared to EC, i.e., the significance of BrC regarding global warming (Fig. 6). It is noteworthy that the fraction of solar energy absorbed by BrC relative to EC, i.e., $f_{BrC/EC}$, reached ~ 150 and 275% (based on $EC_{refined}$) for Cases *B* and *C*, respectively, in the ultraviolet wavelength range of 300–400 nm (Fig. 6). To our knowledge, such high $f_{BrC/EC}$ was rarely seen in other regions, highlighting the importance of BrC in main agricultural regions in Northeast China.

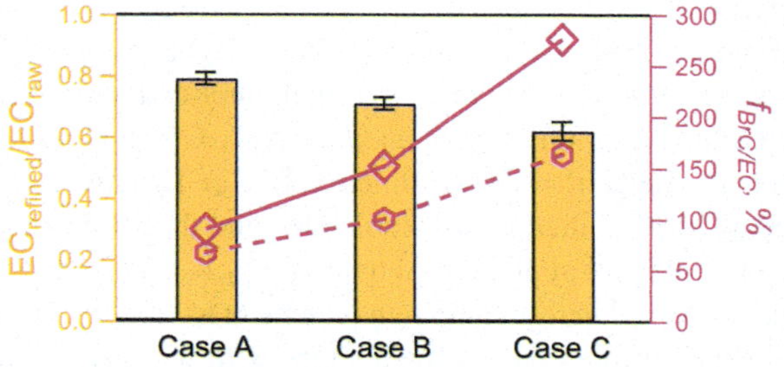

Fig. 6 Variations of $EC_{refined}$ to EC_{raw} ratio (orange solid bars, determined based on linear regression of $EC_{refined}$ on EC_{raw}) and the fraction of solar energy absorbed by BrC relative to EC, i.e., $f_{BrC/EC}$ (solid and dashed lines for the fractions computed using $EC_{refined}$ and EC_{raw}, respectively) from Cases *A* through *C*, with increasing biomass burning impacts. $f_{BrC/EC}$ are computed for the wavelength range of 300–400 nm

If we recall the temporal variations of BrC absorption and particulate components in this campaign, the three months from December 2018 to February 2019 were occupied with higher concentrations. This is in response to the release of the local policy *Interim Provisions of Heilongjiang Province on Reward and Punishment for Straw Open Burning Management*, which approved a window of approximately three months (from December 11, 2018, through March 9, 2019) for the open burning of crop residues. The low ambient temperatures during these three months and crop residues buried under snow likely resulted in the emissions in smoldering conditions. This suggested that the sustainable use of crop residues and the effectiveness of residue burning policy need to be re-considered carefully, especially in major agricultural regions such as Northeast China.

5 Conclusion

Based on a six-month field campaign conducted in Harbin (i.e., the central city in HC), we show that the increase of OC mainly drove the formation of heavily polluted $PM_{2.5}$ episodes during the heating season. Furthermore, based on a synthesis of various source-related information, such as the PMF results and different chemical signatures, we suggest that biomass burning was the dominant contributor to OC (with an average contribution of ~ 50%). Agricultural fires with relatively low combustion efficiencies could considerably enhance the contribution of biomass burning (e.g., ~ 40 versus 80% between Cases *A* and *C*). This study revealed the distinct influence of crop residue combustion on regional-level haze pollution in the HC area. However, the model simulation could not correctly reproduce the influence of agricultural field fires. Significant under-prediction of $PM_{2.5}$ concentration was identified when biomass burning impacts were substantial, likely due to the underestimation of open burning emissions by FINN. The intensive agricultural fire emissions in Northeast China introduced substantial difficulties to the accuracy of $PM_{2.5}$ simulation.

In addition to the characteristics of agricultural fires, we also discussed the climate impacts of low-efficiency biomass burning in this study. On the one hand, smoldering combustion of crop residues emitted a large amount of BrC; in contrast, the agricultural fire emissions also led to an overestimation of EC mass. Both resulted in a substantial underestimation of the relative importance of BrC in global warming. A refined estimation suggested that the contribution of BrC radiative forcing was comparable with or even overwhelmed that of EC, highlighting the importance of BrC in main agricultural regions in Northeast China.

Agricultural production in the Northeast China Plain is of great importance for food security in China, but sustainable use of crop residues remains a challenge; hence open burning, although prohibited, is difficult to be eliminated. Our study indicates that the massive agricultural sector in Northeast China deserved our urgent attention from various aspects, e.g., air quality improvement, model simulation accuracy, climate change mitigation, etc., and a new roadmap toward sustainable use of crop residues should be re-designed.

Acknowledgements This work was supported by the National Natural Science Foundation of China (42222706, 41805097), the Natural Science Foundation of Heilongjiang Province (YQ2019D004), the State Key Laboratory of Urban Water Resource and Environment (2020DX14), and the Heilongjiang Touyan Team.

References

Andreae, M.O., and A. Gelencsér. 2006. Black carbon or brown carbon? The nature of light-absorbing carbonaceous aerosols. *Atmospheric and Chemistry Physics* 6: 3131–3148.

Badarinath, K.V.S., S.K. Kharol, K.M. Latha, T.K. Chand, V.K. Prasad, A.N. Jyothsna and K. Samatha. 2007. Multiyear ground-based and satellite observations of aerosol properties over a tropical urban area in India. *Atmospheric Science Letters* 8(1): 7–13.

Badarinath, K.V.S., S.K. Kharol, A.R. Sharma and V.K. Prasad. 2009. Analysis of aerosol and carbon monoxide characteristics over Arabian Sea during crop residue burning period in the Indo-Gangetic plains using multi-satellite remote sensing datasets. *Journal of Atmospheric and Solar-Terrestrial Physics* 71(12): 1267–1276.

Badarinath, K.V.S., S. Kumar Kharol, V. Krishna Prasad, A. Rani Sharma, E.U.B. Reddi, H.D. Kambezidis, and D.G. Kaskaoutis. 2008. Influence of natural and anthropogenic activities on UV Index variations–a study over tropical urban region using ground based observations and satellite data. *Journal of Atmospheric Chemistry* 59(3): 219–236.

Cheng, Y., X.B. Cao, J.M. Liu, Q.Q. Yu, Y.J. Zhong, G.N. Geng, Q. Zhang, and K.B. He. 2022. New open burning policy reshaped the aerosol characteristics of agricultural fire episodes in Northeast China. *Science of the Total Environment* 810: 152272.

Cheng, Y., G. Engling, K.B. He, F.K. Duan, Y.L. Ma, Z.Y. Du, J.M. Liu, M. Zheng, and R.J. Weber. 2013. Biomass burning contribution to Beijing aerosol. *Atmospheric and Chemistry Physics* 13: 7765–7781.

Cheng, Y., Q. Yu, J. Liu, X. Cao, Y. Zhong, Z. Du, L. Liang, G. Geng, W. Ma, H. Qi, Q. Zhang, and K. He. 2021a. Dramatic changes in Harbin aerosol during 2018–2020: The roles of open burning policy and secondary aerosol formation. *Atmospheric and Chemistry Physics* 21: 15199–15211.

Cheng, Y., Q.Q. Yu, X.B. Cao, C.Q. Yan, Y.J. Zhong, Z.Y. Du, L.L. Liang, W.L. Ma, H. Qi, M. Zheng, and J.M. Liu. 2021b. Strong impacts of legitimate open burning on brown carbon aerosol in Northeast China. *Environmental Science & Technology Letters* 8: 732–738.

Cheng, Y., Q.Q. Yu, J.M. Liu, Z.Y. Du, L.L. Liang, G.N. Geng, B. Zheng, W.L. Ma, H. Qi, Q. Zhang, and K.B. He. 2021c. Strong biomass burning contribution to ambient aerosol during heating season in a megacity in Northeast China: Effectiveness of agricultural fire bans? *Science of the Total Environment* 754: 142144.

Cheng, Y., Q.Q. Yu, J.M. Liu, Y.W. Sun, L.L. Liang, Z.Y. Du, G.N. Geng, W.L. Ma, H. Qi, Q. Zhang, and K.B. He. 2021d. Formation of secondary inorganic aerosol in a frigid urban atmosphere. *Frontiers of Environmental Science & Engineering* 16: 18.

Cheng, Y., Q.Q. Yu, J.M. Liu, S. Zhu, M. Zhang, H. Zhang, B. Zheng, and K.B. He. 2021e. Model vs. observation discrepancy in aerosol characteristics during a half-year long campaign in Northeast China: the role of biomass burning. *Environmental Pollution* 269:116167.

Gupta, P.K., V.K. Prasad, C. Sharma, A.K. Sarkar, Y. Kant, K.V.S. Badarinath, and A.P. Mitra. 2001a. CH_4 emissions from biomass burning of shifting cultivation areas of tropical deciduous forests–experimental results from ground-based measurements. *Chemosphere-Global Change Science* 3 (2): 133–143.

Gupta, P.K., V.K. Prasad, Y. Kant, C. Sharma, A.B. Ghosh, M.C. Sharma, A.K. Sarkar, S.L. Jain, O.P. Tripathi, R.C. Sharma, and K.V.S. Badarinath. 2001b. Study of trace gases and aerosol emissions due to biomass burning at shifting cultivation sites in East Godavari District (Andhra Pradesh) during INDOEX IFP-99. *Current Science* 186–196.

Hodgson, A.K., W.T. Morgan, S. O'Shea, S. Bauguitte, J.D. Allan, E. Darbyshire, M.J. Flynn, D.T. Liu, J. Lee, B. Johnson, J.M. Haywood, K.M. Longo, P.E. Artaxo, and H. Coe. 2018. Near-field emission profiling of tropical forest and Cerrado fires in Brazil during SAMBBA 2012. *Atmospheric and Chemistry Physics* 18: 5619–5638.

Hu, W.W., M. Hu, W. Hu, J.L. Jimenez, B. Yuan, W.T. Chen, M. Wang, Y.S. Wu, C. Chen, Z.B. Wang, J.F. Peng, L.M. Zeng, and M. Shao. 2016. Chemical composition, sources, and aging process of submicron aerosols in Beijing: Contrast between summer and winter. *Journal of Geophysical Research Atmosphere* 121: 1955–1977.

Kharol, S.K., K.V.S. Badarinath, A.R. Sharma, D.V. Mahalakshmi, D. Singh and V.K. Prasad. 2012. Black carbon aerosol variations over Patiala city, Punjab, India—a study during agriculture crop residue burning period using ground measurements and satellite data. *Journal of Atmospheric and Solar-Terrestrial Physics* 84: 45–51.

Laskin, A., J. Laskin, and S.A. Nizkorodov. 2015. Chemistry of atmospheric brown carbon. *Chemical Reviews* 115: 4335–4382.

Lasko, K. and K. Vadrevu. 2018. Improved rice residue burning emissions estimates: accounting for practice-specific emission factors in air pollution assessments of Vietnam. *Environmental Pollution* 236: 795–806.

Lasko, K., K.P. Vadrevu. and T.T.N. Nguyen. 2018. Analysis of air pollution over Hanoi, Vietnam using multi-satellite and MERRA reanalysis datasets. *PLoS One* 13(5): e0196629.

Lasko, K., K.P. Vadrevu, V.T. Tran, E. Ellicott, T.T. Nguyen, H.Q. Bui. and C. Justice. 2017. Satellites may underestimate rice residue and associated burning emissions in Vietnam. *Environmental Research Letters* 12(8): 085006.

Li, Y.C., J. Liu, H. Han, T.L. Zhao, X. Zhang, B.L. Zhuang, T.J. Wang, H.M. Chen, Y. Wu, and M.M. Li. 2019. Collective impacts of biomass burning and synoptic weather on surface $PM_{2.5}$ and CO in Northeast China. *Atmospheric Environment* 213: 64–80.

Liang, L.L., G. Engling, C. Liu, W.Y. Xu, X.Y. Liu, Y. Cheng, Z.Y. Du, G. Zhang, J.Y. Sun, and X.Y. Zhang. 2021. Measurement report: chemical characteristics of $PM_{2.5}$ during typical biomass burning season at an agricultural site of the North China Plain. *Atmospheric and Chemistry Physics* 21: 3181–3192.

Oanh, K., N.T. Permadi, D.A. Dong, N.P. and D.A Nguyet. 2018. Emission of toxic air pollutants and greenhouse gases from crop residue open burning in Southeast Asia. In *Land-atmospheric research applications in South and Southeast Asia*, 47–66. Cham: Springer.

Pan, X.L., Y. Kanaya, Z.F. Wang, F. Taketani, H. Tanimoto, H. Irie, H. Takashima, and S. Inomata. 2012. Emission ratio of carbonaceous aerosols observed near crop residual burning sources in a rural area of the Yangtze River Delta Region. *China. Journal of Geophysical Research* 117: D22304.

Sullivan A.P., A.S. Holden, L.A. Patterson, G.R. McMeeking, S.M. Kreidenweis, W.C. Malm, W.M. Hao, C.E. Wold, and J.L. Collett. 2008. A method for smoke marker measurements and its potential application for determining the contribution of biomass burning from wildfires and prescribed fires to ambient $PM_{2.5}$ organic carbon. *Journal of Geophysical Research* 113: D22302.

Uranishi, K., F. Ikemori, H. Shimadera, A. Kondo, and S. Sugata. 2019. Impact of field biomass burning on local pollution and long-range transport of $PM_{2.5}$ in Northeast Asia. *Environmental Pollution* 244: 414–422.

Vadrevu, K., and K. Lasko. 2015. Fire regimes and potential bioenergy loss from agricultural lands in the Indo-Gangetic plains. *Journal of Environmental Management 148*:10–20.

Vadrevu K., T. Ohara and C. Justice. 2017. Land cover, land use changes and air pollution in Asia: a synthesis. *Environmental Research Letters* 12(12): 120201.

Vadrevu, K.P., T. Ohara, and C. Justice, eds. 2018. Land-atmospheric research applications in South and Southeast Asia. Cham: Springer.

Vadrevu, K.P., T. Ohara, and C. Justice, eds. 2021b. Biomass burning in South and Southeast Asia: Mapping and monitoring, vol. 1. CRC Press.

Vadrevu, K.P., T. Ohara, and C. Justice. 2014. Air pollution in Asia. *Environmental Pollution (Barking, Essex: 1987)* 195:233–235

Vadrevu Le Toan, T., S.S. Ray, and C. Justice. 2022a. Remote sensing of agriculture and land cover/land use changes in South and Southeast Asian Countries. Cham: Springer.

Vadrevu, K.P., A. Eaturu, E. Casadaban, and S. Biswas. 2022b. Agricultural fires in South Asian Countries and implications. In *Remote sensing of agriculture and land cover/land use changes in South and Southeast Asian Countries*, 501–516. Cham: Springer.

Wiedinmyer, C., S.K. Akagi, R.J. Yokelson, L.K. Emmons, J.A. Al-Saadi, J.J. Orlando, and A.J. Soja. 2011. The fire INventory from NCAR (FINN): a high resolution global model to estimate the emissions from open burning. *Geoscientific Model Development* 4: 625–641.

Xu, W.Q., Y.L. Sun, Q.Q. Wang, J. Zhao, J.F. Wang, X.L. Ge, C.H. Xie, W. Zhou, W. Du, J. Li, P.Q. Fu, Z.F. Wang, D.R. Worsnop, and H. Coe. 2019. Changes in aerosol chemistry from 2014 to 2016 in winter in Beijing: Insights from high-resolution aerosol mass spectrometry. *Journal of Geophysical Research Atmosphere* 124: 1132–1147.

Air Pollution Modeling in Southeast Asia—An Overview

Teerachai Amnuaylojaroen

Abstract Rapid economic growth and industrial development, including traditional practices such as slash-and-burn agriculture, have increased trace gas emissions in Southeast Asia (SEA) with impacts such as acid deposition, regional haze, air quality degradation, and climate change. In particular, air quality degradation has become an increasing concern in Southeast Asian countries. Various air pollution models were developed to understand the sources, transport, and chemical transformation of air pollutants to address this concern. In addition, the models serve as critical technical tools for air quality management because they provide helpful information to understand the interactions between emissions, their sources, and mitigation options. This study reviews some important air quality models, their potential, and their limitations, including applications in SEA. An air quality modeling case study over northern Thailand is also presented.

Keywords Biomass burning · Air pollution model · Air pollution · Air quality · Southeast Asia

1 Introduction

Air pollution in Southeast Asia (SEA) is a major concern throughout the year, particularly from January to April. Of the different sources, biomass burning, in-particular vegetation fires (Fig. 1), is a substantial contributor to air pollution on a regional to local scale in SEA (Amnuaylojaroen et al. 2010, 2014; Albar et al. 2018; Arvelyna et al. 2021; Hayasaka et al. 2014; Biswas et al. 2015; Lasko et al. 2017; 2018; 2021; Yin et al. 2019). Further, the increase in particulate pollution and significant haze episodes in this region have become more frequent and severe in recent years. Widespread biomass burning and airborne pollutants from human activities contribute significantly to Southeast Asia's air pollution (Lee et al. 2018; 2019; Amnuaylojaroen et al. 2019). Furthermore, several meteorological and topographic

---

T. Amnuaylojaroen (✉)
School of Energy and Environment, University of Phayao, Phayao 56000, Thailand
e-mail: teerachai4@gmail.com

K. P. Vadrevu et al. (eds.), *Vegetation Fires and Pollution in Asia*,
https://doi.org/10.1007/978-3-031-29916-2_31

characteristics contribute to Southeast Asia's air pollution (Vadrevu et al. 2014a,b; Eaturu and Vadrevu 2021). Air pollutant emissions or intercontinental pollution transport were discovered to be the primary drivers of future air quality (Colette et al. 2013; Nguyen et al. 2019; Zhang et al. 2017; Vadrevu et al. 2018). While air pollutant-related emissions were effectively decoupled from economic growth in many parts of the world, it is not the case in both South and Southeast Asia (Badarinath et al. 2007; 2008, 2009; Vadrevu and Justice 2011; Kharol et al. 2012). In S/SEA, increased economic activities are closely related to air pollution in continental Southeast Asia, and they will continue to increase if no additional policy efforts are made (Amann et al. 2013; Justice et al. 2015; Lasko and Vadrevu 2018; Vadrevu et al. 2017; 2021a,b; 2022a, 2022b). Controlling air pollutant emissions can positively impact air quality and human health (Fiore et al. 2015). High levels of air pollution can have several harmful health impacts. PM2.5 is a major air pollutant, and it has a substantial influence on human health (Lasko et al. 2021). According to a Pollution Control Department (PCD) report, at the beginning of 2020, PM2.5 measurements of air pollution in Chiang Mai, one of the largest cities in northern Thailand, reached the highest level of PM2.5 concentrations of up to 360 g/m^3 over the weekend, making the northern city the most polluted city in the world.

Air pollution modeling is a numerical tool for understanding the interaction between emissions, meteorology, atmospheric concentrations, deposition, and other parameters. Air pollution measurements provide valuable quantitative information regarding ambient concentrations and deposition; however, they have limitations. The measurements can only provide information on the air quality at specific locations and times if they provide clear guidance on identifying the origins of the air quality problem. Instead, air pollution modeling can provide a more comprehensive and prescriptive description of the air quality problem, including an analysis of various factors such as emission sources, meteorological processes, and physical and chemical changes and processes) including some guidance on mitigation measures implementation. Modeling results on air pollution can also help assess the influence of air pollutants on human health. Because of its ability to determine the relative relevance of critical processes, air pollution models play an essential role in science. Air pollution models are the only tools helpful in quantifying the deterministic relationship between emissions and concentrations or depositions, including understanding the effects of past and future scenarios for successful abatement strategies. As a result, air pollution models are critical in regulatory, research, and forensic applications. They are essential for estimating the relative contributions of various sources, monitoring compliance with air quality rules, and arriving at policy decisions.

2 Type of Air Pollution Modeling

Air pollution models utilize mathematical and numerical techniques to simulate the physical and chemical processes of air pollutant behavior as they distribute and react in the atmosphere. Based on meteorological data and source information such as

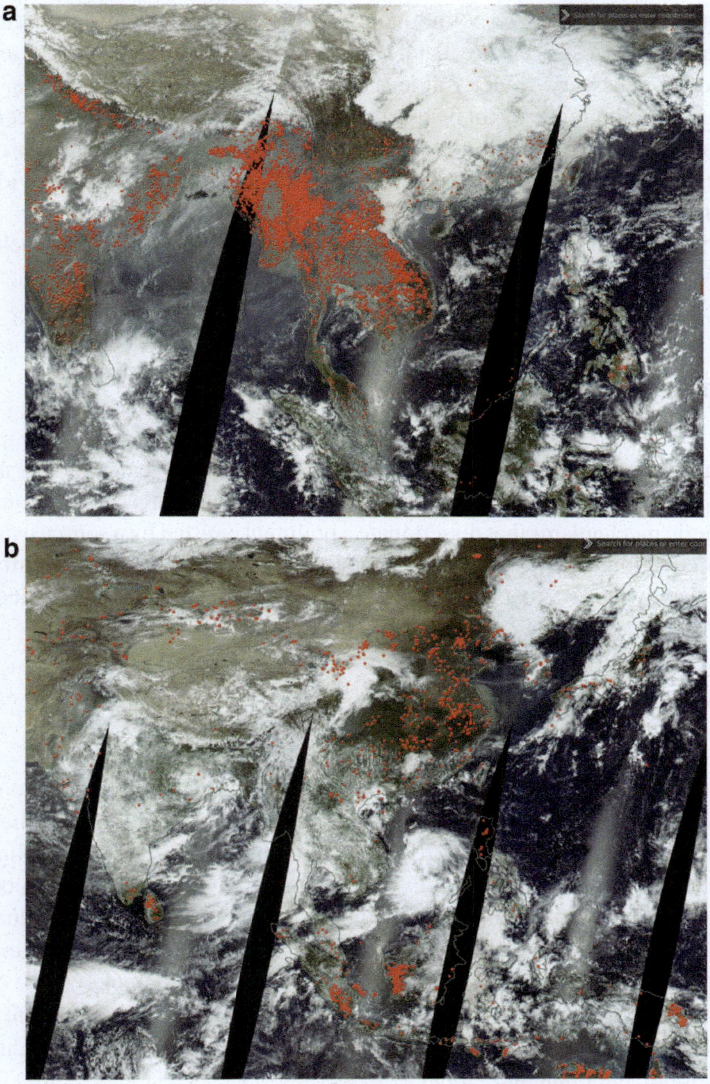

Fig. 1 Fires and thermal anomalies retrieved using Suomi NPP/VIIRS 375 m (day and night) data. **a** Shows peak fires in Southeast Asian countries of Myanmar, Thailand, Vietnam, and Laos on March 20, 2021, and **b** shows peak fires in Indonesia on September 23, 2021

emission rates and stack heights, the models characterize primary pollutants that are emitted directly into the atmosphere and, in some cases, secondary pollutants that are formed as a result of complex chemical reactions within the atmosphere. The models are critical to air quality management systems because they are frequently utilized by organizations tasked with regulating air pollution to identify source contributions to air quality problems and assist in creating successful measures to eliminate dangerous air pollutants. Air pollution models, for example, can be used throughout the permitting process to ensure that a new source will not exceed ambient air quality criteria or, if necessary, to determine acceptable additional control needs. Furthermore, air pollution models can be used to anticipate future pollutant concentrations from many sources following the implementation of a new regulatory program to measure the program's efficacy in decreasing hazardous exposures to individuals and the environment.

The most commonly used air pollution models based on EPA's approach include the following:

2.1 Dispersion Modeling

These models are typically used in the permitting process to calculate the air pollution at specific ground-level receptors near an emissions source. It characterizes the atmospheric mechanisms that disseminate a pollutant emitted by a source using mathematical equations. For example, a dispersion model can forecast concentrations at selected downwind receptor locations based on emissions and meteorological inputs.

2.2 Photochemical Modeling

These models are commonly used in regulatory or policy assessments to simulate the effects of all sources by calculating pollutant concentrations and the deposition of both inert and chemically reactive pollutants over broad geographic scales. Photochemical air quality models have become widely recognized and commonly used tools for regulatory analysis and attainment demonstrations by evaluating the efficacy of management techniques. These photochemical models are large-scale air quality models that use a set of mathematical equations to characterize the chemical and physical processes in the atmosphere to simulate changes in pollutant concentrations in the atmosphere. These models are used at various spatial scales, including local, regional, national, and global.

The Lagrangian trajectory model, which employs a moving frame of reference, and the Eulerian grid model, which employs a fixed coordinate system about the ground, are the two types of photochemical air quality models widely employed in air quality assessments. Because of its computational simplicity, earlier generation modeling projects frequently used the Lagrangian technique to simulate the production of contaminants. On the other hand, the Lagrangian approach needs to be more complete in terms of the physical processes it can represent. Therefore, most contemporary operational photochemical air quality models use three-dimensional Eulerian grid modeling, owing to its capacity to define physical processes in the atmosphere and predict species concentrations across the whole model domain.

2.3 Receptor Modeling

These models are observational techniques that employ the chemical and physical properties of gases and particles measured at the source and receptor to detect and quantify the presence of source contributions to receptor concentrations. Receptor models are mathematical or statistical processes that are used to detect and quantify the sources of air pollutants at a receptor location. Unlike photochemical and dispersion air quality models, receptor models do not assess the contribution of sources to receptor concentrations using pollutant emissions, meteorological data, or chemical transformation mechanisms. Instead, receptor models use the chemical and physical properties of gases and particles measured at the source and receptor to detect and quantify the presence of source contributions to receptor concentrations. As a result, these models are a natural complement to other air quality models. In addition, they are used as part of State Implementation Plans (SIPs) to identify sources of air quality concerns.

3 Review of Air Pollution Modeling Studies in Southeast Asia

For many decades, air pollution models have been widely used to examine the quality of air contaminants and the spread of harmful pollutants in Southeast Asia. For example, in Thailand, Amnuaylojaroen et al. (2022) utilized the Nested Regional Climate and Chemistry Model (NRCM-Chem) to predict PM2.5 concentrations over Southeast Asia's northern peninsula from 2020 to 2029 using the Representative Concentration Pathway (RCP) 8.5. PM2.5 concentrations tend to increase over the region in the range of (+ 1)–(+ 35) $\mu g/m^3$ during the dry season (November to April) and decrease in the range of (3)–(30) $\mu g/m^3$ during the wet season in response to variations in meteorological conditions and the emission of PM2.5 precursors (May to October). Amnuaylojaroen et al. 2020, used a coupled atmospheric and air pollution model based on the Weather Research and Forecasting Model (WRF) and a Hybrid Single-particle Lagrangian Integrated Trajectory Model (HYSPLIT). The finding showed that biomass burning from neighboring countries has a greater potential to contribute to air pollution in northern Thailand than national emissions, as evidenced by the number of hotspot locations in Burma being twice that of Thailand under the influence of two major Asian Monsoon channels, including easterly and northwesterly winds that bring pollution. Using the Weather Research and Forecasting Model with Chemistry, Khodmanee and Amnuaylojaroen (2021) investigated the impact of biomass burning on surface O_3, CO, and NO_2 levels in Northern Thailand (WRF-Chem). According to the findings, biomass burning increased O_3, CO, and NO_2 levels by 9%, 51%, and 96%, respectively. Amnuaylojaroen et al. 2019 investigated the main volatile organic compound (VOC) ozone precursors during high levels of biomass burning emissions in March 2014 over upper Southeast Asia using a Weather Research and Forecasting Model with Chemistry (WRF-CHEM) model that included anthropogenic emissions from EDGAR-HTAP, biomass burning from FINN, and biogenic emissions from MEGAN. According to the model results, CO and VOCs like BIGENE play an important role in atmospheric oxidation to surface O_3. Furthermore, biomass burning emissions increase surface O_3 by one ppmv and the reaction rate of CO and BIGENE by roughly 0.5106 and 1106 molecules/cm^3/s, respectively, in upper Southeast Asia. Amnuaylojaroen et al. (2014) used the Model for Ozone and Related Chemical Tracers (MOZART) gas-phase chemistry and the Global Ozone Chemistry Aerosol Radiation and Transport (GOCART) aerosols to investigate differences in predicted carbon monoxide (CO) and ozone (O_3) surface mixing ratios for Southeast Asia in March and December 2008. When comparing the March biomass burning period to the December period with low biomass burning emissions, biomass burning emissions cause a significant rise in both O_3 and CO by 29% and 16%, respectively. The simulations reveal that none of the anthropogenic emission inventories forecast O_3 surface mixing ratios better than the others. However, simulations with different anthropogenic emission inventories disagree in their predictions of CO surface mixing ratios, with differences of 30% for March and 10–20% for December at Thai surface monitoring locations. Amnuaylojaroen and Kreasuwun (2012) investigated PM2.5 distributions and the relative contributions of a fine fraction (PM2.5) and a coarse fraction (PM10–2.5) to the PM10 fraction from forest fires in the Chiang Mai basin in March 2007. They employed the WRF/CALPUFF modeling system. The significant atmospheric stability and low-level light breeze over the Chiang Mai basin provided a favorable environment for the buildup of particulate matter. The simulated PM2.5 distributions were largely concentrated near the burning zones. The model results show that PM2.5 and PM10–2.5 contributions account for 74% and 26% of the PM10, respectively. Because of the greater contribution of PM2.5, PM10 concentrations are more strongly related to PM2.5 than PM10–2.5. Amnuaylojaroen et al. (2018) investigated the sensitivity of air quality in Southeast Asia regarding ozone, future emissions, and climate change. According to model results, climate change raises ozone concentrations in Southeast Asia by roughly 30%, while the combination of future climate and emission related changes increases ozone by another 10% compared to the simulation with only future climate change. In March 2012, Amnuaylojaroen and Anuma (2021) used the coupled regional atmospheric model Weather Research and Forecasting (WRF) and air quality trajectory model HYSPLIT to determine the source contribution of PM10 in upper Southeast Asia and nearby regions. The Emissions Database for Global Atmospheric Research (EDGAR) and the NCAR Fire INventory (FINN) were employed for anthropogenic and biomass burning emission input data. According to the model results, biomass burning was responsible for approximately 56% of the PM10 concentration, while anthropogenic emissions accounted for approximately 44%. Pimonsree and Vongruang (2018) used Fire Radiative Power (FRP) to adapt FINN to solve the uncertainty of biomass burning emissions through an air quality modeling system. The WRF-CMAQ modeling system was used to simulate PMs in mainland of Southeast Asia, with a focus on Thailand, during a haze occurrence in March 2012. The simulation results were compared to satellite and ground-based data. The comparison of simulated PMs with modified FINN by FRP (PMFINN-FRP) generally revealed good agreement, with the modeling system capturing most of the significant observable traits. Comparing PMs in the source location revealed a significant improvement in simulated PMs. The simulated PMFINN-FRP has a factor of two of the observations greater than 70%, and the spatial correlation with the observations is greater than 0.8. Vongruang and Pimonsree 2020 explored the characteristics of biomass burning and their effects on PM10 concentrations during a haze outbreak between the March 1 and 31, 2012, inside and outside Phayao city, Thailand. During a smog episode, the authors assessed the emission control measures using Weather Research and Forecasting (WRF) and Community Multi-scale Air Quality (CMAQ) modeling systems. The study found a strong impact from biomass burning outside of the city on PM concentrations within the city, with contributions of approximately 85% and 89% for PM10 and PM2.5, respectively. Control of biomass burning in the city significantly affects local PM concentrations. According to our research, PM10 was emitted by biomass burning and accounted for 72% of all sources. Using the WRF-CMAQ modeling technique, we found that severe PM10 levels in Thailand surpassed the World Health Organization's (WHO) recommended safe limit for air quality by 51%.

In another research, Nguyen et al. (2019) employed an online coupled meteorology and chemistry WRF-CMAQ model to predict future O_3 and PM2.5 air quality over Continental Southeast Asia. According to the RCP4.5 scenario, the future atmosphere seems to have lower O_3 and PM2.5 concentrations, implying a potential "climate advantage" for air quality. During the dry season, significant increases were seen in northern Vietnam (for O_3) and southern Vietnam (for PM2.5). Xing et al. (2021) investigated the effects of biomass burning in peninsular Southeast Asia on PM2.5 concentrations and O_3 production in southern China. They tested an air pollution episode from March 21 to March 25, 2015, using a source-oriented WRF-Chem model. Sensitivity experiments demonstrate that biomass burning in Southeast Asia increases regional average PM2.5 concentrations by 39.3 $\mu g/m^3$ (68.0%) in Yunnan Province (YNP) and 8.4 $\mu g/m^3$ (24.1%) in other downwind areas (ODAs) in southern China, including the provinces of Guizhou, Guangxi, Hunan, Guangdong, Jiangxi, Fujian, and Zhejiang. Lin et al. (2014) investigated how biomass burning generates aerosols and air pollutants in Southeast Asia during the spring season. According to simulation studies, increased concentrations throughout the spring are associated with biomass burning plumes transported from Southeast Asia's Indochinese peninsula. During NASA's 2006 Biomass Burning Aerosols in Southeast Asia: Smoke Impact Assessment, Fu et al. (2012) ran numerical simulations (BASE-ASIA). The study demonstrates that biomass burning in Southeast Asia considerably impacts air quality in both local and downwind locations, particularly during biomass burning episodes. Li et al. (2017) investigated the regional influence of biomass burning (BB) on aerosols and source–receptor associations in Southeast Asia during March–April 2013 using a nested air quality prediction modeling system (NAQPMS) with an online tracer-tagged module. Aerosols from biomass burning were blown northward from the Indochina peninsula to China's southwestern provinces via the first pathway. The average contribution of biomass burning decreased from 70 to 80% in the source regions to 10–40% in southwestern China. Myanmar was the most important exporter. PM2.5 released by biomass burning was raised into the Indochina peninsula's mid-altitudes (2000 m) and transported eastward to the western Pacific at altitudes of 2500–4000 m, passing across the South China Sea, southern China, and western Pacific. Huang et al. (2013) evaluated the impact of biomass burning originating in SEA using a combination of numerical simulation, ground-based measurement, and satellite observation within the framework of NASA's 2006 Biomass Burning Aerosols in Southeast Asia: Smoke Impact Assessment (BASE-ASIA). In the spring of 2006, biomass burning emissions peaked in March, when the most extensive biomass burning took place in Myanmar, northern Thailand, Laos, and parts of Vietnam and Cambodia. Comparing AERONET aerosol optical characteristics with simulations at different sites indicated the impact of biomass burning via long-distance transmission. The contribution of biomass burning to AOD in the source region was calculated to be more than 56%. While the contribution was still significant in downwind zones, ranging from 26 to 62%. Yin et al. 2019 examined the temporal distribution of biomass burning in mainland Southeast Asia and its implications for local ambient air quality from 2001 to 2016 using remote sensing data, modeling data, and emission inventories. According to the findings, the monthly fire hotspots peaked at 34,512 in March. The monthly fluctuation followed a seasonal pattern strongly tied to precipitation and farming operations. Reddington et al. (2021) investigated the influence of forest and vegetation fires on air quality degradation and public health in Southeast Asia using a combination of regional and global air quality models and observations (including Mainland Southeast Asia and south-eastern China). They discovered that eliminating fire related emissions might significantly enhance regional air quality across Southeast Asia by reducing population exposure to PM2.5 concentrations by 7% and surface ozone concentrations by 5%. Lee et al. (2018) conducted numerical simulations using the Weather Research and Forecasting (WRF) model coupled with a chemistry component (WRF-Chem) to quantify the contributions of aerosols emitted from fire (i.e., biomass burning) versus non-fire (including fossil fuel combustion, road dust, and so on) sources to the degradation of air quality and visibility over Southeast Asia. According to the model results, 39% of observed low-visibility days (LVDs) can be explained by either fossil fuel burning or biomass burning emissions alone, 20% by fossil fuel burning alone, 8% by biomass burning alone, and 3% by a combination of fossil fuel burning and biomass burning. Permadi et al. (2018) used the WRF-CHIMERE model to investigate the potential benefits of specified black carbon (BC) emission reduction policies on air pollution and climate forcing in Southeast Asia (SEA). All the statistical criteria for PM assessment provided for the modeled PM10 and BC were met satisfactorily. The model results also revealed that the BC AOD contributed 7.5–12% AOD, which was consistent with prior research for areas with high emissions.

In all the above studies, several factors, such as emission input data accuracy and spatial resolution, could affect the air pollution model's performance. The observations were point-based at individual observation sites and might be affected by local emissions nearby, whereas the modeled concentrations were average grid values. However, to address these discrepancies between simulations and observations, further studies on the region's emissions, mechanisms, and meteorology are required (Amnuaylojaroen et al. 2022). Furthermore, Georgiou et al. (2018) postulated that the chemical mechanism processes overestimate the air pollutant's simulation. For example, mechanistic aerosol models slightly underestimate nitrate aerosols. These changes are due to the different handling of gas-to-particle partitioning from nitric acid to ammonium nitrate as a function of humidity (Balzarini et al. 2015), which employs the Zaveri et al. (2008) method where the diurnal cycle of relevant chemical components such as NO_3 and N_2O_5 is crucial during the nighttime. The discrepancy between biomass burning emissions and PM2.5 simulations was revealed by Pimonsree and Vongruang (2018).

4 Case Study of Air Pollution Modeling

We present a case study for predicting monthly PM2.5 concentrations in northern Thailand using the Nested Regional Climate Model with Chemistry (NRCM-Chem) between 2020 and 2029. The northern part of Thailand is mountainous and dominated by croplands. Most areas are considerably affected by traffic issues; however, farmers also burn crop residues in preparation for the upcoming rain and rice planting. The narrow valleys provide suitable basins for this smog and smoke problem. Furthermore, forest fires are a common occurrence in northern Thailand, occurring throughout the dry season. The model setup was described in Amnuaylojaroen et al. (2022), with three domains covering the bulk of Southeast Asia and parts of China. The first domain has a grid spacing of 60 km, the second domain in Thailand has a grid spacing of ten kilometers, and the finest domain covers northern Thailand with a grid spacing of one kilometer. Figure 2 depicts the monthly mean PM2.5 concentrations compared to Pollution Control Department observations in 2020. The pattern of monthly simulated PM2.5 concentrations is comparable to observations. The model achieved a satisfactory simulation of PM2.5, with an overestimation during the dry season (January–April) and an underestimation during the wet season (June–December). By comparing with the previous year's data from 1990–1999, the modeled average PM2.5 concentration increased in the range from (1) to (35) $\mu g/m^3$ during the dry season (November to April) and decreases in the ranges of (3)–(30) $\mu g/m^3$ during the wet season (May to October) (Fig. 2). The maximum increase was found in March, with concentrations of $> 40\ \mu g/m^3$ (Fig. 3c).

5 Conclusion

The problem of air quality deterioration in Southeast Asia has been increasing with trace gas and aerosol emissions. Air pollution models have been widely employed to understand better the sources, transport, and chemical changes of air pollutants. This study reviewed existing studies on air quality and modeling in Southeast Asia.

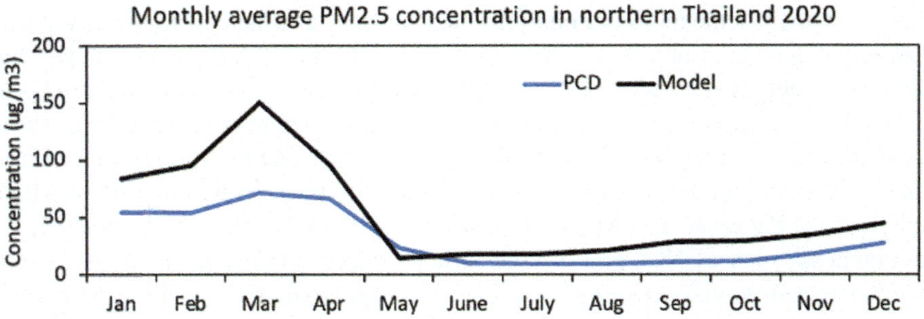

Fig. 2 Monthly average of PM2.5 concentration in the year 2020 between the model (black line) and ground-based measurement (blue line) in northern Thailand

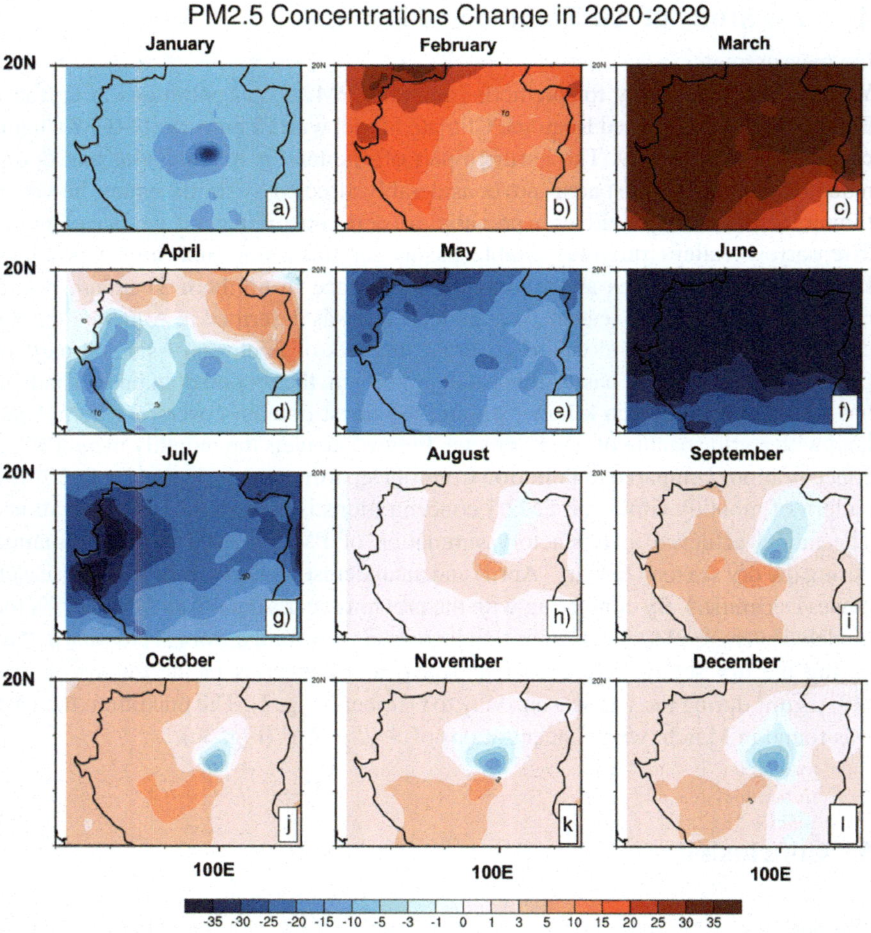

Fig. 3 Difference of PM2.5 concentrations from January–December in 2020–2029 (shown below) and 1990–1999 over the northern Thailand

Most air quality models gather quantitative information on the amounts of chemical species in the atmosphere, the distributions of which are regulated by four governing processes: emission, transport, chemistry, and deposition. For effective results, the air quality models need good input data, i.e., emission inventories which are not well developed in most of the countries in Southeast Asia, thus, restricting the models' capabilities. In addition, biomass burning is a major contributor to Southeast Asia's poor air quality. Several air pollutants, including PM10, PM2.5, O_3, and CO, are predicted to increase due to biomass burning activities, particularly during the peak of the dry season. Over the last few years, O_3 concentrations in SEA have gradually increased, and several studies relate the increase in O_3 emissions to an increase in biomass burning emissions. On the other hand, the PMs trend, particularly PM2.5, is expected to rise due to rapid industrial development and biomass burning. The link between BC and PMs was also established, which increased the AOD in Southeast Asia. With the advancement in monitoring capability via in situ, airplane, and satellite measurements, it has become vital to integrate observations and models to gain a greater knowledge of the sources, movement, and transformation of air pollutants. By adding observations into models, inverse modeling has been shown to be an efficient technique for providing observational constraints on a priori information and reducing simulation errors. Inverse modeling combined with atmospheric observations of air pollutant concentrations can efficiently reduce uncertainty in the emission inventory. The author believes that inverse modeling and its technical advancement should be stressed further in air quality models for Southeast Asia.

References

Albar, I. I. Jaya, B.H. Saharjo, B. Kuncahyo and K.P. Vadrevu. 2018. Spatio-temporal analysis of land and forest fires in Indonesia using MODIS active fire dataset. In *Land-Atmospheric Research Applications in South and Southeast Asia*, 105–127. Cham: Springer.

Amann, M., Z. Klimont, and F. Wagner. 2013. Regional and global emissions of air pollutants: recent trends and future scenarios. *Annual Review of Environment and Resources* 38: 31–55.

Amnuaylojaroen, T. and N. Anuma. 2021. Estimation of PM10 concentrations from biomass burning and anthropogenic emissions using the WRF–HYSPLIT modeling system. In *Biomass Burning in South and Southeast Asia*, 297–308. CRC Press.

Amnuaylojaroen, T., J. Kreasuwun, S. Towta, and K. Siriwitayakorn. 2010. Dispersion of particulate matter (PM10) from forest fires in Chiang Mai province. *Thailand. Chiang Mai J Sci* 37 (1): 39–47.

Amnuaylojaroen, T., and J. Kreasuwun. 2012. Investigation of fine and coarse particulate matter from burning areas in Chiang Mai, Thailand using the WRF/CALPUFF. *Chiang Mai Journal of Science* 39 (2): 311–326.

Amnuaylojaroen, T., M.C. Barth, L.K. Emmons, G.R. Carmichael, J. Kreasuwun, S. Prasitwattanaseree, and S. Chantara. 2014. Effect of different emission inventories on modeled ozone and carbon monoxide in Southeast Asia. *Atmospheric Chemistry and Physics* 14 (23): 12983–13012.

Amnuaylojaroen, T., M.C. Barth, G. Pfister, and C. Bruyere. 2018. Simulations of emissions, air quality, and climate contribution in Southeast Asia for March and December. In *Land-atmospheric research applications in South and Southeast Asia*, 233–254. Cham: Springer.

Amnuaylojaroen, T., J. Inkom, R. Janta, and V. Surapipith. 2020. Long range transport of southeast asian PM2. 5 pollution to northern Thailand during high biomass burning episodes. *Sustainability* 12(23):10049.
Amnuaylojaroen, T., R.C. Macatangay, and S. Khodmanee. 2019. Modeling the effect of VOCs from biomass burning emissions on ozone pollution in upper Southeast Asia. *Heliyon* 5 (10): e02661.
Amnuaylojaroen, T., V. Surapipith, and R.C. Macatangay. 2022. Projection of the near-future PM2.5 in Northern Peninsular Southeast Asia under RCP8.5. *Atmosphere* 13(2): 305.
Arvelyna, Y., H. Takahashi, L. Indrayanti, H. Hayasaka, K.P. Vadrevu, R. Maryani, M. Osaki, and H. Kazuyo. 2021. Peatland surface loss due to fires in central Kalimantan, Indonesia–a case study using differential interferometry SAR (DInSAR). In *Biomass Burning in South and Southeast Asia*, 77–90. CRC Press.
Badarinath, K.V.S., S.K. Kharol, K.M. Latha, T.K. Chand, V.K. Prasad, A.N. Jyothsna, and K. Samatha. 2007. Multiyear ground-based and satellite observations of aerosol properties over a tropical urban area in India. *Atmospheric Science Letters* 8 (1): 7–13.
Badarinath, K.V.S., S.K. Kharol, A.R. Sharma, and V.K. Prasad. 2009. Analysis of aerosol and carbon monoxide characteristics over Arabian Sea during crop residue burning period in the Indo-Gangetic Plains using multi-satellite remote sensing datasets. *Journal of Atmospheric and Solar-Terrestrial Physics* 71 (12): 1267–1276.
Badarinath, K.V.S., S. Kumar Kharol, V. Krishna Prasad, A. Rani Sharma, E.U.B. Reddi, H.D. Kambezidis, and D.G. Kaskaoutis. 2008. Influence of natural and anthropogenic activities on UV Index variations–a study over tropical urban region using ground based observations and satellite data. *Journal of Atmospheric Chemistry* 59 (3): 219–236.
Balzarini, A., G. Pirovano, L. Honzak, R. Žabkar, G. Curci, R. Forkel, M. Hirtl, R. San José, P. Tuccella, and G.A. Grell. 2015. WRF-Chem model sensitivity to chemical mechanisms choice in reconstructing aerosol optical properties. *Atmospheric Environment* 115: 604–619.
Biswas, S., K.D. Lasko, and K.P. Vadrevu. 2015. Fire disturbance in tropical forests of Myanmar—Analysis using MODIS satellite datasets. *IEEE Journal of Selected Topics in Applied Earth Observations and Remote Sensing* 8 (5): 2273–2281.
Colette, A., B. Bessagnet, R. Vautard, S. Szopa, S. Rao, S. Schucht, Z. Klimont, L. Menut, G. Clain, F. Meleux, and G. Curci. 2013. European atmosphere in 2050, a regional air quality and climate perspective under CMIP5 scenarios. *Atmospheric Chemistry and Physics* 13 (15): 7451–7471.
Eaturu, A., K.P. Vadrevu. 2021. Evaluation of sentinel-3 SLSTR data for mapping fires in forests, peatlands, and croplands–a case study over Australia, Indonesia, and India. In *Biomass burning in South and Southeast Asia*, 39–59. CRC Press.
Fiore, A.M., V. Naik, and E.M. Leibensperger. 2015. Air quality and climate connections. *Journal of the Air and Waste Management Association* 65 (6): 645–685.
Fu, J.S., N.C. Hsu, Y. Gao, K. Huang, C. Li, N.H. Lin, and S.C. Tsay. 2012. Evaluating the influences of biomass burning during 2006 BASE-ASIA: a regional chemical transport modeling. *Atmospheric Chemistry and Physics* 12 (9): 3837–3855.
Georgiou, G.K., T. Christoudias, Y. Proestos, J. Kushta, P. Hadjinicolaou, and J. Lelieveld. 2018. Air quality modelling in the summer over the eastern Mediterranean using WRF-Chem: chemistry and aerosol mechanism intercomparison. *Atmospheric Chemistry and Physics* 18 (3): 1555–1571.
Hayasaka, H., I. Noguchi, E.I. Putra, N. Yulianti, and K. Vadrevu. 2014. Peat-fire-related air pollution in Central Kalimantan, Indonesia. *Environmental Pollution* 195: 257–266.
Huang, K., J.S. Fu, N.C. Hsu, Y. Gao, X. Dong, S.C. Tsay, and Y.F. Lam. 2013. Impact assessment of biomass burning on air quality in Southeast and East Asia during BASE-ASIA. *Atmospheric Environment* 78: 291–302.
Justice, C., G. Gutman, and K.P. Vadrevu. 2015. NASA land cover and land use change (LCLUC): an interdisciplinary research program. *Journal of Environmental Management* 148: 4–9.
Kharol, S.K., K.V.S. Badarinath, A.R. Sharma, D.V. Mahalakshmi, D. Singh, and V.K. Prasad. 2012. Black carbon aerosol variations over Patiala city, Punjab, India—a study during agriculture crop

residue burning period using ground measurements and satellite data. *Journal of Atmospheric and Solar-Terrestrial Physics* 84: 45–51.

Khodmanee, S., and T. Amnuaylojaroen. 2021. Impact of biomass burning on Ozone, Carbon Monoxide, and Nitrogen Dioxide in Northern Thailand. *Frontiers in Environmental Science* 9: 27.

Lasko, K., and K. Vadrevu. 2018. Improved rice residue burning emissions estimates: accounting for practice-specific emission factors in air pollution assessments of Vietnam. *Environmental Pollution* 236: 795–806.

Lasko, K., K.P. Vadrevu, and T.T.N. Nguyen. 2018. Analysis of air pollution over Hanoi, Vietnam using multi-satellite and MERRA reanalysis datasets. *PLoS ONE* 13 (5): e0196629.

Lasko, K., K.P. Vadrevu, V. Bandaru, T.N.T. Nguyen, and H.Q. Bui.2021. PM2. 5 emissions from biomass burning in South/Southeast Asia–Uncertainties and trade-offs. In *Biomass burning in south and Southeast Asia*, 149–169. CRC Press.

Lasko, K., K.P. Vadrevu, V.T. Tran, E. Ellicott, T.T. Nguyen, H.Q. Bui, and C. Justice. 2017. Satellites may underestimate rice residue and associated burning emissions in Vietnam. *Environmental Research Letters* 12 (8): 085006.

Lee, H.H., O. Iraqui, and C. Wang. 2019. The impact of future fuel consumption on regional air quality in Southeast Asia. *Scientific Reports* 9 (1): 1–20.

Lee, H.H., O. Iraqui, Y. Gu, S.H.L. Yim, A. Chulakadabba, A.Y.M. Tonks, Z. Yang, and C. Wang. 2018. Impacts of air pollutants from fire and non-fire emissions on the regional air quality in Southeast Asia. *Atmospheric Chemistry and Physics* 18 (9): 6141–6156.

Li, J., Y. Zhang, Z. Wang, Y. Sun, P. Fu, Y. Yang, H. Huang, J. Li, Q. Zhang, C. Lin, and N.H. Lin. 2017. Regional impact of biomass burning in Southeast Asia on atmospheric aerosols during the 2013 Seven South-East Asian Studies project. *Aerosol Air Qual. Res* 17 (12): 2924–2941.

Lin, C.Y., C. Zhao, X. Liu, N.H. Lin, and W.N. Chen. 2014. Modelling of long-range transport of Southeast Asia biomass-burning aerosols to Taiwan and their radiative forcings over East Asia. *Tellus B: Chemical and Physical Meteorology* 66 (1): 23733.

Nguyen, G.T.H., H. Shimadera, K. Uranishi, T. Matsuo, and A. Kondo. 2019. Numerical assessment of PM2.5 and O_3 air quality in Continental Southeast Asia: impacts of potential future climate change. *Atmospheric Environment* 215: 116901.

Permadi, D.A., N.T. Kim Oanh, and R. Vautard. 2018. Integrated emission inventory and modeling to assess distribution of particulate matter mass and black carbon composition in Southeast Asia. *Atmospheric Chemistry and Physics* 18 (4): 2725–2747.

Pimonsree, S., and P. Vongruang. 2018. Impact of biomass burning and its control on particulate matter over a city in mainland Southeast Asia during a smog episode. *Atmospheric Environment* 195: 196–209.

Reddington, C.L. L. Conibear, S. Robinson, C. Knote, S.R. Arnold and D.V. Spracklen. 2021. Air pollution from forest and vegetation fires in Southeast Asia disproportionately impacts the poor. *GeoHealth* 5(9): p.e2021GH000418.

Vadrevu, K., A. Eaturu, E. Casadaban, K. Lasko, W. Schroeder, S. Biswas, L. Giglio, and C. Justice. 2022. Spatial variations in vegetation fires and emissions in South and Southeast Asia during COVID-19 and pre-pandemic. *Scientific Reports* 12 (1): 1–21.

Vadrevu, K., T. Ohara, and C. Justice. 2017. Land cover, land use changes and air pollution in Asia: a synthesis. *Environmental Research Letters* 12 (12): 120201.

Vadrevu, K.P., and C.O. Justice. 2011. Vegetation fires in the Asian region: Satellite observational needs and priorities. *Global Environmental Research* 15 (1): 65–76.

Vadrevu, K.P., K. Lasko, L. Giglio, and C. Justice. 2014a. Analysis of Southeast Asian pollution episode during June 2013 using satellite remote sensing datasets. *Environmental Pollution* 195: 245–256.

Vadrevu, K.P., T. Ohara, and C. Justice, eds. 2018. *Land-atmospheric research applications in South and Southeast Asia.* Cham: Springer.

Vadrevu, K.P., T. Ohara, and C. Justice. eds. 2021a. *Biomass burning in south and Southeast Asia: Impacts on the biosphere*, vol. 2. CRC Press.

Vadrevu, K.P., T. Ohara, and C. Justice. eds. 2021b. *Biomass burning in South and Southeast Asia: Mapping and monitoring,* vol. 1. CRC Press.

Vadrevu, K.P., T. Ohara, and C. Justice. 2014a. Air pollution in Asia. *Environmental Pollution (Barking, Essex: 1987)* 195: 233–235

Vadrevu Le Toan, T., S.S. Ray and C. Justice. 2022a. Remote sensing of agriculture and land cover/land use changes in South and Southeast Asian Countries. Cham: Springer

Vongruang, P., and S. Pimonsree. 2020. Biomass burning sources and their contributions to PM10 concentrations over countries in mainland Southeast Asia during a smog episode. *Atmospheric Environment* 228: 117414.

Xing, L., N. Bei, J. Guo, Q. Wang, S. Liu, Y. Han, S. Pongpiachan, and G. Li. 2021. Impacts of biomass burning in Peninsular Southeast Asia on PM2. 5 concentration and ozone formation in Southern China during Springtime—a case study. *Journal of Geophysical Research: Atmospheres* 126(22): p.e2021JD034908

Yin, S., X. Wang, X. Zhang, M. Guo, M. Miura, and Y. Xiao. 2019. Influence of biomass burning on local air pollution in mainland Southeast Asia from 2001 to 2016. *Environmental Pollution* 254: 112949.

Zaveri, R.A., R.C. Easter, J.D Fast, and L.K. Peters. 2008. Model for simulating aerosol interactions and chemistry (MOSAIC). *Journal of Geophysical Research: Atmospheres* 113(D13)

Zhang, Y., K. Wang, and C. Jena. 2017. Impact of projected emission and climate changes on air quality in the US: from national to state level. *Procedia Computer Science* 110: 167–173.

Trace Gases and Air Quality in Northwestern Vietnam During Recurrent Biomass Burning on the Indochina Peninsula Since 2014—Field Observations and Atmospheric Simulations

Simone M. Pieber, Stephan Henne, Nhat Anh Nguyen, Dac-Loc Nguyen, and Martin Steinbacher

Abstract Biomass burning (BB), including forest wildfires and agricultural waste burning, is a significant source of atmospheric trace gases and aerosols that are harmful to human health and alter the Earth's radiative balance. Here, we study the influence of BB on trace gas levels and air quality on the Indochinese Peninsula in Southeast Asia, where the dry season from mid-December until mid-April leads to regularly recurring large-scale BB until the onset of the Asian summer monsoon. Since 2014, trace gases and air quality parameters (carbon dioxide (CO_2), methane (CH_4), carbon monoxide (CO), and ozone (O_3)) and aerosol optical properties have been continuously monitored in situ at the regional Global Atmosphere Watch (GAW) station Pha Din (PDI) in rural Northwestern Vietnam. The station is well suited to study the large-scale fires on the Indochinese Peninsula, as emission plumes are frequently transported toward the site. Our investigations indicate that the annually recurrent large-scale BB leads to CO mixing ratios at PDI that exceed 1000 ppb (24 h means), typically during the February to April dry season. Instead, mixing ratios well below 100 ppb are observed in summer, when precipitation reaches a maximum. Further, we compare CO mixing ratios from two atmospheric transport and chemistry simulations (Copernicus Atmospheric Monitoring Service (CAMS)

S. M. Pieber (✉) · S. Henne · M. Steinbacher
Empa, Swiss Federal Laboratories for Materials Science and Technology, Dübendorf, Switzerland
e-mail: simone.pieber@empa.ch

S. M. Pieber
AirUCI (Atmospheric Integrated Research), University of California, Irvine, USA

N. A. Nguyen
Hydro-meteorological Observation Center, Vietnam Meteorological and Hydrological Administration, Ministry of Natural Resources and Environment, Ha Noi, Vietnam

D.-L. Nguyen
Institut Für Umweltmedizin, Helmholtz Zentrum München, Munich, Germany

Chair of Analytical Chemistry, University of Rostock, 18059 Rostock, Germany

K. P. Vadrevu et al. (eds.), *Vegetation Fires and Pollution in Asia*,
https://doi.org/10.1007/978-3-031-29916-2_32

global reanalysis and dedicated FLEXPART backward simulations) to the observations at PDI. Although the simulations tend to overestimate CO values during the periods when the highest CO mixing ratios are observed, they generally confirm BB's large impact on PDI observations. In addition, the results indicate that mostly very fresh BB emissions are sampled in the most prominent plumes (age 24 to 72 h).

Keywords Air quality · Biomass burning · Trace gases · Aerosol · Vietnam · GAW

1 Introduction

Biomass burning (BB) is a substantial source of air pollutants (trace gases and aerosols) in the atmosphere (Andreae and Merlet 2001; Andreae 2019). Globally, BB is a significant contributor to greenhouse gas emissions and, thus, to climate change (Kato et al. 2013). At smaller spatial scales, BB can be responsible for poor regional air quality and can cause adverse health effects (Yadav et al. 2017; Cusworth et al. 2018). In the tropics, the recurrent appearance of BB is strongly driven by rainfall and monsoon patterns (Yen et al. 2013; Chin et al. 2017; Yang et al. 2022), and large-scale oscillations like ENSO-El Nino and its influence (e.g., Van Der Werf et al. 2004, 2006). On a continental scale, the largest fire emissions are usually observed in South America and Sub-Sahara Africa (Schultz et al. 2008; van der Werf et al. 2017; Van Marle et al. 2017). However, BB in Southeast Asia is often of particular concern as it impacts highly populated areas (Betha et al. 2014; Vadrevu et al. 2017; 2021a,b). Carbon monoxide (CO) is a common tracer for BB plumes (Mu et al. 2011; Saito et al. 2022) because CO emissions from wildfires such as BB are typically rather large due to a low combustion efficiency (e.g., Akagi et al. 2011). Moreover, the lifetime of CO in the atmosphere is sufficiently long to allow tracking of the plumes downwind of the fires. The Indochina Peninsula, including Myanmar, Thailand, Cambodia, Laos, as well as Vietnam, is known for recurrent and intense BB, especially during the dry season in spring (Streets et al. 2003; Yen et al. 2013; Le et al. 2014; Biswas et al. 2015; Oanh et al. 2018; Lasko and Vadrevu 2018; Adam and Balasubramanian 2021) and previous work on the 7-SEAS campaign (Chuang et al. 2013; Li et al. 2013; Lin et al. 2013; Tsai et al. 2013; Chuesaard et al. 2014). However, as fires are often not spotted appropriately by satellites due to a thick aerosol-cloud system, particularly over Northern Vietnam, ground-based measurements offer an invaluable validation possibility. Here, we report on long-term observations of CO, ozone (O_3), and carbonaceous greenhouse gases (GHG), including carbon dioxide (CO_2) and methane (CH_4), based on measurements performed at the regional Global Atmosphere Watch (GAW) station Pha Din (PDI) in Northwestern Vietnam. Our study on trace gases complements previous studies at PDI, which focused on aerosol optical properties (Bukowiecki et al. 2019) and carbonaceous aerosol composition (Nguyen et al. 2021).

2 Study Area

Pha Din (PDI) (21.573°N, 103.516°E, 1466 m a.s.l.) is a meteorological station of the Vietnam Meteorological and Hydrological Administration (VNMHA), which started its operation in January 2012 after the observations were moved from a nearby site to this site in September 2011. Measurements of trace gases and aerosol optical properties were added in March 2014 in the context of a Swiss-Vietnamese collaboration called Capacity Building and Twinning for Climate Observing Systems (CATCOS) supported by the Swiss Agency for Development and Cooperation SDC. CATCOS aimed at initiating and resuming long-term observations of the essential climate variables in data sparse regions. In 2014, the Pha Din was accepted as a regional station of the World Meteorological Organization's (WMO) Global Atmosphere Watch (GAW) programme. The station, described in detail in Bukowiecki et al. (2019), is located 360 km northwest of Hanoi, 200 km south of the border with China, and 120 km east of the border with Laos (see Fig. 1). The observation site is located on the top of a hill (see Fig. 2) about 1 km north of the Pha Din mountain pass, which connects Son-La city (to the southeast) and Dien-Bien-Phu city (to the west) via national highway AH13. The Pha Din pass represents the border of the Dien-Bien and the Son-La provinces. Population density near the station is low and well below the average of the neighboring provinces (64 and 90 persons per km2 in the Dien-Bien and Son-La provinces in 2020, respectively (http://www.gso.gov.vn/) and even. No relevant residential or industrial areas exist within the surrounding 10 to 20 km, except for sparse individual farmhouses. The closest farmhouse is located a 1 km distance in the NE direction. A few ethnic H'mong households using wood logs and debris for residential cooking and heating are located within 5 km of the sampling site. The laboratory building provides air-conditioned laboratory space and basic accommodations for station operators and guest researchers. Two custodians constantly look after the station, and the presence of other staff is rare.

3 Data and Methods

3.1 Observations

The PDI measurement site consists of a meteorological tower and a laboratory building (see Fig. 2) and represents a level 3 meteorological station of the VNMHA. Meteorological data of temperature, relative humidity, wind speed, and direction in 6 h intervals are provided since January 2012. Continuous measurements of greenhouse gases (CO_2, CH_4, CO), surface ozone (O_3), and aerosol optical properties (scattering and absorption coefficients), operated in an air-conditioned laboratory, are provided since March 2014. The inlet height for the trace gas and aerosol observations is 12 m above ground and roughly 6 m above the top of the roof (see Fig. 3). CO_2, CH_4, and CO are measured with a cavity ring-down spectrometer (CRDS)

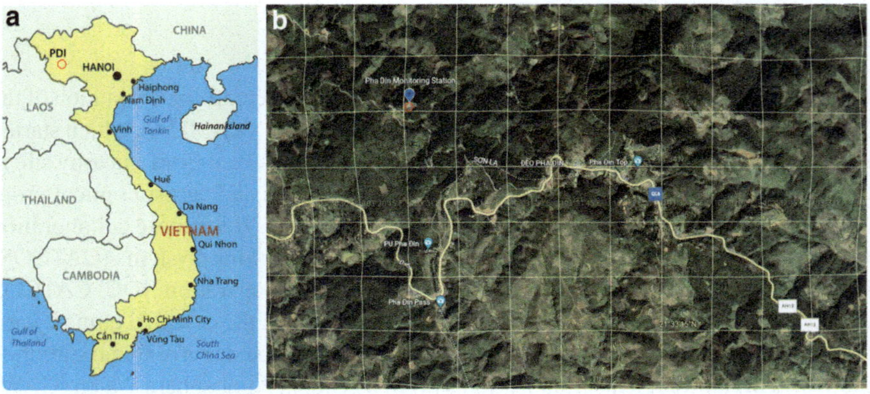

Fig. 1 Maps showing the PDI monitoring station (indicated by red circle) **a** within Vietnam in relation to Hanoi and **b** detailing the mountain pass

Fig. 2 **a** PDI measurement site: location on top of a mountain hill (view of the station from below), **b** meteorological tower, **c** laboratory building with inlet mast, and **d** close up of the trace gas and aerosol (particulate matter, PM) inlet configuration

(Picarro Inc., G2401); O_3 is measured by UV absorption (Thermo Scientific, 49i). The CRDS analyzer is regularly (every 3 to 6 days) calibrated with four reference gases. The quality of the calibration is verified with a fifth reference gas (target cylinder). Reference gases were either purchased from the Global Atmosphere Watch Central Calibration Laboratory (CCL) hosted by NOAA in Boulder (USA) or were prepared by the GAW World Calibration Centre for CH_4, CO_2, CO and surface O3 (WCC-Empa). Data are reported on the recent GAW reference scales, i.e., WMO CO2 X2019, WMO CH4 X2004A, and WMO CO X2014A. Quality assurance procedures involve time series plots, target tanks (i.e., cylinders containing natural air with assigned gas mole fractions that are treated as a (unknown) sample in a sequence of analyses) measurements, evaluation of the evolution of the instrument sensitivity, and consistency checks. The measurements were made without sample drying, and a water vapor correction was applied. See Rella et al. (2013) for details of the empirical determination of the correction parameters. Traceability of the O_3 data to the primary ozone standard was ensured through the calibration of the ozone

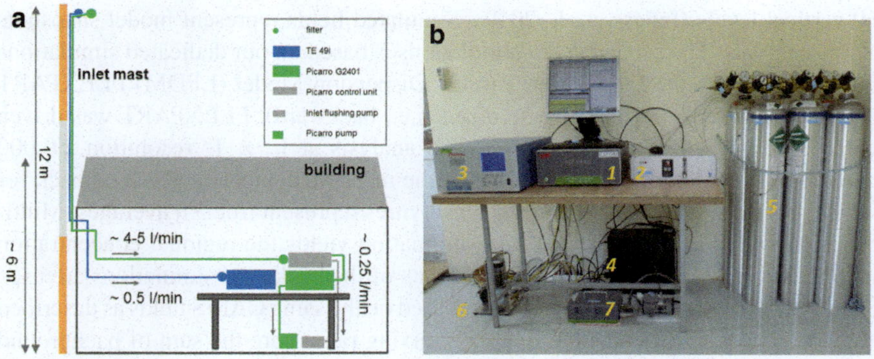

Picture Credit: M. Steinbacher and CATCOS Team

Fig. 3 **a** Laboratory set-up for trace gases and **b** picture of trace gas instrumentation; the numbered items are described in the text. The numbered items in **b** correspond to the CRDS Picarro laser spectrometer for carbon dioxide (CO_2), methane (CH_4), carbon monoxide (CO) measurements(#1), the calibration unit for spectrometer (#2), the UV absorption analyzer for measuring ozone (O_3) (#3), the computer with data acquisition software (#4), a set of six cylinders with calibration gases (#5), a zero air unit for O_3 calibration (#6), and two pumps (#7)

analyzer with a transfer standard, which was calibrated against a Standard Reference Photometer (SRP) of the CCL-NIST-SRP family at WCC-Empa. Trace gas observations are publically available at the World Data Centre for Greenhouse (CO_2, CH_4, CO; https://gaw.kishou.go.jp) and the World Data Centre for Reactive Gases (O_3; https://www.gaw-wdcrg.org/). Details on the in-situ aerosol observations can be found in Bukowiecki et al. (2019). Details on quartz filter collection for chemical analysis of carbonaceous aerosol can be found in Nguyen et al. (2021). Trace gas data have been collected continuously since 2014 and are available throughout most of 2014–2020. Several gaps in the data series are due to various typical known and unknown malfunctions of the system (e.g., a lightning strike in April 2014, failures of pumps and ventilation systems, temperature stabilization, and hard disk issues). Data collection was halted in 2021, but the full operation has been achieved since spring 2022 again.

3.2 *Simulations*

Two sets of air quality and greenhouse gas simulations were evaluated against the observations at PDI. The first is global reanalysis products from the Copernicus Climate Change and Atmospheric Monitoring (CAMS) service provided by the European Centre for Medium range Weather Forecast (ECMWF). Both air quality (ECC4, CO, O_3) and greenhouse gas reanalysis (EGG4, CO2, CH4) are based on ECMWF's integrated forecasting system (IFS), applying data assimilation of satellite-observed concentrations and are available globally at $0.75° \times 0.75°$ horizontal resolution and

60 vertical levels (Inness et al., 2019). Simulated fields represent model snapshots at 3 h intervals. The second set of simulations is based on our dedicated simulations for PDI applying the Lagrangian Particle Dispersion Model (LPDM) FLEXPART (Pisso et al., 2019) in time-reverse mode (i.e., backward). FLEXPART was driven with ECMWF operational meteorological analysis at $1^\circ \times 1^\circ$ resolution. 50,000 particles were released in 3 h intervals at the measurement site and traced back for ten days. The model-derived surface sensitivities represent true 3 h averages. Multiplication with surface emissions and summation yields the regional concentration increments for the target area. The endpoints of the backward simulations are used to calculate background concentrations (based on the same CAMS fields as described above). Final model concentrations (i.e., mixing ratios) are the sum of background and regional increments. The Lagrangian approach offers additional information on regional emissions. It can be used to determine when and where emissions were picked up by an air mass that was later sampled at the receptor site. As such, it is also possible to calculate the plume age (time since emission) of regional BB emissions that reach the measurement site. Two different emission datasets were treated in the Lagrangian approach: (i) emissions from BB as obtained from the ECMWF Global Fire Assimilation System (GFAS; (Kaiser et al., 2012)) and (ii) anthropogenic emissions available from the EDGAR database (version 5.0; (Crippa et al., 2020)). In comparison to real topography, the topography of both models is considerably smoothed. Hence, simulations were not evaluated in the lowermost model layer (particles not released at the surface in the case of FLEXPART) but at a level that represents the middle between the model and real topography (level 55 for CAMS products, release at 270 m above ground level for FLEXPART). Here, we evaluate data for three consecutive years (2014–2016).

4 Results and Discussion

4.1 Long-Term Records of Trace Gases and Meteorological Parameters

Figure 4 shows the observed precipitation (as weekly accumulated sums) and daily temperature and relative humidity (daily averages) at PDI from 2014 to 2020. The observations confirm the typical climate conditions expected for Northern Vietnam: dry (precipitation and relative humidity), relatively cold winters, and moist and hot summers with the onset of the Asian summer monsoon. Interestingly, the onset of warmer temperatures precedes that of wetter conditions by one to two months, leading to favorable conditions for BB. Figure 5 illustrates the recorded time series of the trace gases (O_3, CO_2, CH_4, CO) mixing ratios. Trace gas data have been collected continuously since 2014 and are available throughout most of 2014 to 2020. From June to August, conditions at PDI typically allow background observations when hourly CO mixing ratios reach values sometimes below 60 ppb and rarely above

130 ppb. Monthly CO and O_3 averages are usually between 80 and 120 ppb and 15 and 25 ppb, respectively. This is also when monthly CH_4 mixing ratios are around 1900 ppb, about 100 ppb lower than the rest of the year. CO_2 mixing ratios are also low during this period but usually reach their minimum in early fall (September) when the ecosystem uptake declines. This illustrates the influence of the ample vegetation in the station's vicinity. In spring, a clear BB signal is seen in the CO time series as the station is regularly exposed to pollution from recurrent fires in the area (discussed in further detail in 4.2). Such BB signal is less obvious in the other trace gas observations (CH_4, CO_2, or O_3). This is likely due to the relatively small contribution of recent BB emissions to those trace gases compared to their larger atmospheric background levels (e.g., for CH_4), the complex photochemistry producing and destroying O_3, and the dominant role of other sources and sinks (and a large background burden) for CO_2.

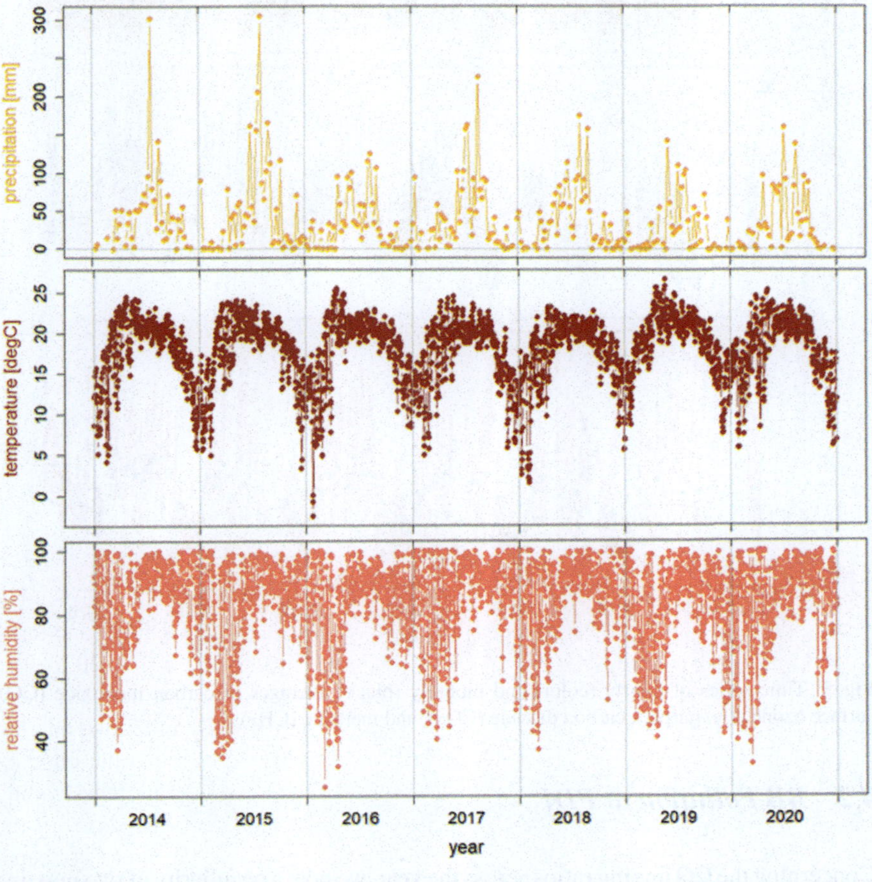

Fig. 4 Time series weekly sums of precipitation, daily averages of temperature, and relative humidity at PDI

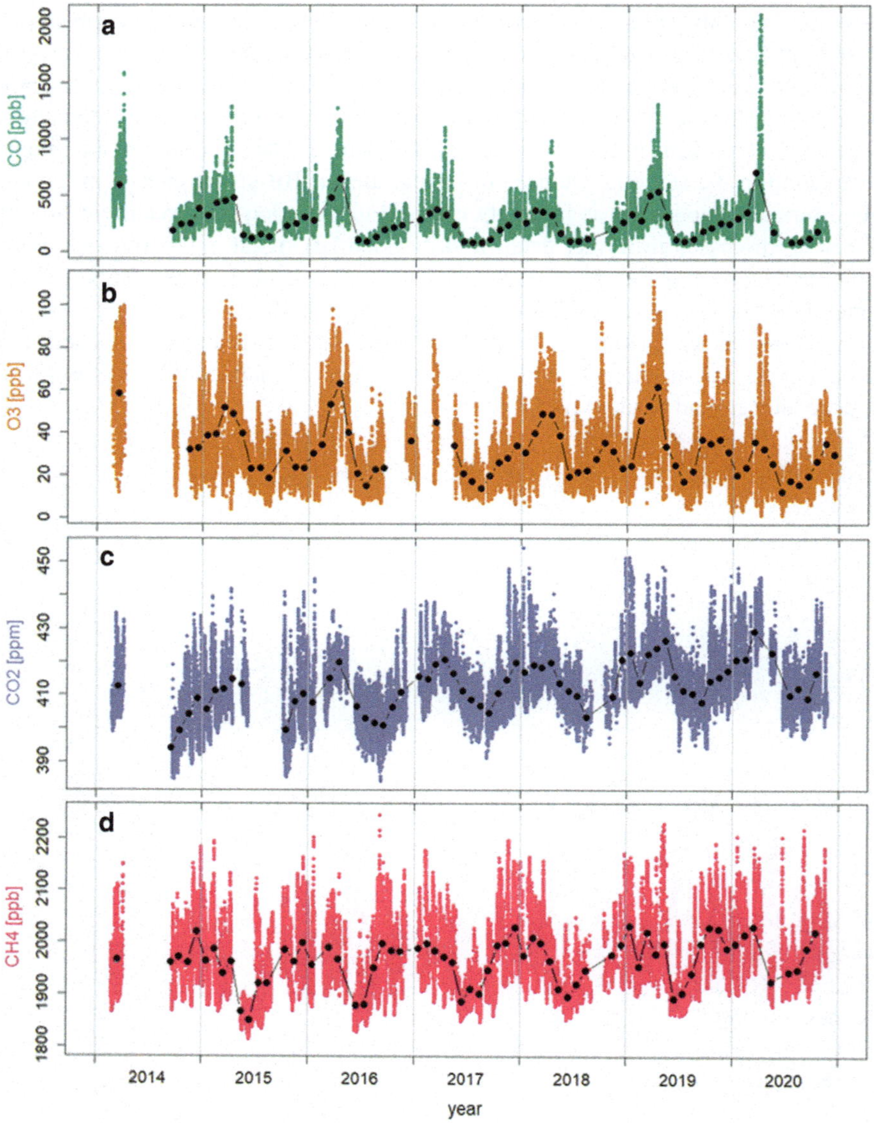

Fig. 5 Time series of hourly (color) and monthly (black) averages of carbon monoxide (CO), surface ozone (O_3, purple), carbon dioxide (CO_2), and methane (CH_4)

4.2 BB Pollution at PDI

Concerning the CO mixing ratios across the year, we note a peculiarity of two maxima visible in the time series (one typically in October to November and a second typically in March to April). While the first one seems to be associated with air masses from north–east (perhaps non-BB, anthropogenic), the second one appears associated with air masses from the south–west (likely BB on the Indochinese Peninsula). Hourly averages of CO mixing ratios reach the highest values (> 1000 ppb) during the dry season in March and April before the mixing ratios drop to values typical for rural background stations with the onset of persistent precipitation in May. The regular recurrent pattern is also apparent in the sawtooth-like feature of the monthly CO averages. Aside from trace gases, BB releases incomplete combustion products in the particle phase, particularly carbonaceous aerosols. These are composed of elemental carbon (or black carbon) and a complex mix of organic carbon, including, for instance, sugars, PAHs, and a mix of light-absorbing compounds termed "brown carbon." Bukowiecki et al. (2019) used the light-absorbing properties of PM2.5 to quantify BB and traffic contributions at PDI and identified BB as a relevant and periodically dominant component. Correlations of these data with fire hotspots on the Indochinese Peninsula were high. In addition, Nguyen et al. (2021) identified distinct episodes of a low, medium, and high BB pollution based on clustering 51 organic markers in PM2.5 samples and EC at PDI. Samples were collected during an intensive campaign in March and April 2015. Receptor modeling confirmed that cleaner air masses arrived from the northeast during this 3-week measurement period. Instead, the polluted periods included more continental recirculation and advection from the southwest. In these areas, fire count density was highest during this period. The high BB pollution during the campaign also affected the visibility, as presented in Fig. 6, showing the station under clean and polluted conditions (Nguyen et al., 2021).

4.3 *Simulations of CO Mixing Ratios at PDI*

Figures 7 and 8 present simulations of CO mixing ratios in comparison to observations. Figure 7 indicates that CO mixing ratios are overestimated by both models (FLEXPART and CAMS) during the BB seasons, which are associated with the highest observed CO values in spring. Instead, they nicely reproduce the observed CO mixing ratios during less polluted and unpolluted conditions.

Figure 8 evaluates the simulated CO mixing ratios concerning the age of emissions (i.e., plume age) for the main BB period in 2015. The arrival of intermittent BB plumes with different plume ages is discernable. In early March 2015, at relatively small CO mixing ratios, arriving plumes were generally young (below 36 h). Similarly, with intensified BB starting in mid-March, with the exception of 23rd March, when the average plume age was determined to be around 72 h. Again, the largest observed mixing ratios at the beginning of April were connected with recent (< 24 h) emissions. For the final BB event around 18th April, the largest CO levels with very recent (< 18 h) emissions were predicted. However, no observations were available to confirm these model estimates. For the whole period, the timing of the CO signal is captured quite well by the simulation, whereas the CO magnitude, especially during the intense phase at the beginning of April, tends to be overestimated by the model.

Fig. 6 Pictures showing the view from the study site PDI under clean (**a, c**) and polluted (**b, d**) conditions in south and west direction. Clean and polluted conditions refer to the low and high BB periods on April 12–13, 2015, and April 6–7, 2015, respectively, as identified in Nguyen et al. (2021). Pictures are taken at 4 PM local time

Interestingly, both model simulations (CAMS and FLEXPART) overestimate CO during the time with the highest observed CO and lowest plume age before arrival at the station. Since both models use the same BB emission dataset, it seems likely that the overestimation originates from an overestimation in the emissions. Another possibility could be the proximity of the measurement site to the sources and the limited resolution of the transport models and emission data. In the CAMS model, emissions occurring in the grid cell, in which PDI is located, get immediately spread throughout this whole grid cell (size ca. 70 km × 70 km), although the real plumes may have been much finer scale and "missed" PDI. Although the same argument does not apply to the Lagrangian simulation, it is still limited by the resolution of the emission dataset (ca. 10 km × 10 km). Another potential factor for overestimating FLEXPART could be the assumption of BB emissions near the surface. In reality, these are lifted by the heat released in the fires. The CAMS model considers this convective lifting, potentially explaining lower values in CAMS than in FLEXPART simulations.

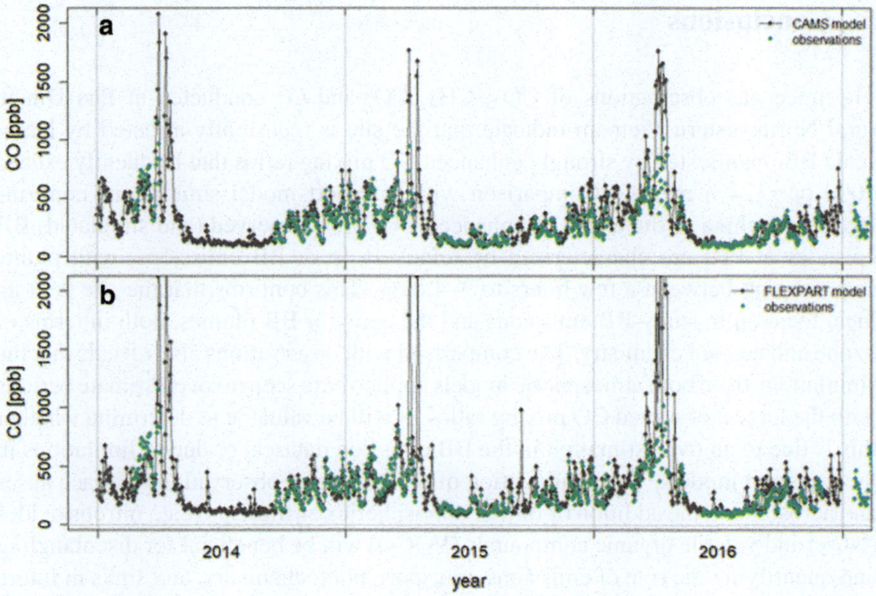

Fig. 7 Time series of daily averaged CO observations and **a** simulated CO with the CAMS model, **b** simulated CO with the FLEXPART model, for the years 2014 to 2016 at Pha Din

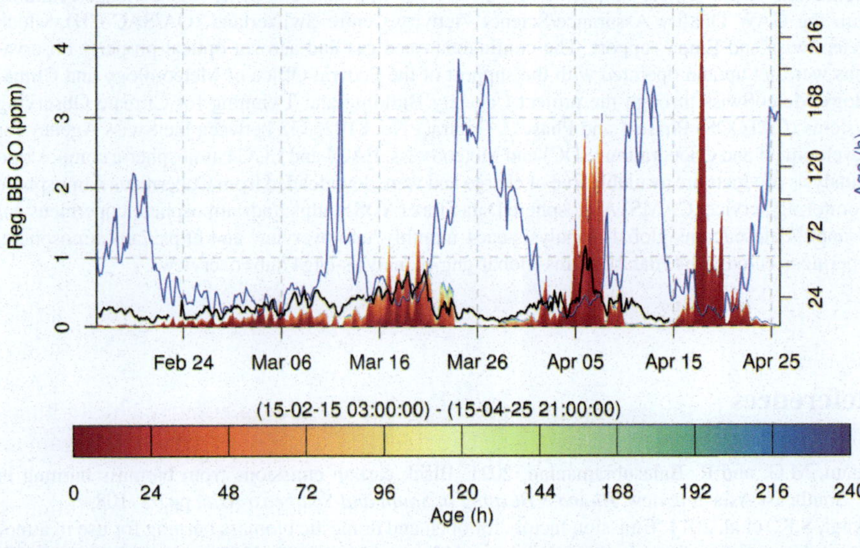

Fig. 8 Regional contribution to CO mixing ratios simulated by FLEXPART (color according to age/time since release), average age (blue line), and observed regional signal (black line). The latter was derived by subtracting the mixing ratios of the simulated baseline (i.e., background conditions) from the observations

5 Conclusions

The trace gas observations of CO_2, CH_4, CO, and O_3 conducted at Pha Din in rural Northwestern Vietnam indicate that the site is recurrently affected by large-scale BB, manifested by strongly enhanced CO mixing ratios that frequently exceed 1000 ppb (24 h means). Comparison with transport model simulations confirms BB in Indochina as the origin of enhanced CO. Most observed (and simulated) BB episodes at PDI are characterized by relatively fresh BB emissions, with plume ages ranging between a few hours to 3–4 days. This confirms that the site is at an ideal location to study BB emissions and the aging of BB plumes, both in terms of ozone and aerosol chemistry. The comparison with observations also reveals that the simulations from both atmospheric models applied here seem to overestimate periods with the largest observed CO mixing ratios. It will be valuable to determine whether this is due to an overestimation in the BB emission datasets or due to limitations in the transport models. The continuation of the long-term observations of trace gases and aerosols and the addition of further atmospheric constituents (e.g., nitrous oxides (NOx) and volatile organic compounds (VOCs)) will be beneficial for disentangling and quantifying the role of emissions, transport, photochemistry, and sinks in future studies.

Acknowledgements Simone M. Pieber acknowledges funding from the Swiss National Science Foundation under project number P400P2_194390. Martin Steinbacher acknowledges funding from the GAW Quality Assurance/Science Activity Centre Switzerland (QA/SAC-CH), which MeteoSwiss and Empa support. The continuous trace gas and aerosol optical property observations were set up and operated with the support of the Federal Office of Meteorology and Climatology MeteoSwiss through the project Capacity Building and Twinning for Climate Observing Systems (CATCOS) Phase 1 and Phase 2, Contract No. 81025332 between the Swiss Agency for Development and Cooperation (SDC) and MeteoSwiss. EAC4 and EGG4 atmospheric composition reanalysis products are available free of charge and were downloaded from Copernicus Atmosphere Monitoring Service (CAMS) Atmosphere Data Store (ADS) (https://ads.atmosphere.copernicus.eu/cdsapp#!/dataset/cams-global-reanalysis-eac4-monthly?tab=overview and https://ads.atmosphere.copernicus.eu/cdsapp#!/dataset/cams-global-ghg-reanalysis-egg4?tab=overview).

References

Adam, M.G. and R. Balasubramanian. 2021. Black carbon emissions from biomass burning in Southeast Asia–a review. *Biomass Burning in South and Southeast Asia*, pp.95–108.

Akagi, S.K., et al. 2011. Emission factors for open and domestic biomass burning for use in atmospheric models. *Atmospheric Chemistry and Physics* 11 (9): 4039–4072. https://doi.org/10.5194/acp-11-4039-2011.

Andreae, M.O. 2019. Emission of trace gases and aerosols from biomass burning—an updated assessment. *Atmospheric Chemistry and Physics Discussions* 19: 8523–8546. https://doi.org/10.5194/acp-2019-303.

Andreae, M.O., and P. Merlet. 2001. Emission of trace gases and aerosols from biomass burning. *Global Biogeochemical Cycles* 15 (4): 955–966. https://doi.org/10.1029/2000GB001382.

Betha, R., S.N. Behera, and R. Balasubramanian. 2014. 2013 Southeast Asian smoke haze: fractionation of particulate-bound elements and associated health risk. *Environmental Science and Technology* 48 (8): 4327–4335. https://doi.org/10.1021/es405533d.

Biswas, S., K.P. Vadrevu, Z.M. Lwin, K. Lasko, and C.O. Justice. 2015. Factors controlling vegetation fires in protected and non-protected areas of Myanmar. *PLoS ONE* 10 (4): e0124346.

Bukowiecki, N., et al. 2019. Effect of large-scale biomass burning on aerosol optical properties at the GAW regional station Pha Din, Vietnam. *Aerosol Air Qual Philosophy and Phenomenological Research* 19: 1172–1187. https://doi.org/10.4209/aaqr.2018.11.0406.

Chin, M. et al. (2017). Connection between East Asian air pollution and monsoon system. In *Air pollution in Eastern Asia: an integrated perspective*. Cham: Springer International Publishing, 87–103. https://doi.org/10.1007/978-3-319-59489-7_5.

Chuang, M.T., et al. 2013. Characterization of aerosol chemical properties from near-source biomass burning in the northern Indochina during 7-SEAS/Dongsha experiment. *Atmospheric Environment Elsevier Ltd* 78: 72–81. https://doi.org/10.1016/j.atmosenv.2012.06.056.

Chuesaard, T., et al. 2014. Influence of biomass burning on the levels of atmospheric polycyclic aromatic hydrocarbons and their nitro derivatives in Chiang Mai, Thailand. *Aerosol and Air Quality Research* 14 (4): 1247–1257. https://doi.org/10.4209/aaqr.2013.05.0161.

Crippa, M., et al. 2020. High resolution temporal profiles in the emissions database for global atmospheric research. *Scientific Data Springer, US* 7 (1): 1–17. https://doi.org/10.1038/s41597-020-0462-2.

Cusworth, D. H. et al. (2018). Quantifying the influence of agricultural fires in northwest India on urban air pollution in Delhi, India. *Environmental Research Letters*, 13(4). https://doi.org/10.1088/1748-9326/aab303.

Inness, A. et al. (2019). CAMS global reanalysis (EAC4). Copernicus atmosphere monitoring service (CAMS) atmosphere data store (ADS). https://ads.atmosphere.copernicus.eu/cdsapp#!/dataset/cams-global-reanalysis-eac4?tab=overview (Accessed 31 May 2022).

Kaiser, J.W., et al. 2012. Biomass burning emissions estimated with a global fire assimilation system based on observed fire radiative power. *Biogeosciences* 9 (1): 527–554. https://doi.org/10.5194/bg-9-527-2012.

Kato, E., T. Kinoshita, A. Ito, M. Kawamiya, and Y. Yamagata. 2013. Evaluation of spatially explicit emission scenario of land-use change and biomass burning using a process-based biogeochemical model. *Journal of Land Use Science* 8 (1): 104–122.

Lasko, K., and K. Vadrevu. 2018. Improved rice residue burning emissions estimates: accounting for practice-specific emission factors in air pollution assessments of Vietnam. *Environmental Pollution* 236: 795–806.

Le, T. H. et al. (2014). Vegetation fires and air pollution in Vietnam'. *Environmental Pollution (Barking, Essex : 1987)* 195(December 2008), 267–275. https://doi.org/10.1016/j.envpol.2014.07.023.

Li, C., et al. 2013. Characteristics and composition of atmospheric aerosols in Phimai, central Thailand during BASE-ASIA'. *Atmospheric Environment Elsevier Ltd* 78: 60–71. https://doi.org/10.1016/j.atmosenv.2012.04.003.

Lin, N.H., et al. 2013. An overview of regional experiments on biomass burning aerosols and related pollutants in Southeast Asia: from BASE-ASIA and the Dongsha Experiment to 7-SEAS. *Atmospheric Environment Elsevier Ltd* 78: 1–19. https://doi.org/10.1016/j.atmosenv.2013.04.066.

Mu, M., et al. 2011. Daily and 3-hourly variability in global fire emissions and consequences for atmospheric model predictions of carbon monoxide. *Journal of Geophysical Research Atmospheres* 116 (24): 1–19. https://doi.org/10.1029/2011JD016245.

Nguyen, D.-L., et al. 2021. Carbonaceous aerosol composition in air masses influenced by large-scale biomass burning: A case study in Northwestern Vietnam. *Atmospheric Chemistry and Physics* 21 (10): 8293–8312. https://doi.org/10.5194/acp-21-8293-2021.

Oanh, K. N. T., Permadi, D. A., Dong, N. P., and Nguyet, D. A. (2018). Emission of toxic air pollutants and greenhouse gases from crop residue open burning in Southeast Asia. *Land-atmospheric research applications in South and Southeast Asia,* 47–66. Springer, Cham.

Pisso, I., et al. 2019. The Lagrangian particle dispersion model FLEXPART version 10.4. *Geoscientific Model Development* 12 (12): 4955–4997. https://doi.org/10.5194/gmd-12-4955-2019.

Rella, C.W., et al. 2013. High accuracy measurements of dry mole fractions of carbon dioxide and methane in humid air. *Atmospheric Measurement Techniques* 6 (3): 837–860. https://doi.org/10.5194/amt-6-837-2013.

Saito, M., et al. 2022. Sensitivity of biomass burning emissions estimates to land surface information. *Biogeosciences* 19 (7): 2059–2078. https://doi.org/10.5194/bg-19-2059-2022.

Schultz, M.G., et al. 2008. Global wildland fire emissions from 1960 to 2000. *Global Biogeochemical Cycles* 22 (2): 1–17. https://doi.org/10.1029/2007GB003031.

Streets, D.G. et al. (2003). Biomass burning in Asia: Annual and seasonal estimates and atmospheric emissions. *Global Biogeochemical Cycles* 17(4). https://doi.org/10.1029/2003gb002040.

Tsai, Y.I., et al. 2013. Source indicators of biomass burning associated with inorganic salts and carboxylates in dry season ambient aerosol in Chiang Mai Basin, Thailand. *Atmospheric Environment Elsevier Ltd* 78: 93–104. https://doi.org/10.1016/j.atmosenv.2012.09.040.

Vadrevu, K., T. Ohara, and C. Justice. (2017) Land cover, land use changes and air pollution in Asia: a synthesis. *Environmental Research Letters*, 12(12). https://doi.org/10.1088/1748-9326/aa9c5d.

Vadrevu, K., T. Ohara, and C. Justice. (Eds.). (2021a). Biomass burning in south and Southeast Asia: impacts on the biosphere, Volume Two. CRC Press.

Vadrevu, K., T. Ohara, and C. Justice. (Eds.). (2021b). Biomass burning in South and Southeast Asia: mapping and monitoring, volume one. CRC Press.

Van Der Werf, G.R., et al. 2004. Continental-scale partitioning of fire emissions during the 1997 to 2001 El Niño/La Niña period. *Science* 303 (5654): 73–76. https://doi.org/10.1126/science.1090753.

Van Der Werf, G.R., et al. 2006. Interannual variability in global biomass burning emissions from 1997 to 2004. *Atmospheric Chemistry and Physics* 6 (11): 3423–3441. https://doi.org/10.5194/acp-6-3423-2006.

van der Werf, G.R., et al. 2017. Global fire emissions estimates during 1997–2016. *Earth System Science Data* 9 (2): 697–720. https://doi.org/10.5194/essd-9-697-2017.

Van Marle, M.J.E., et al. 2017. Historic global biomass burning emissions for CMIP6 (BB4CMIP) based on merging satellite observations with proxies and fire models (1750–2015). *Geoscientific Model Development* 10 (9): 3329–3357. https://doi.org/10.5194/gmd-10-3329-2017.

Yadav, I.C., et al. 2017. Biomass burning in Indo-China peninsula and its impacts on regional air quality and global climate change-a review. *Environmental Pollution Elsevier Ltd* 227: 414–427. https://doi.org/10.1016/j.envpol.2017.04.085.

Yang, Sixiao et al. (2022). 'Impacts of radiative effect of pre-monsoon biomass burning aerosols on atmospheric circulation and rainfall over Southeast Asia and southern China'. *Climate Dynamics.* Springer Berlin Heidelberg, (0123456789). https://doi.org/10.1007/s00382-021-06135-7.

Yen, M.C. et al. (2013). 'Climate and weather characteristics in association with the active fires in northern Southeast Asia and spring air pollution in Taiwan during 2010 7-SEAS/Dongsha Experiment'. *Atmospheric Environment.* Elsevier Ltd 78(x), 35–50. https://doi.org/10.1016/j.atmosenv.2012.11.015.

Southeast Asian Transboundary Haze in the Southern Philippines, 2019 and Meteorological Drivers

Krishna E. Santos, Mylene G. Cayetano, and Prisco D. Nilo

Abstract This study investigated the effects of meteorological systems on Southeast Asian (SEA) forest fires and the transport of particulate matter leading to the transboundary haze episode in the Philippines. The main goal is to investigate the prevailing meteorological patterns and the behavior of the pollutants during the haze episodes. Data were obtained from the NASA-Aerosol Robotic Network (NASA-AERONET, Philippine Atmospheric, Geophysical, and Astronomical Services Administration (PAGASA) ground-based stations, ECMWF-CAMS, and HIMAWARI-8 satellite data. The data gathered were divided into three periods, namely pre-haze (August 01, 2019), during the haze (September 13–21, 2019), and post-haze (October 01, 2019). In addition, backward and forward air trajectories using the NOAA HYSPLIT and wind vectors from MERRA-2 were plotted to find the sources of biomass burning to the recurring smoke haze in this region. In conclusion, air trajectory analysis and the results of aerosol sample analysis using ground-based data indicate that the transboundary air pollution from SEA influenced the haze event experienced in the Southern Philippines.

Keywords Biomass burning · Transboundary haze · Southeast Asia · Air quality

1 Introduction

Vegetation fires are common in several South/Southeast Asian (S/SEA) countries as they are often used as land clearing tool. For example, fires are used to clear the forests through slash and burn agriculture as in north India, Myanmar, northern Thailand, Cambodia, Laos, Philippines, etc. (Prasad et al. 2008; Albar et al. 2018; Perez et al. 2021; Hayasaka et al. 2014; Inoue 2018; Tariq and Ul-Haq 2018). In Indonesia, fires are mostly used to clear forests and peatlands for oil palm planting (Syaufina and Maulana (2021); Saharjo and Yungan 2018). In addition to forest biomass burning,

K. E. Santos (✉) · M. G. Cayetano · P. D. Nilo
Institute of Environmental Science and Meteorology, University of the Philippines-Diliman, Quezon City, Philippines
e-mail: kesantos2@up.edu.ph

K. P. Vadrevu et al. (eds.), *Vegetation Fires and Pollution in Asia*,
https://doi.org/10.1007/978-3-031-29916-2_33

crop residue burning too is most common in several S/SEA countries such as India, Pakistan, Myanmar, Vietnam, Thailand, Laos and Cambodia (Biswas et al. 2015a, b; Gadde et al. 2009; Lasko et al. 2017, 2018a,b; Lasko and Vadrevu 2018; Vadrevu and Lasko 2018). Although there are some advantages of using fires as a management tool, such as saving labor costs and controlling insects, diseases, and invasive weeds, most often, biomass burning results in negative impacts to the environment (Yulianti et al. 2020). For example, most of these biomass burning activities have been reported to emit large amounts of greenhouse gases such as CO_2, CO, non-methane hydrocarbons (NMHC), NOx, SO_2, particulate matter, and other gases (Kant et al. 2000; Gupta et al. 2001; Prasad et al. 2000, 2001; Kharol et al. 2012) which may significantly deteriorate the ambient air quality and contribute to the regional transport of air pollution (Phairuang 2021; Uranishi et al. 2021; Vadrevu et al. 2021a,b). Also, biomass burning is known to cause severe health effects such as chronic obstructive pulmonary disease, pneumoconiosis, bronchitis, cataract, corneal opacity, and blindness (Salvi and Barnes 2010). Further, the biomass burning activities release a significant amount of smoke which can reduce visibility, thus causing road accidents (Adam and Balasubramanian 2021). Most importantly, the pollutants and smoke can get transported to longer distances impacting regional air quality and people's health.

In August 2019, a vast biomass burning event broke out in Indonesia's peatlands. The burned area was much bigger than the previous forest fire in 2018 and 2015, with a total of 857,756 hectares (2.12 million acres) burned (Albar et al., 2018; Reuters, 2019). During this time, Indonesia dealt with forest fires in multiple areas: Central, West, North, and South Kalimantan, Riau, Jambi, and South Sumatra. The 1997 Indonesian Peatland Fire showed that using ground measurements, the estimated amount of carbon released into the atmosphere was 0.19 to 0.23 gigatons (Gt) via peat combustion (Page et al. 2002). The strong El Niño phenomenon experienced during the same year is associated with forest fire occurrences (Kita et al. 2000; Siegert et al. 2001). The Indonesian Peatland Fire that occurred in September 1997 and 2015 also reportedly came from biomass burning, which has caused transboundary air pollution (Fang and Huang 1998). The continuous activities connected to land conversions in Indonesia and the El Niño phenomenon that led to the Indonesian Peatland Fire are becoming a recurring problem in the Southeast Asian (SEA) Region.

One of the most affected areas in the Philippines during the transboundary haze event is the Southern Philippines region. The Philippines comprises three (3) main islands, namely Luzon, Visayas, and Mindanao. The southern part of the Philippines is composed of urban and rural areas; however, local biomass burning methods include field burning, wherein in rice and sugarcane cultivation, the fire was utilized as a rapid and labor-saving technique for crop waste removal (Mendoza and Samson, 1999; Rolph et al. 2017; Stein et al. 2015) and at the same time in preparation for agricultural planting (Mabalay et al. 2022), residue burning (Gadde et al. 2009) is still evident in the region, causing significant air pollution.

Several studies have investigated the influence of various pollutants from Southeast Asian biomass burning on the atmosphere of the Philippines, including the size of resolved-aerosol composition in Metro Manila (Braun et al. 2020; Cruz et al. 2019) and carbon emissions (Shi et al. 2014). However, no comprehensive studies have explored the long-range atmospheric transport (LRT) behavior of anthropogenic emissions from the Indonesian Peatland Fire through the transboundary haze.

To investigate the characteristics of the anthropogenic emissions from Indonesian biomass burning to their receptors and the active synoptic weather systems, the satellite datasets and in-situ measurements were interpreted according to the non-haze and haze events during the biomass burning periods in Indonesia.

2 Study Area

The air particulate samples of various filter types (PM10 and PM2.5) used for the chemical analysis were collected from five regions in the Philippines, namely Region 4B, 8, 9, 12, 13, and Autonomous Region in Muslim Mindanao (ARMM).

The chosen regions such as Regions 9, 12, 13, and ARMM were based on the closeness of the region to Indonesia, and Regions 4b, and 8 were based on the number of reports received during the event. It should also be noted that the Air Quality Monitoring Stations (AQMS) were located near roadside and agricultural areas; hence, there would also be other emissions such as roadside emissions and local biomass burnings that will be recorded during the study period (Fig. 1 and Table 1)

3 Data and Methods

3.1 Data

The visibility data were gathered by sending an electronic request form through the Google Forms prepared by the Climatology and Agrometeorology Division (CADS) of the Philippine Atmospheric, Geophysical, and Astronomical Services Administration (PAGASA). The stations used for the study were either airport stations or synoptic stations.

Additionally, the study used upper air radiosonde data from September 13 to 20, 2019, from the three stations of PAGASA in Southern Philippines: (i) Mactan Station, (ii) Davao Station, and (iii) Puerto Princesa Station. These data were accessed using the upper air sounding data archive website of Wyoming University. The upper air sounding data show the selected stations' wind speed and direction.

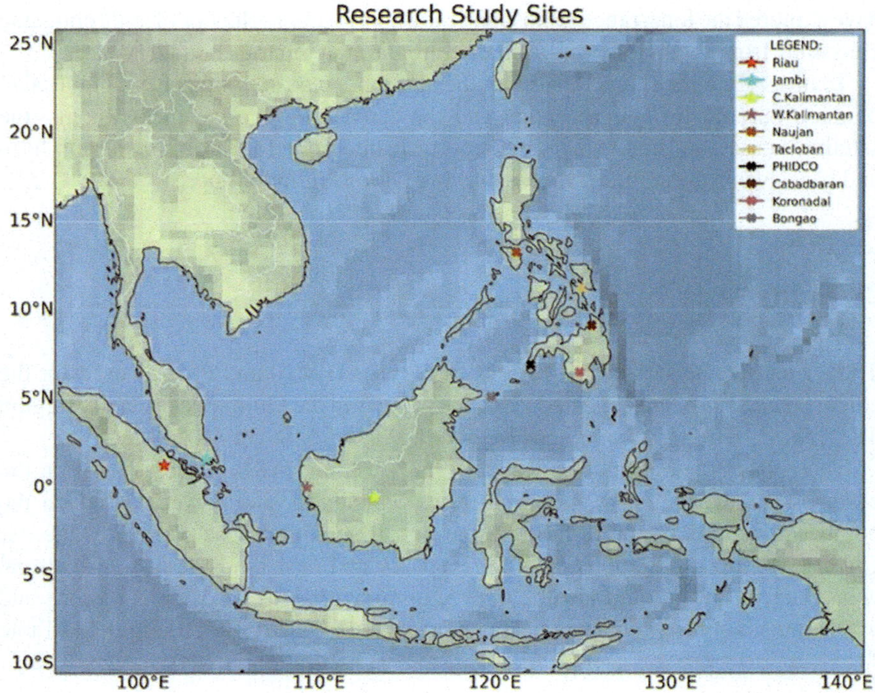

Fig. 1 Study sites showing the locations in the Philippines and the hotspots of Indonesia's 2019 peatlands fire plotted using Python

Table 1 Study sites in the Philippines

Region	Station	Geographical coordinates
4B	Naujan	13.308889, 121.29333300000007
	Baco	13.358889, 121.09638900000004
	Calapan	13.3950058, 121.16710899999998
8	Tacloban	11.207542167784, 125.00854202118
9	EMB-9 compound	6.9104444444444, 122.06958333333
	Philippine international development incorporated (PHIDCO)	6.9121666666667, 122.05736111111003
	Zamboanga city medical center	6.90727777778, 122.08088888889006
12	Koronadal city	6.4974° N, 124.8472° E
13	R13-Cabadbaran city hall compound	9.121752, 125.54647499999999
ARMM	Bongao, Tawi-Tawi	5.1042° N, 119.8121° E

3.2 Satellite Data

The trajectory from the models of NASA-MERRA-2 and NOAA HYSPLIT is utilized in this study. Using the NOAA HYSPLIT, the forward and backward trajectory plot is generated. The following settings were set into a total run time of three hundred fifteen (315) hours and starting a new trajectory per one (1) hour with a maximum number of trajectories of seventy-two (72) runs (Rolph et al. 2017; Stein et al. 2015).

The satellite data imagery of an active fire from the National Aeronautics and Space Administration Land Atmosphere Near real-time Capability for EOS (NASA LANCE) is used as a secondary dataset. The Moderate Resolution Imaging Spectroradiometer (MODIS) hotspot data (Collection 6 active fire products) from August to November 2019 are also used in this study. The hotspot selection is focused on the Indonesian Area and Southern Philippines. This was also coupled with data from NASA-CALIPSO to measure the height of the smoke and plume from the ground.

The same method will be used for the ground-based AOD/AOD values using the data gathered from NASA-Aerosol Robotic Network (NASA-AERONET) and Japan Aerospace Exploration Agency (JAXA) Himawari-8. The full disk data for the HIMAWARI-8 satellite dataset were provided through the network provided by JAXA (http://www.eorc.jaxa.jp/ptree/index.html) (eorc.jaxa.jp, 2022). The aerosol products such as the AOD were gathered using the JAXA polar-orbiting satellite and averaged daily Resolution Imaging Spectroradiometer (MODIS) NASA Terra (EOS AM) and Aqua (EOS PM) satellites. NASA Terra and Aqua data were used for interpreting AOD variations during the morning and evening respectively.

Furthermore, all the datasets used for this study were summarized below (Table 2).

3.3 Approach

A back-trajectory analysis from the National Oceanic and Atmospheric Administration (NOAA) Hybrid Single-Particle is utilized to investigate the effects of the transboundary haze induced by the Indonesian Peatland fire using the Lagrangian Integrated Trajectory (HYSPLIT) model. The HYSPLIT model is used for Hazardous Materials (HazMat) events (Rolph et al., 2017; Stein et al., 2015) which would then project the air mass pathway for the characteristics of PM2.5 and PM10 particles. We also used Pearson correlation coefficient to identify the strength (a) between the AOD from two different AERONET stations; Palangkaraya, Indonesia, and Koronadal City, Philippines, (b) PM2.5 Values from Region 4b, 9, 12, and 13 against the MODIS AOD Values and Himawari-8 AOD Values.

Table 2 Datasets used in the study

Variable	Date	Instrument	Resolution	Information
Visibility	August 1 to October 30	PAGASA stations records	3 h	
Upper-air sounding	September 13 to 21	PAGASA stations records	00 and 12 UTC	
Wind speed and wind direction	September	PAGASA stations records	3 h	
PM 2.5 and PM 10	August to October	DENR-EMB stations	Daily averaged	
AOD	August to November	NASA AERONET	All points level 2 daily averaged	
AOD	September	Himawari-8	10 min, 0.05° 1 day, 0.05°	Full DiskLevel 2
Active fire	August to November	MODIS Aqua/Terra	1 km	Daily
Wind speed and wind direction	August to September	MERRA-2	2 km 0.5° × 0.625°	Hourly
Particulate matter	September	ECMWFCAMS	80 km	Daily

4 Results and Discussion

4.1 Wind Patterns and Wind Trajectory

During this period, the prevailing weather system in the study area is the southwest monsoon. The effect of the monsoon explains the northward movement of air coming from Indonesia toward the Philippines. Tropical cyclones and the low-pressure regions also formed in the Philippine Area of Responsibility (PAR) during September 5–7, 17–22, 29, and October 01, 2019, which intensified the southwest monsoon (Fig. 2) (Ellrod, 2022).

The air pathway results using the forward trajectory of HYSPLIT showed that the air came from the Indonesian Peninsula, where 76% of the hotspots are located. The time set for the NOAA HYSPLIT model is 315 h; however, the model approached the maximum limit of data display which is why it only shows five days. The forward trajectories of the air move in the northwest direction. The air mass started at a level of 300–500 m above ground level (AGL); as it approaches the Philippines' receptor sites, it moves southwest from the source locations (Indonesia).

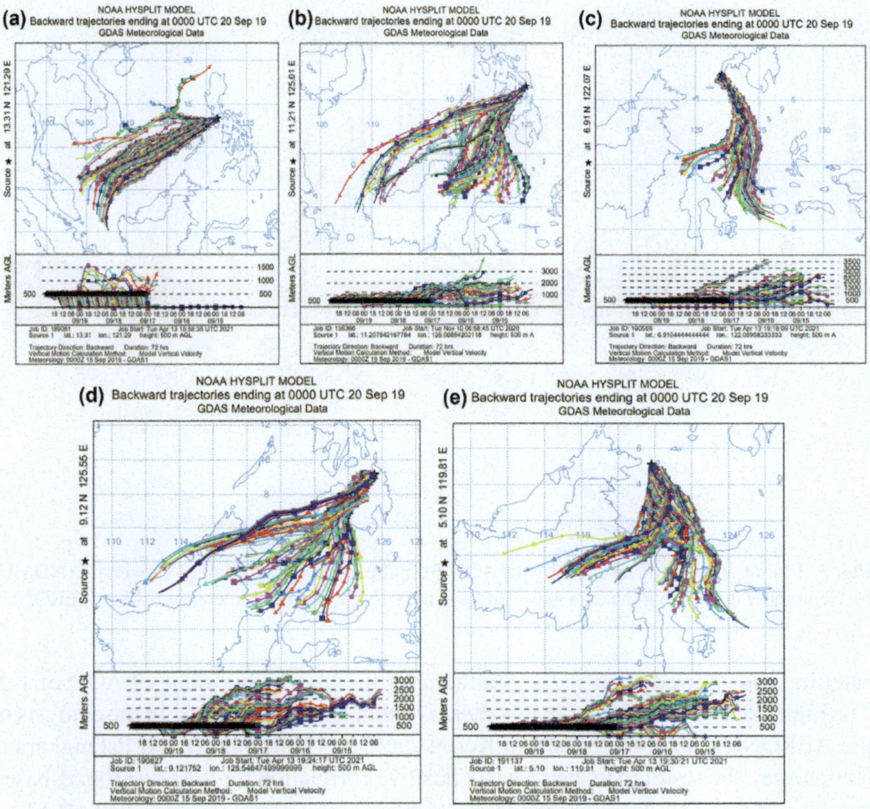

Fig. 2 The backward trajectory with wind barbs for the different stations in the Philippines for September 13–20, 2019, with 315 h running time. The ending heights range from 1000 to 3500 m AGL. **a** Naujan, Oriental Mindoro for Region 4b; **b** Tacloban for Region 8; **c** Philippine International Development Incorporated for Region 9 **d** Cabadbaran City Hall for Region 13; **e** Bongao, Tawi-Tawi in ARMM

4.2 *Impact of Indonesian Peatland Fire on the Air Quality in the Southern Philippines*

4.2.1 Aerosol Optical Depth

The ground-based data using the NASA-AERONET show that during the Indonesian Peatland Fire, the highest value of AOD was on September 14, 2019. It should also be noted that the AERONET instrument is installed at Notre Dame Mabel University in Koronadal City, South Cotabato, located in Region 12. During this date, PAGASA's Synoptic Station located in Awang Airport, Cotabato City, observed hazy conditions with the brief occurrence of showers. In October and November, the AOD was slowly decreasing. Thus, signifying the decline of the effect of the SEA transboundary haze in the Southern Philippines. The generated plot using the AOD of Himawari-8

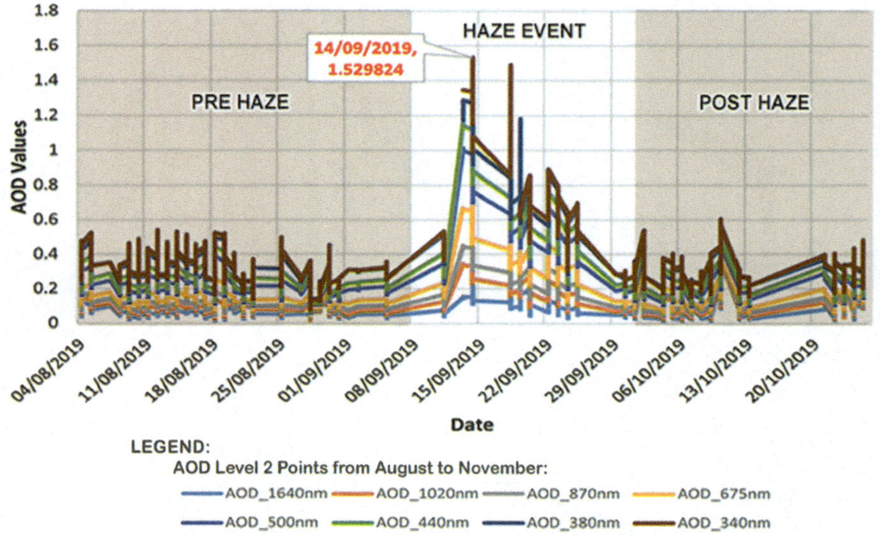

Fig. 3 Plot of all-point level 2 quality of AOD from the Notre Dame Marbel University (NDMU) for September 2019 using the data gathered from the CIMEL sun photometer of NASA-AERONET

satellite data showed areas in the Southern Philippines with a value of AOD of 1-2. The high values of AOD denote that there is a thin haze in the areas. In the analysis of the AERONET Stations located in Koronadal City (Philippines) and Palangkaraya (Indonesia), the result indicated a weak positive correlation for the two ground-based stations (R= 0.34). Negative moderate to high correlations (R = – 0.86, 0.12, – 0.66, and – 0.66) were found using the Pearson-R correlation values for Region 9. Lastly, the Himawari-8 satellite AOD provides higher geographical and temporal resolution than MODIS Terra and Aqua AOD (Figs. 3 and 4).

4.2.2 Visibility Values

The visibility data gathered from the PAGASA airport stations show that before the Indonesian Peatland Fire, the visibility in some locations was in the range of 18–20 km, which are considered the average values for the stations. During the event, the visibility decreased from 12 to 6 km.

However, showers and thunderstorms present in the area also play a role in reducing visibility. The earliest low visibility and high PM values were recorded on September 16, 2019, in Regions 9 and 12. The visibility went back to its average value shortly after the haze event (Table 3).

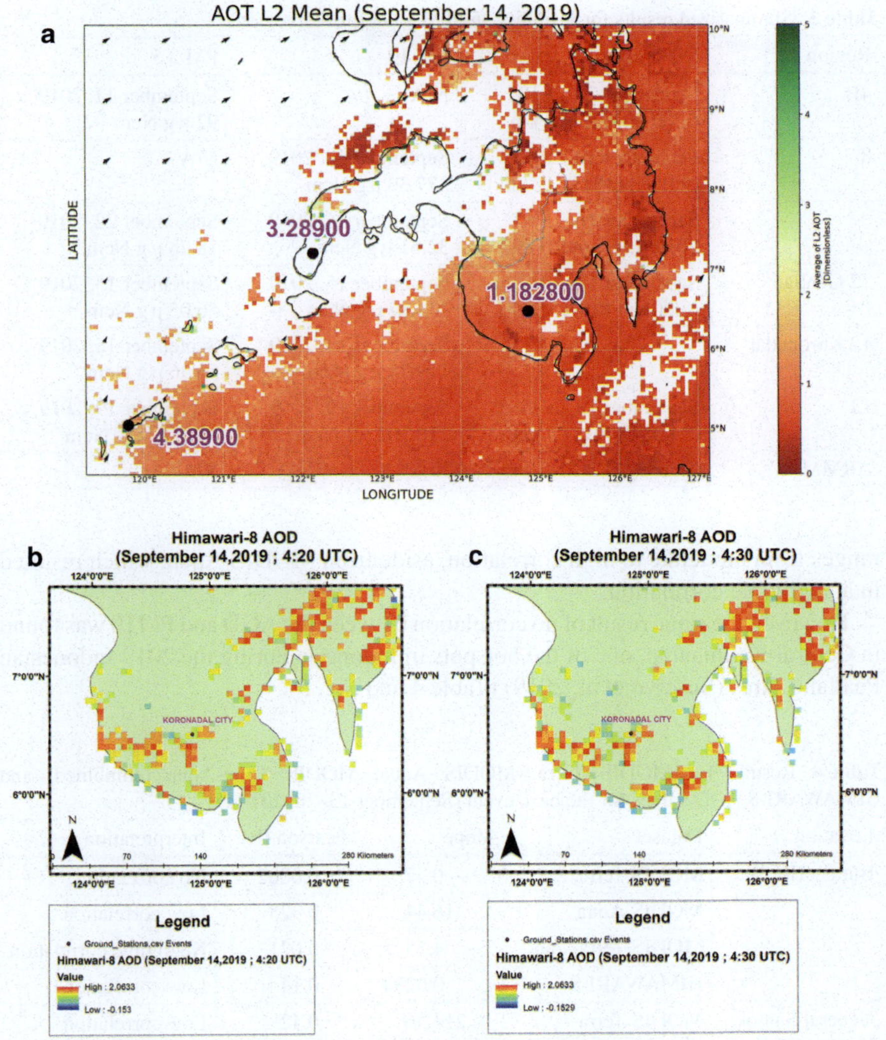

Fig. 4 **a** Mean AOD for September 14, 2019, from Himawari-8 retrievals and the wind (quiver) movement from MERRA-2. **b** 4:20 UTC, and **c** 4:30 UTC AOD in Region 12

4.2.3 Validation of PM 2.5 and AOD

All the analyses conducted for Himawari-8, MODIS Terra, and Aqua are summarized in Table 4. Due to the limited amount of data and the differences in terms of resolution (spatial and temporal), the correlation came out poorly.

Region 13 and Region 12 analysis shows little to no correlation between the AOD and $PM_{2.5}$ values. However, for Region 9, the analysis for the Pearson R-value

Table 3 Summarized results for visibility and PM values for all regions

Region	Visibility	PM 10	PM 2.5
4B	September 20, 2019 (2:00 pm PHT) 6 km	N/A	September 17, 2019 92 μg Ncm
8	September 19, 2019 (5:00 pm PHT) 12 km	September 18, 2019 377.36 μg Ncm	N/A
9	September 16, 2019(5:00 pm PHT) 7 km	September 16, 2019 32.61 μg Ncm	September 22, 2019 18.28 μg Ncm
12 GenSan	September 16, 2019 (2:00 pm PHT) 10 km	September 15, 2019 114.87 μg Ncm	September 15, 2019 56.05 μg Ncm
12 Koronadal		September 14, 2019 119.35 μg Ncm	September 16, 2019 56.96 μg Ncm
13	September 19, 2019 (8:00 pm PHT) 8 km	September 19, 2019 210.709 μg Ncm	September 19, 2019 149.104 μg Ncm
ARMM	N/A	N/A	N/A

ranges from moderate to high correlation, aside from MODIS Aqua, which resulted in a negligible correlation.

However, the same result of no correlation between the AOD and PM10 was found in Central Kalimantan, one of the hotspots in Indonesia during the 2019 Indonesian Peatland Fires (Susetyo et al. 2019) (Table 4 and Fig. 5).

Table 4 Results for MODIS Terra, MODIS Aqua, MODIS Terra-Aqua (combined) and HIMAWARI-8 AOD values for the haze event (September 13–20, 2019)

Location	Dataset	Slope	Pearson (R)	Interpretation
Butuan (R13)	MODIS Terra	− 0.47	0.0002	No correlation
	MODIS Aqua	16.44	0.325	Low correlation
	MODIS TerraAqua	4.55	0.021	Negligible correlation
	HIMAWARI-8	0.0234	0.14	Low correlation
General Santos Koronadal (R12)	MODIS Terra	214.36	0.42	Low correlation
	MODIS Aqua	69.885	0.064	Negligible correlation
	MODIS TerraAqua	115.69	0.20	Negligible correlation
	HIMAWARI-8	0.0032	0.44	Low correlation
Zamboanga (R9)	MODIS Terra	− 53.32	− 0.86	High (negative) correlation
	MODIS Aqua	25.04	0.12	Negligible correlation
	MODIS TerraAqua	− 108.23	− 0.66	Moderate (negative) correlation
	HIMAWARI-8	− 0.05	− 0.66	Moderate (negative) correlation

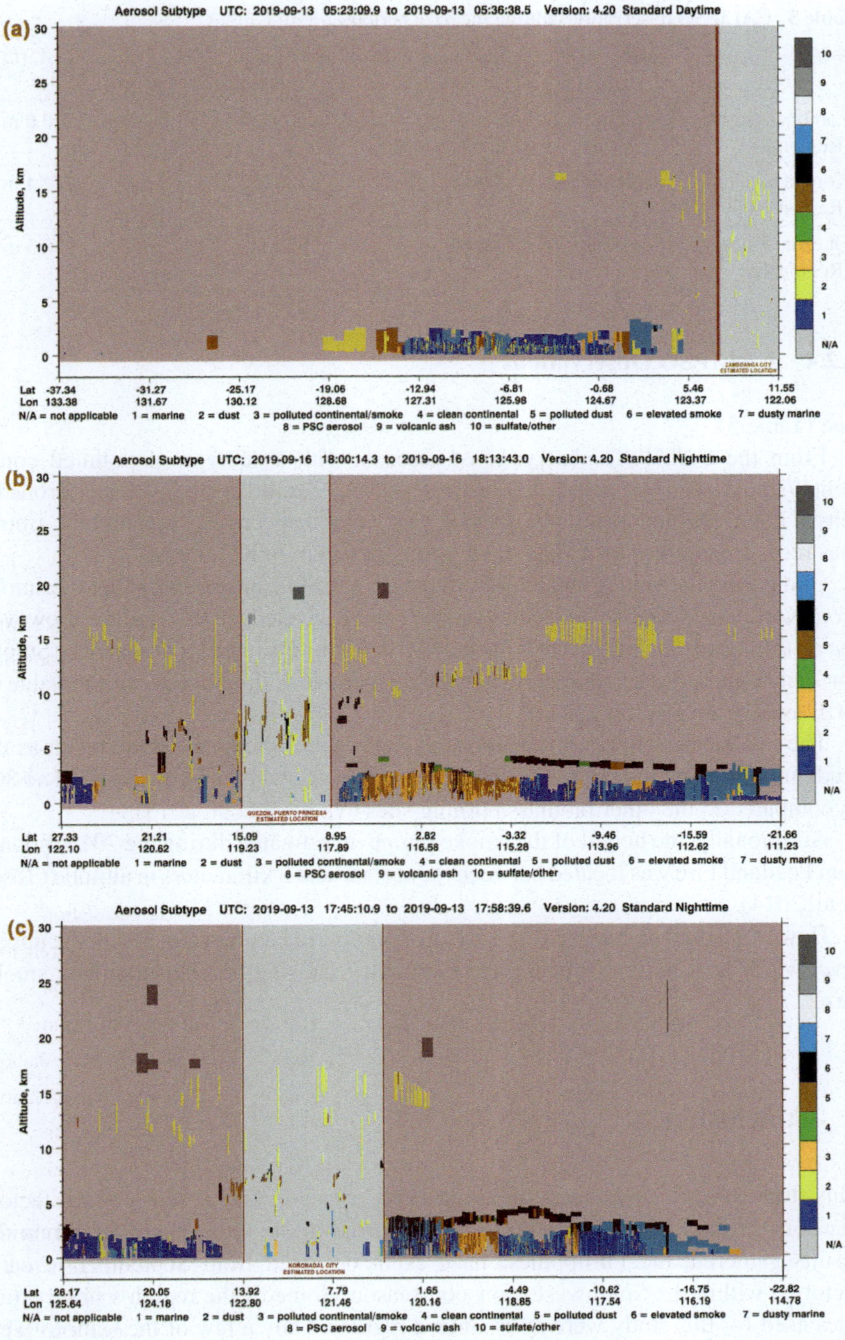

Fig. 5 Compiled CALIPSO passthrough results in **a** Zamboanga City, **b** Quezon, Puerto Princesa City, and **c** Koronadal City

Table 5 CALIPSO observations during the haze period

Site	Date	Smoke altitude (m)	Polluted continental/smoke (m)	LCL (m)
Zamboanga city (Region 9)	September 13	N/A	500 m	770.6 m
Koronadal city (Region 12)	September 13	2000 m	1000 m	772.4 m
Quezon, Palawan (Region 4B)	September 16	5000 m	900 m	986.3 m

4.2.4 CALIPSO Observations

See (Table 5).

From the CALIPSO observations, aerosols with a subtype of polluted continental/smoke were recorded at Quezon, Palawan, Zamboanga City, and Koronadal City with an altitude of 900 m, 0.5 km, and 1 km, respectively. Meanwhile, smoke was recorded at 5 km for Quezon, Palawan, and 2 km in Koronadal.

Comparing the values obtained from the CALIPSO dataset shows that the smoke from Koronadal (2000 m) and Palawan (5000 m) exceeded the PBL height. However, the result shows that 500 m, 1000 m, and 900 m for the Polluted Continental or Smoke for Zamboanga, Koronadal, and Palawan, respectively, did not exceed the value of LCL aside from Region 12.

The low smoke plume in the Northern Kalimantan shows the same result as the study investigating the biomass burnings in Indonesia, wherein they recorded an 800 m compared to the other biomass burning sites (Vadrevu et al. 2015).

Additionally, the height of the smoke plume investigated during the 2015 Indonesian Peatland Fire was located between the surface and 3 kilometers in altitude (Tosca et al. 2014).

However, some of the smoke plumes from Indonesia tend to move with the upper-level winds as it moves from country to country, causing the transboundary smoke haze.

5 Conclusion

This study aims to characterize the long-range transport and meteorological factors that formed the transboundary haze that occurred in the Philippines. Based on the results gathered, the Philippines' haze event occurred from September to early October, within the Southwest Monsoon season. Some of the region's observation sites used for this study were in urbanized regions. Only a few of the stations were in rural areas. Looking at synoptic-scale weather in this study, we investigated the impact of the southwest monsoon and other weather systems on aerosol transport from Indonesia toward the Southern Philippines.

The anthropogenic aerosol transport originated from the Indonesian Peninsula heading toward the Philippines using air mass back trajectories from the NOAA HYSPLIT model. Using the satellite models' Copernicus and MERRA-2, we also identified the prevailing meteorological system as the Southwest Monsoon or Habagat. The influence on aerosols in different regions was shown through enhancements in AOD values and visibility decrease.

The characteristic of the receptor sites, whether rural or urban, has a massive influence on the amount of aerosols detected in the area. The anthropogenic emissions from the transboundary haze enhanced the amount of aerosols in the receptor sites. Moreover, the weather systems heightened the strength of the monsoon, which advected the emissions in the Southern Philippines.

Throughout the observation, the thunderstorm's effect was considered the main challenge in using the airport and synoptic stations' visibility. Heavy rain showers associated with thunderstorms reduce aerosol concentrations including the visibility. Another challenge is the homogeneity of the data. Some data were obtained using manual methods, while the other stations used automated methods. The different datasets lead to the difference in the results per region.

Acknowledgements This work is supported by the GIST Research Institute (GRI) grant funded by the Gwangju Institute of Science and Technology (GIST) in 2021.

We also thank PAGASA, JMA, NASA, and NOAA for the datasets. Furthermore, we would like to extend sincere thanks to Mr. Bernard Alan Racoma of UP-IESM, Mr. Sam Lillo of NOAA and CIRES, Ms. Nina Yulianti of the University of Palangkaraya, and Mr. Krishna Prasad Vadrevu of NASA for providing important comments and suggestions.

References

Adam, M.G. and R. Balasubramanian. (2021). Black carbon emissions from biomass burning in Southeast Asia–a review. *Biomass Burning in South and Southeast Asia*, pp.95–108.

Albar, I., I. Jaya, B.H. Saharjo, B. Kuncahyo, and K.P. Vadrevu. (2018). Spatio-temporal analysis of land and forest fires in Indonesia using MODIS active fire dataset. In Land-atmospheric research applications in South and Southeast Asia (pp. 105–127). Springer, Cham.

Biswas, S., K.P. Vadrevu, Z.M. Lwin, K. Lasko, and C.O. Justice. 2015a. Factors controlling vegetation fires in protected and non-protected areas of Myanmar. *PLoS ONE* 10 (4): e0124346.

Biswas, S., K.D. Lasko, and K.P. Vadrevu. 2015b. Fire disturbance in tropical forests of Myanmar—analysis using MODIS satellite datasets. *IEEE Journal of Selected Topics in Applied Earth Observations and Remote Sensing* 8 (5): 2273–2281.

Braun, R., M. Aghdam, P. Bañaga, G. Betito, M. Cambaliza, M. Cruz, G. Lorenzo, A. MacDonald, J. Simpas, C. Stahl, and A. Sorooshian. 2020. Long-range aerosol transport and impacts on size-resolved aerosol composition in Metro Manila Philippines. *Atmospheric Chemistry and Physics* 20 (4): 2387–2405.

Cruz, M., P. Bañaga, G. Betito, R. Braun, C. Stahl, M. Aghdam, M. Cambaliza, H. Dadashazar, M. Hilario, G. Lorenzo, L. Ma, A. MacDonald, P. Pabroa, J. Yee, J. Simpas, and A. Sorooshian. 2019. Size-resolved composition and morphology of particulate matter during the southwest monsoon in Metro Manila Philippines. *Atmospheric Chemistry and Physics* 19 (16): 10675–10696.

Ellrod, G. (2022). Use of the NOAA ARL HYSPLIT trajectory model for the short range prediction of coastal stratus and fog. In 17th AMS aviation, range, and aerospace meteorology conference. [online] Available at <https://www.researchgate.net/publication/273679533_Use_of_the_NOAA_ARL_HYSPLIT_Trajectory_Model_For_the_Short_Range_Prediction_of_Coastal_Stratus_and_Fog>.

Elvidge, C.D., M. Zhizhin, K. Baugh, and F.C. Hsu. (2021). Identification of smoldering peatland Fires in Indonesia via triple-phase temperature analysis of VIIRS nighttime data. In Biomass burning in South and Southeast Asia (pp. 25–38). CRC Press.

Eorc.jaxa.jp. (2022). JAXA Himawari Monitor (P-Tree System). [online] Available at <http://www.eorc.jaxa.jp/ptree/index.html>.

Fang, M., and W. Huang. 1998. Technical note—Tracking the Indonesian forest fire using NOAA/AVHRR images. *International Journal of Remote Sensing* 19 (3): 387–390.

Gadde, B., S. Bonnet, C. Menke, and S. Garivait. 2009. Air pollutant emissions from rice straw open field burning in India Thailand and the Philippines. *Environmental Pollution* 157 (5): 1554–1558.

Gupta, P.K., V.K. Prasad, C. Sharma, A.K. Sarkar, Y. Kant, K.V.S. Badarinath, and A.P. Mitra. 2001. CH4 emissions from biomass burning of shifting cultivation areas of tropical deciduous forests–experimental results from ground-based measurements. *Chemosphere-Global Change Science.* 3 (2): 133–143.

Hayasaka, H., I. Noguchi, E.I. Putra, N. Yulianti, and K. Vadrevu. 2014. Peat-fire-related air pollution in Central Kalimantan Indonesia. *Environmental Pollution* 195: 257–266. https://doi.org/10.1016/j.envpol.2014.06.031.

Inoue, Y. (2018). Ecosystem carbon stock, atmosphere, and food security in slash-and-burn land use: a geospatial study in mountainous region of Laos. In Land-atmospheric research applications in South and Southeast Asia (pp. 641–665). Springer, Cham.

Kant, Y., A.B. Ghosh, M.C. Sharma, P.K. Gupta, V.K. Prasad, K.V.S. Badarinath, and A.P. Mitra. 2000. Studies on aerosol optical depth in biomass burning areas using satellite and ground-based observations. *Infrared Physics and Technology.* 41 (1): 21–28.

Kharol, S.K., K.V.S. Badarinath, A.R. Sharma, D.V. Mahalakshmi, D. Singh, and V.K. Prasad. 2012. Black carbon aerosol variations over Patiala city, Punjab, India—a study during agriculture crop residue burning period using ground measurements and satellite data. *Journal of Atmospheric and Solar-Terrestrial Physics.* 84: 45–51.

Kita, K., M. Fujiwara, and S. Kawakami. 2000. Total ozone increase associated with forest fires over the Indonesian region and its relation to the El Niño-Southern oscillation. *Atmospheric Environment* 34 (17): 2681–2690.

Lasko, K., and K. Vadrevu. 2018. Improved rice residue burning emissions estimates: accounting for practice-specific emission factors in air pollution assessments of Vietnam. *Environmental Pollution* 236: 795–806.

Lasko, K., K.P. Vadrevu, V.T. Tran, E. Ellicott, T.T. Nguyen, H.Q. Bui, C. Justice. (2017). Satellites may underestimate rice residue and associated burning emissions in Vietnam. *Environmental Research Letters.* 12(8):085006.

Lasko, K., Vadrevu, K.P. and Nguyen, T.T.N. 2018a. Analysis of air pollution over Hanoi, Vietnam using multi-satellite and MERRA reanalysis datasets. PloS one. 13(5):e0196629.

Lasko, K., K.P. Vadrevu, V.T. Tran, and C. Justice. 2018b. Mapping double and single crop paddy rice with Sentinel-1A at varying spatial scales and polarizations in Hanoi Vietnam. *IEEE Journal of Selected Topics in Applied Earth Observations and Remote Sensing* 11 (2): 498–512.

Mabalay, M.R., J. Raviz, E. Alosnos, M. Barbieri, E. Quicho, J.E.A. Bibar, M. Barroga, M. Coñado, P. Mabalot, J.R. Mirandilla, and A. Arocena Jr. (2022). The Philippine rice information system (PRiSM): an operational monitoring and information system on rice. *Remote Sensing of Agriculture and Land Cover/Land Use Changes in South and Southeast Asian Countries*, p. 133.

Mendoza, T. and R. Samson. (1999). Strategies to avoid crop residue burning in the Philippine context. Reap-canada.com. Available at <http://reap-canada.com/library/IntDev/8%20Strategies%20to%20Avoid%20Crop%20Residue%20Burning%20in%20the%20Philippine%20Context%20(1999).htm>.

Page, S., F. Siegert, J. Rieley, H. Boehm, A. Jaya, and S. Limin. 2002. The amount of carbon released from peat and forest fires in Indonesia during 1997. *Nature* 420 (6911): 61–65.

Perez, G.J., J.C. Comiso, M.G. Cayetano. (2021). Swidden agriculture and biomass burning in the Philippines. In Biomass burning in South and Southeast Asia (pp. 183–198). CRC Press.

Phairuang, W. (2021). Biomass burning and their impacts on air quality in Thailand. In Biomass burning in South and Southeast Asia (pp. 21–38). CRC Press.

Prasad, V.K., K.V.S. Badarinath, and A. Eaturu. 2008. Biophysical and anthropogenic controls of forest fires in the Deccan Plateau India. *Journal of Environmental Management* 86 (1): 1–13.

Prasad, V.K., P.K. Gupta, C. Sharma, A.K. Sarkar, Y. Kant, K.V.S. Badarinath, T. Rajagopal, and A.P. Mitra. 2000. NOx emissions from biomass burning of shifting cultivation areas from tropical deciduous forests of India–estimates from ground-based measurements. *Atmospheric Environment* 34 (20): 3271–3280.

Prasad, V.K., Y. Kant, and K.V.S. Badarinath. 2001. CENTURY ecosystem model application for quantifying vegetation dynamics in shifting cultivation areas: a case study from Rampa Forests, Eastern Ghats (India). *Ecological Research.* 16 (3): 497–507.

Reuters. (2019). Area burned in 2019 forest fires in Indonesia exceeds 2018 - official. Available at <https://www.reuters.com/article/us-southeast-asia-haze-idUSKBN1X00VU>

Rolph, G., A. Stein, and B. Stunder. 2017. Real-time environmental applications and display system: READY. *Environmental Modelling and Software* 95: 210–228.

Saharjo, B.H. and A. Yungan. (2018). Forest and land fires in Riau Province: a case study in fire prevention policy implementation with local concession holders. In Land-atmospheric research applications in South and Southeast Asia (pp. 143–169). Springer, Cham.

Salvi, S., and P.J. Barnes. 2010. Is exposure to biomass smoke the biggest risk factor for COPD globally? *Chest* 138 (1): 3–6.

Shi, Y., T. Sasai, and Y. Yamaguchi. 2014. Spatio-temporal evaluation of carbon emissions from biomass burning in Southeast Asia during the period 2001–2010. *Ecological Modelling* 272: 98–115.

Siegert, F., G. Ruecker, A. Hinrichs, and A. Hoffmann. 2001. Increased damage from fires in logged forests during droughts caused by El Niño. *Nature* 414 (6862): 437–440.

Stein, A., R. Draxler, G. Rolph, B. Stunder, M. Cohen, and F. Ngan. 2015. NOAA's HYSPLIT atmospheric transport and dispersion modeling system. *Bulletin of the American Meteorological Society* 96 (12): 2059–2077.

Susetyo, K.E., K. Kusin, Y. Nina, Y. Jagau, M. Kawasaki, and D. Naito. (2019). Peatland and forest fires in Central Kalimantan, Indonesia. Tropical Peatland Society Project, [online] Available at <https://www.chikyu.ac.jp/peatlands/img/newsletter/vol8.pdf>

Syaufina, L. and S.I. Maulana. (2021). Biomass burning emissions in Indonesia and policy measures–an overview. *Biomass Burning in South and Southeast Asia*, pp. 135–148

Tariq, S., and Z. Ul-Haq. (2018). Satellite remote sensing of aerosols and gaseous pollution over Pakistan 2018. In: Vadrevu, KP, T. Ohara, and C. Justice (Eds). Land-atmospheric research applications in South/Southeast Asia. Springer, pp.523–552.

Tosca, M., D. Diner, M. Garay, and O. Kalashnikova. 2014. Observational evidence of fire-driven reduction of cloud fraction in tropical Africa. *Journal of Geophysical Research: Atmospheres* 119 (13): 8418–8432.

Uranishi, K., H. Shimadera, and A. Kondo. (2021). Biomass burning influence on PM2. 5 regional and long-range transport in Northeast Asia. In Biomass Burning in South and Southeast Asia (pp. 281–295). CRC Press.

Vadrevu, K., K. Lasko, L. Giglio, and C. Justice. 2015. Vegetation fires, absorbing aerosols and smoke plume characteristics in diverse biomass burning regions of Asia. *Environmental Research Letters* 10 (10): 105003.

Vadrevu, K.P., and K. Lasko. (2018). Intercomparison of MODIS AQUA and VIIRS I-Band fires and emissions in an agricultural landscape—implications for air pollution research. Remote sensing. 10(7):978. https://doi.org/10.3390/rs10070978

Vadrevu, K.P., T. Ohara, and C. Justice (eds.). (2021a). Biomass burning in south and Southeast Asia: Impacts on the biosphere, Volume Two. CRC Press.
Vadrevu, K.P., T. Ohara, and C. Justice (eds.). (2021b). Biomass burning in South and Southeast Asia: Mapping and monitoring, volume one. CRC Press.
Yulianti, N., K. Kusin, D. Naito, M. Kawasaki, O. Kozan, and K. Susatyo. 2020. The Linkage of El Niño-induced peat fires and its relation to current haze condition in Central Kalimantan. *Journal of Wetlands Environmental Management* 8 (2): 100.

An Operational Fire Danger Rating System for Thailand and Lower Mekong Region: Development, Utilization, and Experiences

Veerachai Tanpipat, Kasemsan Manomaiphiboon, Robert D. Field, William J. deGroot, Prayoonyong Nhuchaiya, Narin Jaroonrattanapak, Chatchaya Buaniam, and Jittisak Yodcum

Abstract This study describes development, utilization, and experiences of implementing a fire danger rating forecast system for Thailand and Upper ASEAN, or the Lower Mekong River Region, which is referred to as "FR-Mek (short for Fire Danger Rating System for Lower Mekong River Region)." It was developed within the widely used framework of the Canadian Fire Danger Rating System which has been adopted in equatorial Southeast Asia. Following its development, FR-MeK has been transferred to and officially operated since 2015 by the Department of National Parks, Wildlife, and Plant Conservation and since 2020 by the Royal Forest Department. The system uses two fire danger indices: the Fine Fuel Moisture Code (FFMC) and Fire Weather Index (FWI) representing, how easy a fire would be ignited and how dangerous an ignited fire would be in terms of spread and intensity, respectively. To adapt the FR-MeK for the regional fire environment, fires over Upper ASEAN or the Lower Mekong River Region were studied using Terra/Aqua Moderate Resolution Imaging Spectroradiometer (MODIS) and their relationship with fire weather was investigated. The FFMC and FWI were calibrated to four fire danger levels from

V. Tanpipat (✉)
Upper ASEAN Wildland Fire Special Research Unit, Forestry Research Center, Faculty of Forestry, Kasetsart University, Bangkok, Thailand
e-mail: fforvrc@ku.ac.th

K. Manomaiphiboon
The Joint Graduate School of Energy and Environment, King Mongkut's University of Technology Thonburi, Bangkok, Thailand

R. D. Field
Department of Applied Physics and Applied Mathematics, NASA Goddard Institute for Space Studies, Columbia University, New York, USA

W. J. deGroot
Natural Resources Canada, Ottawa, Canada

P. Nhuchaiya · N. Jaroonrattanapak · C. Buaniam
Department of National Parks, Wildlife, and Plant Conservation, Ministry of Natural Resources and Environment, Bangkok, Thailand

J. Yodcum
Royal Forest Department, Ministry of Natural Resources and Environment, Bangkok, Thailand

K. P. Vadrevu et al. (eds.), *Vegetation Fires and Pollution in Asia*,
https://doi.org/10.1007/978-3-031-29916-2_34

low to high. A suite of computer programs were developed as FR-Mek for a daily forecast operation capable of forecasting fire danger at grid resolutions of 3–9 km over six days (current day and next five days). The daily FWI value is used to adjust FWI signs placed in front of 305 forest fire suppression stations throughout Thailand. Beyond this direct public data sharing, more intensive operational use is still evolving. During the 2021 fire season, its use has increased and become a part of daily operation of forest fire, open burning and smoke haze control and management for 17 northern provinces of Thailand.

Keywords Forest fire · Fire danger rating · Weather · Forecast · Thailand and upper ASEAN or lower Mekong River Region

1 Introduction

Fires and smoke haze have become an annual hazard in Thailand and the Upper ASEAN (or Lower Mekong River Region) during the dry season, particularly in Upper North Thailand (UNT) and most of the Lower Mekong River Region (LMR), which is the study area of interest. The LMR geographically includes Thailand, Myanmar, Lao PDR, Cambodia, and Vietnam. For the region, the dry season typically lasts six months (November–April), while the wet season includes the remaining months. In the early part of the dry season, the northeast monsoon prevails and brings cool, dry continental air from mid-latitudes. In the later part, the monsoon weakens and is replaced by warm lows. Although the air becomes warmer and more humid, the moisture availability needs to be higher to produce considerable rain. These conditions are favorable to vegetation fires. In addition to climate, most of the fires in Southeast Asia are attributed to human factors (Biswas et al. 2015; Alabar et al. 2018; Hayasaka and Sepriando 2018; Lasko and Vadrevu 2018; Saharjo and Yungan 2018; Syaufina and Maulana 2021; Vadrevu et al. 2021a,b). Agricultural burning usually intensifies in this period due to land clearing and preparation for crop farming in the forthcoming wet season (Oanh et al. 2018). Fires can cause many impacts. They are an environmental disturbance that may alter the richness of forest ecology and biodiversity. Severe fires can seriously threaten firefighters and people within affected areas. It also emits a large amount of air pollutants and greenhouse gasses into the atmosphere, potentially resulting in degraded air quality due to smoke haze and, in turn, negatively affecting tourism, business, air, and land transportation, and, importantly, public health (Manomaiphiboon et al. 2009 and 2017; Phairuang 2021; Lam and Roy 2021; Adam and Balasubramanian 2021; Park and Takeuchi 2021; Lee and Wang 2021; Uranishi et al. 2021).

In Thailand, several agencies are responsible for forest fire, open burning, and smoke haze control, management, and mitigation, namely: the Department of National Parks, Wildlife, and Plant Conservation (DNP), the Royal Forest Department (RFD), and the Pollution Control Department (PCD). All are under the Ministry of Natural Resources and Environment (MNRE). At the national level, the Forest Fire

Control Division (FFCD) in both DNP and RFD has utilized several technical tools to assist their fire control, manage and mitigate related activities (e.g., applications of satellite-detected active fire hotspots to fire monitoring by NASA Fire Information for Resources Management System, FIRMS (Tanpipat et al. 2009), School of Mines in Colorado's VIIRS NightFire Alert System (Elvidge et al. 2019), VIIRS NOAA-the National Environmental Satellite, Data, and Information Service (NESDIS) Active Fire Alert System, and use of rapid communication channels between fire suppression stations and commanding centers. The earlier researchers conducted field experiments to understand fire characteristics and behavior (e.g., Akaakara 1996; Akaakara et al. 2005). Fire Danger Rating Systems (FDRS), which is used to indicate conditions under which fires can start and spread (Van Wagner 1987), can also be extended as early warning or forecast information to support fire control, management, and mitigation in terms of resources preparation, logistics and for awareness among the public (Moore 2019; Vadrevu 2021, 2021a, 2021b). Different FDRSs have been developed, and they have been used across different regions of the world (Dimitrakopoulos et al. 2011; de Groot et al. 2015). The Canadian Forest FDRS is widely used (Van Wagner 1987; Field et al. 2015). However, such a forecast technology has not yet been calibrated and established for the LMR, thus implementing this study. In contrast to Lower Southeast Asia, Malaysia, and Indonesia, the FDRS technology adopted from the Canadian FDRS has been since 2004 (de Groot et al. 2007). Although located nearby, the fire environment of those countries is distinct from the LMR. Continuous evaluation of the products from the Indonesian and Malaysian FDRS products for ASEAN since 2010 suggested that, in their current form, they do not apply to the LMR, having been calibrated for a wet tropical environment and with a focus on peat fires. During informal cooperation in 2012–2013 and with a formal effort in 2014 supported by the National Research Council of Thailand (NRCT), we have worked on calibrating FDRS for Thailand and Upper ASEAN and laying the groundwork for its operational use.

2 Data and Methods

Deciduous forests and agricultural regions with distinct dry and rainy seasons dominate the LMR study area. Over the deciduous forests, leaves shed during the dry season provide the primary fuel source when burning for non-timber products in a forest fire. Over the agricultural regions, residue burning from rice and maize cash crops mainly contributes to open burning. Our goal was to calibrate several components of the Canadian Fire Weather Index (FWI) System (a sub-component of the Canadian FDRS) for use in the LMR fire environment. The three fuel moisture codes and fire behavior indices of the FWI System are as follows.

1. The Fine Fuel Moisture Code (FFMC) captures variations in the moisture content of fine fuels and leaf litter on the forest floor, driven by temperature, relative humidity, wind speed, and rainfall

2. The Duff Moisture Code (DMC) is driven by temperature, relative humidity, and rainfall. It represents the moisture content of loosely compacted forest floor organic matter and relates to the likelihood of lightning ignition, whose value has no limit
3. The Drought Code (DC) represents the moisture content of deep, compacted organic soils and heavy surface fuels driven by temperature and rainfall. Similar to DMC, its value also has no upper limit
4. The Initial Spread Index (ISI) reflects how quickly a fire spreads after ignition, which is determined by wind speed and FFMC
5. The Buildup Index (BUI) reflects the fuel available for fire and is driven by DMC and DC
6. The Fire Weather Index (FWI), which takes ISI and BUI into account, aggregately gives a rating of fireline intensity in a reference fuel type and level terrain.

For every index, the higher the value, the more extreme the fire danger. The same computer program as Field et al. (2015) was used to calculate these indices over the LMR, based on that of Van Wagner and Pickett (1985). The calculation uses four simple daily surface weather variables (observed at noon): air temperature at 2 m (above ground level), relative humidity at 2 m, wind speed at 10 m, and 24 h accumulated rainfall.

In developing the FDRS forecast system specifically for the LMR, there are two main tasks: (1) to calibrate the above-mentioned system to the forest ecology or conditions in the region, and (2) to develop a suite of computer programs that can predict the desired fire danger indices (here, FFMC and FWI) ahead of time and with low latency. The system calibration determines the number of threshold values for a particular index corresponding to different fire danger levels. Since one region may have different forest types from another, it is thus necessary to find such threshold values suitable for the region of interest. Here, a fire-prone area was defined as the area over which a large number of fires occur, compared to other areas (Fig. 1), using active fire hotspots detected by the MODIS sensors on board of the National Aeronautics and Space Administration (NASA)'s satellite Terra and Aqua (Justice et al. 2002). The active fire hotspots were considered over 10 years (November 2002–October 2012) in the two fire-prone areas combined and only found in MODIS-based vegetative areas (Friedl et al. 2010). As for historical FFMC and FWI data used in the system calibration were directly extracted from Global Fire Weather Database (GFWED) (Field et al. 2015; https://data.giss.nasa.gov/impacts/gfwed/). The GFWED data, derived from Modern Era Retrospective-Analysis for Research and Applications (MERRA) data (Rienecker et al. 2011; https://gmao.gsfc.nasa.gov/reanalysis/MERRA), cover a long-term period from 1980 to the present and have a moderate grid resolution of $0.5° \times 0.667°$. The FFMC and FWI values in the GFWED data were evaluated against those computed using surface weather data at several stations located in the region. The surface weather data were obtained from the Thai Meteorological Department (TMD, https://www.tmd.go.th/) and the National Centers for Environmental Information (NCEI, https://www.ncei.noaa.gov/). The GFWED data quality was found to be fair, well-capturing monthly mean variation, and in reasonable agreement with daily values statistically (not shown). The original MODIS active fire hotspots were arranged as fire groups using a geometric aggregation technique to represent fire occurrences better and burned areas, which were later coupled with FFMC and FWI, respectively. The relationship between FFMC and total fire group count and FWI versus total burned area was established by statistical fitting using nonlinear log-logistic functions. Changes in the slope of the estimated functions were used to suggest threshold values.

For the second task, a suite of computer programs was developed as an operational fire danger forecast system, which is referred to as short for Fire Danger Rating System for Lower Mekong River Region (FR-Mek). It integrates data from various sources as input to FR-Mek implementation (Fig. 2). They are National Centers for Environmental Prediction (NCEP) Final Reanalysis (FNL) data (https://rda.ucar.edu/datasets/ds083.2) for fair-resolution temperature, relative humidity, and wind speed data, Climate Prediction Center (CPC) unified gauge-based analysis data (Xie et al. 2007; https://climatedataguide.ucar.edu/climate-data/cpc-unified-gauge-based-analysis-global-daily-precipitation) for fair-resolution rainfall data, the Hydro-Informatics Institute (HII) for high-resolution weather data from its official weather forecast operation (Torsri et al. 2014), and Global Forecast System (GFS) data for fair-resolution weather data. The FNL and CPC data are used together to

Fig. 1 Fire-prone area (shaded)

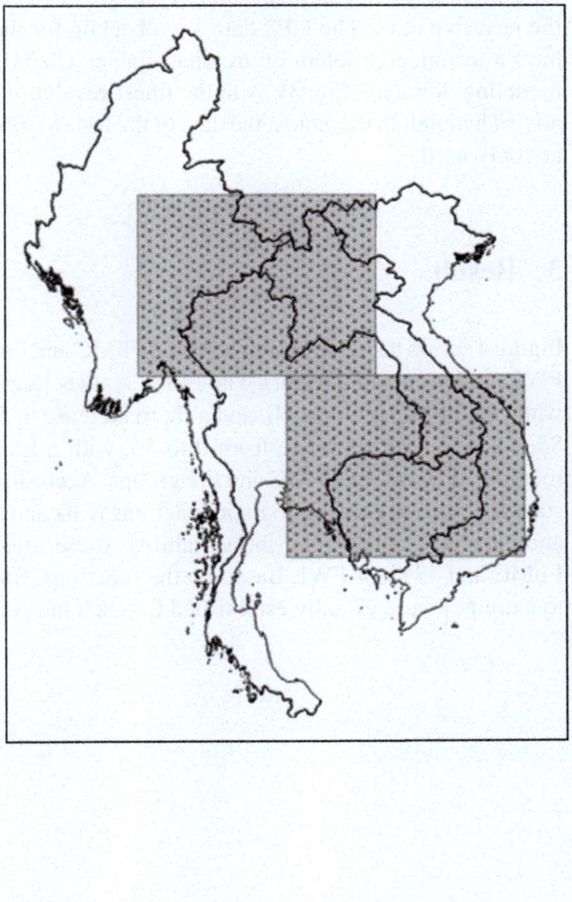

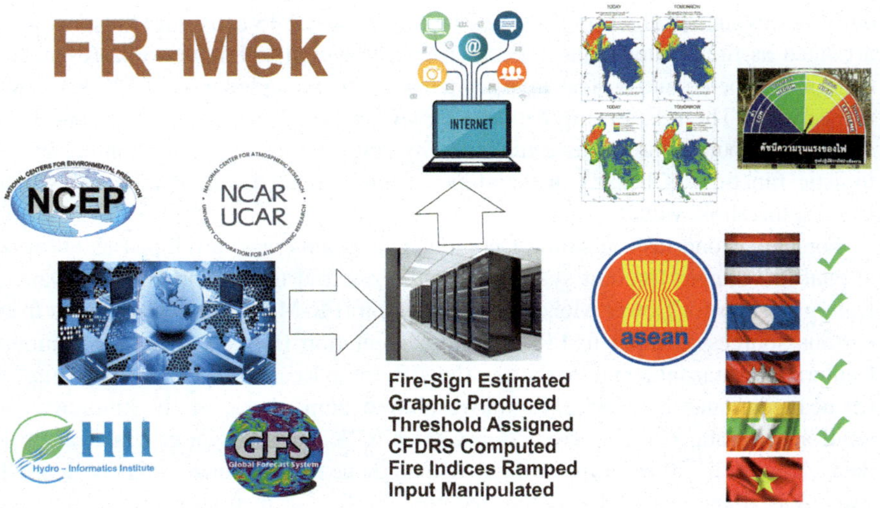

Fig. 2 Overall schematic of FR-Mek

spin up the system over a historical period (at least one year) until the day before the current day of operation. The HII data drive calculations for the current day and the next five days. The GFS data are a backup for the HII data in case the HII data have a technical problem or are unavailable. The HII forecast operation uses three modeling domains (Fig. 3), with the finest-resolution (3 km) domain encompassing all of Thailand. In this study, the data of the last two domains (3 and 9 km resolutions) are only used.

3 Results

Figure 4 shows the relationship between FFMC and fire groups (left) and that between FWI and burned area (right). The FFMC values range from 20 to 95, with little fire when the FFMC is below 60, and a sharp increase in fire count when FFMC is above 85. The FWI values range from 0 to 50, with a less clear distinction than for the relationship between FFMC and fire groups. According to the system calibration, the standard 3-parameter log-logistic functions (Ritz and Streibig, 2005) were employed and capable of explaining the variability; these fitted data reasonably by 45% for FFMC and 47% for FWI. Based on the functions, four fire danger levels (from low to extreme) were visually established for each index as follows:

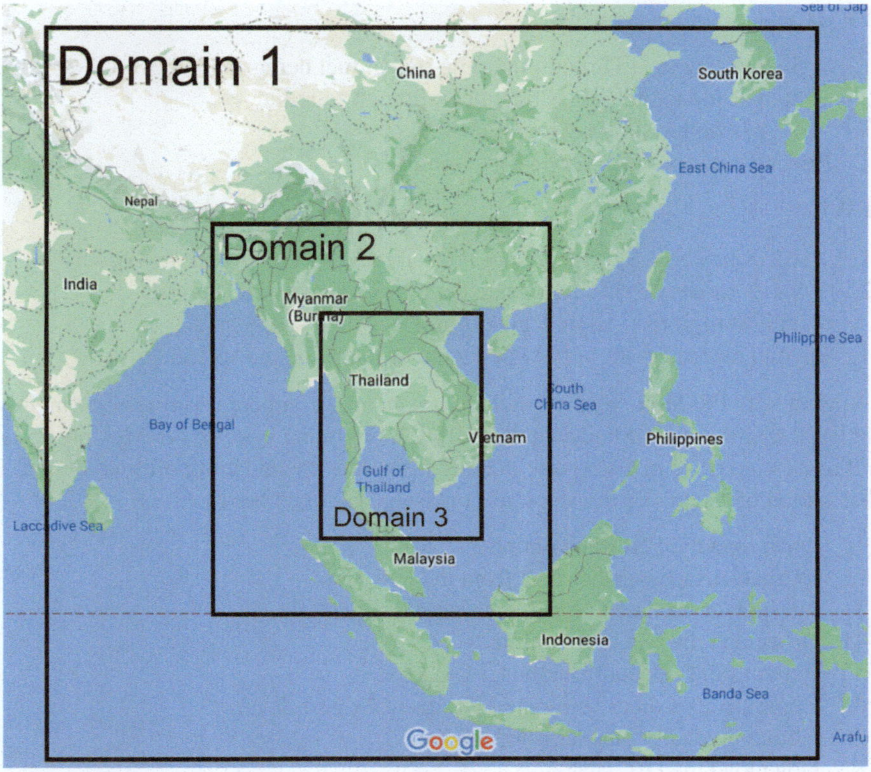

Fig. 3 Modeling domains used by the HII. The above three domains have the grid resolutions of 27 km, 9 km, and 3 km, respectively

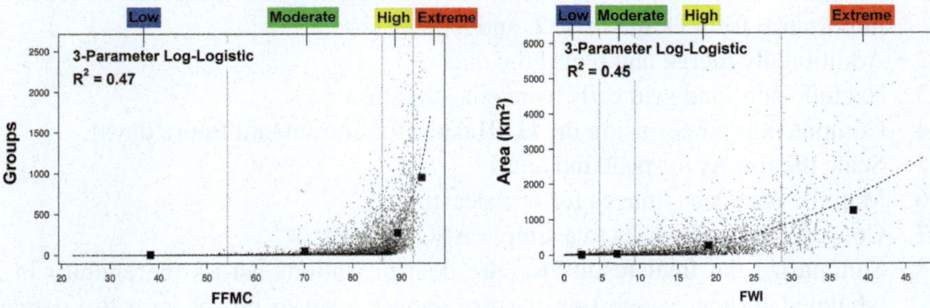

Fig. 4 FFMC and FWI fitting, specified thresholds, and their corresponding fire danger levels. For FFMC (left), the X-axis is daily GFWED FFMC (averaged over each fire-prone area, then pooled into the fitting), and the Y-axis is daily fire groups (total over each fire-prone area, then pooled into the fitting). For FWI (right), the X-axis is daily GFWED FWI (averaged over each fire-prone area, then pooled into the fitting), and the Y-axis is summed area of all daily fire groups (summed over each fire-prone area, then pooled into the fitting)

FFMC:

1. < 54.2 as low risk: Difficult to ignite (given fuel being available)
2. < 86.0 as moderate risk: Moderately difficult to ignite
3. < 92.4 as high risk: Easy to ignite
4. Beyond extreme risk: Very easy to ignite.

FWI:

1. < 4.7 as low risk: Low degree of intensity and spread (when fires occur)
2. < 9.4 as moderate risk: Moderate degree of intensity and spread
3. < 28.7 as high risk: High degree of intensity and spread and
4. Beyond extreme risk: Very high degree of intensity and spread.

Most of FR-Mek was written in standard computer languages (mainly, FORTRAN, R, and C-Shell) within a LINUX operating system. FR-Mek processes all associated tasks in a stepwise manner from data downloading till final output. It has a total of 19 processing steps, as concisely described below:

1. Download the FNL data online
2. Extract surface weather data from the FNL data
3. Interpolate the data to local noon time
4. Download the CPC data online
5. Extract rainfall data from the CPC data
6. Compute all indices using the FNL data (for past days)
7. Climatologically bias-adjusted the data
8. Download the HII data
9. Extract and arrange the data in a flat binary format
10. Extract the data from
11. Merge and regrid the data of the three modeling domains altogether in order of importance from Domains 1, 2, and 3
12. Additionally merge and regrid the data
13. Exclude non-land grid cells from consideration
14. Compute all indices using the HAII data (for current and future days)
15. Scale the results for peatland areas
16. Map the results as images for broadcasting
17. Convert the final results to a simple ASCII format
18. Summarize the final results for fire control stations across the country in a tabulated format, suggesting the fire danger level to be set on a fire danger warning sign at each station (Fig. 5) and
19. Finalize maps for online broadcasting at http://www2.dnp.go.th/gis/FDRS/FDRS.php and https://wildfire.forest.go.th/fdrs/FDRS.php. Figure 6 shows an example of daily FFMC and FWI forecasts posted at the website.

Fig. 5 Example of fire danger signs at local fire suppression stations

4 Discussion

FR-Mek has been successfully developed in support of daily fire danger forecast operation after calibration for the LMR fire environment. FR-Mek was tested, enhanced, and then transferred to the Department of National Parks, Wildlife, and Plant Conservation and Royal Forest Department. It has been officially adopted and operational daily since 2015. The parallel forecast system at the Royal Forest Department began in 2020. The system is automatically initialized at 7 am and runs to completion in about 30 min. In terms of its prediction performance, an evaluation test was preliminarily performed, comparing forecast results and MODIS active fire hotspots for March–May 2015. It was found to be acceptable but tends to underestimate FFMC and FWI for severe fire events (Fig. 7). Over the years, more data are being collected for further calibration. Moreover, there is over a decade of fire field experimental data which can be used. A comprehensive performance evaluation can better understand the developed system's strengths and weaknesses, guiding further development. Bias correction on the modeled weather input data should be explored to increase the existing system's reliability, given automated telemetry weather stations installed

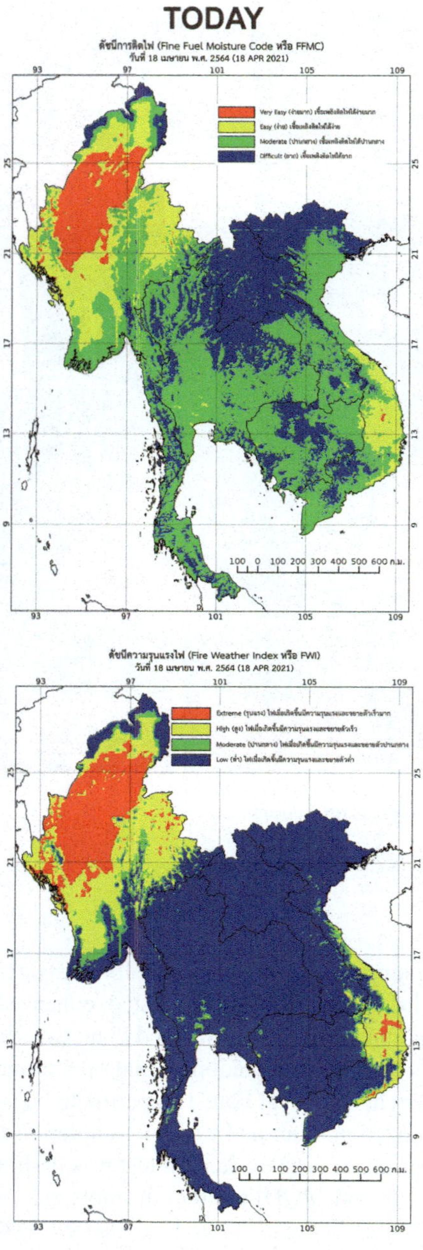

Fig. 6 Example of daily FFMC and FWI forecasts (http://www2.dnp.go.th/gis/FDRS/FDRS.php and https://wildfire.forest.go.th/fdrs/FDRS.php)

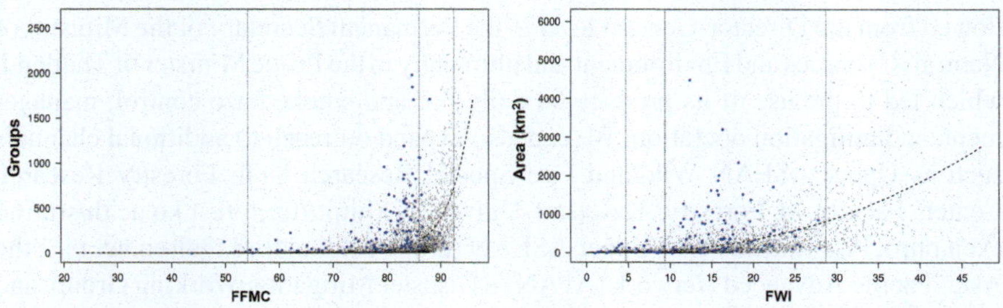

Fig. 7 Evaluation of daily forecast FFMC and FWI for March–May 2015. The blue dots are of forecast while the small black dots and the fitted dashed lines are of calibration

across the country. The use of a higher spatial resolution and a longer leading forecast time will increase its applicability.

5 Utilization

At the beginning of Thailand's FDRS operational effort, many official letters went out from DNP to other government departments to let them know about the availability of Fire Weather Index (FWI) and Fire Fuel Moisture Content (FFMC) information from the Forecast Fire Danger Rating System (FDRS) that can be used in order to control, management and mitigate their fires and smoke haze related activities more efficiently. There was also considerable effort and follow-up to explain the FDRS within the DNP and the RFD, and many follow-up internal letters, too; for example, consistent fire danger signs in front of over 305 fire control stations in both DNP and RFD were placed throughout the country. Still, there are misunderstandings about the FDRS; thus, need for more outreach. For example, some users still think that the FDRS outputs predict where fires will occur rather than the *potential* for fires to occur according to the fuel's estimated moisture content.

Delivering and communicating these findings to stakeholders at an appropriate level of scientific detail are crucial. Transforming Science to Policy Makers (SOP) is essential since high-level government officers and politicians usually need help understanding the scientific facts that would act as their background knowledge when making decisions. There is also a need for an effective mechanism to communicate the science of fire weather and fire danger rating to operational government departments and agencies. When making fire-related policies, they should have the correct information to support their decision process. For Thailand, the first step is to put all findings and facts together and communicate through the existing operational channels, namely the Forest Fire Division in both the Department of National Parks, Wildlife, and Plant Conservation (http://www2.dnp.go.th/gis/FDRS/FDRS.php) and the Royal Forest Department (https://wildfire.forest.go.th/fdrs/FDRS.php) under the Ministry of Natural Resources and Environment. Institutionally, the information has flowed from the Director-General level to the Permanent Secretary of the Ministry of Natural Resources and Environment and ultimately to the Prime Minister of Thailand, which led to the use of its products in daily fire and smoke haze control, management, and mitigation operation. We can also expand outreach to additional channels such as Upper ASEAN Wildland Fire Special Research Unit, Forestry Research Center, Faculty of Forestry, Kasetsart University (http://frc.forest.ku.ac.th/sru/indexen.php), Fire Emissions in Upper ASEAN (http://wildlandfire.thairen.net.th/), the Asia–Pacific Advanced Network (APAN)'s Disaster Mitigation Working Group, and newly its Open and Sharing Data Working Group (https://apan.net/wg). Through collaborations with the Asian Forest Cooperation Organization (AFoCO, http://afocosec.org/), the possibility exists to improve the system with the more advanced FDRS of the Korean Forest Institute of Forest Science, Korean Forest Service (https://eng-nifos.forest.go.kr/). Regionally, the operational Forecast Thailand and Upper ASEAN FDRS began running in 2015 at the DNP. Based on requests from Lao PDR, Cambodia, and Myanmar, we have created separate web pages for each country, e.g., https://wildfire.forest.go.th/fdrs/FDRS_Laos.php, https://wildfire.forest.go.th/fdrs/FDRS_Cambodia.php, and https://wildfire.forest.go.th/fdrs/FDRS_Myanmar.php. In 2020, a backup or parallel system was set up and running at the RFD. In the 2021 fire season, it became a primary supporting information system in daily fires and smoke haze control, management, and mitigation under the Ministry of Natural Resources and Management.

The FR-Mek has become a part of daily forest fire, open burning, and smoke haze control, management, and mitigation national operation system after nearly a decade of research, system development, and outreach. The improvement of the system is still needed to make daily operations and seasonal planning more efficient. Currently, only web-based broadcasting is offered. Mobile application development may facilitate broader use of its predicted results. Collaboration with different groups is needed to advance the FDRS technology. The fundamental education and understanding of FDRS to the public, government agencies, and NGOs should be continued. Further recalibration by using 30 years of historical weather and a decade of fire field experiment data with Natasha Jurko of Great Lakes Forestry Center, Natural Resources Canada – Government of Canada and Mike Wotton of University of Toronto is on going.

Acknowledgements The authors sincerely thank the DNP and RFD for all technical assistance and support given to the study, the HII for the weather forecast data, the TMD and the NCEI for the observed surface weather data, NASA's Fire Information for Resources Management (FIRMS) for the satellite active fire hotspot data and Land Processes Distributed Active Archive Center (LP DAAC) for land cover data. We thank Retired Dir. Siri Akaakara (DNP) for his pioneering fire danger work in the region that has inspired us, and Dr. Kobsak Wanthongchai, Dean of the Faculty of Forestry, Kasetsart University, for his valuable suggestions. This study was supported by the Hydro-Informatics Institute, the Joint Graduate School of Energy and Environment, the King Mongkut's University of Technology Thonburi, the National Research Council of Thailand, and the Biodiversity-Based Economy Development Office.

References

Adam, M.G. and R. Balasubramanian. (2021). Black carbon emissions from biomass burning in Southeast Asia–a review. *Biomass Burning in South and Southeast Asia*, pp.95–108.

Akaakara, S. (1996). Fire danger rating in dry dipterocarp forest at Doi Suthep-Pui National Park. Technical Report, Royal Forest Department. (in Thai)

Akaakara, S., K. Wiriya, and P. Nhuchaiya. (2005). Fuel characteristics in mixed deciduous forest at Huai Kha Khaeng Wildlife Sanctuary. Technical Report, Department of National Parks, Wildlife and Plant Conservation. (in Thai)

Albar, I., I. Jaya, B.H. Saharjo, B. Kuncahyo, and K.P. Vadrevu. (2018). Spatio-temporal analysis of land and forest fires in Indonesia using MODIS active fire dataset. In *Land-Atmospheric Research Applications in South and Southeast Asia* (pp. 105–127). Springer, Cham.

Biswas, S., K.P. Vadrevu, Z.M. Lwin, K. Lasko, and C.O. Justice. (2015). Factors controlling vegetation fires in protected and non-protected areas of Myanmar. *PLoS ONE* 10 (4): e0124346.

de Groot, W.J., R.D. Field, M.A. Brady, R. Orbita, and M. Mohamad. (2007). Development of the Indonesian and Malaysian fire danger rating systems. *Mitigation and Adaptation Strategies for Global Change.* 12: 165–180.

de Groot, W.J., M.B. Wotton, and M.D. Flannigan. (2015). Chapter 11: Wildland fire danger rating and early warning systems. *Wildfire Hazards, Risks, and Disasters.* Elsevier. https://doi.org/10.1016/B978-0-12-410434-1.00011-7.

Dimitrakopoulos, A.P., A.M. Bemmerzouk, and I.D. Mitsopoulos. (2011). Evaluation of the Canadian fire weather index system in an eastern Mediterranean environment. *Meteorological Applications.* 18: 83–93.

Elvidge, C.D., M. Zhizhin, K. Baugh, F.C. Hsu, and T. Ghosh. 2019. Extending nighttime combustion source detection limits with short wavelength VIIRS data. *Remote Sensing* 11 (4): 395.

Field, R.D., A.C. Spessa, N.A. Aziz, A. Camia, A. Cantin, R. Carr, W.J. de Groot, A.J. Dowdy, M.D. Flannigan, K. Manomaiphiboon, F. Pappenberger, V. Tanpipat, and X. Wang. (2015). Development of a global fire weather database. *Natural Hazards and Earth System Sciences.* 15: 1407–1423.

Friedl, M.A., D. Sulla-Menashe, B. Tan, A. Schneider, N. Ramankutty, A. Sibley, and X. Huang. 2010. MODIS collection 5 global land cover: algorithm refinements and characterization of new datasets. *Remote Sensing of Environment* 114: 168–182.

Hayasaka, H., and A. Sepriando. (2018). Severe air pollution due to peat fires during 2015 super El Nino in Central Kalimantan, Indonesia. In *Land-atmospheric research applications in South and Southeast Asia* (pp. 105–127). Springer, Cham.

Justice, C.O., L. Giglio, S. Korontzi, J. Owens, J.T. Morisette, D. Roy, J. Descloitres, S. Alleaume, F. Petitcolin, and Y. Kaufman. (2002). The MODIS fire products. *Remote Sensing of Environment.* 83: 244–262.

Kim Oanh, N.T., D.A. Permadi, N.P. Dong, and D.A. Nguyet. (2018). Emission of toxic air pollutants and greenhouse gases from crop residue open burning in Southeast Asia. In *Land-atmospheric research applications in South and Southeast Asia* (pp. 47–66). Springer, Cham.

LAM, Y.F. and S. Roy. (2021). Impact of biomass burning on local air quality in South China. In *Biomass Burning in South and Southeast Asia* (pp. 39–56). CRC Press.

Lasko, K., and K. Vadrevu. (2018). Improved rice residue burning emissions estimates: accounting for practice-specific emission factors in air pollution assessments of Vietnam. *Environmental Pollution* 236: 795–806.

Lee, H.H. and C. Wang. (2021). Impacts of biomass burning aerosols on air quality and convective systems in Southeast Asia. In *Biomass Burning in South and Southeast Asia* (pp. 267–280). CRC Press.

Manomaiphiboon, K., V. Surapipith, N. Wiwatwattana, P. Pengchai, S. Thepanondh, R. Onchang, C. Chusai, and M. Octaviani. (2009). Development of a prototype of smoke-haze forecast modeling system for Upper Northern Thailand and its application to drought fire seasons. Final report,

The Joint Graduate School of Energy and Environment, funded by National Research Council of Thailand. (in Thai)
Manomaiphiboon, K., V. Tanpipat, P. Nhuchaiya, N. Jaroonrattanapak, and C. Buaniam. (2017). Development of a regional fire danger forecast system for Upper Thailand and Lower Mekong River Basin areas in support of forest fire management and control. Final report, conducted by The Joint Graduate School of Energy and Environment, funded by Biodiversity-Based Economy Development Office and National Research Council of Thailand. (in Thai)
Moore, P.F. (2019). Global wildland fire management research needs. *Current Forestry Reports* 5: 210–225. https://doi.org/10.1007/s40725-019-00099-y.
Park, H. and W. Takeuchi. (2021). Satellite-based estimation of global CO2 emissions from biomass burning. In *Biomass Burning in South and Southeast Asia* (pp. 109–125). CRC Press.
Phairuang, W. (2021). Biomass burning and their impacts on air quality in Thailand. In *Biomass Burning in South and Southeast Asia* (pp. 21–38). CRC Press.
Rienecker, M.M., M.J. Suarez, R. Gelaro, and R. Todling. (2011). MERRA: NASA's modern-era retrospective analysis for research and applications. *Journal of Climate.* 24: 3624–3648.
Ritz, C., and J.C. Streibig. (2005). Bioassay analysis using R. *Journal of Statistical Software* 12: 1–22.
Saharjo, B.H. and A. Yungan. (2018). Forest and land fires in Riau Province: a case study in fire prevention policy implementation with local concession holders. In *Land-Atmospheric Research Applications in South and Southeast Asia* (pp. 143–169). Springer, Cham.
Syaufina, L. and S.I. Maulana. (2021). Biomass burning emissions in Indonesia and policy measures–an overview. *Biomass Burning in South and Southeast Asia*, pp.135–148.
Tanpipat, V., K. Honda, and P. Nuchaiya. (2009). MODIS hotspot validation over Thailand. *Remote Sensing.* 1: 1043–1054.
Torsri, K., W. Wannawong, K. Sarinnapakorn, S. Boonya-aroonnet, and R. Chiradon. (2014). An application of air-sea model components in the Coupled Ocean Atmosphere Wave Sediment Transport (COAWST) modelling system over an Indochina Peninsular sub-region: impact of high spatiotemporal SST on WRF model in precipitation prediction. 11th Annual Meeting of Asia Oceania Geosciences Society, Sapporo, Hokkaido, Japan.
Uranishi, K., H. Shimadera, and A. Kondo. (2021). Biomass burning influence on PM2. 5 regional and long-range transport in Northeast Asia. In *Biomass Burning in South and Southeast Asia* (pp. 281–295). CRC Press.
Vadrevu, K.P. (2021). Fire danger indices and methods: An appraisal. In *Biomass Burning in South and Southeast Asia* (pp. 201–232). CRC Press.
Vadrevu, K.P., T. Ohara, and C. Justice, (eds.). (2021a). *Biomass burning in south and Southeast Asia: impacts on the biosphere*, Volume Two. CRC Press.
Vadrevu, K.P., T. Ohara, and C. Justice, (eds.). (2021b). *Biomass burning in South and Southeast Asia: Mapping and monitoring,* Volume one. CRC Press.
Van Wagner, C.E. and Pickett, T.L. (1985). Equations and FORTRAN program for the Canadian forest fire weather index system. Forestry Technical Report 33. Canadian Forest Service.
Van Wagner, C. E. (1987). Development and structure of the Canadian forest fire weather index system. Forestry Technical Report 35. Canadian Forest Service.
Xie, P., A. Yatagai, M. Chen, T. Hayasaka, Y. Fukushima, C. Liu, and S. Yang. (2007). A gauge-based analysis of daily precipitation over East Asia. *Journal of Hydrometeorology.* 8: 607–626.

Fires Hotspot Forecasting in Indonesia Using Long Short-Term Memory Algorithm and MODIS Datasets

Evizal Abdul Kadir, Hsiang Tsung Kung, Arbi Haza Nasution, Hanita Daud, Amal Abdullah AlMansour, Mahmod Othman, and Sri Listia Rosa

Abstract Vegetation fires are most common in South and Southeast Asian countries, including Indonesia. In addition to anthropogenic causes, climate change in the form of droughts is the biggest driver of fires in Indonesia. In particular, the peatlands in Indonesia are highly vulnerable to droughts with recurrent fires. In this study, we used a long short-term memory (LSTM) algorithm to predict the fire hotspots based on the 2010 to 2021 fire data. More than 700,000 fire hotspots from 2010 to 2021 have been collected and used as a training dataset to forecast fires for the year 2022. The LSTM algorithm successfully predicted 2022 fires with the minimum root mean squared error and high accuracy. Furthermore, the results of the 2022 prediction year matched the previous year's fire data seasonally, with increasing fires from August to November. The study highlights the potential use of the LSTM algorithm for forecasting fires in Indonesia.

Keywords Fires hotspot · Forecasting · Indonesia · LSTM · MODIS

1 Introduction

Fires are one of the biggest natural threats to forests, woodlands, and grasslands in many countries, including Indonesia (Albar et al. 2018; Akther and Hassan, 2011; Goldammer 2012; Hayasaka et al. 2014; Petropoulos et al. 2013; Justice et al. 2015; Kadir et al. 2019, 2020, 2021). In several South/Southeast Asian countries, fire is used to clear the forests for agriculture through slash and burn (Albar et al. 2018;

E. A. Kadir (✉) · A. H. Nasution · S. L. Rosa
Department of Informatics Engineering, Universitas Islam Riau, Pekanbaru 28284, Indonesia
e-mail: evizal@eng.uir.ac.id

E. A. Kadir · H. T. Kung
Department of Computer Science, Harvard University, Cambridge, MA 02134, USA

H. Daud · M. Othman
Department of Applied Mathematics, Universiti Teknologi Petronas, 86400 Perak, Malaysia

A. A. AlMansour
Department of Computer Science, King Abdul Aziz University, Jeddah 80200, Saudi Arabia

K. P. Vadrevu et al. (eds.), *Vegetation Fires and Pollution in Asia*,
https://doi.org/10.1007/978-3-031-29916-2_35

Badarinath et al. 2007; 2008, 2009, 2015b; Badarinath and Prasad 2011; Biswas et al. 2015a); Biswas et al. 2021; Kant et al. 2000; Lasko and Vadrevu 2018; Petropoulos et al. 2013; Prasad et al. 2001a,b; Prasad et al. 2002a, 2002b;2003; 2004; Prasad and Badarinath 2004; Prasad et al. 2005; Prasad and Badarinth 2006; Vadrevu 2008; 2021a,b; Wooster et al. 2021; (Biswas et al. 2015a,b; Prasad et al. 2001a, b, 2002a, b;) agricultural residues after crop harvest to clear the land for the next crop (Lasko et al. 2017; 2018a,b; 2021; Vadrevu and Lasko 2015), to clear the forested lands for plantations (Albar et al. 2018), promoting the growth of grass in pasture lands for cattle (Thapa et al. 2022), etc., in addition to intentional or accidental human activities. While most of these fires are anthropogenic, the drivers of fires can also be natural such as lightning and extreme and prolonged drought conditions. Especially in tropical regions, there are usually two alternating rainy and dry seasons, and forests and grassland fires are highly vulnerable to fires during the dry season. Indonesia is one of the tropical countries with major fire issues, especially in Kalimantan and Sumatra Islands with recurrent fires (Hayasaka et al. 2014). Regardless of the ignition source, in forested areas, the fires can spread rapidly and become uncontrollable due to the local meteorological and environmental conditions. Further, fires are a major important source of air pollution which results in the release of greenhouse gas emissions and aerosols (Ito and Penner 2005; Gupta et al. 2001; Lasko and Vadrevu 2018; Vadrevu and Badarinath 2009; Vadrevu and Justice 2011; Kharol et al. 2012; Vadrevu and Lasko 2015; Vadrevu 2015; Vadrevu et al. 2008; 2013;). The smoke particles released from fires can interact with the cloud droplets and alter Earth's radiation budget (Martins and Dias 2009). The GHG emissions from biomass burning represent the largest source of inter-annual variability, in particular, CO_2 fluxes (Szopa et al. 2007; Kant et al. 2000;). Biomass burning is estimated to contribute to 7600 ± 359 Tg CO2eq year $-$ 1 (FAOSTAT 2020). In addition, biomass burning has been shown to influence various land-atmospheric interactions at different scales, such as vegetation transpiration, soil erosion, albedo (Crutzen and Andreae 1990). Smoke-borne aerosols from fires disrupt normal hydrological processes and reduce rainfall, potentially contributing to regional drought. In addition to these effects on Earth's radiation, atmosphere, climate, and ecosystems, the pollutants released from the fires (Vadrevu et al. 2014a,b, 2017, 2018 2019) can impact health resulting in asthma, acute respiratory illness, eye irritation, cardiovascular mortality, thrombosis, etc. (Sigsgaard et al. 2015). Thus, fires can become a disaster for humans and the environment due to their severity and intensity. Considering these effects, mapping and monitoring of fires, including forecasting, can not only help in understanding land-atmospheric interactions useful for climate change studies but also protecting human lives, ecosystems, and related functions (Goldammer 2012; Eaturu and Vadrevu 2021; Vadrevu and Justice 2011; Vadrevu et al. 2020; 2021a,b, 2022a; b).

Several techniques have been proposed to forecast fires, such as fire danger indices combining climate data with site characteristics and fire data records (Akhter and Hassan 2011; Vadrevu et al. 2021a, b). In addition, multiple machine learning algorithms were also used to characterize fire patterns and predict fires. For most algorithms, previous fire data is essential for calibration and prediction (Liang et al. 2019;

Omar et al. 2021; Lamjiak et al. 2021; Abdul Kadir et al. 2022; Mohan et al. 2021). These studies considered both the climate and environmental factors in predicting the fires. Including meteorological factors in the prediction of fires is important as they can drive accuracy. A comprehensive data analysis of fire hotspot occurrences, their fire size, intensity, and how they can potentially spread into new areas, including forecasting methods, were given in earlier works (Khabarov et al. 2008; Han et al. 2019; Kadir et al. 2019; Kukuk and Kilimci 2021; Prapas et al. 2021). Recently, deep learning algorithms are gaining popularity in various fields, such as pattern recognition, including forecasting (Benzekri et al., 2020). In this study, we use the popular long short-term memory (LSTM) algorithm to forecast fires in Indonesia for 2022. We used the fire spots data derived from the Moderate Resolution Imaging Spectroradiometer (MODIS) from 2010 to 2021 and tested the algorithm's robustness in predicting the fires for 2022.

2 Datasets and Methodology

We used the NASA MODIS fire hotspots data from 2010–2021 for our study. Table 1 shows the sample fires dataset for Indonesia. The data has been normalized and grouped into a single date of fire occurrence. The data has been split into training and testing for fire forecasting. In the field of deep learning, the LSTM algorithm is an artificial recurrent neural network (RNN) architecture and was first introduced by Hochreiter and Schmidhuber (1997). LSTM is a special model of RNN that capable of learning in long-term dependencies and remembering information for prolonged periods as a default. Figure 1 shows the RNN-LSTM model's architecture, consisting of several main blocks called cells with input, output, and forget gates. The sigmoid activation function classifies the values in probabilities for the two predefined classes in the dense output layer.

The LSTM model can be explained as short-term memory, which acts when the information is being acquired, retains for a few seconds, and then destines it to be kept for more extended periods or discards it. Long-term memory permanently retains information, allowing its recovery or recall. It contains all our autobiographical data and all our knowledge. LSTM model can handle the problem with long-term dependencies of RNN in which the RNN algorithm cannot do in the prediction of the information stored in the long-term memory but can give more accurate prediction from the recent information. LSTM can use by default to retain the data for a long-term period. The algorithm can predict, process, and classify based on time series data (Le et al. 2019). The LSTM model has an incredible way of forecasting and works well in time series data. Furthermore, this model can organize in the form of a chain structure and has four interacting layers with a unique method of communication in data processing. Figure 2 shows an analysis block diagram of how the forecasting process of the fire hotspot is done in our study.

Table 1 Detail of fires hotspot dataset from year 2010 to 2021 (NASA 2021)

	Latitude	Longitude	Brightness	Scan	Track	acq_data	acq_time	Satellite	Instrument	Confidence	Version	Bright_t31	Frp	Daynight	Type
0	0.02110	116.87390	315.30	1.10	1.10	2010-01-01	251	terra	MODIS	42	6.2	295.60	8.70	D	0.0
1	0.48080	116.08060	312.30	1.00	1.00	2010-01-01	251	terra	MODIS	66	6.2	295.00	6.90	D	0.0
11	2.15090	117.49680	320.60	1.00	1.00	2010-01-01	550	Aqua	MODIS	0	6.2	297.50	10.60	D	0.0
10	– 8.10890	118.07430	319.30	1.00	1.00	2010-01-01	547	Aqua	MODIS	0	6.2	300.80	9.10	D	0.0
8	– 8.15960	117.58570	319.60	1.00	1.00	2010-01-01	547	Aqua	MODIS	43	6.2	297.70	9.10	D	0.0
…	…	…	…	…	…	…	…	…	…	…	…	…	…	…	…
14208	– 7.22331	110.42920	316.80	1.00	1.00	2021-12-31	300	terra	MODIS	67	6.1NRT	293.096	8.56	D	NaN
14209	– 6.96059	110.45844	318.84	1.00	1.00	2021-12-31	300	terra	MODIS	55	6.1NRT	291.66	6.42	D	NaN
14210	– 5.80178	139.61118	309.54	1.00	1.00	2021-12-31	418	Aqua	MODIS	65	6.1NRT	283.90	5.50	D	NaN
14211	– 4.51654	136.84802	313.68	1.21	1.09	2021-12-31	418	Aqua	MODIS	56	6.1NRT	291.74	7.84	D	NaN
14213	– 4.54666	136.77507	309.21	1.22	1.10	2021-12-31	418	Aqua	MODIS	52	6.1NRT	287.79	5.59	D	NaN

703116 rows × 15 columns

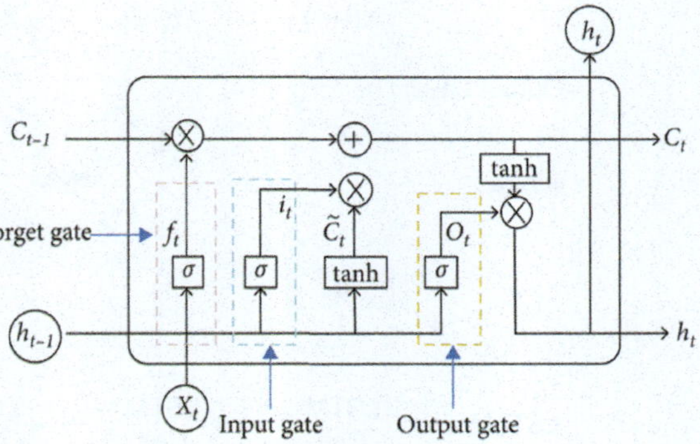

Fig. 1 Structure of RNN-LSTM algorithm

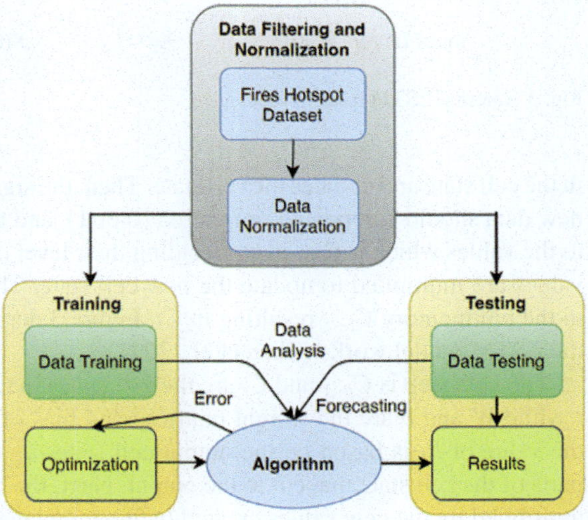

Fig. 2 Approach followed for fire forecasting using LSTM

The first step in data processing in forecasting is to construct an LSTM network model to identify the inputs and eliminate the information that is not necessary for the cell structure of LSTM (Fig. 1). The process of identifying and excluding data is governed by the sigmoid function, which takes the output of the last LSTM unit h_{t-1} at time $t - 1$ and the current input X_t at time t. Additionally, the sigmoid function determines which part from the old output should be eliminated. This gate is called the forget gate f_t; where f is a vector with values ranging from 0 to 1, corresponding to each number in the cell state, C_{t-1}. . Our collected data had more than 700,000 fire hotspots within 12 years and, after normalization, became 4365 datasets of fires grouped in each day. The data was divided into training and testing data (Fig. 2). The optimization process was followed to evaluate results, increase the performance and enhance accuracy to minimize the error and final forecasting. The LSTM cell with sigmoid function W_f and b_f are the weight matrices and bias, respectively, of the forget gate. This step decides and stores the input data from the new information X_t

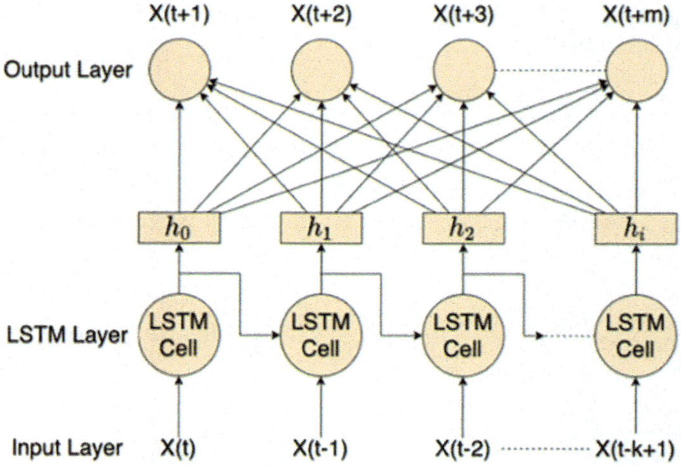

Fig. 3 Internal LSTM model process

in the cell state and updates the cell state. Then, the sigmoid layer decides whether the new data should be updated or ignored (0 or 1), and the tan *h* function gives weight to the values which is passed by deciding their level of importance (1 to 1). The two values are multiplied to update the new cell state. This new memory is then added to the old memory C_{t-1} resulting in C_t. Figure 3 depicts how the neuron process of the LSTM model works (Chen et al., 2021).

The next step is C_{t-1} and C_t are the cell states in the LSTM cell at time C_{t-1} and t while W and b are the weight matrices and bias of the cell state. In the last step, the value of h_t is based on the output cell state o_t, a sigmoid layer decides which parts of the cell state make it to the output. Next, the output of the sigmoid gate o_t is multiplied by the new values created by the *tanh* layer from the cell state C_t, with a value ranging between 1 and 1. Finally, the performance of the fire forecasting was done using the root mean square error (RMSE) with the prediction and actual data values using the below equation (1).

$$\text{RMSE} = \sqrt{\frac{\sum_{i=1}^{n}\left(X_i - \hat{X}_i\right)^2}{n}} \tag{1}$$

In the equation, X_i and X_i' are the actual fires hotspot data compared to forecasting fires data at the time t; X_i is the mean of actual values fires data and n is the total number of data. The smaller the RMSE values, the better the prediction.

3 Results and Discussion

Our fire dataset consisted of several parameters such as coordinate or location of fire occurrence, date and time, confidence level (probability of becoming a big fire and spreading out), brightness, day or nighttime, etc. (Table 1). In addition, we specifically used parameters that have a major impact and are essential to forecasting, which includes coordinates (latitude and longitude), acquisition date (acq_date), and confidence level. Figure 4a shows the mapping of fire hotspot distribution in Indonesia for 2021 and Fig. 4b for 2020. The fire hotspots were classified into five confidence levels, starting with the lowest from 0, low impact, and less potential to spread till 100, with high impact and high probability spread potential to become a big fire. The five-level classifications with confidence levels are shown in different colors (0–20 blue dot; level 21–40 green; 41–60 yellow; 61–80 as orange and 81–100 red with the highest).

The month-wise fire distribution is shown in Fig. 5a, b for the years 2021 and 2020, respectively. Classification based on confidence level and the distribution matched well with the total number of hotspots. Mostly, the map showed yellow and orange colors with confidence levels varying from 41–60 and 61–80, respectively. While red color is the highest potential of fire hotspots spread, they showed less in number in the predicted map.

Results from the LSTM suggested a similar pattern and number of daily fire hotspot incidents, with a maximum of 600 to 700 from the September to November dry or summer season. The daily average number of hotspots is 87. Although this number is insignificant for the entire of Indonesia, the number might increase drastically due to the prevailing weather and other fire-favorable factors. Another issue is the type of land that gets affected due to fires. For example, the Sumatra and Kalimantan Islands peatlands are easily ignited when dry land and fires are difficult to control. The LSTM algorithm for forecasting fire hotspots in Indonesia has been tested preliminarily to the 2121 data before 2022. Figure 6 compares actual fire hotspot data and forecasting results for the year 2021; the results showed a good agreement between the graphs. Preliminary forecasting suggested an RMSE error of 4.56%. We then fine-tuned the LSTM forecasting algorithm for 2022 by training more than 4000 datasets using the filtered data from 2010 to 2021; in essence, 30% of the total data was used for training and the rest 70% for testing.

Figure 7 shows a good agreement and similar normal distribution patterns for all the years, i.e., 2020, 2021, and 2022. The high occurrence of fire hotspots detected in the early part of the year, i.e., March, and lesser in the middle of the year, then increasing from September to November, is a typical pattern reflected in the figures. The spikes during the few days in late September are attributed to seasonal fires in Sumatra Island.

Overall, the LSTM RNN algorithm showed successful results with minimum error. The results of the 2022 prediction year matched the previous 2021-year fire data. Forecasting results in 2022 show good agreement and a similar pattern of fires with increasing fires from August to November. By comparing the predicted data

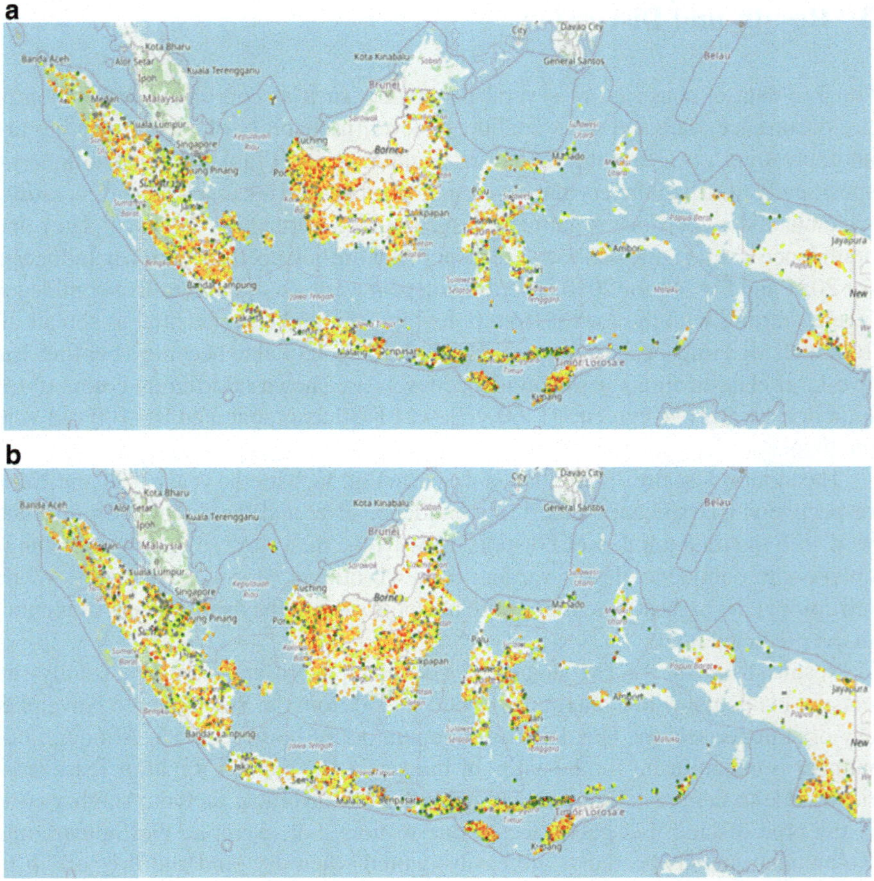

Fig. 4 Mapping of fires hotspots in Indonesia **a** year 2021 **b** year 2020

with the previous year's data, we could achieve an accuracy of up to 95% with an RMSE error of 4.56%. More robust data is required on the local conditions to achieve further high accuracy at specific locations. Our future studies will focus on the same, i.e., collecting and analyzing the data at a much higher spatial resolution for different regions in Indonesia.

4 Conclusion

We demonstrated the long short-term memory (LSTM) algorithm's potential in predicting and forecasting fire hotspots in Indonesia. A fire hotspots dataset from 2010 to 2021 obtained from the NASA MODIS data has been used to train and forecast fires for 2022. By comparing the predicted data with the previous year's

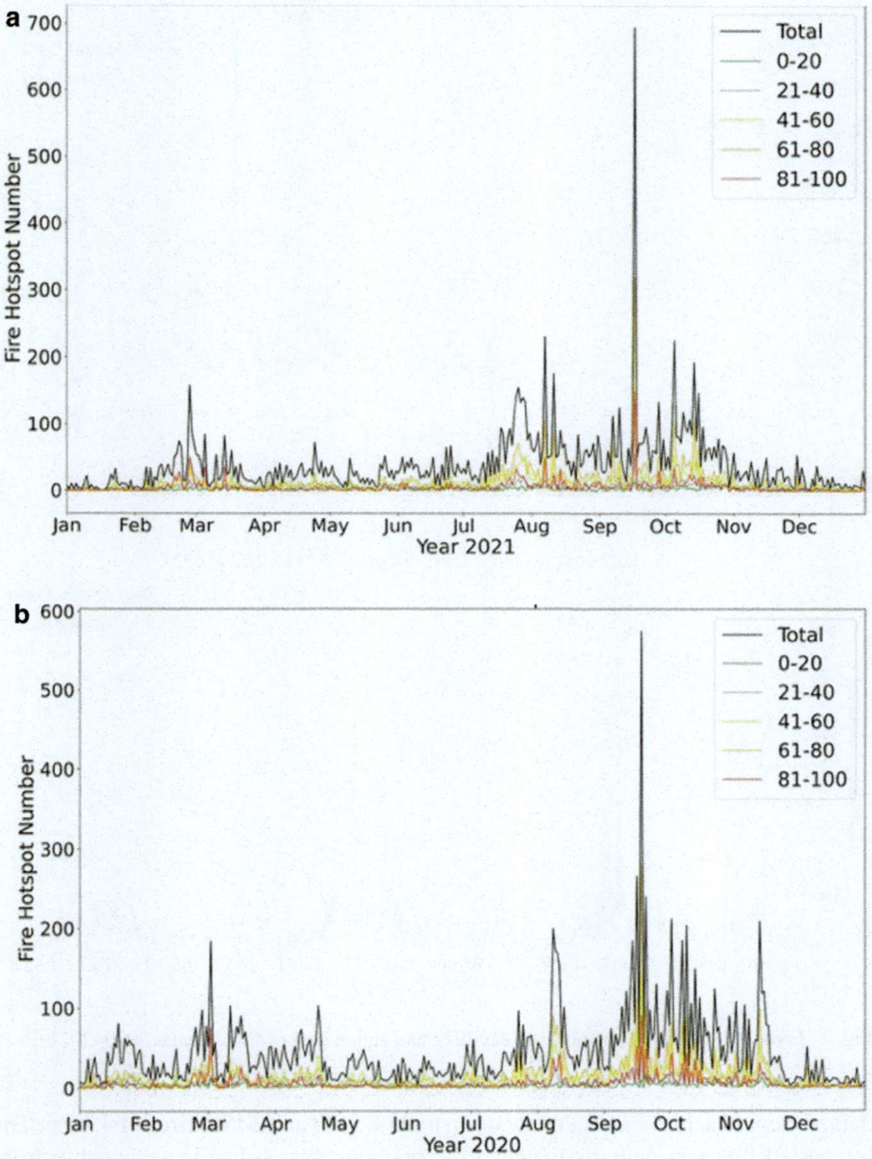

Fig. 5 Distribution of fires hotspots in Indonesia for the year from January to December **a** year 2021 **b** year 2020

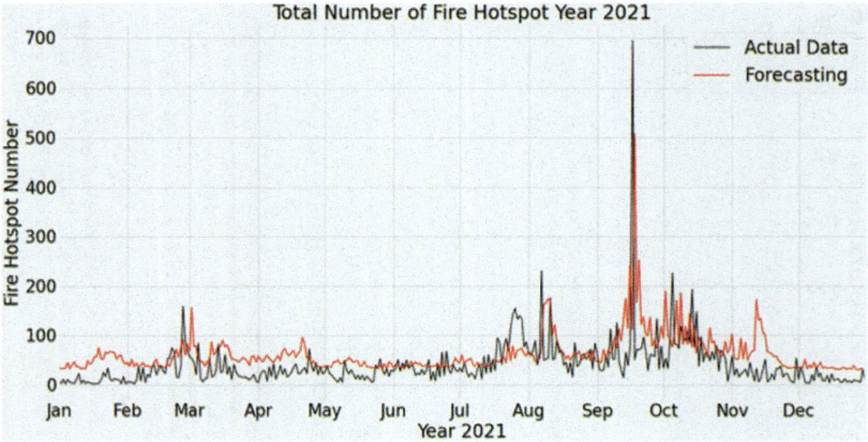

Fig. 6 Comparison of actual and fire forecasting data for year 2021

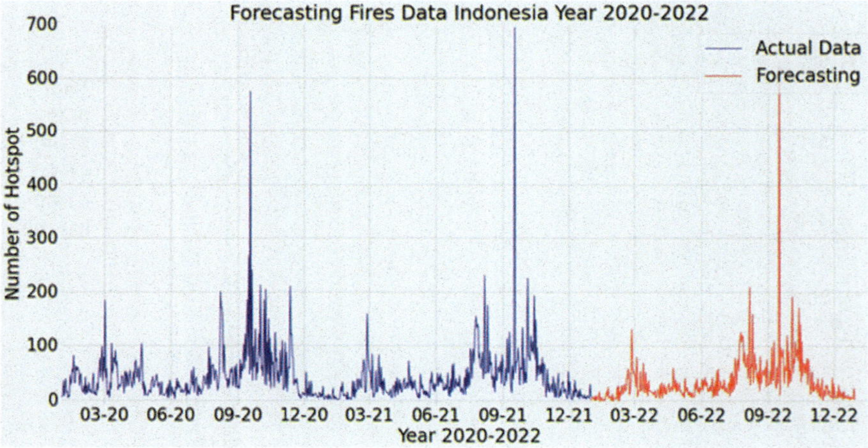

Fig. 7 Forecasting of fires hotspots in year 2022 and actual data of fires in year 2020–2021

data, we could achieve an accuracy of up to 95% with an RMSE error of 4.56%. The forecasted fire data patterns matched the previous year's data in seasonality from January to December. It is noted that the number of hotspots increase by the end of each year due to the dry season in Indonesia.

Acknowledgements We thank the Ministry of Education, Culture, Research, and Technology of Indonesia for funding the research and American Indonesia Exchange Foundation (AMINEF) and the Fulbright fellowship. We also acknowledge Harvard University, Universiti Teknologi Petronas, Universitas Islam Riau, and King Abdul Aziz University for their research facilities.

References

Abdul Kadir, E., S. Listia Rosa, A. Syukur, M. Othman, and H. Daud. 2022. Forest fire spreading and carbon concentration identification in tropical region Indonesia. *Alexandria Engineering Journal* 61: 1551–1561.

Akther, M.S., and Q.K. Hassan. 2011. Remote sensing-based assessment of fire danger conditions over boreal forest. *IEEE Journal of Selected Topics in Applied Earth Observations and Remote Sensing* 4: 992–999.

Albar, I., I. Jaya, B.H. Saharjo, B. Kuncahyo, K.P. Vadrevu. (2018). Spatio-temporal analysis of land and forest fires in Indonesia using MODIS active fire dataset. In Land-atmospheric research applications in South and Southeast Asia, (pp. 105–127). Springer, Cham.

Badarinath, K.V.S., and K.V. Prasad. 2011. Carbon dioxide emissions from forest biomass burning in India. *Global Environmental Research* 15: 45–52.

Badarinath, K.V.S., S.K. Kharol, K.M. Latha, T.K. Chand, V.K. Prasad, A.N. Jyothsna, and K. Samatha. 2007. Multiyear ground-based and satellite observations of aerosol properties over a tropical urban area in India. *Atmospheric Science Letters* 8 (1): 7–13.

Badarinath, K.V.S., S.K. Kharol, A.R. Sharma, and V.K. Prasad. 2009. Analysis of aerosol and carbon monoxide characteristics over Arabian Sea during crop residue burning period in the Indo-Gangetic Plains using multi-satellite remote sensing datasets. *Journal of Atmospheric and Solar-Terrestrial Physics* 71 (12): 1267–1276.

Badarinath, K.V.S., S. Kumar Kharol, V. Krishna Prasad, A. Rani Sharma, E.U.B. Reddi, H.D. Kambezidis, and D.G. Kaskaoutis. 2008. Influence of natural and anthropogenic activities on UV Index variations–a study over tropical urban region using ground based observations and satellite data. *Journal of Atmospheric Chemistry* 59 (3): 219–236.

Benzekri, W., A.E. Moussati, O. Moussaoui, and M. Berrajaa. 2020. Early forest fire detection system using wireless sensor network and deep learning. *International Journal of Advanced Computer Science and Applications (IJACSA)* 11: 496–503.

Biswas, S., K.D. Lasko, and K.P. Vadrevu. 2015a. Fire disturbance in tropical forests of Myanmar—analysis using MODIS satellite datasets. *IEEE Journal of Selected Topics in Applied Earth Observations and Remote Sensing* 8 (5): 2273–2281.

Biswas, S., K.P. Vadrevu, Z.M. Lwin, K. Lasko, and C.O. Justice. 2015b. Factors controlling vegetation fires in protected and non-protected areas of Myanmar. *PLoS ONE* 10 (4): e0124346.

Biswas, S., K.P. Vadrevu, M.S. Mon, and C. Justice. 2021. Contemporary forest loss in Myanmar: effect of democratic transition and subsequent timber bans on landscape structure and composition. *Ambio* 50 (4): 914–928.

Chen, H., M. Guan, and H. Li. 2021. Air quality prediction based on integrated dual LSTM model. *IEEE Access* 9: 93285–93297.

Crutzen, P.J., and M.O. Andreae. 1990. Biomass burning in the tropics: impact on atmospheric chemistry and biogeochemical cycles. *Science* 250 (4988): 1669–1678.

Eaturu, A. and K.P. Vadrevu. (2021). Evaluation of sentinel-3 SLSTR data for mapping fires in forests, peatlands, and croplands–a case study over Australia, Indonesia, and India. In Biomass burning in South and Southeast Asia (pp. 39–59). CRC Press.

FAOSTAT. (2020). http://www.fao.org/faostat/en/

Goldammer, J.G. (ed.). (2012). Fire in the tropical biota: Ecosystem processes and global challenges (Vol. 84). Springer Science & Business Media.

Gupta, P.K., V.K. Prasad, C. Sharma, A.K. Sarkar, Y. Kant, K.V.S. Badarinath, and A.P. Mitra. 2001. CH_4 emissions from biomass burning of shifting cultivation areas of tropical deciduous forests–experimental results from ground-based measurements. *Chemosphere-Global Change Science* 3 (2): 133–143.

Han, J., G. Kim, C. Lee, Y. Han, U. Hwang, S. Kim. (2019). Predictive models of fire via deep learning exploiting colorific variation. In 2019 international conference on artificial intelligence in information and communication (ICAIIC), 11–13 Feb. 2019. 579–581.

Hayasaka, H., I. Noguchi, E.I. Putra, N. Yulianti, and K. Vadrevu. 2014. Peat-fire-related air pollution in Central Kalimantan, Indonesia. *Environmental Pollution* 195: 257–266.

Hochreiter, S., and J. Schmidhuber. 1997. Long short-term memory. *Neural Computation* 9: 1735–1780.

Ito, A., and J.E. Penner. 2005. Historical emissions of carbonaceous aerosols from biomass and fossil fuel burning for the period 1870–2000[J]. *Global Biogeochemical Cycles* 19 (2): 1–14.

Justice, C., G. Gutman, and K.P. Vadrevu. 2015. NASA land cover and land use change (LCLUC): an interdisciplinary research program. *Journal of Environmental Management* 148: 4–9.

Kadir, E. A., H. Irie, and S.L. Rosa. (2019). Modeling of wireless sensor networks for detection land and forest fire hotspot. In 2019 international conference on electronics, information, and communication (ICEIC), 22–25 Jan 2019 2019. 1–5.

Kadir, E.A., M. Othman, and S.L. Rosa. (2021). Smart sensor system for detection and forecasting forest fire hotspot in Riau province Indonesia. In 2021 international congress of advanced technology and engineering (ICOTEN), 4–5 July 2021 2021. 1–6.

Kadir, E.A., S.L. Rosa, and R. A. Ramadhan. (2020). Detection of forest fire used multi sensors system for peatland area in Riau Province. In: W. Sutopo (ed.). 6th international conference on industrial mechanical electrical chemical engineering 2020, 2020 Surakarta, Indonesia. AIP Conference Proceedings.

Kant, Y., V.K. Prasad, and K.V.S. Badarinath. 2000. Algorithm for detection of active fire zones using NOAA AVHRR data. *Infrared Physics and Technology* 41 (1): 29–34.

Khabarov, N., E. Moltchanova, and M. Obersteiner. 2008. Valuing weather observation systems for forest fire management. *IEEE Systems Journal* 2: 349–357.

Kharol, S.K., K.V.S. Badarinath, A.R. Sharma, D.V. Mahalakshmi, D. Singh, and V.K. Prasad. 2012. Black carbon aerosol variations over Patiala city, Punjab, India—a study during agriculture crop residue burning period using ground measurements and satellite data. *Journal of Atmospheric and Solar-Terrestrial Physics* 84: 45–51.

Kukuk, S.H.B., and Z.H. Kiliminci. (2021). Comprehensive analysis of forest fire detection using deep learning models and conventional machine learning algorithms. International Journal of Computational and Experimental Science and Engineering (IJCESEN). Turkey: İskender Akkurt.

Lamjiak, T., R. Kaewthongrach, B. Sirinaovakul, P. Hanpattanakit, A. Chithaisong, and J. Polvichai. 2021. Characterizing and forecasting the responses of tropical forest leaf phenology to El Nino by machine learning algorithms. *PLoS ONE* 16: e0255962.

Lasko, K., and K. Vadrevu. 2018. Improved rice residue burning emissions estimates: accounting for practice-specific emission factors in air pollution assessments of Vietnam. *Environmental Pollution* 236: 795–806.

Lasko, K., K.P. Vadrevu, and T.T.N. Nguyen. 2018. Analysis of air pollution over Hanoi, Vietnam using multi-satellite and MERRA reanalysis datasets. *PLoS ONE* 13 (5): e0196629.

Lasko, K., K.P. Vadrevu, V. Bandaru, T.N.T. Nguyen, and H.Q. Bui. (2021). PM2. 5 emissions from biomass burning in South/Southeast Asia–Uncertainties and trade-offs. In Biomass burning in south and Southeast Asia (pp. 149–169). CRC Press.

Lasko, K., K.P. Vadrevu, V.T. Tran, E. Ellicott, T.T. Nguyen, H.Q. Bui, and C. Justice. 2017. Satellites may underestimate rice residue and associated burning emissions in Vietnam. *Environmental Research Letters* 12 (8): 085006.

Le, X.-H., H.V. Ho, G. Lee, and S. Jung. 2019. Application of long short-term memory (LSTM) neural network for flood forecasting. *Water* 11: 1387.

Liang, H., M. Zhang, and H. Wang. 2019. A neural network model for wildfire scale prediction using meteorological factors. *IEEE Access* 7: 176746–176755.

Martins, J.A., and M.S. Dias. 2009. The impact of smoke from forest fires on the spectral dispersion of cloud droplet size distributions in the Amazonian region. *Environmental Research Letters* 4 (1): 015002.

Mohan, K.V.M., A.R. Satish, K.M. Rao, R.K. Yarava, and G.C. Babu. (2021). Leveraging machine learning to predict wild fires. In 2021 2nd international conference on smart electronics and communication (ICOSEC), 7–9 Oct. 2021. 1393–1400.

NASA 2021. Fires Hotspot Information. 2010–2021. USA.

Omar, N., A. Al-Zebari, and A. Sengur. (2021). Deep learning approach to predict forest fires using meteorological measurements. In 2021 2nd international informatics and software engineering conference (IISEC), 16–17 Dec. 2021 2021. 1–4.

Petropoulos, G.P., K.P. Vadrevu, and C. Kalaitzidis. 2013. Spectral angle mapper and object-based classification combined with hyperspectral remote sensing imagery for obtaining land use/cover mapping in a Mediterranean region. *Geocarto International* 28 (2): 114–129.

Prapas, I., S. Kondylatos, I. Papoutsis, G. Camps-Valls, M. Ronco, M.-N.F. Ndez-Torres, M.P. Guillem, and N. Carvalhais. (2021). Deep learning methods for daily wildfire danger forecasting. In 35th conference on neural information processing systems (NeurIPS 2021). Sydney, Australia.

Prasad, K.V., and K.V.S. Badarinath. 2006. Soil surface nitrogen losses from agriculture in India: a regional inventory within agroecological zones (2000–2001). *The International Journal of Sustainable Development and World Ecology* 13 (3): 173–182.

Prasad, V.K., and K.V.S. Badarinth. 2004. Land use changes and trends in human appropriation of above ground net primary production (HANPP) in India (1961–98). *Geographical Journal* 170 (1): 51–63.

Prasad, V.K., E. Anuradha, and K.V.S. Badarinath. 2005. Climatic controls of vegetation vigor in four contrasting forest types of India—evaluation from National Oceanic and atmospheric administration's advanced very high resolution radiometer datasets (1990–2000). *International Journal of Biometeorology* 50 (1): 6–16.

Prasad, V.K., K.V.S. Badarinath, S. Yonemura, and H. Tsuruta. 2004. Regional inventory of soil surface nitrogen balances in Indian agriculture (2000–2001). *Journal of Environmental Management* 73 (3): 209–218.

Prasad, V.K., Y. Kant, and K.V.S. Badarinath. 2001a. CENTURY ecosystem model application for quantifying vegetation dynamics in shifting cultivation areas: a case study from Rampa forests, Eastern Ghats (India). *Ecological Research* 16 (3): 497–507.

Prasad, V.K., Y. Kant, and K.V.S. Badarinath. 2002a. Land use changes and modeling carbon fluxes from satellite data. *Advances in Space Research* 30 (11): 2511–2516.

Prasad, V.K., Y. Kant, P.K. Gupta, C. Elvidge, and K.V.S. Badarinath. 2002b. Biomass burning and related trace gas emissions from tropical dry deciduous forests of India: a study using DMSP-OLS data and ground-based measurements. *International Journal of Remote Sensing* 23 (14): 2837–2851.

Prasad, V.K., Y. Kant, P.K. Gupta, C. Sharma, A.A. Mitra, and K.V.S. Badarinath. 2001b. Biomass and combustion characteristics of secondary mixed deciduous forests in Eastern Ghats of India. *Atmospheric Environment* 35 (18): 3085–3095.

Prasad, V.K., M. Lata, and K.V.S. Badarinath. 2003. Trace gas emissions from biomass burning from northeast region in India—estimates from satellite remote sensing data and GIS. *The Environmentalist* 23 (3): 229–236.

Sigsgaard, T., B. Forsberg, I. Annesi-Maesano, A. Blomberg, A. Bølling, C. Boman, J. Bønløkke, M. Brauer, N. Bruce, M.E. Héroux, and M.R. Hirvonen. 2015. Health impacts of anthropogenic biomass burning in the developed world. *European Respiratory Journal.* 46 (6): 1577–1588.

Thapa, S.K., J.F. de Jong, A.R. Hof, N. Subedi, L.R. Joshi, and H.H. Prins. 2022. Fire and forage quality: postfire regrowth quality and pyric herbivory in subtropical grasslands of Nepal. *Ecology and Evolution* 12 (4): e8794.

Szopa, S., D.A. Hauglustaine, and P. Ciais. (2007). Relative contributions of biomass burning emissions and atmospheric transport to carbon monoxide interannual variability. Geophysical Research Letters. 34(18).

Vadrevu, K., and K. Lasko. 2015. Fire regimes and potential bioenergy loss from agricultural lands in the Indo-Gangetic Plains. *Journal of Environmental Management* 148: 10–20.

Vadrevu, K. 2015. International regional science meeting of NASA-LCLUC: land cover/land use change (LC/LUC) and environmental impacts in South Asia, Coimbatore, India, 19–23 January 2013. *Journal of Environmental Management* 148: 1–163.
Vadrevu, K., A. Eaturu, E. Casadaban, K. Lasko, W. Schroeder, S. Biswas, L. Giglio, and C. Justice. 2022a. Spatial variations in vegetation fires and emissions in South and Southeast Asia during COVID-19 and pre-pandemic. *Scientific Reports* 12 (1): 1–21.
Vadrevu, K., A. Heinimann, G. Gutman, and C. Justice. 2019. Remote sensing of land use/cover changes in South and Southeast Asian Countries. *International Journal of Digital Earth* 12 (10): 1099–1102.
Vadrevu, K., T. Ohara, and C. Justice. 2017. Land cover, land use changes and air pollution in Asia: a synthesis. *Environmental Research Letters* 12 (12): 120201.
Vadrevu, K.P., and K.V.S. Badarinath. 2009. Spatial pattern analysis of fire events in Central India–a case study. *Geocarto International* 24 (2): 115–131.
Vadrevu, K.P., and C.O. Justice. 2011. Vegetation fires in the Asian region: satellite observational needs and priorities. *Global Environmental Research* 15 (1): 65–76.
Vadrevu, K.P. 2008. Analysis of fire events and controlling factors in eastern India using spatial scan and multivariate statistics. *Geografiska Annaler: Series a, Physical Geography* 90 (4): 315–328.
Vadrevu, K.P., K.V.S. Badarinath, and E. Anuradha. 2008. Spatial patterns in vegetation fires in the Indian region. *Environmental Monitoring and Assessment* 147 (1): 1–13.
Vadrevu, K.P., A. Eaturu, S. Biswas, K. Lasko, S. Sahu, J.K. Garg, and C. Justice. 2020. Spatial and temporal variations of air pollution over 41 cities of India during the COVID-19 lockdown period. *Scientific Reports* 10 (1): 1–15.
Vadrevu, K.P., L. Giglio, and C. Justice. 2013. Satellite based analysis of fire–carbon monoxide relationships from forest and agricultural residue burning (2003–2011). *Atmospheric Environment* 64: 179–191.
Vadrevu, K.P., K. Lasko, L. Giglio, and C. Justice. 2014a. Analysis of Southeast Asian pollution episode during June 2013 using satellite remote sensing datasets. *Environmental Pollution* 195: 245–256.
Vadrevu, K.P., T. Ohara, and C. Justice, (eds.). (2018). Land-atmospheric research applications in South and Southeast Asia.
Vadrevu, K.P., T. Ohara, and C. Justice (eds.). (2021a). Biomass burning in south and Southeast Asia: impacts on the biosphere, Volume Two. CRC Press.
Vadrevu, K.P., T. Ohara, and C. Justice (eds.). (2021b). Biomass burning in South and Southeast Asia: mapping and monitoring, volume one. CRC Press.
Vadrevu, K.P., T. Ohara, and C. Justice (eds.). (2014a). Air pollution in Asia. Environmental Pollution (Barking, Essex: 1987), 195, pp.233–235.
Vadrevu, T. Le Toan, S.S. Ray, and C. Justice. (2022a). Remote sensing of agriculture and land cover/land use changes in South and Southeast Asian Countries. Springer Cham.
Vadrevu, K.P., A. Eaturu, E. Casadaban, and S. Biswas. (2022b). Agricultural fires in South Asian countries and implications. In Remote sensing of agriculture and land cover/land use changes in South and Southeast Asian Countries (pp. 501–516). Springer, Cham.
Wooster, M.J., G.J. Roberts, L. Giglio, D.P. Roy, P.H. Freeborn, L. Boschetti, C. Justice, C. Ichoku, W. Schroeder, D. Davies, and A.M. Smith. 2021. Satellite remote sensing of active fires: History and current status, applications and future requirements. *Remote Sensing of Environment* 267: 112694.

The manufacturer's authorised representative in the EU is Springer Nature Customer Service Centre GmbH, Europaplatz 3, 69115 Heidelberg, Germany. If you have any concerns regarding our products, please contact ProductSafety@springernature.com

Printed and bound by CPI Group (UK) Ltd, Croydon, CR0 4YY
01/10/2026
02225902-0020